$$[p_e]_{11} = \frac{1}{30l_e}(\ 37p_i + 36p_{i+1} - 3p_{i+2})$$

$$[p_e]_{12} = \frac{1}{30l_e}(-44p_i - 32p_{i+1} - 4p_{i+2})$$

$$[p_e]_{13} = \frac{1}{30l_e}(\ 7pi - 4p_{i+1} + 7p_{i+2})$$

$$[p_e]_{22} = \frac{1}{30l_e}(\ 48p_i + 64p_{i+1} + 48p_{i+2})$$

$$[p_e]_{23} = \frac{1}{30l_e}(\ -4p_i - 32p_{i+1} - 44p_{i+2})$$

$$[p_e]_{33} = \frac{1}{30l_e}(\ -3p_i + 36p_{i+1} + 37p_{i+2})$$

with $[p_e]_{ij} = [p_e]_{ji}$,

$$[q_e]_{11} = \frac{l_e}{420}(\ 39q_i + 20q_{i+1} - 3q_{i+2})$$

$$[q_e]_{12} = \frac{l_e}{420}(\ 20q_i + 16q_{i+1} - 8q_{i+2})$$

$$[q_e]_{13} = \frac{l_e}{420}(-3q_i - 8q_{i+1} - 3q_{i+2})$$

$$[q_e]_{22} = \frac{l_e}{420}(\ 16q_i + 192q_{i+1} + 16q_{i+2})$$

$$[q_e]_{23} = \frac{l_e}{420}(-8q_i + 16q_{i+1} + 20q_{i+2})$$

$$[q_e]_{33} = \frac{l_e}{420}(-3q_i + 20q_{i+1} + 39q_{i+2})$$

with $[q_e]_{ij} = [q_e]_{ji}$, and

$$[f_e]_1 = \frac{l_e}{30}(\ 4f_i + 2f_{i+1} - f_{i+2})$$

$$[f_e]_2 = \frac{l_e}{30}(\ 2f_i + 16f_{i+1} + 2f_{i+2})$$

$$[f_e]_3 = \frac{l_e}{30}(-f_i + 2f_{i+1} + 4f_{i+2})$$

Approximate elemental matrices for quadratic interpolation

A First Course in the Finite Element Method

A First Course in the Finite Element Method

Second Edition

William B. Bickford
Arizona State University

IRWIN
Burr Ridge, Illinois
Boston, Massachusetts
Sydney, Australia

This symbol indicates that the paper in this book is made of recycled paper. Its fiber content exceeds the recommended minimum of 50% waste paper fibers as specified by the EPA.

Associate editor: *Kelley Butcher*
Editorial coordinator: *Christine Bara*
Project editor: *Stephanie M. Britt*
Production manager: *Laurie Kersch*
Cover designer: *Nikki Life*
Art manager: *Kim Meriwether*
Compositor: *Publication Services, Inc.*
Typeface: *10/12 Times Roman*
Printer: *R. R. Donnelley & Sons Company*

Library of Congress Cataloging-in-Publication Data

Bickford, William (William B.)
A first course in the finite element method / William B. Bickford.
p. cm.
Includes bibliographical references and index.
ISBN 0-256-14472-9
1. Finite element method. I. Title.
TA347.F5B53 1994
620'.001'51535—dc20 93-20874

Printed in the United States of America
1 2 3 4 5 6 7 8 9 0 DOC 0 9 8 7 6 5 4 3

To my parents,
for unlimited horizons

Preface to the Second Edition

The main purpose of the second edition is still systematically introducing the reader to the methodology of the finite element method, independent of the many specific areas of application. The seven basic ingredients, or steps, (discretization, interpolation, elemental formulation, assembly, constraints, solution, and computation of derived variables) remain firmly in focus throughout the text. Applications to solid mechanics, heat transfer, fluid mechanics, and vibrations continue to play a central role in the text.

Chapter 2 has been significantly rearranged. After discussing the class of problems to be considered in the chapter, the chapter presents the Galerkin method for deriving the finite element model for one-dimensional boundary value problems. The example applications from solid mechanics, fluid mechanics, and heat transfer, as well as the eigenvalue problems, immediately follow. If you prefer variational formulation, the Ritz method for deriving the finite element model appears next. Quadratic interpolation and its applications to solid mechanics, fluid mechanics, heat transfer, and eigenvalue problems round out the chapter. Two other important changes include moving the material on weighted residuals to Appendix A and increasing the number of end-of-chapter exercises by approximately 80 percent.

Chapter 3 has been adjusted to reflect user preferences by presenting the Galerkin approach to developing the finite element model for scalar two-dimensional boundary value problems before presenting the corresponding variational finite element model. Again, end-of-chapter exercises have been expanded.

Chapters 4 and 5 are largely unchanged.

Consistent with strong user input, the source listings for the codes in Appendixes B and C have been converted to FORTRAN. The codes available to the instructor, on the diskette included with the solutions manual, are also written in FORTRAN.

KEY FEATURES

- Reiteration of the seven basic ingredients throughout the text.
- Increased number of end-of-chapter exercises.
- Solutions manual containing worked-out solutions and numerical results to most of the end-of-chapter exercises.
- FORTRAN listing of codes in appendices and solutions manual.

ACKNOWLEDGMENTS

I gratefully acknowledge the contributions of the following individuals to the second edition: Jim H. Akin, University of Arkansas–Fayetteville; Timothy K. Hight, Santa Clara University; John E. Jackson, Jr., University of Alabama–Tuscaloosa; Jesa H. Kreiner, California State University–Fullerton; V. Dakshina Murty, University of Portland; Mehrdad Negahban, University of Nebraska–Lincoln; Virgil W. Snyder, Michigan Technological University; John M. Sullivan, Jr., Worcester Polytechnic Institute; David G. Taggart, University of Rhode Island; Robert F. Tucker, Texas A&I University; and John D. Whitcomb, Texas A&M University–College Station.

William B. Bickford

Preface to the First Edition

The finite element method is an analytical tool that can be applied very effectively to the analysis of many of the physical and mathematical models of interest to engineers, scientists, and applied mathematicians. These physical and mathematical models usually arise in the process of modeling problems in application areas such as solid and fluid mechanics, heat transfer, vibrations, and electrical potentials, and magnetic fields. The finite element method has been successfully applied and is regularly used as an analysis tool in virtually all areas of engineering, both in academic and industrial circles. The finite element method is also frequently employed as a numerical method for solving differential equations.

As normally used, the finite element method can be considered as a definite set of basic ingredients or steps. In order, these are

1. Discretization
2. Interpolation
3. Elemental formulation
4. Assembly
5. Constraints
6. Solution
7. Computation of derived variables

These basic steps can be very useful in systematically applying the finite element method to problems in engineering, physics, and applied mathematics.

This book is intended to provide a thorough introduction to the basic ideas employed in the application of the finite element method to the problems of interest to scientists, engineering students, and practicing engineers. A student who has mastered the content of this book should be able to perform hand analyses of small-sized problems using the finite element method, and to write medium-sized codes for analyzing classes of problems of interest. Additionally, and more to the point in many cases, mastery of the material contained in this text should enable the student to more intelligently use commercially available codes such as ANSYS and NASTRAN, which are based on the finite element method.

Background for the course includes a working knowledge of matrix algebra and differential equations in addition to elementary courses in solid mechanics and fluid mechanics. Numerous examples are also taken from heat transfer and vibrations as well as from the areas of ordinary and partial differential equations.

The level of the text is appropriate for study by seniors or first-year graduate students. With proper background it is also quite suitable for self study.

In order, the chapters treat successively more complex classes of problems, beginning with discrete systems and ending with some of the partial differential equations occurring in solid mechanics. Chapters 2 through 5 are each laid out in the same way: The seven basic steps are discussed and illustrated for the class of problems being considered. This will help the student see that the implementation of the finite element method is in a sense "invariant" in that it can generally be applied to many different classes of physical and mathematical problems in exactly the same fashion.

Experience has shown that several of the basic steps of the finite element method are most easily understood through the use of matrices as opposed to the more commonly used index notation. For this reason, the matrix notation is used throughout the book for representing the basic ideas at both the elemental and global levels.

Chapter 1 begins with an overview of the finite element method and contains a general discussion of the process of converting a physical problem that is modeled as continuous into a corresponding discrete problem. This process is approached toward seeing that, although it may not be apparent, a typical problem considered to be discrete implicitly contains the basic steps of discretization, interpolation, and elemental formulation. Chapter 1 also reviews the formulation and analysis of typical discrete problems in solid mechanics, heat transfer, and fluid mechanics. The emphasis is on having the student begin to consider the relationships between familiar classes of discrete problems and the basic steps or building blocks of the finite element method.

Chapter 2 is devoted to the application of the finite element method to second- and fourth-order ordinary differential equations in general and to several application areas for which ordinary differential equations appearing as boundary value problems or eigenvalue problems are the appropriate mathematical model. These application areas include solid mechanics, heat transfer, vibrations, and fluid mechanics. Approximate methods for the solution of ordinary differential equations are discussed in general to enable the student to see the place of the finite element method in the context of approximate methods in general. Certain basic concepts of the calculus of variations, which are indispensable to the understanding of the relationship between the Ritz and Galerkin finite element models, are introduced and used regularly. Both the Variational and Galerkin approaches to formulating the finite element model for ordinary differential equations are introduced and illustrated.

Chapter 3 covers situations in which two-dimensional elliptic boundary value problems are the appropriate mathematical model. This is a natural next step from Chapter 2, where the steady-state one-dimensional problems considered are essentially one-dimensional elliptic problems. Important physical problems that fall within this category are steady-state heat conduction-convection, two-dimensional inviscid irrotational fluid flow, electric potential problems, and the torsion of prismatic bars. Several isoparametric elements are discussed and used in developing the finite element models for typical problems in heat transfer, solid and fluid mechanics, and vibrations.

In Chapter 4, time-dependent diffusion (parabolic) and wave propagation (hyperbolic) type problems are covered. Semidiscrete interpolation for time-dependent problems is introduced and discussed. The finite element model for a typical diffusion problem consists of a set of first-order ordinary differential equations with time as the independent variable, whereas for a typical wave propagation problem the finite element model consists of a set of second-order ordinary differential equations. Analytical approaches and numerical algorithms for solving these sets of ordinary differential equations are introduced and demonstrated.

Chapter 5 is devoted to linear elasticity. The governing differential equations of equilibrium for linear elasticity are coupled partial differential equations with the displacements as dependent variables, thus introducing the student to a typical situation where there is more than one dependent variable. The finite element models for plane and axisymmetric elasticity are developed and applied to problems of classical and technical importance.

There are numerous exercises at the ends of the chapters. They are generally of two basic types:

1. Those intended for drill to help the student become proficient in applying the ideas and techniques presented, or that ask the student to fill in some details of the developments in the chapter.
2. Those that extend to some degree the material presented in the chapter, or that introduce additional material. These exercises are starred.

The student should endeavor to at least *read through* all the exercises, as some of the information may be useful, even if all the details of the exercises are not investigated or verified.

Also included at the end of the exercises are computer projects. These computer projects suggest and outline codes that can be written to solve the particular class of problems covered in the chapter. It has been found that the students' understanding of the basic ingredients of the finite element method is greatly enhanced when they are required to complete several of these codes. Successfully completing the first code assigned tends to be somewhat painful for the student, with subsequent assignments following much more easily as the student masters and is able to reuse—with suitable minor modifications—some portions of previously developed codes.

Appendix B contains a general discussion of the Gaussian elimination procedure for solving sets of linear algebraic equations. A listing of the code BANDSOLV for the solution of positive-definite symmetric sets of equations stored in banded form is given. Also included is a listing of the popular code DECOMP-SOLVE, which is based on Gaussian elimination with partial pivoting. Source files for both BANDSOLV and DECOMP-SOLVE are available on diskette in the solutions manual.

Appendix C contains the listing for several codes to determine the eigenvalues and eigenvectors of the generalized eigenvalue problem $\mathbf{Ax} - \lambda\mathbf{Bx} = \mathbf{0}$. FWDT and INVIT are routines for forward and inverse iteration, respectively. FWDT is an iteration algorithm that will yield the eigenvalue-eigenvector pair corresponding to the eigenvalue of maximum modulus. INVIT is an iteration algorithm that will yield the eigenvalue-eigenvector pair corresponding to the eigenvalue of minimum modulus. The code GENJAC is an algorithm that uses rotations to transform both the mass and stiffness matrices to diagonal form from which the eigenvalues and eigenvectors can be determined. It is suitable for relatively small- or medium-sized problems. The code SUBSPACE uses an algorithm that is usually used to determine the first $q \ll n$ eigenvalue-eigenvector pairs of $\mathbf{Ax} - \lambda\mathbf{Bx} = \mathbf{0}$. Source files for each of these algorithms are available on diskette in the solutions manual.

Appendix D contains a discussion of the relationship between boundary value problems and their corresponding finite element models, as it pertains to solvability. In particular, the difficulties associated with the Neumann problem for both one- and two-dimensional boundary value problems are discussed and demonstrated in terms of both the boundary value problems and their corresponding finite element models.

A solutions manual containing solutions to most of the end-of-chapter exercises is available. Also included in the solutions manual is a diskette containing executable files, and the corresponding FORTRAN source code, which solve the finite element models for many of the classes of problems considered in the text. A utility for obtaining hardcopies of the source files is included. In this way, flexibility is provided for those instructors who desire to provide the students with part or all of the codes, as well as for the instructor who wishes to have the students do all the programming.

For a one-semester senior-level course, the following coverage has been found workable.

Reading	*Sections*	*Programming*
Chapter 1	All	First computer assignment
Chapter 2	2.1–2.4.1 and 2.5–2.10	Second computer assignment
Chapter 3	3.1–3.5, 3.8.1, and 3.8.2 or 3.8.3	First or second computer assignment
Chapter 4	4.1 and 4.2 or 4.3	None

For a one-semester beginning graduate-level course, the following is suggested. Chapter 1 should be assigned as self study and review.

Reading	*Sections*	*Programming*
Chapter 2	All	Third or fourth computer assignment
Chapter 3	All	Fourth or fifth computer assignment
Chapter 4	All	First or second computer assignment
Chapter 5	All	Computer assignment of choice

KEY FEATURES

- Reiteration of the seven basic ingredients throughout the text.
- Numerous end-of-chapter exercises for reinforcing material in the text.
- Consistent use of the easy-to-follow matrix notation.
- Solutions manual containing worked-out solutions and numerical results to most of the end-of-chapter exercises.
- Maximum flexibility in handling computer assignments provided by the inclusion in the solutions manual of a diskette containing executable and FORTRAN source files for codes that solve many of the classes of problems discussed in the text.

ACKNOWLEDGMENTS

I am indebted to many individuals who have contributed to the development of this textbook. I would like to thank the reviewers whose constructive criticism and suggestions were very helpful. Whenever possible, I tried to accommodate reviewer suggestions, and I hope that the final product will meet the needs of instructors teaching the course.

Reviewers for the text were Nicholas Alterio, Michigan State University; Robert Archer, University of Massachusetts; Eduardo Bayo, University of California–Santa Barbara; George E. Blandford, University of Kentucky; Linda Hayes, University of Texas at Austin; Edward Hensel, New Mexico State University; John Jackson, Clemson University; Timothy Kennedy, Oregon State University; Subramanian Rajan, Arizona State University; Robert Rankin, Arizona State University; Thomas R. Rogge, Iowa State University; Thomas Rudolphi, Iowa State University; Stephen Swanson, University of Utah; I-Chih Wang, University of Cincinnati; and Kenneth M. Will, Georgia Institute of Technology.

My special thanks to Bob Rankin. His assistance with many aspects of "the infernal machine" over the years has been most valuable and appreciated.

William B. Bickford

To the Student

During the course of the four or five years spent in pursuit of an undergraduate engineering education, a student studies many different areas: solid mechanics, fluid mechanics, heat transfer, vibrations, and perhaps many others. Given that the basic principles governing the physics of a particular area are understood, the goal of these studies is to develop a mathematical model based on realistic assumptions regarding the physical model. The mathematical model is then analyzed with the idea that the results can potentially be used to predict the behavior of the physical model.

Specifically, the steps in carrying out this process are:

1. Identification of the basic physical principles that are assumed to govern the behavior of a system.
2. Application of the physical principle in developing the governing equation(s) or mathematical model on which the analysis or design is predicated.
3. Selection of the appropriate tool for the analysis of the governing equations or mathematical model.
4. Solution of the governing equations.
5. Interpretation of the results of the analysis with subsequent reanalysis if, for instance, the design criteria are not met.

Frequently, the resulting mathematical model consists of differential equations together with appropriate auxiliary conditions in the form of initial and/or boundary conditions. Depending on the complexity of the differential equations and auxiliary conditions, it may be possible to attack the problem successfully with one of several classical analytical tools such as those studied in elementary courses in ordinary and partial differential equations. In most real-life situations, however, a numerical approach such as the finite element method (FE method) is realistically the only way to obtain the desired information about the physical problem in question.

In one very important respect, the finite element method is an ideal tool for the analysis of physically based problems: A correctly developed finite element model, at either the elemental or system level, has built into it—at least approximately—the basic physics of the original problem (i.e., the solution from the mathematical model should satisfy the basic physics at least approximately). You are strongly encouraged to develop the habit of regularly determining the physical significance of the matrices appearing at the elemental and system levels, and checking that the output from the final system of equations does in fact satisfy the basic physics.

Students often tend to associate a particular tool for analyzing a mathematical model with the corresponding physical problem itself, instead of attempting to identify and make use of the essential structure of the mathematical model. It is hoped that as a result of this introduction to the finite element method, you will be better able to see some of the mathematical structure that exists in many classes of problems of interest, and to benefit from this awareness by being able to get the most out of the analysis of the mathematical model and its connections with the corresponding physical model.

A word of caution! Be sure you understand the basic physics of the problem being analyzed. The connections between the physical and mathematical models must be understood well enough so that the results obtained from an analysis of the mathematical model using the finite element method can be checked in the appropriate sense. That is, is equilibrium (of the physical model) really satisfied? Or is energy (of the physical model) really balanced?

Many of the large commercially available codes you may be asked to use are capable of analyzing physical problems whose mathematical models are quite complex. In situations such as these, it is very tempting to try one's hand at "analysis by input" when there can be some question as to whether the user can recognize incorrect output when it occurs. Before such a code is used in an application, it is highly advisable that a set of appropriate test problems—ones that possess known solutions for the class of problems in question and that are understood by the user—be input to the code for the purpose of debugging.

W. B. B.

Contents

Chapter 1 Introduction and Discrete Problems **1**

1.1 Introduction, 2
1.2 Typical applications, 2
 1.2.1 Solid mechanics—structures
 1.2.2 Fluid mechanics
 1.2.3 Heat transfer
 1.2.4 Groundwater seepage
1.3 Basic ingredients—discrete problems, 4
1.4 Examples of discrete problems, 10
 1.4.1 Equilibrium of a spring-mass system—vectorial approach
 1.4.2 Equilibrium of a spring-mass system—variational approach
 1.4.3 Steady-state heat transfer
 1.4.4 Piping networks
 1.4.5 Plane truss structures
1.5 Closure, 34
References, 36
Computer projects, 37
Exercises, 40

Chapter 2 One-Dimensional Boundary Value Problems **49**

2.1 Introduction, 50
2.2 The general problem, 57
2.3 Galerkin finite element models, 61
2.4 Example applications, 75
 2.4.1 Solid mechanics—axial deformation
 2.4.2 Fluid mechanics—thin-film lubrication
 2.4.3 One-dimensional heat conduction with convection
2.5 Eigenvalue problems, 91
 2.5.1 Torsional vibrations
2.6 Evaluation of elemental matrices, 97
2.7 Variational finite element models, 114
 2.7.1 The weak formulation
 2.7.2 Calculus of variations
 2.7.3 Ritz finite element models
2.8 Higher-order interpolations, 138

2.9 Quadratic interpolation examples, 146
2.9.1 Solid mechanics—axial deformation
2.9.2 Fluid mechanics—thin-film lubrication
2.9.3 One-dimensional heat conduction with convection
2.9.4 Torsional vibrations
2.10 Transverse deflections of beams, 158
2.11 Errors and convergence, 168
2.12 Closure, 174
References, 176
Computer projects, 178
Exercises, 179

Chapter 3 Two-Dimensional Boundary Value Problems **215**

3.1 Introduction, 216
3.1.1 Problem statement
3.2 The Galerkin finite element model, 220
3.3 Variational finite element models, 237
3.3.1 The weak formulation and variational principles
3.3.2 The Ritz finite element model
3.4 Evaluation of matrices—linearly interpolated triangular elements, 249
3.5 Example applications, 257
3.5.1 Torsion of a rectangular section
3.5.2 Conduction heat transfer with boundary convection
3.6 Rectangular elements, 272
3.6.1 Evaluation of matrices
3.6.2 Torsion of a rectangular section
3.6.3 Conduction heat transfer with boundary convection
3.7 Eigenvalue problems, 286
3.7.1 Introduction
3.7.2 Finite element models for the Helmholtz equation
3.7.3 Examples for the Helmholtz equation
3.8 Isoparametric elements and numerical integration, 295
3.8.1 Four-node quadrilateral (Q4) elements
3.8.1.1 Transformations and shape functions
3.8.1.2 Evaluation of Q4 elemental matrices
3.8.1.3 Application—stress concentration for the torsion of an angle section
3.8.1.4 Application—steady-state heat transfer
3.8.2 Eight-node quadrilateral (Q8) elements
3.8.2.1 Transformations and shape functions
3.8.2.2 Evaluation of Q8 elemental matrices
3.8.2.3 Application—stress concentration for the torsion of an angle section
3.8.2.4 Application—steady-state heat transfer
3.8.3 Six-node triangular (T6) elements
3.8.3.1 Transformations and shape functions
3.8.3.2 Evaluation of T6 elemental matrices

3.8.3.3 Application—stress concentration for the torsion of an angle section
3.8.3.4 Application—steady-state heat transfer
3.9 Axisymmetric problems, 364
3.10 Closure, 374
References, 378
Computer projects, 379
Exercises, 381

Chapter 4 Time-Dependent Problems **407**

4.1 Introduction, 408
4.2 One-dimensional diffusion or parabolic equations, 408
4.2.1 The Galerkin finite element model
4.2.2 Example of one-dimensional diffusion
4.2.3 Analytical integration techniques
4.2.4 Time domain integration techniques—first-order systems
4.2.4.1 The Euler method
4.2.4.2 Improved Euler or Crank-Nicolson method
4.2.4.3 Analysis of algorithms
4.3 One-dimensional wave or hyperbolic equations, 434
4.3.1 The Galerkin finite element model
4.3.2 One-dimensional wave example
4.3.3 Analytical integration techniques
4.3.4 Time domain integration techniques—second-order systems
4.3.4.1 Central difference method
4.3.4.2 Newmark's method
4.3.4.3 Analysis of algorithms
4.4 Two-dimensional diffusion, 464
4.5 Two-dimensional wave equations, 467
4.6 Closure, 471
References, 471
Computer projects, 472
Exercises, 475

Chapter 5 Elasticity **483**

5.1 Introduction, 484
5.1.1 Kinetics
5.1.2 Kinematics
5.1.3 Constitution
5.1.4 Combination—boundary value problems
5.1.4.1 Plane stress
5.1.4.2 Plane strain
5.2 Weak formulation and variational principles, 491
5.3 The Ritz finite element model, 495
5.4 Evaluation of elemental matrices, 503
5.5 Applications, 510
5.5.1 Uniform tension of a thin rectangular plate
5.5.2 Thin rectangular plate—moment loading

5.5.3 Thin rectangular plate—moment loading, meshes 2 and 3
5.5.4 Stress concentration—plate with circular hole
5.6 Rectangular elements, 535
5.6.1 Thin rectangular plate—moment loading
5.6.2 Thin rectangular plate—moment loading, meshes 2 and 3
5.7 Axisymmetric problems, 550
5.7.1 The Ritz finite element models—triangular elements
5.7.2 Evaluation of elemental matrices
5.7.2.1 Thick-walled pressure vessel
5.7.2.2 Notched shaft under tension
5.8 Isoparametric formulations, 572
5.8.1 Applications
5.8.1.1 Thin rectangular plate—moment loading
5.8.1.2 Stress concentration—plate with circular hole
5.9 Closure, 589
References, 591
Computer projects, 592
Exercises, 594

Appendix A Approximate Methods of Analysis **601**

Appendix B Linear Algebraic Equations **611**

Appendix C Eigenvalue Problems **623**

Appendix D Solvability of Finite Element Equations **644**

Index **649**

CHAPTER 1

Introduction and Discrete Problems

Chapter Contents

1.1 Introduction
1.2 Typical applications
  1.2.1 Solid mechanics—structures
  1.2.2 Fluid mechanics
  1.2.3 Heat transfer
  1.2.4 Groundwater seepage
1.3 Basic ingredients—discrete problems
1.4 Examples of discrete problems
  1.4.1 Equilibrium of a spring-mass system—vectorial approach
  1.4.2 Equilibrium of a spring-mass system—variational approach
  1.4.3 Steady-state heat transfer
  1.4.4 Piping networks
  1.4.5 Plane truss structures
1.5 Closure
References
Computer projects
Exercises

1.1 INTRODUCTION

Two of the main goals of engineering analysis are to be able to identify the basic physical principle(s) that govern the behavior of a system and to translate those principles into a mathematical model involving an equation or equations that can be solved to accurately predict the qualitative and quantitative behavior of the system. The resulting mathematical model is frequently a single differential equation or a set of differential equations whose solution should be consistent with and accurately represent the basic physics of the system.

In situations where the system is relatively simple, it may be possible to analyze the problem by using some of the classical methods learned in elementary courses in ordinary and partial differential equations. Far more frequently, however, the governing differential equation(s) or the regions in which the solution is sought are such that it is necessary to use some sort of approximate or numerical method for extracting the desired information regarding the behavior of the system. Until relatively recently the primary approximate method used by engineers, physicists, and applied mathematicians for analyzing complex problems involving ordinary and partial differential equations was the finite difference method in one of its many forms.

Over roughly the last quarter of a century, the finite element method (FEM) has become a viable alternative that in many applications has certain advantages over the finite difference methods. The FEM has evolved as a generalization of ideas in structural analysis. Among the first efforts along these lines were the presentations of Turner, Clough, Martin, and Topp [1] and of Argyris and Kelsey [2]. Clough [3] was apparently responsible for coining the phrase "finite element method." Prior to that, some of the essential ideas of the FEM could be found in the works of Hrenikoff [4], Courant [5], McHenry [6], and Levy [7]. Today, the FEM is regularly applied as a successful analysis tool in virtually every area of modern engineering.

1.2 TYPICAL APPLICATIONS

In this section a few typical examples of the wide range of applications of the finite element method are indicated. The areas chosen are meant to be representative, but by no means exhaustive, of the wide range of application of the finite element method.

1.2.1 Solid Mechanics—Structures

The aerospace industry regularly uses the finite element method to determine the static and dynamic response of aircraft and spacecraft to the wide range of environments which they encounter. The Columbia spacecraft and an indication of the corresponding finite element model are shown in Fig. 1–1. The subregions shown are often chosen so as to correspond to components of the spacecraft. These subregions are then usually further subdivided into suitable finite elements for the analysis.

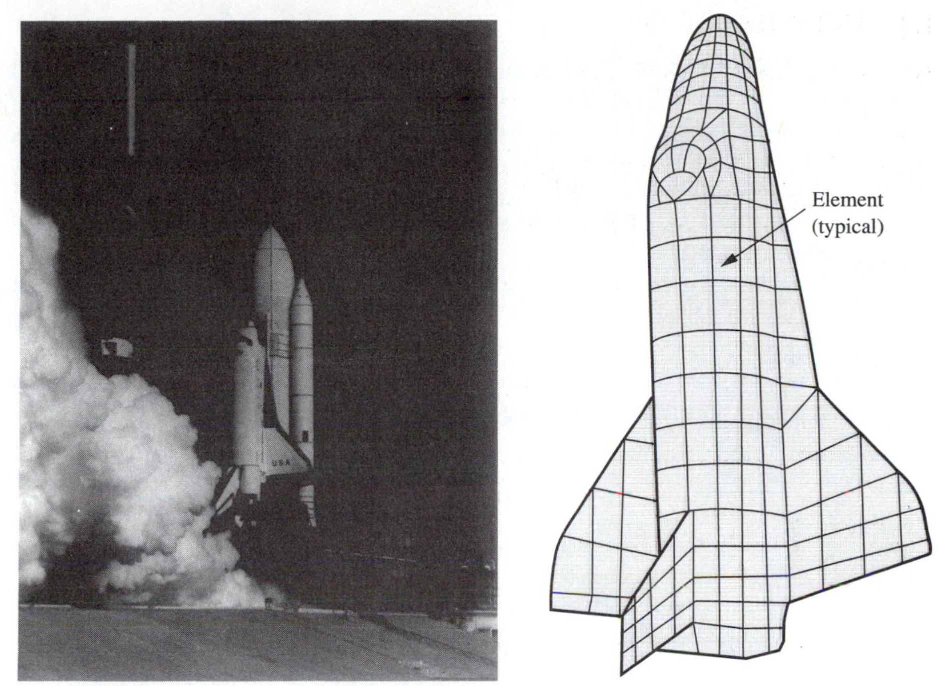

FIGURE 1–1 The Columbia and its corresponding finite element model

1.2.2 Fluid Mechanics

An example of an application of the finite element method to fluid mechanics is the problem of the flow of air past an airfoil, such as shown in the cross section of an aircraft wing in Fig. 1–2. Of interest are the lift and drag forces on the wing arising from the flow. The flow region near the wing is subdivided into elements

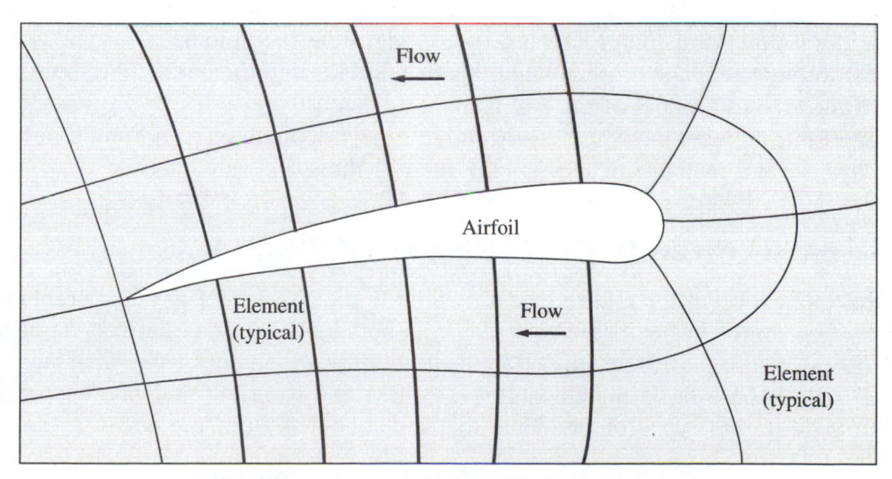

FIGURE 1–2 Cross section of aircraft wing

as indicated. The solution of the finite element model allows for the computation of the desired lift and drag forces.

1.2.3 Heat Transfer

An example of the application of the finite element method to the general problem of heat transfer is that of a gas turbine engine. Temperatures reach very high levels in the engine and certain components in the interior must be cooled in order to survive. The blades in a typical rotating part generally contain cavities so as to be able to pass cooler air through the interior of the blade for cooling purposes. A typical finite element model of a cross section of a blade is shown in Fig. 1–3. The model is used to determine the proper number, size and location of the holes that are necessary to properly cool the blade.

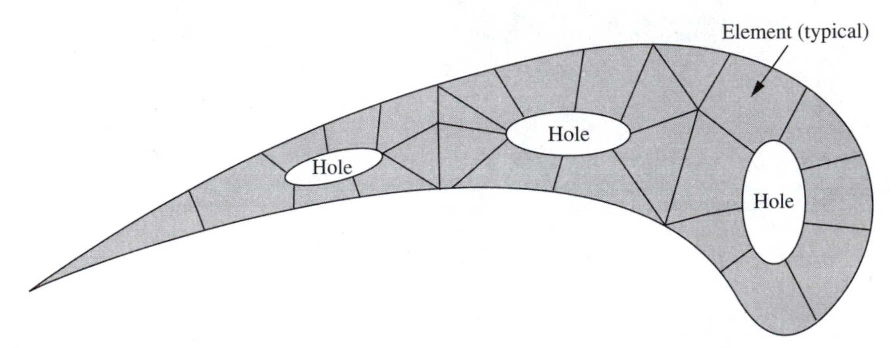

FIGURE 1–3 Finite element model of engine blade

1.2.4 Groundwater Seepage

An important application in the area of geomechanics is the problem of groundwater seepage. A typical situation is indicated in Fig. 1–4, where water is impounded behind an impervious dam. The task is to determine the amount of water which is lost through seepage or percolation under the dam into the earth. The mesh for the finite element model of the soil is also indicated.

The finite element method is easily and routinely applied to an enormous variety of physical and mathematical problems such as those indicated above.

1.3 BASIC INGREDIENTS—DISCRETE PROBLEMS

In the application of the finite element method to the analysis of physical problems there are certain basic ingredients that can be identified and routinely applied regardless of the particular area of application. It is the intent of this first section to introduce these basic ingredients of the FEM and to briefly indicate how they relate to different types of problems that are modeled from the outset as being discrete.

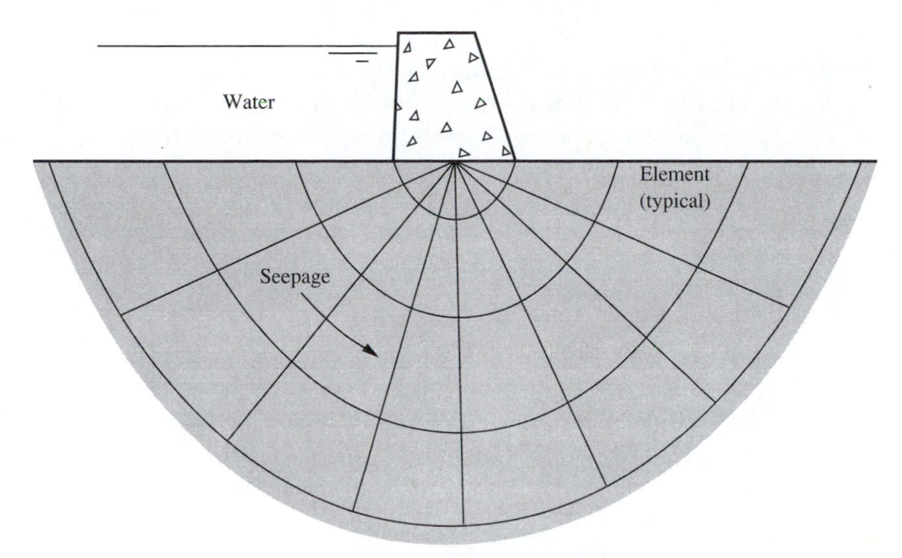

FIGURE 1–4 Water behind a dam

Any application of the finite element method to a mathematical or physical problem involves, either implicitly or explicitly, the use of several distinct steps or building blocks. These can be listed in order as:

1. Discretization
2. Interpolation
3. Elemental description or formulation
4. Assembly
5. Constraints
6. Solution
7. Computation of derived variables

These will be discussed in detail in subsequent chapters. Whether the finite element method is being used for (1) setting up and solving a problem by hand, (2) using an existing code to solve a problem, or (3) generating or writing a code to solve a class of problems, these basic ingredients can be of value in establishing and following a rational procedure in setting up and solving the finite element model. This text is set up with this in mind, so that it will be natural for the student to proceed along these lines in his or her subsequent use of the finite element method.

Discrete models are approximations of corresponding continuous models. They are generally designed to be simpler as far as the mathematics is concerned. The process of converting a continuous model into a discrete model is referred to as **discretization**. As a result of this process, a mathematical model involving an ordinary differential equation is generally converted into a mathematical model involving algebraic equations. A mathematical model involving a partial differential equation is generally converted into a mathematical model involving ordinary differential equations or algebraic equations.

The process of discretization is the basis for many of the developments the student has seen in elementary courses in solids, fluids, heat transfer, and vibrations where a lumping of one sort or another is performed. In many of these situations it is possible to formulate and solve the problem without any reference whatsoever to the finite element method. In this section, we will investigate just what transpires in the process of the nonfinite element conversion of a continuous system into a discrete system, in hopes of seeing how steps of the process fit in with the basic ingredients of the finite element method. Specifically, we will consider the conversion of a continuous model of a one-dimensional solid mechanics problem into a corresponding discrete model with the intention of identifying several of the basic ingredients previously enumerated.

When the finite element method is applied to discrete systems, the discretization step has already taken place. The steps of **interpolation** and **elemental description** are generally implicit in the discrete model. It is the purpose of this section to carry through the details of converting the continuous model of a typical physical problem into the corresponding discrete model and to illustrate that the steps of interpolation and elemental description are in fact present in the discrete model.

Consider then the problem of the axial deformation of a linearly elastic bar as indicated in Fig. 1–5, where $u(x)$ is the axial displacement, P is the external resultant axial force applied at $x = L$, and $q(x)$ represents any distributed external loadings. $A(x)$ is the area, E is Young's modulus, and $\rho(x)$ is the mass density.

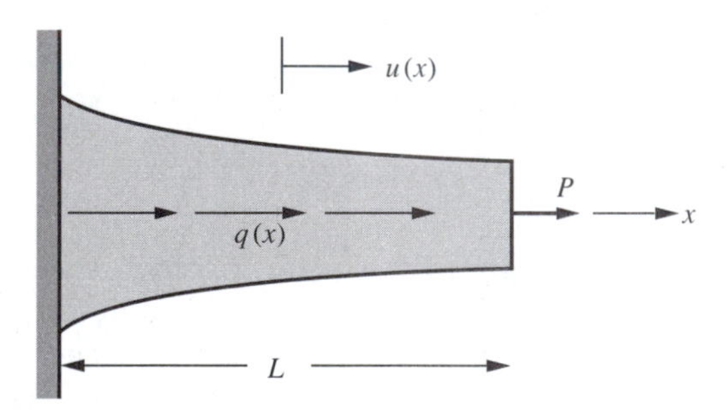

FIGURE 1–5 Axial deformation of a linearly elastic bar

This is an example of a continuous problem in that the dependent variable $u(x)$ and the geometrical and material constants A, E, and ρ are considered to be continuously varying functions on the interval $0 \leq x \leq L$. Although the exact analytical solution to the static case could be obtained in this case, it is instructive to investigate the process of converting the continuous problem into a corresponding discrete problem. **Discrete** means essentially that we are willing to accept a model that will yield information about the dependent variable(s) at a finite number of points, referred to as **nodes**, within the interval $0 \leq x \leq L$. Fig. 1–6 shows the results of this discretization process as applied to the axial deformation problem where the nodes, numbered sequentially as shown, are chosen at the ends of the bar and at three equally spaced interior points.

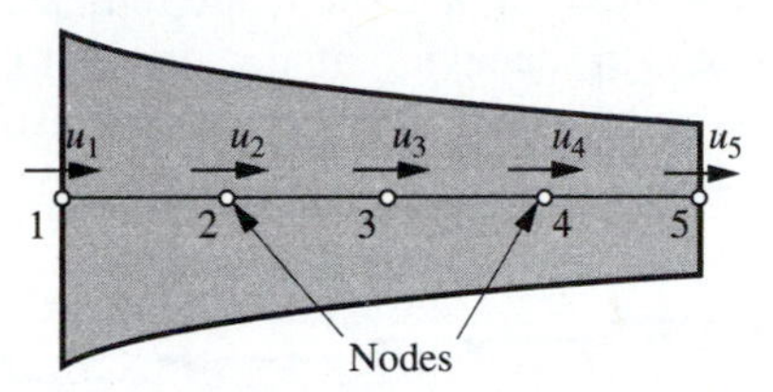

FIGURE 1–6 Discretization of region $0 \leq x \leq L$

Each node is assigned a displacement u_i, $i = 1$ through 5, as indicated. This discretization begins the process of converting the continuous problem, with an infinite number of degrees of freedom (the values of $u(x)$ at all the points $0 \leq x \leq L$), into a discrete problem with a finite number of degrees of freedom—the five nodal displacements in this case.

The next step in developing a discrete model is to replace the discrete portions of the bar (i.e., the portions between the nodes) by linear springs, thus producing the model shown in Fig. 1–7. These discrete portions between the nodes are called **elements**.

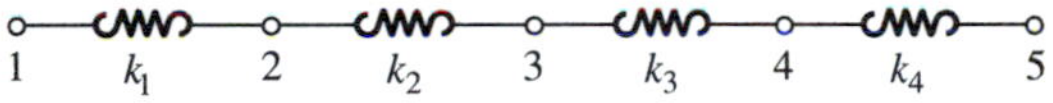

1 k_1 2 k_2 3 k_3 4 k_4 5

FIGURE 1–7 Replacement springs for the axial deformation problem

The basis for this step is the knowledge that an elastic bar of length l under the action of an axial load P elongates according to

$$e = \frac{Pl}{A_{\text{avg}}E} \tag{1.1}$$

where A_{avg} is a suitably defined average area. Verification of this result is left to the Exercises. Eq. (1.1) can be rewritten as

$$P = \left(\frac{A_{\text{avg}}E}{l}\right)e = ke$$

where $k = A_{\text{avg}}E/l$ is the stiffness of the bar. Thus an elastic bar of length l is seen to be equivalent to a simple linear spring of stiffness $k = A_{\text{avg}}E/l$. It is clear that each of the springs indicated in Fig. 1–7 will have a different stiffness due to a different A_{avg}. We will approximate each of these stiffnesses by taking

$$A_{\text{avg}} = \frac{A_i + A_{i+1}}{2}$$

with the resulting springs of stiffnesses

$$k_i = \frac{(A_i + A_{i+1})E}{2l_i}$$

where $A_i = A(x_i)$, $A_{i+1} = A(x_{i+1})$, and the l_i are indicated in Fig. 1–8. Thus, the continuously distributed elastic behavior of the original continuous problem has been replaced by a problem with piecewise constant elastic behavior as indicated in Fig. 1–8.

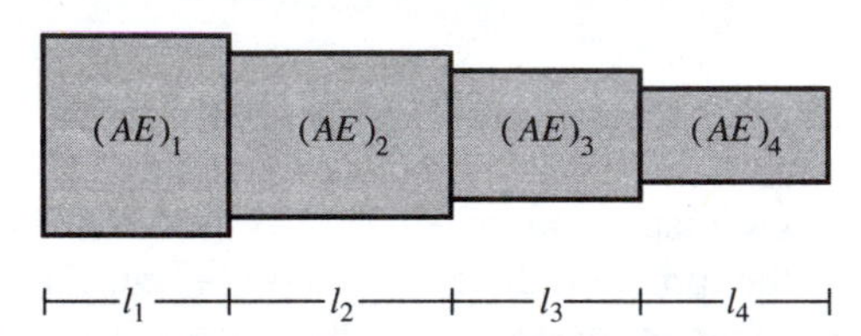

FIGURE 1–8 Equivalent springs for the discrete problem

The basic physical properties of each of the elements are contained in the expression

$$F_i = k_i e_i = k_i(u_{i+1} - u_i)$$

where F_i is the force transmitted by the element and $e_i = u_{i+1} - u_i$ is the elongation. This is essentially the **elemental description**, and will be further discussed in the next two sections.

Also part of discretization is the process of lumping of the masses, that is, the process of assigning masses to the nodes. There are several options for handling the distributed mass. In line with the choice of the elements as outlined, we proceed as follows. Consider the portion of the bar between the ith and $(i + 1)$st nodes, that is, the ith element, as indicated in Fig. 1–9.

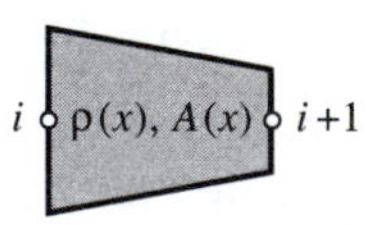

FIGURE 1–9 Distribution of masses

One method of lumping the distributed mass is simply to take an average intensity $m* = [\rho(x_i)A(x_i) + \rho(x_{i+1})A(x_{i+1})]/2 = (m_i + m_{i+1})/2$, multiply by the length l_i to get an approximate total mass associated with the element, and then distribute the total equally between the ith and $(i + 1)$st nodes. Performing these steps for each element and adding yields the replacement or lumped masses shown in Fig. 1–10.

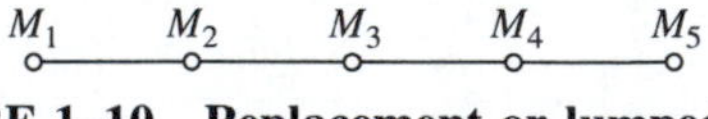

M_1 M_2 M_3 M_4 M_5

FIGURE 1–10 Replacement or lumped masses

In Fig. 1–10,

$$M_1 = (m_1 + m_2)\frac{l_1}{4}$$

$$M_2 = (m_1 + m_2)\frac{l_1}{4} + (m_2 + m_3)\frac{l_2}{4}$$

$$M_3 = (m_2 + m_3)\frac{l_2}{4} + (m_3 + m_4)\frac{l_3}{4}$$

$$M_4 = (m_3 + m_4)\frac{l_3}{4} + (m_4 + m_5)\frac{l_4}{4}$$

$$M_5 = (m_4 + m_5)\frac{l_4}{4}$$

This method of lumping is generally satisfactory when the variables ρ and A do not change rapidly or significantly on the interval $0 \leq x \leq L$. Note that the sum of the masses should satisfy at least approximately

$$\sum M_i = \int_0^L \rho(x)A(x)\,dx$$

The last step in the process of discretization is the conversion of the continuously distributed loading $q(x)$ and the concentrated load P into nodal loads. The load P should clearly be applied at node 5. There are several options for handling the distributed loads. In line with the choice of the springs and masses as previously outlined, consider the loading of a typical element located between nodes i and $(i + 1)$ as indicated in Fig. 1–11.

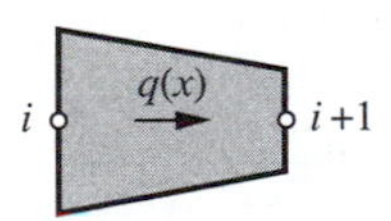

FIGURE 1–11 Loading of a typical element

Consistent with the approach used for lumping the masses, take an average intensity $q* = (q(x_i)+q(x_{i+1}))/2 = (q_i+q_{i+1})/2$, multiply by the length l_i to get an approximate total load associated with the element, and then distribute the total equally between the ith and $(i + 1)$st nodes. Performing these steps for each element and adding the load P at node 5 yields the replacement loads shown in Fig. 1–12.

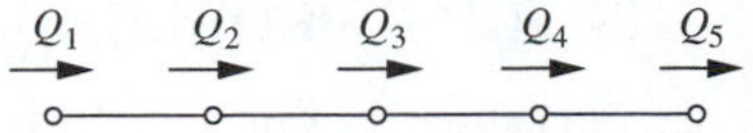

Q_1 Q_2 Q_3 Q_4 Q_5

FIGURE 1–12 Replacement or lumped nodal loads

In Fig. 1–12,

$$Q_1 = (q_1 + q_2)\frac{l_1}{4}$$

$$Q_2 = (q_1 + q_2)\frac{l_1}{4} + (q_2 + q_3)\frac{l_2}{4}$$

$$Q_3 = (q_2 + q_3)\frac{l_2}{4} + (q_3 + q_4)\frac{l_3}{4}$$

$$Q_4 = (q_3 + q_4)\frac{l_3}{4} + (q_4 + q_5)\frac{l_4}{4}$$

$$Q_5 = (q_4 + q_5)\frac{l_4}{4} + P$$

The sum of the loads should satisfy at least approximately

$$\sum Q_i = \int_0^L q(x)\ dx + P$$

The final discrete model appears as in Fig. 1–13, where all the springs, masses, and loads are indicated.

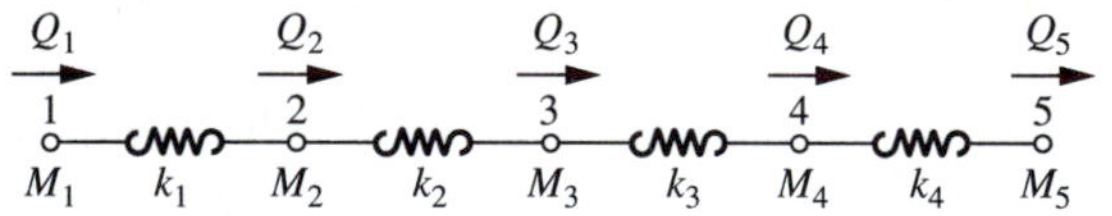

FIGURE 1–13 Discrete springs, masses, and loads

This completes the process of converting the continuous system into what is hoped will be an equivalent discrete system. The basic ingredient of **discretization** is quite apparent. For this example, **interpolation** is implicit in the step of replacing the continuously distributed elastic properties by the linear spring. Demonstration of this fact is straightforward and is left to the Exercises. The **elemental description** is essentially the relationship $P = ke$.

Whether the axial deformation problem is modeled as continuous or discrete, the basic physical principle that must be satisfied is Newton's second law. For a static problem the equations of equilibrium must be developed and solved, whereas for a dynamic situation, equations of motion are required. The remaining steps of **assembly**, **constraints**, **solution**, and **computation of derived variables** can best be demonstrated after deciding on what method to use for generating the equations of equilibrium, and will be discussed in the next two sections.

1.4 EXAMPLES OF DISCRETE PROBLEMS

The purpose of the examples presented in this section is to introduce the student to some of the basic ingredients of the FEM as they relate to problems considered from the outset to be discrete. The student will likely have encountered problems of this sort in courses in solid mechanics, fluid mechanics, heat transfer, and electric circuits as well as in other disciplines. In such a system, the decision has already been made to approximate the original physical problem by subdividing it, in some sense, into a number of pieces that, when recombined and analyzed as a system, give accurate estimates as to the behavior of the original continuous system. The pieces, or elements as they are often called, should contain the basic physics of the continuous problem being replaced. The emphasis in this section will be on the manner in which the basic physical principle being enforced (balance of momentum, balance of energy, balance of mass as examples) is a part of the elemental and system equations, both of which reflect, at least approximately, the basic physics of the problem.

1.4.1 Equilibrium of a Spring-Mass System—Vectorial Approach

Consider a typical spring-mass system as indicated in Fig. 1–14. Such a system could represent the result of the discretization of a continuous model of a physical problem as discussed in the previous section. As such the discretization is implicit.

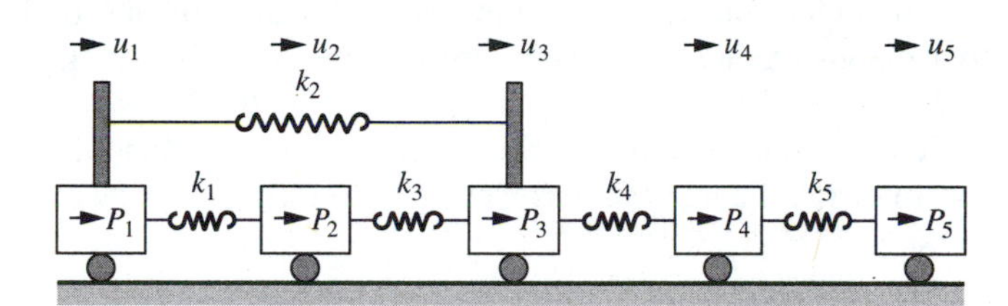

FIGURE 1–14 Discrete spring-mass system

Each of the springs k_i is assumed to behave in a linear fashion ($F = kx$) and the loads P_i to be applied slowly enough to the masses so that the problem is essentially static in nature. As discussed in the previous section, both the interpolation and elemental formulation are contained within the basic assumption as to the linear character of the spring element. The nodal displacements u_i and u_{i+1} along with the corresponding internally transmitted forces f_i and f_{i+1} are shown for a typical spring or element in Fig. 1–15.

u_i → u_{i+1} →

f_i → k_i → f_{i+1}

FIGURE 1–15 Elemental displacements and forces

These four variables are related according to

$$f_i = k_i(u_i - u_{i+1})$$
$$f_{i+1} = k_i(u_{i+1} - u_i)$$

where k_i is the spring constant. The basic physical principle of equilibrium is clearly satisfied for the individual spring or element by virtue of the fact that $f_i + f_{i+1} = 0$. Expressed as a matrix equation, the force-displacement relations are

$$\begin{bmatrix} f_i \\ f_{i+1} \end{bmatrix} = \begin{bmatrix} k_i & -k_i \\ -k_i & k_i \end{bmatrix} \begin{bmatrix} u_i \\ u_{i+1} \end{bmatrix}$$

or

$$\mathbf{f_e} = \mathbf{k_e u_e} \tag{1.2}$$

where $\mathbf{k_e}$ is referred to as the **elemental stiffness matrix**, with $\mathbf{f_e}$ and $\mathbf{u_e}$ the **elemental force** and **elemental displacement vectors** respectively. Eq. (1.2) is the statement at the *elemental level* that $F = kx$. As will be seen in what follows, the individual $\mathbf{k_e}$'s can be considered as being **assembled** or **loaded** to form the **global stiffness matrix**, which represents the physics of the entire system.

Referring again to Fig. 1–14, the external loads P_i are taken to be the inputs with the displacements u_i as the outputs or unknowns. The basic physical idea in developing the system or global equations is that each of the masses must be in equilibrium under the action of the loads P_i and the internal forces in the springs. All displacements and loads are assumed positive in the positive x-direction as indicated. A free-body diagram of each of the masses is drawn and the appropriate equilibrium equation written for each. In particular consider the free-body diagram of mass 2 and the corresponding equilibrium equation shown in Fig. 1–16.

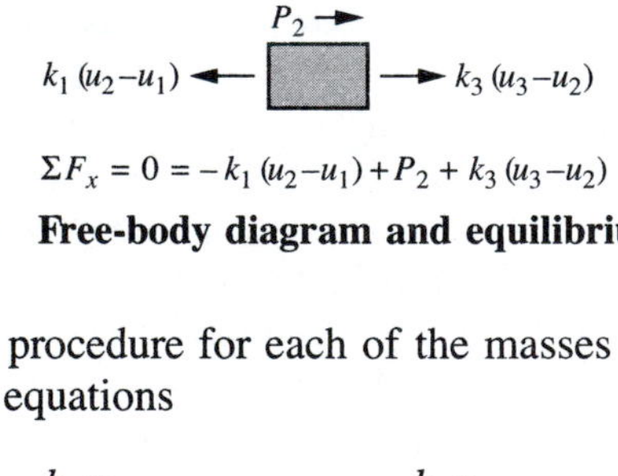

FIGURE 1–16 Free-body diagram and equilibrium for mass m_2

Following the same procedure for each of the masses yields the set of equilibrium system or global equations

$$\begin{aligned}
(k_1 + k_2)u_1 \quad - k_1u_2 \quad - k_2u_3 \quad &= P_1 \\
-k_1u_1 + (k_1 + k_3)u_2 \quad - k_3u_3 \quad &= P_2 \\
-k_2u_1 \quad - k_3u_2 + (k_2 + k_3 + k_4)u_3 \quad - k_4u_4 &= P_3 \\
- k_4u_3 + (k_4 + k_5)u_4 - k_5u_5 &= P_4 \\
- k_5u_4 + k_5u_5 &= P_5
\end{aligned}$$

These can be written in matrix notation as

$$\mathbf{K_G u_G} = \mathbf{P_G} \tag{1.3}$$

where

$$\mathbf{K_G} = \begin{bmatrix} k_1 + k_2 & -k_1 & -k_2 & 0 & 0 \\ -k_1 & k_1 + k_3 & -k_3 & 0 & 0 \\ -k_2 & -k_3 & k_2 + k_3 + k_4 & -k_4 & 0 \\ 0 & 0 & -k_4 & k_4 + k_5 & -k_5 \\ 0 & 0 & 0 & -k_5 & k_5 \end{bmatrix}$$

$$\mathbf{u}_\mathbf{G}^\mathrm{T} = [\,u_1 \quad u_2 \quad u_3 \quad u_4 \quad u_5\,]$$

$$\mathbf{P}_\mathbf{G}^\mathrm{T} = [\,P_1 \quad P_2 \quad P_3 \quad P_4 \quad P_5\,]$$

$\mathbf{K_G}$ is referred to as the **global stiffness matrix**, $\mathbf{u_G}$ as the **global displacement vector**, and $\mathbf{P_G}$ as the **global load vector**. Eq. (1.3) is the statement at the global level that $F = kx$.

A careful inspection of the global equilibrium equations reveals that each of the elemental stiffness matrices $\mathbf{k_{ei}}$ is present with its location in the global stiffness matrix determined by the node numbers assigned to the nodes at the ends of the springs. To see this we rewrite $\mathbf{K_G}$ as

$$\mathbf{K_G} = \mathbf{k_{G1}} + \mathbf{k_{G2}} + \mathbf{k_{G3}} + \mathbf{k_{G4}} + \mathbf{k_{G5}}$$

where

$$\mathbf{k_{G1}} = \begin{bmatrix} k_1 & -k_1 & 0 & 0 & 0 \\ -k_1 & k_1 & 0 & 0 & 0 \\ 0 & 0 & 0 & 0 & 0 \\ 0 & 0 & 0 & 0 & 0 \\ 0 & 0 & 0 & 0 & 0 \end{bmatrix} \qquad \mathbf{k_{G2}} = \begin{bmatrix} k_2 & 0 & -k_2 & 0 & 0 \\ 0 & 0 & 0 & 0 & 0 \\ -k_2 & 0 & k_2 & 0 & 0 \\ 0 & 0 & 0 & 0 & 0 \\ 0 & 0 & 0 & 0 & 0 \end{bmatrix}$$

$$\mathbf{k_{G3}} = \begin{bmatrix} 0 & 0 & 0 & 0 & 0 \\ 0 & k_3 & -k_3 & 0 & 0 \\ 0 & -k_3 & k_3 & 0 & 0 \\ 0 & 0 & 0 & 0 & 0 \\ 0 & 0 & 0 & 0 & 0 \end{bmatrix} \qquad \mathbf{k_{G4}} = \begin{bmatrix} 0 & 0 & 0 & 0 & 0 \\ 0 & 0 & 0 & 0 & 0 \\ 0 & 0 & k_4 & -k_4 & 0 \\ 0 & 0 & -k_4 & k_4 & 0 \\ 0 & 0 & 0 & 0 & 0 \end{bmatrix}$$

$$\mathbf{k_{G5}} = \begin{bmatrix} 0 & 0 & 0 & 0 & 0 \\ 0 & 0 & 0 & 0 & 0 \\ 0 & 0 & 0 & 0 & 0 \\ 0 & 0 & 0 & k_5 & -k_5 \\ 0 & 0 & 0 & -k_5 & k_5 \end{bmatrix}$$

where each of the $\mathbf{k_{Gi}}$ represents the corresponding $\mathbf{k_{ei}}$ expanded to match the size of the global array $\mathbf{K_G}$. Throughout the text we will write

$$\mathbf{K_G} = \sum_e \mathbf{k_{Ge}} = \sum_e \mathbf{k_G}$$

to indicate the process of assembly. The small **k** denotes an elemental matrix and the capital **G** subscript indicates the appropriate expansion to global size. The small e indicates that the sum is to be taken over the elements. Implicit in the assembly step is what will be referred to generally throughout the text as interelement continuity, where, the displacement at the right end of the ith element is identical to the displacement at the left end of the $(i + 1)$st element.

For a specific example, assume that mass 1 is fixed against displacement, that is, $u_1 = 0$, and that a displacement U_0 is imposed on the mass 5. These constraints are accomplished very simply by replacing the u_1 equation by $u_1 = 0$ and the u_5 equation by $u_5 = U_0$, after which the equilibrium equations appear (in augmented form) as

$$\left[\begin{array}{ccccc|c} 1 & 0 & 0 & 0 & 0 & 0 \\ -k_1 & k_1 + k_3 & -k_3 & 0 & 0 & P_2 \\ -k_2 & -k_3 & k_2 + k_3 + k_4 & -k_4 & 0 & P_3 \\ 0 & 0 & -k_4 & k_4 + k_5 & -k_5 & P_4 \\ 0 & 0 & 0 & 0 & 1 & U_0 \end{array}\right]$$

For many systems the unconstrained set of equations is symmetric, a property which has computational advantages. Thus before beginning the solution procedure, it is advantageous to perform the elementary row operations necessary to convert the set of equations to symmetric form. For this case, these elementary row operations result in

$$\left[\begin{array}{ccccc|c} 1 & 0 & 0 & 0 & 0 & 0 \\ 0 & k_1 + k_3 & -k_3 & 0 & 0 & P_2 \\ 0 & -k_3 & k_2 + k_3 + k_4 & -k_4 & 0 & P_3 \\ 0 & 0 & -k_4 & k_4 + k_5 & 0 & P_4 + k_5 U_0 \\ 0 & 0 & 0 & 0 & 1 & U_0 \end{array}\right]$$

Note the additional effective force $+k_5 U_0$ introduced at node 4 by the nonzero constraint.

If, rather than imposing a displacement constraint on u_5, a force P_5 had been applied at mass 5, the final symmetric set of equations would have been

$$\left[\begin{array}{ccccc|c} 1 & 0 & 0 & 0 & 0 & 0 \\ 0 & k_1 + k_3 & -k_3 & 0 & 0 & P_2 \\ 0 & -k_3 & k_2 + k_3 + k_4 & -k_4 & 0 & P_3 \\ 0 & 0 & -k_4 & k_4 + k_5 & -k_5 & P_4 \\ 0 & 0 & 0 & -k_5 & k_5 & P_5 \end{array}\right]$$

These two specific situations, regarding the type of condition specified at mass 5, are typical of boundary value problems in general, in that when other external influences are absent at a boundary, either the displacement or the force (the derived variable) can be prescribed. This idea will be discussed in more detail in later chapters after introduction to some of the basic tools of the calculus of variations. In either of the two cases presented, the symmetric set of equations is then sent to the appropriate equation solver, that is, the solution phase.

After having determined the unknown displacements u_i from the solution phase, the internal forces in the springs can be computed by applying Eq. (1.2) to each of the elements:

$$\mathbf{f}_{ei} = \mathbf{k}_{ei}\mathbf{u}_{ei}$$

As a specific example, consider the case where $k_1 = k_3 = 2k, k_2 = k_4 = k_5 = k, P_2 = P_3 = P_4/2 = P$, with $U_0 = 0$. The final constrained symmetric set of equations becomes

$$\left[\begin{array}{ccccc|c} 1 & 0 & 0 & 0 & 0 & 0 \\ 0 & 4 & -2 & 0 & 0 & \alpha \\ 0 & -2 & 4 & -1 & 0 & \alpha \\ 0 & 0 & -1 & 2 & 0 & 2\alpha \\ 0 & 0 & 0 & 0 & 1 & 0 \end{array}\right]$$

where $\alpha = P/k$. The solution is $u_1 = u_5 = 0, u_2 = 3\alpha/4, u_3 = \alpha$, and $u_4 = 3\alpha/2$. Using the elemental equations, the forces in the springs are computed to be (computation of derived variables)

$$f_1 = 2k(u_2 - u_1) = \frac{3P}{2}$$

$$f_2 = k(u_3 - u_1) = P$$

$$f_3 = 2k(u_3 - u_2) = \frac{P}{2}$$

$$f_4 = k(u_4 - u_3) = \frac{P}{2}$$

$$f_5 = k(u_5 - u_4) = \frac{-3P}{2}$$

These forces can easily be seen to satisfy the equilibrium equations developed on the basis of the free-body diagrams drawn previously. Checks of this sort on the basic physics, which can and should be made, are extremely valuable in checking a hand calculation or in checking and debugging a computer program before relying on the output.

For this discrete problem, several of the basic ingredients enumerated previously are not explicitly present. The *discretization* has already taken place on the basis that the problem is initially discrete. The *interpolation* and *elemental formulation* are buried in the basic relationship $F = kx$. *Assembly* in a formal sense is absent but is seen to be implicit in writing the global equations of equilibrium. The steps involving *constraint, solution,* and *computation of derived variables* are executed in a recognizable manner.

1.4.2 Equilibrium of a Spring-Mass System—Variational Approach

An alternate formulation of the equilibrium problem considered in the previous section is in terms of the principle of stationary potential energy. As discussed in Meriam and Kraige [8], the equations of equilibrium for a conservative mechanical system can be obtained by requiring that the total potential energy of the system be stationary. For a system with a finite number of degrees of freedom, the total potential energy can be represented as

$$V = V(u_1, u_2, \ldots, u_N)$$

where $u_1, u_2, \ldots, u_N$ are the degrees of freedom of the system. The stationary value of V is then determined by equating to zero each of the partial derivatives; that is,

$$\frac{\partial V}{\partial u_k} = 0 \qquad k = 1, 2, \ \ldots \ , N$$

The solution(s) u_i to these N equations represent equilibrium configurations of the system. For the systems considered in this text, V is a quadratic function of the N variables u_k, resulting in equilibrium equations that are represented as a symmetric set of linear algebraic equations.

Consider again the system discussed in the previous article, shown in Fig. 1–17. Discretization is again implicit.

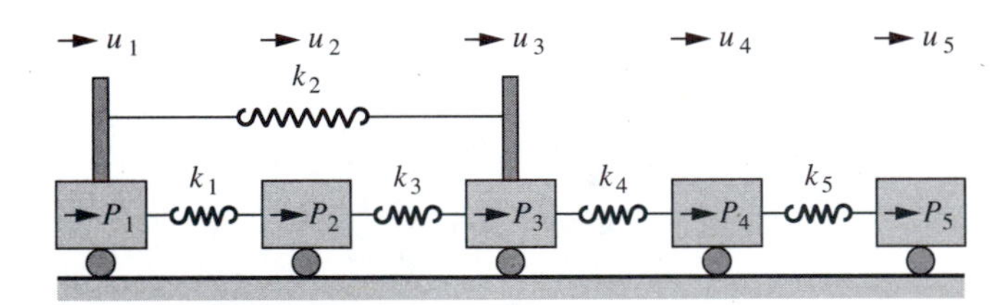

FIGURE 1–17 Discrete spring-mass system

Each of the linear elastic springs has a potential energy of the form

$$U_i = \frac{1}{2} k_i e_i^2$$

where k_i is the spring stiffness and e_i is the elongation of the spring. This energy is obtained by integrating the force-displacement relation $F = kx$ and thus contains the basic physics. The steps of interpolation and elemental formulation are now contained in the equation for the potential energy, $U = ke^2/2$.

Each of the loads P_i has a potential energy Ω_i given by

$$\Omega_i = -P_i u_i$$

The total potential energy V of the system can then be written as

$$V = \sum_1^M U_i + \sum_1^N \Omega_i = \sum_1^M \frac{k_i(e_i)^2}{2} - \sum_1^N P_i u_i$$

where for the example of Fig. 1–17, both M and $N = 5$. The spring elongations are easily seen to be

$$\begin{aligned} e_1 &= u_2 - u_1 \\ e_2 &= u_3 - u_1 \\ e_3 &= u_3 - u_2 \\ e_4 &= u_4 - u_3 \\ e_5 &= u_5 - u_4 \end{aligned}$$

so that the total potential energy becomes

$$\begin{aligned} V = {} & \frac{k_1}{2}(u_2 - u_1)^2 + \frac{k_2}{2}(u_3 - u_1)^2 + \frac{k_3}{2}(u_3 - u_2)^2 + \frac{k_4}{2}(u_4 - u_3)^2 \\ & + \frac{k_5}{2}(u_5 - u_4)^2 - P_1 u_1 - P_2 u_2 - P_3 u_3 - P_4 u_4 - P_5 u_5 \end{aligned} \tag{1.4}$$

The reader should show that using Eq. (1.4) and computing the partial derivatives as outlined above yields exactly the same set of equations as in the previous section.

It is very instructive, however, to view Eq. (1.4) in the following light. Consider the typical term

$$\begin{aligned} \frac{k_1}{2}(u_2 - u_1)^2 &= [u_1 \;\; u_2]\begin{bmatrix} k_1 & -k_1 \\ -k_1 & k_1 \end{bmatrix}\begin{bmatrix} u_1 \\ u_2 \end{bmatrix}\frac{1}{2} \\ &= \mathbf{u}_{\mathbf{e1}}^{\mathrm{T}} \mathbf{k}_{\mathbf{e1}} \frac{\mathbf{u}_{\mathbf{e1}}}{2} \end{aligned}$$

where

$$\mathbf{k}_{\mathbf{e1}} = \begin{bmatrix} k_1 & -k_1 \\ -k_1 & k_1 \end{bmatrix}$$

is referred to as the elemental stiffness matrix for element 1. The potential energies for each of the other springs can be written in an analogous fashion.

It is a simple matter to see that the potential energy for spring 1 can be expressed as

$$\begin{aligned} \frac{k_1}{2}(u_2 - u_1)^2 &= \mathbf{u}_{\mathbf{G}}^{\mathrm{T}} \begin{bmatrix} k_1 & -k_1 & 0 & 0 & 0 \\ -k_1 & k_1 & 0 & 0 & 0 \\ 0 & 0 & 0 & 0 & 0 \\ 0 & 0 & 0 & 0 & 0 \\ 0 & 0 & 0 & 0 & 0 \end{bmatrix} \frac{\mathbf{u}_{\mathbf{G}}}{2} \\ &= \mathbf{u}_{\mathbf{G}}^{\mathrm{T}} \mathbf{k}_{\mathbf{G1}} \frac{\mathbf{u}_{\mathbf{G}}}{2} \end{aligned}$$

where $\mathbf{u}_{\mathbf{G}}^{\mathrm{T}} = [u_1\ u_2\ u_3\ u_4\ u_5]$ and $\mathbf{k}_{\mathbf{G1}}$ is the elemental stiffness matrix for spring or element 1 expanded to the global level. In a similar fashion the potential energy of spring 2 can be written as

$$\frac{k_2}{2}(u_3 - u_1)^2 = \mathbf{u}_{\mathbf{G}}^{\mathrm{T}} \begin{bmatrix} k_2 & 0 & -k_2 & 0 & 0 \\ 0 & 0 & 0 & 0 & 0 \\ -k_2 & 0 & k_2 & 0 & 0 \\ 0 & 0 & 0 & 0 & 0 \\ 0 & 0 & 0 & 0 & 0 \end{bmatrix} \frac{\mathbf{u}_{\mathbf{G}}}{2}$$

$$= \mathbf{u}_{\mathbf{G}}^{\mathrm{T}} \mathbf{k}_{\mathbf{G2}} \frac{\mathbf{u}_{\mathbf{G}}}{2}$$

Each of the spring potential energies can be written in a similar form so that the total potential energy can be expressed as

$$V = \frac{1}{2}\mathbf{u}_{\mathbf{G}}^{\mathrm{T}}(\mathbf{k}_{\mathbf{G1}} + \mathbf{k}_{\mathbf{G2}} + \mathbf{k}_{\mathbf{G3}} + \mathbf{k}_{\mathbf{G4}} + \mathbf{k}_{\mathbf{G5}})\mathbf{u}_{\mathbf{G}} - \mathbf{u}_{\mathbf{G}}^{\mathrm{T}}\mathbf{F}_{\mathbf{G}}$$

$$= \frac{1}{2}\mathbf{u}_{\mathbf{G}}^{\mathrm{T}} \sum \mathbf{k}_{\mathbf{Ge}}\mathbf{u}_{\mathbf{G}} - \mathbf{u}_{\mathbf{G}}^{\mathrm{T}}\mathbf{F}_{\mathbf{G}}$$

$$= \frac{1}{2}\mathbf{u}_{\mathbf{G}}^{\mathrm{T}}\mathbf{K}_{\mathbf{G}}\mathbf{u}_{\mathbf{G}} - \mathbf{u}_{\mathbf{G}}^{\mathrm{T}}\mathbf{F}_{\mathbf{G}}$$

where

$$\mathbf{K}_{\mathbf{G}} = \sum \mathbf{k}_{\mathbf{Ge}} = \begin{bmatrix} k_1 + k_2 & -k_1 & -k_2 & 0 & 0 \\ -k_1 & k_1 + k_3 & -k_3 & 0 & 0 \\ -k_2 & -k_3 & k_2 + k_3 + k_4 & -k_4 & 0 \\ 0 & 0 & -k_4 & k_4 + k_5 & -k_5 \\ 0 & 0 & 0 & -k_5 & k_5 \end{bmatrix}$$

is the global stiffness matrix and where

$$\mathbf{F}_{\mathbf{G}} = [P_1\ P_2\ P_3\ P_4\ P_5]^{\mathrm{T}}$$

is the global load vector. The manner in which the elemental stiffness matrices contribute to the global stiffness matrix in terms of the assembly process is made quite clear using the variational formulation. After generating the equations of equilibrium by computing the partial derivatives

$$\frac{\partial V}{\partial u_k} = 0 \qquad k = 1, 2, \ldots, 5$$

the steps of constraints, solution, and computation of derived variables are carried out exactly as in the previous section.

1.4.3 Steady-State Heat Transfer

A practical problem of frequent interest in heat transfer involves the situation where several different materials are placed in contact as shown in Fig. 1–18.

The temperatures T_L and T_R at the boundaries are assumed to be known. It is then desired to determine the rate at which energy (as heat) flows because of a temperature difference $T_L - T_R$ across the region.

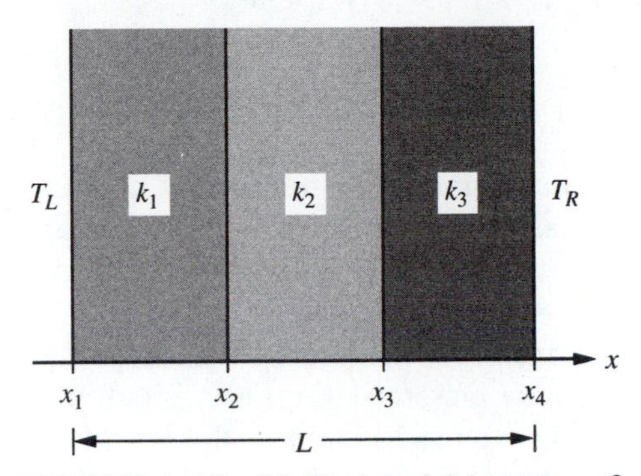

FIGURE 1–18 Multimaterial heat transfer

As indicated by Reynolds [9], Fourier's law of heat conduction states that the local heat flux or flow of energy q is given by

$$q = -k(x)A(x)\frac{dT}{dx}$$

where k is a material property called the thermal conductivity, A is the area, and dT/dx is the temperature gradient. For a problem that is being modeled as discrete, this general expression is replaced in each of the regions $x_i \leq x \leq x_{i+1}$ (i.e., the elements) by

$$q_i = -\frac{k_i A_i (T_{i+1} - T_i)}{l_i}$$

where k_i and A_i are appropriately chosen average values, and T_i and T_{i+1} are the temperatures at the ends of the segment.

Figure 1–19 depicts the situation regarding a typical segment or element, where

$$q_i = \frac{k_i A_i (T_i - T_{i+1})}{l_i}$$

$$q_{i+1} = \frac{k_i A_i (T_{i+1} - T_i)}{l_i}$$

FIGURE 1–19 Element temperatures and fluxes

These can be written in matrix form as

$$\mathbf{q} = \mathbf{kT} \tag{1.5}$$

where $q = [q_i \; q_{i+1}]^{\mathrm{T}}, \mathbf{T} = [T_i \; T_{i+1}]^{\mathrm{T}}$, and

$$\mathbf{k} = \begin{bmatrix} 1 & -1 \\ -1 & 1 \end{bmatrix} \frac{k_i A_i}{l_i}$$

where $\mathbf{k}$ will be called the **elemental conductivity matrix, q** the **elemental flux vector**, and **T** the **elemental temperature vector**. The appearance of elemental conductivity matrices in the governing equations for the present discrete problem will now be illustrated.

The basic physical principle which must be satisfied is *conservation of energy* or *balance of energy*. In the present setting, this is enforced by requiring that the amount of energy leaving a typical region to the left of x_i must, in the absence of any energy input at node i, be equal to the amount of energy entering the next region to the right of x_i.

Returning to the problem of Fig. 1–18, balance of energy is enforced at each of the internal material boundaries x_2 and x_3, yielding

$$\frac{-k_1 A(T_2 - T_1)}{l_1} = \frac{-k_2 A(T_3 - T_2)}{l_2}$$

$$\frac{-k_2 A(T_3 - T_2)}{l_2} = \frac{-k_3 A(T_4 - T_3)}{l_3}$$

or with $K_i = k_i / l_i$,

$$\begin{aligned} -K_1 T_1 + (K_1 + K_2) T_2 - K_2 T_3 \qquad\qquad &= 0 \\ -K_2 T_2 + (K_2 + K_3) T_3 - K_4 T_4 &= 0 \end{aligned}$$

that is, two equations for four unknowns. The additional equations for T_1, T_2, T_3, and T_4 come from the boundary conditions. For this type of heat transfer problem, there are three legitimate options for correctly specifying boundary conditions. Which option to select depends on the physics of the problem.

The first possibility is that the temperature at a boundary is prescribed; that is, the temperature is known at the boundary. At the left boundary, for instance, the corresponding equation would appear as

$$T_1 = T_L$$

This equation is simply appended to the balance equations, which for the problem of Fig. 1–18 would then appear as

$$\begin{aligned} T_1 \qquad\qquad\qquad\qquad\qquad &= T_L \\ -K_1 T_1 + (K_1 + K_2) T_2 - K_2 T_3 \qquad\qquad &= 0 \\ - K_2 T_2 + (K_2 + K_3) T_3 - K_4 T_4 &= 0 \end{aligned}$$

A prescribed temperature at the right boundary would be handled in precisely the same fashion.

The second possibility is that the amount of flux or energy entering at the boundary is known. At the left boundary, this is equivalent to stating that

$$Q_L = -\frac{k_1 A(T_2 - T_1)}{l_1}$$

where Q_L is the known flux *into* the region. In this instance the balance equations would be appended to read

$$\begin{aligned} K_1T_1 - K_1T_2 \qquad\qquad\qquad\qquad\qquad\qquad &= Q_L \\ -K_1T_1 + (K_1 + K_2)T_2 - K_2T_3 \qquad\qquad\qquad &= 0 \\ - K_2T_2 + (K_2 + K_3)T_3 - K_3T_4 &= 0 \end{aligned}$$

The corresponding condition at the right boundary would appear as

$$Q_R = \frac{k_3 A(T_4 - T_3)}{l_3}$$

with Q_R the prescribed flux *into* the region.

As discussed in Holman [10], the third possibility is that, at a boundary, there is a *local balance of energy* between convection exterior to the region and conduction interior to the region. This condition is pictured at the left boundary in Fig. 1–20.

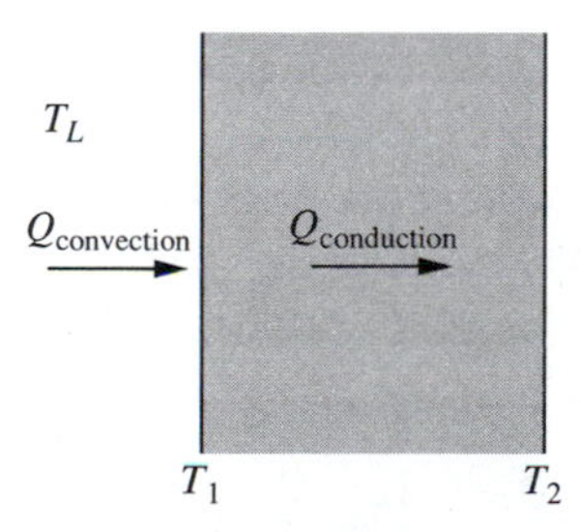

FIGURE 1–20 Local energy balance at the left boundary

In equation form this condition would be expressed as

$$Q_{\text{convection}} = Q_{\text{conduction}}$$

or

$$Ah_L(T_L - T_1) = -\frac{k_1 A(T_2 - T_1)}{l_1}$$

where T_L is known as the free stream temperature at the left boundary, and h_L is the convective heat transfer coefficient. A similar equation can be written to express the local energy balance at the right boundary.

For a specific example, assume that there is a convective type boundary condition at the left boundary resulting in the additional equation, just presented. Also assume that the temperature prescribed at the right boundary is $T_4 = T_R$.

The four equations for the four unknowns T_1, T_2, T_3, and T_4 can then be written in augmented form as

$$\left[\begin{array}{cccc|c} h_L + K_1 & -K_1 & 0 & 0 & h_L T_L \\ -K_1 & K_1 + K_2 & -K_2 & 0 & 0 \\ 0 & -K_2 & K_2 + K_3 & -K_3 & 0 \\ 0 & 0 & 0 & 1 & T_R \end{array}\right] \tag{1.6}$$

where the common area A has been canceled and $K_i = k_i/l_i, i = 1, 2, 3$. The positions in which the elemental conductivity matrices for elements 1 and 2 occur are indicated by the dotted boxes, and represent the contributions to the global or system equations from the elements. The assembly process is buried in the process of writing the balance equations and, in terms of this formulation, can only be seen in retrospect. The constraint is essentially the fourth equation $T_4 = T_R$.

We could proceed directly to the solution phase but for reasons of economy mentioned previously, choose to first alter the form of Eq. (1.6) as follows. The fourth row is multiplied by K_3 and added to the third row to produce the symmetric set of equations

$$\left[\begin{array}{cccc|c} h_L + K_1 & -K_1 & 0 & 0 & h_L T_L \\ -K_1 & K_1 + K_2 & -K_2 & 0 & 0 \\ 0 & -K_2 & K_2 + K_3 & 0 & K_3 T_R \\ 0 & 0 & 0 & 1 & T_R \end{array}\right]$$

As a specific numerical example, take $K_1 = K, K_2 = 2K$, $K_3 = K = h_L$ and $T_R = 2T_L$. The three equations for remaining unknowns T_1, T_2, and T_3 are

$$\left[\begin{array}{ccc|c} 2 & -1 & 0 & T_R/2 \\ -1 & 3 & -2 & 0 \\ 0 & -2 & 3 & T_R \end{array}\right]$$

with solution $T_1 = 9T_R/14$, $T_2 = 11T_R/14$, and $T_3 = 12T_R/14$. Using Fourier's law, the fluxes within the elements are easily calculated (computation of derived variables) to be

$$\frac{q_1}{A} = -K_1(T_2 - T_1) = -K\frac{2T_R}{14}$$

$$\frac{q_2}{A} = -K_2(T_3 - T_2) = -2K\frac{T_R}{14}$$

$$\frac{q_3}{A} = -K_3(T_4 - T_3) = -K\frac{2T_R}{14}$$

each of which represents the total energy flow across the corresponding element. Equality of the q's clearly indicates that the energy is conserved.

For this example *discretization* is implicit in the characterization of the problem. *Interpolation* and *elemental description* are essentially contained in the specialization of Fourier's law to a finite region. *Assembly* is accounted for in generating the governing equations by requiring balance of energy across each of the interfaces. Each of the elemental **k** matrices is loaded into the global or system equations according to the numbers of the temperatures defined at the ends of the element. The *constraint(s)* (boundary conditions) must then be imposed, after which the *solution* is obtained in the usual manner. The *derived variables* (fluxes) are computed by using the basic Eq. (1.5) for each element.

1.4.4 Piping Networks

Consider a network of interconnected pipes that distribute fluid as indicated in Fig. 1–21.

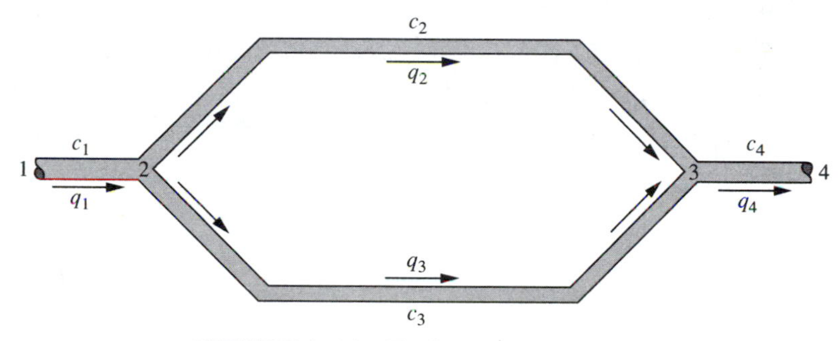

FIGURE 1–21 Typical network of pipes

Each segment of the network is characterized by a length L, a diameter d, and a suitable dynamic viscosity μ for the fluid flow. Using basic principles of fluid mechanics, an approximate relationship between the flow and the pressure drop in a segment can be expressed by

$$q = \frac{\pi d^4 \Delta p}{128 \mu L} = c \Delta p$$

where q is the flow rate, d is the diameter of the pipe, μ is the kinematic viscosity, L is the length, Δp is the pressure drop between the ends and $c = \pi d^4/128\mu L$. For simplicity of the model, we will assume that no fluid is introduced or removed between points 1 and 4 in Fig. 1–21. For a discussion of the physics of the problem see White [11].

The basic physical principle governing the distribution of the flow in the network is *conservation of mass* or *balance of mass;* that is, the amount of mass entering the junction of any two or more pipes must be the amount of mass leaving that same junction. In terms of the flows and pressures indicated in Fig. 1–21, the flow rate equations for the segments can be written as

$$\begin{aligned} q_1 &= C_1(p_1 - p_2) \\ q_2 &= C_2(p_2 - p_3) \\ q_3 &= C_3(p_2 - p_3) \\ q_4 &= C_4(p_3 - p_4) \end{aligned} \tag{1.7}$$

Note that there are four unknown flow rates and four unknown pressures. The necessary balance equations are enforced at junctions 2 and 3 to yield

$$\begin{aligned} q_1 &= q_2 + q_3 \\ q_2 + q_3 &= q_4 \end{aligned}$$

Eliminating the flow rates using Eq. (1.7) yields

$$\begin{aligned} -C_1 p_1 + (C_1 + C_2 + C_3) p_2 - (C_2 + C_3) p_3 \qquad\qquad &= 0 \\ -(C_2 + C_3) p_2 + (C_2 + C_3 + C_4) p_3 - C_4 p_4 &= 0 \end{aligned}$$

that is, two equations for four unknowns. The additional equations are supplied from the boundary conditions: a condition at point 1, the inlet, and a condition at point 4, the outlet. An appropriate condition at a boundary is either on the pressure or the flow rate.

If for instance the pressures are specified as $p_1 = P_1$ and $p_4 = P_4$, these conditions are simply appended to the balance equations to yield

$$\begin{aligned} p_1 \qquad\qquad\qquad\qquad &= P_1 \\ -C_1 p_1 + (C_1 + C_2 + C_3) p_2 - (C_2 + C_3) p_3 \qquad\qquad &= 0 \\ -(C_2 + C_3) p_2 + (C_2 + C_3 + C_4) p_3 - C_4 p_4 &= 0 \\ p_4 &= P_4 \end{aligned}$$

which can then be solved for p_2 and p_3. The flow rates can then be determined using Eq. (1.7).

The other type of permissible boundary consists of specifying a flow rate. At point 4 for example this would take the form $Q_4 = C_4(p_3 - p_4)$, with Q_4 representing the prescribed flow out. This equation is appended to the balance equations to yield

$$\begin{aligned} -C_1 p_1 + (C_1 + C_2 + C_3)\, p_2 - (C_2 + C_3) p_3 \qquad\qquad &= 0 \\ -(C_2 + C_3)\, p_2 + (C_2 + C_3 + C_4) p_3 - C_4 p_4 &= 0 \\ -C_4 p_3 + C_4 p_4 &= -Q_4 \end{aligned}$$

An equation representing the boundary at the left end would also be necessary.

As a specific example take $C_1 = C$, $C_2 = 2C$, $C_3 = 3C$, $C_4 = 2C$, $p_1 = P_1$, and $q_4 = Q$; that is, specify the pressure at the inlet and the flow at the outlet. The resulting equations can be expressed in augmented form as

$$\left[\begin{array}{cccc|c} 1 & 0 & 0 & 0 & P_1 \\ -1 & 6 & -5 & 0 & 0 \\ 0 & -5 & 7 & -2 & 0 \\ 0 & 0 & -2 & 2 & -Q/C \end{array}\right]$$

which can be solved to yield

$$p_2 = P_1 - \frac{Q}{C}$$

$$p_3 = P_1 - \frac{6Q}{5C}$$

The outlet pressure p_4 can then be computed as $p_4 = P_1 - 17Q/10C$. If the pressure at the outlet is not to be negative, that is $p_4 \geq 0$, (in order to prevent undesirable behavior of the flow in the pipe), we see that

$$P_1 \geq \frac{17Q}{10C}$$

If P_1 is then assumed to be the minimum allowable value $P_1 = 17Q/10C$, the pressures become

$$p_1 = P_1 = \frac{17Q}{10C}$$

$$p_2 = \frac{7Q}{10C} = \frac{7P_1}{17}$$

$$p_3 = \frac{5Q}{10C} = \frac{5P_1}{17}$$

$$p_4 = 0$$

Using Eq. (1.7) the corresponding flows in the pipes can be computed as

$$q_1 = C_1(p_1 - p_2) = Q!!! \text{ (Global conservation)}$$

$$q_2 = C_2(p_2 - p_3) = 0.4Q$$

$$q_3 = C_3(p_2 - p_3) = 0.6Q$$

$$q_4 = C_4(p_3 - p_4) = Q$$

and can easily be shown to satisfy the continuity equations.

As in the previous examples it is possible to identify elemental matrices relating the basic and derived variables of the problem.

FIGURE 1–22 Element pressures and flows

From the basic relationship

$$q = \frac{\pi d^4 \Delta p}{128 \mu L}$$

and Fig. 1–22, it is easily seen that

$$\begin{aligned} q_i &= c(p_i - p_{i+1}) \\ q_{i+1} &= c(p_{i+1} - p_i) \end{aligned}$$

or in matrix notation

$$\mathbf{q} = \mathbf{cp} \tag{1.8}$$

where

$$\mathbf{q} = \begin{bmatrix} q_i \\ q_{i+1} \end{bmatrix}$$

is the **flow vector** for the element,

$$\mathbf{c} = \begin{bmatrix} c & -c \\ -c & c \end{bmatrix}$$

is the **elemental flow resistance matrix,** and

$$\mathbf{p} = \begin{bmatrix} p_i \\ p_{i+1} \end{bmatrix}$$

is the **elemental pressure vector**.

The student should observe that the system flow resistance matrix can be considered to be assembled from 2 × 2 elemental flow resistance matrices in positions according to the names of the pressures defined at the ends of the elements: *assembly* of the elemental flow resistance matrices is implicit in the governing mass balance equations for this discrete problem. *Discretization* is implicit in the original model. *Interpolation* and *elemental formulation* are contained in the basic relationship $q = c\Delta p$, with *assembly* accomplished by enforcing balance of mass at all the network junctions. *Constraints* are enforced by specifying the pressure at a boundary, with the *solution* and *computation of derived variables* steps being accomplished in the usual fashion.

From the chapters that follow, it should be clear that, regardless of the complexity of the physical model, the final system or global equations resulting from the application of the FEM can be considered as the appropriate superposition or assembly of contributions from the elements.

1.4.5 Plane Truss Structures

Many structures are constructed so that it is appropriate to assume that the individual members are connected by pins. Each element of the structure is a so-called two-force member, or truss member, meaning that it transmits only axial forces.

There are no shear, bending, or torsional loads transmitted by the member of such a structure. If the members can be assumed to be linear we have a system of interconnected stiff linear springs, each of which behaves according to

$$P_i = k_i(u_i - u_{i+1})$$
$$P_{i+1} = k_i(u_{i+1} - u_i)$$

where P_i, P_{i+1}, u_i, and u_{i+1} are as indicated in Fig. 1–23.

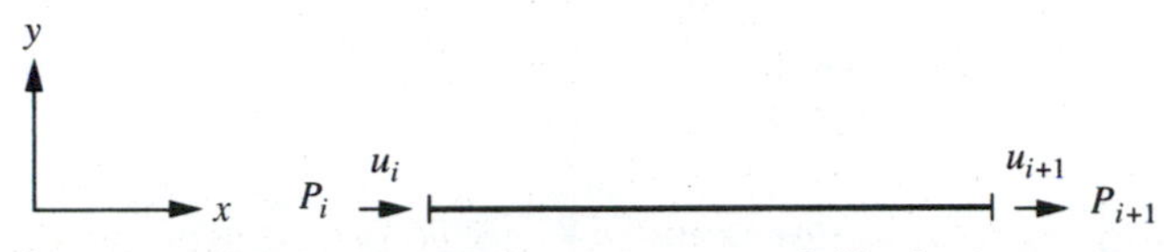

FIGURE 1–23 Element forces and displacements

The spring stiffness constant k_i is $(AE/l)_i$, where A is an area, E is Young's modulus, and l is the length of the bar. The displacements u and the forces P can be thought of as being expressed in the local xy coordinate system indicated in Fig. 1–23. For a general discussion of local and global coordinates see Martin [12].

A typical plane truss with four bars or elements is indicated in Fig. 1–24.

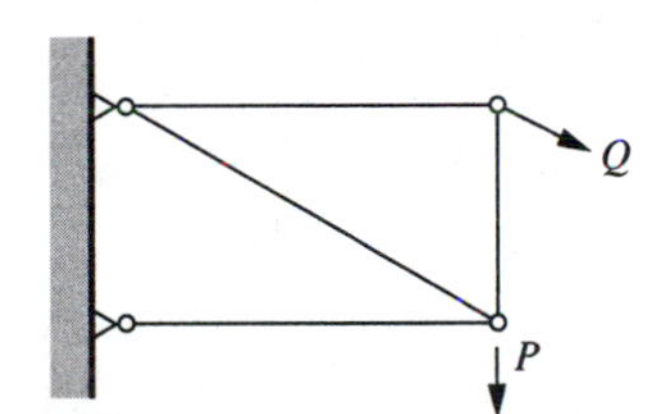

FIGURE 1–24 Typical plane truss

It is obvious that the individual members of the truss are not all lined up; that is, although each member in general will elongate (or shorten) and sustain a tensile (or compressive) load, the forces in and elongations of the members are in different directions. What is needed is the ability to refer all the forces and elongations in the members of the truss to a common reference frame such as the XY frame indicated in Fig. 1–25.

Consider then a typical member inclined at an angle θ with respect to the global XY system as shown in Fig. 1–25. The displacement and force components referred to both the elemental or local xy system and the global XY system are indicated in Fig. 1–25. The force components in the two coordinate systems at end 1 of the member are related according to

$$F_{X1} = f_{x1} \cos\theta - f_{y1} \sin\theta$$
$$F_{Y1} = f_{x1} \sin\theta + f_{y1} \cos\theta$$

and the displacement components as

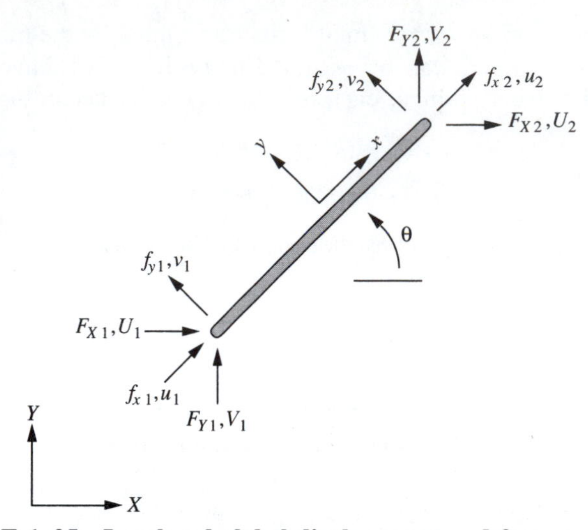

FIGURE 1–25 Local and global displacement and force components

$$
\begin{aligned}
U_1 &= u_1 \cos\,\theta - v_1 \sin\,\theta \\
V_1 &= u_1 \sin\,\theta + v_1 \cos\,\theta
\end{aligned}
$$

or when expressed in matrix notation

$$
\begin{aligned}
\mathbf{F_1} &= \mathbf{Rf_1} \\
\mathbf{U_1} &= \mathbf{Ru_1}
\end{aligned}
\tag{1.9}
$$

where

$$
\mathbf{F_1} = \begin{bmatrix} F_{X1} \\ F_{Y1} \end{bmatrix} \qquad \mathbf{f_1} = \begin{bmatrix} f_{x1} \\ f_{y1} \end{bmatrix} \qquad \mathbf{U_1} = \begin{bmatrix} U_1 \\ V_1 \end{bmatrix} \qquad \mathbf{u_1} = \begin{bmatrix} u_1 \\ v_1 \end{bmatrix}
$$

are the force and displacement vectors referred to the global (X, Y) and local (x, y) axes, respectively, at end 1.

R, given by

$$
\mathbf{R} = \begin{bmatrix} \cos\,\theta & -\sin\,\theta \\ \sin\,\theta & \cos\,\theta \end{bmatrix}
$$

is the frequently encountered transformation matrix for a rotation of axes. It is an example of an orthogonal matrix having the easily verified property that $\mathbf{R}^{-1} = \mathbf{R}^{\mathrm{T}}$.

In a completely similar fashion we can write

$$
\begin{aligned}
\mathbf{F_2} &= \mathbf{Rf_2} \\
\mathbf{U_2} &= \mathbf{Ru_2}
\end{aligned}
\tag{1.10}
$$

where

$$\mathbf{F}_2 = \begin{bmatrix} F_{X2} \\ F_{Y2} \end{bmatrix} \qquad \mathbf{f}_2 = \begin{bmatrix} f_{x2} \\ f_{y2} \end{bmatrix} \qquad \mathbf{U}_2 = \begin{bmatrix} U_2 \\ V_2 \end{bmatrix} \qquad \mathbf{u}_2 = \begin{bmatrix} u_2 \\ v_2 \end{bmatrix}$$

are the force and displacement vectors referred to the global and local axes, respectively, at end 2.

The stiffness matrix for the axial element in the elemental or local coordinate system can be written as

$$\begin{bmatrix} f_{x1} \\ f_{x2} \end{bmatrix} = \begin{bmatrix} k & -k \\ -k & k \end{bmatrix} \begin{bmatrix} u_1 \\ u_2 \end{bmatrix}$$

When expanded to include the y components of the forces and displacements, there results

$$\begin{bmatrix} f_{x1} \\ f_{y1} \\ f_{x2} \\ f_{y2} \end{bmatrix} = \begin{bmatrix} k & 0 & -k & 0 \\ 0 & 0 & 0 & 0 \\ -k & 0 & k & 0 \\ 0 & 0 & 0 & 0 \end{bmatrix} \begin{bmatrix} u_1 \\ v_1 \\ u_2 \\ v_2 \end{bmatrix} \tag{1.11}$$

where the second and fourth equations reflect the fact that only axial loads, which are in the x-direction locally, are possible in the absence of bending, shear, and torsion. Write Eq. (1.11) in partitioned form as

$$\begin{bmatrix} \mathbf{f}_1 \\ \hline \mathbf{f}_2 \end{bmatrix} = \left[\begin{array}{c|c} \mathbf{k}_{11} & \mathbf{k}_{12} \\ \hline \mathbf{k}_{21} & \mathbf{k}_{22} \end{array}\right] \begin{bmatrix} \mathbf{u}_1 \\ \hline \mathbf{u}_2 \end{bmatrix}$$

or

$$\begin{aligned} \mathbf{f}_1 &= \mathbf{k}_{11}\mathbf{u}_1 + \mathbf{k}_{12}\mathbf{u}_2 \\ \mathbf{f}_2 &= \mathbf{k}_{21}\mathbf{u}_1 + \mathbf{k}_{22}\mathbf{u}_2 \end{aligned}$$

In order to convert these relations to global form, we eliminate all the quantities referred to the elemental system through the use of Eqns. (1.9) and (1.10) to obtain

$$\begin{aligned} \mathbf{R}^{-1}\mathbf{F}_1 &= \mathbf{k}_{11}\mathbf{R}^{-1}\mathbf{U}_1 + \mathbf{k}_{12}\mathbf{R}^{-1}\mathbf{U}_2 \\ \mathbf{R}^{-1}\mathbf{F}_2 &= \mathbf{k}_{21}\mathbf{R}^{-1}\mathbf{U}_1 + \mathbf{k}_{22}\mathbf{R}^{-1}\mathbf{U}_2 \end{aligned}$$

or, since $\mathbf{R}^{-1} = \mathbf{R}^{\mathrm{T}}$

$$\begin{aligned} \mathbf{F}_1 &= \mathbf{R}\mathbf{k}_1 1\mathbf{R}^{\mathrm{T}}\mathbf{U}_1 + \mathbf{R}\mathbf{k}_1 2\mathbf{R}^{\mathrm{T}}\mathbf{U}_2 \\ \mathbf{F}_2 &= \mathbf{R}\mathbf{k}_2 1\mathbf{R}^{\mathrm{T}}\mathbf{U}_1 + \mathbf{R}\mathbf{k}_2 2\mathbf{R}^{\mathrm{T}}\mathbf{U}_2 \end{aligned}$$

Putting this equation in partitioned form leads to

$$\begin{bmatrix} \mathbf{F}_1 \\ \hline \mathbf{F}_2 \end{bmatrix} = \left[\begin{array}{c|c} \mathbf{R}\mathbf{k}_{11}\mathbf{R}^{\mathrm{T}} & \mathbf{R}\mathbf{k}_{12}\mathbf{R}^{\mathrm{T}} \\ \hline \mathbf{R}\mathbf{k}_{21}\mathbf{R}^{\mathrm{T}} & \mathbf{R}\mathbf{k}_{22}\mathbf{R}^{\mathrm{T}} \end{array}\right] \begin{bmatrix} \mathbf{U}_1 \\ \hline \mathbf{U}_2 \end{bmatrix}$$

or

$$\begin{bmatrix} \mathbf{F}_1 \\ \hline \mathbf{F}_2 \end{bmatrix} = \left[\begin{array}{c|c} \mathbf{R} & \mathbf{O} \\ \hline \mathbf{O} & \mathbf{R} \end{array}\right] \left[\begin{array}{c|c} \mathbf{k}_{11} & \mathbf{k}_{12} \\ \hline \mathbf{k}_{21} & \mathbf{k}_{22} \end{array}\right] \left[\begin{array}{c|c} \mathbf{R}^{\mathrm{T}} & \mathbf{O} \\ \hline \mathbf{O} & \mathbf{R}^{\mathrm{T}} \end{array}\right] \begin{bmatrix} \mathbf{U}_1 \\ \hline \mathbf{U}_2 \end{bmatrix}$$

and finally, with $\mathbf{T}$ defined in the obvious way, as

$$\mathbf{F} = \mathbf{T}\mathbf{k}\mathbf{T}^{\mathrm{T}}\mathbf{U} = \mathbf{K}\mathbf{U}$$

where $\mathbf{K}$ is the global stiffness matrix for the arbitrarily located axial element. The multiplications can be easily carried out to produce

$$\mathbf{K} = k \begin{bmatrix} \lambda^2 & \lambda\mu & -\lambda^2 & -\lambda\mu \\ \lambda\mu & \mu^2 & -\lambda\mu & -\mu^2 \\ -\lambda^2 & -\lambda\mu & \lambda^2 & \lambda\mu \\ -\lambda\mu & -\mu^2 & \lambda\mu & \mu^2 \end{bmatrix}$$

where $\lambda = \cos\theta$ and $\mu = \sin\theta$.

The final set of equilibrium equations for a truss structure can be obtained by either a vectorial or a variational approach in a manner entirely analogous to what was presented in Sections 1.4.1 and 1.4.2, respectively.

For this class of problems, *discretization* has already taken place in deciding that the physical model should be a truss. *Interpolation* and *elemental formulation* are contained in the basic force-displacement relationship for the bar, or in the expression for the internal potential energy of the bar. *Assembly* is quite visible when the variational approach is used as for the example of Section 1.4.2. The steps of *constraints*, *solution*, and *computation of derived variables* are accomplished in a straightforward fashion and are quite evident.

Generally, a truss structure can be analyzed by the following steps:

1. Number the nodes, keeping in mind that the bandwidth of the final system of equations will depend on the maximum difference over all the elements of the node numbers defining each of the elements. Consider for example two of the logical choices for numbering the nodes of the truss as indicated in Figs. 1–26*a* and 1–26*b*. The corresponding global stiffness matrices for the two structures

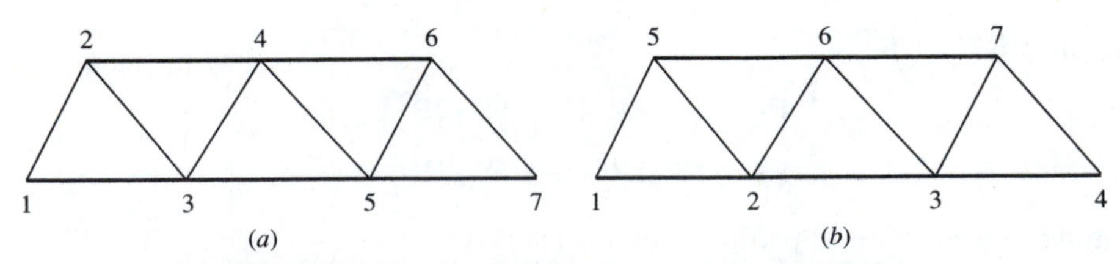

FIGURE 1–26 **(*a*) Numbering 1; (*b*) Numbering 2**

will appear as in Figs. 1–27*a* and 1–27*b*, respectively, with the positions of the x's an indication of the location of the potentially nonzero entries.

$$\begin{bmatrix} x & x & x & 0 & 0 & 0 & 0 \\ x & x & x & x & 0 & 0 & 0 \\ x & x & x & x & x & 0 & 0 \\ 0 & x & x & x & x & x & 0 \\ 0 & 0 & x & x & x & x & x \\ 0 & 0 & 0 & x & x & x & x \\ 0 & 0 & 0 & 0 & x & x & x \end{bmatrix} \qquad \begin{bmatrix} x & x & 0 & 0 & x & 0 & 0 \\ x & x & x & 0 & x & x & 0 \\ 0 & x & x & x & 0 & x & x \\ 0 & 0 & x & x & 0 & 0 & x \\ x & x & 0 & 0 & x & x & 0 \\ 0 & x & x & 0 & x & x & x \\ 0 & 0 & x & x & 0 & x & x \end{bmatrix}$$

(*a*) (*b*)

FIGURE 1–27 (*a*) Global stiffness matrix 1; (*b*) Global stiffness matrix 2

These figures clearly show that the first numbering scheme produces a global stiffness matrix that has a smaller bandwidth, generally resulting in savings during the solution phase. The half bandwidth of a symmetric set of equations is defined as follows. For row i determine the column number j of the last nonzero entry. Compute $(nb)_i$ as

$$(nb)_i = 1 + (j - i)$$

NB, the half bandwidth, is the maximum of the $(nb)_i$ over all the rows.

2. Assemble the elemental stiffness and load matrices. Each of the elemental stiffness matrices must have been transformed to the global coordinate system as previously outlined. This produces the unconstrained global set of equilibrium equations for the structure.
3. Apply any constraints.
4. Solve the equations.
5. Compute the forces in the members. The forces in the members can be computed in terms of components in either the global XY system or the appropriate xy system for the element or member in question. In terms of the local xy system,

$$\mathbf{f} = \mathbf{ku} = \mathbf{kR}^{\mathrm{T}}\mathbf{U}$$

using the transformation between the displacements. This can be expanded to yield

$$\begin{aligned} f_{x1} &= k[(U_1 - U_2)\cos\theta + (V_1 - V_2)\sin\theta] \\ f_{y1} &= 0 \\ f_{x2} &= -k[(U_1 - U_2)\cos\theta + (V_1 - V_2)\sin\theta] \\ f_{y2} &= 0 \end{aligned} \tag{1.12}$$

Each of these steps is carried out in detail in the following example.

Example 1.1.

Consider the very simple two-element truss in Fig. 1–28.

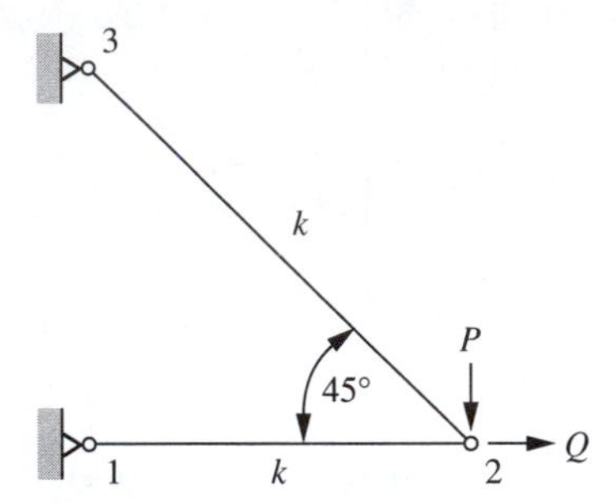

FIGURE 1–28 Two-element truss

Step 1. The node numbers are defined as indicated in Fig. 1–28.

Step 2. Table 1.1 contains the information necessary for developing and assembling the elemental stiffness matrices.

TABLE 1.1 Element information

Member	Node 1	Node 2	Elemental stiffness	θ
1	1	2	k	0
2	2	3	k	$3\pi/4$

The stiffness matrices expanded to the global size are

$$
\mathbf{k}_{\mathbf{G1}} = k \begin{bmatrix} 1 & 0 & -1 & 0 & 0 & 0 \\ 0 & 0 & 0 & 0 & 0 & 0 \\ -1 & 0 & 1 & 0 & 0 & 0 \\ 0 & 0 & 0 & 0 & 0 & 0 \\ 0 & 0 & 0 & 0 & 0 & 0 \\ 0 & 0 & 0 & 0 & 0 & 0 \end{bmatrix} \begin{matrix} U_1 \\ V_1 \\ U_2 \\ V_2 \\ U_3 \\ V_3 \end{matrix} \quad \text{(columns: } U_1\ V_1\ U_2\ V_2\ U_3\ V_3\text{)}
$$

$$
\mathbf{k}_{\mathbf{G2}} = \frac{k}{2} \begin{bmatrix} 0 & 0 & 0 & 0 & 0 & 0 \\ 0 & 0 & 0 & 0 & 0 & 0 \\ 0 & 0 & 1 & -1 & -1 & 1 \\ 0 & 0 & -1 & 1 & 1 & -1 \\ 0 & 0 & -1 & 1 & 1 & -1 \\ 0 & 0 & 1 & -1 & -1 & 1 \end{bmatrix} \begin{matrix} U_1 \\ V_1 \\ U_2 \\ V_2 \\ U_3 \\ V_3 \end{matrix} \quad \text{(columns: } U_1\ V_1\ U_2\ V_2\ U_3\ V_3\text{)}
$$

Assembly of the global stiffness matrix yields

$$
\mathbf{k_G} = \frac{k}{2}
\begin{array}{c}
\begin{array}{cccccc} U_1 & V_1 & U_2 & V_2 & U_3 & V_3 \end{array} \\
\left[\begin{array}{cccccc}
2 & 0 & -2 & 0 & 0 & 0 \\
0 & 0 & 0 & 0 & 0 & 0 \\
-2 & 0 & 3 & -1 & -1 & 1 \\
0 & 0 & -1 & 1 & 1 & -1 \\
0 & 0 & -1 & 1 & 1 & -1 \\
0 & 0 & 1 & -1 & -1 & 1
\end{array}\right]
\end{array}
\begin{array}{c} U_1 \\ V_1 \\ U_2 \\ V_2 \\ U_3 \\ V_3 \end{array}
$$

The global load vector is composed of the external applied loads and is $\mathbf{P}_\mathbf{G}^\mathrm{T} = [0\ 0\ Q\ \ -P\ 0\ 0]$.

The final assembled unconstrained set of equilibrium equations in augmented form is

$$
\left[\begin{array}{cccccc|c}
2 & 0 & -2 & 0 & 0 & 0 & 0 \\
0 & 0 & 0 & 0 & 0 & 0 & 0 \\
-2 & 0 & 3 & -1 & -1 & 1 & 2Q/k \\
0 & 0 & -1 & 1 & 1 & -1 & -2P/k \\
0 & 0 & -1 & 1 & 1 & -1 & 0 \\
0 & 0 & 1 & -1 & -1 & 1 & 0
\end{array}\right]
$$

Step 3. The structure is supported at nodes 1 and 3 so as to prevent motion in all directions. The constraints are $U_1 = V_1 = U_3 = V_3 = 0$, leading to the global constrained set of equations in augmented form

$$
\left[\begin{array}{cccccc|c}
1 & 0 & 0 & 0 & 0 & 0 & 0 \\
0 & 1 & 0 & 0 & 0 & 0 & 0 \\
0 & 0 & 3 & -1 & 0 & 0 & 2Q/k \\
0 & 0 & -1 & 1 & 0 & 0 & -2P/k \\
0 & 0 & 0 & 0 & 1 & 0 & 0 \\
0 & 0 & 0 & 0 & 0 & 1 & 0
\end{array}\right]
$$

Step 4. The final constrained set is easily solved for the two remaining unknowns U_2 and V_2 as $U_2 = Q/k - P/k$, and $V_2 = -3P/k + Q/k$.

Step 5. Using Eqns. (1.12) the force in each member is computed.

Member 1. Referring to Table 1.1, the node numbers are 1 and 2 in that order and $\theta = 0$. There results

$$
f_{x1} = k\left[-\left(\frac{Q}{k} - \frac{P}{k}\right)\right] = -Q + P = -f_{x2}
$$

$$
f_{y1} = f_{y2} = 0
$$

which appears as

$(P-Q)$ → 1 ———————— 2 ← $(P-Q)$

Member 2. For member 2, the node numbers are 2 and 3 and $\theta = 3\pi/4$ so that

$$f_{x1} = k\left[\left(\frac{Q}{k} - \frac{P}{k}\right)\left(-\frac{1}{\sqrt{2}}\right) + \left(\frac{-3P}{k} + \frac{Q}{k}\right)\left(\frac{1}{\sqrt{2}}\right)\right]$$

$$= -\sqrt{2}P = -f_{x2}$$

$$f_{y1} = f_{y2} = 0$$

which appears as

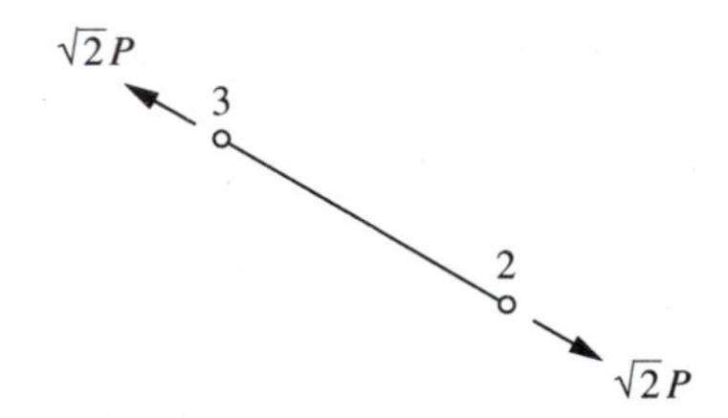

The student should verify that the forces so computed do satisfy equilibrium, the basic physical principle for this problem. This can be easily accomplished by drawing a free-body diagram of node 2.

This example illustrates a situation that occurs frequently in the analysis of two- and three-dimensional structures: the degrees of freedom that are appropriate for the description of an individual element may have different orientations from element to element. This necessitates transforming each of the elemental coordinate systems or degrees of freedom to a common global coordinate system. These transformations would be invisible to the user of a computer code but would certainly have to be included explicitly if the object were to generate such a code or to solve the problem by hand.

1.5 CLOSURE

The main purpose of this chapter was to introduce the student to the basic ingredients of

1. Discretization
2. Interpolation
3. Elemental formulation
4. Assembly
5. Constraints
6. Solution
7. Computation of derived variables

and how they relate to several practical physical problems that are modeled as discrete. In particular, the student should have the assembly process clearly in mind as a result of the examples presented. Similarly, the method of applying the constraints to the global equations should be clear, and the use of the elemental matrices in computation of derived variables should be a familiar step. The basic ingredients introduced in this chapter will be reiterated in the remaining chapters. The student will benefit by following these basic steps in setting up and solving finite element models by hand and also in writing and using codes for implementing the finite element method.

In all three classes of problems discussed in this chapter, there is a remarkable connection between the physics and the mathematics that is well worth discussing. We choose the example of the springs shown in the figure.

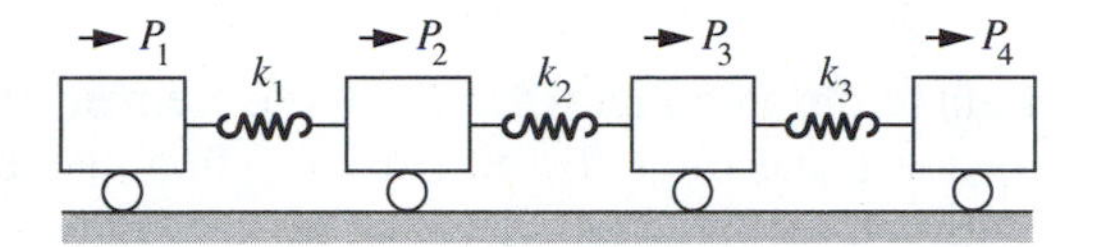

The augmented global equations are

$$\left[\begin{array}{cccc|c} k_1 & -k_1 & 0 & 0 & P_1 \\ -k_1 & k_1+k_2 & -k_2 & 0 & P_2 \\ 0 & -k_2 & k_2+k_3 & -k_3 & P_3 \\ 0 & 0 & -k_3 & k_3 & P_4 \end{array}\right]$$

If we attempt to solve this set of equations, forward reduction results in

$$\left[\begin{array}{cccc|c} k_1 & -k_1 & 0 & 0 & P_1 \\ 0 & k_2 & -k_2 & 0 & P_1+P_2 \\ 0 & 0 & k_3 & -k_3 & P_1+P_2+P_3 \\ 0 & 0 & 0 & 0 & P_1+P_2+P_3+P_4 \end{array}\right]$$

This states that unless $P_1 + P_2 + P_3 + P_4 = 0$ the equations are inconsistent; there is no solution. If $P_1 + P_2 + P_3 + P_4 = 0$, that is, *if the resultant external force is zero*, there are infinite solutions all of which differ by a constant. This constant represents a rigid-body motion of the system of masses and springs.

Physically, this is just what we would expect. If the resultant external force is zero, the springs will deform relative to each other with the position of any one mass being arbitrary. If the resultant external force is not zero there is no equilibrium. Once a constraint is enforced—say, the left mass is required to be stationary—the stiffness matrix is no longer singular, and a unique solution is possible.

A parallel discussion can take place with respect to the heat transfer and piping network problems. When there is no convection at either end for the heat transfer problem and no constraint is enforced on the assembled set of elemental matrices, the conclusion from the theory of $\mathbf{Ax} = \mathbf{b}$ is exactly the same: either there is no solution, or an infinity of solutions results. A physical interpretation is possible in both instances. The unique solution results when an appropriate constraint is imposed. An entirely similar situation arises when no constraint is enforced on the assembled set of elemental matrices for the piping problem. The student is referred to Appendix D for a discussion of these ideas as they relate to the classes of continuous problems covered in Chapters 2 and 3.

It is possible to write a single computer code that will create and analyze a finite element model of all of the problems treated in Sections 1.4.1 through 1.4.4 of this chapter. The student is strongly encouraged to generate this code for a better understanding of the ideas involved and for future use in solving appropriate discrete problems. This code is outlined in the Computer Projects section. Also outlined in that section is a computer code for solving the plane truss type of problem.

Finally, it is important to observe that when the finite element method is applied to a physical problem, the usual rules for dimensional homogeneity must be observed. In particular, any inputs must be such that all the units are consistent.

REFERENCES

1. Turner, M. R., R. W. Clough, H. Martin, and L. Topp: "Stiffness and Deflection Analysis of Complex Structures," *Journal of Aeronautical Sciences,* vol. 23, pp. 805–823, 1956.
2. Argyris, J. H., and S. Kelsey: *Energy Theorems and Structural Analysis,* Butterworth Scientific Publications, London, 1960.
3. Clough, R. W.: "The Finite Element Method in Plane Stress Analysis," *Proceedings of the American Society of Civil Engineers,* vol. 87, pp. 345–378, 1960.
4. Hrenikoff, H.: "Solution of Problems in Elasticity by the Framework Method," *Journal of Applied Mechanics, Transactions ASME*, vol. 8, pp. 169–175, 1941.
5. Courant, R.: "Variational Methods for the Solution of Problems of Equilibrium and Vibration," *Bulletin of the American Mathematical Society,* vol. 49, pp. 1–43, 1943.
6. McHenry, D.: "A Lattice Analogy for the Solution of Plane Stress Problems," *Journal of the Institute of Civil Engineers,* vol. 21, pp. 59–82, 1943.
7. Levy, S.: "Structural Analysis and Influence Coefficients for Delta Wings," *Journal of Aeronautical Sciences,* vol. 20, no. 7, pp. 449–454, 1953.
8. Meriam, J.L., and L.G. Kraige: *Engineering Mechanics*, vol. I, 3rd ed., John Wiley, New York, 1992.
9. Reynolds, W. C.: *Thermodynamics*, 2d ed., McGraw-Hill, New York, 1965.
10. Holman, J. P.: *Heat Transfer*, McGraw-Hill, New York, 1976.
11. White, F. M.: *Fluid Mechanics*, McGraw-Hill, New York, 1979.
12. Martin, H. C.: *Introduction to Matrix Methods of Structural Analysis*, McGraw-Hill, New York, 1966.

GENERAL REFERENCES

Akin, J. E.: *Finite Element Analysis for Undergraduates,* Academic Press, London, 1986.

Allaire, P. E.: *Basics of the Finite Element Method—Solid Mechanics, Heat Transfer and Fluid Mechanics*, W. C. Brown, Dubuque, Iowa, 1985.

Bathe, K. J.: *Finite Element Procedures in Engineering Analysis,* Prentice-Hall, Englewood Cliffs, New Jersey, 1982.

Becker, E. B., G. F. Carey, and J. T. Oden: *Finite Elements—An Introduction,* vol. I, Prentice-Hall, Englewood Cliffs, New Jersey, 1981.

Burnett, D. S.: *Finite Element Analysis, From Concepts to Applications*, Addison-Wesley, Reading, Massachusetts, 1987.
Grandin, H., Jr.: *Fundamentals of the Finite Element Method*, Macmillan, New York, 1986.
Livesley, R. K.: *Finite Elements: An Introduction for Engineers,* Cambridge University Press, Cambridge, 1983.
Norrie, D. H., and G. de Vries: *The Finite Element Method: Fundamentals and Applications*, Academic Press, New York, 1973.
Rao, S. S.: *The Finite Element Method in Engineering*, Pergamon, Oxford, 1982.
Reddy, J. N.: *An Introduction to the Finite Element Method*, McGraw-Hill, New York, 1984.

COMPUTER PROJECTS

When writing a finite element code, it is possible and desirable to establish a correspondence between the basic ingredients and the different subroutines or procedures that will constitute the code. This can be helpful in deciding upon the actual structure of the code.

Additionally, when using complex finite element codes it can be very helpful to divide what may be a large amount of input into more manageable chunks that correspond roughly to the basic ingredients of the finite element method. Such an approach can assist in making sure that all the necessary inputs are in fact provided.

The two codes discussed below are intended to enable the student to begin the process of writing relatively simple codes to obtain numerical results for classes of problems of interest. The student is strongly encouraged to complete at least the first computer project.

Computer Project #1

Develop a computer code for solving the one-dimensional linear problems such as systems of springs, heat transfer, and pipe flow which are governed by elemental statements of the form

$$\mathbf{q} = \mathbf{c}\boldsymbol{\phi}$$

Set up the code in approximately the following form.

INPUT. Input the data required to form elemental matrices, load the elemental matrices into the global matrix, apply the loads, and enforce the constraints.

1.	NN, NE		Number of nodes and elements.
2.	SK(I), NI(I), NJ(I)	I = 1,NE	Values of the elemental stiffnesses SK and the corresponding nodes which define the element.
3.	NL		Number of nodes at which loads are applied.
4.	NLDF(I), FD(I)	I = 1,NL	Nodes NDLF at which the loads FD are applied.
5.	CVL, CVR		When appropriate, the convections that are added to the beginning and end.
6.	NC		Number of nodes at which constraints are applied.
7.	NCDF(I), C(I)	I = 1,NC	Nodes NCDF at which the constraints C are enforced.

OUTPUT OF INPUT DATA. Output all the input data for checking purposes.

FORM AND ASSEMBLE THE ELEMENTAL STIFFNESS MATRICES. Using the information available from 1 and 2 above, the elemental matrices can be formed and assembled or loaded into their respective positions in the global matrix.

FORM THE GLOBAL LOAD MATRIX. Using the information provided in 3 and 4, the loads FD can be positioned properly in the global load matrix. When appropriate, the additional inputs CVL and CVR (item 5) corresponding to the convective type boundary conditions are added to the correct position(s) in the global stiffness matrix.

CONSTRAIN THE ASSEMBLED EQUATIONS. Using the information provided in 6 and 7, the equations can be constrained.

SOLVE THE EQUATIONS. Call an appropriate linear equation solver to determine the ϕ's. Routines BDecomp and BSolve are discussed in Appendix B. Source listings for both are also included.

COMPUTE THE "FORCES." Reuse the information provided in 2 to determine the q's within each of the elements.

OUTPUT THE SOLUTION. Output the ϕ's and q's.

Shown in the Fig. CP1–1 is a flowchart indicating the construction of a code to analyze the type of discrete problems discussed in Sections 1.4.1 through 1.4.4. Note carefully that there is a general correspondence between the basic ingredients of the finite element method and the blocks of the code. The blocks can correspond to subroutines in FORTRAN or to procedures in Pascal and to functions in C.

Computer Project #2

Develop a computer code for solving the one-dimensional truss problem. Set up the code in the following form.

INPUT. Input all the information necessary to form and load the elemental stiffnesses and loads, apply the loads and enforce the constraints.

1.	NN, NE	Number of nodes and elements.
2.	X(I) I = 1,NN	x-coordinates of nodes.
	Y(I) I = 1,NN	y-coordinates of nodes.
3.	NI(I), NJ(I), A(I), E(I)	I = 1,NE Node numbers, area, and material property for each element.
4.	NL	Number of nodes at which loads are applied.
5.	NK(I), FX(I), FY(I)	I = 1,NL Node numbers (NK) at which loads are applied with the x (FX) and y(FY) components of the load.
6.	NC	Number of nodes at which constraints are applied.
7.	ND(I), DX(I), DY(I)	I = 1,NC Node numbers (ND) at which constraints are applied with x (DX) and y (DY) components of the displacement.

OUTPUT. Output all the input data for checking purposes.

FORM AND ASSEMBLE THE ELEMENTAL STIFFNESS MATRICES. Using the information available from 2 and 3 above, the elemental matrices can be formed and assembled or loaded into their respective positions in the global matrix.

SUBROUTINES	*BASIC INGREDIENT*
Initialize arrays	
Input Block Input required data: NN, NE, NI, NJ, SK, NL, NLDF, FD, NC, NCDF, C, CVL, CVR	For the discrete system, the basic ingredients of discretization, interpolation and elemental formulation are implicit in several of the inputs.
Reflect the data for checking purposes.	
Stiffness Assembly Block Do on NE (# of elements) Using NI & NJ, form and load each of the elemental 2 by 2 stiffnesses.	Blocks 4 and 5 clearly correspond to *assembly*.
Load Assembly Block Do on NL (# of loads) Use NDLF to locate the FD properly in the global load vector.	
Add the convections to the global stiffness.	Effect of "convective" boundary conditions. Also part of *assembly* .
Constraint Block Do on NC (# of constraints) Use NCDF to constrain nodal degrees of freedom to the C values. If appropriate, restore the set of equations to symmetry.	
Solution Block Send the constrained set of equations to the appropriate equation solver. Output solution.	*Solution.*
Derived Variables Block Do on NE (# of elements) Using NI and NJ, select the appropriate subset of the degrees of freedom for computing the derived variable in each element. Output same.	Computation of derived variables.

FIGURE CP1–1

FORM THE GLOBAL LOAD MATRIX. Using the information provided in 4 and 5 the loads FX and FY can be positioned properly in the global load matrix.

CONSTRAIN THE ASSEMBLED EQUATIONS. Using the information provided in 6 and 7 the equations can be constrained. The equations should be put in symmetric form before beginning the solution phase.

SOLVE THE EQUATIONS. Call an appropriate linear equation solver to determine the displacements. Routines BDecomp and BSolve are discussed in Appendix B. Source listings for both are also included.

COMPUTE THE FORCES. Reuse the information generated in forming and loading the elemental stiffness matrices to determine the internal forces in each of the elements.

OUTPUT THE SOLUTION. Output the nodal displacements and elemental forces. The computer code DISCRETE contained on the diskette available from your instructor can be used to solve the classes of problems considered in Sections 1.4.1 through 1.4.4.

EXERCISES

Section 1.3

1.1. For constant E, integrate the governing differential equation of equilibrium for the axial deformation of an elastic bar; that is,

$$(AEu')' = 0$$
$$u(0) = 0$$
$$AEu'(L) = P$$

to show that there is an A_{avg} given by

$$\frac{1}{A_{\text{avg}}} = \frac{1}{L}\int_0^L \frac{dx}{A(x)}$$

such that

$$P = \frac{A_{\text{avg}}Eu(L)}{L} = ku(L)$$

that is, an equivalent linear spring.

Section 1.4.1

For each of the Exercises 1.2–1.5, set up and solve the global or system equations. In the process, identify each of the elemental stiffness matrices and where it contributes to the global stiffness matrix. For Exercises 1.2 and 1.3 the inputs are provided by gravity as indicated. Take $k_1 = k_3 = k_6 = k$, $k_2 = k_4 = k_5 = 2k$, and $k_7 = 3k$. Take all masses to be m.

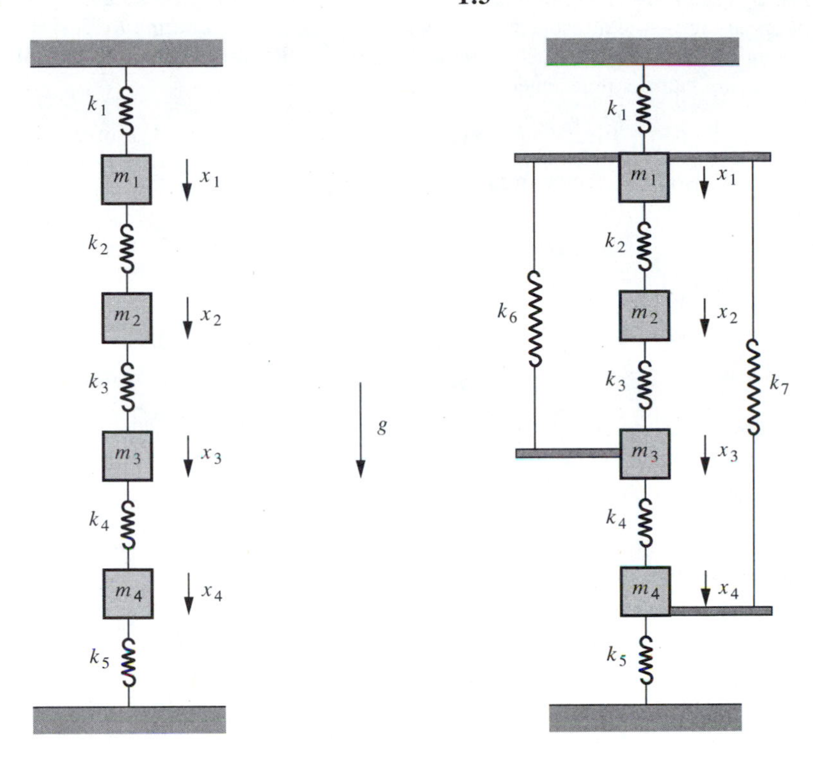

1.4. Repeat Exercise 1.2 with $g = 0$ for the case where a displacement X_2 is imposed at mass 2.

1.5. Repeat Exercise 1.3 with $g = 0$ for the case where a displacement X_3 is imposed at mass 3.

Section 1.4.2

1.6. Show that requiring the potential energy

$$V(x_1, x_2, x_3, x_4) = \frac{\mathbf{x}^T\mathbf{K}\mathbf{x}}{2} - \mathbf{x}^T\mathbf{F}$$

to be stationary by computing

$$\frac{\partial V}{\partial x_i} = 0 \qquad i = 1, 2, 3, 4$$

leads in general to

$$\left(\frac{\mathbf{K} + \mathbf{K}^T}{2}\right)\mathbf{x} = \mathbf{F}$$

so that when $\mathbf{K}$ is symmetric, $\mathbf{Kx} = \mathbf{F}$.

1.7. When springs are arranged in a noncolinear fashion such as in the figure below, it is necessary to define a displacement vector for each node according to $X_i = [x_i \ \ y_i]$. For an inclined member such as member 2 connecting nodes 1 and 4, show that the elongation can be represented as

$$\delta_{14} = \mathbf{n_{14}} \cdot (\mathbf{X_4} - \mathbf{X_1}) = (x_4 - x_1)\cos\alpha + (y_4 - y_1)\sin\alpha$$

where $\mathbf{n_{14}}$ is a unit vector from node 1 to node 4.

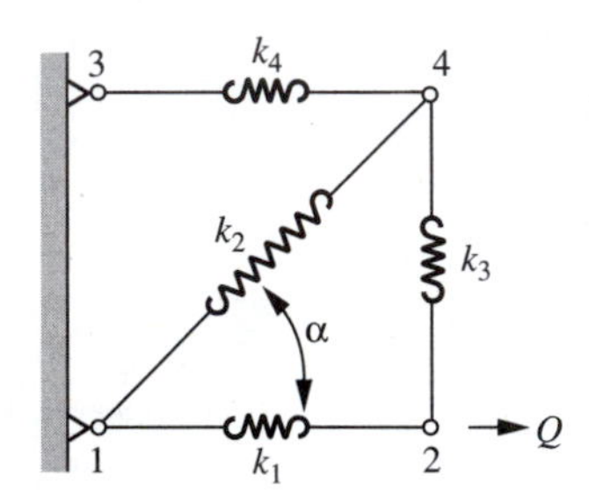

1.8. Use the result of Exercise 1.7, with $l = \cos\alpha$ and $m = \sin\alpha$, to show that the potential energy of the spring can be written as

$$\begin{aligned} U_{e2} &= \frac{1}{2}k_2\left((x_4 - x_1)\cos\alpha + (y_4 - y_1)\sin\alpha\right)^2 \\ &= \frac{k_2}{2}[x_1 \ \ y_1 \ \ x_4 \ \ y_4]\begin{bmatrix} l^2 & lm & -l^2 & -lm \\ lm & m^2 & -lm & -m^2 \\ -l^2 & -lm & l^2 & lm \\ -lm & -m^2 & lm & m^2 \end{bmatrix}\begin{bmatrix} x_1 \\ y_1 \\ x_4 \\ y_4 \end{bmatrix} \\ &= \frac{1}{2}\mathbf{x}_{\mathbf{e2}}^{\mathrm{T}}\mathbf{k}_{\mathbf{e2}}\mathbf{x}_{\mathbf{e2}} \end{aligned}$$

which is the energy of the spring expressed in terms of the elemental displacement vector $\mathbf{x}_{\mathbf{e2}}$ and the elemental stiffness matrix $\mathbf{k}_{\mathbf{e2}}$. Note that this is the variational approach to the formulation of the elemental stiffness for the truss element covered in Section 1.4.5.

1.9. Show that the actual tensile or compressive force in the spring of Exercises 1.7 and 1.8 can be computed as

$$F_2 = k_2\delta_{14} = k_2\left((x_4 - x_1)\cos\alpha + (y_4 - y_1)\sin\alpha\right)$$

Also show that

$$\frac{\partial U_{e2}}{\partial \mathbf{x}_{\mathbf{e2}}} = \mathbf{F}_{\mathbf{e2}} = \mathbf{k}_{\mathbf{e2}}\mathbf{x}_{\mathbf{e2}}$$

represents the nodal components in the x and y directions of the force in the bar.

1.10. Show that the potential energy U_{e2} of Exercise 1.8 can be further expanded to yield

$$U_{e2} = \frac{1}{2}\mathbf{x}_{\mathbf{G}}^{\mathrm{T}}\mathbf{k}_{\mathbf{G2}}\mathbf{x}_{\mathbf{G}}$$

where

$$\mathbf{k}_{G2} = \begin{bmatrix} l^2 & lm & 0 & 0 & 0 & 0 & -l^2 & -lm \\ lm & m^2 & 0 & 0 & 0 & 0 & -lm & -m^2 \\ 0 & 0 & 0 & 0 & 0 & 0 & 0 & 0 \\ 0 & 0 & 0 & 0 & 0 & 0 & 0 & 0 \\ 0 & 0 & 0 & 0 & 0 & 0 & 0 & 0 \\ 0 & 0 & 0 & 0 & 0 & 0 & 0 & 0 \\ l^2 & -lm & 0 & 0 & 0 & 0 & l^2 & lm \\ lm & -m^2 & 0 & 0 & 0 & 0 & lm & m^2 \end{bmatrix}$$

and

$$\mathbf{x}_G^T = [x_1 \ y_1 \ x_2 \ y_2 \ x_3 \ y_3 \ x_4 \ y_4]$$

$\mathbf{k}_{G2}$ is the stiffness matrix for spring or element 2 expressed in terms of the global size of the system.

Since the total internal energy can be represented as $\sum k\delta^2/2$, it follows that for the four-member example, the total internal potential energy can be represented as

$$U = \frac{1}{2}\sum_1^4 \mathbf{x}_{ei}^T \mathbf{k}_{ei} \mathbf{x}_{ei}$$

$$= \frac{1}{2}\sum_1^4 \mathbf{x}_G^T \mathbf{k}_{Gi} \mathbf{x}_G$$

$$= \frac{1}{2}\mathbf{x}_G^T \left(\sum_1^4 \mathbf{k}_{Gi}\right) \mathbf{x}_G$$

$$= \frac{1}{2}\mathbf{x}_G^T \mathbf{K}_G \mathbf{x}_G$$

where $\mathbf{K}_G = \sum \mathbf{k}_{Gi}$ is the global stiffness matrix. With the external potential energy expressed as

$$\Omega = -\mathbf{x}_G^T \mathbf{P}_G$$

with $\mathbf{P}_G$ as the 8×1 vector of nodal loads, the total potential energy can be expressed as

$$V = \frac{1}{2}\mathbf{x}_G^T \mathbf{K}_G \mathbf{x}_G - \mathbf{x}_G^T \mathbf{P}_G$$

and the global equilibrium equations can be generated in the usual way according to

$$\frac{\partial V}{\partial \mathbf{u}_G} = 0 = \mathbf{K}_G \mathbf{u}_G - \mathbf{P}_G$$

or

$$\mathbf{K}_G \mathbf{u}_G = \mathbf{P}_G$$

The constraints are enforced, the equations solved, and the internal member forces determined according to Exercise 8.

For each of the following exercises and figures, determine the nodal displacements and member forces by going through the following steps:

1. Number the nodes and springs.
2. Determine k_i and the inclination α for each spring.
3. Determine the elemental stiffness matrices $\mathbf{k_{ei}}$ for each spring.
4. Expand each of the $\mathbf{k_{ei}}$ to their respective $\mathbf{k_{Gi}}$.
5. Add the $\mathbf{k_{Gi}}$ to determine $\mathbf{K_G}$.
6. Form $\mathbf{P_G}$.
7. Form the augmented global equations $[\mathbf{K_G}|\mathbf{P_G}]$.
8. Apply the constraints.
9. Solve the equations.
10. Determine the internal member forces.
11. Verify that each of the nodes is in equilibrium under the action of the internal member forces and any external loads.

1.11.

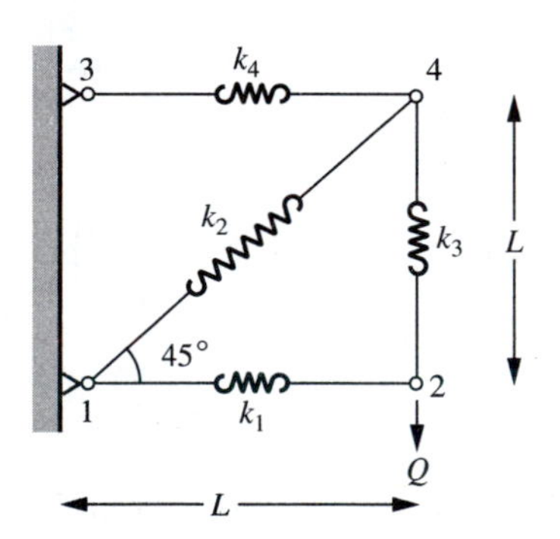

Take $k_1 = k_3 = k_4 = k$, and $k_2 = 2k$. Determine the displacements of nodes 2 and 4 and the forces in each of the members. What observations can be made about the validity of the solution?

1.12.

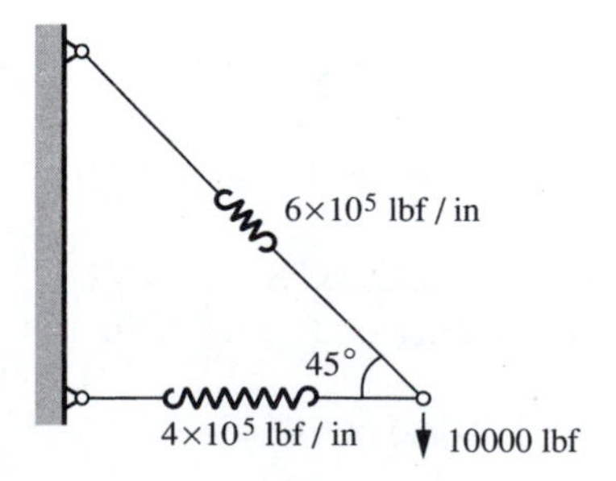

1.13.

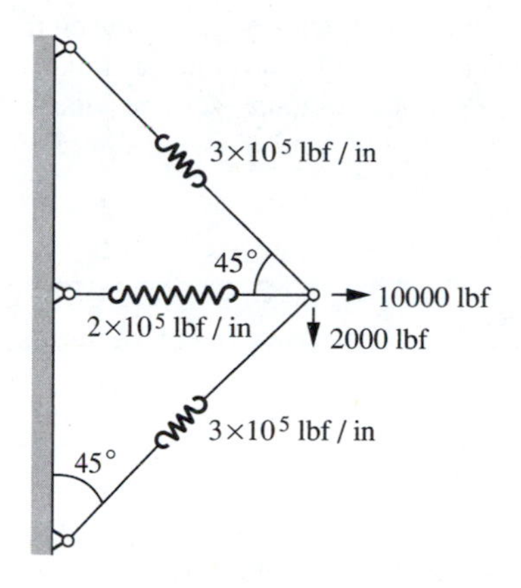

Section 1.4.3

For each of the following exercises and figures, set up and solve the equations necessary to determine the temperatures at the boundaries and interfaces. Also determine the total energy transfer across the region. Take the area to be 80 ft^2.

1.14.

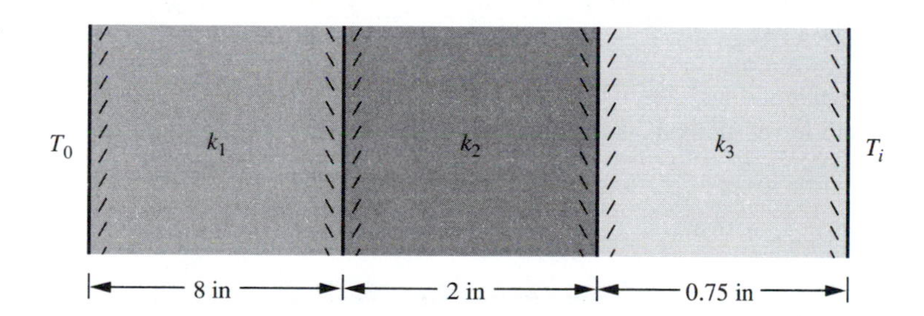

The temperatures T_0 and T_i are prescribed at the outer and inner surfaces of the wall as 110°F and 72°F respectively. Let $k_1 = 0.44$ BTU/hr-ft-°F, $k_2 = 0.03$ BTU/hr-ft-°F, and $k_3 = 0.06$ BTU/hr-ft-°F.

1.15. Repeat Exercise 1.14 with the thickness of the middle region doubled.

1.16. Rework Exercise 1.14 with the condition at the left boundary such that there is convection according to

$$q = Ah_L(T_0 - T_w)$$

where T_w is the outside wall temperature. Take $h_L = 0.1$ BTU/hr-ft^2-°F and the area of the wall to be 80 ft^2. Take $T_0 = 110$°F.

1.17. Rework Exercise 1.14 with the condition at both boundaries convective as in Exercise 1.16. Take the convective temperatures to be 110°F and 72°F, respectively.

1.18. An annular region $r_i \leq r \leq r_0$ is insulated on the top and bottom surfaces as indicated. The temperature at the inner radius is held at T_i and the outer radius at T_0. Break up the region $r_i \leq r \leq r_0$ into two equal-length segments and treat each as an element with area $a_{\text{avg}} = 2\pi t(r_i + r_{i+1})/2$ and length $\Delta r = r_{i+1} - r_i$, where r_i and

r_{i+1} are the radii at the inner and outer edges of the element. Solve for the unknown temperature at the interior point. Determine the flux $q = -ka_{\text{avg}}(T_{i+1} - T_i)/\Delta r$. Take $r_0 = 4r_i$ and $T_i = 3T_0$. Compare your results with the exact solution $T = (T_0 - T_i)\ln(r/r_i)/\ln(r_0/r_i) + T_i$ and $q = -2\pi krtT'$.

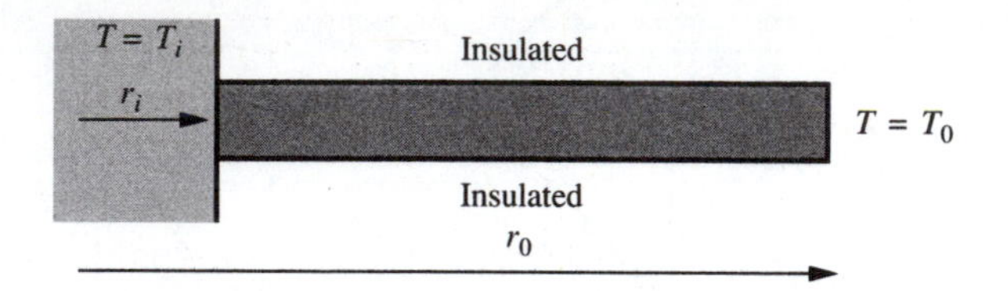

1.19. Repeat Exercise 1.18 for four and eight equal-length subdivisions. Compare the results with those from Exercise 1.18 and with the exact solution.

Section 1.4.4

For each of the next three exercises follow the steps indicated.

1. Number the junctions.
2. Set up the elemental $\mathbf{C_e}$ matrices.
3. Assemble into the global $\mathbf{C}$ array.
4. Assemble the loads.
5. Apply the constraints.
6. Solve the equations.
7. Compute the flows (derived variables).

1.20.

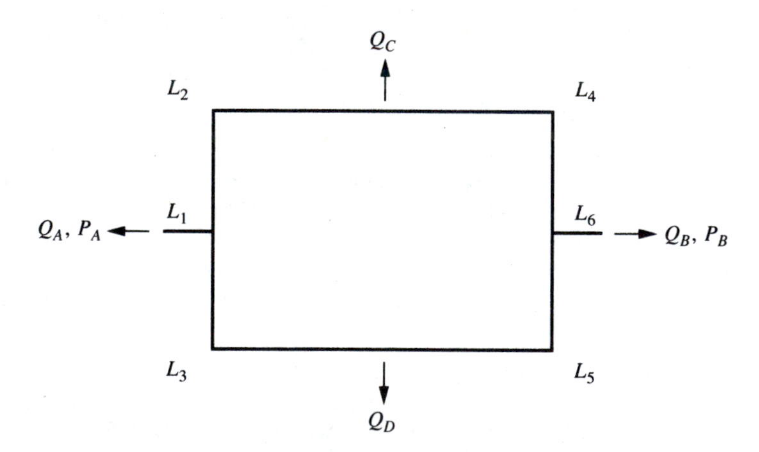

Take $L_1 = L_6 = 200$ ft, $L_2 = 600$ ft, $L_3 = 1000$ ft, $L_4 = 400$ ft, and $L_5 = 1200$ ft, $D_1 = D_6 = 3$ in. and $D_2 = D_3 = D_4 = D_5 = 2$ in. Consider P_A to be known and specify $Q_B = 4000$ in^3/s, $Q_C = Q_D = 0$ and determine the pressures at the junctions and the corresponding flows in the individual segments. What is the minimum required P_A? Take $\mu = 1.5 \times 10^{-7}$ lbf s/in^2.

1.21. Repeat Exercise 1.20 when $Q_B = 4000$ in^3/s, $Q_C = 1600$ in^3/s, and $Q_D = 0$.

1.22. Repeat Exercise 1.20 when $Q_B = 4000$ in^3/s, $Q_C = 1600$ in^3/s, and $Q_D = 2000$ in^3/s.

1.23. A manifold pictured below accepts flows from several sources and ejects them at zero pressure as shown.

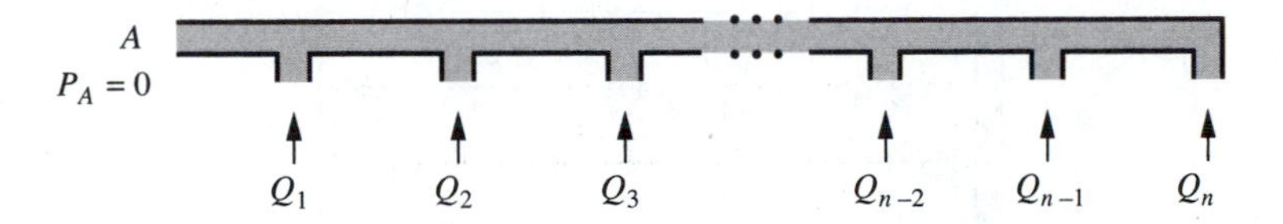

Assuming a $C_i = \pi D_4/128\mu L$ for each portion of the pipe (i.e., between flow inputs), set up the general equations to determine the pressures along the pipe. Note that the continuity equation at each junction must account for the two flows in the pipe and the input.

1.24. In Exercise 1.23, take $n = 2$, $Q_1 = Q_2 = Q_0$, and $C_1 = C_2 = C_0$ and determine the pressures at the nodes and flows within the elements. Plot the results.

1.25. For equal Q's and C's, repeat Exercise 1.23 for $n = 4$, and then $n = 8$. What do you think the pressure and flows would look like for a continuously distributed input along the length?

1.26. A simple flow distribution system can appear as in Exercise 1.23, except that the pressure P_A is some prescribed value and there are known or required flows being ejected rather than input at the locations along the pipe.

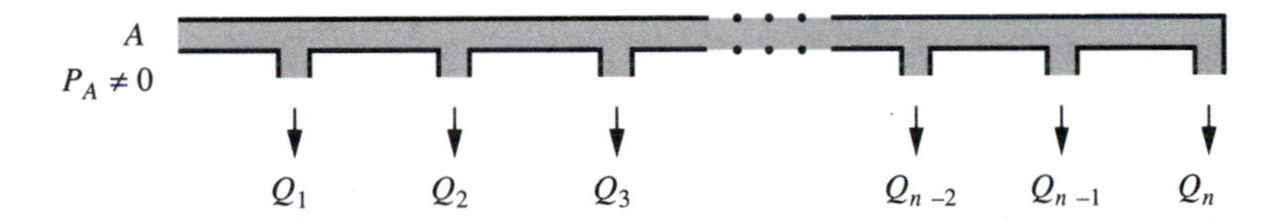

Set up the general equations for determining the pressures along the pipe. Solve these general equations for $n = 2$. Are there any restrictions on any of the variables for the solution to be meaningful? Specialize these restrictions when the Q's and C's are equal.

1.27. For equal C's and Q's, solve the equations for $n = 4$ and $n = 8$. Plot the results for the pressure and flow.

Section 1.4.5

For each of the following exercises and figures, set up and solve for the displacements at the nodes and for the forces transmitted by each of the elements. Follow this sequence of steps:

1. Number the nodes.
2. Determine the (AE/l) and θ for each element.
3. Determine the 4×4 elemental stiffness $\mathbf{k_e}$ for each element.
4. Expand each $\mathbf{k_e}$ to the global level and add to the global stiffness matrix.
5. Form the global load matrix.
6. Enforce the constraints.
7. Solve the equations.

8. Calculate the forces transmitted in each member.
9. Verify nodal equilibrium and calculate the reactions.

1.28. All areas = 1 in^2, all E's = 30×10^6 psi.

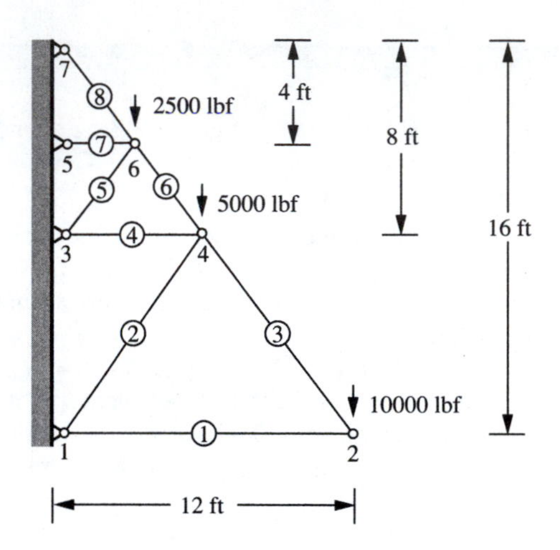

1.29. All areas = 1 in^2, all E's = 30×10^6 psi.

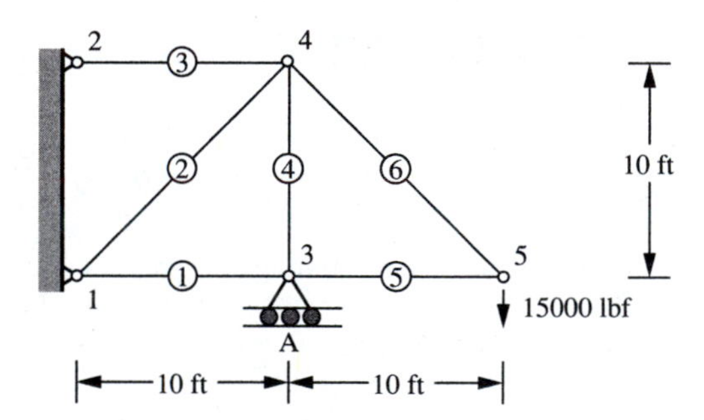

1.30. Repeat Exercise 1–29 with the roller at A replaced by a pin support, that is, complete constraint.

CHAPTER 2

One-Dimensional Boundary Value Problems

Chapter Contents

2.1 Introduction
2.2 The general problem
2.3 Galerkin finite element models
2.4 Example applications
    2.4.1 Solid mechanics—axial deformation
    2.4.2 Fluid mechanics—thin-film lubrication
    2.4.3 One-dimensional heat conduction with convection
2.5 Eigenvalue problems
    2.5.1 Torsional vibrations
2.6 Evaluation of elemental matrices
2.7 Variational finite element models
    2.7.1 The weak formulation
    2.7.2 Calculus of variations
    2.7.3 Ritz finite element models
2.8 Higher-order interpolations
2.9 Quadratic interpolation examples
    2.9.1 Solid mechanics—axial deformation
    2.9.2 Fluid mechanics—thin-film lubrication
    2.9.3 One-dimensional heat conduction with convection
    2.9.4 Torsional vibrations
2.10 Transverse deflections of beams
2.11 Errors and convergence
2.12 Closure
References
Computer projects
Exercises

2.1 INTRODUCTION

In disciplines such as the mechanics of deformable bodies, heat transfer, fluid mechanics, and vibration of elastic bodies, the analysis of a typical problem often involves isolating a differential element of the region in which the phenomenon is occurring. This differential element is then used to construct a free-body diagram, a control mass, or a control volume, and the appropriate basic physical principle (balance of momentum, balance of energy, balance of mass, etc.) is applied, with the result generally being a differential equation. It is frequently possible to make reasonable, simplifying assumptions about the physical behavior of the model such that the differential equation involves only one independent and one dependent variable, and is thus an ordinary differential equation. It is the purpose of this chapter to discuss and develop the finite element method as applied to boundary value and eigenvalue problems associated with linear ordinary differential equations. The physics of the problem will be kept clearly in focus from the standpoint of checking how well the predictions of the finite element model satisfy the basic physical principle. In this section we outline several important engineering application areas; we will use these to demonstrate the finite element method throughout the remainder of the chapter.

SOLID MECHANICS APPLICATION. Consider the class of solid mechanics problems shown in Fig. 2–1. The bar occupying the region shown is loaded by forces producing an internal axial force that must be transmitted along the length of the bar. The task at hand is to determine the displacements and the internal forces.

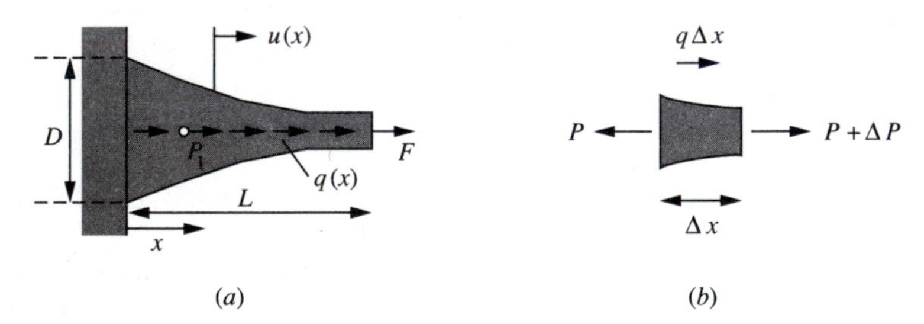

FIGURE 2–1 Classical axial deformation problem

The figure shows a variable-cross-section bar supported at one end against axial motion and loaded axially by distributed loads $q(x)$ and by typical concentrated loads P_1 and F, P_1 applied at an interior location and F at the end of the bar. The variation of the cross section is gradual so as to avoid stress concentrations; it is assumed that $D << L$. The bar is assumed to be composed of a linearly elastic material. The development of the governing equations is based on kinetics or balance of momentum (Newton's second law), kinematics or strain-displacement relations, and constitution or stress-strain relations. We consider each of these three basic ideas in turn, to develop the governing equations necessary for analyzing the relations between the displacement and the internal axial force.

Equilibrium. The free-body diagram shown in Fig. 2–1*b* is constructed to expose the internal force $P(x)$, after which, upon summoning forces, dividing by Δx and passing to the limit, we have

$$P' + q(x) = 0 \tag{a}$$

Also part of equilibrium is the relation between P, the internal force transmitted, and the corresponding axial stress σ, namely

$$P = \int_{\text{Area}} \sigma \, dA \tag{b}$$

Strain-displacement. With $u(x)$ as the axial displacement the corresponding strain-displacement relation is

$$\epsilon = u' \tag{c}$$

Stress-strain. For the one-dimensional elastic solid, the stress-strain relation is

$$\sigma = E\epsilon \tag{d}$$

where E is Young's modulus or the modulus of elasticity.

Combining (*b*), (*c*), and (*d*) gives the force-displacement relation $P = AEu'$, which can be eliminated using (*a*) to obtain the governing differential equation

$$(AEu')' + q = 0 \tag{e}$$

This is a second-order linear ordinary differential equation; it requires two boundary conditions, one at each end, to complete the formulation of the problem. In the present case these two boundary conditions are

$$u(0) = 0 \qquad \text{and} \qquad P(L) = AEu'(L) = F \tag{f}$$

The displacements are determined by solving the governing equation (*e*) and satisfying the boundary conditions (*f*). The internal force can then be computed using the force-displacement relation $P = AEu'$.

Note in passing that for a statically determinate problem the equilibrium equation $P' + q = 0$ can be integrated to obtain the internal force independent of the deformations, but that for a statically indeterminate problem $(AEu')' + q = 0$ must be integrated subjected to two appropriate boundary conditions.

FLUID MECHANICS APPLICATION. Consider the class of fluid mechanics problems pictured in Fig. 2–2*a*, frequently referred to as thin-film lubrication.

The basic physical problem involves two relatively flat surfaces moving relative to each other and prevented from contact by the presence of a viscous fluid between the surfaces. The geometry of the "converging wedge" is such that substantial pressure and viscous forces are developed in the fluid, allowing a load to be transferred through the fluid. The analysis is based upon the application of kinematics or conservation of mass, kinetics or balance of momentum (Newton's second law) and constitution or stress-strain rate relations. Each of these three

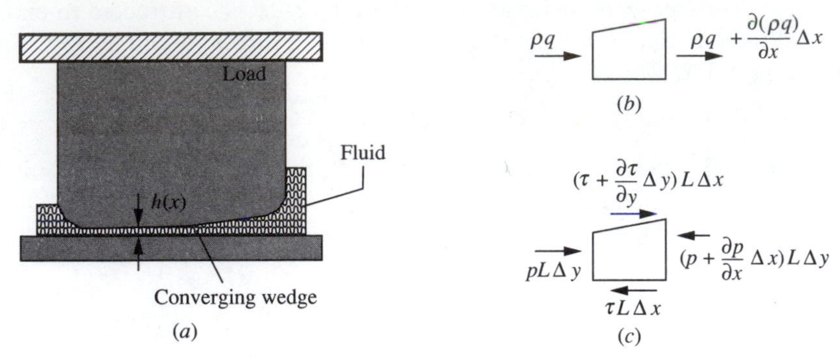

FIGURE 2–2 Thin-film lubrication

basic ideas is considered in turn in developing the governing equations necessary to analyze the relations between the film thickness $h(x)$ and the pressure.

Balance of mass. By requiring that mass be conserved for the differential element shown in Fig. 2–2*b* it follows, assuming incompressible flow, that

$$\frac{\partial(\rho q)}{\partial x} = \rho\frac{\partial q}{\partial x} = 0 \quad \Rightarrow \quad \frac{\partial q}{\partial x} = 0 \tag{a}$$

where q is the volume flow rate across the thickness of the film, related to u, the x-component of the velocity, according to $q = \int u \, dy$. ρ is the mass density.

Balance of momentum. Neglecting acceleration forces compared to pressure forces p and viscous forces τ, balance of momentum in the x-direction (using Fig. 2–2*c*) yields

$$\frac{\partial \tau}{\partial y} = \frac{\partial p}{\partial x} \tag{b}$$

Stress-strain rate. With η the kinematic viscosity, Newton's viscosity relation for this class of problem is

$$\tau = \eta\frac{\partial u}{\partial y} \tag{c}$$

Combining (*b*) and (*c*) yields

$$\frac{\partial}{\partial y}\left(\eta\frac{\partial u}{\partial y}\right) = \frac{\partial p}{\partial x}$$

Assuming that the pressure is independent of y, this can be integrated to yield

$$\eta\frac{\partial u}{\partial y} = y\frac{\partial p}{\partial x} + f(x)$$

Assuming η to be constant, another integration yields

$$u = \left(\frac{y^2}{2\eta}\right)\frac{\partial p}{\partial x} + yF(x) + G(x)$$

Satisfying nonslip boundary conditions on u, namely $u(0) = -U_1$ and $u(h) = U_2$, yields

$$u = \frac{y(y-h)}{2\eta}\frac{\partial p}{\partial x} - U_1\left(1 - \frac{y}{h}\right) + \frac{y}{h}U_2$$

The flow rate is then the integral of the velocity over the thickness, yielding

$$q = \int_0^h u\,dy = -\frac{h^3}{12\eta}\frac{\partial p}{\partial x} + \frac{h}{2}(U_2 - U_1)$$

Substituting into (a) yields

$$\frac{d}{dx}\left(\frac{h^3}{12\eta}\frac{\partial p}{\partial x}\right) = \frac{d}{dx}\left[\frac{h}{2}(U_2 - U_1)\right] \qquad (d)$$

for the pressure p. Eq. (d) is a second-order linear differential equation requiring two boundary conditions. As indicated in Fig. 2–3, we assume that the pressures are known at the ends of the thin film; that is,

$$p(0) = p_1 \qquad \text{and} \qquad p(L) = p_2 \qquad (e)$$

where p_1 and p_2 are frequently taken to be zero, that is, atmospheric pressure.

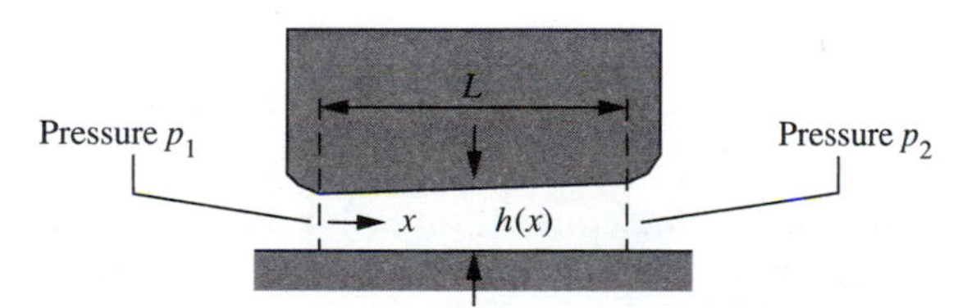

FIGURE 2–3 Boundary conditions for thin-film lubrication

Given $h(x)$ for the shape of the wedge, the pressure is determined by solving the governing equation (d) and satisfying the boundary conditions (f).

HEAT TRANSFER APPLICATION. Consider a classical heat conduction problem as indicated in Fig. 2–4a.

The bar is assumed to have a length L that is large compared to the transverse dimensions B and H, both of which can vary gradually with x. Energy in the form of heat is transferred (a) along the bar due to temperature gradients and (b) from the lateral surfaces due to convection. The basic physical principle that must be satisfied is balance of energy. Coupled with constitution (Fourier's law of heat conduction), the principle of balance of energy will yield the equation governing the temperature distribution in the bar.

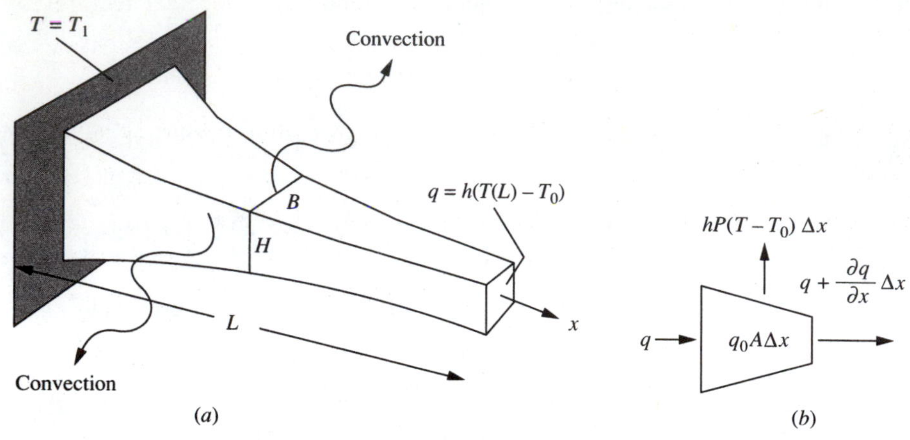

FIGURE 2–4 One-dimensional heat transfer

Balance of energy. Consulting Fig. 2–4*b*, a balance of energy in terms of

$$\text{energy in} - \text{energy out} + \text{internal energy sources} = 0$$

yields

$$q + q_0 A \Delta x - \left(q + \frac{\partial q}{\partial x}\Delta x\right) - hP(T - T_0)\Delta x = 0$$

or

$$\frac{\partial q}{\partial x} + hP(T - T_0) - q_0 A = 0 \tag{a}$$

where P is the perimeter, h is the convection heat transfer coefficient and T_0 is the "free-stream" temperature of the convection medium. For the present geometry, $P = 2(B + H)$.

Constitution. Fourier's law for one-dimensional heat conduction can be expressed as

$$q = -kA\frac{dT}{dx} \tag{b}$$

where k, referred to as the thermal conductivity, is a positive physical property of the material from which the bar is constituted. The assumption that heat is conducted only along the axis of the bar is justified on the basis of the fact that B and H have been taken to be small compared to L.

Combining (*a*) and (*b*), the governing equation for the temperature is

$$\frac{d}{dx}\left(kA\frac{dT}{dx}\right) - hP(T - T_0) + q_0 A = 0 \tag{c}$$

This is a second-order ordinary linear differential equation that requires two boundary conditions. For the situation shown in Fig. 2–4*a*, these are

$$T(0) = T_1 \qquad \text{and} \qquad -kAT'|_L = h(T(L) - T_0) \tag{d}$$

The solution is obtained by solving the governing equation (*c*) and satisfying the boundary conditions (*d*).

TORSIONAL VIBRATIONS APPLICATION. Consider the torsional oscillations of circular-cross-section shaft as pictured in Fig. 2–5*a*. The radius a, assumed to be small compared to the length L, can be a gradually varying function of position.

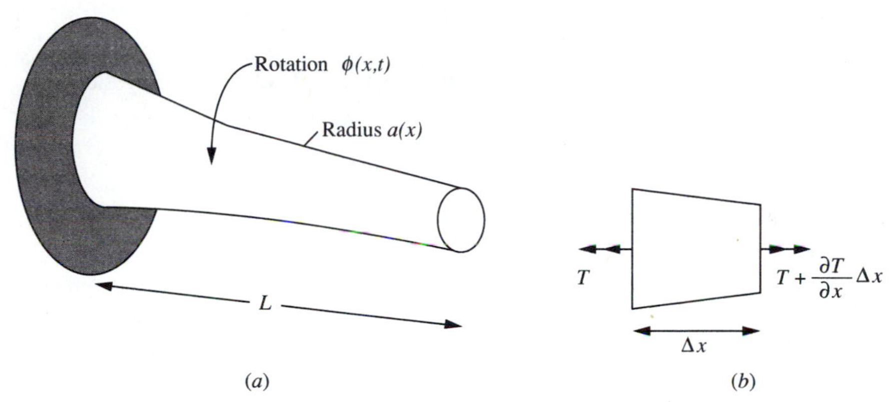

FIGURE 2–5 Torsional vibrations of a circular bar

The three basic ideas of solid mechanics—equations of motion, strain-displacement and stress-strain—will be discussed and combined to arrive at the final equation governing the problem.

Balance of angular momentum. Balance of angular momentum is enforced by virtue of Newton's second law. The internal torque T shown in the free-body diagram of Fig. 2–5*b* must satisfy

$$\sum T_x = -T + \left(T + \frac{\partial T}{\partial x}\Delta x\right) = \rho J \Delta x \frac{\partial^2 \phi}{\partial t^2} \tag{a}$$

or

$$\frac{\partial T}{\partial x} = \rho J \frac{\partial^2 \phi}{\partial t^2}$$

where $J = \pi a^4/2$ is the polar moment of the area and ρ is the mass density. Also from equilibrium, the internal torque is related to the stress according to

$$T = \int r\tau_{x\theta} dA \tag{b}$$

where $\tau_{x\theta}$ is the tangential component of the shear stress on the exposed face.

Strain-displacement. The shear strain associated with the rotation $\phi(x, t)$ shown in Fig. 2–5*a* is given by

$$\gamma_{x\theta}(r, x, t) = r\frac{\partial \phi(x, t)}{\partial x} \tag{c}$$

Stress-strain. For an isotropic elastic solid, the relation between the shear stress $\tau_{x\theta}$ and the shear strain $\gamma_{x\theta}$ is taken as

$$\tau_{x\theta} = G\gamma_{x\theta} \tag{d}$$

Combining Eqs. (*c*) and (*d*) and then substituting into Eq. (*b*) yields

$$T = GJ\frac{\partial \phi}{\partial x} \tag{e}$$

which is the torque-twist relation. Substituting Eq. (*e*) into Eq. (*a*) yields

$$\frac{\partial}{\partial x}\left(GJ\frac{\partial \phi}{\partial x}\right) = PJ\frac{\partial^2 \phi}{\partial t^2} \tag{f}$$

for the equation of motion governing the rotation ϕ. Note that we are now dealing with a partial differential equation. In the most general case we would append the equation of motion (*f*) with two boundary conditions and two initial conditions. For the treatment of this general problem the reader is referred to Chapter 4.

For the class of problems considered in this chapter we are primarily concerned with the free vibrations, determined by taking $\phi(x, t) = \psi(x)\exp(i\omega t)$, leading to

$$(JG\psi')' + P\omega^2 J\psi = 0 \tag{g}$$

Consulting Fig. 2–5*a*, the boundary conditions to be imposed on the variable $\psi(x)$ are

$$\psi(0) = 0 \qquad \text{and} \qquad T(L) = JG\psi'(L) = 0 \tag{h}$$

The mode shapes ψ and the natural frequencies ω are determined by solving the differential equation (*g*) and satisfying the boundary conditions (*h*).

In each of the example applications discussed above, the governing equations consist of a second-order linear ordinary differential equation and two boundary conditions. The task in each case is to determine a function $f(x)$ satisfying the appropriate differential equation and boundary conditions. We expect that in general, the function we are seeking is a relatively smooth function $f(x)$ such as indicated in Fig. 2–6*a*.

Shown in Fig. 2–6*b* is $F(x)$, an approximation to $f(x)$, consisting of piecewise linear segments passing through values $F(x_n) \approx f(x_n)$ at regularly spaced points along the interval from $x = 0$ to $x = L$. The finite element method, presented in the remainder of this chapter, is capable of generating just such an approximation. We will see in the following sections that as the number of regularly spaced points x_n is increased, the values $F(x_n)$ of the approximate piecewise linear functions more and more closely approximate the corresponding exact values $f(x_n)$, and that for all intents and purposes, the approximate solution F approaches the exact solution f.

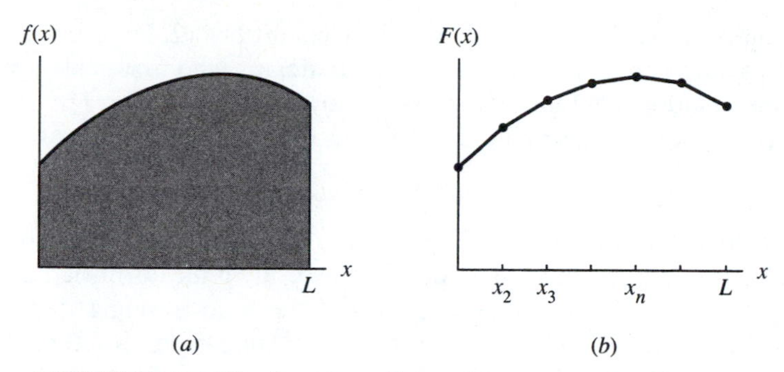

FIGURE 2–6 The function $f(x)$ and an approximation $F(x)$

When the finite element method is applied to any of the example applications presented above, the basic physics present in the governing equations is transferred directly to the equations of the finite element model. It is then possible to check that the basic physical principle is in fact satisfied, at least in a manner consistent with the physics of the resulting finite element model.

Two basic approaches are used to generate the finite element model for the type of problem discussed in the above examples. One approach, presented in Section 2.3, uses the method of Galerkin to attack the differential equation directly. This approach is essentially equivalent to the principle of virtual displacements or virtual work used frequently in mechanics, and leads to a system of linear algebraic equations that can be solved for approximate values of the solution to the original differential equation. For the other approach, presented in Section 2.7, it is necessary to develop a corresponding "energy," represented as an intregral over the interval. This energy, which for a conservative mechanical system is essentially the total potential energy, is then used as the basis for constructing a Ritz finite element model—also a system of linear algebraic equations. For the type of problems considered in this chapter the two finite element models will turn out to be identical.

2.2 THE GENERAL PROBLEM

Many of the ordinary differential equations mentioned in the previous section can be considered as specific examples of the class of boundary value problems defined by the differential equation

$$(pu')' + (\lambda r - q)u + f = 0 \qquad a < x < b \tag{2.1a}$$

and the two boundary conditions

$$\begin{aligned} -p(a)u'(a) + \alpha u(a) &= A \\ p(b)u'(b) + \beta u(b) &= B \end{aligned} \tag{2.1b}$$

We assume that q, r, and f are piecewise continuous functions of x, that p has a continuous derivative, and that p, q, and r are positive on the basic interval $a < x < b$. λ is a parameter with α, β, A, and B constants. The collection of the differential equation (2.1*a*) and boundary conditions (2.1*b*) is called a regular Sturm-Liouville system, after the two mathematicians who first made a systematic study of the solutions and their properties [Boyce and DiPrima, 1].

Occasionally a linear second-order differential equation appears as

$$a_0(x)u'' + a_1(x)u' + a_2(x)u = a_3(x) \tag{2.2}$$

rather than in the form given in Eq. (2.1*a*). From the standpoints of both the physics and the mathematics, it is preferable to alter the form of Eq. (2.2) by collapsing the first two terms $a_0(x)u'' + a_1(x)u'$ into a single term as in the standard form given by Eq. (2.1*a*). For a linear second-order differential equation this is always possible in principle by seeking an integrating factor $\mu(x)$, satisfying the first-order linear differential equation

$$\mu a_1 = (\mu a_0)'$$

which can be solved for μ to yield

$$\mu = \frac{1}{a_0} \exp\left(\int_0^x \frac{a_1}{a_0} dt\right).$$

Multiplying the differential equation (2.2) by the integrating factor μ and defining

$$\mu a_0 = p$$

$$\mu a_2 = \lambda r - q$$

$$\mu a_3 = -f$$

allows the differential equation to be put in the standard form (2.1*a*). This form is referred to in the literature as the **formal self-adjoint** form of the differential equation. The boundary value problem given by the differential equation (2.1*a*) and the specific boundary conditions (2.1*b*) is an example of a **self-adjoint boundary value problem**. We will assume in what follows that any linear second-order differential equation and boundary conditions under consideration have been put in the standard form so that the functions p, q, r, and f, as well as the constants α, A, β, and B, can be identified.

In this regard, it is important to mention that the general form of the boundary condition at $x = a$, namely,

$$-p(a)u'(a) + \alpha u(a) = A$$

can be specialized to the case $-p(a)u'(a) = A$ (i.e., $\alpha = 0$), but *not* to the case where u is prescribed. A boundary condition that contains the u' term specifically is called **natural boundary condition.** For a boundary condition of the form

$$u(a) = U_a$$

called an **essential boundary condition,** both α and A are taken as zero; a special case not obtainable from the general boundary condition by specializing α and A. Similar statements apply for the boundary condition at $x = b$.

Example 2.1.

Consider the equidimensional equation

$$x^2u'' + xu' + u + 1 = 0$$
$$u(1) = 1$$
$$u'(2) + u(2) = 0$$

The integrating factor is computed as

$$\mu = \frac{\exp(\int_0^x t\,dt/t^2)}{x^2} = \frac{1}{x}$$

Multiplying the differential equation by μ yields the standard form

$$xu'' + u' + \frac{u}{x} + \frac{1}{x} = (xu')' + \frac{u}{x} + \frac{1}{x} = 0$$

so that $p = x$, $q = -1/x$, and $f = 1/x$. The u' term is absent in the boundary condition at $x = 1$ so that $\alpha = A = 0$. The boundary condition at $x = 2$ must be multiplied by $p(2) = 2$ for it to be in standard form. This produces

$$p(2)u'(2) + p(2)u(2) = 2u'(2) + 2u(2) = 0$$

so that by comparison, $\beta = 2$ and $B = 0$.

It is worth mentioning that the purpose of the integrating factor is to collapse the first two terms into one. For this to be possible, the coefficient of u' must be the derivative of the coefficient of u''. On this basis it may be possible to ascertain the integrating factor by inspection.

The character of the solution to Eq. (2.1*a*) is strongly influenced by the λru term. When the λru term is absent, f, A, and B are generally not all zero and we have a simple boundary value problem. Solving the boundary value problem can be approached in principle by determining two linearly independent solutions to the homogeneous differential equation together with a particular solution, the sum of which must satisfy the two boundary conditions. A simple case where this approach is possible is presented in Example 2.2.

Example 2.2.

$$x^2u'' - 2xu' + 2u = 1$$
$$u(1) = 1$$
$$u(2) = 0$$

This is also a Euler or equidimensional equation with the general solution given by

$$u = c_1x + c_2x^2 + \frac{1}{2}$$

Satisfying the boundary conditions leads to the set of linear equations from which $c_1 = 5/4$ and $c_2 = -3/4$. The solution is then

$$u = \frac{5x - 3x^2 + 2}{4}$$

When the $\lambda r u$ term is present, f, A, and B are usually all zero and we have what is generally referred to as an eigenvalue problem. Solving the eigenvalue problem can be approached again in principle by determining two linearly independent solutions to the homogeneous differential equation and subsequently satisfying the two homogeneous boundary conditions. This leads to a characteristic equation from which the eigenvalues and eigenfunctions can be determined.

Example 2.3.

$$\begin{aligned} u'' + \lambda u &= 0 \\ u(0) &= 0 \\ u'(L) &= 0 \end{aligned}$$

The homogeneous solution is $u = c_1 \cos \mu x + c_2 \sin \mu x$ where $\mu^2 = \lambda$. Satisfying the boundary conditions leads to

$$\begin{aligned} c_1 1 + c_2 0 &= 0 \\ -\mu c_1 \sin \mu L + \mu c_2 \cos \mu L &= 0 \end{aligned}$$

or

$$\begin{bmatrix} 1 & 0 \\ -\mu \sin \mu L & \mu \cos \mu L \end{bmatrix} \begin{bmatrix} c_1 \\ c_2 \end{bmatrix} = \begin{bmatrix} 0 \\ 0 \end{bmatrix}$$

For nontrivial c_1 and c_2, the determinant must vanish, leading to

$$\mu \cos \mu L = 0$$

The squares of the roots to this characteristic equation are the eigenvalues λ_n given by

$$\lambda_n = \mu_n^2 = \left(\frac{(2n-1)\pi}{2L} \right)^2$$

with the corresponding eigenfunctions given by

$$u_n(x) = \sin\left(\frac{(2n-1)\pi x}{2L} \right)$$

For a vibrations problem, the eigenvalues and eigenfunctions would be related to the natural frequencies of vibration and the mode shapes, respectively.

In these two simple examples, the solutions were easily obtained due to the special character of the functions p, q, and r; that is, the differential equations were Euler and constant coefficient equations, respectively. When the differential equation is not constant coefficient or Euler, the two linearly independent solutions of the homogeneous equation as well as the particular solution cannot generally be expressed in terms of elementary functions. If an analytical solution is desired, one can determine the two linearly independent solutions of the homogeneous equation in terms of power series, or attempt to express the solution in terms of solutions to special, regularly occurring differential equations such as Bessel's or Legendre's equations, for which the solutions and their properties are widely tabulated and discussed.

Solutions generated in this manner—by determining two linearly independent functions that satisfy the differential equation, adding a particular solution, and then satisfying the boundary conditions—are termed **classical solutions.** They are solutions that, with the possible exception of a finite number of points in the interior of the interval, possess derivatives of all orders. When it is possible to proceed effectively in this manner, it is the method of choice since one has essentially generated the "exact" solution to the boundary value problem. Alternatively, in the absence of any of the special circumstances just mentioned, a numerical method is generally used for extracting the necessary information about the solution of the problem.

In the remainder of this chapter, we will investigate several approaches for generating approximate solutions to the general linear one-dimensional second-order boundary value problem. We will investigate in detail the finite element method, a numerical method that can be used for obtaining solutions to the general boundary value problem given by Eqs. (2.1).

2.3 GALERKIN FINITE ELEMENT MODELS

In this section, the method of Galerkin will be used to develop finite element models for applications introduced in Section 2.1. In general, the steps to be followed in generating a finite element model are, again,

1. Discretization
2. Interpolation
3. Elemental formulation
4. Assembly
5. Constraints
6. Solution
7. Computation of derived variables

Each of these basic steps will be discussed in detail in what follows.

Discretization. As indicated in Section 2.1, we expect that the solution $u(x)$ to many of the applications of interest is a relatively smooth function as shown in Fig. 2–7*a*, where for generality the range of the independent variable is chosen to be $a \leq x \leq b$.

The first step in developing the finite element model, discretization, consists of choosing a set of points x_i, $i = 1, 2, \ldots, N+1$, as shown in Fig. 2–7*b*. These points are referred to as the **nodes**. The intervals between the nodes are referred

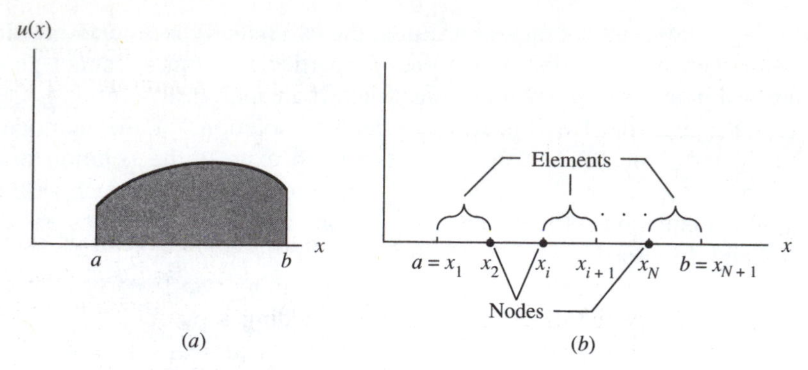

FIGURE 2–7 Discretization—nodes and elements

to as the **elements**. The nodes and elements are commonly referred to collectively as the **mesh**. Note that there are N elements, $N+1$ nodes, and that $a \equiv x_1$ and $b \equiv x_{N+1}$. The nodes are frequently chosen to be uniformly spaced so that all the elements have equal lengths. In regions where rapid changes in the solution $u(x)$ are expected, the nodes can be chosen so that elemental lengths are smaller, thus enabling the approximate finite element solution $v(x)$ to more easily represent the rapid changes. In such instances the nodes should be chosen so that there is a gradual change in the lengths of the elements.

Interpolation. After the locations of the nodes have been selected, an approximate solution $v(x)$ is assumed to be represented in a piecewise linear fashion as indicated in Fig. 2–8*a*. As discussed briefly in the introduction, we expect that if we construct the finite element model correctly the piecewise linear approximation $v(x)$ will approach the corresponding exact solution $u(x)$ as the number of nodes is increased.

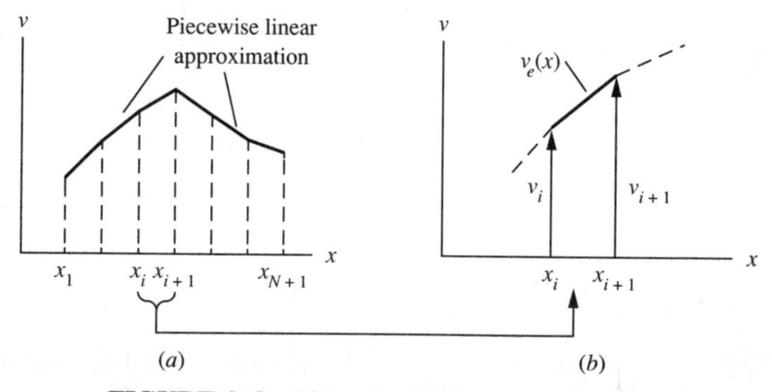

FIGURE 2–8 Piecewise linear approximation

On a typical element $x_i \leq x \leq x_{i_1}$, shown in Fig. 2–8*b*, the form for v is clearly

$$v_e(x) = \alpha + \beta x$$

where α and β are determined by requiring that $v_e(x_i) = v_i$ and $v_e(x_{i+1}) = v_{i+1}$, leading to

$$v_e(x) = \frac{x_{i+1} - x}{x_{i+1} - x_i} v_i + \frac{x - x_i}{x_{i+1} - x_i} v_{i+1} = N_i v_i + N_{i+1} v_{i+1}$$

N_i and N_{i+1}, shown in Fig. 2–9, are referred to as **elemental** linear interpolation functions; $\ell_e = x_{i+1} - x_i$ is the corresponding element length.

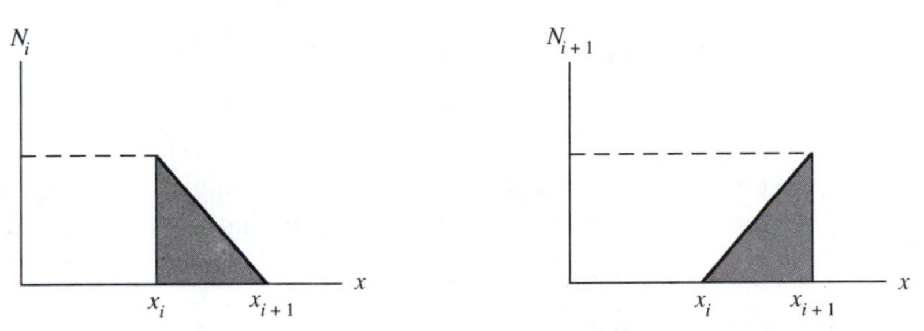

FIGURE 2–9 Elemental linear interpolation functions

On the entire interval $a \leq x \leq b$ the approximate function $v(x)$, shown in Fig. 2–8a, is representing according to

$$v(x) = \sum_{i=1}^{N+1} v_i n_i(x) \tag{2.3}$$

where the $n_i(x)$, shown in Fig. 2–10, are referred to as the **nodal** linear interpolation functions. The character of the corresponding $n_i'(x)$, which are needed in constructing the finite element model, are also shown in Fig. 2–10.

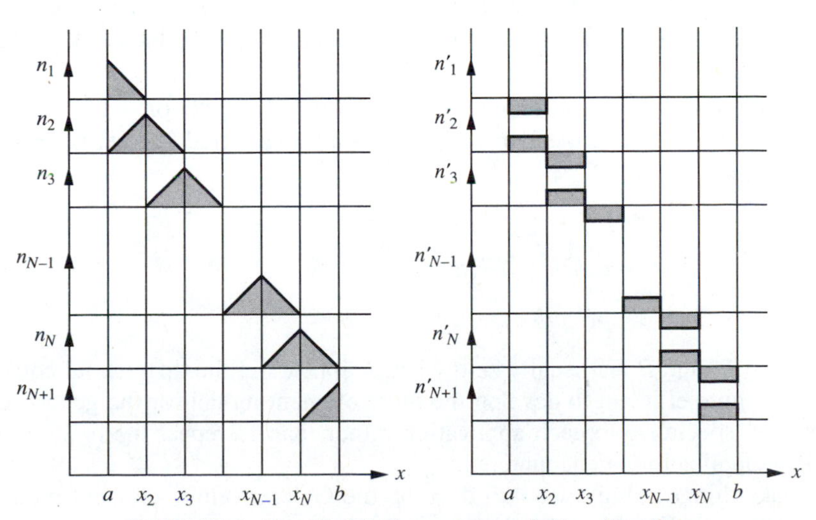

FIGURE 2–10 Nodal linear interpolation functions

Each of the nodal $n_i(x)$ satisfies

$$n_i(x_j) = 1 \qquad \text{if } i = j$$
$$n_i(x_j) = 0 \qquad \text{if } i \neq j$$

In other words, n_i has the value 1 at the node located at x_i and n_i *has the value* 0 *at all the other nodes*.

Evaluating the approximate solution at a typical node located at x_j there results

$$v(x_j) = \sum_{i=1}^{N+1} v_i n_i(x_j) = v_j$$

showing that the *coefficients* v_i *are the nodal values of the approximate solution* $v(x)$. Further, each of the nodal n_i is clearly a linear combination of the appropriate elemental N_i of Fig. 2–9 for the two elements neighboring the node x_i, thus,

$$n_i = N_{i+1} \text{ (from element } i-1) \qquad \text{for} \qquad x_{i-1} \le x \le x_i$$
$$n_i = N_i \text{ (from element } i) \qquad \text{for} \qquad x_i \le x \le x_{i+1}$$

For obvious reason, the n_i are frequently referred to as *hat* or *roof* functions. In what follows, it is important to remember that at an interior node x_i the corresponding n_i is nonzero only in the two elements adjacent to the node located at x_i. At a boundary node the corresponding n_i is nonzero only in the adjacent element.

The process of representing the approximate solution $v(x)$ in terms of the n_i (or the N_i) is generally referred to as **interpolation**, one of the basic steps in the derivation of a finite element model. The nodal interpolation functions n_i are used in developing the finite element model using Galerkin's method, whereas the elemental interpolation functions N_i are used in developing the finite element model using the Ritz method. Other types of interpolation used for developing both the Galerkin and Ritz finite element models are discussed later in the chapter.

Elemental formulation and assembly. As discussed in Section 2.2, each of the applications presented in the introduction can be considered special cases of the general problem

$$(pu')' + (\lambda r - q)u + f = 0 \qquad a \le x \le b$$
$$-p(a)u'(a) + \alpha u(a) = A$$
$$p(b)u'(b) + \beta u(b) = B$$

where $f(x)$, A and B are usually zero when λ appears in the differential equation. It is much more efficient to develop the finite element model for the general cases and to then specialize to each application rather than to repeat the process when each new application is encountered.

To make things definite we will develop the Galerkin finite element model for the noneigenvalue problem

$$(pu')' - qu + f = 0 \qquad a \leq x \leq b$$
$$-p(a)u'(a) + \alpha u(a) = A$$
$$p(b)u'(b) + \beta u(b) = B$$

which can be used to analyze the solids, the fluids, and the heat transfer application. We will develop the model for the case where both boundary conditions are natural boundary conditions and discuss later in this section the case of an essential boundary condition.

As outlined above, we take an approximate solution of the form

$$v(x) = \sum_{i=1}^{N+1} v_i n_i(x)$$

Substitution of the approximate solution $u = v$ into the differential equation yields

$$(pv')' - qv + f = \left(p \sum_{i=1}^{N+1} v_i n_i'(x) \right)' - q \sum_{i=1}^{N+1} v_i n_i(x) + f$$
$$= E(x, v_1, v_2, \ldots, v_{N+1})$$

where E is the error arising from the fact that the approximate solution v does not (in general) satisfy the differential equation. Galerkin's method [2], one of the collection of the so-called methods of weighted residuals (see Appendix A), requires that the error function E be integrated over the interval from a to b against each member of the set of $N + 1$ independent nodal interpolation functions $n_k(x)$ used in representing the solution $v(x)$, or

$$\int_a^b E n_k(x)\, dx = 0 \qquad k = 1, 2, \ldots, N + 1$$

This process generates $N + 1$ linear algebraic equations for the determination of the $N + 1$ unknown coefficients v_i. For large N this effectively forces the error E to approach zero and the approximate piecewise linear solution $v(x)$ to approach the corresponding exact solution $u(x)$.

Substituting the error E into the above integrals yields

$$\int_a^b [(pv')' - qv + f] n_k(x)\, dx = 0 \qquad k = 1, 2, \ldots, N + 1$$

We observe that E contains second derivatives ($(pv')' = p'v' + pv''$) of the approximate solution v. Since v is piecewise linear as shown typically in Fig. 2–11a, v' has a constant value within each element as indicated in Fig. 2–11b.

Thus we see that the second derivatives of v don't exist at the nodes. Fortunately, this difficulty can be overcome if the first term of the error E is integrated by parts to yield

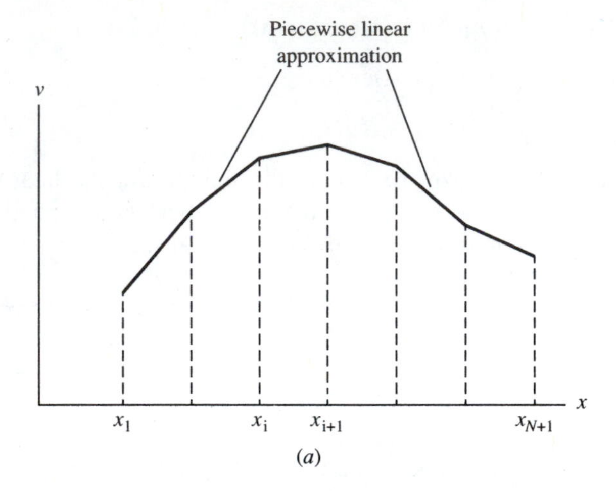

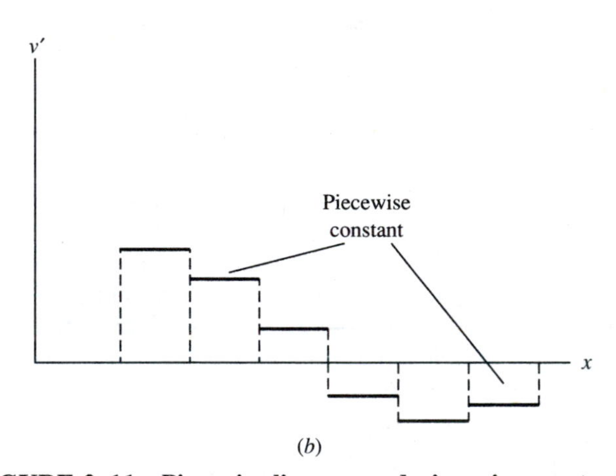

FIGURE 2–11 Piecewise linear v and piecewise constant v'

$$n_k pv'\Big|_a^b - \int_a^b [n_k' pv' + n_k qv + n_k f]\,dx = 0 \qquad k = 1, 2, \ldots, N+1$$

The pv' terms evaluated at the boundary are replaced using the boundary conditions satisfied by the approximate solution v to yield

$$n_k(b)[B - \beta v(b)] - n_k(a)[\alpha v(a) - A] - \int_a^b [n_k' pv' + n_k qv + n_k f]\,dx = 0$$

$$k = 1, 2, \ldots, N+1$$

Only the first derivatives of the interpolation functions derivative $n_k(x)$ appear in this equation. Additionally and equally important is the fact that the integration by parts has entered the boundary conditions to be satisfied by v into the equations.

Finally substituting for $v = \sum v_i n_i$ and accounting for the fact that $n_k(b)$ is zero except when $k = N + 1$ and that $n_k(a)$ is zero except when $k = 1$, these equations can be expressed as

$$\sum_{i=1}^{N+1}\left[\int_a^b (n_k' p n_i' + n_k q n_i)\,dx\right]v_i + \beta v_{N+1}\delta_{kN+1} + \alpha v_1 \delta_{k1}$$

$$= \int_a^b n_k f\,dx + B\delta_{kN+1} + A\delta_{k1} \qquad k = 1, 2, \ldots, N+1 \tag{2.4}$$

where δ_{ij} is the ***Kronecker delta*** given by

$$\delta_{ij} = 0 \quad \text{if} \quad i \neq j$$

$$\delta_{ij} = 1 \quad \text{if} \quad i = j$$

The term δ_{kN+1} indicates the occurrence of a term in only the $(N + 1)$st equation and δ_{k1} indicates the occurrence of a term in only the first equation. Eq. (2.4) can be written as

$$\sum_{i=1}^{N+1} K_{ki} v_i = F_k \qquad k = 1, 2, \ldots, N+1$$

or

$$\mathbf{K}\mathbf{v} = \mathbf{F} \tag{2.5}$$

where

$$K_{ki} = \int_a^b (n_k' p n_i' + n_k q n_i)dx + \beta\delta_{ki}\delta_{iN+1} + \alpha\delta_{ki}\delta_{i1}$$

are the coefficients of the *global stiffness matrix*. Note that when i and k are interchanged on the right-hand side of the equation there is no change in the value of the coefficient K_{ki} so that the matrix $\mathbf{K}$ is *symmetric*. The F_k given by

$$F_k = \int_a^b n_k f\,dx + B\delta_{kN+1} + A\delta_{k1}$$

are the coefficients of the *global load matrix*.

It is instructive to investigate in more detail the nature of this set of equations. For $k \neq 1$ or $k \neq N + 1$, the corresponding n_k is nonzero only in the intervals $x_{k-1} \leq x \leq x_k$ and $x_k \leq x \leq x_{k+1}$, so that the kth equation can be expressed as

$$\sum_{i=1}^{N+1} \int_{x_{k-1}}^{x_{k+1}} (n_k' p n_i' + n_k q n_i)\,dx\,v_i = \int_{x_{k-1}}^{x_{k+1}} n_k f\,dx$$

Then with only the n_i for $i = k-1, k$, and $k+1$ being nonzero in the intervals $x_{k-1} \leq x \leq x_k$ and $x_k \leq x \leq x_{k+1}$, the kth equation can be expressed as

$$
\begin{aligned}
&\int_{x_{k-1}}^{x_k} (n_k' p n_{k-1}' + n_k q n_{k-1})\, dx \; v_{k-1} \\
&+ \int_{x_{k-1}}^{x_{k+1}} (n_k' p n_k' + n_k q n_k)\, dx \; v_k \\
&+ \int_{x_k}^{x_{k+1}} (n_k' p n_{k+1}' + n_k q n_{k+1})\, dx \; v_{k+1} = \int_{x_{k-1}}^{x_{k+1}} n_k f \, dx
\end{aligned}
$$

that is, *only the unknowns* v_{k-1}, v_k, *and* v_{k+1} *appear in the* k^{th} *equation.* In a completely similar fashion it follows that only the unknowns v_1 and v_2 appear in the first equation and that only the unknowns v_N and v_{N+1} appear in the $(N+1)^{\text{st}}$ equation. Thus the global stiffness matrix **K** has the form

$$
\mathbf{K} = \begin{bmatrix}
K_{11} & K_{12} & \dots & \dots & 0 \\
K_{21} & K_{22} & K_{23} & \dots & 0 \\
0 & K_{32} & K_{33} & K_{34} & \\
\vdots & & & & \\
0 & \dots & k_{N,N-1} & k_{N,N} & k_{N,N+1} \\
0 & \dots & & k_{N+1,N} & k_{N+1,N+1}
\end{bmatrix}
$$

This is commonly referred to in the literature as a **tri-diagonal matrix.** This property, for the linearly interpolated finite element model, can be exploited to reduce the effort required to solve the final set of equations.

It turns out that there are distinct advantages in being able to view the final set of equations from an elemental rather that a nodal perspective. To this end, consider a specific case where $N = 2$, that is, two elements and three nodes. Taking account of the regions in which each of the n_k is nonzero, the final set of three unconstrained equations can be written out in full as

$$
\begin{aligned}
&\boxed{\int_a^{x_2} F_{11} dx\, v_1 + \int_a^{x_2} F_{12} dx\, v_2} && + \alpha v_1 = \boxed{\int_a^{x_2} n_1 f\, dx} + A \\
&\boxed{\int_a^{x_2} F_{21} dx\, v_1 + \int_a^{x_2} F_{22} dx\, v_2} + \boxed{\int_{x_2}^{b} F_{22} dx\, v_2 + \int_{x_2}^{b} F_{23} dx\, v_3} && = \boxed{\int_a^{x_2} n_2 f\, dx} + \boxed{\int_{x_2}^{b} n_2 f\, dx} \\
&\boxed{\int_{x_2}^{b} F_{32} dx\, v_2 + \int_{x_2}^{b} F_{33} dx\, v_3} + \beta v_3 = B && + \boxed{\int_{x_2}^{b} n_3 f\, dx}
\end{aligned}
$$

where $F_{ij} = n_i' p n_j' + n_i q n_j$. The boxed terms indicated in the first and second equations arise from the first element, $a \le x \le x_2$, whereas the boxed terms indicated in the second and third equations arise from the second element, $x_2 \le x \le b$. In terms of the elemental interpolation functions $N_1 \equiv N_i$ and $N_2 \equiv N_{i+1}$ defined previously, we define the *elemental stiffness matrices* as

$$
\mathbf{k_{e1}} = \begin{bmatrix}
\int_a^{x_2} F_{11} dx & \int_a^{x_2} F_{12} dx \\
\int_a^{x_2} F_{21} dx & \int_a^{x_2} F_{22} dx
\end{bmatrix}
\qquad
\mathbf{k_{e2}} = \begin{bmatrix}
\int_{x_2}^{b} F_{11} dx & \int_{x_2}^{b} F_{12} dx \\
\int_{x_2}^{b} F_{21} dx & \int_{x_2}^{b} F_{22} dx
\end{bmatrix}
$$

where

$$\begin{aligned} F_{11} &= N_1' p N_1' + N_1 q N_1 \\ F_{12} &= N_1' p N_2' + N_1 q N_2 \\ F_{21} &= N_2' p N_1' + N_2 q N_1 \\ F_{22} &= N_2' p N_2' + N_2 q N_2 \end{aligned}$$

and clearly $F_{12} = F_{21}$; that is, the elemental stiffness matrices are symmetric. If we define an elemental interpolation vector $\mathbf{N}$ as

$$\mathbf{N} = \begin{bmatrix} N_1 \\ N_2 \end{bmatrix} \qquad \text{with} \qquad \mathbf{N}' = \begin{bmatrix} N_1' \\ N_2' \end{bmatrix}$$

it follows that

$$\mathbf{k_{e1}} = \int_a^{x_2} (\mathbf{N}' p \mathbf{N}'^{\mathrm{T}} + \mathbf{N} q \mathbf{N}^{\mathrm{T}})\, dx$$

$$\mathbf{k_{e2}} = \int_{x_2}^{b} (\mathbf{N}' p \mathbf{N}'^{\mathrm{T}} + \mathbf{N} q \mathbf{N}^{\mathrm{T}})\, dx$$

For equal length elements and for p and q constants, $\mathbf{k_{e1}} = \mathbf{k_{e2}}$. For variable p and q, care must be taken to use the correct values in evaluating the integrals.

For $N = 2$ each of the elemental stiffness matrices is then expanded to the global or system size according to

$$\mathbf{k_{g1}} = \begin{bmatrix} \int_a^{x_2} F_{11}\, dx & \int_a^{x_2} F_{12}\, dx & 0 \\ \int_a^{x_2} F_{21}\, dx & \int_a^{x_2} F_{22}\, dx & 0 \\ 0 & 0 & 0 \end{bmatrix} \qquad \mathbf{k_{g2}} = \begin{bmatrix} 0 & 0 & 0 \\ 0 & \int_{x_2}^{b} F_{11}\, dx & \int_{x_2}^{b} F_{12}\, dx \\ 0 & \int_{x_2}^{b} F_{21}\, dx & \int_{x_2}^{b} F_{22}\, dx \end{bmatrix}$$

with

$$\mathbf{k_G} = \mathbf{k_{g1}} + \mathbf{k_{g2}}$$

In a completely similar fashion, we define the *elemental load matrices* as

$$\mathbf{f_{e1}} = \begin{bmatrix} \int_a^{x_2} N_1 f\, dx \\ \int_a^{x_2} N_2 f\, dx \end{bmatrix} \qquad \mathbf{f_{e2}} = \begin{bmatrix} \int_{x_2}^{b} N_1 f\, dx \\ \int_{x_2}^{b} N_2 f\, dx \end{bmatrix}$$

which can be expressed in terms of the elemental interpolation vector $\mathbf{N}$ as

$$\mathbf{f_{e1}} = \int_a^{x_2} \mathbf{N} f\, dx \qquad \text{and} \qquad \mathbf{f_{e2}} = \int_{x_2}^{b} \mathbf{N} f\, dx$$

Each of the elemental load matrices is then expanded to the global or system level according to

$$\mathbf{f_{g1}} = \begin{bmatrix} \displaystyle\int_a^{x_2} N_1 f\,dx \\ \displaystyle\int_a^{x_2} N_2 f\,dx \\ 0 \end{bmatrix} \qquad \mathbf{f_{g2}} = \begin{bmatrix} 0 \\ \displaystyle\int_{x_2}^{b} N_1 f\,dx \\ \displaystyle\int_{x_2}^{b} N_2 f\,dx \end{bmatrix}$$

with

$$\mathbf{f_G} = \mathbf{f_{g1}} + \mathbf{f_{g2}}$$

The boundary terms α and β can be represented at the global level as

$$\mathbf{BT_G} = \begin{bmatrix} \alpha & 0 & 0 \\ 0 & 0 & 0 \\ 0 & 0 & \beta \end{bmatrix}$$

whereas the boundary terms A and B contribute as

$$\mathbf{bt_G} = \begin{bmatrix} A \\ 0 \\ B \end{bmatrix}$$

so that the final set of equations can be represented as $\mathbf{K_G u_G} = \mathbf{F_G}$ where the global stiffness matrix $\mathbf{K_G}$ is given by

$$\mathbf{K_G} = \mathbf{k_G} + \mathbf{BT_G}$$

and the global load matrix $\mathbf{F_G}$ is given by

$$\mathbf{F_G} = \mathbf{f_G} + \mathbf{bt_G}$$

Note that in the general case where there are N elements, the elemental stiffness and load matrices are

$$\mathbf{k_{ei}} = \int_{x_i}^{x_{i+1}} (\mathbf{N}'\mathrm{p}\mathbf{N}'^{\mathrm{T}} + \mathbf{N}\mathrm{q}\mathbf{N}^{\mathrm{T}})\,dx \qquad \text{and} \qquad \mathbf{f_{ei}} = \int_{x_i}^{x_{i+1}} \mathbf{N} f\,dx \tag{2.6}$$

and that for a problem with N elements it is possible to consider the final set of equations from an elemental perspective, thus,

$$\mathbf{K_G} = \sum_{i=1}^{N} \mathbf{k_{gi}} + \mathbf{BT_G} \tag{2.7}$$

and

$$\mathbf{F_G} = \sum_{i=1}^{N} \mathbf{f_{gi}} + \mathbf{bt_G} \tag{2.8}$$

where the $\mathbf{k_{gi}}$ and $\mathbf{f_{gi}}$ correspond to the ith element.

At this juncture it is possible to outline the steps of elemental formulation and assembly. The *elemental formulation* is the part of the process that transfers the basic physics contained in the original differential equation and boundary conditions into the elemental stiffness and load matrices and, hence, into the final set of global equations. As we will see several times later in this chapter, it is possible to interpret the results of the finite element model in a manner indicating that the basic physics is being satisfied approximately. *Assembly* consists of positioning the elemental matrices at the proper locations in the global matrices. For the type of one-dimensional problems being discussed in this chapter, assembly consists of identifying the node numbers associated with a particular element and then adding the elements of the elemental matrices $\mathbf{k_{gi}}$ and $\mathbf{f_{gi}}$ to the corresponding locations in the global stiffness matrices $\mathbf{K_G}$ and $\mathbf{F_G}$. This process is shown in Fig. 2–12, where the i and $i+1$ locations in the global stiffness matrix are clearly indicated.

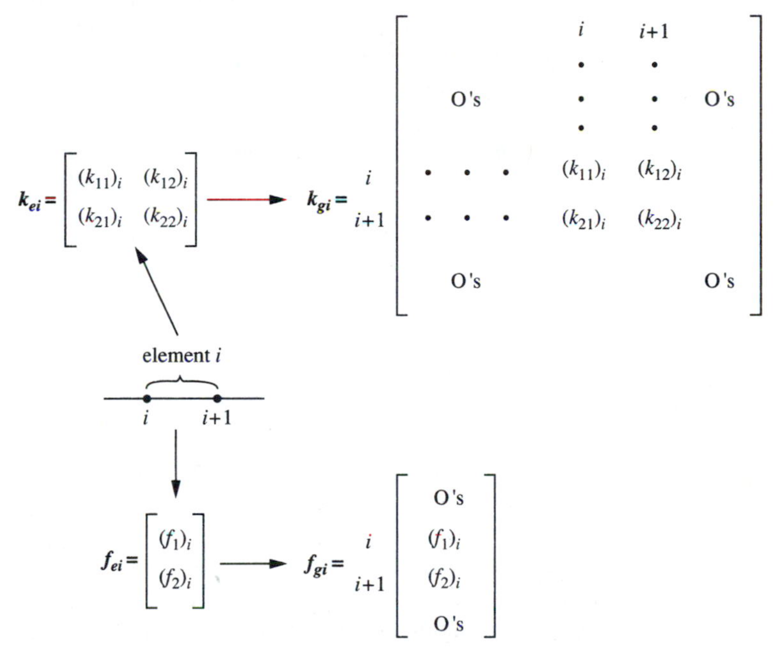

FIGURE 2–12 Contributions at the global level of $\mathbf{k_{ei}}$ and $\mathbf{f_{ei}}$

Constraints. For the case where both boundary conditions are natural, the $N + 1$ by $N + 1$ set of symmetric linear algebraic equations $\mathbf{Kv} = \mathbf{F}$ is then solved for the $N + 1$ unknown nodal values v_i. If, however, the boundary condition at a is of the form $u(a) = U_1$ (i.e., not of the form of the general boundary condition), we require the approximate solution to also satisfy the same equation so that $v_1 = U_1$. Such a condition on the dependent variable is referred to as a *constraint*. The equations $Kv = F$ must be altered to reflect this constraint. To illustrate the process of enforcing the constraint, we represent the unconstrained equation (2.5) in augmented form as

$$\left[\begin{array}{cccccc|c} K_{11} & K_{12} & 0 & \dots & 0 & 0 & F_1 \\ K_{21} & K_{22} & K_{23} & \dots & 0 & 0 & F_2 \\ & & & \dots & & & \\ 0 & 0 & 0 & 0 & K_{N+1,N} & K_{N+1,N+1} & F_{N+1} \end{array}\right]$$

The constraint is enforced by replacing the first equation, the equation corresponding to the unknown v_1, by the equation $v_1 = U_1$, namely,

$$\left[\begin{array}{cccccc|c} 1 & 0 & 0 & \dots & 0 & 0 & U_1 \\ K_{21} & K_{22} & K_{23} & \dots & 0 & 0 & F_2 \\ & & & \dots & & & \\ 0 & 0 & 0 & 0 & K_{N+1,N} & K_{N+1,N+1} & F_{N+1} \end{array}\right]$$

As derived, the unconstrained set of linear algebraic equations is symmetric. There are definite advantages associated with solving a symmetric set of equations as opposed to solving an arbitrary set of equations. Specifically, the amount of storage required for a symmetric array is slightly more than half that required for storing an asymmetric array. In addition, the coding required for the solution of a symmetric set of equations is substantially less than for an arbitrary set of equations. For these reasons we first perform the necessary elementary row operations to convert the constrained set of equations to symmetric form. For the constraint at $x = a$ this consists of multiplying the first row by $-K_{21}$ and then adding row 1 to row 2. This yields the constrained symmetric set of equations

$$\left[\begin{array}{cccccc|c} 1 & 0 & 0 & \dots & 0 & 0 & U_1 \\ 0 & K_{22} & K_{23} & \dots & 0 & 0 & F_2 - K_{21}U_1 \\ & & & \dots & & & \\ 0 & 0 & 0 & 0 & K_{N+1,N} & K_{N+1,N+1} & F_{N+1} \end{array}\right]$$

These equations can then be solved using a solver designed for symmetric sets of linear algebraic equations. A boundary condition at the boundary $x = b$ can be handled in a completely similar fashion.

In summary then, if the boundary condition at $x = a$ does not contain a u' term, the boundary condition is of the form $u(a) = U_1$ and the final system of equations must be constrained as outlined above. If the u' term is present, the values of α and A must be determined and used in the formation of the **BT** matrix and the **bt** vector. Similarly, if the boundary condition at $x = b$ does not contain a u' term the boundary condition is of the form $u(b) = U_2$ and the final system of equations must be properly constrained. If the u' term is present the values of β and B must be determined and added to the proper location in the **BT** matrix and the **bt** vector, respectively.

Another method that is frequently and easily used for enforcing constraints is the so-called **penalty method.** We will demonstrate the ideas associated with the penalty method by means of example. Consider a typical finite element model represented in augmented form as

$$[\mathbf{K} : \mathbf{F}] = \left[\begin{array}{ccccc|c} 1 & -1 & 0 & 0 & 0 & 1 \\ -1 & 2 & -1 & 0 & 0 & 2 \\ 0 & -1 & 2 & -1 & 0 & 2 \\ 0 & 0 & -1 & 2 & -1 & 2 \\ 0 & 0 & 0 & -1 & 1 & 1 \end{array}\right]$$

Suppose that we wish to impose the constraints $u_1 = 0$. Using the method outlined above it is easily verified that the solution of the constrained set of equations is $\mathbf{u}_\mathbf{G}^\mathrm{T} = [0\ 7\ 12\ 15\ 16]$. Using the penalty method, the constraint can be enforced approximately by replacing the first equation by

$$Cu_1 - u_2 + (0) \cdot u_3 + (0) \cdot u_4 + (0) \cdot u_5 = 0$$

where C is some relatively large number, for example $C = 10^n$, where we will temporarily consider n to be a parameter that we choose to make the approximation sufficiently accurate. The system of equations would appear as

$$[\mathbf{K} : \mathbf{F}] = \left[\begin{array}{ccccc|c} 10^n & -1 & 0 & 0 & 0 & 1 \\ -1 & 2 & -1 & 0 & 0 & 2 \\ 0 & -1 & 2 & -1 & 0 & 2 \\ 0 & 0 & -1 & 2 & -1 & 2 \\ 0 & 0 & 0 & -1 & 1 & 1 \end{array}\right]$$

Below are the corresponding solutions for several values of n.

$n = 2: \quad \mathbf{u}_\mathbf{G}^\mathrm{T} = [0.080808\ 7.080808\ 12.080808\ 15.080808\ 16.080808]$

$n = 4: \quad \mathbf{u}_\mathbf{G}^\mathrm{T} = [0.000800\ 7.000800\ 12.000800\ 15.000800\ 16.000800]$

$n = 6: \quad \mathbf{u}_\mathbf{G}^\mathrm{T} = [0.000008\ 7.000008\ 12.000008\ 15.000008\ 16.000008]$

Clearly, as n becomes large (i.e., $n \geq 4$), the approximate penalty method solution approaches the corresponding exact solution with an error of the order of 10^{-n}. The idea is that by taking the coefficient of the main diagonal element (corresponding to the degree of freedom to be constrained) to be several orders of magnitude larger than any other element in the coefficient matrix, the value of the constrained degree of freedom can be forced to zero without significantly altering the values of the other degrees of freedom. The main advantage of the penalty method is that symmetry (when originally present in the model) is not destroyed. There are no additional elementary row operations necessary as in the first approach to enforcing constraints. As shown in what follows, the penalty method can also be easily applied to enforce nonzero constraints.

In an actual computer application, where the stiffness matrix has been formed internally by a code, C can be chosen by (1) testing all of the main diagonal elements, (2) choosing the largest main diagonal element, say $d_{kk} \approx 10^M$, and (3) setting $C \approx 10^{n+M}$, with n chosen so as to produce the desired approximation (i.e., so that the error is sufficiently small). A nonzero constraint $u_i = \Delta$ is enforced in a completely similar manner, that is, by replacing the corresponding k_{ii} by 10^{n+M} and the corresponding F_i by $10^{n+M}\Delta$.

Example 2.4.

Consider the augmented set of equations

$$[\mathbf{K} : \mathbf{F}] = \left[\begin{array}{rrrrr|r} 17000 & -17000 & 0 & 0 & 0 & 16 \\ -17000 & 32000 & -15000 & 0 & 0 & 12 \\ 0 & -15000 & 24000 & -9000 & 0 & 10 \\ 0 & 0 & -9000 & 15000 & -6000 & 8 \\ 0 & 0 & 0 & -6000 & 6000 & 6 \end{array}\right]$$

Take the constraints to be $u_1 = 0$ and $u_5 = 0.005$. The exact solution is $\mathbf{u}_\mathbf{G}^\mathrm{T} = [0 \;\; 0.001931 \;\; 0.003321 \;\; 0.004526 \;\; 0.005000]$. The largest main diagonal element is of the order of 10^4, so we will take $M = n = 4$ for the penalty method. The corresponding augmented equations can be written as

$$[\mathbf{K} : \mathbf{F}] = \left[\begin{array}{rrrrr|r} 1.0E8 & -17000 & 0 & 0 & 0 & 16 \\ -17000 & 32000 & -15000 & 0 & 0 & 12 \\ 0 & -15000 & 24000 & -9000 & 0 & 10 \\ 0 & 0 & -9000 & 15000 & -6000 & 8 \\ 0 & 0 & 0 & -6000 & 1.0E8 & 5.0E5 \end{array}\right]$$

The solution is $\mathbf{u}_\mathbf{G}^\mathrm{T} = [0.000000 \;\; 0.001932 \;\; 0.003321 \;\; 0.004526 \;\; 0.005000]$; this shows the accuracy of the penalty method.

Note that an essential boundary condition is automatically satisfied when the corresponding constraint is enforced on the global set of equations. However, a natural boundary condition is not satisfied identically. Although the constants α, A or β, B appear in the final assembled set of equations, the natural boundary condition is satisfied only approximately, not identically. The accuracy with which the natural boundary condition is satisfied depends upon the mesh, and is such that as the number of elements is increased the natural boundary condition is more closely satisfied. This idea can be used as a practical means for deciding upon the accuracy of the finite element solution when a natural boundary condition is present.

Solution. Using either of the above approaches for enforcing the constraints, the equations are then ready for solution by hand or ready to be sent to appropriate software for extracting the solution. The user should take some care in selecting the software or in generating the code for determining the solution. The subroutine or procedure should be able to detect any singularities or near singularities that may inadvertently be present in the equations. A routine that has at least partial pivoting should be used. An estimate of the condition number is also useful. And the routine should be selected to take advantage of properties such as symmetry and bandedness. See Appendix B for descriptions and source listings of suitable routines.

Computation of derived variables. For the class of problems under consideration, the quantity pu', not simply the derivitive u', is the appropriate derived variable. The derived variable pu' is computed for each element. The computation will depend upon the function $p(x)$, the type of interpolation used, and the procedure used for evaluating the elemental stiffness matrices. If for instance $p(x)$ is a constant, say p_0, and linear interpolation is used, there will be a single constant value of pu' within each element, namely $p_0(u_{i+1} - u_i)/\ell_e$. If p is not a constant, the derived variable is

$$pu' = \frac{p(x)(u_{i+1} - u_i)}{\ell_e}$$

which is a function of x within the element in question. The obvious question arises: At what point(s) within the element should pu' be evaluated? Moan [3] showed that for constant p, or in the limit as $\ell_e \to 0$, evaluating pu' at the midpoint of a linearly interpolated element is generally more accurate than at other locations within the element. This again produces a single constant value for the derived variable within the element. In any event, the constant value computed will in general differ from element to element resulting in interelement discontinuities in pu', as indicated in Fig. 2–13.

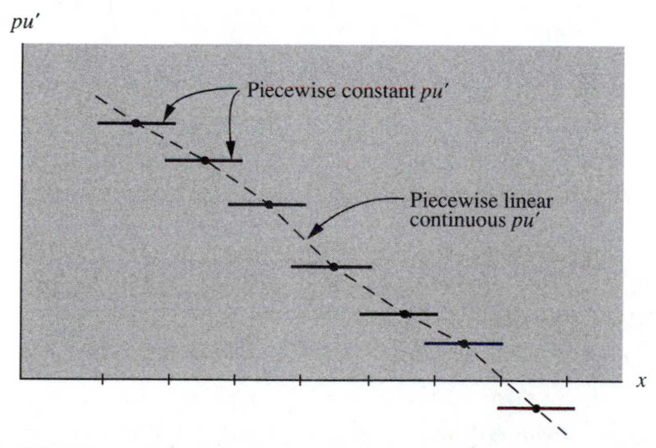

FIGURE 2–13 Piecewise constant $-pu'$ and a piecewise linear approximation

In situations where the derived variable $-pu'$ is known to be continuous, ad hoc procedures are sometimes used to generate a suitable continuous derived variable. One such procedure involves simply connecting the midpoint values of the derived variables in adjacent elements to produce a piecewise linear continuous approximation for pu'. The result of this procedure for the piecewise constant pu' of Fig. 2–13 is indicated by the dotted line.

2.4 EXAMPLE APPLICATIONS

The introduction to this chapter presented the governing equations and boundary conditions for the application areas of solid mechanics, fluid mechanics, heat transfer, and vibrations. Each of these can be considered as special cases of the general one-dimensional boundary value problem discussed in Section 2.2. In this section, we will show how the Galerkin finite element model developed in Section 2.3 for the general problem can be specialized to solve typical physical problems. In view of the fact that many of the examples considered have exact solutions, immediate feedback on the validity and accuracy of the finite element method will be available.

2.4.1 Solid Mechanics—Axial Deformation

Consider the problem of a straight uniform-cross-section bar of specific weight γ and area A under the action of its own weight as indicated in Fig. 2–14*a*.

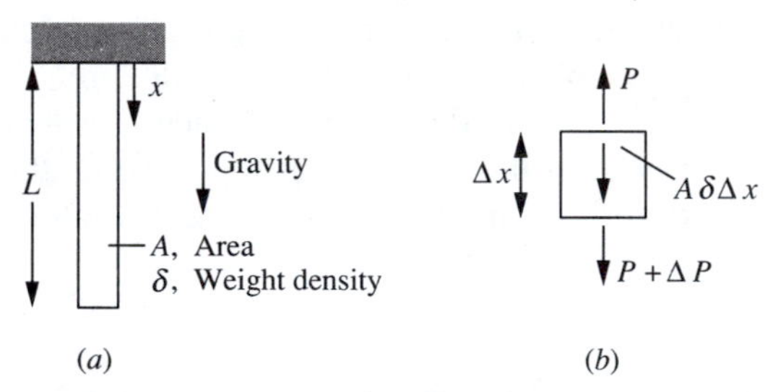

FIGURE 2–14 Hanging bar problem

By consulting the discussion of equilibrium in the introduction and Fig. 2–14*b*, we can see that the loading $q(x)$ is $A\gamma$, where γ is the weight density $[FL^{-3}]$ of the bar. The governing equation of equilibrium in terms of the displacements is then

$$(AEu')' + A\gamma = 0$$

The boundary conditions, one at each end, are

$$u(0) = 0 \qquad \text{and} \qquad P(L) = AEu'(L) = 0$$

Note that the first boundary condition is essential whereas the second boundary condition is natural. Comparing with the standard form of the one-dimensional boundary value problem yields $p = AE, q = 0, f = A\gamma$, and $\alpha = \beta = A = B = 0$.

The finite element model for this problem will be set up and solved following the seven basic steps.

Discretization, interpolation and elemental formulation. Consider first a two-element model consisting of three nodes and two equal-length elements as indicated in Fig. 2–15.

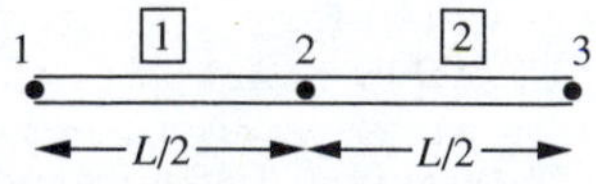

1 [1] 2 [2] 3

← L/2 → ← L/2 →

FIGURE 2–15 Nodes and elements—two-element model

Using linear interpolation, the elemental matrices are then

$$\mathbf{k_e} = \int_{x_i}^{x_{i+1}} (\mathbf{N}' p \mathbf{N}'^{\mathrm{T}} + \mathbf{N} q \mathbf{N}^{\mathrm{T}})\, dx = \int_{x_i}^{x_{i+1}} \mathbf{N}' AE \mathbf{N}'^{\mathrm{T}} dx$$

and

$$\mathbf{f_e} = \int_{x_i}^{x_{i+1}} \mathbf{N} A\gamma\, dx$$

where

$$\mathbf{N} = \begin{bmatrix} \dfrac{x_{i+1} - x}{x_{i+1} - x_i} \\ \dfrac{x - x_i}{x_{i+1} - x_i} \end{bmatrix} \quad \text{and} \quad \mathbf{N}' = \frac{1}{\ell_e}\begin{bmatrix} -1 \\ 1 \end{bmatrix}$$

with $x_i = 0, x_{i+1} = L/2$ for element 1 and $x_i = L/2, x_{i+1} = L$ for element 2, and with $\ell_e = x_{i+1} - x_i = L/2$ for each element. With AE constant,

$$\mathbf{k_{e1}} = \frac{4}{L^2}\int_0^{L/2} \begin{bmatrix} -1 \\ 1 \end{bmatrix} AE \begin{bmatrix} -1 & 1 \end{bmatrix} dx = \frac{2AE}{L}\begin{bmatrix} 1 & -1 \\ -1 & 1 \end{bmatrix}$$

$$\mathbf{k_{e2}} = \frac{4}{L^2}\int_{L/2}^{L} \begin{bmatrix} -1 \\ 1 \end{bmatrix} AE \begin{bmatrix} -1 & 1 \end{bmatrix} dx = \frac{2AE}{L}\begin{bmatrix} 1 & -1 \\ -1 & 1 \end{bmatrix}$$

Similarly, with $A\gamma$ constant,

$$\mathbf{f_{e1}} = A\gamma \int_0^{L/2} \begin{bmatrix} 1 - \frac{2x}{L} \\ \frac{2x}{L} \end{bmatrix} dx = \frac{A\gamma L}{4}\begin{bmatrix} 1 \\ 1 \end{bmatrix}$$

$$\mathbf{f_{e2}} = A\gamma \int_{L/2}^{L} \begin{bmatrix} 2(1 - \frac{x}{L}) \\ \frac{2x}{L} - 1 \end{bmatrix} dx = \frac{A\gamma L}{4}\begin{bmatrix} 1 \\ 1 \end{bmatrix}$$

Assembly. Expanding the elemental matrices to the global level yields

$$\mathbf{k_{g1}} = \frac{2AE}{L}\begin{bmatrix} 1 & -1 & 0 \\ -1 & 1 & 0 \\ 0 & 0 & 0 \end{bmatrix} \qquad \mathbf{k_{g2}} = \frac{2AE}{L}\begin{bmatrix} 0 & 0 & 0 \\ 0 & 1 & -1 \\ 0 & -1 & 1 \end{bmatrix}$$

and

$$\mathbf{f_{g1}} = \frac{A\gamma L}{4}\begin{bmatrix} 1 \\ 1 \\ 0 \end{bmatrix} \qquad \mathbf{f_{g2}} = \frac{A\gamma L}{4}\begin{bmatrix} 0 \\ 1 \\ 1 \end{bmatrix}$$

Assembly then yields

$$\mathbf{K_G} = \mathbf{k_{g1}} + \mathbf{k_{g2}} = \frac{2AE}{L}\begin{bmatrix} 1 & -1 & 0 \\ -1 & 2 & -1 \\ 0 & -1 & 1 \end{bmatrix}$$

which is equivalent to two linear springs each with stiffness $k = AE/(L/2)$ in series as discussed in Chapter 2. Similarly,

$$\mathbf{F_G} = \mathbf{f_{g1}} + \mathbf{f_{g2}} = \frac{A\gamma L}{4}\begin{bmatrix} 1 \\ 2 \\ 1 \end{bmatrix}$$

The total weight $A\gamma L$, which in the original continuous problem was uniformly distributed along the length, is now applied as discrete or concentrated forces at the nodes. The specific nodal distribution indicated for $\mathbf{F_G}$ is a direct result of having used the linear interpolation. As we will see in later sections, other interpolations produce other "lumpings."

With $\alpha = \beta = A = B = 0$, $\mathbf{BT_G} = 0$ and $\mathbf{bt_G} = 0$, so that the augmented unconstrained global set of equations can be expressed as

$$\left[\begin{array}{ccc|c} 1 & -1 & 0 & \phi \\ -1 & 2 & -1 & 2\phi \\ 0 & -1 & 1 & \phi \end{array}\right]$$

where $\phi = \gamma L^2/8E$.

Constraints. The essential boundary condition at $x = 0$ translates into the finite element model as $u_1 = 0$, which when imposed on the augmented set of equations yields

$$\left[\begin{array}{ccc|c} 1 & 0 & 0 & 0 \\ -1 & 2 & -1 & 2\phi \\ 0 & -1 & 1 & \phi \end{array}\right]$$

Performing the elementary row operations necessary to restore symmetry yields

$$\left[\begin{array}{ccc|c} 1 & 0 & 0 & 0 \\ 0 & 2 & -1 & 2\phi \\ 0 & -1 & 1 & \phi \end{array}\right]$$

Solution. This system is easily solved to yield

$$\mathbf{u_G} = [0 \quad 3\phi \quad 4\phi]^{\mathrm{T}}$$

for the nodal values at $x = L/2$ and $x = L$. The results are shown in Fig. 2–16.

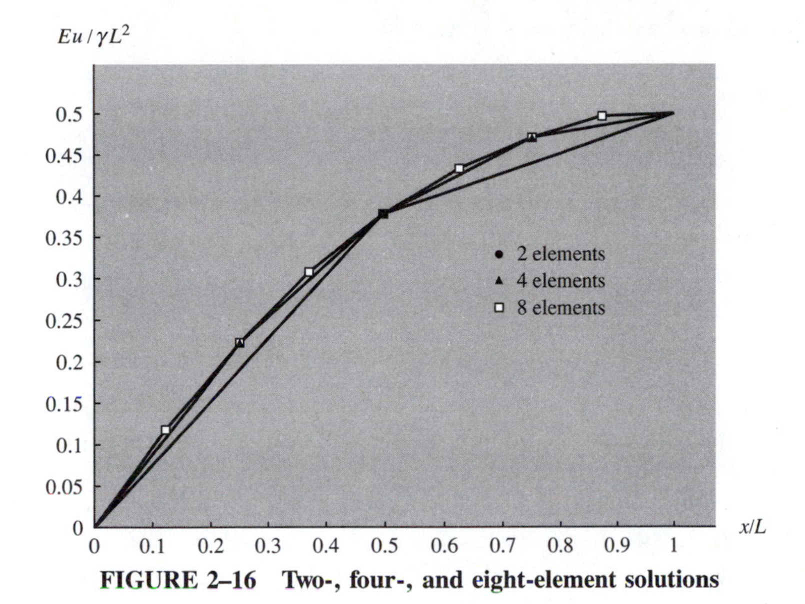

FIGURE 2–16 Two-, four-, and eight-element solutions

Computation of derived varibles. For this problem, the derived variable is the internal force transmitted, that is, $P = pu' = AEu'$, and can be computed for each element.

Element 1. We can easily see u' to be

$$u' = \frac{(u_2 - u_1)}{l_e} = \frac{2(u_2 - u_1)}{L}$$

so that

$$P_1 = AEu_1' = \left(\frac{2}{L}\right)\left(\frac{3A\gamma L^2}{8}\right) = \frac{3A\gamma L}{4}$$

Element 2. Again, we easily see u' to be

$$u' = \frac{(u_3 - u_2)}{l_e} = \frac{2(u_3 - u_2)}{L}$$

so that

$$P_2 = AEu_2' = \left(\frac{2}{L}\right)\left(\frac{A\gamma L^2}{8}\right) = \frac{A\gamma L}{4}$$

as shown in Fig. 2–17. These results will be discussed in detail after investigating successively finer meshes for this problem.

For a linearly interpolated four-element model there are five degrees of freedom. The length l_e of each equal-length element is $L/4$. The assembled unconstrained global equations can easily be verified to be

$$\left[\begin{array}{ccccc|c} 1 & -1 & 0 & 0 & 0 & \phi \\ -1 & 2 & -1 & 0 & 0 & 2\phi \\ 0 & -1 & 2 & -1 & 0 & 2\phi \\ 0 & 0 & -1 & 2 & -1 & 2\phi \\ 0 & 0 & 0 & -1 & 1 & \phi \end{array}\right]$$

where $\phi = \gamma L^2/32E$. Enforcing the constraints and solving the constrained equations yields

$$\mathbf{u}_G^T = [0 \quad 7 \quad 12 \quad 15 \quad 16] \frac{\gamma L^2}{32E}$$

The elemental internal transmitted forces are easily computed to be $P_1 = 7A\gamma L/8E, P_2 = 5A\gamma L/8E, P_3 = 3A\gamma L/8E$, and $P_4 = A\gamma L/8E$.

An eight-element linearly interpolated model with equal lengths of $L/8$ yields the results

$$\mathbf{u}_G^T = [0 \quad 15 \quad 28 \quad 39 \quad 48 \quad 55 \quad 60 \quad 63 \quad 64] \frac{\gamma L^2}{128E}$$

and with an obvious notation

$$\mathbf{P} = [15 \quad 13 \quad 11 \quad 9 \quad 7 \quad 5 \quad 3 \quad 1]^T \frac{A\gamma L}{16}$$

Comparison of these three models, which have resulted from successive halving of the mesh size, indicate a definite pattern for the displacements and for the force transmitted which the student should be able to discern. By studying the pattern, the student should be able to deduce what is going to happen in the limit as the mesh is further refined by these successive halvings.

Discussion of results. The results for the displacements are plotted in Fig. 2–16. The exact solution, given by $u(x) = \gamma(2Lx - x^2)/2E$, is a parabola that passes through all of the nodal values predicted by each of the finite element models. The results for the internal force transmitted are plotted in Fig. 2–17, again along with the exact results $P(x) = A\gamma(L - x)$.

The internal forces predicted by the finite element model are constants within an element. This is an inescapable consequence of having used linear interpolation for this problem. From Fig. 2–17 we easily see that employing the ad hoc procedure discussed in Section 2.3 for representing the derived variable $P = AEu'$ results in the exact curve for the force transmitted along the length of the bar.

It is very instructive to draw the free-body diagrams of each of the elements. For the four-element model, the free-body diagrams are indicated in Fig. 2–18, where $f = A\gamma L/8$. Fig. 2–19 indicates the situation that results when the elements are "put back together." Equilibrium is clearly satisfied.

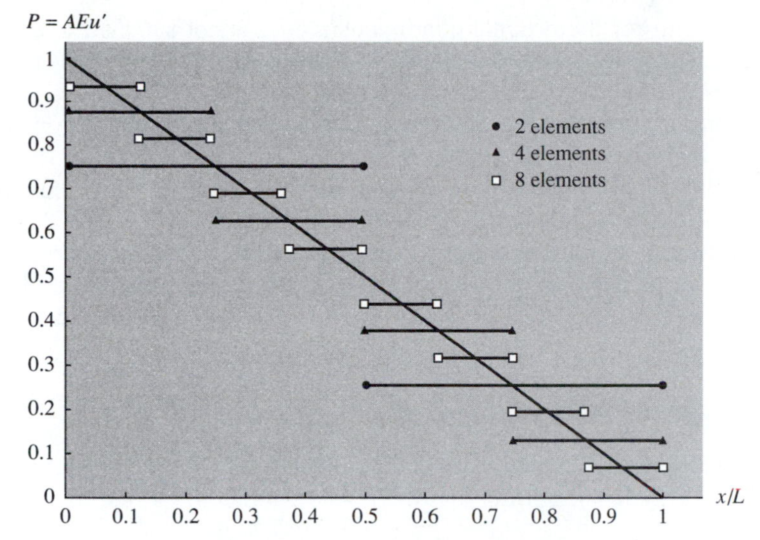

FIGURE 2–17 Finite element and exact internal forces

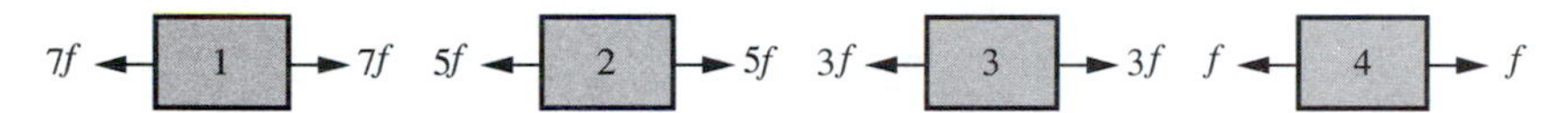

FIGURE 2–18 Free-body diagrams of elements—four-element model

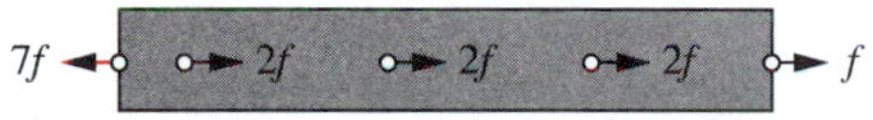

7f 2f 2f 2f f

FIGURE 2–19 Reassembled bar showing external nodal forces

The discontinuity in the internal force between the elements is exactly equal to and in this instance caused by the external "lumped" force applied at each node. Consider in this regard a free-body diagram of a small slice of the bar including node 3 as shown in Fig. 2–20. The discontinuity in the internal force transmitted is clearly in equilibrium with the external force at that node.

5f 2f 3f

FIGURE 2–20 Free-body diagram of a portion of the bar containing node 3

Additionally, we can see from Fig. 2–20 that the reaction at node 1 is equal to the sum of the external loads applied at the nodes below the support. Finally, the $7A\gamma L/8$ reaction at the top of the bar can be considered to be the sum of $(A\gamma L - A\gamma L/8)$, which is the true reaction minus the amount of load lost when the constraint was applied. Thus in a sense, equilibrium, the basic principle involved for this problem, is satisfied in terms of the above discussions regarding the internal forces and applied external loads for the finite element model.

That equilibrium for the original continuous model is not satisfied is easily seen to be the case by virtue of the fact that in a given element the force is constant in the presence of continuous nonzero external loading.

As mentioned in Section 2.3, the natural boundary condition $P(L) = AEu'(L) = 0$ is not satisfied identically. As more elements are taken, the value of the force transmitted by the last element is the load at the last node. With N the number of equal-length elements, it can be seen from induction that the value of the force transmitted in the last element is $f = A\gamma L/2N$, which approaches zero for large N. Or, if one uses the ad hoc procedure of connecting the midpoints of the pu' segments, the natural boundary condition is satisfied identically for this application.

2.4.2 Fluid Mechanics—Thin-Film Lubrication

Consider the specific problem where the slider bearing has a parabolic shape $h(x) = h_0(1 + (x/L)^2)$, with $U_2 = 0, U_1 = -U$, and $p_L = p_0$, as indicated in Fig. 2–21.

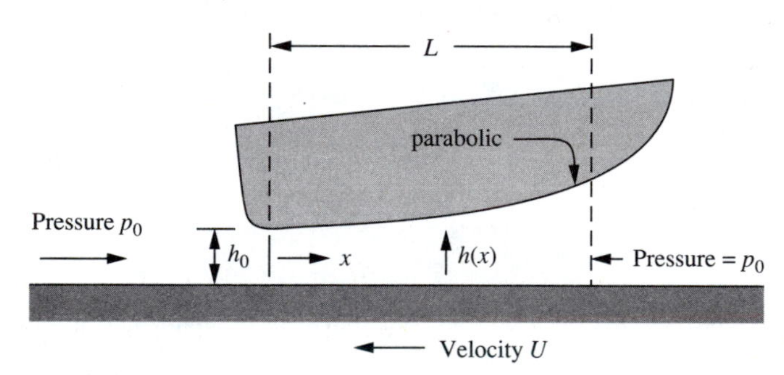

FIGURE 2–21 Slider bearing with parabolic profile

The boundary value problem here is

$$(h(x)^3 p')' = 6\eta(U_2 - U_1)h'$$

$$p(0) = p_0$$

$$p(L) = p_0$$

We will nondimensionalize the problem according to

$$\frac{(p - p_0)h_0{}^2}{12\eta UL} = P \qquad \text{and} \qquad \frac{x}{L} = s$$

resulting in

$$((1 + s^2)^3 P')' + s = 0$$

$$P(0) = 0$$

$$P(1) = 0$$

where $' \equiv d/ds$. With the independent variable now s, comparison with the standard form indicates that $p(s) = (1 + s^2)^3, q = 0$, and $f = s$. Since both boundary conditions are essential, A, B, α, and β are all zero. The finite element model for this problem will be set up and solved following the seven basic steps.

Discretization and interpolation. Consider first a two-element model consisting of three nodes and two equal-length elements as indicated in Fig. 2–22.

FIGURE 2–22 Nodes and elements—two-element model

Elemental formulation. Using linear interpolation, the elemental matrices are then

$$\mathbf{k_e} = \int_{s_i}^{s_{i+1}} (\mathbf{N}' p \mathbf{N}'^{\mathrm{T}} + \mathbf{N} q \mathbf{N}^{\mathrm{T}}) ds = \int_{s_i}^{s_{i+1}} \mathbf{N}'(1 + s^2)^3 \mathbf{N}'^{\mathrm{T}} ds$$

$$\mathbf{f_e} = \int_{s_i}^{s_{i+1}} \mathbf{N} s \, ds$$

where

$$\mathbf{N} = \begin{bmatrix} \dfrac{s_{i+1} - s}{s_{i+1} - s_i} \\ \dfrac{s - s_i}{s_{i+1} - s_i} \end{bmatrix} \qquad \text{and} \qquad \mathbf{N}' = \frac{1}{\ell_e} \begin{bmatrix} -1 \\ 1 \end{bmatrix}$$

For element 1 $s_i = 0$, $s_{i+1} = 0.5$ and for element 2 $s_i = 0.5$, $s_{i+1} = 1.0$, with $\ell_e = s_{i+1} - s_i = 0.5$ for each element. Specifically, the elemental matrices are

$$\mathbf{k_{e1}} = 4 \int_0^{1/2} \begin{bmatrix} -1 \\ 1 \end{bmatrix} (1 + s^2)^3 [-1 \quad 1] \, ds = 2.579464 \begin{bmatrix} 1 & -1 \\ -1 & 1 \end{bmatrix}$$

$$\mathbf{k_{e2}} = 4 \int_{1/2}^{1} \begin{bmatrix} -1 \\ 1 \end{bmatrix} (1 + s^2)^3 [-1 \quad 1] \, ds = 8.391964 \begin{bmatrix} 1 & -1 \\ -1 & 1 \end{bmatrix}$$

$$\mathbf{f_{e1}} = \int_0^{1/2} \begin{bmatrix} 1 - \frac{2s}{L} \\ \frac{2s}{L} \end{bmatrix} s \, ds = \frac{1}{24} \begin{bmatrix} 1 \\ 2 \end{bmatrix}$$

$$\mathbf{f_{e2}} = \int_{1/2}^{1} \begin{bmatrix} 2(1 - \frac{s}{L}) \\ \frac{2s}{L} - 1 \end{bmatrix} s \, ds = \frac{1}{24} \begin{bmatrix} 4 \\ 5 \end{bmatrix}$$

Assembly. Expanding the elemental matrices to the global level yields

$$\mathbf{k_{g1}} = \begin{bmatrix} 2.579464 & -2.579464 & 0 \\ -2.579464 & 2.579464 & 0 \\ 0 & 0 & 0 \end{bmatrix} \qquad \mathbf{k_{g2}} = \begin{bmatrix} 0 & 0 & 0 \\ 0 & 8.391964 & -8.391964 \\ 0 & -8.391964 & 8.391964 \end{bmatrix}$$

and

$$\mathbf{f_{g1}} = \frac{1}{24}\begin{bmatrix} 1 \\ 2 \\ 0 \end{bmatrix} \qquad \mathbf{f_{g2}} = \frac{1}{24}\begin{bmatrix} 0 \\ 4 \\ 5 \end{bmatrix}$$

Assembly then yields

$$\mathbf{K_G} = \mathbf{k_{g1}} + \mathbf{k_{g2}} = \begin{bmatrix} 2.579464 & -2.579464 & 0 \\ -2.579464 & 10.971428 & -8.391964 \\ 0 & -8.391964 & 8.391964 \end{bmatrix}$$

and

$$\mathbf{F_G} = \mathbf{f_{g1}} + \mathbf{f_{g2}} = \frac{1}{24}\begin{bmatrix} 1 \\ 6 \\ 5 \end{bmatrix}$$

With $\alpha = \beta = A = B = 0$, $\mathbf{BT_G} = 0$ and $\mathbf{bt_G} = 0$ so that the augmented unconstrained global set of equations can be expressed as

$$\left[\begin{array}{ccc|c} 2.579464 & -2.579464 & 0 & 0.041667 \\ -2.579464 & 10.971428 & -8.391964 & 0.250000 \\ 0 & -8.391964 & 8.391964 & 0.208333 \end{array}\right]$$

Constraints. The essential boundary conditions at $s = 0$ and $s = 1$ translate into the finite element model as $P_1 = P_3 = 0$. When these constraints are imposed on the augmented set of equations, there results

$$\left[\begin{array}{ccc|c} 1 & 0 & 0 & 0 \\ -2.579464 & 10.971428 & -8.391964 & 0.25 \\ 0 & 0 & 1 & 0 \end{array}\right]$$

Performing the elementary row operations necessary to restore symmetry yields

$$\left[\begin{array}{ccc|c} 1 & 0 & 0 & 0 \\ 0 & 10.971428 & 0 & 0.25 \\ 0 & 0 & 1 & 0 \end{array}\right]$$

Solution. This system is easily solved to yield

$$\mathbf{P_G} = [0 \quad 0.022786 \quad 0]^{\mathrm{T}}$$

for the pressures at the nodes. The results are shown in Fig. 2–23.

For a four-element model the nodal pressures are

$$\mathbf{P_G^T} = [0.000000 \quad 0.018558 \quad 0.0210593 \quad 0.012415 \quad 0.000000]$$

and for an eight-element model,

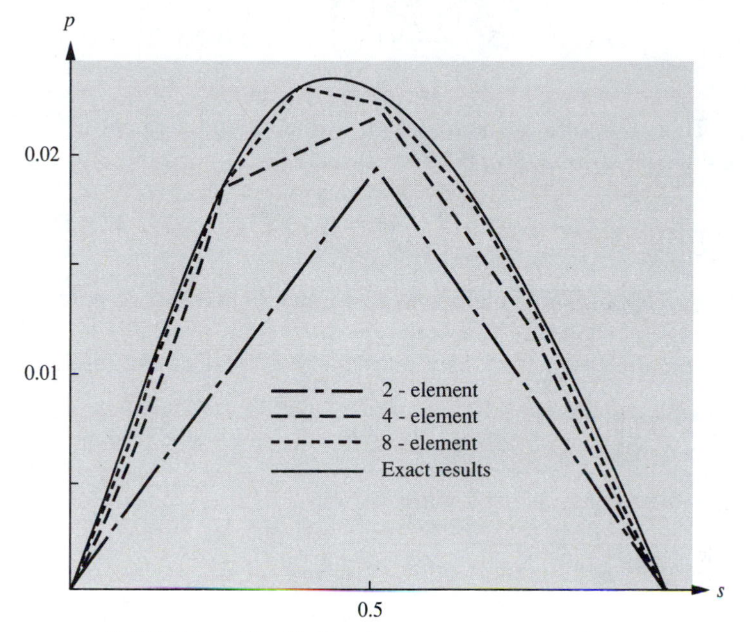

FIGURE 2–23 Finite element and exact solutions

$$\mathbf{P}_G^T = [0.000000 \quad 0.010848 \quad 0.018987 \quad 0.022870 \quad 0.022458 \quad 0.018749 \quad 0.013060 \quad 0.006544 \quad 0.000000]$$

The pressure results for the two-, four-, and eight-element models are displayed in Fig. 2–23, along with the corresponding exact solution.

Computation of derived variables. For this example, the derived variable is the volume flow rate q given by

$$q = -\left(\frac{h^3}{12\eta}\right)\frac{dp}{dx} + \frac{(U_2 - U_1)h}{2}$$

which can be expressed in terms of the dimensionless variables as

$$-\frac{q}{Uh_0} = (1 + s^2)\left[\frac{1}{2} + (1 + s^2)^2 \frac{dP}{ds}\right]$$

The finite element solution should demonstrate, at least approximately, that the basic principle of conservation of mass is being satisfied: q should be approximately constant from element to element. As discussed in Section 2.3, these flow rates should be evaluated at the midpoints of the elements.

Element 1. The value of s at the midpoint of the first element is 0.25, so that with $dP/ds = (0.022786 - 0)/0.5 = 0.045572$,

$$-\left(\frac{q}{Uh_0}\right)_1 = (1+0.25^2)\left[\frac{1}{2} + (1+0.25^2)^2 0.045572\right] = 0.5859$$

Element 2. The value of s at the midpoint of the second element is 0.75, so that with $dP/ds = (0 - 0.022786)/0.5 = -0.045572$,

$$-\left(\frac{q}{Uh_0}\right)_2 = (1+0.75^2)\left[\frac{1}{2} - (1+0.75^2)^2 0.045572\right] = 0.6074$$

Each of the two approximate elemental flow rates is less than 2 percent different from the average of the two elemental values.

The corresponding elemental flow rates for the four-element model are

$$-\frac{\mathbf{q}}{\mathbf{Uh_0}} = [0.05887 \quad 0.5896 \quad 0.5925 \quad 0.5969]^T$$

The corresponding eight-element solutions are

$$-\frac{\mathbf{q}}{\mathbf{Uh_0}} = [0.05898 \quad 0.5898 \quad 0.5899 \quad 0.5901 \quad 0.5905 \quad 0.5910 \quad 0.5916 \quad 0.5922]^T$$

These results clearly indicate that q, as predicted by the results from both the four- and eight-element models, is approaching a constant value.

2.4.3 One-Dimensional Heat Conduction with Convection

The general problem of one-dimensional heat transfer situation was discussed in Section 2.1. Now we will investigate the specific problem of a circular-cross-section bar conducting heat along the axis of the bar as shown in Fig. 2–24. We will assume that convection occurs along the length of the bar and at the end $x = L$.

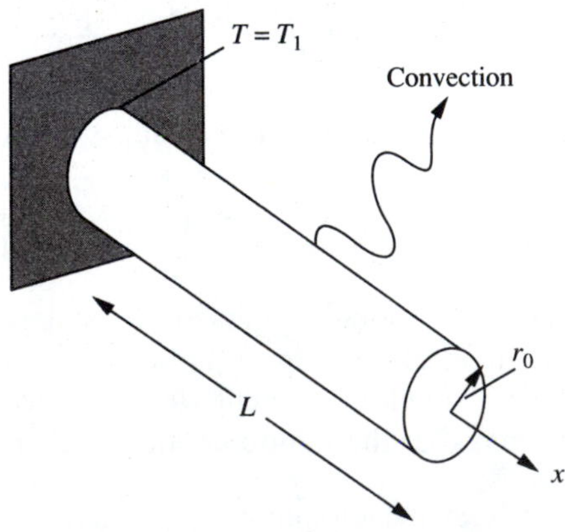

FIGURE 2–24 Geometry—one-dimensional diffusion with convection

With T the temperature, the governing equation and boundary conditions can be expressed as

$$(k\pi r_0^2 T')' - h2\pi r_0(T - T_0) = 0$$

$$T(0) = T_1$$

$$k(L)T'(L) + h_L T(L) = 0$$

Nondimensionalize by taking $u = (T - T_0)/T_1 - T_0)$ and $x/L = s$, after which the problem can be restated as

$$(u')' - \phi^2 u = 0$$

$$u(0) = 1$$

$$u'(L) + \psi u(L) = 0$$

where $\psi = hL/k$ and $\phi^2 = 2\psi L/r_0$. Comparing with the standard form discussed in Section 2.2, $p = 1, q = \phi^2, f = 0, A = \alpha = B = 0$, and $\beta = \psi$. The finite element model for this problem will be set up and solved following the seven basic steps. The numerical results will be based on the specific values $\psi = 1$ and $\phi^2 = 10$.

Discretization and interpolation. Consider first a two-element model consisting of three nodes and two equal-length elements as indicated in Fig. 2–25.

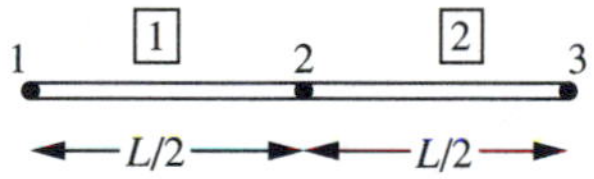

FIGURE 2–25 Nodes and elements—two-element model

Elemental formulation. Using linear interpolation, the elemental matrices are

$$\mathbf{k_e} = \int_{x_i}^{x_{i+1}} (\mathbf{N}'1\mathbf{N}'^{\mathrm{T}} + \mathbf{N}10\mathbf{N}^{\mathrm{T}})\,dx = \mathbf{p_e} + \mathbf{q_e}$$

where

$$\mathbf{N} = \begin{bmatrix} \dfrac{x_{i+1} - x}{x_{i+1} - x_i} \\ \dfrac{x - x_i}{x_{i+1} - x_i} \end{bmatrix} \qquad \text{and} \qquad \mathbf{N}' = \frac{1}{\ell_e}\begin{bmatrix} -1 \\ 1 \end{bmatrix}$$

For element 1 $x_i = 0, x_{i+1} = 0.5$ and for element 2 $x_i = 0.5, x_{i+1} = 1.0$, with $\ell_e = x_{i+1} - x_i = L/2$ for each element. Specifically,

$$\mathbf{p_{e1}} = 4\int_0^{1/2} \begin{bmatrix} -1 \\ 1 \end{bmatrix} 1 \begin{bmatrix} -1 & 1 \end{bmatrix} ds = 2\begin{bmatrix} 1 & -1 \\ -1 & 1 \end{bmatrix} = \mathbf{p_{e2}}$$

and

$$\mathbf{q}_{\mathbf{e1}} = 10\int_0^{1/2}\begin{bmatrix} 1-\frac{2s}{L} \\ \frac{2s}{L} \end{bmatrix}\begin{bmatrix} 1-\frac{2s}{L} & \frac{2s}{L} \end{bmatrix} ds = \frac{5}{6}\begin{bmatrix} 2 & 1 \\ 1 & 2 \end{bmatrix}$$

$$\mathbf{q}_{\mathbf{e2}} = 10\int_{1/2}^{1}\begin{bmatrix} 2(1-\frac{s}{L}) \\ \frac{2s}{L}-1 \end{bmatrix}\begin{bmatrix} 2(1-\frac{s}{L}) & \frac{2s}{L}-1 \end{bmatrix} ds = \frac{5}{6}\begin{bmatrix} 2 & 1 \\ 1 & 2 \end{bmatrix}$$

Thus

$$\mathbf{k}_{\mathbf{e1}} = \mathbf{p}_{\mathbf{e1}} + \mathbf{q}_{\mathbf{e1}} = 2\begin{bmatrix} 1 & -1 \\ -1 & 1 \end{bmatrix} + \frac{5}{6}\begin{bmatrix} 2 & 1 \\ 1 & 2 \end{bmatrix} = \frac{1}{6}\begin{bmatrix} 22 & -7 \\ -7 & 22 \end{bmatrix} = \mathbf{k}_{\mathbf{e2}}$$

Assembly. Expanding the elemental matrices to the global level yields

$$\mathbf{k}_{\mathbf{g1}} = \frac{1}{6}\begin{bmatrix} 22 & -7 & 0 \\ -7 & 22 & 0 \\ 0 & 0 & 0 \end{bmatrix} \qquad \mathbf{k}_{\mathbf{g2}} = \frac{1}{6}\begin{bmatrix} 0 & 0 & 0 \\ 0 & 22 & -7 \\ 0 & -7 & 22 \end{bmatrix}$$

With $\alpha = A = B = 0$ and $\beta = 1$, $\mathbf{bt_G} = 0$ and

$$\mathbf{BT_G} = \begin{bmatrix} 0 & 0 & 0 \\ 0 & 0 & 0 \\ 0 & 0 & 1 \end{bmatrix}$$

Assembly then yields

$$\mathbf{K_G} = \mathbf{k}_{\mathbf{g1}} + \mathbf{k}_{\mathbf{g2}} + \mathbf{BT_G} = \frac{1}{6}\begin{bmatrix} 22 & -7 & 0 \\ -7 & 44 & -7 \\ 0 & -7 & 28 \end{bmatrix}$$

Thus the unconstrained global equations can be expressed in augmented form as

$$\left[\begin{array}{ccc|c} 22 & -7 & 0 & 0 \\ -7 & 44 & -7 & 0 \\ 0 & -7 & 28 & 0 \end{array}\right]$$

The essential boundary condition $u(0) = 1$ is enforced as the constraint $u_1 = 1$, leading to

$$\left[\begin{array}{ccc|c} 1 & 0 & 0 & 1 \\ -7 & 44 & -7 & 0 \\ 0 & -7 & 28 & 0 \end{array}\right]$$

Performing the elementary row operations necessary to restore symmetry yields

$$\left[\begin{array}{ccc|c} 1 & 0 & 0 & 1 \\ 0 & 44 & -7 & 7 \\ 0 & -7 & 28 & 0 \end{array}\right]$$

Solution. Solving the constrained equations yields

$$\mathbf{u_G} = [\,1.000000 \quad 0.165680 \quad 0.041420\,]^T$$

Four- and eight-element models yield, respectively,

$$\mathbf{u_G} = [\,1.000000 \quad 0.445087 \quad 0.200699 \quad 0.096333 \quad 0.059176\,]^T$$

and

$$\mathbf{u_G} = [\,1.000000 \quad 0.672458 \quad 0.452797 \quad 0.305778 \quad 0.207814 \quad 0.143189 \quad 0.101535 \quad 0.076171 \quad 0.063026\,]^T$$

The nodal temperatures for each of the three models are shown in Fig. 2–26.

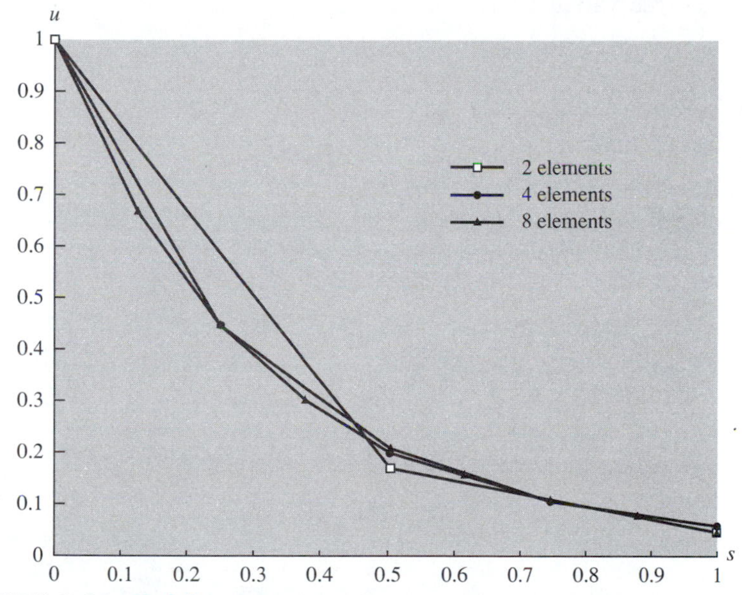

FIGURE 2–26 Nodal temperatures for two-, four-, and eight-element models

Computation of derived variables. For the heat transfer problem the derived variable is given by $q = -kAdT/dx$ or $q = -(kA/L)(T_1 - T_0)du/ds$ in terms of the nondimensionalized variables. Since $(kA/L)(T_1 - T_0)$ is constant, the derived variable will be constant within each element. For the two-element model the two derived variables are

$$-\frac{q_1 L}{kA(T_1 - T_0)} = \frac{0.165680 - 1}{0.5} = 1.6686$$

and

$$-\frac{q_2 L}{kA(T_1 - T_0)} = \frac{0.041420 - 0.165680}{0.5} = 0.2485$$

For the four- and eight-element models the corresponding fluxes **Q** are, respectively,

$$\mathbf{Q_4} = [2.2197 \quad 0.9776 \quad 0.4175 \quad 0.1486]^{\mathrm{T}}$$

and

$$\mathbf{Q_8} = [2.6203 \quad 1.7573 \quad 1.1762 \quad 0.7837 \quad 0.5170 \quad 0.3332 \quad 0.2029 \quad 0.1052]^{\mathrm{T}}$$

Results for the derived variable from each of the three models computed in this manner are displayed in Fig. 2–27. Also shown is the piecewise linear representation generated by connecting the midpoint values of the $-pu'$ segments. This ad hoc result would appear to be a reasonable representation for the exact continuous derived variable.

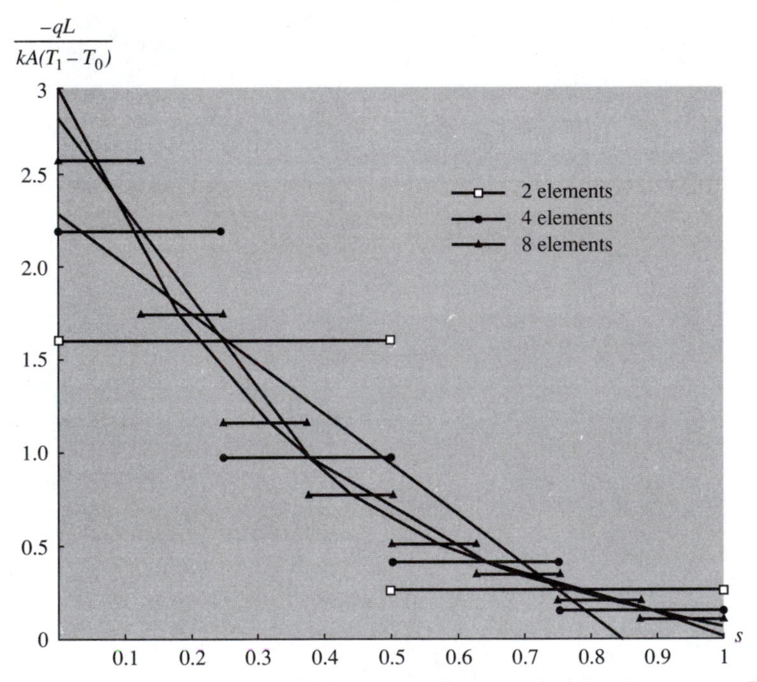

FIGURE 2–27 Derived variable—two-, four-, and eight-element models

In order to check that the basic physics is being satisfied, at least approximately, we consider an energy balance at a typical node. The energies consist of the fluxes $-pu'$ from each of the neighboring elements, the convections $\mathbf{h_e}$, and the sources $\mathbf{f_e}$. The fluxes are computed as outlined above. The nodal convective energy terms $\mathbf{h_e}$ are calculated according to

$$\mathbf{h_e} = \mathbf{q_e u_e}$$

An energy balance can then be performed at the node in question to check that the basic physics is satisfied in the form

$$\text{energy out} - \text{energy in} = \text{source energy}$$

For the results of the two-element model above, the nodal convective terms are computed respectively to be

$$\mathbf{h}_{e1} = \frac{5}{6}\begin{bmatrix} 2 & 1 \\ 1 & 2 \end{bmatrix}\begin{bmatrix} 1.000000 \\ 0.165680 \end{bmatrix} = \begin{bmatrix} 1.8047 \\ 1.1095 \end{bmatrix}$$

$$\mathbf{h}_{e2} = \frac{5}{6}\begin{bmatrix} 2 & 1 \\ 1 & 2 \end{bmatrix}\begin{bmatrix} 0.165680 \\ 0.041420 \end{bmatrix} = \begin{bmatrix} 0.3107 \\ 0.2071 \end{bmatrix}$$

The energy balance can be written as

$$\text{flux out} + \text{convection out} = \text{flux in}$$

Then with flux out $= 0.2485$ (from the second element) and flux in $= 1.6686$ (from the first element), and

$$\text{convection out} = (\mathbf{h}_{e1})_2 + (\mathbf{h}_{e2})_1 = 1.1095 + 0.3107 = 1.4202$$

there results

$$0.2485 + 1.4202 \stackrel{?}{=} 1.6686$$

$$1.6687 \stackrel{?}{=} 1.6686$$

It is left to the student to show that this is essentially the second of the assembled global equations, that is, the Galerkin equation for node 2. These checks indicate that in terms of the different nodal conduction and convection energies, the basic principle of conservation of energy is satisfied.

2.5 EIGENVALUE PROBLEMS

As mentioned in Section 2.2, Sturm-Liouville problems arise that are eigenvalue problems rather than inhomogeneous boundary value problems such as we have studied thus far in this chapter. The development and application of finite element models to these eigenvalue problems will be discussed in this section.

The standard form of the eigenvalue problem associated with the Sturm–Liouville problem can be expressed as

$$(pu')' + (\lambda r - q)u = 0 \tag{2.9a}$$

$$-p(a)u'(a) + \alpha u(a) = 0 \tag{2.9b}$$

$$p(b)u'(b) + \beta u(b) = 0 \tag{2.9c}$$

The task confronting us is to determine the special values of the parameter λ for which there are corresponding nontrivial solutions u. The λ's and corresponding u's are termed *eigenvalues* and *eigenfunctions*, respectively.

To this end we assume an approximate solution of the form

$$v(x) = \sum_{i=1}^{N+1} v_i n_i(x)$$

and substitute into Eq. (2.9*a*). The $n_i(x)$ are the linear nodal interpolation functions introduced and discussed in Section 2.3. The result is

$$(pv')' + (\lambda r - q)v = \left(p \sum_{i=1}^{N+1} v_i n_i'(x)\right)' + (\lambda r - q) \sum_{i=1}^{N+1} v_i n_i(x)$$

$$= E(x, v_1, v_2, \ldots, v_{N+1})$$

where E is the error arising from the fact that the approximate solution v does not (in general) satisfy the differential equation. It is left for the student to show that carrying through the integration by parts and the subsequent development as in Section 2.3 leads to

$$\mathbf{A_G u_G} - \lambda \mathbf{B_G u_G} = \mathbf{0}$$

where

$$\mathbf{A_G} = \Sigma(\mathbf{p_G} + \mathbf{q_G}) + \mathbf{BT_G} \tag{2.10}$$

$$\mathbf{B_G} = \Sigma \mathbf{r_G} \tag{2.11}$$

with

$$\mathbf{p_e} = \int_{x_i}^{x_{i+1}} \mathbf{N}' p \mathbf{N}'^{\mathrm{T}} dx \tag{2.12}$$

$$\mathbf{q_e} = \int_{x_i}^{x_{i+1}} \mathbf{N} q \mathbf{N}^{\mathrm{T}} dx \tag{2.13}$$

$$\mathbf{r_e} = \int_{x_i}^{x_{i+1}} \mathbf{N} r \mathbf{N}^{\mathrm{T}} dx \tag{2.14}$$

$\mathbf{BT_G}$ is the $N + 1$ by $N + 1$ diagonal matrix

$$\mathbf{BT_G} = \begin{bmatrix} \alpha & 0 & \cdots & & 0 \\ 0 & 0 & 0 & 0 & 0 \\ 0 & 0 & 0 & \cdots & 0 \\ 0 & 0 & 0 & 0 & \beta \end{bmatrix}$$

Constraints arising from essential boundary conditions are enforced by deleting from both **A** and **B** the row and column corresponding to the constrained variable. We write the constrained set of equations as

$$(\mathbf{A} - \lambda \mathbf{B})u = 0 \tag{2.15}$$

where **A** and **B** are now reduced $M \times M$ matrices with $M = N + 1 - m$, m being the number of essential boundary conditions that have been imposed.

Eq. (2.15) is an example of the *generalized linear algebraic eigenvalue problem*. It is very similar in character to the *algebraic eigenvalue problem*

$$(\mathbf{A} - \lambda I)u = 0$$

which students encounter in a linear algebra course. The scalars λ_i are the eigenvalues and the corresponding nontrivial vectors u_i satisfying

$$(\mathbf{A} - \lambda_i \mathrm{B})u_{\mathrm{i}} = 0$$

are the eigenvectors. For small hand calculation finite element models, the λ_i and u_i are frequently obtained in the classical manner by expanding the determinant

$$DET(\mathbf{A} - \lambda\mathbf{B}) = 0$$

to obtain an Mth order polynomial

$$P_M(\lambda) = 0$$

whose roots are the approximate eigenvalues. These M roots are then substituted one at a time into the equations

$$(\mathbf{A} - \lambda_i \mathbf{B})\mathbf{u_i} = 0$$

to determine the corresponding approximate eigenvectors.

With the matrices $\mathbf{p_G}$, $\mathbf{q_G}$, $\mathbf{r_G}$, and $\mathbf{BT_G}$ symmetric, **A** and **B** are also symmetric. In such a case the theory can be used to show that all the eigenvalues λ_i are real and that eigenvectors $\mathbf{u_i}$ and $\mathbf{u_j}$ corresponding to distinct eigenvalues λ_i and λ_j satisfy a bi-orthogonality relationship given by

$$\mathbf{u_i B u_j^T} = 0 \qquad i \neq j$$

These general results can be used as checks on the calculations when determining the eigenvalues and eigenvectors.

For eigenvalue problems of dimension larger than three or four, it is essential to have available a reliable computer code for extracting the eigenvalues and eigenvectors. Appendix C contains a discussion of and listings for several routines appropriate for this task. In addition, EISPACK [4, 5] contains several appropriate codes.

2.5.1 Torsional Vibrations

Consider the problem of the torsional vibrations of a uniform circular-cross-section bar as indicated in Fig. 2–28.

As developed in the introduction, the differential equation and boundary conditions that must be investigated in order to determine the mode shapes and natural frequencies are

$$(JG\psi')' + \omega^2 \rho J \psi = 0$$

$$\psi(0) = 0$$

$$\psi'(L) = 0$$

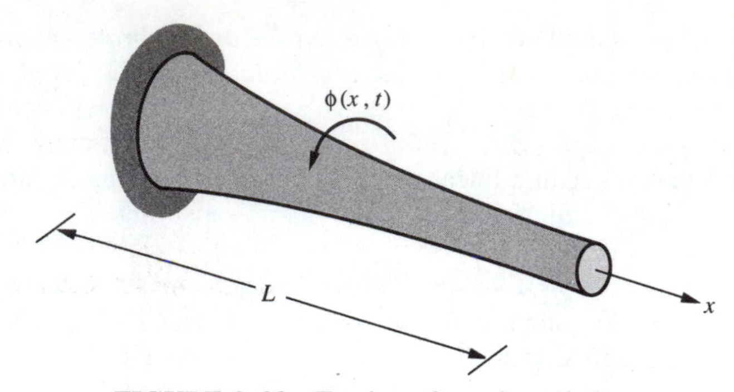

FIGURE 2–28 Torsion of a prismatic bar

With $\lambda = \omega^2$, comparison with the standard form shows that $p = JG, r = \rho J$, $q = 0$, and $\alpha = \beta = 0$.

Two-element solution. Consider a two-element model with equal-length elements. The elemental matrices are

$$\mathbf{p_e} = \int_{x_i}^{x_{i+1}} \mathbf{N}' JG \mathbf{N}'^{\mathrm{T}} dx \qquad \text{and} \qquad \mathbf{r_e} = \int_{x_i}^{x_{i+1}} \mathbf{N} \rho J \mathbf{N}^{\mathrm{T}} dx$$

For the present physical problem with $q = 0, \mathbf{k_e} = \mathbf{p_e}$ are the elemental **mechanical stiffness matrices**. The $\mathbf{r_e}$ are the corresponding elemental **mass matrices** and will be denoted by $\mathbf{m_e}$ in what follows. In an analogous fashion we will use **K** rather than **A**, and **M** rather than **B** at the global level.

Then with

$$\mathbf{N} = \begin{bmatrix} \dfrac{x_{i+1} - x}{x_{i+1} - x_i} \\ \dfrac{x - x_i}{x_{i+1} - x_i} \end{bmatrix} \qquad \text{and} \qquad \mathbf{N}' = \frac{1}{\ell_e}\begin{bmatrix} -1 \\ 1 \end{bmatrix}$$

and $\ell_e = x_{i+1} - x_i = L/2$ for each element,

$$\mathbf{k_{e1}} = \frac{4JG}{L^2}\int_0^{L/2} \begin{bmatrix} -1 \\ 1 \end{bmatrix} 1 \, [-1 \quad 1] \, dx = \frac{2JG}{L}\begin{bmatrix} 1 & -1 \\ -1 & 1 \end{bmatrix} = \mathbf{k_{e2}}$$

and

$$\mathbf{m_{e1}} = \int_0^{L/2} \begin{bmatrix} 1 - \frac{2x}{L} \\ \frac{2x}{L} \end{bmatrix} \rho J \, [1 - \tfrac{2x}{L} \quad \tfrac{2x}{L}] \, dx = \frac{\rho J L}{12}\begin{bmatrix} 2 & 1 \\ 1 & 2 \end{bmatrix}$$

$$\mathbf{m_{e2}} = \int_{L/2}^{L} \begin{bmatrix} 2(1 - \frac{x}{L}) \\ \frac{2x}{L} - 1 \end{bmatrix} \rho J \, [2(1 - \tfrac{x}{L}) \quad \tfrac{2x}{L} - 1] \, dx = \frac{\rho J L}{12}\begin{bmatrix} 2 & 1 \\ 1 & 2 \end{bmatrix}$$

Expanded to the global level,

$$\mathbf{k}_{G1} = \frac{2JG}{L}\begin{bmatrix} 1 & -1 & 0 \\ -1 & 1 & 0 \\ 0 & 0 & 0 \end{bmatrix} \qquad \mathbf{k}_{G2} = \frac{2JG}{L}\begin{bmatrix} 0 & 0 & 0 \\ 0 & 1 & -1 \\ 0 & -1 & 1 \end{bmatrix}$$

and

$$\mathbf{m}_{G1} = \frac{\rho JL}{12}\begin{bmatrix} 2 & 1 & 0 \\ 1 & 2 & 0 \\ 0 & 0 & 0 \end{bmatrix} \qquad \mathbf{m}_{G2} = \frac{\rho JL}{12}\begin{bmatrix} 0 & 0 & 0 \\ 0 & 2 & 1 \\ 0 & 1 & 2 \end{bmatrix}$$

so that with $\mathbf{BT}_G = 0$ the assembled $\mathbf{K}_G \equiv \mathbf{A}_G$ and $\mathbf{M}_G \equiv \mathbf{B}_G$ matrices are easily seen to be

$$\mathbf{K}_G = \frac{2JG}{L}\begin{bmatrix} 1 & -1 & 0 \\ -1 & 2 & -1 \\ 0 & -1 & 1 \end{bmatrix} \qquad \text{and} \qquad \mathbf{B}_G = \frac{\rho JL}{12}\begin{bmatrix} 2 & 1 & 0 \\ 1 & 4 & 1 \\ 0 & 1 & 2 \end{bmatrix}$$

The constraint $\psi_1 = 0$ arises from the essential boundary condition $\psi(0) = 0$. Denoting $\mathbf{K}$ and $\mathbf{M}$ as $\mathbf{K}_G$ and $\mathbf{M}_G$ with the first row and column deleted, there results

$$(\mathbf{K} - \phi\mathbf{M})\psi = \begin{bmatrix} 2 - 4\phi & -1 - \phi \\ -1 - \phi & 1 - 2\phi \end{bmatrix}\begin{bmatrix} \psi_1 \\ \psi_2 \end{bmatrix} = 0$$

where $\phi = \omega^2 L^2 \rho / 24G$. Requiring the determinant of $\mathbf{K} - \phi\mathbf{M}$ to vanish yields

$$2(1 - 2\phi)^2 = (1 + \phi)^2$$

with roots $\phi_1 = 0.1082$ and $\phi_2 = 1.3204$. The corresponding frequencies are given by

$$\omega_1^2 = \frac{24G\phi_1}{\rho L^2} = \frac{2.5968G}{\rho L^2}$$

$$\omega_2^2 = \frac{24G\phi_2}{\rho L^2} = \frac{31.690G}{\rho L^2}$$

These are to be compared to the exact eigenvalues

$$(\omega_1^2)_{\text{exact}} = \left(\frac{\pi}{2L}\right)^2 \frac{G}{\rho} = \frac{2.4674G}{\rho L^2}$$

$$(\omega_2^2)_{\text{exact}} = \left(\frac{3\pi}{2L}\right)^2 \frac{G}{\rho} = \frac{22.207G}{\rho L^2}$$

The estimate of the lowest eigenvalue is quite acceptable (5.1% error), whereas the estimate of the second eigenvalue is much less satisfactory (42.7% error).

The eigenvectors are obtained by substituting the ϕ_i's, one at a time, back into the constrained equations. For the first eigenvalue-eigenvector pair, the first equation becomes

$$(2 - 4\phi_1)\psi_{12} - (1 + \phi_1)\psi_{13} = 0$$

where $\boldsymbol{\psi}^{\mathrm{T}} = [\psi_{12} \quad \psi_{13}]$ is the constrained first eigenvector. Solving for ψ_{12} yields

$$\psi_{12} = 0.707\psi_{13}$$

Repeating for the ϕ_2 yields

$$\psi_{22} = -0.707\psi_{23}$$

for the second eigenvector. The corresponding approximate eigenfunctions or mode shapes appear as in Fig. 2–29.

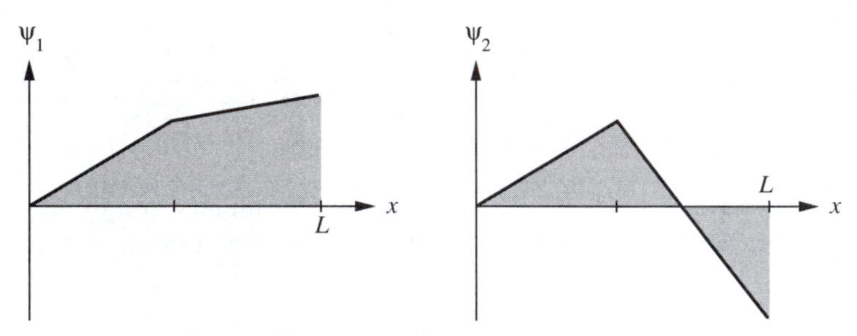

FIGURE 2–29 Eigenfunctions from the two-element model

These two eigenfunctions or mode shapes are approximations to the exact eigenfunctions $\psi_n(x) = \sin(2n - 1)\pi x/2L$ of the original continuous problem. Note that the approximate eigenvectors can be made, by the appropriate choice of the arbitrary constant arising in the solution, to coincide at the nodes with the eigenfunctions they are trying to represent, and that they have the correct number of interior zeros $(n - 1)$ as required by the theory.

Four-element solution. The elemental $\mathbf{k_e}$ and $\mathbf{m_e}$ matrices have exactly the same form as in the two-element model with $\mathbf{l_e}$ now taken as $L/4$. Omitting some of the details, the constrained 4×4 eigenvalue problem is

$$(\mathbf{K} - \phi\mathbf{M})\boldsymbol{\psi} = \mathbf{0}$$

where

$$\mathbf{K} = \begin{bmatrix} 2 & -1 & 0 & 0 \\ -1 & 2 & -1 & 0 \\ 0 & -1 & 2 & -1 \\ 0 & 0 & -1 & 1 \end{bmatrix} \qquad \mathbf{M} = \begin{bmatrix} 4 & 1 & 0 & 0 \\ 1 & 4 & 1 & 0 \\ 0 & 1 & 4 & 1 \\ 0 & 0 & 1 & 2 \end{bmatrix}$$

$\boldsymbol{\psi}^T = [\psi_2 \quad \psi_3 \quad \psi_4 \quad \psi_5]$, and $\phi = \omega^2 L^2 \rho/96G$. Requiring the determinant of $\mathbf{K} - \phi\mathbf{M}$ to vanish yields four roots, which are displayed in Table 2.1 along with the corresponding eigenvalues, exact eigenvalues, and percent errors.

TABLE 2.1 Eigenvalues from four-element model

i	ϕ_i	$\omega_i^2 L^2 \rho/G$	$(\omega_i^2 L^2 \rho/G)_{exact}$	% error
1	0.026034	2.4993	2.4674	1.30
2	0.259085	24.872	22.207	12.0
3	0.854924	82.073	61.685	33.1
4	1.787792	171.63	120.90	42.0

The corresponding eigenfunctions are shown in Fig. 2–30. Again note the correct number of interior zeros of each successive approximate eigenfunction.

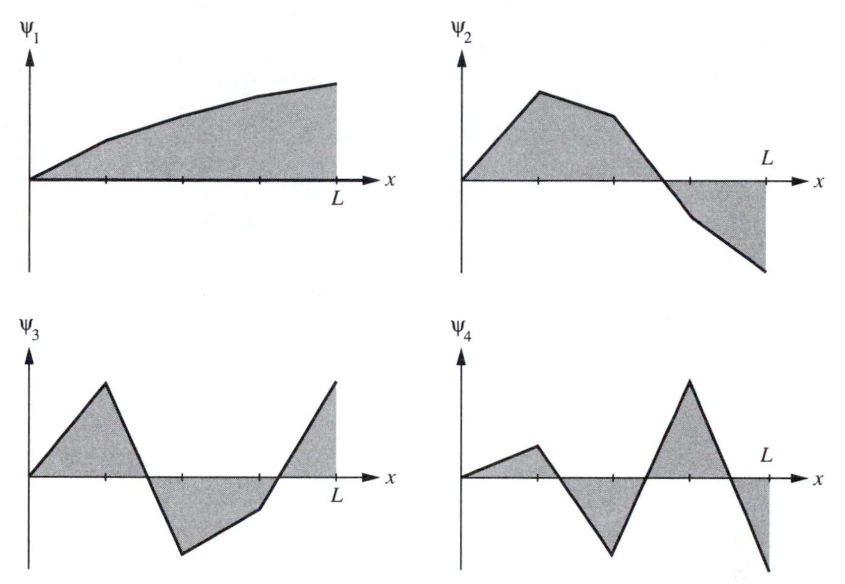

FIGURE 2–30 Eigenfunctions from the four-element model

For the two-element model, the number of constrained degrees of freedom is two. The lowest eigenvalue predicted by that model is 5.2 percent in error, a good estimate. For the four-element model, with four degrees of freedom, the two lowest eigenvalue estimates are 1.3 and 12.0 percent in error respectively, again quite reasonable. In both cases the estimates of the higher eigenvalues are poor. This observation can be stated as a rule of thumb: *For an algebraic eigenvalue problem of the type considered in this section, a model with 2N constrained degrees of freedom is necessary to obtain good estimates for the first N eigenvalues*. This rule of thumb will be further demonstrated in the Exercises.

2.6 EVALUATION OF ELEMENTAL MATRICES

The elemental matrices developed in Section 2.3 are

$$\mathbf{k_e} = \int_{x_i}^{x_{i+1}} (\mathbf{N}' p \mathbf{N}'^{\mathrm{T}} + \mathbf{N} q \mathbf{N}^{\mathrm{T}}) dx$$

$$\mathbf{f_e} = \int_{x_i}^{x_{i+1}} \mathbf{N} f \, dx$$

It is convenient to further delineate the elemental stiffness $\mathbf{k_e}$ according to $\mathbf{k_e} = \mathbf{p_e} + \mathbf{q_e}$, where

$$\mathbf{p_e} = \int_{x_i}^{x_{i+1}} \mathbf{N}' p \mathbf{N}'^{\mathrm{T}} dx$$

$$\mathbf{q_e} = \int_{x_i}^{x_{i+1}} \mathbf{N} q \mathbf{N}^{\mathrm{T}} dx$$

clearly corresponding to the p and q functions. The matrix $\mathbf{r_e}$ encountered in Section 2.5 is given by

$$\mathbf{r_e} = \int_{x_i}^{x_{i+1}} \mathbf{N} r \mathbf{N}^{\mathrm{T}} dx$$

This has precisely the same *form* as $\mathbf{q_e}$. In this section we will outline the several approaches for evaluating the elemental matrices $\mathbf{p_e}$, $\mathbf{q_e}$, $\mathbf{r_e}$, and $\mathbf{f_e}$, corresponding to the functions $p(x)$, $q(x)$, $r(x)$, and $f(x)$, respectively, appearing in the original differential equation.

Recall that the elemental interpolation vector $\mathbf{N}$ appearing in $\mathbf{f_e}$ and $\mathbf{q_e}$ is given by

$$\mathbf{N} = \begin{bmatrix} \dfrac{x_{i+1} - x}{x_{i+1} - x_i} \\ \dfrac{x - x_i}{x_{i+1} - x_i} \end{bmatrix}$$

$\mathbf{N}'$, appearing in $\mathbf{p_e}$, is then easily computed to be

$$\mathbf{N}' = \begin{bmatrix} \dfrac{-1}{x_{i+1} - x_i} \\ \dfrac{1}{x_{i+1} - x_i} \end{bmatrix} = \frac{1}{x_{i+1} - x_i} \begin{bmatrix} -1 \\ 1 \end{bmatrix} = \frac{1}{\ell_e} \begin{bmatrix} -1 \\ 1 \end{bmatrix}$$

These will be used extensively in what follows for the evaluation of the elemental matrices.

Consider first the evaluation of $\mathbf{p_e}$. Expressed in terms of the global coordinate x, $\mathbf{p_e}$ can be written as

$$\mathbf{p_e} = \int_{x_i}^{x_{i+1}} \begin{bmatrix} \dfrac{-1}{l_e} \\ \dfrac{1}{l_e} \end{bmatrix} p(x) \begin{bmatrix} \dfrac{-1}{l_e} & \dfrac{1}{l_e} \end{bmatrix} dx$$

$$= \begin{bmatrix} \left(\frac{1}{l_e}\right)^2 \int_{x_i}^{x_{i+1}} p(x)dx & -\left(\frac{1}{l_e}\right)^2 \int_{x_i}^{x_{i+1}} p(x)dx \\ -\left(\frac{1}{l_e}\right)^2 \int_{x_i}^{x_{i+1}} p(x)dx & \left(\frac{1}{l_e}\right)^2 \int_{x_i}^{x_{i+1}} p(x)dx \end{bmatrix}$$

$$= \left(\frac{1}{l_e}\right)^2 \int_{x_i}^{x_{i+1}} p(x)dx \begin{bmatrix} 1 & -1 \\ -1 & 1 \end{bmatrix}$$

In terms of the local coordinate ξ given by $\xi = x - x_i$,

$$\mathbf{p_e} = \left(\frac{1}{l_e}\right)^2 \int_0^{l_e} p(x_i + \xi)d\xi \begin{bmatrix} 1 & -1 \\ -1 & 1 \end{bmatrix}$$

Whether the choice is made for evaluating the integrals in terms of the global or local coordinates, care must be taken to insert the values of the function p in the interval (x_i, x_{i+1}) when evaluating the integral.

The elemental stiffness matrix $\mathbf{q_e}$ corresponding to the function q was determined to be

$$\mathbf{q_e} = \int_{x_i}^{x_{i+1}} \mathbf{N}q\mathbf{N}^{\mathrm{T}} dx$$

When expressed in terms of the global coordinate x,

$$\mathbf{q_e} = \int_{x_i}^{x_{i+1}} \begin{bmatrix} N_i(x) \\ N_{i+1}(x) \end{bmatrix} q(x)[N_i(x) \;\; N_{i+1}(x)]dx$$

where N_i and N_{i+1} are the elemental interpolation functions. When expressed in terms of the local coordinates

$$\mathbf{q_e} = \int_0^{l_e} \begin{bmatrix} 1 - \dfrac{\xi}{l_e} \\ \dfrac{\xi}{l_e} \end{bmatrix} q(x_i + \xi) \begin{bmatrix} 1 - \dfrac{\xi}{l_e} & \dfrac{\xi}{l_e} \end{bmatrix} d\xi$$

Here again, special care must be taken to insert the correct values for the function $q(x_i + \xi)$, on the interval (x_i, x_{i+1}), in evaluating the integrals. Again, the *form* of the $\mathbf{q_e}$ and $\mathbf{r_e}$ are exactly the same.

The elemental matrices $\mathbf{f_e}$ are given by

$$\mathbf{f_e} = \int_{x_i}^{x_{i+1}} \mathbf{N}f\,dx = \int_{x_i}^{x_{i+1}} \begin{bmatrix} N_i(x) \\ N_{i+1}(x) \end{bmatrix} f(x)dx$$

when expressed in terms of the global coordinates. When expressed in terms of the local coordinate ξ, $\mathbf{f_e}$ appears as

$$\mathbf{f_e} = \int_0^{l_e} \begin{bmatrix} N_i(\xi) \\ N_{i+1}(\xi) \end{bmatrix} f(x_i + \xi)d\xi$$

We have seen that the dummy variable appearing in the elementary matrices $\mathbf{p_e}$, $\mathbf{q_e}$, and $\mathbf{f_e}$ can be either the global variable x or the local variable ξ, both indicated in Fig. 2–31.

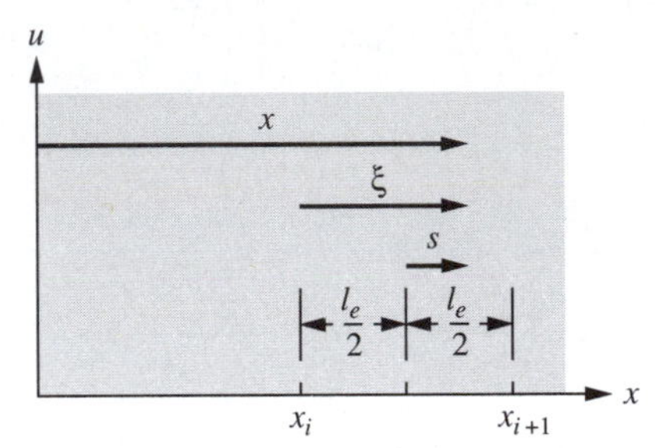

FIGURE 2–31 Global, local, and natural coordinates

In what follows, it will be convenient to consider yet a third coordinate s defined by the relation

$$\frac{\xi}{l_e} = \frac{(1+s)}{2}$$

also indicated in Fig. 2–31. As ξ varies from 0 to l_e, s varies from -1 to 1. It follows that integrals from 0 to l_e with respect to ξ can be converted according to

$$\int_0^{l_e} f(\xi)d\xi = \int_{-1}^{1} f\left(\frac{l_e(1+s)}{2}\right)\left(\frac{l_e}{2}\right)ds = \int_{-1}^{1} F(s)ds$$

In the first equality, the relation between the differentials $d\xi = (l_e/2)ds = J\,ds$ where $J = l_e/2$ is the *Jacobian* of the transformation $\xi = \xi(s)$. For a one-dimensional transformation of this sort, J is clearly a ratio of the differential lengths in ξ and s systems. Often, s is referred to as a natural or intrinsic coordinate.

The integrals $\mathbf{p_e}$, $\mathbf{q_e}$, and $\mathbf{f_e}$ representing the elemental contributions of p, q, and f can be evaluated using the fundamental theorem of integral calculus or by consulting a table of integrals. Alternatively, one can use numerical integration procedures which are now discussed.

NUMERICAL INTEGRATION. Generally we are concerned with the evaluation of integrals of the form

$$I = \int_{x_i}^{x_{i+1}} f(x)dx$$

or

$$I = \int_0^{l_e} f(x_i + \xi)d\xi$$

or

$$I = \int_{-1}^{1} F(s)ds$$

The basic idea of any of the commonly used approximate integration or **quadrature methods,** as they are sometimes called, is to replace the actual integrand by an approximation in terms of elementary functions, usually polynomials. This approximate integrand should be such that the integral is readily integrated with the result having a small error compared with the corresponding exact evaluation. Three different approaches to this problem are presented.

Newton-Cotes quadratures. The family of Newton-Cotes formulas for quadrature are based on subdividing the basic interval (a, b) into a number of suitably chosen subintervals (x_i, x_{i+1}). The integrand is then approximated by a polynomial on each of the subintervals with a resulting estimate for the value of the integral. The first two members of the family are presented as follows.

Trapezoidal rule. For the trapezoidal rule the integrand f is replaced by a linear function F on each of the subintervals as indicated in Fig. 2–32. This clearly corresponds to linear interpolation.

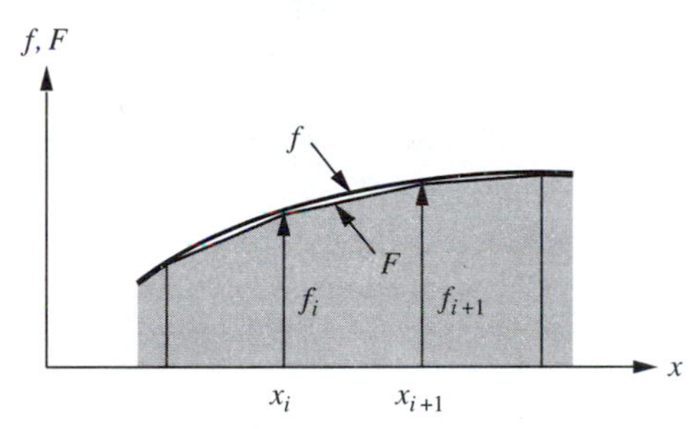

FIGURE 2–32 Linear interpolation for the trapezoidal rule

The value of the integral is approximated by

$$I = \int_a^b f(x)dx \approx \sum_1^N \int_{x_i}^{x_{i+1}} F(x)dx$$

where F is the linear approximation on the interval (x_i, x_{i+1}) given in terms of the linear interpolation functions N_i and N_{i+1} by $F = f_iN_i + f_{i+1}N_{i+1}$ with f_i and f_{i+1} usually taken as the nodal values of $f(x)$. The integrals on the subintervals are evaluated to yield

$$I = \sum_1^N \left(F(x_i) + F(x_{i+1})\right)\frac{h_i}{2}$$

where $h_i = x_{i+1} - x_i$. If all the subintervals are taken to be of equal length h, the result can be written as

$$I = \sum_1^{N+1} w_i F(x_i) = \sum_1^{N+1} w_i f(x_i)$$

in view of the fact that F and f coincide at the x_i. The so-called *weights* w_i for the trapezoidal rule are thus determined to be

$$w_i = \frac{h[1\ 2\ 2\ \ldots\ 2\ 1]}{2}$$

a result that should be familiar to the student who has taken an elementary numerical methods class.

Simpson's rule. For Simpson's rule the interval (a, b) is divided into N subintervals of length $2h_i = (x_{i+1} - x_{i-1})$. In each of these subintervals, the integrand f is replaced by a quadratic function F which is constructed by passing a parabola through the three successive f values, $f_{i-1} = f(x_{i-1})$, $f_i = f(x_i)$, and $f_{i+1} = f(x_{i+1})$ as indicated in Fig. 2–33.

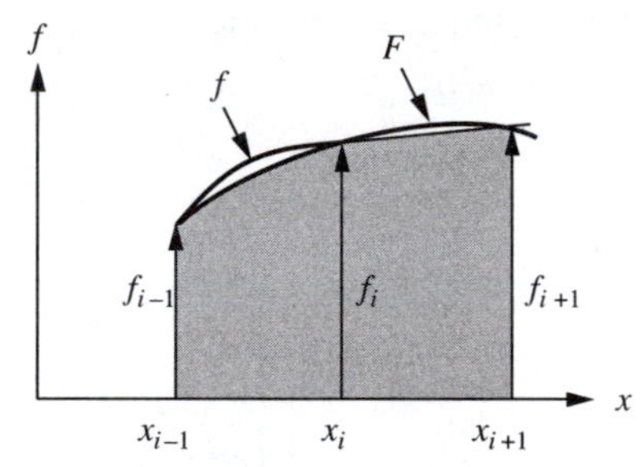

FIGURE 2–33 Quadratic interpolation for Simpson's rule

F has the representation $F = f_{i-1}N_{i-1} + f_iN_i + f_{i+1}N_{i+1}$, where N_{i-1}, N_i, and N_{i+1} are the quadratic interpolation functions which will be discussed later in this chapter. The value of the integral is approximated by

$$I \approx \sum_1^N \int_{x_{i-1}}^{x_{i+1}} F(x)dx$$

The integrals on the subintervals are easily evaluated to yield

$$I \approx \sum_1^N \frac{(F(x_{i-1}) + 4F(x_i) + F(x_{i+1}))\,h_i}{3}$$

where $h_i = (x_{i+1} - x_{i-1})/2$. If all the subintervals are taken to be of equal length $2h$, the result can be written as

$$I = \sum_1^{2N+1} w_iF(x_i) = \sum_1^{2N+1} w_if(x_i)$$

Thus the so-called *weights* w_i for Simpson's rule are collectively

$$w = \frac{h}{3}[1\ 4\ 2\ 4\ 2\ \ldots\ 2\ 4\ 2\ 4\ 1]$$

Higher-order integration formulas are possible but will not be presented or discussed here.

Gauss-Legendre quadrature. In the Newton-Cotes methods, sets of $n + 1$ equally spaced interior points are used to define nth degree polynomials with the property that any function of degree n or less can be integrated exactly using the quadrature rule. The weights and integration points are fixed for a given quadrature rule. No attempt is made to optimize the position of the sampling points or the weights in order to achieve the best accuracy. In Gaussian quadrature a different strategy is used, namely, the N weights and the locations of the N integration points (a total of $2N$ parameters) are chosen so as to integrate a polynomial of degree $2N - 1$ exactly within a certain interval. The general form of the quadrature is

$$\int_{-1}^{1} F(u)du = \sum_{1}^{N} W_i F(u_i)$$

For instance, in a one-term Gauss quadrature, the weight and the integration point are selected so as to ensure that any polynomial of degree one or less is integrated exactly. A two-term Gauss scheme should be such that any polynomial of degree three or less is integrated exactly.

Given the integral

$$I = \int_{a}^{b} f(x)dx$$

it is first necessary to transform the variable of integration to the natural coordinate described previously. This transformation, namely

$$x = x(u) = \frac{b + a}{2} + \frac{(b - a)u}{2}$$

is used to convert the integral to the form

$$I = \int_{-1}^{1} f\left(\frac{b + a}{2} + \frac{(b - a)u}{2}\right)\frac{b - a}{2}du = \int_{-1}^{1} F(u)du$$

Usually Gauss quadrature is based on having first transformed the integral in question to this standard form.

Gauss $N = 1$ scheme. In the $N = 1$ Gauss scheme

$$\int_{-1}^{1} F(u)du = W_1 F(u_1)$$

the two quantities W_1 and u_1 are to be determined such that any linear function of u be integrated exactly. We accomplish this as follows.

First take $F = 1$, a constant. It readily follows that

$$\int_{-1}^{1} 1du = 2 = W_1 F(u_1) = W_1$$

from which $W_1 = 2$. Then with $F(u) = u$,

$$\int_{-1}^{1} u\,du = 0 = W_1 F(u_1) = 2u_1$$

so that $u_1 = 0$. Thus the one-term Gauss quadrature takes the form

$$\int_{-1}^{1} F(u)du = 2F(0)$$

On the basis of Fig. 2–34, this quadrature clearly integrates any linear function exactly.

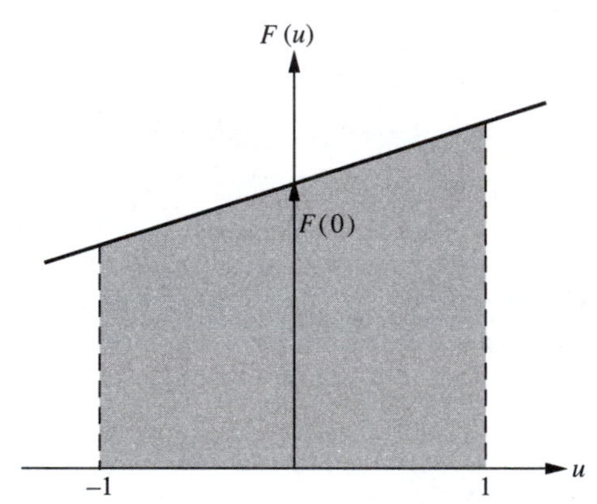

FIGURE 2–34 Gauss $N = 1$, for linear functions $F(u)$

Gauss $N = 2$ scheme. In the $N = 2$ Gauss scheme

$$\int_{-1}^{1} F(u)du = W_1 F(u_1) + W_2 F(u_2)$$

the four quantities W_1, W_2, u_1, and u_2 are to be determined so that any polynomial of degree three or less is integrated exactly. This is accomplished by requiring that each of the monomials 1, u, u^2, and u^3 is integrated exactly by the $N = 2$ Gauss quadrature.

$$F = 1 \qquad \int_{-1}^{1} 1\,du = 2 = W_1 F(u_1) + W_2 F(u_2) = W_1 + W_2$$

$$F = u \qquad \int_{-1}^{1} u\,du = 0 = W_1 F(u_1) + W_2 F(u_2) = W_1 u_1 + W_2 u_2$$

$$F = u^2 \qquad \int_{-1}^{1} u^2 du = \frac{2}{3} = W_1 F(u_1) + W_2 F(u_2) = W_1 u_1^2 + W_2 u_2^2$$

$$F = u^3 \qquad \int_{-1}^{1} u^3 du = 0 = W_1 F(u_1) + W_2 F(u_2) = W_1 u_1^3 + W_2 u_2^3$$

It is left to the Exercises to show that

$$W_1 = W_2 = 1$$

$$u_1 = \frac{-1}{\sqrt{3}} = -u_2$$

so that

$$\int_{-1}^{1} F(u)du = F\left(\frac{-1}{\sqrt{3}}\right) + F\left(\frac{1}{\sqrt{3}}\right)$$

integrates up to and including a third-degree polynomial exactly.

It is known by other approaches to Gauss quadrature that the integration points for the N-point Gauss quadrature are the roots of the Nth degree Legendre polynomial $P_n(x)$. The first few of the Legendre polynomials are $P_0(x) = 1$, $P_1(x) = x$, $P_2(x) = (3x^2 - 1)/2$, and $P_3(x) = (5x^3 - 3x)/2$. For this reason one often sees the term *Gauss-Legendre quadrature* describing the previous quadrature rule.

The known weights W_i and the integration points u_i for the first few values of N are given in Table 2.2 for the Gauss-Legendre quadrature. As mentioned previously, Gauss-Legendre quadrature has the property that an Nth order scheme integrates exactly a polynomial of degree $2N - 1$ or less.

TABLE 2.2 Data for Gauss-Legendre quadrature

Order N	Points u_i	Weights W_i
1	0	2.00000000
2	±0.577350269	1.00000000
3	0	0.88888889
	±0.774596669	0.55555556
4	±0.339981044	0.65214515
	±0.861136312	0.34785485

Example 2.5.

For the purposes of illustration and comparison, we consider the application of the trapezoidal, Simpson's, and Gauss-Legendre quadrature formulas to the evaluation of the integral

$$I = \int_{1}^{3} \frac{dx}{(3x+4)} = \frac{1}{3}\log_e\left(\frac{13}{7}\right) = 0.206346$$

Trapezoidal rule.

One interval:

$$I \approx \frac{2}{2}\left(\frac{1}{7} + \frac{1}{13}\right) = 0.219780$$

Two intervals:

$$I \approx \frac{1}{2}\left(\frac{1}{7} + \frac{2}{10} + \frac{1}{13}\right) = 0.209890$$

Four intervals:

$$I \approx \frac{1}{4}\left(\frac{1}{7} + 2\left(\frac{1}{8.5} + \frac{1}{10} + \frac{1}{11.5}\right) + \frac{1}{13}\right) = 0.207247$$

Simpson's rule.

One interval:

$$I \approx \frac{1}{3}\left(\frac{1}{7} + \frac{4}{10} + \frac{1}{13}\right) = 0.206593$$

Two intervals:

$$I \approx \frac{1}{6}\left(\frac{1}{7} + 4\left(\frac{1}{8.5} + \frac{1}{11.5}\right) + \frac{2}{10} + \frac{1}{13}\right) = 0.206366$$

Gauss-Legendre. The transformation necessary for reduction to standard form is $x = 2 + u$, leading to

$$I = \int_{-1}^{1} \frac{du}{(3u + 10)}$$

$N = 1$

$$I \approx 2\left(\frac{1}{10}\right) = 0.200000$$

$N = 2$

$$I \approx 1\left(\frac{1}{10 - 3(0.557350)} + \frac{1}{10 + 3(0.557350)}\right) = 0.206186$$

$N = 3$

$$I \approx \left(\frac{5}{10 - 3(0.774600)} + \frac{5}{10 + 3(0.774600)} + \frac{8}{10}\right) \Big/ 9 = 0.206342$$

Comparing the results of each of the three methods for three function evaluations indicates the relative accuracy of the Gauss-Legendre quadrature for this example. In later sections we will see that Gauss-Legendre quadrature is ideally suited for the evaluation of many of the integrals appearing in the elemental matrices associated with two- and three-dimensional isoparametric elements.

SPECIAL INTEGRATION FORMULAS. When generating a finite element model by hand, it is reasonable to assume that the functions p, q, and f are known explicitly so that at least in principle the elemental matrices $\mathbf{p_e}$, $\mathbf{q_e}$, and $\mathbf{f_e}$ can be evaluated either exactly or with high accuracy using one of the quadrature rules. In a situation where a computer code is being used, it is much less likely that the functions p, q, and f can be specified explicitly as functions of the interval (a, b), but rather that they are known or specified at discrete points x_i of the interval. In a one-dimensional situation, inputs to the code can include the locations of the nodes and the corresponding nodal values of p, q, and f.

The special integration formulas presented below are based on the idea that the functions p, q, and f are sufficiently smooth so that on a subinterval or element (x_i, x_{i+1}), the functions can be replaced by their linear interpolants as indicated in Fig. 2–35.

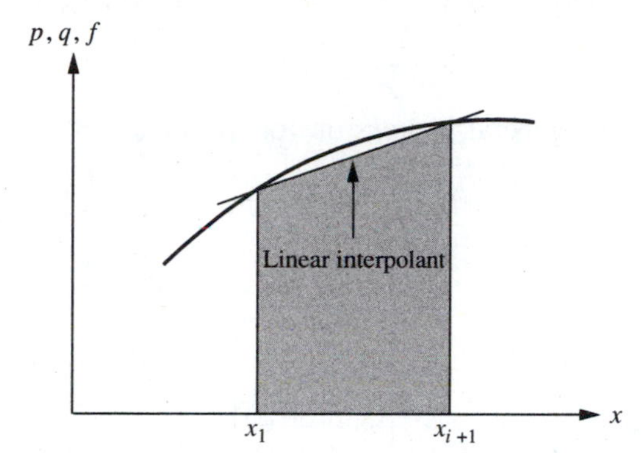

FIGURE 2–35 Replacement by linear interpolant

The results will represent exact evaluations of the integrals if the functions are of degree one or less and will be approximate otherwise. This approach essentially says that coefficients in the differential equation are going to be represented or approximated in terms of the same interpolation functions that are used to represent the solution.

Evaluation of $\mathbf{p_e}$. The general expression for $\mathbf{p_e}$ is

$$\mathbf{p_e} = \left(\frac{1}{l_e^2}\right)\int_{x_i}^{x_{i+1}} p(x)dx \begin{bmatrix} 1 & -1 \\ -1 & 1 \end{bmatrix}$$

or

$$\mathbf{p_e} = \left(\frac{1}{l_e^2}\right)\int_{0}^{l_e} p(x_i + \xi)d\xi \begin{bmatrix} 1 & -1 \\ -1 & 1 \end{bmatrix}$$

Represent or approximate $p(x)$ as a linear function on an interval or element (x_i, x_{i+1}) according to

$$p(x) = p_i N_i + p_{i+1} N_{i+1}$$

where p_i and p_{i+1} are suitably chosen values of $p(x)$. Frequently p_i and p_{i+1} are chosen to be the values of p at x_i and x_{i+1}, respectively. For the linear interpolation functions,

$$\int_{x_i}^{x_{i+1}} N_i dx = \int_0^{l_e} N_i d\xi = \int_{x_i}^{x_{i+1}} N_{i+1} dx = \int_0^{l_e} N_{i+1} d\xi = \frac{l_e}{2}$$

so that

$$\mathbf{p_e} \approx \frac{(p_i + p_{i+1})}{2l_e}\begin{bmatrix} 1 & -1 \\ -1 & 1 \end{bmatrix} = \frac{p_{\text{avg}}}{l_e}\begin{bmatrix} 1 & -1 \\ -1 & 1 \end{bmatrix} \tag{2.16}$$

Generally, $p_{\text{avg}} = (p_i + p_{i+1})/2$ is not the true average unless $p(x)$ is actually linear on the interval, in which case the approximate evaluation for $\mathbf{p_e}$ is exact.

Evaluation of $\mathbf{q_e}$. The general expression for $\mathbf{q_e}$ is

$$\mathbf{q_e} = \int_{x_i}^{x_{i+1}} \begin{bmatrix} N_i(x) \\ N_{i+1}(x) \end{bmatrix} q(x)[N_i(x) \quad N_{i+1}(x)]dx$$

or

$$\mathbf{q_e} = \int_0^{l_e} \begin{bmatrix} 1 - \dfrac{\xi}{l_e} \\ \dfrac{\xi}{l_e} \end{bmatrix} q(x_i + \xi)\begin{bmatrix} 1 - \dfrac{\xi}{l_e} & \dfrac{\xi}{l_e} \end{bmatrix} d\xi$$

Approximate q as

$$q(x) = q_i N_i + q_{i+1} N_{i+1}$$

where q_i and q_{i+1} are suitably chosen values of q. The q_i and q_{i+1} are frequently taken to be the values of q at nodes x_i and x_{i+1}, respectively. With this representation for the function q, the $\mathbf{q_e}$ when expressed in terms of the local coordinate ξ becomes

$$\mathbf{q_e} \approx \begin{bmatrix} \int_0^{l_e} (N_i^3 q_i + N_i^2 N_{i+1} q_{i+1}) d\xi & \int_0^{l_e} (N_i^2 N_{i+1} q_i + N_i N_{i+1}^2 q_{i+1}) d\xi \\ \int_0^{l_e} (N_i^2 N_{i+1} q_i + N_i N_{i+1}^2 q_{i+1}) d\xi & \int_0^{l_e} (N_i N_{i+1}^2 q_i + N_{i+1}^3 q_{i+1}) d\xi \end{bmatrix}$$

The integrals are evaluated to yield

$$\mathbf{q_e} \approx \frac{l_e}{12}\begin{bmatrix} 3q_i + q_{i+1} & q_i + q_{i+1} \\ q_i + q_{i+1} & q_i + 3q_{i+1} \end{bmatrix} \tag{2.17}$$

which is exact if q is constant or a linear function of x. For the special case where q is a constant q_0, $\mathbf{q_e}$ reduces to

$$\mathbf{q_e} = \frac{q_0 l_e}{6}\begin{bmatrix} 2 & 1 \\ 1 & 2 \end{bmatrix}$$

Evaluation of $\mathbf{r_e}$**.** The general and approximate expressions for $\mathbf{r_e}$ have precisely the same form as those for $\mathbf{q_e}$ with the nodal q values replaced by the nodal r values.

Evaluation of $\mathbf{f_e}$**.** The general expression for $\mathbf{f_e}$ is

$$\mathbf{f_e} = \int_{x_i}^{x_{i+1}} \begin{bmatrix} N_i(x) \\ N_{i+1}(x) \end{bmatrix} f(x)dx$$

or

$$\mathbf{f_e} = \int_0^{l_e} \begin{bmatrix} N_i(\xi) \\ N_{i+1}(\xi) \end{bmatrix} f(x_i + \xi)d\xi$$

When f is approximated as

$$f(x) = f_i N_i + f_{i+1} N_{i+1}$$

there results

$$\mathbf{f_e} \approx \begin{bmatrix} \int_0^{l_e} (N_i^2 f_i + N_i N_{i+1} f_{i+1})d\xi \\ \int_0^{l_e} (N_i N_{i+1} f_i + N_{i+1}^2 f_{i+1})d\xi \end{bmatrix}$$

The integrals are easily evaluated to yield

$$\mathbf{f_e} \approx \frac{l_e}{6} \begin{bmatrix} 2f_i + f_{i+1} \\ f_i + 2f_{i+1} \end{bmatrix} \tag{2.18}$$

which is exact if f is constant or a linear function of x. If $f = f_0$, a constant, $\mathbf{f_e}$ reduces to

$$\mathbf{f_e} \approx \frac{f_0 l_e}{2} \begin{bmatrix} 1 \\ 1 \end{bmatrix}$$

In other words, one half of the total load $f_0 l_e$ for the element is lumped at each of the two nodes for that element.

The utility and accuracy of these special integration formulas will be demonstrated in the examples that follow.

Example 2.6

It is the purpose of this example to investigate the effect on the solution of the choice of method for evaluating the elemental matrices. We will investigate two-, four-, and eight-element models with the elemental matrices evaluated using the special integration formulas, $N = 1$ Gauss quadrature, and $N = 2$ Gauss quadrature.

As a vehicle, consider the boundary value problem

$$(x^2u')' - 2u + x^2 = 0$$
$$u(1) = 0$$
$$u'(2) = 0$$

The exact classical solution is

$$u(x) = \frac{17x - 12/x^2 - 5x^2}{20}$$

First consider two elements, using special integration for the evaluation of the integrals. The required parameters are $p = x^2$, $q = 2$, and $f = x^2$. Note that the special integration approach evaluates the $\mathbf{q_e}$ exactly and the $\mathbf{p_e}$ and $\mathbf{f_e}$ approximately. The elemental matrices are

$$\mathbf{p_{e1}} = \begin{bmatrix} 3.2500 & -3.2500 \\ -3.2500 & 3.2500 \end{bmatrix} \qquad \mathbf{p_{e2}} = \begin{bmatrix} 6.2500 & -6.2500 \\ -6.2500 & 6.2500 \end{bmatrix}$$

$$\mathbf{q_{e1}} = \begin{bmatrix} 0.3333 & 0.1667 \\ 0.1667 & 0.3333 \end{bmatrix} \qquad \mathbf{q_{e2}} = \begin{bmatrix} 0.3333 & 0.1667 \\ 0.1667 & 0.3333 \end{bmatrix}$$

$$\mathbf{f_{e1}} = \begin{bmatrix} 0.3542 \\ 0.4583 \end{bmatrix} \qquad \mathbf{f_{e2}} = \begin{bmatrix} 0.7083 \\ 0.8542 \end{bmatrix}$$

Assembling, enforcing the constraint $u_1 = 0$, and solving yields $u_2 = 0.4303$ and $u_3 = 0.5274$. The derived variables, pu', are computed to be

Element 1.

$$pu' = \frac{(1.25)^2(0.4303 - 0.0000)}{0.5} = 1.3448$$

Element 2.

$$pu' = \frac{(1.75)^2(0.5274 - 0.4303)}{0.5} = 0.5945$$

Repeating all these steps using $N = 1$ Gauss quadrature for evaluating the integral yields

$$\mathbf{p_{e1}} = \begin{bmatrix} 3.1250 & -3.1250 \\ -3.1250 & 3.1250 \end{bmatrix} \qquad \mathbf{p_{e2}} = \begin{bmatrix} 6.1250 & -6.1250 \\ -6.1250 & 6.1250 \end{bmatrix}$$

$$\mathbf{q_{e1}} = \begin{bmatrix} 0.2500 & 0.2500 \\ 0.2500 & 0.2500 \end{bmatrix} \qquad \mathbf{q_{e2}} = \begin{bmatrix} 0.2500 & 0.2500 \\ 0.2500 & 0.2500 \end{bmatrix}$$

$$\mathbf{f_{e1}} = \begin{bmatrix} 0.3906 \\ 0.3906 \end{bmatrix} \qquad \mathbf{f_{e2}} = \begin{bmatrix} 0.7656 \\ 0.7656 \end{bmatrix}$$

all of which are approximate. The corresponding solution is $u_2 = 0.4294$ and $u_3 = 0.5158$. The derived variables are $(pu')_1 = 1.3419$ and $(pu')_2 = 0.5293$.

Finally, using $N = 2$ Gauss quadrature for evaluating the integrals results in

$$\mathbf{p}_{e1} = \begin{bmatrix} 3.1667 & -3.1667 \\ -3.1667 & 3.1667 \end{bmatrix} \qquad \mathbf{p}_{e2} = \begin{bmatrix} 6.1667 & -6.1667 \\ -6.1667 & 6.1667 \end{bmatrix}$$

$$\mathbf{q}_{e1} = \begin{bmatrix} 0.3333 & 0.1667 \\ 0.1667 & 0.3333 \end{bmatrix} \qquad \mathbf{q}_{e2} = \begin{bmatrix} 0.3333 & 0.1667 \\ 0.1667 & 0.3333 \end{bmatrix}$$

$$\mathbf{f}_{e1} = \begin{bmatrix} 0.3437 \\ 0.4479 \end{bmatrix} \qquad \mathbf{f}_{e2} = \begin{bmatrix} 0.6979 \\ 0.8438 \end{bmatrix}$$

all of which are exact. The corresponding solution is $u_2 = 0.4314$ and $u_3 = 0.5280$. The derived variables are $(pu')_1 = 1.3481$ and $(pu')_2 = 0.5918$.

These same three options are used in generating four- and eight-element models for this problem. The results are indicated in Figs. 2–36 through 2–41. It is clear from these figures that as the number of elements is increased, the differences in the numerical results from the different models tend to zero. In particular, for the eight-element cases, the $N = 2$ Gauss results are essentially the same as the results based on using the special integration technique.

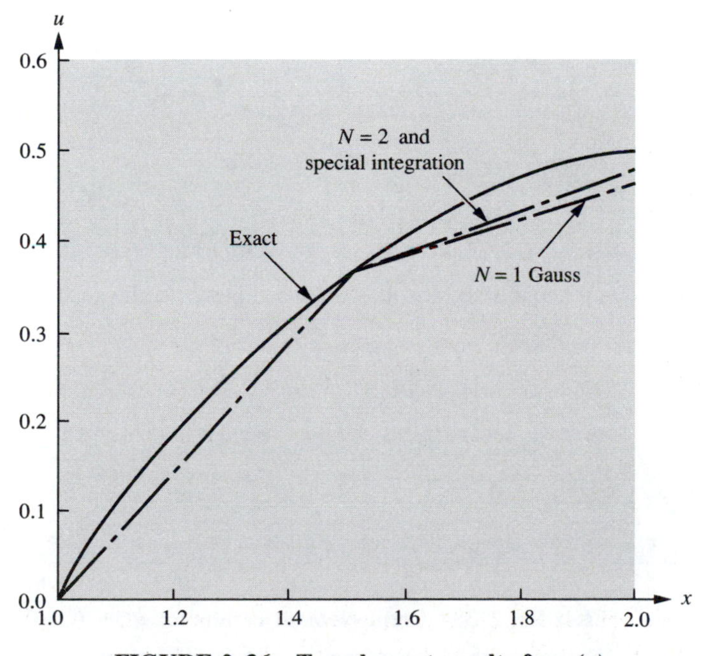

FIGURE 2–36 Two-element results for *u(x)*

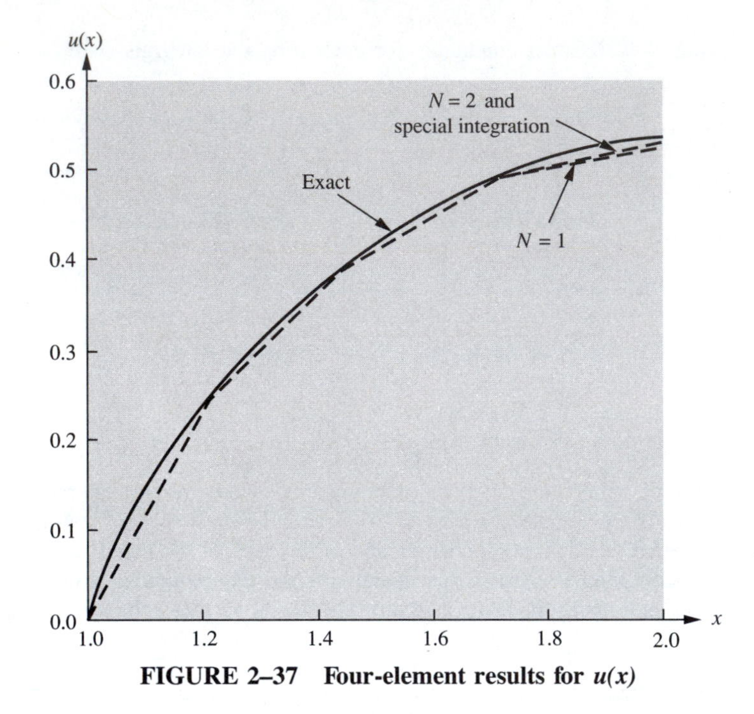

FIGURE 2–37 Four-element results for $u(x)$

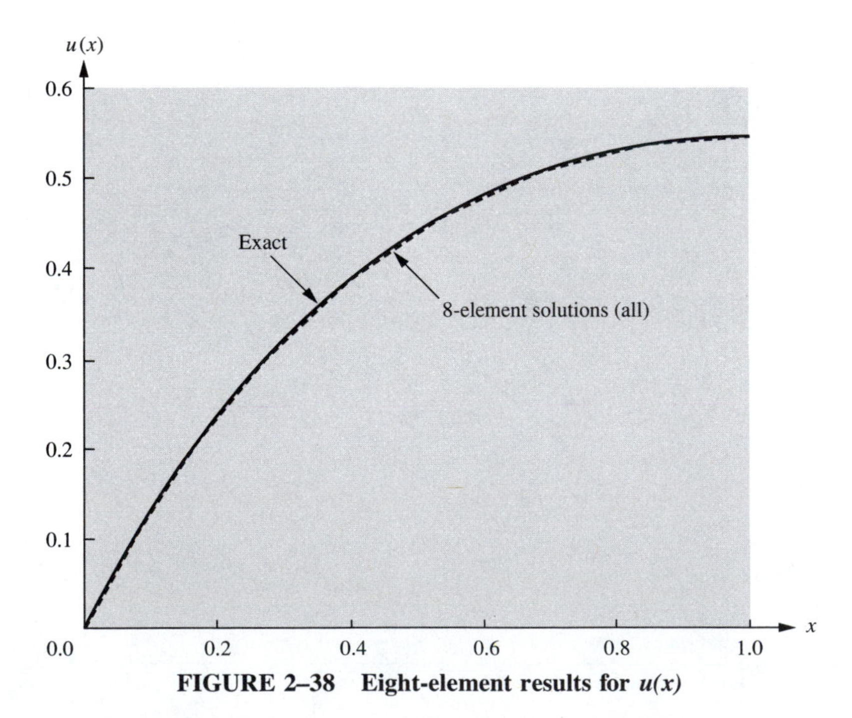

FIGURE 2–38 Eight-element results for $u(x)$

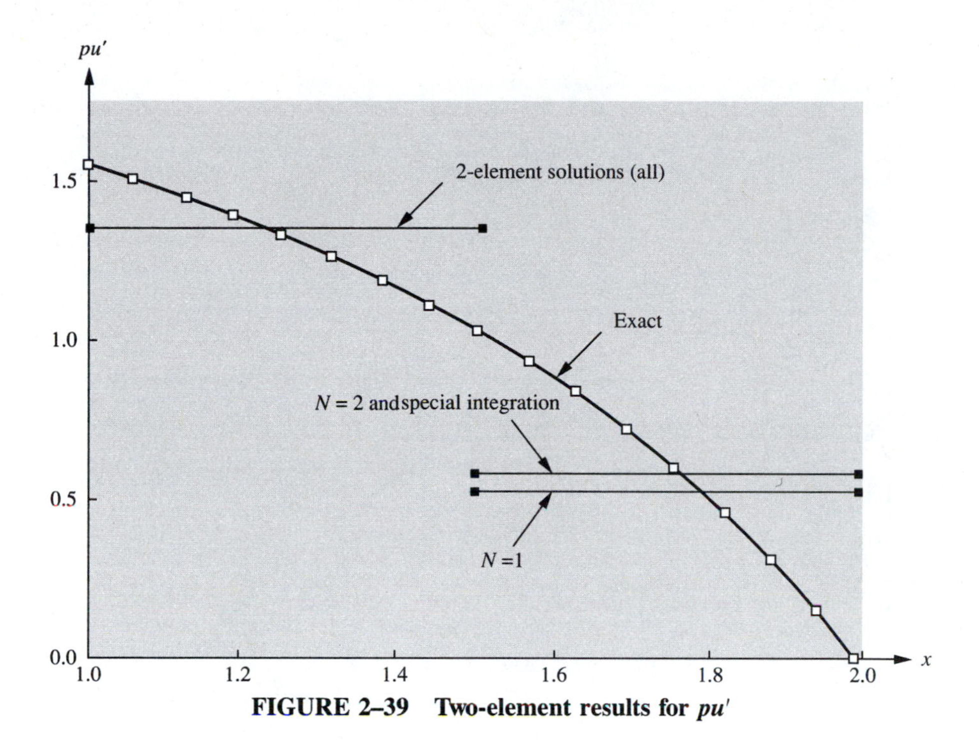

FIGURE 2–39 Two-element results for *pu'*

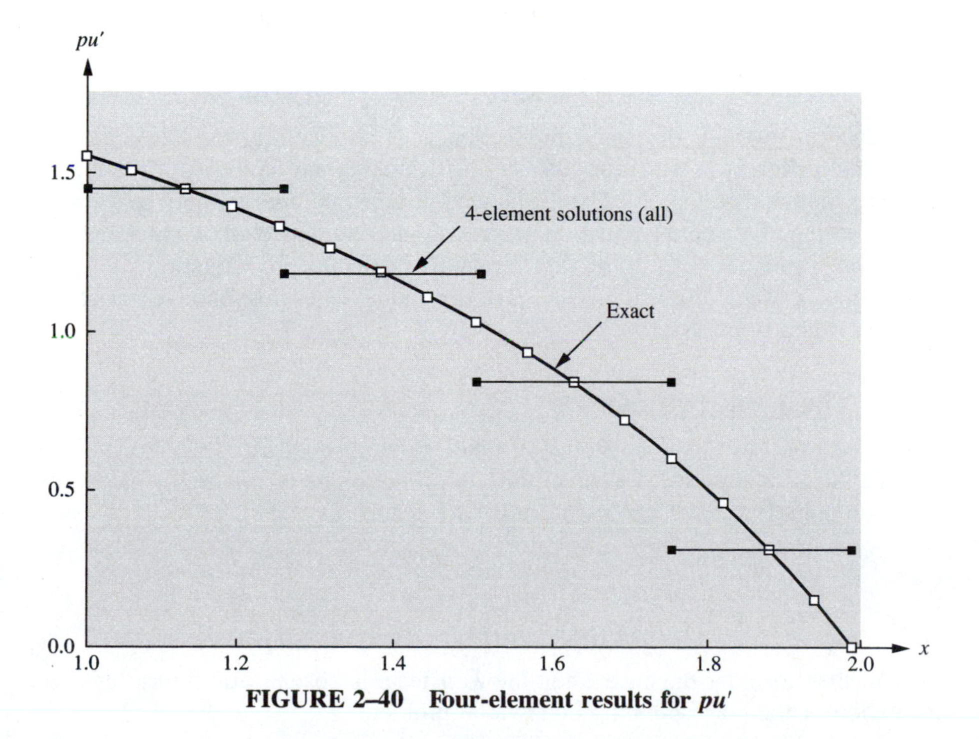

FIGURE 2–40 Four-element results for *pu'*

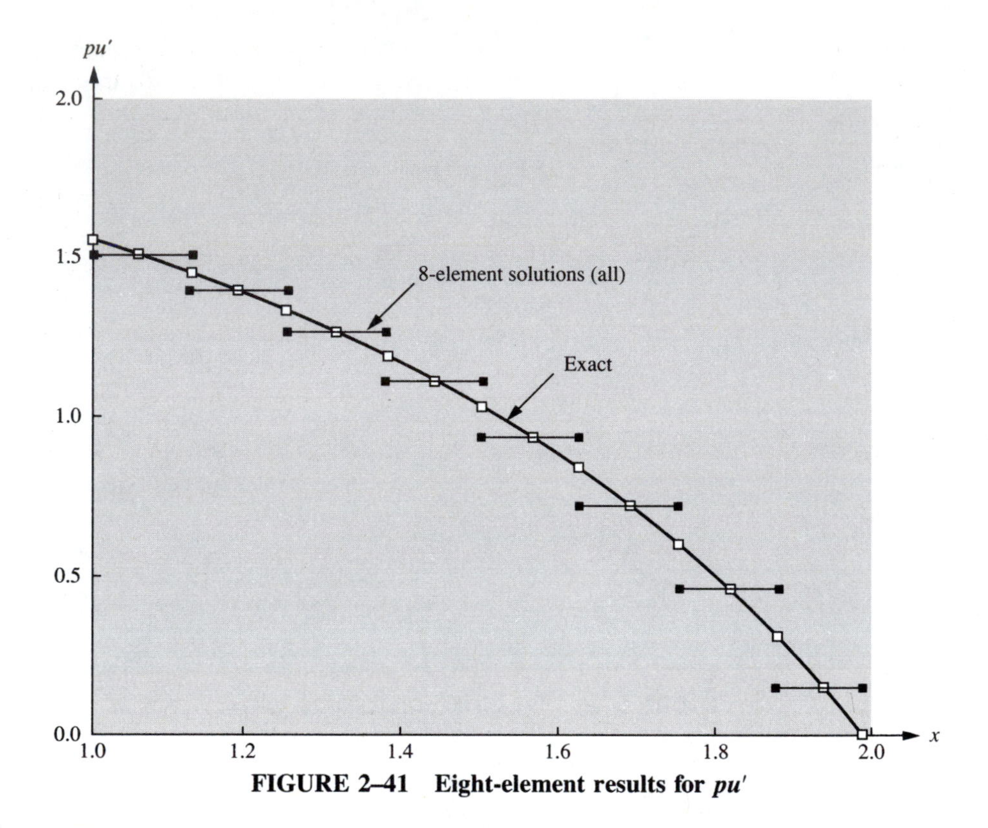

FIGURE 2–41 Eight-element results for *pu'*

2.7 VARIATIONAL FINITE ELEMENT MODELS

The Galerkin method, discussed in Section 2.3, is generally used to attack the governing differential equation directly to generate an approximate solution. The Ritz method, on the other hand, is used to generate approximate solutions to the governing differential equation based on the existence of a corresponding variational principle. In this section, we will begin the investigation of the relationships between the differential equation and the corresponding variational principle when it exists.

2.7.1 The Weak Formulation

In this section, we will develop the **weak formulation** as a preliminary step towards the Ritz approach for generating the finite element model for the Sturm-Liouville boundary value problem, which we restate as

$$5(pu')' + (\lambda r - q)u + f = 0 \qquad a \leq x \leq b \tag{2.19a}$$

$$-p(a)u'(a) + \alpha u(a) = A \tag{2.19b}$$

$$p(b)u'(b) + \beta u(b) = B \tag{2.19c}$$

We will first consider the case when the $\lambda r u$ term is absent, and return later to a discussion of the corresponding eigenvalue problem.

The approach taken to develop the weak formulation is based on satisfying the differential equation in an average sense on the interval $a \le x \le b$. To this end we multiply the differential equation (2.19a) by a suitable "test" function $v(x)$, and integrate from a to b to obtain

$$\int_a^b [(pu')' - qu + f]v\,dx = \int_a^b L(u)v\,dx = 0 \qquad (2.20)$$

Note the similarity with the Galerkin method, where the v was there chosen to be one of the nodal interpolation functions in the approximate solution. For this development, the test function v is taken to be an arbitrary continuous function satisfying the homogeneous form of any essential boundary conditions imposed on u. (For the Sturm-Liouville problem, the essential boundary conditions are boundary conditions imposed on the dependent variable u. The reader is referred to Section 2.7.2 on the calculus of variations for further details.) The use of the word "test" refers to the concept of testing the differential equation in the following sense. The arbitrary continuous test function $v(x)$ can be considered to be a function that vanishes outside a small neighborhood of an interior point $x*$ contained in $[a, b]$ as shown in Fig. 2–42.

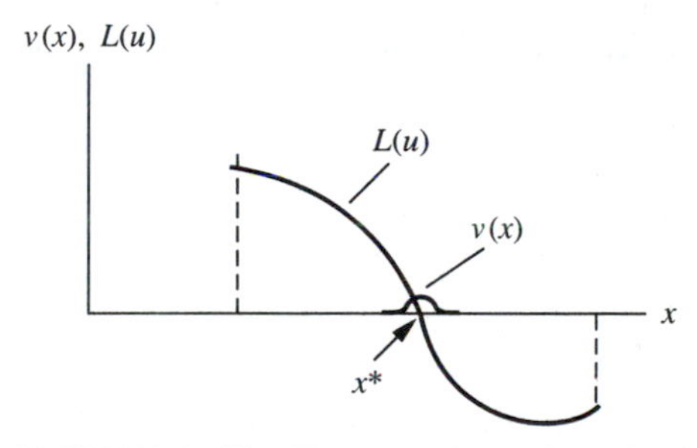

FIGURE 2–42 Error and test functions

As seen in the figure it is then necessary for $L(u)$, considered to be continuous, to vanish at some point in the neighborhood of $x*$. Roughly, the idea is that if Eq. (2.20) is satisfied for all possible test functions v with this character, the only logical possibility for satisfying the equation is that the quantity multiplying the test function must vanish; that is, the differential equation is satisfied, at least on average, on the interval $[a, b]$.

Eq. (2.20) constitutes what is referred to as the *weak statement* of the problem. By *weak* we mean that rather than requiring that the differential equation be satisfied exactly or identically at every point in the interval $a \le x \le b$ as was the case for the previously described classical solution (Section 2.2), we instead require that it be satisfied in the average sense described above, over any small subinterval of $a \le x \le b$.

We note that in Eq. (2.20), u must have a second derivative, $(pu')' = pu'' + p'u'$, whereas v is only required to be continuous. Having different continuity requirements on u and v results in a certain lack of symmetry in the formulation which it is desirable to avoid if possible. We can easily remedy the situation by integrating Eq. (2.20) by parts to obtain

$$v p u'\big|_a^b - \int_a^b [v' p u' + v q u - v f]\, dx = 0$$

or

$$v(b) p(b) u'(b) - v(a) p(a) u'(a) - \int_a^b [v' p u' + v q u - v f]\, dx = 0$$

From the boundary conditions, Eqs. (2.19),

$$p(b)u'(b) = B - \beta u(b)$$

and

$$p(a)u'(a) = -A + \alpha u(a)$$

so that upon elimination of the derivative terms at the boundary, the integrated form of the weak statement can be expressed as

$$\int_a^b [v' p u' + v q u] dx + \beta v(b) u(b) + \alpha v(a) u(a) = \int_a^b f v dx + B v(b) + A v(a) \tag{2.21}$$

We note that the dependent variable u is now only required to have a first derivative; that is, the continuity requirements on u have been weakened. The requirements on the continuity of u and v are clearly now the same.

Eq. (2.21) can be written as

$$B(v, u) = \ell(v)$$

where

$$B(v, u) = \int_a^b [v' p u' + v q u] dx + \beta v(b) u(b) + \alpha v(a) u(a)$$

and

$$\ell(v) = \int_a^b f v dx + B v(b) + A v(a)$$

$B(v, u)$ is an example of a **bilinear functional** and $\ell(v)$ is an example of a **linear functional.** A bilinear functional $B(v, u)$ is linear in each of the variables u and v, according to

$$B(\alpha_1 v_1 + \alpha_2 v_2, u) = \alpha_1 B(v_1, u) + \alpha_2 B(v_2, u)$$

and

$$B(v, \beta_1 u_1 + \beta_2 u_2) = \beta_1 B(v, u_1) + \beta_2 B(v, u_2)$$

where v, v_1, v_2, u, u_1, u_2 are functions, with α_1, α_2, β_1, and β_2 scalars or constants. We also note that B is a symmetric bilinear functional: $B(v, u) = B(u, v)$. By this we mean that if u and v are interchanged the form of B is unchanged. For this reason, Eq.(2.21) is referred to as the ***symmetric weak form*** of the boundary value problem given by Eqs. (2.19).

Similarly, the linear functional $\ell(v)$ satisfies

$$\ell(\gamma_1 v_1 + \gamma_2 v_2) = \gamma_1 \ell(v_1) + \gamma_2 \ell(v_2)$$

where v_1 and v_2 are functions and γ_1 and γ_2 are scalars or constants.

It can be shown [Reddy and Rasmussen, 6] that when the weak form can be expressed as in (2.21), namely that

$$B(v, u) = \ell(v)$$

with $B(v, u) = B(u, v)$, there exists a corresponding quadratic functional given by

$$I(u) = \frac{B(u, u)}{2} - \ell(u) \tag{2.22}$$

and that requiring this functional to be stationary yields the corresponding original boundary value problem. For the Sturm-Liouville problem, the functional in question can be written as

$$I(u) = \frac{1}{2}\int_a^b [p(u')^2 + qu^2]dx + \frac{\beta u(b)^2}{2} + \frac{\alpha u(a)^2}{2} - \int_a^b f u dx - Bu(b) - Au(a) \tag{2.23}$$

whose stationary value, given by $\delta I = 0$, yields the statement of the boundary value problem in Eqs. (2.19). The concept of a stationary value of a functional involves some of the basic ideas of the calculus of variations, discussed in the next section.

Example 2.7.

Consider the classical problem of the axial deformation of a prismatic elastic bar pictured in Fig. 2–43.

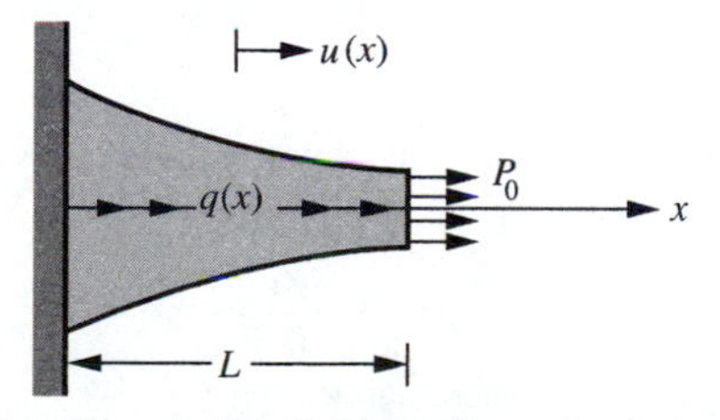

FIGURE 2–43 Axial deformation of a prismatic bar

The boundary value problem for the case indicated in the figure is

$$(AEu')' + q(x) = 0 \qquad 0 \le x \le L$$
$$u(0) = 0$$
$$AEu'(L) = P_0$$

A is the area of the bar and E is Young's modulus.

Multiplying the differential equation by a test function $v(x)$ that satisfies the essential boundary condition $v(0) = 0$ and integrating over the domain $(0, L)$ yields

$$\int_0^L v[(AEu')' + q(x)]dx = 0$$

Integrating by parts results in

$$vAEu'|_0^L - \int_0^L [v'AEu' - vq]dx = 0$$

The boundary term AEu' at $x = L$ is eliminated to produce

$$\int_0^L v'AEu' \, dx = \int_0^L vq \, dx + v(L)P_0$$

This is the symmetric weak form, with

$$B(v, u) = \int_0^L v'AEu' \, dx$$

and

$$l(v) = \int_0^L vq \, dx + v(L)P_0$$

B is clearly symmetric, so that the quadratic functional corresponding to the problem of the axial deformation of an elastic bar becomes

$$I(u) = \int_0^L \left[\frac{AE(u')^2}{2} - qu\right]dx - u(L)P_0$$

This functional represents the *total potential energy* of the system, with $B(u, u)/2$ the internal or strain energy and $l(u)$ the negative of the potential energy of the external forces. It is also important to point out that by letting v be a virtual displacement δu, equation (a) can be interpreted as a statement of the *principle of virtual work*.

Example 2.8.

Consider a bar conducting heat along its axis with a convective condition on the lateral surface as shown in Fig. 2–44.

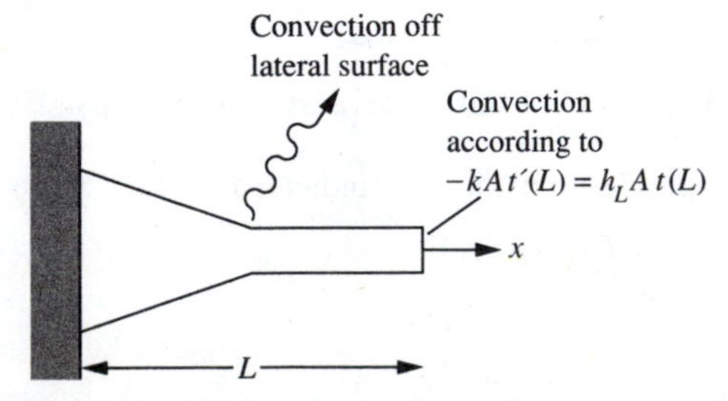

FIGURE 2–44 Heat conduction with lateral convection

With $t(x)$ as the temperature, the boundary value problem is taken to be

$$(kAt')' - hPt + q_0A = 0$$
$$t(0) = 0$$
$$kAt'(L) + h_LAt(L) = 0$$

A is the area, k the thermal conductivity, h and h_L are convective heat transfer coefficients, P is the perimeter, and q_0 an internal energy term.

To develop the weak form, the differential equation is multiplied by a test function $s(x)$ satisfying $s(0) = 0$ and integrated from 0 to L to obtain

$$\int_0^L s[(kAt')' - hPt + q_0A]\,dx = 0$$

Integrating by parts and eliminating the kAt' term using the second of the boundary conditions yields

$$\int_0^L [s'(kAt') + shPt]\,dx + s(L)h_LAt(L) = \int_0^L sq_0\,A\,dx$$

This is the *weak form* of the boundary value problem, with

$$B(s,t) = \int_0^L [s'kAt' + shPt]\,dx + s(L)h_LAt(L)$$

and

$$l(s) = \int_0^L sq_0A\,dx$$

B is obviously symmetric, that is, $B(s,t) = B(t,s)$, so that the functional $I(t)$ can be expressed as

$$I(t) = \int_0^L \frac{kA(t')^2 + hPt^2}{2}\,dx + \frac{h_LAt^2(L)}{2} - \int_0^L q_0At\,dx$$

We shall see in what follows later in this section, that the weak statement and corresponding variational principle can be used to produce a Ritz finite element model that, for the Sturm-Liouville class of problems, will be identical to the Galerkin finite element model developed in Section 2.3.

2.7.2 Calculus of Variations

In Chapter 1, we saw that the energies of some discrete systems could be represented in terms of **functions**. By a function $f(x_1, x_2, \ldots, x_N)$ we mean that the specification of the set of real variables $x_1, x_2, \ldots$ is necessary for the determination of the single real variable f. In the modern literature f is frequently called a mapping. An example is the potential energy function for the system of springs and masses studied in Sections 1.4.1 and 1.4.2

$$V(x_1, x_2, \ldots, x_N) = \frac{\mathbf{x}^T\mathbf{K}\mathbf{x}}{2} - \mathbf{x}^T\mathbf{F}$$

It was further shown in Chapter 1 that in some instances there exist stationary principles such that the equations resulting from requiring that the function be stationary are identical to those derived on the basis of a balance law. With reference to the spring-mass example above, the stationary value of the total potential energy function

$$\frac{\partial V}{\partial \mathbf{x}} = 0$$

leads to

$$\mathbf{Kx} = \mathbf{F}$$

which are the equilibrium equations that can also be obtained on the basis of a force balance.

In Section 2.7.1, the examples presented indicated that for certain systems considered to be continuous, the corresponding energies are appropriately expressed in terms of **functionals**. By a functional we mean that an entire function is mapped into a real number. In Example 2.7, for instance, the functional was

$$I(u) = \int_0^L \left(\frac{AEu'^2}{2} - qu \right) dx - P_0 u(L)$$

Here, in a sense, the independent variable is the function $u(x)$ defined on the interval $0 \leq x \leq L$. The entire function u must be specified on the interval $0 \leq x \leq L$ in order to compute the value of the functional I. Generally, functionals represent energies or generalized energies for the physical problems in question.

In a conservative mechanical system we expect the *principle* of stationary potential energy to be valid regardless of whether the system being studied is considered to be discrete, as was the case in Chapter 1, or continuous, as in Example 2.6. Thus we expect that, by requiring the functional $I(u)$ to be stationary, the appropriate balance equations for the continuous system, which are the equilibrium equations for the axial deformation problem, will result. Our study of the calculus of variations will indicate what is meant by and how to determine the conditions that must be satisfied in order that a functional be stationary. For further reading consult Weinstock [7], Gelfand and Fomin [8], Lanczos [9], and Hildebrand [10].

Consider then the following functional, which is typical of some we have been encountering in our discussions:

$$I(u) = \int_a^b F(x, u, u')dx \tag{2.24}$$

We will assume that F, x, and u are real, so that $I(u)$ is also a real number. We wish here to investigate certain elementary tools from the calculus of variations that are intended to answer questions about what it means to seek the stationary value of an integral of the type represented in Eq. (2.24). When we speak of the stationary value of I, we must keep in mind that $I = I(u)$, so in a sense the function u is the independent variable and we are attempting to determine an entire function on the interval in question. We must consider varying in some suitable fashion the entire function u over the interval, in order to determine the effect of the changes in u on the value of the functional I. Ultimately, we must determine some means for deciding with respect to which function(s) u the value of I is stationary. Generally we do this by varying the function u over all functions with continuous first derivatives (since u' appears in F) which also satisfy any so-called essential boundary conditions imposed on the solution and determine which, if any, of these functions produce vanishingly small changes in I for correspondingly vanishingly small changes in the function u.

To make this more definite, consider the integral

$$I(u) = \int_a^b F(x, u, u')dx$$

with the essential boundary conditions $u(a) = u_a$, and $u(b) = u_b$. Such a problem is known as a **fixed endpoint problem**. It is one of the simplest of the problems of the calculus of variations and serves as a means for beginning our study. The geometry of the situation is indicated in Fig. 2–45.

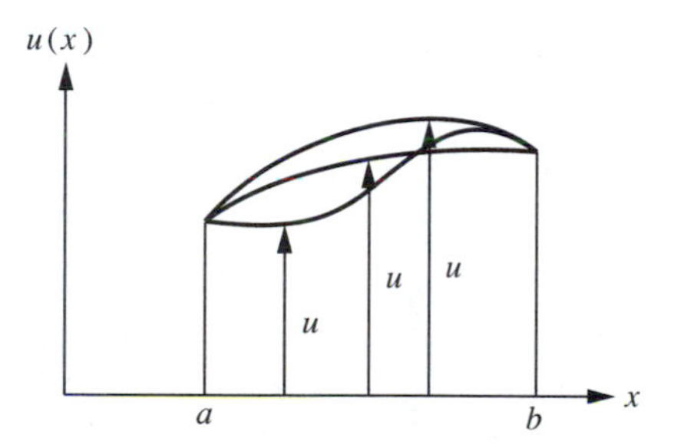

FIGURE 2–45 Admissible functions

Shown are several possible continuous functions that satisfy the two boundary conditions. Each of these functions is called an admissible function. Our task is to determine which of these admissible functions or curves provides the stationary value to the integral I. It is conceptually advantageous to think of the process of determining the correct curve as a sequence of experiments, which we now describe.

First, assume or guess a possible solution u such as one of the curves pictured in Fig. 2–45. Second, compute the corresponding value of $I(u)$. Next, perturb the initial guess for the function $u(x)$ in an arbitrary infinitesimal fashion and see if the change in the integral I is vanishingly small. If the answer is yes, *for all possible infinitesimal changes* in the function $u(x)$, we have determined our solution! If not we must guess another possible u and repeat the process. Although this would seem to be a hopelessly involved sequence of experiments, it turns out that we can set up a scheme that will perform all the experiments simultaneously and provide us with an algorithm to determine the possible function(s) that render the functional stationary.

To this end, assume that $u(x)$ is the desired solution, yielding the stationary value. Define a one-parameter family of comparison curves $U(x) = u(x)+\epsilon\eta(x)$ as indicated in Fig. 2–46.

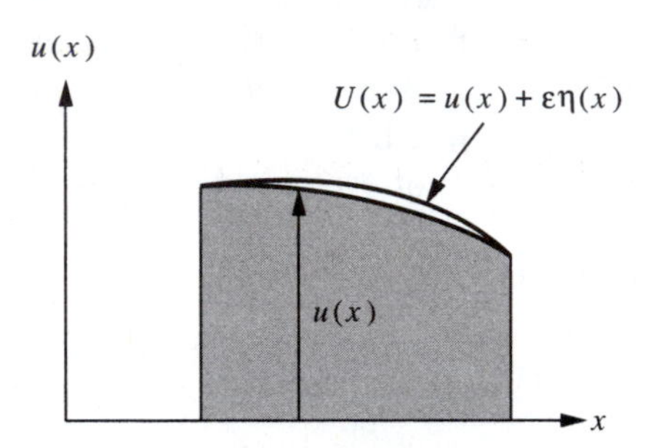

FIGURE 2-46 One-parameter family of curves

In the equation defining U, ϵ is a small parameter and $\eta(x)$ is an *arbitrary* continuously differentiable function which must also vanish at the ends of the interval in order for all the comparison curves $U = u+\epsilon\eta$ to satisfy the boundary conditions. Thus $\eta(a) = \eta(b) = 0$. With u known and η at our disposal, we can consider I to be a function of ϵ and write

$$I = I(\epsilon) = \int_a^b F(x, u + \epsilon\eta,\ u' + \epsilon\eta')dx$$

Requiring I to be stationary with respect to ϵ as ϵ approaches zero ($U \to u$), we can state that a necessary condition for u to provide a stationary value of I is that

$$\lim_{\epsilon\to 0} \frac{dI}{d\epsilon} = I'(0) = 0$$

for all η. Computing $dI/d\epsilon$ we find

$$\frac{d}{d\epsilon}\int_a^b F(x, u + \epsilon\eta,\ u' + \epsilon\eta')dx = \int \frac{dF}{d\epsilon}dx$$

$$= \int_a^b \left(\frac{\partial F}{\partial(u + \epsilon\eta)} \cdot \frac{d(u + \epsilon\eta)}{d\epsilon} + \frac{\partial F}{\partial(u' + \epsilon\eta')} \cdot \frac{d(u' + \epsilon\eta')}{d\epsilon}\right)dx$$

$$= \int_a^b \left(\frac{\partial F}{\partial(u + \epsilon\eta)}\eta + \frac{\partial F}{\partial(u' + \epsilon\eta')}\eta'\right)dx$$

and thus

$$I'(0) = \int_a^b \left(\eta\frac{\partial F}{\partial u} + \eta'\frac{\partial F}{\partial u'}\right)dx = 0$$

Integrating the second term by parts yields

$$\eta \frac{\partial F}{\partial u'}\bigg|_a^b - \int_a^b \left(\frac{d(\partial F/\partial u')}{dx} - \frac{\partial F}{\partial u} \right) \eta dx = 0$$

or since η vanishes at both endpoints,

$$\int_a^b \left(\frac{d(\partial F/\partial u')}{dx} - \frac{\partial F}{\partial u} \right) \eta dx = 0$$

On the basis that η is an arbitrary function on the interval $a < x < b$, it follows* that the quantity which multiplies η in the integrand must vanish, that is,

$$\frac{d(\partial F/\partial u')}{dx} - \frac{\partial F}{\partial u} = 0 \qquad a < x < b$$

This second-order ordinary differential equation is the so-called *Euler equation*. Solutions of the Euler equation are termed *extremals*. The solution to the original problem, the function giving the integral its stationary value, is the particular extremal that also satisfies the two essential boundary conditions $u(a) = u_a$ and $u(b) = u_b$. Just what is meant by an essential boundary condition will be explained in what follows.

Example 2.9.

Consider a boundary value problem related to the heat conduction problem discussed in Example 2.7.

$$(kT')' - qT + f = 0 \qquad (a)$$

$$T(0) = 0 \qquad (b)$$

$$T(L) = 0 \qquad (c)$$

The corresponding functional is

$$I(T) = \int_0^L \left(\frac{kT'^2}{2} + \frac{qT^2}{2} - Tf \right) dx \qquad (d)$$

subject to the essential boundary conditions

$$T(0) = 0$$

$$T(L) = 0$$

This is easily seen to be of the form

$$I(T) = \int_0^L F(x, T, T')dx$$

*This result is referred to as the Fundamental Lemma of the calculus of variations [Weinstock, 7].

Forming the necessary derivatives for $\delta I = 0$,

$$\frac{\partial F}{\partial T} = qT - f$$

and

$$\frac{\partial F}{\partial T'} = kT'$$

the corresponding Euler equation is

$$\frac{\partial F}{\partial T} - \frac{d(\partial F/\partial T')}{dx} = qT - f - \frac{d(kT')}{dx} = 0$$

With the inclusion of the essential boundary conditions, the boundary value problem becomes

$$(kT')' - qT + f = 0$$
$$T(0) = 0$$
$$T(L) = 0$$

which is seen to be the original boundary value problem. Thus it is clear that the original boundary value problem can be obtained from the functional given by Eq. (*d*) and, as was shown in Example 2.7, that the functional can be developed from the boundary value problem. This indicates that the differential equation and variational formulations are equivalent from the standpoint that each can be obtained from the other. As we shall see in later sections, this duality provides us with alternate means for generating the finite element model for the corresponding physical problem.

An important generalization of the fixed endpoint problem just discussed occurs when the function u is not prescribed at one or both of the boundaries. For instance, consider the problem of determining the stationary value of the integral

$$I(u) = \int_a^b F(x, u, u')dx$$

with only $u(a) = u_a$ prescribed. Again defining a one-parameter family of curves according to

$$U(x) = u(x) + \epsilon\eta(x)$$

where u is the known solution, we follow exactly the same reasoning and sequence of steps as for the fixed end point problem, leading to

$$\eta\frac{\partial F}{\partial u'}\bigg|_a^b - \int_a^b \left(\frac{d(\partial F/\partial u')}{dx} - \frac{\partial F}{\partial u}\right)\eta dx = 0$$

as a result of requiring the integral to be stationary. Again $\eta(a)$ is zero, leading to

$$\eta\frac{\partial F}{\partial u'}\bigg|_b - \int_a^b \left(\frac{d(\partial F/\partial u')}{dx} - \frac{\partial F}{\partial u}\right)\eta dx = 0$$

First we choose an η such that $\eta(b) = 0$, so that

$$\int_a^b \left(\frac{d(\partial F/\partial u')}{dx} - \frac{\partial F}{\partial u} \right) \eta dx = 0$$

which on the basis of the arbitrary character of η on the remainder of the interval again leads to the conclusion that

$$\frac{d(\partial F/\partial u')}{dx} - \frac{\partial F}{\partial u} = 0 \qquad \text{for } a < x < b$$

We then choose an arbitrary η that does not vanish at $x = b$, leading to the conclusion that

$$\left. \frac{\partial F}{\partial u'} \right|_b = 0 \tag{2.25}$$

This is a natural boundary condition. Natural boundary conditions are often associated with specifying a flux such as AEu' or kT' at a boundary.

If u is specified at a boundary, the boundary condition is termed an *essential boundary condition*, whereas if u is not specified, a *natural boundary condition* of the form given in Eq. (2.1) must be satisfied. It can also turn out that a boundary condition formed from a linear combination of the function and its derivative may be appropriate, as is the case for our boundary value problem in standard form,

$$(pu')' - qu + f = 0$$

$$-p(a)u'(a) + \alpha u(a) = A$$

$$p(b)u'(b) + \beta u(b) = B$$

where both of the boundary conditions as stated are natural boundary conditions. Any time a first derivative term appears in a boundary condition associated with the second-order Sturm-Liouville problem, the boundary condition is *natural*. If only the dependent variable itself appears the boundary condition is *essential*.

Generally, for a functional of the form

$$I = \int_a^b F(x, u, u', u'', \ldots, u^{(n)}) dx \tag{2.26}$$

where $u^{(n)}$ indicates the nth derivative, essential boundary conditions are imposed on the variables u, u', u'', ..., $u^{(n-1)}$. Natural boundary conditions *can* involve the variables u, u', u'', ..., $u^{(n-1)}$, but in addition *must* involve at least one of the variables $u^{(n)}$, $u^{(n+1)}$, $u^{(n+2)}$, ..., $u^{(2n-1)}$. For our one-dimensional functional of the form

$$I(u) = \int_a^b F(x, u, u') dx$$

$n = 1$, so that an essential boundary condition is imposed on the dependent variable u and a natural boundary condition can involve u but must also contain a u' term. Recall from the discussion in Section 2.2 that essential boundary conditions such as $u(a) = u_a$ and $u(b) = u_b$ cannot be obtained by specializing the natural boundary conditions of the form

$$-p(a)u'(a) + \alpha u(a) = A$$

and

$$p(b)u'(b) + \beta u(b) = B$$

respectively. For the essential boundary condition $u(a) = u_a$, with α and A taken to be zero. For the essential boundary condition $u(b) = u_b$, with β and B taken to be zero.

Example 2.10.

For the boundary value problem

$$\begin{aligned} (xu')' - u + \sin x &= 0 \\ u(1) &= 0 \\ u'(2) &= 1 \end{aligned}$$

the boundary condition at $x = 1$ is an essential boundary condition, whereas the boundary condition at $x = 2$ is a natural boundary condition. $\alpha = A = 0$, with $\beta = 0$ and $B = 2$.

Example 2.11.

For the boundary value problem

$$\begin{aligned} (x^2u')' + 2u - x &= 0 \\ -u'(1) + 2u(1) &= 0 \\ u(2) &= 1 \end{aligned}$$

the boundary conditions at $x = 1$ and $x = 2$ are natural and essential, respectively. $\alpha = 2$, $A = 0$, $\beta = B = 0$.

Example 2.12.

Consider again the axial deformation problem of Example 2.6, for which the potential energy functional (in the absence of the concentrated load at $x = L$) is

$$I(u) = \int_0^L \left(\frac{AE(u')^2}{2} - qu \right) dx$$

with $u(0) = 0$. Since u is not prescribed at $x = L$, the boundary condition at $x = L$ will be a natural boundary condition. We easily identify the integrand as

$$F = \frac{AE(u')^2}{2} - qu$$

The Euler equation

$$\frac{d(\partial F/\partial u')}{dx} - \frac{\partial F}{\partial u} = 0$$

becomes

$$\frac{d(AEu')}{dx} + q = 0 = (AEu')' + q$$

with the integrated term appearing as

$$AEu'(L)\eta(L) = 0$$

The differential equation is obviously the same as before. One of the boundary conditions which is to be imposed on the extremals is the essential boundary condition $u(0) = 0$. Since u is not prescribed at $x = L$, η is arbitrary there and in order to satisfy the boundary term at $x = L$, the natural boundary condition $AEu'(L) = 0$ must be imposed. This clearly states that the force transmitted at $x = L$ must be zero. Thus the complete statement of the boundary value problem generated by applying the principle of stationary energy to the potential energy functional is

$$\begin{aligned}(AEu')' + q &= 0\\ u(0) &= 0\\ AEu'(L) &= 0\end{aligned}$$

The equation is already in self-adjoint form, enabling us to easily identify $p = AE$, $q = 0$, and $f = q$. The boundary condition at $x = 0$ is essential so that α and A are taken as zero. β and B are both zero for the natural boundary condition at b.

Example 2.13.

Consider the problem of the transverse deformation of a straight prismatic bar as indicated in Figure 2–47.

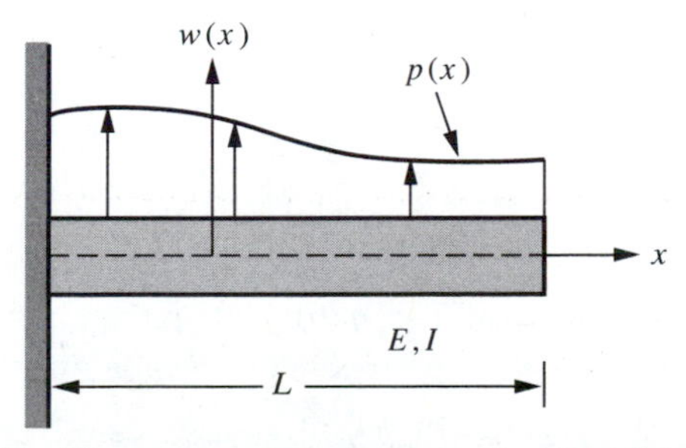

FIGURE 2–47 Transverse deformation of a straight prismatic bar

The total potential energy for the type of loading shown can be expressed as

$$V(w) = \int_o^L \left(\frac{EI(w'')^2}{2} - pw\right)dx$$

with

$$w(0) = w'(0) = 0$$

as the essential boundary conditions. I is the second moment of the area, and E is Young's modulus. The first and second terms of V are the internal and external potential energies, respectively. The functional in question is of the form

$$I(w) = \int_0^L F(x, w, w', w'')dx$$

Requiring the integral to be stationary leads to (see the Exercises at the end of the chapter)

$$I'(0) = \int_0^L \left(\eta \frac{\partial F}{\partial w} + \eta' \frac{\partial F}{\partial w'} + \eta'' \frac{\partial F}{\partial w''}\right) dx = 0$$

which after integrating by parts becomes

$$\eta \left[\frac{\partial F}{\partial w'} - \frac{d(\partial F/\partial w'')}{dx}\right]\Bigg|_0^L + \eta' \left[\frac{\partial F}{\partial w''}\right]\Bigg|_0^L$$

$$+ \int_0^L \eta \left(\frac{d^2(\partial F/\partial w'')}{dx^2} - \frac{d(\partial F/\partial w')}{dx} + \frac{\partial F}{\partial w}\right) dx = 0$$

On the basis of the arbitrary character of η on the interval $0 < x < L$, the Euler equation is

$$\frac{d^2(\partial F/\partial w'')}{dx^2} - \frac{d(\partial F/\partial w')}{dx} + \frac{\partial F}{\partial w} = 0$$

or

$$\frac{d^2(EIw'')}{dx^2} - p = 0$$

which is the governing differential equation of equilibrium. The boundary conditions arise from the integrated terms,

$$\eta \left[\frac{\partial F}{\partial w'} - \frac{d(\partial F/\partial w'')}{dx}\right]\Bigg|_0^L = 0$$

$$\eta' \left[\frac{\partial F}{\partial w''}\right]\Bigg|_0^L = 0$$

Since $w(0) = 0$, $w'(0) = 0$, and $\eta(0) = \eta'(0) = 0$, we are left with the two natural boundary conditions

$$\left[\frac{\partial F}{\partial w'} - \frac{d(\partial F/\partial w'')}{dx}\right]\Bigg|_L = -(EIw'')'\Big|_L = 0$$

and

$$\frac{\partial F}{\partial w''}\Bigg|_L = EIw''\Big|_L = 0$$

stating that the shear given by $-(EIw'')'$ and the moment given by EIw'' must vanish at $x = L$. Comparing with the general form Eq. (2.26) when $n = 2$, we see that essential boundary conditions are imposed on w and w' whereas any boundary conditions involving w'' or w''' are termed natural boundary conditions.

Generally, for a solid or fluid mechanics problem the essential boundary conditions are imposed on the geometry of the deformation or on the velocity, that is, the kinematical variables, whereas the natural boundary conditions are imposed on or involve the force resultants or the kinetic variables.

2.7.3 Ritz Finite Element Models

The starting point for the development of the Galerkin finite element model was the differential equation and boundary conditions. An approximate solution was assumed and used directly in the differential equation. The boundary conditions were entered into the formulation by an integration by parts. In the previous sections we have shown that it is possible to develop a so-called symmetric weak form associated with the differential equation and boundary conditions and, through the calculus of variations, to state a corresponding variational principle based on a functional equivalent to the differential equation and boundary conditions. This functional,

$$I(u) = \frac{1}{2}\int_a^b [p(u')^2 + qu^2]dx + \frac{\beta u(b)^2}{2} + \frac{\alpha u(a)^2}{2} - \int_a^b fu dx - Bu(b) - Au(a)$$

serves as the starting point for the development of the so-called **Ritz finite element model,** outlined in what follows. Assuming that the same discretization and degree of interpolation is used, the Ritz finite element model will turn out to be exactly the same as the Galerkin finite element model for the class of problems considered in this chapter.

In general, the steps to be followed in generating a finite element model are, as before,

1. Discretization
2. Interpolation
3. Elemental formulation
4. Assembly
5. Constraints
6. Solution
7. Computation of derived or secondary variables

Each of these basic steps will be discussed in detail in what follows.

Discretization. The basic interval $a \leq x \leq b$ is broken up into subintervals as indicated in Fig. 2–48.

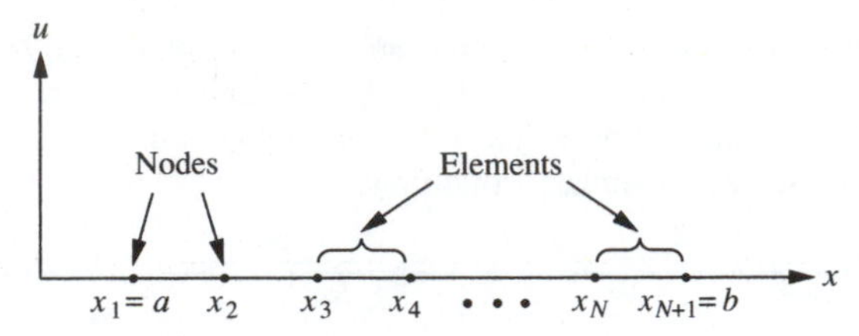

FIGURE 2–48 Nodes and elements

The points that separate the subintervals are called *nodes*. The subdomains between the nodes are called *elements*. Nodes are frequently placed so as to produce equal-length elements. The collection of nodes and elements is referred to as the *mesh*. The choice of a uniform mesh is not necessary and when it is known that there will be larger gradients or a more rapidly changing solution near a boundary, for instance, it is desirable to have the nodes more closely spaced in such regions. The transition between a fine mesh and a crude mesh should be accomplished gradually.

Interpolation. The process of choosing a mesh and the choice for the interpolation in constructing a particular finite element model are not actually independent. Generally the representation for the approximate solution within an element is formed by passing an nth degree polynomial through the $n + 1$ values u_i of the dependent variable at $n + 1$ equally spaced points of the mesh. Thus a two-node element uses straight lines, a three-node element uses parabolas, and so on. In Fig. 2–49 a typical situation for two-node elements is indicated.

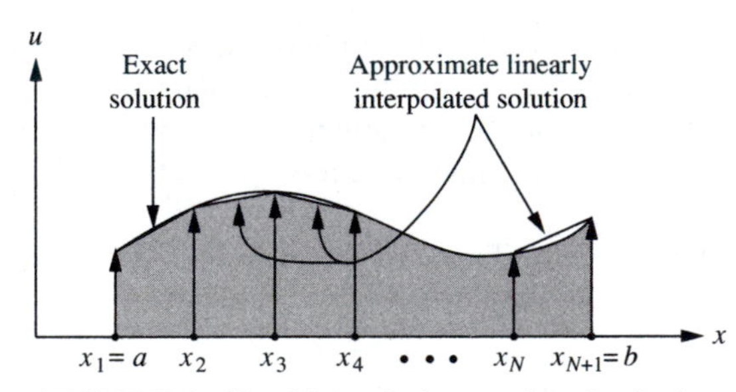

FIGURE 2–49 Linearly interpolated solution

The typical exact solution is shown as a continuous function. In addition, an approximation to the function $u(x)$ consisting of piecewise linear functions in each of the subintervals or elements is shown. This piecewise representation of the approximate solution within the element is an example of *interpolation*. It is called linear interpolation in this case since the approximation within each element is a linear function. Specifically, we write for a typical element

$$u_e(x) = \alpha_e + \beta_e x \tag{2.27}$$

and require that

$$u_e(x_i) = \alpha_e + \beta_e x_i = u_i$$

and

$$u_e(x_{i+1}) = \alpha_e + \beta_e x_{i+1} = u_{i+1}$$

where u_i and u_{i+1} are the unknown nodal values of the solution at nodes i and $i+1$ respectively. Solving these two equations for α_e and β_e and substituting the results into Eq. (2.27) yields

$$u_e(x) = u_i \frac{(x_{i+1} - x)}{(x_{i+1} - x_i)} + u_{i+1} \frac{(x - x_i)}{(x_{i+1} - x_i)} = u_i N_i + u_{i+1} N_{i+1} \tag{2.28}$$

The derivative $u'_e(x)$ is computed to be

$$u'_e(x) = u_i \frac{-1}{(x_{i+1} - x_i)} + u_{i+1} \frac{1}{(x_{i+1} - x_i)} = u_i N'_i + u_{i+1} N'_{i+1}$$

These can be expressed in matrix notation as

$$u_e(x) = \mathbf{N}^{\mathrm{T}} \mathbf{u_e} = \mathbf{u_e^{\mathrm{T}}} \mathbf{N}$$

and (2.29)

$$u'_e(x) = \mathbf{N}'^{\mathrm{T}} \mathbf{u_e} = \mathbf{u_e^{\mathrm{T}}} \mathbf{N}'$$

where

$$\mathbf{N} = \begin{bmatrix} N_i \\ N_{i+1} \end{bmatrix} \qquad \mathbf{N}' = \begin{bmatrix} N'_i \\ N'_{i+1} \end{bmatrix} \qquad \text{and} \qquad \mathbf{u_e} = \begin{bmatrix} u_i \\ u_{i+1} \end{bmatrix}$$

with $\mathbf{N}$ the elemental interpolation vector and $\mathbf{u_e}$ the elemental "displacement vector."

It is frequently convenient to represent the interpolation functions N_i and N_{i+1} in terms of a local coordinate ξ as shown in Fig. 2–50. The transformation between the *global* variable x and the *local* variable ξ is a simple translation given by $x_i + \xi = x$, where x_i is the value of x at the left or beginning end of the element. Represented in terms of the local variable ξ, the linear interpolation functions become $N_i(\xi) = 1 - \xi/l_e$ and $N_{i+1}(\xi) = \xi/l_e$, where $l_e = x_{i+1} - x_i$ is the length of the element.

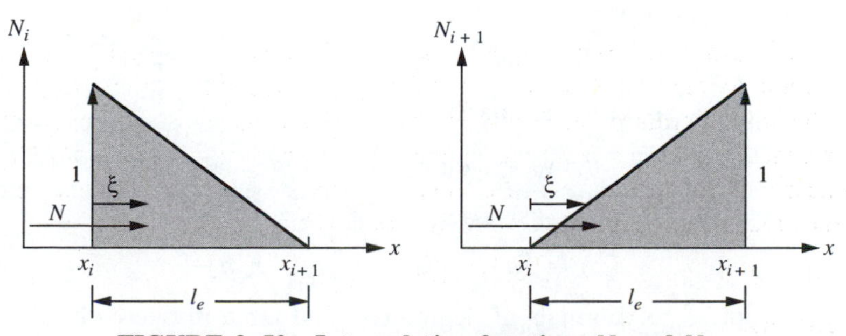

FIGURE 2–50 Interpolation functions N_i and N_{i+1}

The derivatives of the interpolation functions $\mathbf{N}_i'$ and N_{i+1}' are indicated in Fig. 2–51.

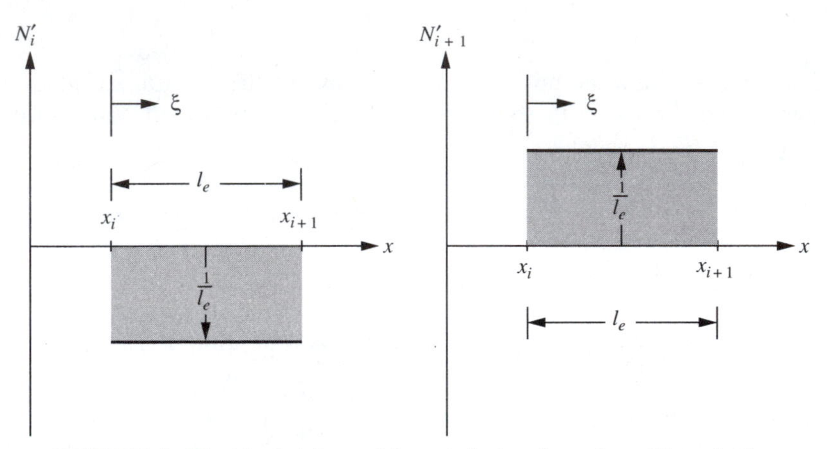

FIGURE 2–51 Derivatives of interpolation functions N_i' and N_{i+1}'

N_i and N_{i+1} form a basis for the space of functions that are linear on the interval (x_i, x_{i+1}), in that any linear function can be represented as a linear combination of N_i and N_{i+1} on the interval (x_i, x_{i+1}).

Before concluding our discussion of interpolation, we note that the selection of the interpolation functions certainly seems arbitrary with respect to the myriad of functions that could be selected. Our main concern, however, is that in the limit as the mesh is refined the choice of the interpolation functions results in convergence of the finite element solution to the exact solution. For the class of one-dimensional problems involving differential equations of second order, sufficient conditions that the interpolation functions must satisfy in order to ensure convergence are:

1. The interpolation functions must be such that continuity at the interelement boundaries is satisfied.
2. The interpolation functions must be capable of representing solutions that are constant and solutions that have constant first derivatives.

The second condition is often referred to as *completeness*. For one-dimensional problems, any polynomial containing both the monomials 1 and x is complete. The linear interpolation just discussed,

$$u_e(x) = 1a_0 + xa_1$$

obviously qualifies, as do representations such as

$$u_e(x) = 1a_0 + xa_1 + x^2a_2$$

and

$$u_e(x) = 1a_0 + xa_1 + x^2a_2 + x^3a_3$$

which are complete polynomials of degree two and three respectively. Representations such as

$$u_e(x) = xa_1 + x^2a_2$$

and

$$u_e(x) = 1a_0 + x^2a_2$$

are not acceptable, because the first cannot assume a constant value within the element and the second cannot assume a constant nonzero derivative within the element. In general, it turns out that for one-dimensional problems governed by differential equations of order $2n$, a complete polynomial of degree $2n - 1$ is required, that is, a polynomial of the form

$$u_e(x) = 1a_0 + xa_1 + x^2a_2 + \cdots + x^{2n-1}a_{2n-1}$$

where all the a_i, $i = 1, 2, \ldots, 2n - 1$, are taken to be nonzero.

Elemental formulation. Recall that the functional for the one-dimensional Sturm-Liouville boundary value problems is

$$I(u) = \int_a^b \frac{\left(p(u')^2 + qu^2\right)dx}{2} + \frac{\beta u(b)^2}{2} + \frac{\alpha u(a)^2}{2} - \int_a^b fu\,dx - Au(a) - Bu(b)$$

As discussed previously, some or all of the boundary terms may be absent depending on whether the boundary conditions are essential or natural. Consistent with the choice for the discretization and interpolation we express the energy as

$$I(u) = \sum_e \int_{x_i}^{x_{i+1}} \frac{\left(p(u')^2 + qu^2\right)dx}{2} + \frac{\beta u_{N+1}^2}{2} + \frac{\alpha u_1^2}{2}$$

$$- \sum_e \int_{x_i}^{x_{i+1}} fu\; dx - Au_1 - Bu_{N+1} \tag{2.30}$$

where $u_1 = u(a)$ and $u_{N+1} = u(b)$ are the unknown values at the boundaries. This is quite simply a use of one of the fundamental properties of the integral that states

$$\int_a^b = \int_a^{x_1} + \int_{x_1}^{x_2} + \int_{x_2}^{x_3} + \cdots + \int_{x_{N-2}}^{x_{N-1}} + \int_{x_{N-1}}^{x_N} + \int_{x_N}^{b}$$

No approximation is involved. The $\sum$ in Eq.(2.30) indicates a sum over all the subintervals or elements.

Temporarily write (2.30) as

$$I(u) = \sum_e \frac{I_{pe} + I_{qe}}{2} + \frac{\beta u_{N+1}^2}{2} + \frac{\alpha u_1{}^2}{2} - \sum_e I_{fe} - Au_1 - Bu_{N+1}$$

Consider then in turn the evaluation of the integrals I_{pe}, I_{qe}, and I_{fe}.

The integral I_{pe},

$$I_{pe} = \int_{x_i}^{x_{i+1}} u'pu'dx$$

can be thought of as the elemental energy associated with the p function. On the typical interval (x_i, x_{i+1}) we replace the first u' by $\mathbf{u_e^T N'}$ and the second u' by $\mathbf{N'^T u_e}$ to obtain

$$I_{pe} \approx \int_{x_i}^{x_{i+1}} \mathbf{u_e^T N'} p(x) \mathbf{N'^T u_e} dx = \mathbf{u_e^T} \left(\int_{x_i}^{x_{i+1}} \mathbf{N'} p(x) \mathbf{N'^T} dx \right) \mathbf{u_e}$$

$$= \mathbf{u_e^T p_e u_e}$$

where

$$\mathbf{p_e} = \int_{x_i}^{x_{i+1}} \mathbf{N'} p(x) \mathbf{N'^T} dx$$

is the 2×2 elemental stiffness matrix arising from the $pu'^2/2$ term in the functional, or alternatively, from the $(pu')'$ term in the differential equation, for the interval (x_i, x_{i+1}).

Expressed in terms of the global coordinate x, $\mathbf{p_e}$ can be written as

$$\mathbf{p_e} = \int_{x_i}^{x_{i+1}} \begin{bmatrix} \dfrac{-1}{l_e} \\ \dfrac{1}{l_e} \end{bmatrix} p(x) \begin{bmatrix} \dfrac{-1}{l_e} & \dfrac{1}{l_e} \end{bmatrix} dx$$

$$= \begin{bmatrix} \left(\dfrac{1}{l_e}\right)^2 \displaystyle\int_{x_i}^{x_{i+1}} p(x)dx & -\left(\dfrac{1}{l_e}\right)^2 \displaystyle\int_{x_i}^{x_{i+1}} p(x)dx \\ -\left(\dfrac{1}{l_e}\right)^2 \displaystyle\int_{x_i}^{x_{i+1}} p(x)dx & \left(\dfrac{1}{l_e}\right)^2 \displaystyle\int_{x_i}^{x_{i+1}} p(x)dx \end{bmatrix}$$

$$= \left(\frac{1}{l_e}\right)^2 \int_{x_i}^{x_i+1} p(x)dx \begin{bmatrix} 1 & -1 \\ -1 & 1 \end{bmatrix}$$

In terms of the local coordinate ξ given by $\xi = x - x_i$,

$$\mathbf{p_e} = \left(\frac{1}{l_e}\right)^2 \int_0^{l_e} p(x_i + \xi) d\xi \begin{bmatrix} 1 & -1 \\ -1 & 1 \end{bmatrix}$$

Note that care must be taken to ensure that the correct values, namely $p(x_i + \xi)$, of the function p are used in the integrand.

The integral I_{qe},

$$I_{qe} = \int_{x_i}^{x_{i+1}} uqu \, dx$$

can be thought of as the elemental energy associated with the $-qu$ term in the differential equation. On the typical interval (x_i, x_{i+1}) we replace the first u by $\mathbf{u_e^T N}$ and the second u by $\mathbf{N^T u_e}$ to obtain

$$I_{qe} \approx \int_{x_i}^{x_{i+1}} \mathbf{u}_\mathbf{e}^\mathrm{T}\mathbf{N}q(x)\mathbf{N}^\mathrm{T}\mathbf{u}_\mathbf{e}\,dx = \mathbf{u}_\mathbf{e}^\mathrm{T}\left(\int_{x_i}^{x_{i+1}} \mathbf{N}q(x)\mathbf{N}^\mathrm{T}dx\right)\mathbf{u}_\mathbf{e}$$
$$= \mathbf{u}_\mathbf{e}^\mathrm{T}\mathbf{q}_\mathbf{e}\mathbf{u}_\mathbf{e}$$

where

$$\mathbf{q}_\mathbf{e} = \int_{x_i}^{x_{i+1}} \mathbf{N}q(x)\mathbf{N}^\mathrm{T}dx$$

is the 2×2 elemental stiffness matrix arising from the $qu^2/2$ term in the functional, or alternatively, from the $-qu$ term in the differential equation, for the interval (x_i, x_{i+1}).

Expressed in terms of the global coordinate x, $\mathbf{q}_\mathbf{e}$ can be written as

$$\mathbf{q}_\mathbf{e} = \int_{x_i}^{x_{i+1}} \begin{bmatrix} N_i \\ N_{i+1} \end{bmatrix} q(x)[N_i \quad N_{i+1}]\,dx$$

$$= \begin{bmatrix} \int_{x_i}^{x_{i+1}} (N_i)^2 q(x)\,dx & \int_{x_i}^{x_{i+1}} N_i N_{i+1} q(x)\,dx \\ \int_{x_i}^{x_{i+1}} N_{i+1} N_i q(x)\,dx & \int_{x_i}^{x_{i+1}} (N_{i+1})^2 q(x)dx \end{bmatrix}$$

where N_i and N_{i+1} are the interpolation functions given in Eq. (2.29). In terms of the local coordinate ξ, $\mathbf{q}_\mathbf{e}$ becomes

$$\mathbf{q}_\mathbf{e} = \int_0^{l_e} q(x_i + \xi) \begin{bmatrix} \left(1 - \dfrac{\xi}{l_e}\right)^2 & \dfrac{\xi(1 - \xi/l_e)}{l_e} \\ \dfrac{\xi(1 - \xi/l_e)}{l_e} & \left(\dfrac{\xi}{l_e}\right)^2 \end{bmatrix} d\xi$$

The integral I_{fe} is

$$I_{fe} = \int_{x_i}^{x_{i+1}} uf\,dx$$

Substituting $u = \mathbf{u}_\mathbf{e}^\mathrm{T}\mathbf{N}$ yields

$$I_{fe} \approx \int_{x_i}^{x_{i+1}} \mathbf{u}_\mathbf{e}^\mathrm{T}\mathbf{N}f\,dx = \mathbf{u}_\mathbf{e}^\mathrm{T}\mathbf{f}_\mathbf{e}$$

where $\mathbf{f}_\mathbf{e}$ is termed the elemental load vector, and has the form

$$\mathbf{f}_\mathbf{e} = \int_{x_i}^{x_{i+1}} \mathbf{N}f\,dx$$

When expressed in terms of the global coordinates $\mathbf{f}_\mathbf{e}$ appears as

$$\mathbf{f}_\mathbf{e} = \int_{x_i}^{x_{i+1}} \begin{bmatrix} N_i(x) \\ N_{i+1}(x) \end{bmatrix} f(x)\,dx$$

or in terms of the local coordinates ξ as

$$\mathbf{f_e} = \int_0^{l_e} \begin{bmatrix} 1 - \dfrac{\xi}{l_e} \\ \dfrac{\xi}{l_e} \end{bmatrix} f(x_i + \xi)d\xi$$

Assembly. The functional $I(u)$, which as a result of the discretization process has been converted into an approximate function $I(u_1, u_2, \ldots, u_{n+1})$, can be written as

$$I(u_1, u_2, u_3, u_4, \ldots, u_{N+1}) = \frac{\sum \mathbf{u_e^T k_e u_e}}{2} - \sum \mathbf{u_e^T f_e} + \frac{\beta u_{N+1}^2}{2} + \frac{\alpha u_1^2}{2} - Bu_{N+1} - Au_1$$

where the $\sum$'s are taken over the elements and where $\mathbf{k_e} = \mathbf{p_e} + \mathbf{q_e}$. Consider the $\mathbf{k_e}$ term arising from the first element. We write for the contribution from the first element,

$$\mathbf{u_e^T k_e u_e} = [u_1 \; u_2] \begin{bmatrix} (k_{11})_1 & (k_{12})_1 \\ (k_{21})_1 & (k_{22})_1 \end{bmatrix} \begin{bmatrix} u_1 \\ u_2 \end{bmatrix}$$

or when considered in the framework of the global system,

$$\mathbf{u_e^T k_e u_e} = [u_1 \; u_2 \; u_3 \ldots u_{N+1}] \begin{bmatrix} (k_{11})_1 & (k_{12})_1 & 0 & \ldots \\ (k_{21})_1 & (k_{22})_1 & 0 & \\ 0 & 0 & & \\ \vdots & & & \end{bmatrix} \begin{bmatrix} u_1 \\ u_2 \\ \vdots \\ u_{N+1} \end{bmatrix}$$

$$= \mathbf{u_G^T k_{G1} u_G}$$

showing clearly the elemental contribution at the global level. In a similar fashion, the contribution from f to the global load vector from the first element can be clearly seen to be

$$\mathbf{u_e^T f_e} = [u_1 u_2] \begin{bmatrix} (f_{e1})_1 \\ (f_{e2})_1 \end{bmatrix}$$

or again when considered at the global level

$$\mathbf{u_e^T f_e} = [u_1 \; u_2 \; u_3 \; \ldots \; u_{N+1}] \begin{bmatrix} (f_{e1})_1 \\ (f_{e2})_1 \\ 0 \\ \vdots \end{bmatrix}$$

$$= \mathbf{u_G^T f_{G1}}$$

With this idea as to the transition from the elemental to the global contribution in mind, the function I can be written as

$$I(u_1, u_2, u_3, u_4, \ldots, u_{N+1}) = \frac{\mathbf{u}_G^T \mathbf{K}_G \mathbf{u}_G}{2} - \mathbf{u}_G^T \mathbf{F}_G$$

where

$$\mathbf{K}_G = \sum_e \mathbf{k}_G + \mathbf{BT}_G$$

and

$$\mathbf{F}_G = \sum_e \mathbf{f}_G + \mathbf{bt}_G$$

The sums clearly indicate the assembly process. The $(N+1) \times (N+1)$ matrix $\mathbf{BT}_G$, given by

$$\mathbf{BT}_G = \begin{bmatrix} \alpha & 0 & & \cdots & & & \\ 0 & 0 & 0 & & & & \\ \vdots & & & & & & \\ & & & & 0 & 0 & 0 \\ & & & & 0 & 0 & \beta \end{bmatrix}$$

arises from the $\alpha u_1^2/2$ and $\beta u_{N+1}^2/2$ terms of the natural boundary condition, and the $(N+1) \times 1$ vector

$$\mathbf{bt}_G^T = [A \ 0 \ 0 \ \ldots \ 0 \ B]$$

arises from the nonhomogeneous terms Au_1 and Bu_{N+1} of the natural boundary conditions.

In view of the fact that we now have a function

$$I = I(u_1, u_2, \ldots, u_{N+1})$$

the stationary value is obtained in the usual manner for a function by requiring each of the partial derivatives to vanish according to

$$\frac{\partial I}{\partial u_i} = 0 \qquad i = 1, 2, \ldots, N+1$$

or in matrix notation as

$$\frac{\partial I}{\partial \mathbf{u}_G} = 0$$

leading to

$$\frac{1}{2}(\mathbf{K}_G + \mathbf{K}_G^T)\overline{\mathbf{u}}_G = \overline{\mathbf{F}}_G$$

or, since $\mathbf{K_G} = \mathbf{K_G^T}$

$$\mathbf{K_G u_G} = \mathbf{F_G}$$

The process followed in converting the original functional into a function of the nodal unknowns and then requiring the resulting function to be stationary with respect to each nodal unknown, is essentially the celebrated Ritz method, further discussed in Appendix A.

At this point, the basic steps of discretization, interpolation, elemental formulation, and assembly for the variational formulation of the finite element model have been carried out. The results are identical to those obtained using the Galerkin method presented in the first part of Section 2.3. The steps of constraints, solution, and computation of derived variables are carried out in precisely the same fashion as in the last part of Section 2.3, to which the reader is referred.

2.8 HIGHER-ORDER INTERPOLATIONS

The use of linear interpolation as the basis for a finite element model has advantages and some definite disadvantages. The main advantage is that the linear interpolation functions and their derivatives are relatively simple analytically, making it possible to concentrate on the basic ingredients of the finite element method rather than having to deal with the complexities associated with higher-order interpolations. The main disadvantages include the fact that the derivative of the solution within an element is constant, resulting in relatively large discontinuities in the derived variable at the interelement boundaries, and the relatively slow convergence exhibited by the solution. In this section we present higher-order polynomial interpolation functions and investigate their use in typical problems.

Quadratic interpolation. When generating a quadratically interpolated finite element model, the discretization must be such that the number of nodes is odd, as indicated in Fig. 2–52. This is necessary for the simple reason that each element is defined by three successive nodes, usually equally spaced, which are used as the basis for constructing a parabolic approximation of the solution within the element. Such a parabolic representation requires three points for its definition. Within the element, whose equally spaced nodes are x_i, x_{i+1}, and x_{i+2}, as shown in Fig. 2–52, the approximate solution is taken as a parabola of the form

$$u_e(x) = \alpha + \beta x + \gamma x^2 \tag{2.31}$$

Note that this assumption satisfies the completeness criterion. Requiring that the parabola pass through the three points (x_i, u_i), (x_{i+1}, u_{i+1}), and (x_{i+2}, u_{i+2}) leads to the three equations

$$\begin{aligned}
u_i &= \alpha + \beta x_i + \gamma x_i^2 \\
u_{i+1} &= \alpha + \beta x_{i+1} + \gamma x_{i+1}^2 \\
u_{i+2} &= \alpha + \beta x_{i+2} + \gamma x_{i+2}^2
\end{aligned}$$

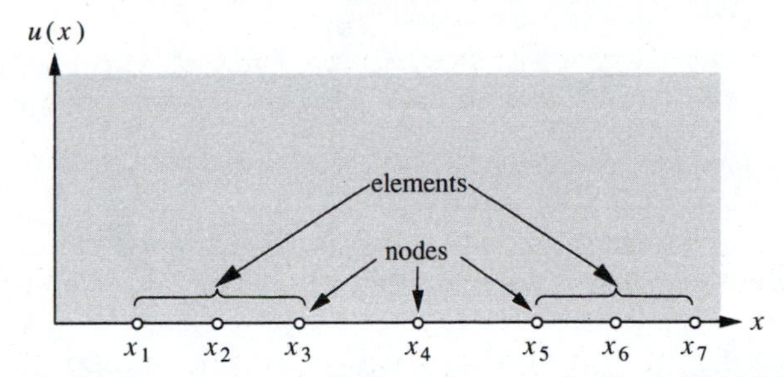

FIGURE 2–52 Nodes and elements—quadratic interpolation

Solving for α, β, and γ and substituting back into Eq. (2.31) yields

$$\begin{aligned} u_e(x) &= N_i(x)u_i + N_{i+1}(x)u_{i+1} + N_{i+2}(x)u_{i+2} \\ &= \mathbf{u}_\mathbf{e}^\mathrm{T}\mathbf{N} = \mathbf{N}^\mathrm{T}\mathbf{u}_\mathbf{e} \end{aligned}$$

where

$$\begin{aligned} N_i(x) &= \frac{(x - x_{i+1})(x - x_{i+2})}{(x_i - x_{i+1})(x_i - x_{i+2})} \\ N_{i+1}(x) &= \frac{(x - x_i)(x - x_{i+2})}{(x_{i+1} - x_i)(x_{i+1} - x_{i+2})} \\ N_{i+2}(x) &= \frac{(x - x_i)(x - x_{i+1})}{(x_{i+2} - x_i)(x_{i+2} - x_{i+1})} \end{aligned} \tag{2.32}$$

The $N_k(x)$ are collectively referred to as the **quadratic interpolation functions** or, as often termed in numerical analysis, the **quadratic Lagrange interpolating polynomials**. They appear as in Fig. 2–53. The derivatives appear in Fig. 2–54 and are clearly linear functions within the element. Interelement continuity of the dependent variable u is again apparent. The derived variable pu' will in general be improved relative to the corresponding quantity using linear interpolation but may, depending upon the character of the problem being modeled, still be discontinuous at the interelement nodes.

Note that quadratic interpolation is the basis of Simpson's rule discussed in Section 2.6 for the numerical evaluation of integrals.

Expressed in terms of the local coordinate $\xi = x - x_i$, the interpolation functions appear as

$$\mathbf{N}(\xi) = \begin{bmatrix} \left(1 - \dfrac{2\xi}{l_e}\right)\left(1 - \dfrac{\xi}{l_e}\right) \\ 4\left(\dfrac{\xi}{l_e}\right)\left(1 - \dfrac{\xi}{l_e}\right) \\ \dfrac{\xi}{l_e}\left(\dfrac{2\xi}{l_e} - 1\right) \end{bmatrix} \tag{2.33}$$

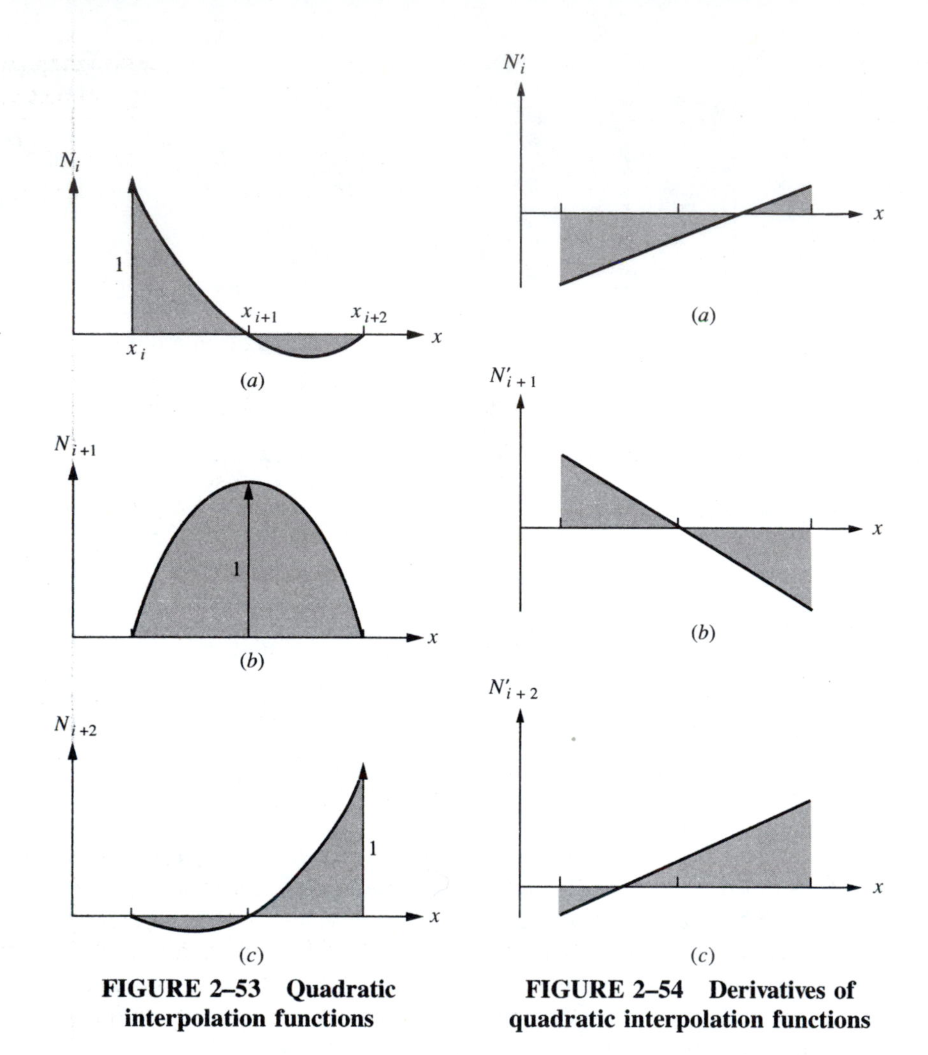

FIGURE 2–53 Quadratic interpolation functions

FIGURE 2–54 Derivatives of quadratic interpolation functions

Since the transformation is a simple translation, $dx = d\xi$ and $dN_k(x)/dx = dN_k(\xi)/d\xi$ and thus

$$\mathbf{N}'(\xi) = \begin{bmatrix} \frac{1}{l_e}\left(4\frac{\xi}{l_e} - 3\right) \\ \frac{1}{l_e}\left(4 - 8\frac{\xi}{l_e}\right) \\ \frac{1}{l_e}\left(4\frac{\xi}{l_e} - 1\right) \end{bmatrix} \tag{2.34}$$

where $l_e = x_{i+2} - x_i$ is the length of the element.

It is sometimes convenient to further transform these interpolation functions into a dimensionless coordinate s whose origin is at the center of the element and whose domain is $(-1,1)$ according to

$$\frac{\xi}{l_e} = \frac{1}{2} + \frac{s}{2}$$

such that

$$N(s) = \begin{bmatrix} \dfrac{s(s-1)}{2} \\ (1+s)(1-s) \\ \dfrac{s(s+1)}{2} \end{bmatrix}$$

Here, since $d/dx = d/d\xi = (2/l_e)d/ds$, it follows that

$$\frac{dN_k(x)}{dx} = \left(\frac{2}{l_e}\right)\frac{dN_k(s)}{ds}$$

and

$$\mathbf{N}' = \frac{2}{l_e}\begin{bmatrix} \dfrac{2s-1}{2} \\ -2s \\ \dfrac{2s+1}{2} \end{bmatrix}$$

The variable s is frequently referred to as an **intrinsic** or **natural coordinate**. Other higher-order interpolations such as cubic interpolation can be defined and used. We will investigate these in the Exercises.

Variational formulation using quadratic interpolation. Recall that the boundary value problem in question is

$$(pu')' - qu + f = 0$$

$$-p(a)u'(a) + \alpha u(a) = A$$

$$p(b)u'(b) + \beta u(b) = B$$

and that the corresponding functional is

$$I(u) = \int_a^b \left(\frac{pu'^2 + qu^2}{2} - fu\right)dx + \frac{\beta u(b)^2}{2} + \frac{\alpha u(a)^2}{2} - Bu(b) - Au(a)$$

With a discretization appropriate for quadratic interpolation in mind, we write

$$I(u) = \sum_e \int_{x_i}^{x_i+2} \left(\frac{pu'^2 + qu^2}{2} - fu\right)dx + \frac{\beta u(b)^2}{2} + \frac{\alpha u(a)^2}{2} - Bu(b) - Au(a)$$

or

$$I(u) = \sum_e \left(\frac{I_{pe} + I_{qe}}{2} - I_{fe} \right) + \frac{\beta u(b)^2}{2} + \frac{\alpha u(a)^2}{2} - Bu(b) - Au(a) \quad (2.35)$$

where

$$I_{pe} = \int_{x_i}^{x_{i+2}} u' p u' dx = \mathbf{u_e^T} \left[\int_{x_i}^{x_{i+2}} \mathbf{N}' p \mathbf{N}'^{\mathrm{T}} dx \right] \mathbf{u_e}$$

$$= \mathbf{u_e^T p_e u_e}$$

$$I_{qe} = \int_{x_i}^{x_{i+2}} uqu\, dx = \mathbf{u_e^T} \left[\int_{x_i}^{x_{i+2}} \mathbf{N} q \mathbf{N}^{\mathrm{T}} dx \right] \mathbf{u_e}$$

$$= \mathbf{u_e^T q_e u_e}$$

and

$$I_{fe} = \int_{x_i}^{x_{i+2}} uf\, dx = \mathbf{u_e^T} \int_{x_i}^{x_{i+2}} \mathbf{N} f\, dx$$

$$= \mathbf{u_e^T f_e}$$

where

$$\mathbf{p_e} = \int_{x_i}^{x_{i+2}} \mathbf{N}' p \mathbf{N}'^{\mathrm{T}} dx$$

$$\mathbf{q_e} = \int_{x_i}^{x_{i+2}} \mathbf{N} q \mathbf{N}^{\mathrm{T}} dx$$

and

$$\mathbf{f_e} = \int_{x_i}^{x_{i+2}} \mathbf{N} f\, dx$$

As mentioned when the elemental formulation was discussed for the linear interpolation, and as is clearly seen here, the *form* of the elemental matrices is independent of the particular interpolation used. The interpolation vectors and their derivatives are of course different in size and character, but we can still state that

$$\mathbf{p_e} = \int_{x_i}^{x_{i+2}} \mathbf{N}'(x) p(x) \mathbf{N}'(x)^{\mathrm{T}} dx = \int_0^{l_e} \mathbf{N}'(\xi) p(x_i + \xi) \mathbf{N}'(\xi)^{\mathrm{T}} d\xi \quad (2.36)$$

$$\mathbf{q_e} = \int_{x_i}^{x_{i+2}} \mathbf{N}(x) q(x) \mathbf{N}(x)^{\mathrm{T}} dx = \int_0^{l_e} \mathbf{N}(\xi) q(x_i + \xi) \mathbf{N}(\xi)^{\mathrm{T}} d\xi \quad (2.37)$$

and

$$\mathbf{f_e} = \int_{x_i}^{x_{i+2}} \mathbf{N}(x) f(x) dx = \int_0^{l_e} \mathbf{N}(\xi) f(x_i + \xi) d\xi \tag{2.38}$$

The boundary terms from the natural boundary conditions are handled in exactly the same fashion as before so that I, given by Eq. (2.35), which is now a function of the nodal values u_i, becomes

$$I(\mathbf{u_G}) = \frac{\mathbf{u_G^T K_G u_G}}{2} - \mathbf{u_G^T F_G}$$

where

$$\mathbf{K_G} = \sum_e (\mathbf{p_G} + \mathbf{q_G}) + \mathbf{BT_G}$$

and (2.39)

$$\mathbf{F_G} = \sum_e \mathbf{f_G} + \mathbf{bt_G}$$

The square matrix $\mathbf{BT_G}$ and the column vector $\mathbf{bt_G}$ are exactly the same as for the linear interpolation, that is,

$$\mathbf{BT_G} = \begin{bmatrix} \alpha & 0 & \cdots & & \\ 0 & 0 & & & \\ \vdots & & & & \\ & & & 0 & 0 \\ & & & 0 & \beta \end{bmatrix}$$

and

$$\mathbf{bt_G^T} = [A \quad 0 \quad \ldots \quad 0 \quad B]$$

As usual, the assembly is implied by the $\sum$ in Eqs. (2.39).

Requiring the function $I(\mathbf{u_G})$ to be stationary yields

$$\frac{\partial I}{\partial \mathbf{u_G}} = \frac{1}{2}(\mathbf{K_G} + \mathbf{K_G^T})\mathbf{u_G} - \mathbf{F_G} = \mathbf{K_G u_G} - \mathbf{F_G} = 0$$

or

$$\mathbf{K_G u_G} = \mathbf{F_G}$$

Constraints are enforced in the usual manner and the **solution** is obtained with a suitable equation solver.

Computation of derived variables. Generally for this class of boundary value problems, the derived or secondary variable is not u', but rather pu', the term appearing in the differential equation and also in the natural boundary conditions. Thus for each element it is appropriate to consider

$$pu'(x) = p(x)\mathbf{u_e^T N'}(x)$$

which is obviously a function of x as indicated in Fig. 2–55, and can be evaluated at any x within the element.

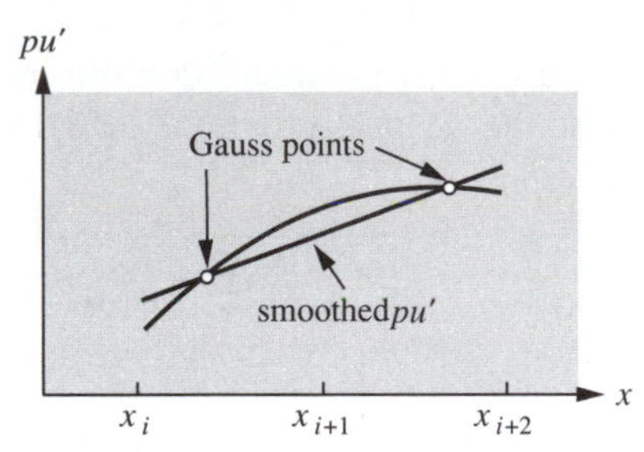

FIGURE 2–55 ***pu'* for the quadratically interpolated model**

Obvious choices for points at which to evaluate pu' would be the three nodes. However, it turns out [Moan, 3] that there are special points within the element at which the flux term pu' is more accurate. For constant p, these points are essentially the Gauss points corresponding to the order of the interpolation that has been used. In particular, if the model has been generated using linear interpolation, the flux pu' is most accurate when computed at the center of the element as pointed out in Section 2.3. For a quadratically interpolated model, the special points for constant p are approximately the Gauss points located at $\pm l_e/2\sqrt{3}$ relative to the center of the element. In practice the derivatives are evaluated at the Gauss points within the element in question whether or not p is a constant.

A frequently used ad hoc procedure for displaying the derived variable is to evaluate pu' at the Gauss points and then use these two values to construct a linear function on the element in question. The results of this process, commonly referred to as *smoothing*, are also indicated in Fig. 2–55. The rationale behind this process is that in the limit as the mesh is refined, p is essentially constant over the element, which, since u' is linear, results in a linear pu'. This ad hoc procedure simply anticipates this result and also uses the idea for meshes with larger element sizes.

Evaluation of integrals. In the last part of Section 2.6, special approximate integration formulas for the evaluation of the elementary matrices associated with the linearly interpolated finite element model were presented. It will again be assumed that p, q, and f are smooth enough so that they can be represented on an element by a second degree polynomial, that is, a polynomial of the same degree as is being used for the interpolation of the solution. In the unlikely event that p, q, and f are not sufficiently smooth, or if it is desired to evaluate the integrals exactly, a Gauss quadrature of sufficiently high order can be used.

Evaluation of $\mathbf{p_e}$. Generally $\mathbf{p_e}$ is given by

$$\mathbf{p_e} = \int_0^{l_e} \mathbf{N}' p(x_i + \xi) \mathbf{N}'^{\mathrm{T}} \, d\xi \tag{2.40}$$

where the $\mathbf{N}'(\xi)$ are given by Eq. (2.34). Representing p approximately by its quadratic interpolant

$$p(x_i + \xi) = p_i N_i(\xi) + p_{i+1} N_{i+1}(\xi) + p_{i+2} N_{i+2}(\xi)$$

the integrals in Eq. (2.40) can be evaluated in a straightforward manner to yield

$$\begin{aligned}
[p_e]_{11} &= \frac{1}{30l_e}(37p_i + 36p_{i+1} - 3p_{i+2}) \\
[p_e]_{12} &= \frac{1}{30l_e}(-44p_i - 32p_{i+1} - 4p_{i+2}) \\
[p_e]_{13} &= \frac{1}{30l_e}(7p_i - 4p_{i+1} + 7p_{i+2}) \\
[p_e]_{22} &= \frac{1}{30l_e}(48p_i + 64p_{i+1} + 48p_{i+2}) \\
[p_e]_{23} &= \frac{1}{30l_e}(-4p_i - 32p_{i+1} - 44p_{i+2}) \\
[p_e]_{33} &= \frac{1}{30l_e}(-3p_i + 36p_{i+1} + 37p_{i+2})
\end{aligned} \tag{2.41}$$

with $(p_e)_{ij} = (p_e)_{ji}$.

Evaluation of $\mathbf{q_e}$. The general expression for $\mathbf{q_e}$ is

$$\mathbf{q_e} = \int_0^{l_e} \mathbf{N} q(x_i + \xi) \mathbf{N}^{\mathrm{T}} d\xi \tag{2.42}$$

where the $\mathbf{N}$ are given by Eq. (2.33). Representing $q(x_i + \xi)$ in terms of its quadratic interpolant as

$$q(x_i + \xi) = q_i N_i(\xi) + q_{i+1} N_{i+1}(\xi) + q_{i+2} N_{i+2}(\xi)$$

the integrals in (2.42) can be evaluated to yield

$$\begin{aligned}
[q_e]_{11} &= \frac{l_e}{420}(39q_i + 20q_{i+1} - 3q_{i+2}) \\
[q_e]_{12} &= \frac{l_e}{420}(20q_i + 16q_{i+1} - 8q_{i+2}) \\
[q_e]_{13} &= \frac{l_e}{420}(-3q_i - 8q_{i+1} - 3q_{i+2}) \\
[q_e]_{22} &= \frac{l_e}{420}(16q_i + 192q_{i+1} + 16q_{i+2})
\end{aligned} \tag{2.43}$$

$$[q_e]_{23} = \frac{l_e}{420}(-8q_i + 16q_{i+1} + 20q_{i+2})$$

$$[q_e]_{33} = \frac{l_e}{420}(-3q_i + 20q_{i+1} + 39q_{i+2})$$

with $[q_e]_{ij} = [q_e]_{ji}$.

Evaluation of $\mathbf{f_e}$. The general expression for $\mathbf{f_e}$ is given by

$$\mathbf{f_e} = \int_0^{l_e} \mathbf{N} f(x_i + \xi)\, d\xi \tag{2.44}$$

Representing f in terms of its quadratic interpolant as

$$f(x_i + \xi) = f_i N_i(\xi) + f_{i+1} N_{i+1}(\xi) + f_{i+2} N_{i+2}(\xi)$$

the integrals in (2.44) can be evaluated to yield

$$\begin{aligned}
[f_e]_1 &= \frac{l_e}{30}(4f_i + 2f_{i+1} - f_{i+2}) \\
[f_e]_2 &= \frac{l_e}{30}(2f_i + 16f_{i+1} + 2f_{i+2}) \\
[f_e]_3 &= \frac{l_e}{30}(-f_i + 2f_{i+1} + 4f_{i+2})
\end{aligned} \tag{2.45}$$

Note carefully that these expressions for the elemental matrices $\mathbf{p_e}$, $\mathbf{q_e}$, and $\mathbf{f_e}$ are exact only if p, q, and f are polynomials of degree two or less. In all other instances, they are approximate expressions whose accuracy depends upon how well the parabolas fit the data on the intervals corresponding to each of the elements.

2.9 QUADRATIC INTERPOLATION EXAMPLES

In this section we repeat each of the examples considered in Section 2.4, setting up and solving finite element models using quadratic interpolation. The student is asked to refer to each of the examples for a description of the physical problems.

2.9.1 Solid Mechanics—Axial Deformation

The boundary value problem is stated as

$$\begin{aligned}
(AEu')' + A\gamma &= 0 \\
u(0) &= 0 \\
P(L) = AEu'(L) &= 0
\end{aligned}$$

so that $p = AE$, $q = 0$, $f = A\gamma$, with $\alpha = \beta = A = B = 0$.

One-element model. With a one-element model, $l_e = L$, and $\mathbf{p_e}$, $\mathbf{q_e}$, and $\mathbf{f_e}$ are easily evaluated using Eqs. (2.41), (2.43), and (2.45), respectively, as

$$\mathbf{p_e} = \frac{AE}{3L}\begin{bmatrix} 7 & -8 & 1 \\ -8 & 16 & -8 \\ 1 & -8 & 7 \end{bmatrix}$$

$$\mathbf{q_e} = \mathbf{0}$$

$$\mathbf{f_e} = \frac{A\gamma L}{6}\begin{bmatrix} 1 \\ 4 \\ 1 \end{bmatrix}$$

For a one-element model, the global stiffness and load matrices are the same as the elemental stiffness and load matrices so that the final unconstrained global equations in augmented form are

$$\left[\begin{array}{ccc|c} 7 & -8 & 1 & \phi \\ -8 & 16 & -8 & 4\phi \\ 1 & -8 & 7 & \phi \end{array}\right]$$

where $\phi = \gamma L^2/2E$. Enforcing the constraint $u_1 = 0$ and solving the resulting set of equations yields

$$\mathbf{u}_\mathbf{G}^\mathrm{T} = \frac{\gamma L^2}{8E}[0 \;\; 3 \;\; 4]$$

Within the element

$$u_e(x) = \mathbf{u}_\mathbf{e}^\mathrm{T}\mathbf{N} = \frac{\gamma L^2}{8E}[0 \quad 3 \quad 4]\begin{bmatrix} N_1 \\ N_2 \\ N_3 \end{bmatrix}$$

which with $N_1(x) = (2x/L - 1)(x/L - 1)$, $N_2(x) = (4x/L)(1 - x/L)$, and $N_3(x) = (x/L)(2x/L - 1)$ becomes

$$u_e(x) = \frac{\gamma L^2}{2E}\frac{x}{L}\left(2 - \frac{x}{L}\right)$$

The derived variable, which is the force transmitted, is

$$P(x) = AEu_e'(x) = AE\mathbf{u}_\mathbf{e}^\mathrm{T}\mathbf{N}' = A\gamma(L - x)$$

These expressions for the displacement and force transmitted are exact, that is, they coincide with the classical solution of the differential equation and boundary conditions.

The emergence of the exact solution from the finite element model is an inherent property of the class of problems considered in this chapter, namely, that if for *some* choice of the parameters u_i, the exact solution is contained within the approximate solution assumed for the finite element model, the method will select the values of the u_i that correspond to that exact solution. When the exact solution is not contained within the approximate finite element solution, the solution will choose the *best* values of the unknowns u_i. For a discussion of *best* the reader is referred to Stakgold [11] and Reddy [12]. It is left to the student to show in the Exercises that a model using two or more elements still produces the exact solution.

2.9.2 Fluid Mechanics—Thin-Film Lubrication

Consider again the specific problem from Section 2.4.2, shown in Fig. 2–56.

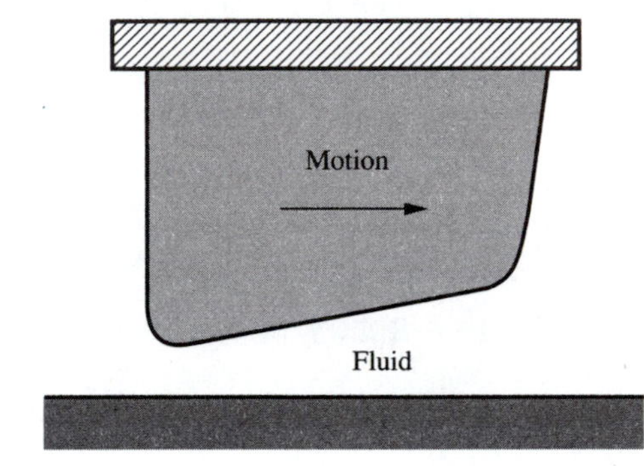

FIGURE 2–56 Parabolic profile slider bearing

The boundary value problem to be solved is

$$\left((1+s^2)^3 P'\right)' + s = 0$$
$$P(0) = 0$$
$$P(1) = 0$$

where $' \equiv d/ds$. Comparison with the standard form indicates that $a = 0$, $b = 1$, $p(s) = (1+s^2)^3$, $q = 0$, and $f = s$. A, B, α, and β are taken as zero.

Generally the elemental matrices are

$$\mathbf{p_e} = \int_0^{l_e} \mathbf{N}'(\xi)\, p(s_i+\xi)\mathbf{N}'^{\mathrm{T}}(\xi)\,d\xi$$

$$\mathbf{q_e} = \mathbf{0}$$

$$\mathbf{f_e} = \int_0^{l_e} \mathbf{N}(\xi) f(s_i+\xi)\,d\xi$$

where $p(s_i+\xi) = (1+(s_i+\xi)^2)^3$, $f(s_i+\xi) = s_i+\xi$ with $\mathbf{N}(\xi)$ and $\mathbf{N}'(\xi)$ given in the previous section.

For the numerical results that follow, two sets of finite element models will be formed, one using the special approximate integration formulas of Section 2.6 for evaluating the elemental matrices and the other by evaluating the matrices exactly. This will provide further calibration as far as to when the approximate evaluations are adequate. Note that the $\mathbf{f_e}$ will be identical for both models.

One-element model. For the approximate evaluation:

$$\mathbf{p}_{e1} = \begin{bmatrix} 2.777083 & -4.616667 & 1.839583 \\ & 18.566667 & -13.950000 \\ \text{symm} & & 12.110417 \end{bmatrix}$$

$$\mathbf{q}_{e1} = \mathbf{0}$$

$$\mathbf{f}_{e1} = \begin{bmatrix} 0.000000 \\ 0.333333 \\ 0.166667 \end{bmatrix}$$

For the exact evaluation:

$$\mathbf{p}_{e1} = \begin{bmatrix} 3.253968 & -5.050794 & 1.796825 \\ & 18.158730 & -13.107937 \\ \text{symm} & & 11.311111 \end{bmatrix}$$

with the $\mathbf{q}_e$ and the $\mathbf{f}_e$ the same as for the approximate evaluation. Assembling, constraining and solving yields

$$\mathbf{u}_G^T = [0.000000 \quad 0.017953 \quad 0.000000]$$

and

$$\mathbf{u}_G^T = [0.000000 \quad 0.018357 \quad 0.000000]$$

respectively, for the approximate and exact evaluations.

Two-element solution. For the approximate evaluation:

$$\mathbf{p}_{e1} = \begin{bmatrix} 4.954753 & -6.013021 & 1.058268 \\ & 14.567708 & -8.554688 \\ \text{symm} & & 7.496419 \end{bmatrix}$$

$$\mathbf{p}_{e2} = \begin{bmatrix} 12.372982 & -16.000521 & 3.627539 \\ & 48.126042 & -32.125521 \\ \text{symm} & & 28.497982 \end{bmatrix}$$

$$\mathbf{q}_{e1} = \mathbf{q}_{e2} = \mathbf{0}$$

$$\mathbf{f}_{e1} = \begin{bmatrix} 0.000000 \\ 0.083333 \\ 0.041667 \end{bmatrix} \qquad \mathbf{f}_{e2} = \begin{bmatrix} 0.041667 \\ 0.250000 \\ 0.083333 \end{bmatrix}$$

For the exact evaluation:

$$\mathbf{p}_{e1} = \begin{bmatrix} 5.000794 & -6.056052 & 1.055258 \\ & 14.538889 & -8.482837 \\ \text{symm} & & 7.427579 \end{bmatrix}$$

$$\mathbf{p}_{e2} = \begin{bmatrix} 12.637698 & -16.256647 & 3.618948 \\ & 48.043651 & -32.787004 \\ \text{symm} & & 28.168056 \end{bmatrix}$$

with the $\mathbf{q}_{e1}$ and the $\mathbf{f}_{e1}$ again the same as for the approximate evaluation.

The solutions using the approximate and exact methods for evaluating the integrals are

$$\mathbf{u}_G^T = [0.000000 \quad 0.018998 \quad 0.022610 \quad 0.012712 \quad 0.000000]$$

and

$$\mathbf{u}_G^T = [0.000000 \quad 0.018872 \quad 0.022522 \quad 0.012824 \quad 0.000000]$$

respectively. The corresponding eight-element solutions are

$$\mathbf{u}_G^T = [0.000000 \quad 0.010786 \quad 0.018888 \quad 0.022783 \quad 0.022438 \quad 0.018730 \quad 0.013092 \quad 0.006544 \quad 0.000000]$$

and

$$\mathbf{u}_G^T = [0.000000 \quad 0.010782 \quad 0.018890 \quad 0.022780 \quad 0.022434 \quad 0.018734 \quad 0.013087 \quad 0.006552 \quad 0.000000]$$

respectively. The results for the one-, two-, and four-element models using the approximate method for evaluation of the integrals in the elemental matrices are displayed in Fig. 2–57.

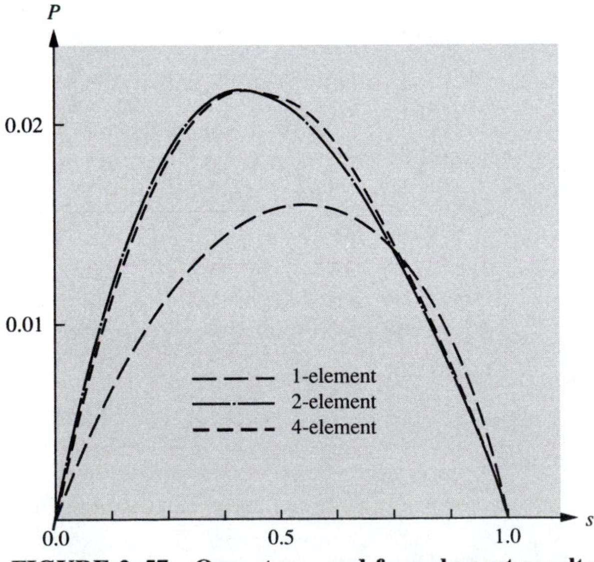

FIGURE 2–57 One-, two-, and four-element results

Computation of derived variables. For this example, the derived variable is the volume flow rate q_x given by

$$q_x = \frac{-h^3}{12\eta}\frac{dp}{dx} + \frac{(U_2 - U_1)h}{2}$$

which can be expressed in terms of the dimensionless variables as

$$-\frac{q_x}{Uh_0} = (1 + s^2)\left(0.5 + (1 + s^2)^2\frac{dP}{ds}\right)$$

For a quadratically interpolated finite element solution, it was shown by Moan [3] that the best points within the element for evaluating the derived variable are the two Gauss points corresponding to $N = 2$, that is, $\pm l_e/2\sqrt{3}$ relative to the midpoint of the element. The results of the calculations are presented in a suggestive form showing the predicted flows within each of the elements for both the approximate and exact evaluation of the elemental matrices. The values quoted are the average of the two values calculated at the Gauss points for the element.

Two-element model. Approximate and exact evaluation of integrals respectively yields

$-(q_x/Uh_0)_1$	$-(q_x/Uh_0)_2$
0.5909	0.5915
0.5908	0.5910

Four-element model.

$-(q_x/Uh_0)_1$	$-(q_x/Uh_0)_2$	$-(q_x/Uh_0)_3$	$-(q_x/Uh_0)_4$
0.5902	0.5902	0.5903	0.5902
0.5902	0.5903	0.5902	0.5902

These results clearly indicate that for both the two- and four-element models q_x is very uniform, so that the physics of the problem is being satisfied to a high degree of accuracy. The results also indicate that the differences in the results arising from the two different methods of evaluating the elemental matrices are negligible for the quadratic interpolation models. The improved accuracy of the quadratically interpolated results relative to the linearly interpolated results is also noted.

As was pointed out in Section 2.4.2, the amount of work involved in evaluating the integrals exactly suggests that a good strategy would be to use the approximate method for evaluating the elemental matrices in setting up and solving a finite element model for an initially fairly crude mesh, and to evaluate the integrals exactly using a Gauss-Legendre scheme when writing a code to perform the analyses for finer meshes.

2.9.3 One-Dimensional Heat Conduction with Convection

From Section 2.4.3, the nondimensional form of the governing equation and boundary conditions can be stated as

$$(u')' - \phi^2 u = 0$$
$$u(0) = 1$$
$$u'(1) + \psi u(1) = 0$$

where $\psi = hL/k$ and $\phi^2 = 2\psi L/r_o$. Comparing with the standard form discussed in Section 2.2, $p = 1$, $q = \phi^2$, $f = 0$, $A = \alpha = B = 0$, and $\beta = \psi$. The numerical results will be based on the specific values $\psi = 1$ and $\phi^2 = 10$.

Consider first a one-element quadratically interpolated model. Using Eqs. (2.41), (2.43), and (2.45), the elemental matrices are then

$$\mathbf{p_e} = \int_0^{\ell_e} \mathbf{N}' 1 \mathbf{N}'^{\mathrm{T}} d\xi = \frac{1}{3}\begin{bmatrix} 7 & -8 & 1 \\ -8 & 16 & -8 \\ 1 & -8 & 7 \end{bmatrix}$$

and

$$\mathbf{q_e} = \int_0^{\ell_e} \mathbf{N} 10 \mathbf{N}^{\mathrm{T}} d\xi = \frac{10}{30}\begin{bmatrix} 4 & 2 & -1 \\ 2 & 16 & 2 \\ -1 & 2 & 4 \end{bmatrix}$$

Thus with $\alpha = A = B = 0$ and $\beta = 1$, $\mathbf{bt_G} = \mathbf{0}$ and

$$\mathbf{BT_G} = \begin{bmatrix} 0 & 0 & 0 \\ 0 & 0 & 0 \\ 0 & 0 & 1 \end{bmatrix}$$

Assembly then yields

$$\mathbf{K_G} = \mathbf{p_{G1}} + \mathbf{q_{G2}} + \mathbf{BT_G} = \frac{1}{3}\begin{bmatrix} 11 & -6 & 0 \\ -6 & 32 & -6 \\ 0 & -6 & 14 \end{bmatrix}$$

Thus the unconstrained global equations can be expressed in augmented form as

$$\left[\begin{array}{ccc|c} 11 & -6 & 0 & 0 \\ -6 & 32 & -6 & 0 \\ 0 & -6 & 14 & 0 \end{array}\right]$$

The essential boundary condition $u(0) = 1$ is enforced as the constraint $u_1 = 1$, which, after restoring symmetry, leads to

$$\begin{bmatrix} 1 & 0 & 0 & | & 1 \\ 0 & 32 & -6 & | & 6 \\ 0 & -6 & 14 & | & 0 \end{bmatrix}$$

Solving the constrained equations yields

$$\mathbf{u_G} = [1.000000 \quad 0.203883 \quad 0.087379]^{\mathrm{T}}$$

Four- and eight-element models yield

$$\mathbf{u_G} = [1.000000 \quad 0.454417 \quad 0.211780 \quad 0.103911 \quad 0.065315]^{\mathrm{T}}$$

and

$$\mathbf{u_G} = [1.000000 \quad 0.674155 \quad 0.455318 \quad 0.308276 \quad 0.210167 \quad 0.145197 \quad 0.103274 \quad 0.077635 \quad 0.064320]^{\mathrm{T}}$$

respectively. The nodal temperatures for each of the three models are shown in Fig. 2–58.

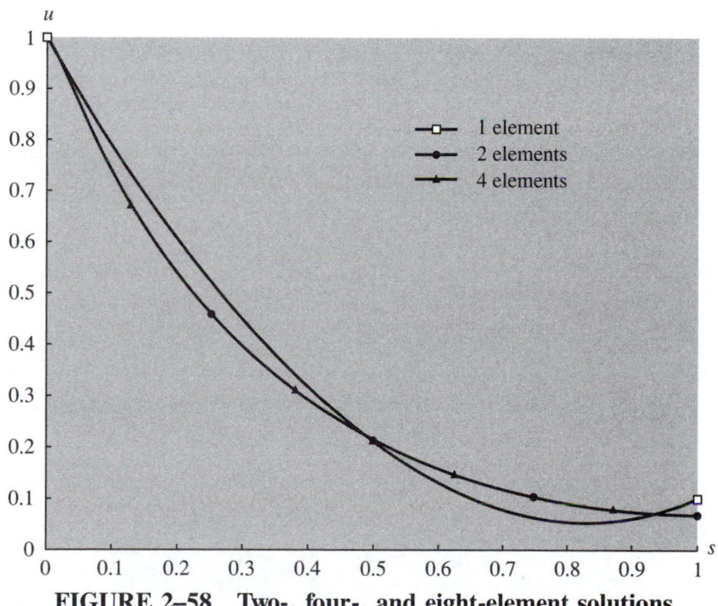

FIGURE 2–58 Two-, four-, and eight-element solutions

For the heat transfer problem the derived variable is given by $q = -kAdT/dx$ or $q = -(kA/L)(T_1 - T_0)du/ds$ in terms of the nondimensionalized variables. For the one-element model, the derived variables $du/ds = -q_1L/(kA(T_1 - T_0))$ at the two Gauss points are computed as

$$\frac{du}{ds} = \mathbf{u}_\mathbf{e}^{\mathrm{T}}\mathbf{N}'|_{gp}$$

At the Gauss point located at $\xi/\ell_e = 0.211325$

$$\frac{du}{ds} = [1.000000 \quad 0.203883 \quad 0.087379]\frac{1}{1}\begin{bmatrix} 4(0.211325) - 3 \\ 4 - 8(0.211325) \\ 4(0.211325) - 1 \end{bmatrix}$$

$$= -1.6974$$

Similarly at the Gauss point located at $\xi/\ell_e = 0.788675$

$$\frac{du}{ds} = [1.000000 \quad 0.203883 \quad 0.087379]\frac{1}{1}\begin{bmatrix} 4(0.788675) - 3 \\ 4 - 8(0.788675) \\ 4(0.788675) - 1 \end{bmatrix}$$

$$= -0.1279$$

For the four- and eight-element models, the corresponding vectors of Gauss point fluxes Q are

$$\mathbf{Q_4} = -[2.2761 \quad 0.8768 \quad 0.4529 \quad 0.1329]^{\mathrm{T}}$$

and

$$\mathbf{Q_8} = -[2.6730 \quad 1.6845 \quad 1.2066 \quad 0.7546 \quad 0.5340 \quad 0.3211 \quad 0.2127 \quad 0.0989]^{\mathrm{T}}$$

Results for the derived variable from each of the three models computed in this manner are displayed in Fig. 2–59. Also shown are the piecewise linear representations generated by connecting the Gauss point values of the pu' segments. The results would clearly seem to indicate a reasonable approximation of the exact continuous pu'.

As mentioned in Section 2.3, an essential boundary condition is satisfied automatically by enforcing the corresponding constraint, whereas a natural boundary condition is only satisfied in the limit as the number of nodes and elements is increased. With $\psi = 1$, the nondimensional form of the boundary condition at $s = 1$ is $u' + u = 0$. In general, the derivative is given by

$$u' = \mathbf{u_e N'} = \mathbf{u_e^T}\frac{1}{\ell_e}\begin{bmatrix} 4\dfrac{\xi}{\ell_e} - 3 \\ 4 - 8\dfrac{\xi}{\ell_e} \\ 4\dfrac{\xi}{\ell_e} - 1 \end{bmatrix}$$

which is to be evaluated at $\xi = \ell_e$ with the $\mathbf{u_e}$ for the last element in the mesh. For the one-element model with $\ell_e = 1$,

$$u' = [1.000000 \quad 0.203883 \quad 0.087379]\frac{1}{1}\begin{bmatrix} 4(1) - 3 \\ 4 - 8(1) \\ 4(1) - 1 \end{bmatrix} = 0.4466$$

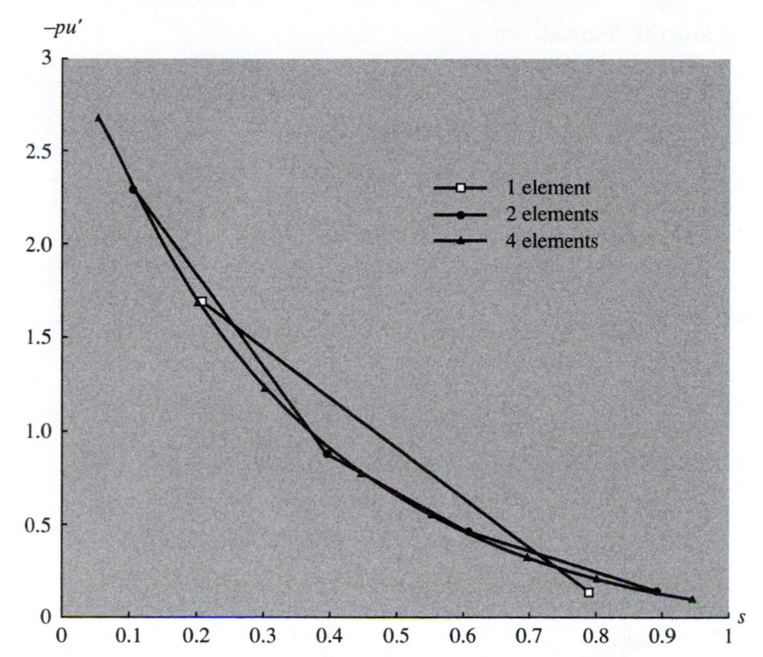

FIGURE 2–59 Derived variable—two-, four-, and eight-element models

so that

$$u'(1) + u(1) = 0.4466 + 0.0874 = 0.5340$$

showing quite a large error in satisfying the boundary condition at $s = 1$. For the two-element model with $\ell_e = 1/2$,

$$u' = [0.211780\ 0.103911\ 0.065315]\frac{2}{1}\begin{bmatrix} 4(1) - 3 \\ 4 - 8(1) \\ 4(1) - 1 \end{bmatrix} = -0.0158$$

so that

$$u'(1) + u(1) = -0.0158 + 0.0653 = 0.0495$$

For the four-element model,

$$u' = [0.103274\ 0.077635\ 0.064320]\frac{4}{1}\begin{bmatrix} 4(1) - 3 \\ 4 - 8(1) \\ 4(1) - 1 \end{bmatrix} = -0.0572$$

so that

$$u'(1) + u(1) = -0.0572 + 0.0643 = 0.0071$$

The natural boundary condition is clearly being satisfied in the limit as the element size is decreased.

2.9.4 Torsional Vibrations

The eigenvalue problem for the torsional vibration problem of Section 2.5.1 is

$$(JG\psi')' + \omega^2 \rho J \psi = 0$$
$$\psi(0) = 0$$
$$\psi'(L) = 0$$

With $\lambda = \omega^2$, the parameters in the standard form for the one-dimensional eigenvalue value problem are $p = JG$, $q = 0$, and $r = \rho J$. Additionally, $\alpha = \beta = 0$. The elemental matrices are $\mathbf{q_e} = 0$

$$\mathbf{P_e} = \int_0^{\ell_e} \mathbf{N}' JG \mathbf{N}'^{\mathrm{T}} d\xi$$

and

$$\mathbf{r_e} = \int_0^{\ell_e} \mathbf{N} \rho J \mathbf{N}^{\mathrm{T}} d\xi$$

For a constant JG and ρJ, these are easily evaluated to yield

$$\mathbf{P_e} = \frac{JG}{3\ell_e}\begin{bmatrix} 7 & -8 & 1 \\ -8 & 16 & -8 \\ 1 & -8 & 7 \end{bmatrix}$$

and

$$\mathbf{r_e} = \frac{\rho J \ell_e}{30}\begin{bmatrix} 4 & 2 & -1 \\ 2 & 16 & 2 \\ -1 & 2 & 4 \end{bmatrix}$$

With the elemental stiffness matrices $\mathbf{k_e} = \mathbf{P_e}$, the final form of the unconstrained matrix eigenvalue problem is

$$(\mathbf{k_G} - \phi \mathbf{M_G})\boldsymbol{\psi} = \mathbf{0}$$

with $\mathbf{K_G} = \Sigma \mathbf{P_G}$ and $\mathbf{M_G} = \Sigma \mathbf{r_G}$.

With $\ell_e = L$ for a one-element model, the assembled unconstrained matrices are simply the $\mathbf{k_e}$ and $\mathbf{r_e}$ above. The constraint is $\psi_1 = 0$, leading to

$$\left\{\begin{bmatrix} 16 & -8 \\ -8 & 7 \end{bmatrix} - \beta \begin{bmatrix} 16 & 2 \\ 2 & 4 \end{bmatrix}\right\}\begin{bmatrix} \psi_2 \\ \psi_3 \end{bmatrix} = \begin{bmatrix} 0 \\ 0 \end{bmatrix} \tag{2.46}$$

where $\beta = \rho\omega^2 L^2/10G$. Requiring the determinant to vanish yields $\beta_1 = 0.2486$ and $\beta_2 = 3.2181$. The corresponding frequencies are given by

$$\omega_1{}^2 = \frac{2.4860G}{\rho L^2} \qquad \omega_2{}^2 = \frac{32.181G}{\rho L^2}$$

compared to the exact eigenvalues

$$(\omega_1^2)_{ex} = \left(\frac{\pi}{2L}\right)^2 \frac{G}{\rho} = \frac{2.4674G}{\rho L^2} \qquad (\omega_2^2)_{ex} = \left(\frac{3\pi}{2L}\right)^2 \frac{G}{\rho} = \frac{22.207G}{\rho L^2}$$

The estimate of the lowest eigenvalue is quite acceptable (0.75 percent error), whereas the estimate of the second eigenvalue is quite crude (44.9 percent error).

The eigenvectors obtained by solving Eq. (2.46) are

$$[\psi_2\psi_3]_1 = [0.707 \quad 1.000]$$

and

$$[\psi_2\psi_3]_2 = [-0.407 \quad 1.000]$$

The eigenfunctions or mode shapes appear as in Fig. 2–60 and clearly have properties similar (the nth eigenfunction has $n - 1$ interior zeroes) to the exact eigenfunctions $\psi_n(x) = \sin(2n - 1)\pi x/2L$.

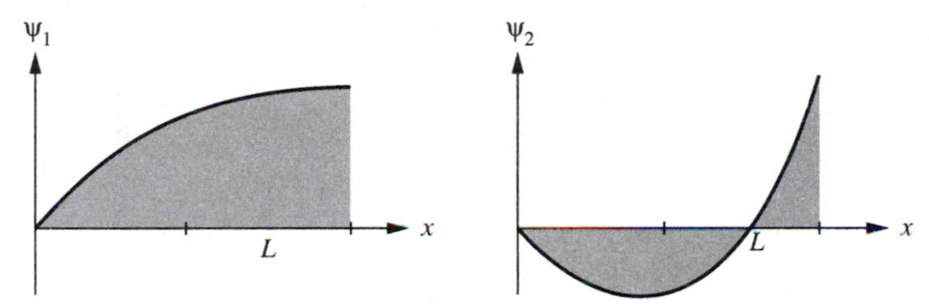

FIGURE 2–60 First and second approximate eigenfunctions

Two-element solution. The elemental $\mathbf{p_e}$ and $\mathbf{r_e}$ matrices have exactly the same form as in the two-element model, with ℓ_e now taken as L/2. Omitting some of the details, the constrained 4 × 4 eigenvalue problem is

$$\left\{\begin{bmatrix} 16 & -8 & 0 & 0 \\ -8 & 14 & -8 & 1 \\ 0 & -8 & 16 & -8 \\ 0 & 1 & -8 & 7 \end{bmatrix} - \beta \begin{bmatrix} 16 & 2 & 0 & 0 \\ 2 & 8 & 2 & -1 \\ 0 & 2 & 16 & 2 \\ 0 & -1 & 2 & 4 \end{bmatrix}\right\} \begin{bmatrix} \psi_2 \\ \psi_3 \\ \psi_4 \\ \psi_5 \end{bmatrix} = \begin{bmatrix} 0 \\ 0 \\ 0 \\ 0 \end{bmatrix}$$

where $\beta = \rho\omega^2 L^2/40G$. Requiring the determinant to vanish yields four roots; these are displayed in Table 2.3 below along with the corresponding eigenvalues, exact eigenvalues, and percent errors.

TABLE 2.3 Approximate eigenvalues—four-element model

i	βi	$\rho\omega^2\beta_i G/L^2$	$(\rho\omega^2\beta_i G/L^2)_{ex}$	% error
1	0.061717	2.4687	2.4674	0.05
2	0.573654	22.946	22.207	3.33
3	0.926578	77.063	61.685	24.9
4	4.967463	198.70	120.90	64.4

These results again bear out the rule of thumb stated in the example in Section 2.5.; namely, that approximately $2N$ unconstrained degrees of freedom are needed for an accurate estimate of the first N eigenvalues. The corresponding eigenfunctions are shown in Fig. 2–61, and again clearly have properties similar to the exact eigenfunctions.

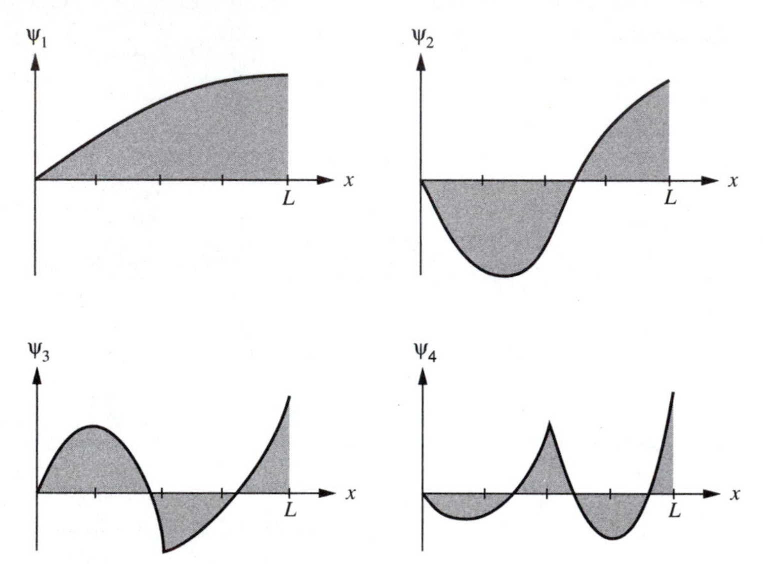

FIGURE 2–61 Approximate eigenfunctions—four-element model

The inability of the two parabolas to accurately represent the third and fourth eigenfunctions is apparent from the figures, resulting in the poor estimates for the corresponding eigenvalues.

2.10 TRANSVERSE DEFLECTIONS OF BEAMS

A very important technical problem in structures is the analysis of beams subjected to transverse loading. In this section we will investigate the application of the finite element method to this class of problems.

As developed in Section 2.7.2, and indicated in Fig. 2–62, the governing differential equation for the transverse deflections of beams is

$$(EIw'')'' = q(x) \qquad 0 \leq x \leq L$$

with *two* boundary conditions at each end. The corresponding functional, representing the potential energy, is

$$V(w) = \int_0^L \left(\frac{EI(w'')^2}{2} - q(x)w(x) \right) dx \tag{2.47}$$

If present, concentrated moments and concentrated transverse loads should be included. It is also occasionally appropriate to include either linear and/or torsional springs in the potential energy. In terms of either the differential equation or the corresponding energy, it is clear that the problem is fourth-order rather than second-order as was the case for the Sturm-Liouville class of problems.

Recall that in Section 2.7.3, the requirements on interpolation functions used to represent solutions to problems involving second-order differential equations were set forward as 1) continuity at the interelement boundaries, and 2) the ability to represent a constant solution and to represent a solution that has a constant first derivative. Functions that are required to be continuous are commonly known as C^0 functions, the 0 superscript referring to the zeroth derivative.

Recall also from Section 2.7.3, that for a fourth-order problem the interpolation functions are required to be at least complete cubic polynomials. An arbitrary cubic polynomial contains four arbitrary constants. For a fourth-order problem, these four constants are used to enforce continuity of the function *and* its first derivative at the interelement boundaries (nodes). Physically this corresponds to requiring the displacement *and* slope to be continuous at the nodes. This is referred to as C^1 continuity.

Consider then a typical problem as indicated in Fig. 2–62.

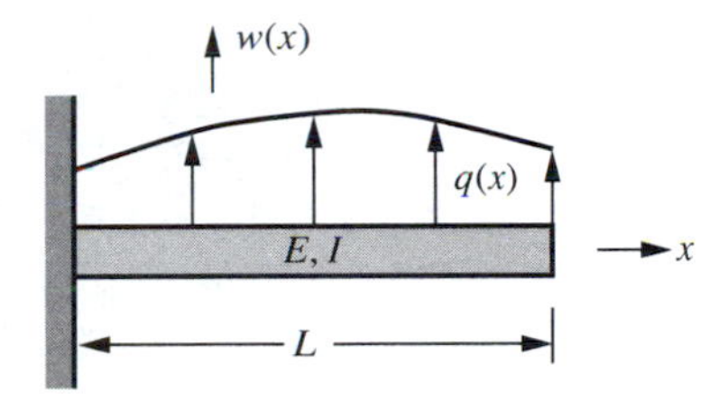

FIGURE 2–62 Typical geometry and loading

Discretization. The discretization is accomplished in a straightforward fashion with the nodes chosen so as to easily accommodate idealized concentrated loads and moments when they occur. Nodes should also be placed at positions corresponding to the beginning and end of a loaded region, as indicated in Fig. 2–63.

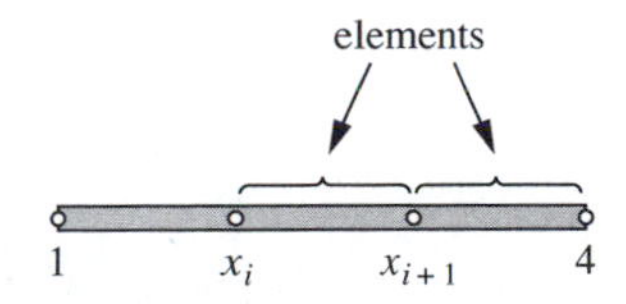

FIGURE 2–63 Nodes and elements

Interpolation. As mentioned above, the basic assumption for the dependent variable $w(x)$ within an element, $x_i \leq x \leq x_{i+1}$, is

$$w(x) = c_0 + c_1x + c_2x^2 + c_3x^3 \tag{2.48}$$

It is much more convenient to develop the interpolation in terms of the local coordinate ξ shown in Fig. 2–64.

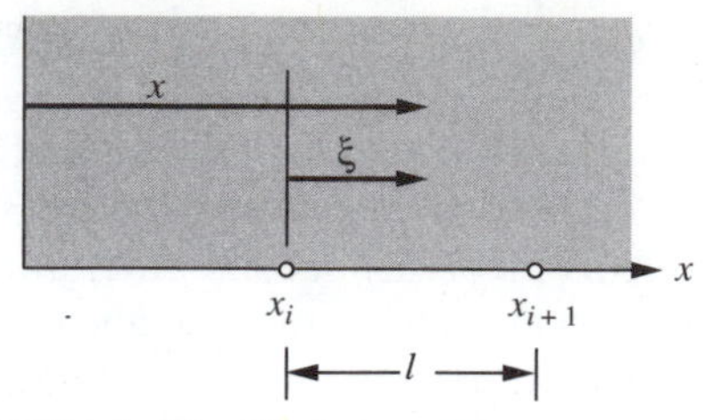

FIGURE 2–64 Global and local coordinates

In terms of the local variable $0 \le \xi \le l_e$, we write

$$w(\xi) = c_0 + c_1\xi + c_2\xi^2 + c_3\xi^3 \tag{2.49}$$

The four constants are determined by satisfying

$$w(0) = w_1 \qquad w(l_e) = w_2$$

$$w'(0) = \theta_1 \qquad w'(l_e) = \theta_2$$

where w_1, w_2, θ_1, and θ_2 are indicated in Fig. 2–65.

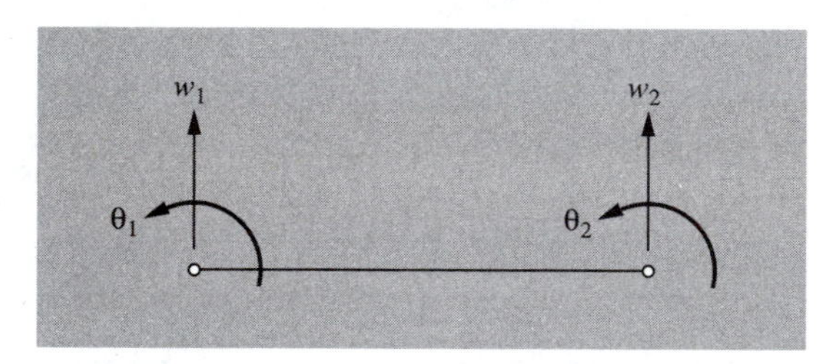

FIGURE 2–65 Nodal degrees of freedom

Solving for the constants and back substituting, Eq. (2.48) can be put in the form

$$\begin{aligned} w(\xi) &= w_1N_1(\xi) + \theta_1N_2(\xi) + w_2N_3(\xi) + \theta_2N_4(\xi) \\ &= \mathbf{w_e^T N}(\xi) \end{aligned} \tag{2.50}$$

where

$$\mathbf{w_e^T} = [w_1 \quad \theta_1 \quad w_2 \quad \theta_2]$$

and

$$N_1(\xi) = 1 - 3\left(\frac{\xi}{l_e}\right)^2 + 2\left(\frac{\xi}{l_e}\right)^3$$

$$N_2(\xi) = \xi\left(1 - \frac{2\xi}{l_e} + \left(\frac{\xi}{l_e}\right)^2\right)$$

$$N_3(\xi) = 3\left(\frac{\xi}{l_e}\right)^2 - 2\left(\frac{\xi}{l_e}\right)^3$$

$$N_4(\xi) = \xi\left(\left(\frac{\xi}{l_e}\right)^2 - \frac{\xi}{l_e}\right)$$

The N_i, frequently referred to as **Hermite** or **cubic interpolating polynomials**, are pictured in Fig. 2–66.

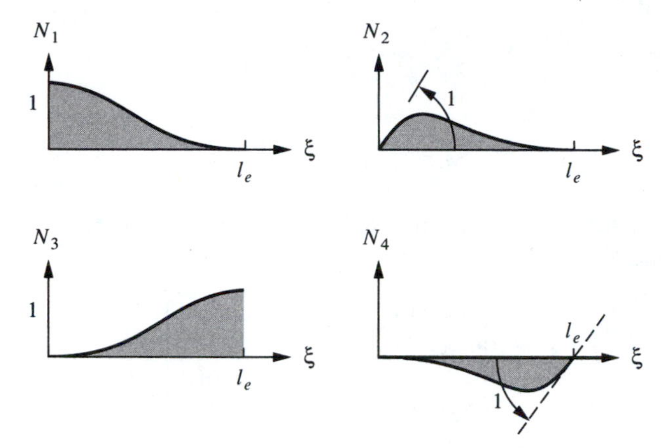

FIGURE 2–66 Cubic or Hermite interpolating functions

Elemental formulation. We will develop the elemental matrices for the beam problem using the potential energy given in Eq. (2.47), that is,

$$V(w) = \int_0^L \left(\frac{EI(w'')^2}{2} - q(x)w(x) \right) dx$$

Consistent with a typical discretization, we write the potential energy as the sum of the potential energies over all the elements according to

$$V(w) = \sum_e \int_{x_i}^{x_{i+1}} \left(\frac{EI(w'')^2}{2} - q(x)w(x) \right) dx$$

which we will express as

$$V(w) = \sum_e U_e + \sum_e \Omega_e$$

where U_e represents an elemental potential energy and Ω_e an elemental external potential energy. It follows that

$$2U_e = \int_{x_i}^{x_{i+1}} EI(w'')^2 dx = \int_0^{l_e} EI(w'')^2 d\xi \tag{2.51}$$

$$-\Omega_e = \int_{x_i}^{x_{i+1}} w(x)q(x)dx = \int_0^{l_e} w(\xi)q(\xi)d\xi \tag{2.52}$$

where ξ is a local variable indicated in Fig. 2–64.

Evaluation of $\mathbf{U_e}$. From Eq. (2.50) it follows that

$$w'' = \mathbf{w}_\mathbf{e}^\mathrm{T}\mathbf{N}'' = \mathbf{N}''^\mathrm{T}\mathbf{w}_\mathbf{e}$$

Substitution into Eq. (2.51) yields

$$2U_e = \int_0^{l_e} EI(w'')^2 d\xi = \int_0^{l_e} \mathbf{w_e^T N''} EI \mathbf{N''^T w_e} d\xi$$
$$= \mathbf{w_e^T k_e w_e}$$

where

$$\mathbf{k_e} = \int_0^{l_e} \mathbf{N''} EI \mathbf{N''^T} d\xi$$

is the elemental stiffness matrix. It is left to the exercises to show that for constant EI, $\mathbf{k_e}$ becomes

$$\mathbf{k_e} = \begin{bmatrix} 12 & 6l_e & -12 & 6l_e \\ & 4l_e^2 & -6l_e & 2l_e^2 \\ & & 12 & -6l_e \\ \text{symm} & & & 4l_e^2 \end{bmatrix} \frac{EI}{l_e^3} \tag{2.53}$$

The elements of $\mathbf{k_e}$ can be interpreted as classical stiffnesses in the structural sense in that $[k_e]_{ij}$ = the force at node i due to a unit displacement at node j with all other nodal displacements constrained to zero.

Evaluation of Ω_e. Substitution of the elemental expression for w into Ω_e yields

$$-\Omega_e = \int_0^{l_e} \mathbf{w_e^T N} q(\xi) d\xi$$
$$= \mathbf{w_e^T q_e}$$

where

$$\mathbf{q_e} = \int_0^{l_e} \mathbf{N} q(\xi) d\xi$$

is called an elemental load vector. Assuming that $q(\xi)$ is a linear function of position within the element according to

$$q(\xi) = \left(1 - \frac{\xi}{l_e}\right) q_L + \left(\frac{\xi}{l_e}\right) q_R$$

$\mathbf{q_e}$ becomes

$$\mathbf{q_e} = \frac{l_e}{20} \begin{bmatrix} 7q_L + 3q_R \\ \dfrac{l_e(3q_L + 2q_R)}{3} \\ 3q_L + 7q_R \\ \dfrac{-l_e(2q_L + 3q_R)}{3} \end{bmatrix} = \begin{bmatrix} Q_1 \\ m_1 \\ Q_2 \\ m_2 \end{bmatrix} \tag{2.54}$$

These nodal loads are indicated in Fig. 2–67.

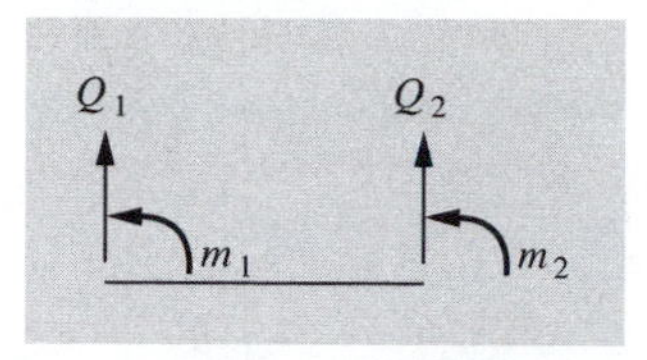

FIGURE 2–67 Nodal loads for linearly varying q

Assembly. Assembly of the elemental stiffness matrices takes place in the expected manner with an overlap of 2×2 matrices corresponding to requiring that both the displacement (w) and slope (θ) be continuous. The situation for a two-element model is indicated in Fig. 2–68. The stiffnesses for the first element are indicated by a's and for the second element by b's.

$$\mathbf{K_G} = \sum_e \mathbf{k_G} = \begin{bmatrix} a_{11} & a_{12} & a_{13} & a_{14} & 0 & 0 \\ a_{21} & a_{22} & a_{23} & a_{24} & 0 & 0 \\ a_{31} & a_{32} & a_{33}+b_{11} & a_{34}+b_{12} & b_{13} & b_{14} \\ a_{41} & a_{42} & a_{43}+b_{21} & a_{44}+b_{22} & b_{23} & b_{24} \\ 0 & 0 & b_{31} & b_{32} & b_{33} & b_{34} \\ 0 & 0 & b_{41} & b_{42} & b_{43} & b_{44} \end{bmatrix}$$

FIGURE 2–68 Stiffness assembly for a two-element model

The global load vector for the same two-element model would appear as

$$\mathbf{F_G} = \sum_e \mathbf{f_G} = \begin{bmatrix} (Q_1)_1 \\ (m_1)_1 \\ ((Q_2)_1 + (Q_1)_2) \\ ((m_2)_1 + (m_1)_2) \\ (Q_2)_2 \\ (m_2)_2 \end{bmatrix}$$

where the subscripts outside the bracket refer to the element number. Accounting for the proper sign, any concentrated loads or moments would need to be included in the global load vector.

Constraints and solution. Constraints would be enforced on the displacement or slope as necessary at any appropriate node and the solution then obtained.

Derived variables. The derived variables for this problem are the two higher-order (($n-2$)nd and ($n-1$)st) derivatives w'' and w'''. They are related to the force variables, the moment M and shear V, given respectively by

$$M = EIw'' \qquad \text{and} \qquad V = -EIw'''$$

The moments and shears at the left and right ends of the element are given by the basic equation

$$\mathbf{F_e} = \mathbf{k_e w_e}$$

or

$$\begin{bmatrix} V_1 \\ M_1 \\ V_2 \\ M_2 \end{bmatrix} = \mathbf{k_e} \begin{bmatrix} w_1 \\ \theta_1 \\ w_2 \\ \theta_2 \end{bmatrix}$$

Alternatively, the approximate expressions for the moment and shear can be computed as functions of position within the element according to

$$M = EIw'' = EI\mathbf{w}_\mathbf{e}^\mathrm{T}\mathbf{N}''$$

and

$$-V = EIw''' = EI\mathbf{w}_\mathbf{e}^\mathrm{T}\mathbf{N}'''$$

Example 2.14.

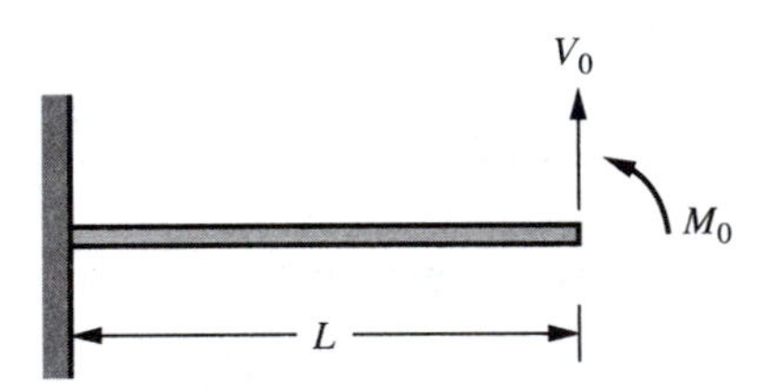

FIGURE 2–69 Cantilever beam with end shear and moment

Consider the elementary problem, indicated in Fig. 2–69, of a cantilever beam loaded by a moment M_0 and a shear force V_0. Note that $l_e = L$. With only one element and the two nodal loads, the unconstrained global equations are easily seen to be

$$\begin{bmatrix} 12 & 6L & -12 & 6L \\ & 4L^2 & -6L & 2L^2 \\ & & 12 & -6L \\ \text{symm} & & & 4L^2 \end{bmatrix} \begin{bmatrix} w_1 \\ \theta_1 \\ w_2 \\ \theta_2 \end{bmatrix} = \begin{bmatrix} 0 \\ 0 \\ V_0 \\ M_0 \end{bmatrix} \frac{L^3}{EI}$$

The *constraints* are $w_1 = \theta_1 = 0$. *Solving* the resulting 2×2 system of equations yields

$$w_2 = \frac{M_0 L^2}{2EI} + \frac{V_0 L^3}{3EI}$$

$$\theta_2 = \frac{M_0 L}{EI} + \frac{V_0 L^2}{2EI}$$

from which it follows using Eq. (2.49) that

$$w(x) = \frac{M_0 x^2}{2EI} + \frac{V_0 x^2 (3L - x)}{6EI}$$

which is the exact solution. This is to be expected for this problem, since when there is no loading interior to the (single) element, the exact solution is a cubic contained within the displacements (Eq. (2.49)) used for the finite element solution.

The nodal forces are computed according to

$$\mathbf{F_e} = \mathbf{k_e w_e}$$

$$= \begin{bmatrix} 12 & 6L & -12 & 6L \\ & 4L^2 & -6L & 2L^2 \\ & & 12 & -6L \\ \text{symm} & & & 4L^2 \end{bmatrix} \begin{bmatrix} 0 \\ 0 \\ w_2 \\ \theta_2 \end{bmatrix} \frac{L^3}{EI}$$

$$= \begin{bmatrix} -V_0 \\ -M_0 - V_0 L \\ V_0 \\ M_0 \end{bmatrix}$$

This can easily be verified to be correct.

Example 2.15.

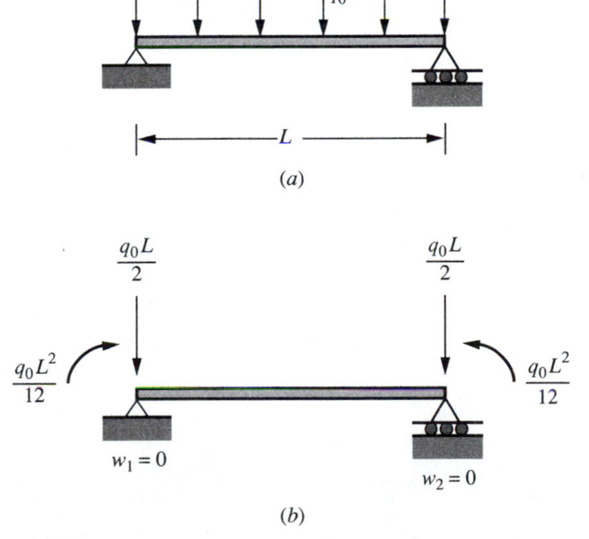

FIGURE 2–70 Beam, nodal loads, and constraints for a uniformly loaded simply supported beam

Consider the elementary problem of the simply supported uniformly loaded beam as indicated in Fig. 2–70*a*. With q_0 negative, the assembled equations for a one-element model are

$$\begin{bmatrix} 12 & 6L & -12 & 6L \\ & 4L^2 & -6L & 2L^2 \\ & & 12 & -6L \\ \text{symm} & & & 4L^2 \end{bmatrix} \begin{bmatrix} w_1 \\ \theta_1 \\ w_2 \\ \theta_2 \end{bmatrix} = \begin{bmatrix} \dfrac{-q_0L}{2} \\ \dfrac{-q_0L^2}{12} \\ \dfrac{-q_0L}{2} \\ \dfrac{q_0L^2}{12} \end{bmatrix} \frac{L^3}{EI}$$

with the nodal loads shown in Fig. 2–70*b*. Applying the constraints indicated in the figure yields

$$\begin{bmatrix} 4 & 2 \\ 2 & 4 \end{bmatrix} \begin{bmatrix} \theta_1 \\ \theta_2 \end{bmatrix} = \begin{bmatrix} -1 \\ 1 \end{bmatrix} \frac{q_0L^3}{12EI}$$

from which

$$\theta_2 = -\theta_1 = \frac{q_0L^3}{24EI}$$

Constructing $w(x)$ yields

$$w(x) = \frac{q_0x}{L}\left(\frac{x}{L} - 1\right)\frac{L^4}{24EI}$$

which has a maximum value at $x = L/2$ of $-4q_0L^4/384EI$ versus the corresponding exact value of $-5q_0L^4/384EI$. The slopes at the ends are $\pm q_0L^3/24EI$, which coincide with the exact values.

The derived variables are easily computed to be

$$M = EIw'' = \frac{q_0L^2}{12}$$

and

$$V = -EIw''' = 0$$

These are plotted versus the corresponding exact values in Figs. 2–71*a* and *b*.

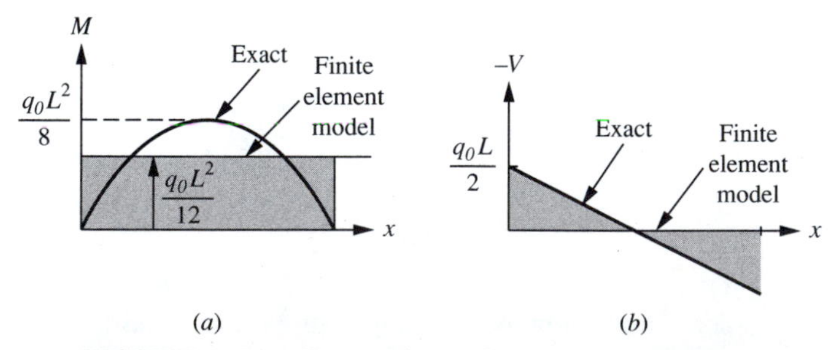

FIGURE 2–71 Comparison of moments and shears

The one-element solution is decidedly unsatisfactory as far as the derived variables are concerned. Even the accuracy of the primary variables w and θ is relatively poor. In the next example we will investigate the improvements possible by using more elements.

Example 2.16.

Consider the problem of Example 2.15 using two elements. With $l_e = L/2$, the assembled unconstrained equations can be written

$$\begin{bmatrix} 12 & 6l_e & -12 & 6l_e & 0 & 0 \\ 6l_e & 4l_e^2 & -6l_e & 2l_e^2 & 0 & 0 \\ -12 & -6l_e & 24 & 0 & -12 & 6l_e \\ 6l_e & 2l_e^2 & 0 & 8l_e^2 & -6l_e & 2l_e^2 \\ 0 & 0 & -12 & -6l_e & 12 & -6l_e \\ 0 & 0 & 6l_e & 2l_e^2 & -6l_e & 4l_e^2 \end{bmatrix} \begin{bmatrix} w_1 \\ \theta_1 \\ w_2 \\ \theta_2 \\ w_3 \\ \theta_3 \end{bmatrix} = \begin{bmatrix} \dfrac{-Q_0 l_e}{2} \\ \dfrac{-Q_0 l_e^2}{12} \\ -Q_0 l_e \\ 0 \\ \dfrac{-Q_0 l_e}{2} \\ \dfrac{Q_0 l_e^2}{12} \end{bmatrix}$$

where $Q_0 = q_0 l_e^3/EI$. The constraints are clearly $w_1 = w_3 = 0$, resulting in

$$\begin{bmatrix} 4l_e^2 & -6l_e & 2l_e^2 & 0 \\ -6l_e & 24 & 0 & 6l_e \\ 2l_e^2 & 0 & 8l_e^2 & 2l_e^2 \\ 0 & 6l_e & 2l_e^2 & 4l_e^2 \end{bmatrix} \begin{bmatrix} \theta_1 \\ w_2 \\ \theta_2 \\ \theta_3 \end{bmatrix} = \begin{bmatrix} \dfrac{-Q_0 l_e^2}{12} \\ Q_0 l_e \\ 0 \\ \dfrac{-Q_0 l_e^2}{12} \end{bmatrix}$$

with solutions

$$\theta_1 = \frac{-q_0 l_e^3}{3EI}$$

$$w_2 = \frac{-5q_0 l_e^4}{24EI}$$

$$\theta_2 = 0$$

$$\theta_3 = \frac{q_0 l_e^3}{3EI}$$

For the first element it follows that

$$w_{e1}(x) = \theta_1 N_2 + w_2 N_3 = -\frac{q_0 L^4}{384EI}\left(16\frac{x}{L} - 4\left(\frac{x}{L}\right)^2 - 16\left(\frac{x}{L}\right)^3\right)$$

from which

$$w_{e1}\left(\frac{L}{2}\right) = \frac{-5q_0 L^4}{384EI}$$

and

$$w'_{e1}(0) = \frac{-q_0 L^3}{24EI}$$

both of which coincide with the corresponding exact values. The finite element solution is symmetric about $x = L/2$, so that any necessary information about the solution and its derivatives is contained in w_{e1}. The derived variables are computed as

$$M = EIw'' = -\frac{q_0L^2}{48}\left(-1 - \frac{12x}{L}\right)$$

$$-V = EIw''' = \frac{q_0 l}{4}$$

Discussion. The displacement (w) predicted by the finite element model coincides with the exact displacement at the middle of the beam. The slopes ($\theta = w'$) given by the finite element solution are also exact at the ends of the beam. The derived variables are plotted versus the exact values for the first element in Figs. 2–72*a* and *b*.

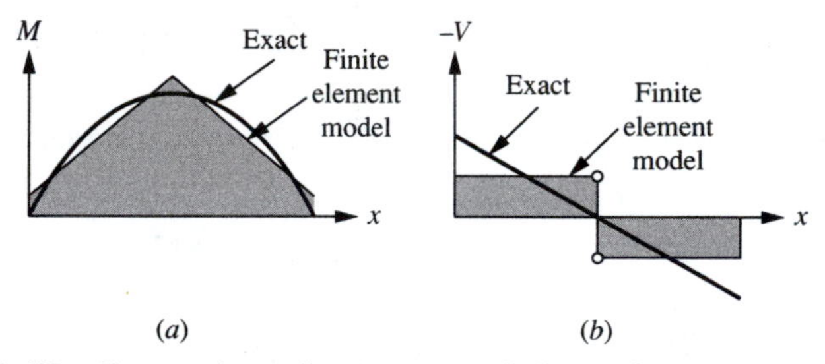

FIGURE 2–72 Comparison of moments and shears for a two-element model

It is clear from Figs. 2–72*a* and *b* that the accuracy of the variables associated with the higher-order derivatives suffers. In particular, many elements may be necessary to sufficiently resolve the shear due to the fact that the shear given by the finite element solution is constant within an element. Smoothing, the process of connecting the midpoint values as discussed previously, can be used to advantage here. Generally as the order of the loading $q(x)$ increases, so also does the number of elements required to obtain accurate estimates of the moments and shears transmitted. This is particularly significant in that the design of a structure is usually dictated by the values of the force resultants (M and V) rather than by the kinematical variables (w and θ), making it necessary to have accurate estimates of M and V.

2.11 ERRORS AND CONVERGENCE

Presented in this section is a brief and heuristic discussion of errors and convergence for the one-dimensional boundary value problem. The results from some of the examples of Sections 2.4 and 2.9 are analyzed using these ideas.

For the one-dimensional boundary value problem, the errors are of two types:

1. Errors due to the numerical evaluation of the integrals and the use of the computer for performing the numerical calculations.
2. Errors due to the approximate character of the finite element formulation.

In what follows we will assume that the integrals have been evaluated exactly, and that there is no error resulting from the use of the computer or from the hand calculations, so that all the error can be considered to be of type 2.

The basic tool for our analysis is the Taylor's series. The exact solution (which is generally unknown) and the approximate finite element solution will both be expanded in Taylor's series, and the difference analyzed on a subinterval corresponding to an element, to assess the character of the error associated with the approximate nature of the finite element method.

To make this definite, consider for simplicity the case where the finite element solution has been interpolated linearly. On an interval (x_i, x_{i+1}), the finite element solution has the representation

$$U_e(x) = c_0 + c_1(x - x_i)$$

that is, the first two terms of the Taylor's series' expansion of the linear function about the point x_i. Also assume that the exact solution has a Taylor's series expansion about the point x_i given by

$$u(x) = \sum_0^{\infty} a_n(x - x_i)^n$$

where $a_n = f^{(n)}(x_i)/n!$. The difference

$$e_{1,0}(x) = u(x) - U_e(x) = \sum_0^{\infty} a_n(x - x_i)^n - c_0 - c_1(x - x_i)$$

is an error associated with the finite element formulation. Generally the notation $e_{m,n}$ is meant to indicate the error in the nth derivative of a finite element solution that has been interpolated using mth order polynomials. Using the remainder associated with the Taylor's series, it follows that

$$\begin{aligned} e_{1,0}(x) &= a_0 + a_1(x - x_1) + \frac{u''(x*)(x_{i+1} - x_i)^2}{2} + \cdots - c_0 - c_1(x - x_i) \\ &= a_0 + a_1(x - x_i) + \frac{u''(x*)h^2}{2} + \cdots - c_0 - c_1(x - x_i) \end{aligned}$$

where $x_i \leq x* \leq x_{i+1}$. It is assumed that for a small enough mesh, $a_0 \approx c_0$ and $a_1 \approx c_1$, so that the constant and linear terms essentially cancel, leaving

$$e_{1,0}(x) = \frac{u''(x*)h^2}{2} = O(h^2)$$

The "big O" symbol, read as "is of the order of" means generally that

$$O(h^2) = (\text{constant})h^2$$

that is, that for $h = x_{i+1} - x_i$ "small enough," the error is proportional to h^2. For example, if h is halved between two meshes, the error $e_{1,0}$ is reduced by roughly a factor of four. In a completely similar and obvious fashion, it can be shown that the error in the derivative u' behaves with respect to the mesh size according to

$$e_{1,1}(x) = O(h)$$

so that halving the step size roughly halves the error of the derivative for the linearly interpolated finite element solution.

It also follows that for a finite element solution using polynomial interpolation of degree p,

$$e_{p,0} = Ch^{p+1}$$
$$e_{p,1} = Dh^{(p+1)-1} = Dh^{p}$$

where C and D are constants, and that the error in the mth derivative behaves according to

$$e_{p,m} = Eh^{(p+1)-m}$$

E being also a constant. General results of this type can be quite useful in estimating errors and assessing the convergence of finite element solutions.

A useful application of results of this type is the so-called Richardson's extrapolation [Atkinson, 13]. The error of a finite element solution interpolated in terms of pth order polynomials is

$$e_{p,0} = u - U_n = O(h^{p+1})$$

or stated differently,

$$u = U_n + O(h^{p+1}) \approx U_n + Ch^{p+1}$$

where U_n represents the solution at a point in the region obtained on the basis of an n-element mesh. For the finite element solution corresponding to the same problem with each of the element lengths halved,

$$u = U_{2n} + C\left(\frac{h}{2}\right)^{p+1}$$

U_{2n} represents the solution at the same point based on halving each of the element lengths of the previous mesh. Eliminating the unknown constant C yields

$$u = \frac{2^{p+1}U_{2n} - U_n}{2^{p+1} - 1}$$

as an improved estimate of the value of the solution u at the point in question. An improved estimate of the value of the mth derivative would be

$$u^{(m)} = \frac{2^{p+1-m}U_{2n}^{(m)} - U_n^{(m)}}{2^{p+1-m} - 1}$$

This idea can be extended and systematized into an algorithm known as the Romberg integration [Yakowitz and Szidarovszky, 14]. For a point common to all of the meshes resulting from successive halving of the element length, a sequence of estimates $U_{1,1}$, $U_{2,1}$, $U_{3,1}$, ... is generated. These estimates are arranged in the first column of a triangular array as indicated in Fig. 2–73.

$$
\begin{array}{llll}
U_{1,1} & & & \\
U_{2,1} & U_{2,2} & & \\
U_{3,1} & U_{3,2} & U_{3,3} & \\
U_{4,1} & U_{4,2} & U_{4,3} & U_{4,4} \\
\cdot & \cdot & \cdot & \cdot
\end{array}
$$

FIGURE 2–73 Triangular array of improved estimates

The first subscript indicates the row and the second subscript the column corresponding to the position of the term in the triangular array. The second and subsequent columns are computed according to

$$U_{m,n} = \frac{2^{p+1}U_{m,n-1} - U_{m-1,n-1}}{2^{p+1} - 1}$$

The calculations are continued until the triangular form is completed. The results towards the bottom right of the triangular array may represent increasingly accurate estimates of the solution at the point in question, and can often indicate the value to which the solution is converging. This indication can be to a different number of significant digits, depending upon the number of entries in the first column, and the character of the problem being examined. Derivatives can also be handled in this way, being careful to use the correct integer $p + 1 - m$ in the construction of the Richardson's extrapolations. It is often more difficult to easily and confidently draw conclusions regarding the derivatives using the Romberg integration.

The Romberg integration is valid in the limit as h approaches zero; results based on large h may have to be discarded in order for the Romberg integration to make sense. Practically, this may mean that one or more of the upper diagonal(s) of the triangular array need to be discarded from the process.

Example 2.17.

For the problem of Section 2.4.3, a one-dimensional heat transfer using linear interpolation, consider the Romberg integration as applied to the results for $u(1/2)$, with the first column representing results from 2-, 4-, 8-, and 16-element models, (i.e., with the element size being halved between successive models). The type of interpolation used is linear, that is, $p = 1$, so that a single computation using Richardson's extrapolation is

$$U_{m,n} = \frac{4U_{m,n-1} - U_{m-1,n-1}}{3}$$

The results appear in Fig. 2–74.

0.165680			
0.200699	0.212372		
0.207814	0.210186	0.209459	
0.209511	0.210077	0.210040	0.210234

FIGURE 2–74 Romberg integration for $u(1/2)$

These numbers indicate fairly strongly that, to three significant figures, an improved estimate of the value of $u(0.5)$ is 0.210. The corresponding exact value is $u(0.5) = 0.210069$.

For the same example, consider also a Romberg integration applied to the value of the derivative at $s = 1$, a point common to all the meshes. The derivatives at $s = 1$ are easily computed from the values at $s = 1$ and the adjacent node to the left; they appear in the first column of the array below. The additional entries in the triangular array are to be computed according to

$$U'_{m,n} = 2U'_{m,n-1} - U'_{m-1,n-1}$$

leading to Fig. 2–75.

−0.248520			
−0.148628	−0.048736		
−0.105160	−0.061692	−0.074648	
−0.084496	−0.063832	−0.065972	−0.057296

FIGURE 2–75 Romberg integration for $u'(1)$

Here the limiting value is not nearly as clear. At best one could assert, on the basis of the middle two entries of the bottom row, that $u'(1) \approx -0.065$, illustrating that the Romberg integration does not always converge strongly to the desired result. The corresponding exact value is $u'(1) = -0.064259$.

Consider also an analysis of the derivative at $s = 0$, easily computed for each of the successive meshes from the values of the solution at $s = 0$ and the adjacent node to the right, which appear in the first column of Fig. 2–76.

−1.668640			
−2.219652	−2.770664		
−2.620336	−3.021020	−3.271376	
−2.867760	−3.115184	−3.209348	−3.147320

FIGURE 2–76 Romberg integration for $u'(0)$

Again, the process does not strongly indicate the value to which $u'(0)$ converges. However, the data do suggest that $u'(0) \approx -3.1$, a better estimate than any of the entries in the first column. The corresponding exact value is −3.156396.

Example 2.18.

Consider the problem from Section 2.9.3, a one-dimensional heat transfer problem. The results of that example were obtained using quadratic interpolation. Any single calculation using the Richardson's extrapolation for results from a quadratically interpolated finite element solution ($p = 2$) is carried out according to

$$U_{m,n} = \frac{8U_{m,n-1} - U_{m-1,n-1}}{7}$$

and the first derivative according to

$$U'_{m,n} = \frac{4U'_{m,n-1} - U'_{m-1,n-1}}{3}$$

Using the $u(0.5)$ results from one-, two-, and four-element models, there results

0.203883			
0.211780	0.212908		
0.210167	0.209937	0.209513	
0.210075	0.210062	0.210080	0.210161

FIGURE 2–77 Romberg integration for $u(0.5)$

These results imply quite strongly that, to four significant digits, the correct value of $u(1/2) = 0.2101$—a bit sharper result than for the linearly interpolated case, as would be expected. Again the exact value is 0.210069.

With each of the entries in the first column calculated from

$$u_e'(l_e) = \mathbf{u}_e^T\mathbf{N}'(l_e) = \frac{u_{N-1} - 4u_N + 3u_{N+1}}{l_e}$$

the Romberg integration for the derivative at $s = 1$ appears as in Fig. 2–78.

+0.446605			
....			
−0.015838	−0.169986		
	....		
−0.057224	−0.071019	−0.038030	
		....	
−0.063000	−0.064925	−0.062894	−0.071182

FIGURE 2–78 Romberg integration for $u'(1)$

Here it is advisable to exclude some of the data for which the element size was not "small enough," mainly the data corresponding to the one-element quadratically interpolated model. Thus, based on the entries below the dotted lines, it is concluded that, to three significant digits, $u'(1) \approx -(0.063 - 0.064)$. Again, the corresponding value is $u'(1) = -0.064259$.

In a similar fashion, with each of the entries in the first column calculated from

$$u_e'(0) = \mathbf{u}_e^T\mathbf{N}'() = \frac{-3u_1 + 4u_2 - u_3}{l_e}$$

the Romberg integration for $u'(0)$ appears as in Fig. 2–79.

−2.271847			
−2.788224	−2.961148		
−3.034792	−3.116981	−3.168925	
−3.121176	−3.149971	−3.160968	−3.158315

FIGURE 2–79 Romberg integration for $u'(0)$

Based upon the bottom row, it is reasonable to conclude that to three significant figures, $u'(0) \approx -(3.15 - 3.16)$, compared to the exact value of $u'(0) = -3.156396$.

The conclusions reached by applying the Romberg integration to the raw data from the linearly and quadratically interpolated finite element models of the one-dimensional heat transfer example are clearly consistent, and in all cases provide valuable additional information regarding the convergence of the finite element method.

2.12 CLOSURE

After the student has mastered the details of the basic ingredients of the finite element method, the actual construction of a finite element model for the solution of linear second-order boundary value problems of the type given in Eqs. (2.1) can be reduced to these steps:

1. Put the equation and boundary conditions in standard form, thus identifying p, q, f, a, b, α, β, A, and B.
2. Decide on an initial mesh and type of interpolation with the resulting definition of the nodes and elements.
3. Evaluate and assemble the matrices $\mathbf{p_e}$, $\mathbf{q_e}$, $\mathbf{f_e}$, **BT**, and **bt**.
4. Constrain and solve the assembled equations.
5. Compute the derived variables.
6. Repeat these steps as needed with finer meshes so as to get an idea of the accuracy and convergence. Perform numerical analyses using the Romberg integration if necessary or desired.

This general approach can be executed by hand or using a code if one is available. Such a code is fairly easy to generate and is certainly desirable if problems of this type arise frequently. An entirely similar sequence of steps can be given for the corresponding eigenvalue type problem.

It is important to note that for the class of boundary value problems considered in this chapter, the *form* of the matrices $\mathbf{p_e}$, $\mathbf{q_e}$, $\mathbf{f_e}$, **BT**, and **bt** is the same regardless of the type of interpolation selected, making it possible to outline a sequence of steps that is essentially independent of the discretization and interpolation. As seen in Fig. 2–80, the FEM acts as a sort of operator, converting the boundary value problem stated on the left into the corresponding set of linear algebraic equations on the right. The composition of the global matrices is clearly indicated by the arrows in Fig. 2–80, which can be thought of as indicating the effect that each of the parameters p, q, f, α, β, A, and B has on the global matrices $\mathbf{K_G}$ and $\mathbf{F_G}$.

For linearly interpolated models, when the functions p, q, and f are replaced on the element by their linear interpolants, the elemental matrices reduce to

$$\mathbf{p_e} \simeq \frac{p_{\text{avg}}}{l_e}\begin{bmatrix} 1 & -1 \\ -1 & 1 \end{bmatrix}$$

$$\mathbf{q_e} \simeq \frac{l_e}{12}\begin{bmatrix} 3q_i + q_{i+1} & q_i + q_{i+1} \\ q_i + q_{i+1} & q_i + 3q_{i+1} \end{bmatrix}$$

$$\mathbf{f_e} \simeq \frac{l_e}{6}\begin{bmatrix} 2f_i + f_{i+1} \\ f_i + 2f_{i+1} \end{bmatrix}$$

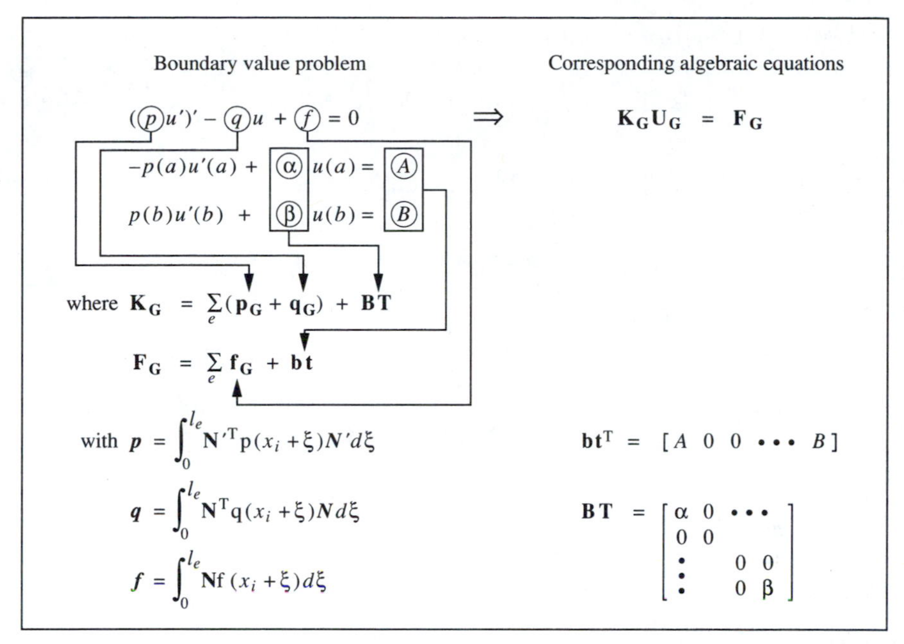

FIGURE 2–80 Conversion of the boundary value problem into the corresponding algebraic equations

These expressions for the elemental matrices are exact only if p, q, and f are polynomials of degree one or less. In all other instances, they are approximate expressions with accuracy that depends on how well the linear approximations fit the data on the intervals corresponding to each of the elements.

For a quadratically interpolated model, with p, q, and f replaced by their quadratic interpolants, the elemental matrices are

$$[p_e]_{11} = \frac{1}{30l_e}(37p_i + 36p_{i+1} - 3p_{i+2})$$

$$[p_e]_{12} = \frac{1}{30l_e}(-44p_i - 32p_{i+1} - 4p_{i+2})$$

$$[p_e]_{13} = \frac{1}{30l_e}(7pi - 4p_{i+1} + 7p_{i+2})$$

$$[p_e]_{22} = \frac{1}{30l_e}(48p_i + 64p_{i+1} + 48p_{i+2})$$

$$[p_e]_{23} = \frac{1}{30l_e}(-4p_i - 32p_{i+1} - 44p_{i+2})$$

$$[p_e]_{33} = \frac{1}{30l_e}(-3p_i + 36p_{i+1} + 37p_{i+2})$$

with $[p_e]_{ij} = [p_e]_{ji}$,

$$[q_e]_{11} = \frac{l_e}{420}(39q_i + 20q_{i+1} - 3q_{i+2})$$

$$[q_e]_{12} = \frac{l_e}{420}(20q_i + 16q_{i+1} - 8q_{i+2})$$

$$[q_e]_{13} = \frac{l_e}{420}(-3q_i - 8q_{i+1} - 3q_{i+2})$$

$$[q_e]_{22} = \frac{l_e}{420}(16q_i + 192q_{i+1} + 16q_{i+2})$$

$$[q_e]_{23} = \frac{l_e}{420}(-8q_i + 16q_{i+1} + 20q_{i+2})$$

$$[q_e]_{33} = \frac{l_e}{420}(-3q_i + 20q_{i+1} + 39q_{i+2})$$

with $[q_e]_{ij} = [q_e]_{ji}$, and

$$[f_e]_1 = \frac{l_e}{30}(4f_i + 2f_{i+1} - f_{i+2})$$

$$[f_e]_2 = \frac{l_e}{30}(2f_i + 16f_{i+1} + 2f_{i+2})$$

$$[f_e]_3 = \frac{l_e}{30}(-f_i + 2f_{i+1} + 4f_{i+2})$$

These expressions for the elemental matrices $\mathbf{p_e}$, $\mathbf{q_e}$, and $\mathbf{f_e}$ are exact only if p, q, and f are polynomials of degree two or less. In all other instances, they are approximate expressions with accuracy that depends on how well the parabolas fit the data on the intervals corresponding to each of the elements.

REFERENCES

1. Boyce, W. E., and R. C. DiPrima: *Elementary Differential Equations and Boundary Value Problems*, John Wiley, New York, 1986.
2. Galerkin, B. G.: "Rods and Plates. Series Occurring in Various Questions Concerning the Elastic Equilibrium of Rods and Plates," *Engineers Bulletin* (*Vestnik inshenerov*), 19, pp. 897–908, 1915.
3. Moan, T.: "On the Local Distribution of Errors by Finite Element Approximations," in *Theory and Practice in Finite Element Structural Analysis*, eds. Y. Yamada and R. H. Gallagher, University of Tokyo Press, Tokyo, 1973.
4. Smith, B. T., et al.: *Lecture Notes in Computer Science, Matrix Eigensystem Routines—EISPACK Guide,* Springer-Verlag, Berlin, 1976.
5. Garbow, G. S., et al.: *Lecture Notes in Computer Science, Matrix Eigensystem Routines—EISPACK Guide Extension*, Springer-Verlag, Berlin, 1977.
6. Reddy, J. N., and M. L. Rasmussen: *Advanced Engineering Analysis*, John Wiley, New York, 1982.

7. Weinstock, R.: *Calculus of Variations*, McGraw-Hill, New York, 1952.
8. Gelfand, I. M., and S. V. Fomin: *Calculus of Variations*, Prentice Hall, Englewood Cliffs, New Jersey, 1963.
9. Lanczos, C.: *The Variational Principles of Mechanics*, University of Toronto Press, Toronto, 1960.
10. Hildebrand, F. B.: *Methods of Applied Mathematics*, Prentice Hall, Englewood Cliffs, New Jersey, 1965.
11. Stakgold, I.: *Boundary Value Problems of Mathematical Physics*, vol. II, Macmillan, New York, 1967.
12. Reddy, J. N.: *Applied Functional Analysis and Variational Methods in Engineering*, McGraw-Hill, New York, 1986.
13. Atkinson, K.: *Elementary Numerical Analysis*, John Wiley, New York, 1985.
14. Yakowitz, S., and F. Szidarovszky: *An Introduction to Numerical Computations*, Macmillan, New York, 1986.

GENERAL REFERENCES

Axelsson, O., and V. A. Barker: *Finite Element Solution of Boundary Value Problems—Theory and Computation*, Academic Press, Orlando, 1984.

Becker, E. B., G. F. Carey, and J. T. Oden: *Finite Elements—An Introduction,* vol. I, Prentice-Hall, Englewood Cliffs, New Jersey, 1981.

Burnett, D. S.: *Finite Element Analysis, From Concepts to Applications*, Addison-Wesley, Reading, Massachusetts, 1987.

Carey, G. F., and J. T. Oden.: *Finite Elements—A Second Course,* vol. II, Prentice Hall, Englewood Cliffs, New Jersey, 1983.

——*Finite Elements—Computational Aspects,* vol. III, Prentice Hall, Englewood Cliffs, New Jersey, 1984.

Chung, T. J.: *Finite Element Analysis in Fluid Dynamics*, McGraw-Hill, New York, 1978.

Connor, J. C., and C. A. Brebbia: *Finite Element Techniques for Fluid Flow*, Butterworths, London, 1976.

Cook, R. D.: *Concepts and Applications of Finite Element Analysis*, John Wiley, New York, 1981.

Desai, C. S.: *Elementary Finite Element Method*, Prentice Hall, Englewood Cliffs, New Jersey, 1979.

Desai, C. S., and J. F. Abel: *Introduction to the Finite Element Method*, Von Nostrand Reinhold, New York, 1972.

Gallagher, R. H.: *Finite Element Analysis, Fundamentals*, Prentice-Hall, Englewood Cliffs, New Jersey, 1975.

Huebner, K. H., and E. A. Thornton: *The Finite Element Method for Engineers*, John Wiley, New York, 1982.

Hughes, T. J. R.: *The Finite Element Method, Linear Static and Dynamic Finite Element Analysis*, Prentice Hall, Englewood Cliffs, New Jersey, 1987.

Kikuchi, N.: *Finite Element Methods in Mechanics*, Cambridge University Press, Cambridge UK, 1986.

Livesley, R. K.: *Finite Elements: An Introduction for Engineers*, Cambridge University Press, Cambridge UK, 1983.

Martin, H. C., and G. F. Carey: *Introduction to Finite Element Analysis—Theory and Application*, McGraw-Hill, New York, 1973.

Norrie, D. H., and G. de Vries: *The Finite Element Method: Fundamentals and Applications*, Academic Press, Orlando, 1973.

——*An Introduction to Finite Element Analysis*, Academic Press, Orlando, 1978.

Rao, S. S.: *The Finite Element Method in Engineering*, Pergamon, Oxford UK, 1982.

Reddy, J. N.: *An Introduction to the Finite Element Method*, McGraw-Hill, New York, 1984.

——*Energy and Variational Methods in Applied Mechanics with an Introduction to the Finite Element Method*, Wiley-Interscience, New York, 1993.

COMPUTER PROJECTS

Develop a computer code for the one-dimensional boundary value problem

$$(pu')' - qu + f = 0$$
$$-p(a)u'(a) + \alpha u(a) = A$$
$$p(b)u'(b) + \beta u(b) = B$$

Structure the program in roughly the following fashion.

INPUT. Input the information about p, q, f, a, b, α, β, A, and B.

1. NE	Number of elements. The number of nodes depends upon the type of interpolation used.
2. a, b	Boundary points.
3. $p(\mathrm{I}), q(\mathrm{I}), f(\mathrm{I})$ I = 1,NN	Nodal p, q, and f values.
4. NA, NB	Boundary code flags at $x = a$ and $x = b$. NA = 1 for an essential boundary condition and NA = 2 for a natural boundary condition. Similarly for NB.
5. UA or AL and AA	UA as the prescribed value of u at a if NA = 1 and AL and AA as α and A if NA = 2.
6. UB or BE and BB	UB as the prescribed value of u at a if NB = 1 and BE and BB as β and B if NB = 2.

Alternatively, p, q, and f can be defined internally by subroutines, procedures, functions, and so on, but would have to be changed for each new application.

DATA REFLECTION. Output the input data for checking purposes.

NODE GENERATION. Generate the additional positions of all the internal nodes using NE, a and b. Usually element lengths are taken to be equal but this is not necessary. An additional input could be used to specify a denser concentration of nodes near a or near b if desired.

ELEMENT FORMULATION AND ASSEMBLY. $\mathbf{p}_e$, $\mathbf{q}_e$, and $\mathbf{f}_e$ need to be formed and assembled for each of the NE elements. Any of the methods discussed in Sections 2.6 and 2.8 can be used for evaluating the integrals. The information supplied in input 3, 4, 5, and 6, above is needed here. If either of the boundary conditions is natural, the appropriate α, β must be added to the global stiffness matrix and the appropriate A, B must be added to the global load matrix.

CONSTRAINTS. If either boundary condition is essential, the first or last equation, respectively, must be altered to reflect the constraint at $x = a$ or $x = b$. The equations should also be put into symmetric form.

SOLUTION. An appropriate equation solver is employed to determine the solution for the nodal u's. There are listings of the routines BDecomp and BSolve in Appendix B for this purpose.

COMPUTATION OF DERIVED VARIABLES. The derived variable pu' is computed for each element. The point(s) at which pu' is evaluated should be the *best* points mentioned in the text.

Specifically, the following codes are easily generated based on the material presented in the chapter:

1. Linearly interpolated elements using the special integration formulas indicated in Section 2.6 for evaluating the $\mathbf{p}_e$, $\mathbf{q}_e$, and $\mathbf{f}_e$.
2. Linearly interpolated elements using a Gauss-Legendre quadrature for evaluating the $\mathbf{p}_e$, $\mathbf{q}_e$, and $\mathbf{f}_e$. The order of the Gauss quadrature can be taken large enough so the $\mathbf{p}_e$, $\mathbf{q}_e$, and $\mathbf{f}_e$ are essentially exact for most of the p, q, and f functions encountered.
3. Quadratically interpolated elements using the special integration formulas indicated in Section 2.8 for evaluating the $\mathbf{p}_e$, $\mathbf{q}_e$, and $\mathbf{f}_e$.
4. Quadratically interpolated elements using a Gauss-Legendre quadrature for evaluating the $\mathbf{p}_e$, $\mathbf{q}_e$, and $\mathbf{f}_e$. Again, the order of the Gauss quadrature can be taken large enough so the $\mathbf{p}_e$, $\mathbf{q}_e$, and $\mathbf{f}_e$ are essentially exact for most any of the p, q, and f functions encountered.

Each of these projects can also be carried out for the eigenvalue type problem. The equation solver becomes a suitable routine for extracting the eigenvalues and eigenvectors. Appendix B contains listings for several appropriate routines in this regard. Derived variables are not usually required.

It is possible to think of each of the above segments, that is, INPUT, DATA REFLECTION, and so on, as corresponding to a subroutine (procedure, function) or a group of subroutines in the code. The reader is referred to the discussion of the computer projects in Chapter 1 in this regard.

GENSL, a computer code for analyzing problems of the type discussed in this chapter, is contained on the disk available from your instructor.

EXERCISES

Section 2.2

For each of the following boundary value problems, put the differential equation in standard form and determine p, q, f, α, A, β, and B.

2.1. $x^2u'' + xu' + u + 1 = 0, u'(1) = 1, u(2) = 1$
2.2. $u'' + u + \sin \pi x = 0, u(0) = 0, u(1) = 0$
2.3. $u'' + u' + 2u = \exp(x), u(0) = 0, u'(1) - u(1) = 2$
2.4. $xu'' + u' + xu = 1, u'(1) - u(1) = 1, u'(2) = 1$
2.5. $x^2u'' + 2xu' + 2u + x = 0, u(1) = 0, u'(2) + u(2) = 1$
2.6. $x^2u'' - xu' + 2xu + x = 0, u(1) = 0, u'(3) + 2u(3) = 1$

For each of the following eigenvalue problems, put the differential equation in standard form and determine p, q, r, α and β.

2.7. $u'' + \lambda u = 0, u(0) = 0, u(L) = 0$
2.8. $u'' + \lambda xu = 0, u(0) = 0, u(L) = 0$
2.9. $u'' + xu' + \lambda u = 0, u'(0) = 0, u(x_0) = 0$
2.10. $x^2u'' + xu' + \lambda x^2u = 0, u'(1) = 0, u(2) = 0$
2.11. $u'' + \lambda u = 0, u(0) = 0, u'(1) - u(1) = 0$

2.12. $x^2u'' + 2xu' + 2\lambda u = 0, u(1) = 0, u'(2) + u(2) = 0$
2.13. $x^2u'' - xu' + 2\lambda xu = 0, u(2) = 0, u'(4) + 2u(4) = 0$

Section 2.3

2.14. For the general one-dimensional problem

$$(pu')' - qu + f = 0$$
$$-p(a)u'(a) + \alpha u(a) = A$$
$$p(b)u'(b) + \beta u(b) = B$$

discretize the interval $a \leq x \leq b$ into four equal-length elements and go through the details of developing the Galerkin finite element model. In particular show that the set of algebraic equations is tridiagonal.

2.15. Show that the final set of equations developed in the previous problem can be interpreted in terms of the elemental matrices $\mathbf{P_{ei}}$, $\mathbf{q_{ei}}$, and $\mathbf{f_{ei}}$, $\mathbf{i} = \mathbf{1, 2, 3, 4}$, arising from the four elements.

2.16. Go through the details to show that the elemental stiffness matrices $\mathbf{k_{ei}}$ can be represented in terms of the elemental interpolation vectors **N** and **N**′ according to Eq. (2.6).

2.17. Go through the details to show that the elemental load matrices $\mathbf{f_{ei}}$ can be represented in terms of the elemental interpolation vector **N** according to Eq. (2.6).

2.18. For $p = p_0$, a constant, show that

$$\mathbf{p_e} = \frac{p_0}{l_e}\begin{bmatrix} 1 & -1 \\ -1 & 1 \end{bmatrix}$$

2.19. For $q = q_0$, a constant, show that

$$\mathbf{q_e} = \frac{q_0 l_e}{6}\begin{bmatrix} 2 & 1 \\ 1 & 2 \end{bmatrix}$$

2.20. For $f = f_0$, a constant, show that

$$\mathbf{f_e} = \frac{f_0 l_e}{2}\begin{bmatrix} 1 \\ 1 \end{bmatrix}$$

Using the results of the previous three exercises, set up and solve each of the following problems. Use the number of equal-length elements indicated and compare your solution with the corresponding exact solution.

2.21. $u'' + u = 0$ $\quad u(0) = 0,\ u(1) = 1$ $\quad$ two elements
2.22. $u'' + u = 0$ $\quad u(0) = 0,\ u(1) = 1$ $\quad$ four elements
2.23. $u'' + u = 0$ $\quad u(0) = 0,\ u(1) = 1$ $\quad$ eight elements
2.24. $u'' - 4u = 2$ $\quad u(0) = 0,\ u(1) = 1$ $\quad$ two elements
2.25. $u'' - 4u = 2$ $\quad u(0) = 0,\ u(1) = 1$ $\quad$ four elements
2.26. $u'' - 4u = 2$ $\quad u(0) = 0,\ u(1) = 1$ $\quad$ eight elements

When p, q, and f are not constant it is shown in Section 2.6 that $\mathbf{p_e}$, $\mathbf{q_e}$, and $\mathbf{f_e}$ can be approximated as

$$\mathbf{p_e} \approx \frac{p_i + p_j}{2\ell_e}\begin{bmatrix} 1 & -1 \\ -1 & 1 \end{bmatrix} \qquad \mathbf{q_e} \approx \frac{\ell_e}{12}\begin{bmatrix} 3q_i + q_j & q_i + q_j \\ q_i + q_j & q_i + 3q_j \end{bmatrix} \qquad \mathbf{f_e} \approx \frac{\ell_e}{6}\begin{bmatrix} 2f_i + f_j \\ f_i + 2f_j \end{bmatrix},$$

where p_i and p_j, q_i and q_j, and f_i and f_j are the nodal value of p, q, and f, respectively, at the nodes for the element in question. After putting each of the following equations in the required standard form, set up and solve the finite element model using the number of equal-length elements indicated, and compare your solution with the corresponding exact solution.

2.27.	$u'' + 2u' + u = 1$	$u(0) = 1,\ u(2) = 2$	two elements
2.28.	$u'' + 2u' + u = 1$	$u(0) = 1,\ u(2) = 2$	four elements
2.29.	$u'' + 2u' + u = 1$	$u(0) = 1,\ u(2) = 2$	eight elements
2.30.	$u'' - 2u' + 2u = 1$	$u(0) = 0,\ u(4) = 0$	two elements
2.31.	$u'' - 2u' + 2u = 1$	$u(0) = 0,\ u(4) = 0$	four elements
2.32.	$u'' - 2u' + 2u = 1$	$u(0) = 0,\ u(4) = 0$	eight elements

Section 2.4.1

2.33. The uniform bar is loaded by a varying intensity distributed loading as shown. Set up and solve two-, four-, and eight-element models for the displacements and internal forces. Make plots for both the displacements and internal forces. Connect the midpoint values for the internal forces and compare the results with the exact solution.

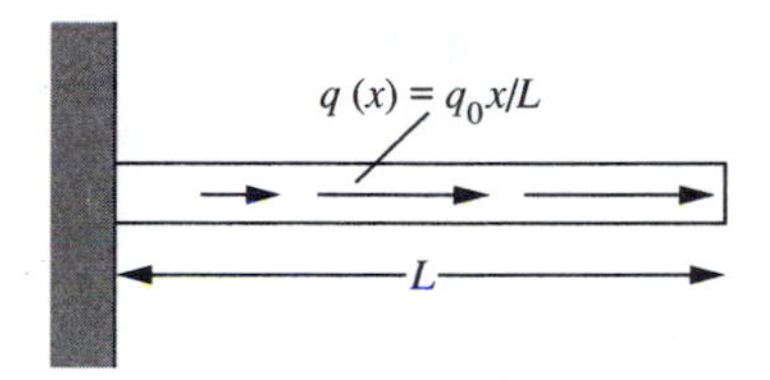

2.34. The uniform bar is loaded by a uniform intensity distributed loading over half of the bar as shown. Set up and solve two-, four-, and eight-element models for the displacements and internal forces. Make plots for both the displacements and internal forces. Connect the midpoint values for the internal forces and compare the results with the exact solution.

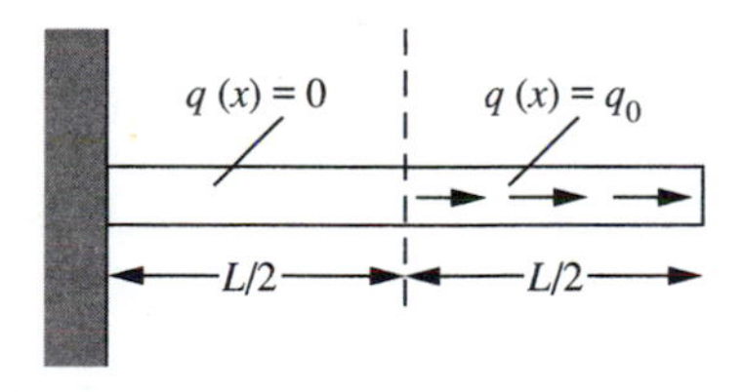

2.35. A variable-cross-section bar hangs under its own weight as shown. Set up and solve two-, four-, and eight-element models for the displacements and internal forces. Make plots for both the displacements and internal forces. Connect the midpoint values for the internal forces and compare the results with the exact solution.

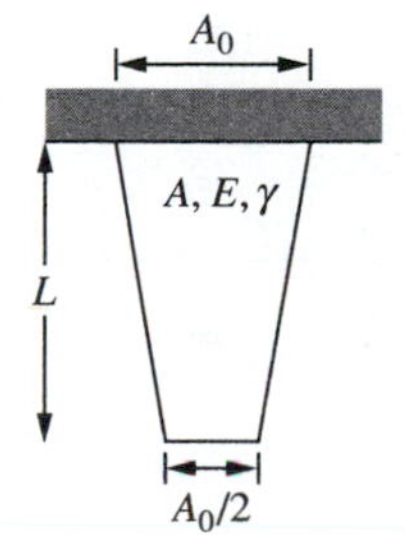

2.36. A variable-cross-section column is supported at the top and bottom and hangs under its own weight as shown. Set up and solve for the displacements and internal forces using models with one-, two-, and four-elements for each of the two sections. Make plots for both the displacements and internal forces. Take $\gamma = 150 \text{ lbf/ft}^2$, $E = 3 \times 10^6$ psi, $D = 6$ ft, $d = 3$ ft, $L_1 = 10$ ft and $L_2 = 10$ ft.

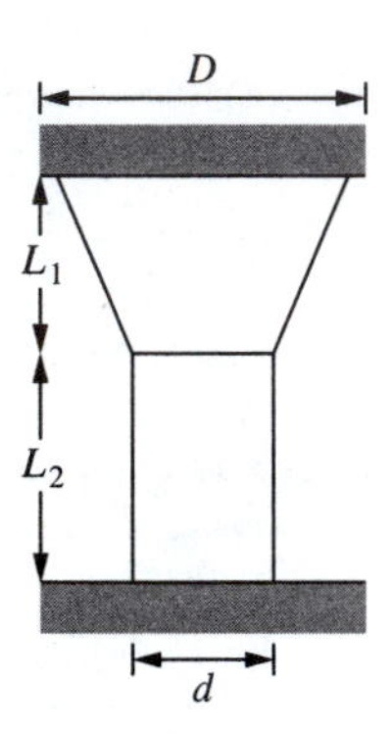

2.37. A uniform-cross-section bar rotates with angular velocity ω about one end as shown. Set up and solve two-, four-, and eight-element models for the displacements and internal forces. Make plots for both the displacements and internal forces. Connect the midpoint values for the internal forces and compare the results with the exact solution.

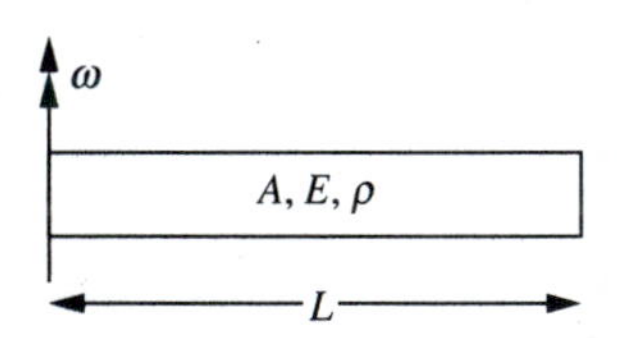

2.38. A uniform-cross-section bar is connected to a shaft of radius r_0 rotating with angular velocity ω as shown. Set up and solve two-, four-, and eight-element models for the displacements and internal forces. Make plots for both the displacements and internal forces. Connect the midpoint values for the internal forces and compare the results with the exact solution.

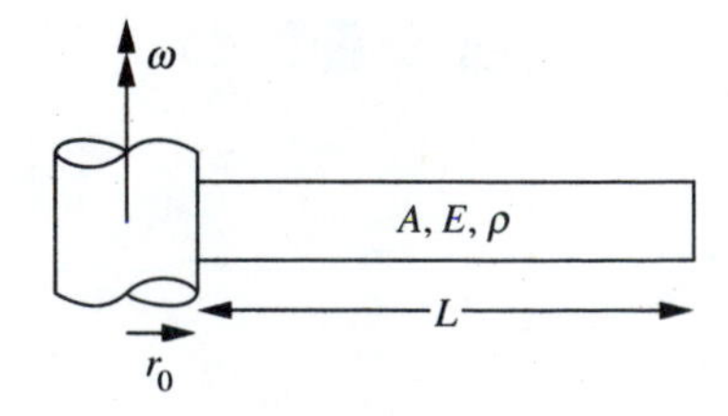

2.39. A variable-diameter, circular-cross-section bar is attached to a rigid shaft of radius R_0 rotating at a constant angular rate ω as indicated. The radius $r(x)$ of the bar is a linear function of position along the axis of the bar. Set up and solve one-, two-, and four-element linearly interpolated finite element models to determine that axial displacement of the bar. Plot the displacement and the force transmitted for each of the models. At what point should the derived variable $P = AEu'$ be evaluated within an element?

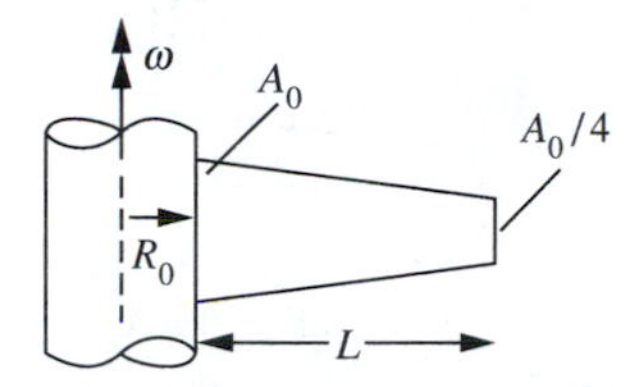

As developed in a mechanics of solids class, the governing equation of equilibrium for the torsion problem is

$$(JG\phi')' + t = 0$$

where $J = \pi a^4/2$ is the polar moment of the area, G is the shear modulus of the elastic material, t represents the distributed torques and ϕ is the rotation. The essential boundary condition is on the rotation ϕ and the natural boundary condition is usually on the torque $T = JG\phi'$. Thus the torsion problem is entirely analogous to the axial problem, with AE replaced by JG and u replaced by ϕ. Use this analogy and the results developed in Section 2.4.1 to solve the following torsional problems.

2.40. The uniform circular-cross-section bar is loaded by a varying intensity distributed torsional loading as shown. Set up and solve two-, four-, and eight-element models for the displacements and internal forces. Make plots for both the rotations and internal torques. Connect the midpoint values for the internal torques and compare the results with the exact solution.

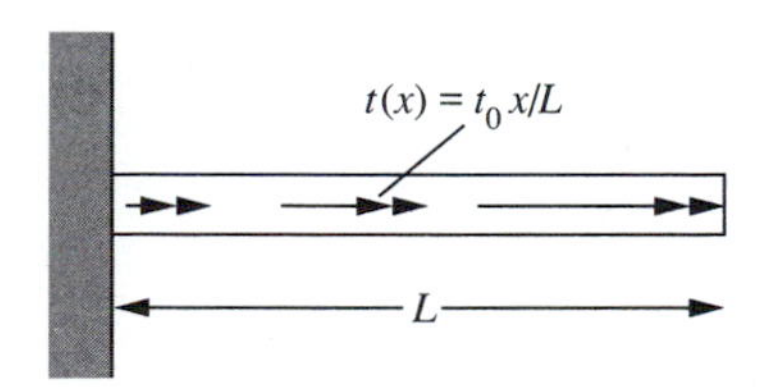

2.41. The uniform bar is loaded by a uniform intensity, distributed torsional loading over half of the bar as shown. Set up and solve two-, four-, and eight-element models for the displacements and internal forces. Make plots for both the displacements and internal torques. Connect the midpoint values for the internal torques and compare the results with the exact solution.

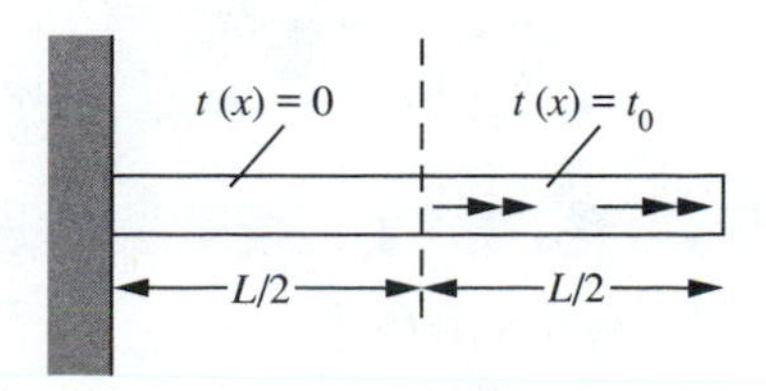

2.42. The variable-radius bar transmits the torque T_0 as shown. Set up and solve two-, four-, and eight-element models for determining the rotations and internal torques. Connect the midpoint values for the internal torques and compare the results with the exact solution.

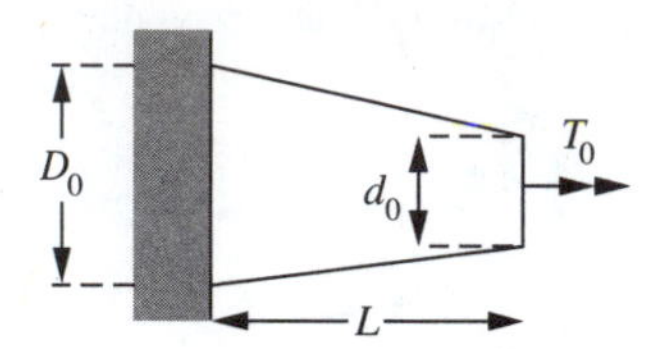

2.43. The composite circular-cross-section bar is loaded by the torque T_0 as indicated. r is a linear function of x for the variable-cross-section portion of the bar. Using one-, two-, and four-elements for the variable-cross-section portion and one element for the constant-cross-section portion, set up and solve two-, three-, and five-element models for determining the rotations and internal torques. Connect the midpoint values for the internal torques. Take $T_0 = 10^4$ in.-lbf, $G = 10^7$ lbf/in^2, $L_1 = L_2 =$ 15 in., $D_0 = 4$ in., and $d_0 = 2$ in.

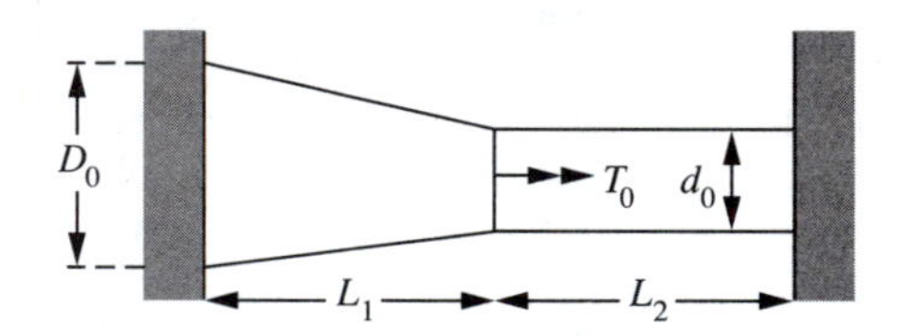

Section 2.4.2

2.44. For the simple slider bearing indicated, set up and solve two, four, and eight equal-length linearly interpolated finite element models for the pressure $p(x)$. For each of the models, plot $p(x)$ and determine what vertical load P can be sustained. Take $L = 12$ in., $h_0 = 10^{-3}$ in., $U = 10$ in./s, $p_{atm} = 14.7$ psi, and $\eta = 10^{-4}$ lbf-s/ft^2.

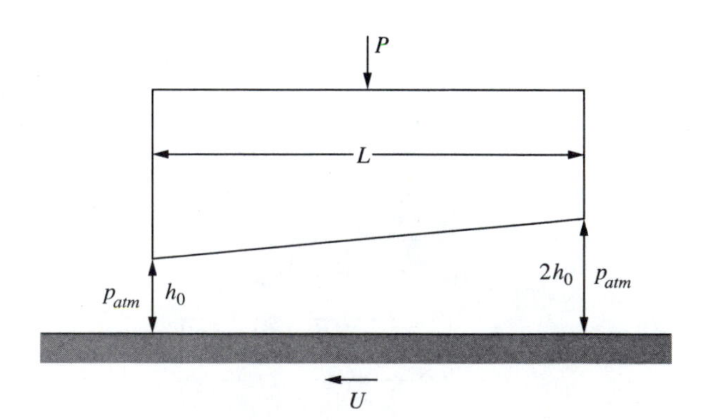

2.45. For the simple stepped wedge slider bearing indicated, set up and solve two, four, and eight equal-length linearly interpolated finite element models for the pressure $p(x)$. For each of the models, plot $p(x)$ and determine what vertical load P can be sustained. Take $L = 6$ in., $h_0 = 10^{-3}$ in., $U = 10$ in./s, $p_{atm} = 14.7$ psi, and $\eta = 10^{-4}$ lbf-s/ft^2.

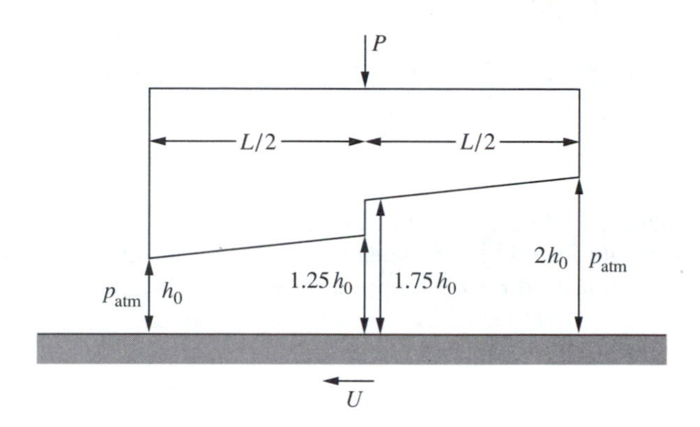

2.46. For the wedge slider bearing indicated, set up and solve 3, 6, and 12 equal-length linearly interpolated finite element models for the pressure $p(x)$. For each of the models, plot $p(x)$ and determine what vertical load P can be sustained. Take $L = 6$ in., $h_0 = 4 \times 10^{-3}$ in., $h_0/h_1 = 4, \delta/h_1 = 0.5, U = 20$ in./s., $p_{atm} = 14.7$ psi, and $\eta = 10^{-4}$ lbf-s/ft^2. What conclusion do you reach about the convergence of the solutions?

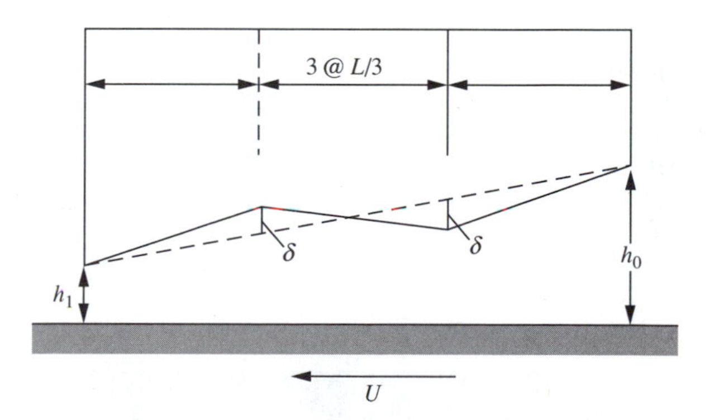

2.47. Repeat problem 2.46 for $\delta/h_1 = 1$.

2.48. For the compound wedge slider bearing indicated, set up and solve two, four, and eight equal-length linearly interpolated finite element models for the pressure $p(x)$. Use equal numbers of elements in both portions. For each of the models, plot $p(x)$ and determine what vertical load P can be sustained. Take $L_1 = L_2 = 12$ in., $h_0 = 10^{-3}$ in., $h_1 = 2 \times 10^{-3}$ in., $U = 10$ in./s., $p_{atm} = 14.7$ psi, and $\eta = 10^{-4}$ lbf-s/ft^2.

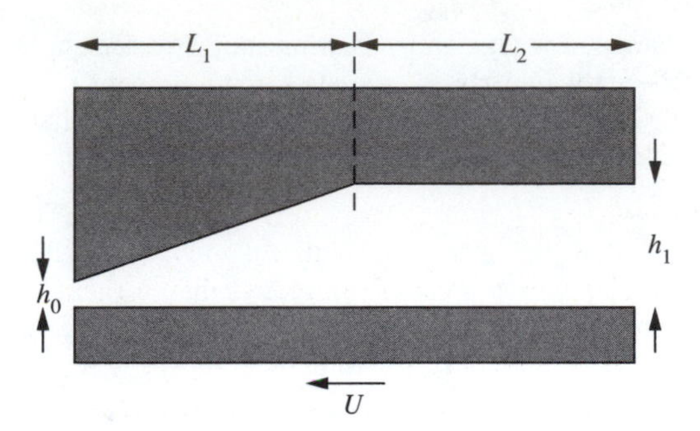

2.49. For the exponential wedge slider bearing indicated, set up and solve two, four, and eight equal-length linearly interpolated finite element models for the pressure $p(x)$. With x as shown take $h(x) = h_0 \exp(-\alpha x)$. For each of the models, plot $p(x)$ and determine what vertical load P can be sustained. Take $L = 12$ in., $h_0 = 10^{-3}$ in., $h_1 = 0.5 \times 10^{-3}$ in., $U = 10$ in./s., $p_{\text{atm}} = 14.7$ psi, and $\eta = 10^{-4}$ lbf-s/ft^2.

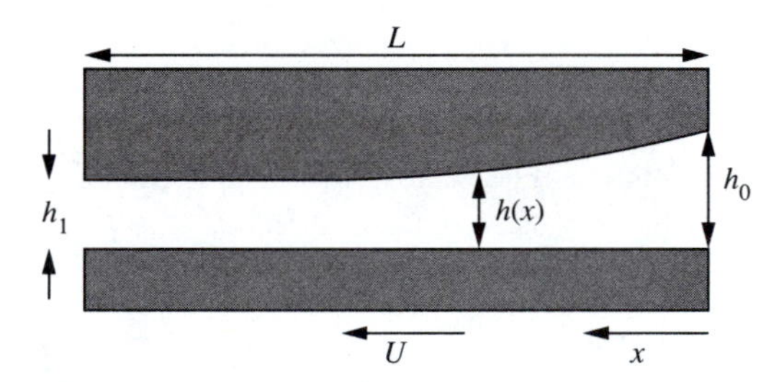

2.50. For the compound wedge slider bearing indicated, set up and solve two, four, and eight equal-length linearly interpolated finite element models for the pressure $p(x)$. Use equal numbers of elements in both portions. For each of the models, plot $p(x)$ and determine what vertical load P can be sustained. Take $L_1 = L_2 = 12$ in., $h_0 = 0.5 \times 10^{-3}$ in., $h_1 = 10^{-3}$ in., $U = 10$ in./s., $p_{\text{atm}} = 14.7$ psi, and $\eta = 10^{-4}$ lbf-s/ft^2.

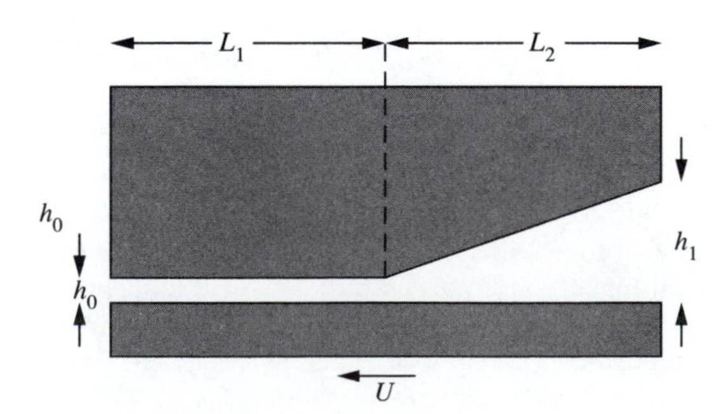

Section 2.4.3

2.51. Repeat the example problem of Section 2.4.3. using two, four, and eight elements when $\phi^2 = 40$. How well does the solution compare with the corresponding exact solution?

2.52. Repeat the example problem of Section 2.4.3. using two, four, and eight elements when $\phi^2 = 100$. How well does the solution compare with the corresponding exact solution? Investigate how a nonuniform mesh might be used to advantage.

2.53. Repeat the example problem of Section 2.4.3. using two, four, and eight elements when $\phi^2 = 400$. How well does the solution compare with the corresponding exact solution? Investigate how a nonuniform mesh might be used to advantage.

2.54. A variable area, rectangular-cross-section fin transmits heat away from a mass as indicated in the figure. The thickness of the fin in the direction perpendicular to the paper is 10 times that shown in the plane of the paper. There is convection on the entire lateral surface.

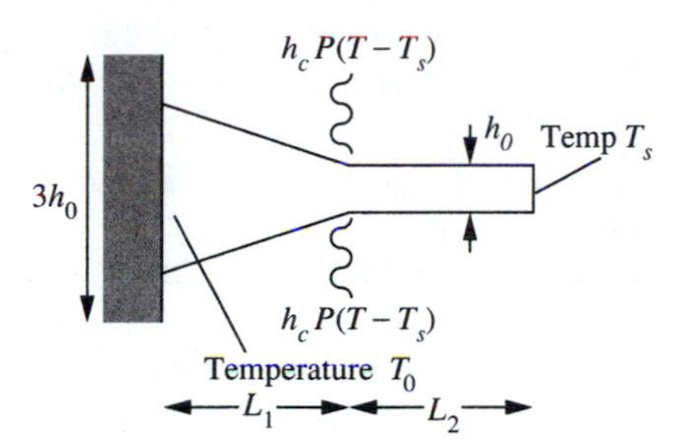

With $h_0 = 5$ cm, $L_1 = L_2 = 10$ cm, $T_0 = 400°\text{C}$, $T_s = 100°\text{C}$, $h_c = 10^{-3}$ W/mm^2−°C, and $k = 0.30$ W/mm−°C, set up two, four, and eight equal-length linearly interpolated finite element models for the temperature T. Plot the results for T and for the derived variable $q_x = -kAT'$.

2.55. Heat is transferred through a thick walled cylinder by conduction as shown in the figure.

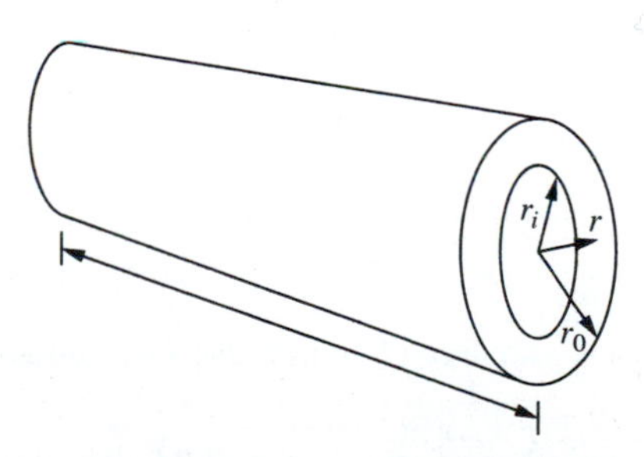

Show that based on an energy balance, the governing differential equation is

$$\frac{d}{dr}\left[(2\pi r L)k\frac{dT}{dr}\right] = 0$$

Appropriate boundary conditions are $T(r_i) = T_i$ and $T(r_0) = T_0$. Set up and solve two-, four-, and eight-element models for the case $T_0/T_i = 3$. Plot the temperature and compare with the exact solution. Show that rT', properly interpreted, is a constant.

2.56. Repeat the previous problem when $T(r_i) = T_i$ and the boundary condition at $r = r_0$ is changed to $T' = 0$. Take $r_0 = 2r_i$.

2.57. Solve the steady-state radial heat conduction in the cylinder when $T(r_i) = T_i$ and the boundary condition at $r = r_0$ is changed to

$$-k\frac{dT}{dr} = h_c(T - T_e)$$

Take the dependent variable to be $u = (T - T_e)/(T_i - T_e)$ and the independent variable to be $s = r/r_i$. Take $r_0 = 2r_i$ and $hr_i/k = 10$.

2.58. When internal sources are present, steady-state radial heat conduction in the cylinder is governed by

$$-\frac{d}{dr}\left[(2\pi rL)k\frac{dT}{dr}\right] = 2\pi rLQ$$

where Q (power/unit volume) is the strength of the internal energy source. Appropriate boundary conditions are $T(r_i) = T_i$ and $T(r_0) = T_0$. Set up and solve two-, four-, and eight-element models for the case $Q = Q_0, T_i = T_0 = 0$. For $r_0 = 2r_i$, plot $u = kT/Q_0r_i^2$ and the flux $q = -rT'$.

2.59. Repeat the previous problem when the boundary at r_0 is replaced by $T'(r_0) = 0$.

2.60. A radial fin whose thickness varies linearly with r transmits heat away from a region as indicated. There is also convection on the surface $r_i \leq r \leq r_0$.

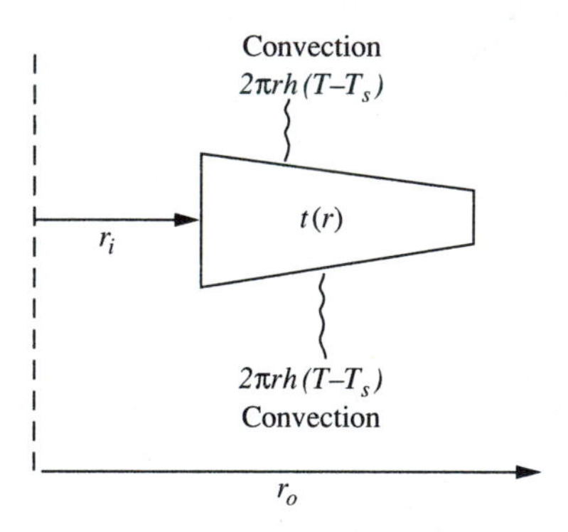

Show that based on an energy balance, the governing differential equation is

$$\frac{d[k(r)rt(r)dT/dr]}{dr} - 2rhT + 2rhT_s = 0$$

Take the boundary conditions to be $T(r_i) = T_i$ and $T(r_0) = T_0$. For $t_i = 5$ cm, $t_0 = 2$ cm, $r_i = 5$ cm, $r_0 = 20$ cm, $T_i = 300°\text{C}$, $T_0 = 80°\text{C}$, and $T_s = 80°\text{C}$. With $k = 0.2$ W/mm−°C, and $h = 10^{-3}$ W/mm^2−°C, set up and solve equal-length linearly interpolated finite element models with two, four, and eight elements.

For each of the elements, plot the temperature and the derived variable $q_r = -krtT$. Where within each of the elements should the derived variable be evaluated?

Section 2.5

2.61. Carry through the details of showing that the results of applying the Galerkin method to the boundary value problem of equations 2.9*a*–2.9*c* leads to equations 2.10 and 2.11.

For each of the following problems set up and solve for the eigenvalues and eigenvectors using models with one, two, and four unconstrained degrees of freedom. How do the results from each of these exercises appear to fit within the "Rule of Thumb" given in Section 2.5?

2.62. $u'' + \lambda u = 0$ $\quad u(0) = 0, u(L) = 0$
2.63. $u'' + \lambda x u = 0$ $\quad u(0) = 0, u(L) = 0$
2.64. $x^2 u'' + x u' + \lambda x^2 u = 0$ $\quad u'(1) = 0, u(2) = 0$
2.65. $x u'' + 2u' + \lambda x^{-1} u = 0$ $\quad u(1) = 0, u'(2) + u(2) = 0$

Section 2.5.1

2.66. Set up and solve a two-element model for the torsional vibrations of the uniform circular shaft fixed at both ends as shown.

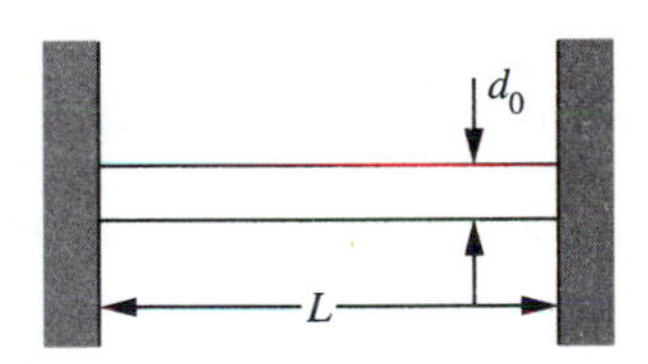

2.67. Repeat the previous problem using three elements.
2.68. Repeat the previous problem using five elements.
2.69. Set up and solve a one-element model for the torsional vibrations of the variable-radius circular shaft fixed at one end as shown. Take $d_0 = D_0/2$.

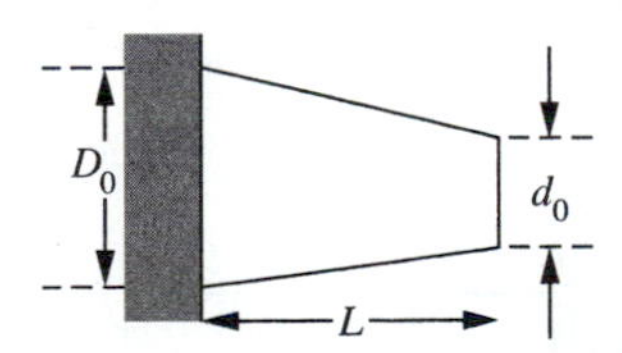

2.70. Repeat the previous problem using two elements.
2.71. Repeat the previous problem using four elements.
2.72. An important technical problem involves systems of shafts connected by gears. The study of such systems can begin by investigating the problem of a circular disk rigidly attached to a uniform shaft as shown.

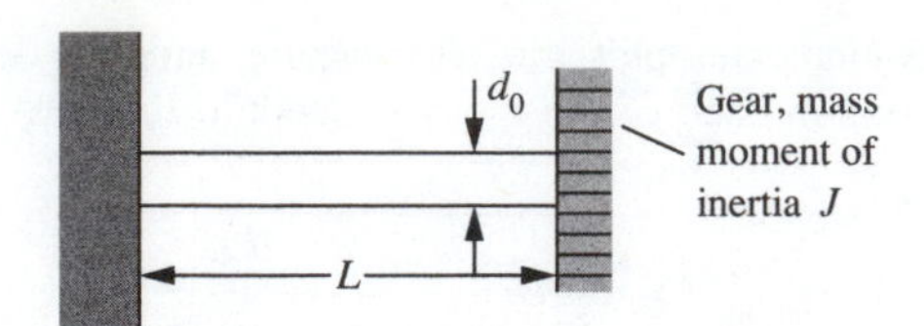

The corresponding finite element model is constructed by simply adding J to the corresponding main diagonal element of the global **M** matrix. The frequencies and mode shapes are quite accurate as long as J is large compared to the mass moment of inertia of the shaft. Set up and solve a one-element model for the torsional vibrations of the circular shaft fixed at one end as shown. Take $J = 2J_{\text{shaft}} = 2(\pi d_0^4 \rho L/32)$.

2.73. Repeat the previous problem using two elements. Take $J = 2J_{\text{shaft}} = 2(\pi d_0^4 \rho L/32)$.

2.74. Repeat the previous problem using four elements. Take $J = 2J_{\text{shaft}} = 2(\pi d_0^4 \rho L/32)$.

2.75. Set up and solve a two-element model for the disk-shaft system shown. Take $J = 2J_{\text{shaft1}} = 2(\pi d_1^4 \rho L_1/32)$, $d_1 = d_2$ and $L_1 = L_2$.

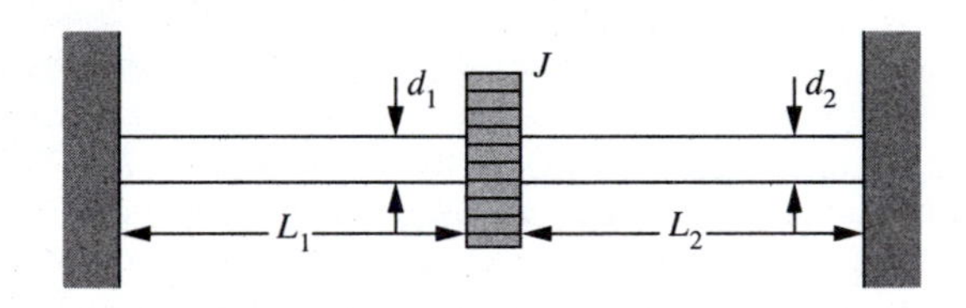

2.76. Repeat the previous problem using four equal-length elements.

Axial vibrations. For axial vibrations the elemental stiffness and mass matrices are given by

$$\mathbf{k_e} = \int_{x_i}^{x_{i+1}} \mathbf{N}' AE \mathbf{N}'^{\mathrm{T}} dx \qquad \text{and} \qquad \mathbf{m_e} = \int_{x_i}^{x_{i+1}} \mathbf{N} A\rho \mathbf{N}^{\mathrm{T}} dx$$

Concentrated masses are handled by simply adding the mass M to the corresponding main diagonal element of the global **M** matrix. The rest of the basic steps in the finite element method are carried out in exactly the same manner as described in the previous section for the corresponding torsion problem.

2.77. Set up and solve a one-element model for the axial motion of the uniform bar shown. Compare your results with those from the corresponding exact solution.

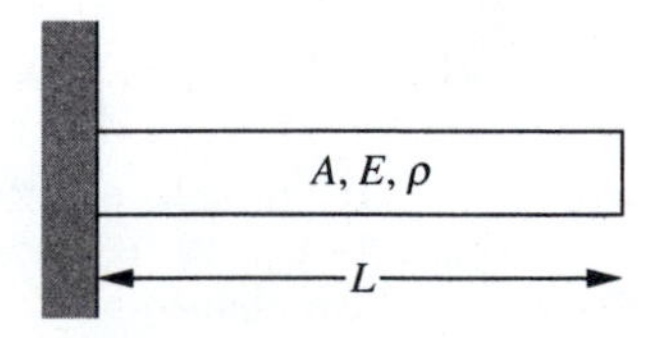

2.78. Repeat the previous problem using two equal-length elements.

2.79. Repeat the previous problem using four equal-length elements.

2.80. For the variable-radius, circular-cross-section bar shown, set up a one-element model to determine the frequencies and mode shapes. Take $d_0 = D_0/2$.

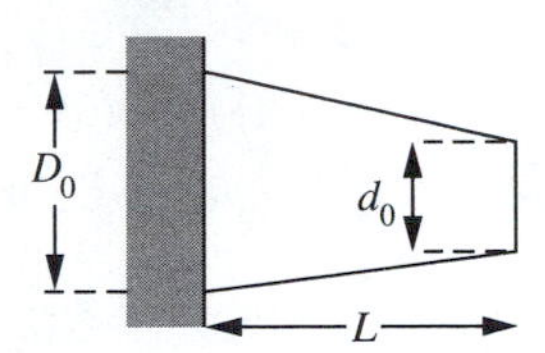

2.81. Repeat the previous problem using two elements.
2.82. Repeat the previous problem using four elements.
2.83. Frequently the axial motion of a bar takes place in the presence of an additional mass M as shown.

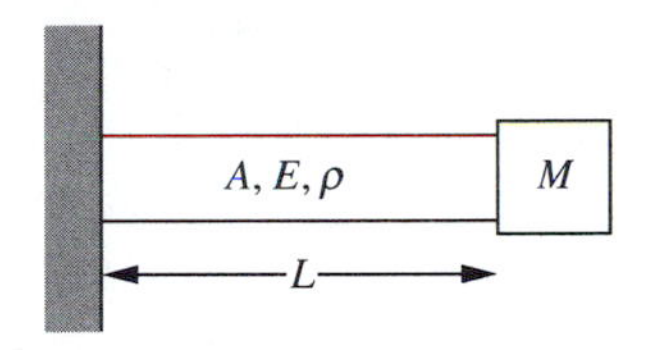

The corresponding finite element model is constructed by simply adding M to the corresponding main diagonal element of the global **M** matrix. The frequencies and mode shapes are quite accurate as long as M is large compared to the mass moment of inertia of the shaft. Set up and solve a one-element model for the axial vibrations of uniform bar fixed at one end as shown. Take $M = 2m_{\text{bar}} = 2\rho AL$.
2.84. Repeat the previous problem using two elements. Take $M = 2m_{\text{bar}}$.
2.85. Repeat the previous problem using four elements. Take $M = 2m_{\text{bar}}$.

Section 2.6

2.86. Verify the special integration formulas developed in Section 2.6 for $\mathbf{p_e}$, $\mathbf{q_e}$, and $\mathbf{f_e}$.
2.87. Verify that the special integration formulas developed in Section 2.6 for $\mathbf{p_e}$, $\mathbf{q_e}$, and $\mathbf{f_e}$ are exact when p, q, and f are constants or linear functions of x.
2.88. In each of the following, evaluate $\mathbf{p_e}$ for $p = x^2$ for each of the two equal-length intervals for a two-element model on the interval (1,3).
- **a.** By using an $N = 1$ Gauss quadrature.
- **b.** By using an $N = 2$ Gauss quadrature.
- **c.** By using an $N = 3$ Gauss quadrature.
- **d.** By evaluating the integrals exactly.
- **e.** By using the special integration formulas.

2.89. Repeat Exercise 2.88 for $\mathbf{q_e}$ with $q(x) = x^2$.
2.90. Repeat Exercise 2.88 for $\mathbf{f_e}$ with $f(x) = x^2$.
2.91–2.93. Repeat Exercises 2.88, 2.89, and 2.90 for four equal-length elements on the interval (1,3).

2.94. Set up and solve the finite element model using two equal-length linearly interpolated elements for

$$(x^2u')' + x^2u + x^2 = 0$$
$$u(1) = 1$$
$$u'(3) + u(3) = 0$$

corresponding to each of the methods a–c above for evaluating the elemental matrices $\mathbf{p_e}$, $\mathbf{q_e}$, and $\mathbf{f_e}$.

2.95. Repeat Exercise 2.94 for four equal-length elements.

2.96. In each of the following, evaluate $\mathbf{p_e}$ for $p = x^3$ for each of the two equal-length intervals for a two-element model on the interval (1,3).

a. By using an $N = 1$ Gauss quadrature.
b. By using an $N = 2$ Gauss quadrature.
c. By using an $N = 3$ Gauss quadrature.
d. By evaluating the integrals exactly.
e. By using the special integration formulas.

2.97. Repeat Exercise 2.96 for $\mathbf{q_e}$ with $q(x) = x^3$.

2.98. Repeat Exercise 2.96 for $\mathbf{f_e}$ with $f(x) = x^3$.

2.99–2.101. Repeat Exercises 2.96, 2.97, and 2.98 for four equal-length elements on the interval (1,3).

2.102. Set up and solve the finite element model using two equal-length linearly interpolated elements for

$$(x^3u')' + x^3u + x^3 = 0$$
$$u(1) = 1$$
$$u'(3) + u(3) = 0$$

corresponding to each of the methods a–c above for evaluating the elemental matrices $\mathbf{p_e}$, $\mathbf{q_e}$, and $\mathbf{f_e}$.

2.103. Repeat Exercise 2.102 for four equal-length elements.

The point of these exercises is to provide some calibration for the applicability of the special integration formulas in Section 2.6 for evaluating the elemental matrices, and to indicate that as the mesh size is decreased, the errors of the special integration formulas decrease rapidly.

Section 2.7.1

In terms of each of the boundary values *as stated* (i.e., do not first put the differential equation in standard form), go through integration by parts to determine the weak form. If either boundary condition specifies u, that is, $u(a) = u_a$ or $u(b) = u_b$, require that the test function satisfy $v(a) = 0$ or $v(b) = 0$, respectively. Identify the bilinear functional $B(v, u)$ and the linear functional $l(v)$. For which of these problems is $B(v, u) = B(u, v)$?

2.104. $u'' + u + \sin \pi x = 0$, $u(0) = 0$, $u(1) = 0$

2.105. $x^2u'' + 2xu' + 2u + x = 0$, $u(1) = 0$, $u'(2) + u(2) = 1$

2.106. $u'' + xu' + u = 1$, $u'(0) - u(0) = 1$, $u'(1) = 1$

2.107. $x^2u'' + xu' + 2xu = x$, $u(1) = 0$, $u'(3) + 2u(3) = 1$

2.108–2.111. Repeat, if necessary, Exercises 2.104–2.107 by first putting the differential equations in standard form. Subsequently, determine the corresponding functional $I(u) = B(u, u)/2 - l(u)$.

It can be seen from Exercises 2.104 through 2.111, that the differential equation must first be put in standard form in order to be able to determine the functional corresponding to the original boundary value problem.

Section 2.7.2

There are certain interchanges of operations routinely performed in the calculus of variations that require verification. One such operation involves the interchange of variation and differentiation, for example,

$$\delta\left(\frac{du}{dx}\right) \stackrel{?}{=} \frac{d(\delta u)}{dx}$$

That these are in fact identical is seen from the following development.

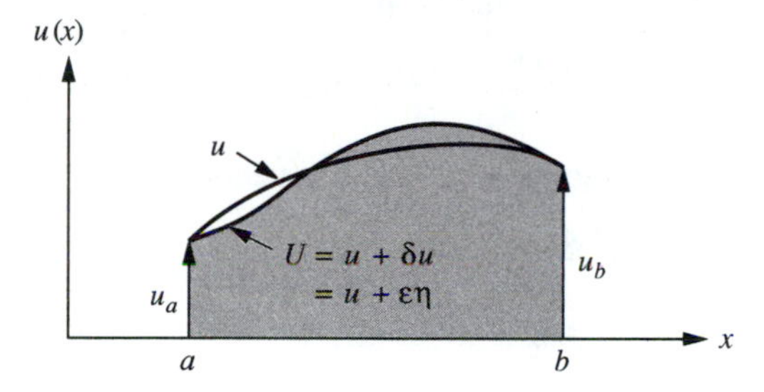

From the figure the variation of u is the difference in the varied curve U and the original curve u; that is,

$$\delta u = U - u = \epsilon\eta$$

so that

$$\frac{d(\delta u)}{dx} = \frac{d(U-u)}{dx} = \frac{d(\epsilon\eta)}{dx} = \epsilon\eta'$$

whereas

$$\delta\left(\frac{du}{dx}\right) = U' - u' = (u + \epsilon\eta)' - u' = \epsilon\eta'$$

showing the equivalence of the derivative of the variation and the variation of the derivative.

In the next three exercises, the student is asked to investigate parallel developments for several other operator interchanges.

***2.112.** Show that

$$\delta\left(\int F\,dx\right) = \int \delta F\,dx$$

***2.113.** Show that

$$\delta(F_1F_2) = F_1\,\delta F_2 + \delta F_1\,F_2$$

*Starred exercises extend the material presented in the chapter, or introduce additional material.

***2.114.** Show that

$$\delta\left(\frac{F_1}{F_2}\right) = \frac{(F_2\,\delta F_1 - F_1\,\delta F_2)}{F_2^2}$$

For each of the following use the formal approach developed in Section 2.7.2 for generating the Euler equations and corresponding boundary conditions.

2.115. $I(u) = \int_1^2 \left(\frac{xu'^2}{2} - \frac{u^2}{2x} - \frac{u}{x}\right)dx, \; u(2) = 1$

2.116. $I(u) = \int_0^L \left(\frac{AEu'^2}{2} - \frac{ku^2}{2} - qu\right)dx, \; u(0) = 0$

***2.117.** Show that for a functional of the form

$$I(u) = \int_a^b F(x, u, u', u'')\,dx$$

requiring $\delta I = 0$ leads to the differential equation

$$\frac{d^2(\partial F/\partial u'')}{dx^2} - \frac{d(\partial F/\partial u')}{dx} + \frac{\partial F}{\partial u} = 0$$

and the boundary conditions

$$\eta\left[\frac{d(\partial F/\partial u'')}{dx} - \frac{\partial F}{\partial u'}\right]\Bigg|_a^b = 0$$

$$\eta'\left[\frac{\partial F}{\partial u''}\right]\Bigg|_a^b = 0$$

* * * * * *

In many instances, the functional whose stationary value is desired is of the form of an integral with an integrand that is a quadratic function of its arguments. An example is given by the problem of the axial deformation problem shown.

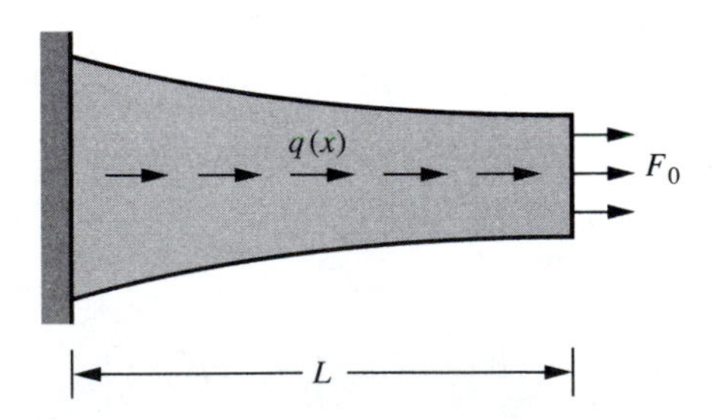

The functional in question is the potential energy given by

$$V(u) = \int_0^L \left(\frac{AEu'^2}{2} - qu\right)dx - F_0u(L) \qquad \text{with} \qquad u(0) = 0$$

Form

$$V(u + \delta u) = \int_0^L \left(\frac{AE(u' + \delta u')^2}{2} - q(u + \delta u) \right) dx - F_0 \left(u(L) + \delta u(L) \right)$$

and

$$\Delta V = V(u + \delta u) - V(u)$$

After some algebra this can be written as

$$\Delta V = \epsilon \left\{ \int_0^L (AEu'\eta' - q\eta)\, dx - F_0\eta(L) \right\} + \left(\frac{\epsilon^2}{2} \right) \int_0^L (AE\eta'^2)\, dx$$

which is expressed as

$$\Delta V = \epsilon\, \delta V + \frac{\epsilon^2\, \delta^2 V}{2!}$$

where δV is the first variation of V encountered in the development in Section 2.7.2, and $\delta^2 V$ is the second variation of V. In the limit as $\epsilon \to 0$,

$$\lim_{\epsilon \to 0} \frac{\Delta V}{\epsilon} = \delta V = \int_0^L \left(AEu'\eta' - q\eta \right) dx - F_0\eta(L) = 0.$$

Integration by parts yields

$$AEu'\eta \Big|_0^L - \int_0^L \left((AEu')' + q \right) \eta\, dx - F_0\eta(L) = 0$$

or since $\eta(0) = 0$,

$$\left(AEu'(L) - F_0 \right) \eta(L) - \int_0^L \left((AEu')' + q \right) \eta\, dx = 0$$

The arbitrariness of η results in the conclusions

$$(AEu')' + q = 0 \qquad 0 \le x \le L$$

$$AEu'(L) - F_0 = 0$$

for the differential equation and natural boundary condition, with the essential boundary condition as

$$u(0) = 0$$

which are exactly the same results as obtained using the formal approach of Section 2.7.2. Note also that this approach allows boundary terms appearing in the functional to be handled easily. In each of the following exercises, use the approach just developed to generate the Euler equations and appropriate boundary conditions.

***2.118.**

$$V(u) = \int_0^L \left(\frac{AEu'^2}{2} + \frac{ku^2}{2} - qu \right) dx + \frac{k_0 u(0)^2}{2} + \frac{k_L u(L)^2}{2} - F_0 u(0) - F_L u(L)$$

***2.119.**

$$V(w) = \int_0^L \left(\frac{EIw''^2}{2} - qw\right)dx - V_L w(L) - M_L w'(L)$$

with $w(0) = w'(0) = 0$.

***2.120.**

$$V(w) = \int_0^L \left(\frac{EIw''^2}{2} - qw\right)dx + \frac{k_0 w(0)^2 + k_L w(L)^2 + C_0 w'(0)^2 + C_L w'(L)^2}{2}$$

****

There are numerous instances of idealizations of inputs to physical problems such as concentrated loads in solid mechanics, point sources in fluid mechanics and heat transfer, and point masses in vibrations. These are idealized from situations where a finite amount of some quantity (force, mass, energy) is input over a relatively very small part of the region. In the limit as the amount of region (line segment, area, volume) is allowed to shrink to zero, the intensity becomes infinite with the total amount of input remaining a constant. Situations of this sort are handled by representing the input according to

$$q(x) = Q\,\delta(x, x_i)$$

where Q is the total amount of input (force, energy, mass) considered to be input at $x = x_i$. The Dirac delta function is $\delta(x, x_i)$, defined as

$$\delta(x, x_i) = 0 \qquad \text{if } x \neq x_i$$

with

$$\int_a^b \delta(x, x_i)\, dx = 1$$

when $a \leq x_i \leq b$. Thus the total area under the δ diagram is unity. Note that the delta function has dimensions of $1/L$ when x has dimensions L. For a two- or three-dimensional situation, the dimensions would be $1/L^2$ and $1/L^3$ respectively. In addition, the important sifting property of the delta function with respect to a continuous function f is expressed as

$$\int_a^b \delta(x, x_i) f(x)\, dx = f(x_i)$$

Some of the applications of this type of idealization are investigated in the next few problems.

***2.121.** In the axial deformation problem indicated, show that the corresponding potential energy can be expressed as

$$V(u) = \int_0^L \left(\frac{AEu'^2}{2} - P\delta(x, a)u - q(x)u\right) dx - P_L u(L)$$

with

$$u(0) = 0$$

and that the Euler equation and boundary conditions are

$$(AEu')' + P\delta(x, a) + q = 0$$
$$u(0) = 0$$
$$AEu'(L) = P_L$$

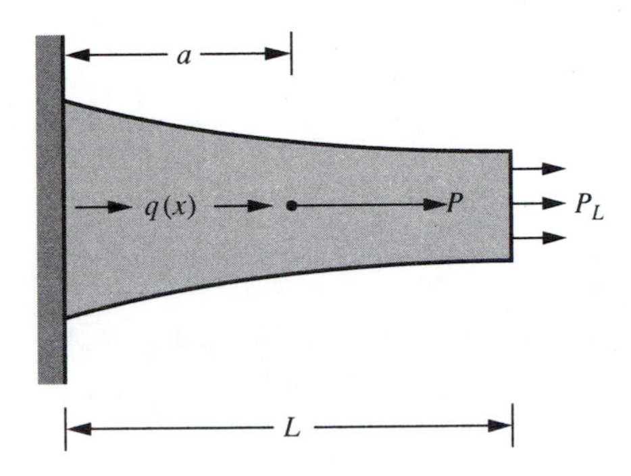

***2.122.** Alternately, in Exercise 2.121, write the first variation as

$$0 = \left\{ \int_0^{a-\epsilon} + \int_{a-\epsilon}^{a+\epsilon} + \int_{a+\epsilon}^{L} \right\} \left(AEu'\eta' - P\delta(x, a)\eta - q(x)\eta \right) dx - P_L\eta(L)$$

and passing to the limit as $\epsilon \to 0$ show that by requiring u, and hence δu, to be continuous at $x = a$, the boundary value problem can be restated:
For $0 \leq x < a$

$$(AEu_1')' + q = 0$$
$$u_1(0) = 0$$

For $a < x \leq L$

$$(AEu_2')' + q = 0$$
$$AEu_2'(L) = P_L$$

and at the interface between the two regions

$$[u]|_a = u_2(a) - u_1(a) = 0$$
$$[AEu']|_a = AEu_1'(a^+) - AEu_2'(a^-) = -P$$

In both of these interface equations, $[\Phi]_a$ represents the jump or discontinuity in the variable Φ at a. The interface conditions state that there is no jump in u at a (u is continuous), whereas the jump in AEu' at a is equal to the force P input at that point. This same result can also be obtained by a force balance on a thin slice of the bar containing $x = a$.

For each of the following, determine

a. The Euler equations and boundary conditions as per the approach indicated by Exercise 2.121.

b. The Euler equations, the boundary, and the jump conditions as per the approach indicated by Exercise 2.122. Interpret the jump conditions physically.

***2.123.** The torsion problem, for which the potential energy associated with the following figure is

$$I(u) = \int_0^L \left(\frac{JG\phi'^2}{2} - T_1\delta(x, a)\phi - t(x)\phi \right) dx - T_L\phi(L)$$

with

$$\phi(0) = 0$$

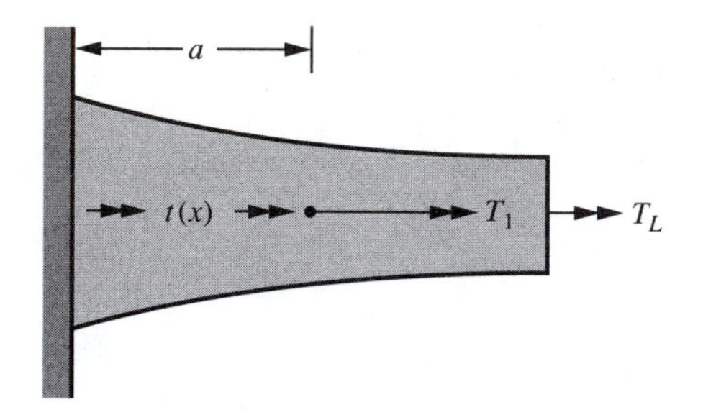

***2.124.** The bending problem, for which the potential energy associated with the following figure is

$$I(w) = \int_0^L \left(\frac{EIw''^2}{2} - P_1\delta(x, a)w - q(x)w \right) dx - V_L w(L) - M_L w'(L)$$

with

$$w(0) = w'(0) = 0$$

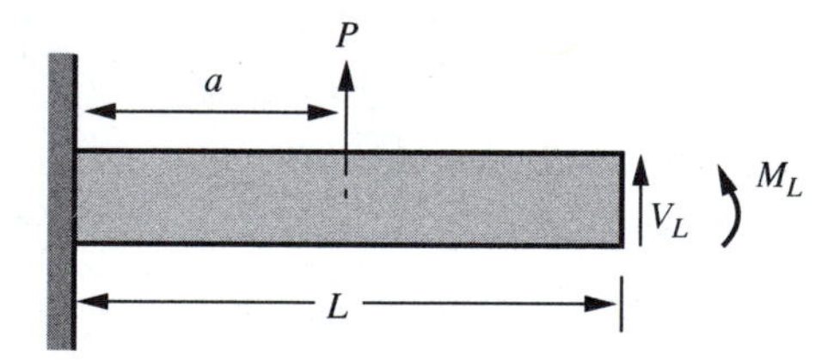

***2.125.** The general Sturm-Liouville functional

$$I(u) = \int_a^b \left(\frac{pu'^2}{2} + \frac{qu^2}{2} - fu - C_1\delta(x, a)u \right) dx$$

with

$$u(a) = u_a$$

Section 2.7.3

***2.126.** For the general one-dimensional problem

$$(pu')' - qu + f = 0$$
$$u(a) = u_a$$
$$u(b) = u_b$$

with the corresponding functional

$$I(u) = \int_a^b \left(\frac{pu'^2 + qu^2}{2} - fu \right) dx$$

go through all the details of showing that the final set of unconstrained global equations can be written as

$$\mathbf{K_G u_G} = \mathbf{F_G}$$

with $\mathbf{K_G} = \sum(\mathbf{p_G} + \mathbf{q_G})$ and $\mathbf{F_G} = \sum \mathbf{f_G}$.

***2.127.** Repeat Exercise 2.126 with

$$I(u) = \int_a^b \left(\frac{pu'^2 + qu^2}{2} - fu \right) dx + \frac{\alpha u(a)^2}{2} + \frac{\beta u(b)^2}{2} - Au(a) - Bu(b)$$

to show that the final set of equations is given by

$$\mathbf{K_G u_G} = \mathbf{F_G}$$

where $\mathbf{K_G} = \sum(\mathbf{p_G} + \mathbf{q_G}) + \mathbf{BT}$ and $\mathbf{F_G} = \sum \mathbf{f_G} + \mathbf{bt}$.

***2.128.** In situations where there are concentrated inputs S_i at points x_i within the interval $a \le x \le b$, the energy functional must be augmented according to

$$I(u) = \int_a^b \left(\frac{pu'^2 + qu^2}{2} - fu \right) dx - \sum S_i u(x_i)$$
$$+ \frac{\alpha u(a)^2}{2} + \frac{\beta u(b)^2}{2} - Au(a) - Bu(b)$$

With a node placed at each x_i in connection with a usual discretization, show that this extra term results in additional loads in $\mathbf{F_G}$ so that $\mathbf{F_G} = \sum \mathbf{f_G} + \mathbf{bt} + \mathbf{S_G}$, where $\mathbf{S_G}$ is an $(N + 1) \times 1$ vector with entries S_i in the ith position.

***2.129.** In certain situations, there may be "spring supports" of strength k_i at interior points x_i within the interval $a \le x \le b$, resulting in an augmented functional of the form

$$I(u) = \int_a^b \left(\frac{pu'^2 + qu^2}{2} - fu \right) dx + \sum \frac{k_i u(x_i)^2}{2}$$
$$+ \frac{\alpha u(a)^2}{2} + \frac{\beta u(b)^2}{2} - Au(a) - Bu(b)$$

With a node placed at each of the x_i in connection with a usual discretization, show that this results in an augmented stiffness matrix

$$\mathbf{K_G} = \sum(\mathbf{p_G} + \mathbf{q_G}) + \mathbf{BT} + \mathbf{K_I}$$

where $\mathbf{K_I}$ is the $(N + 1) \times (N + 1)$ matrix with diagonal terms at the positions corresponding to the nodes at which the springs are located. Note that each of the k_i produces an effect identical to that of the α and β, which are essentially springs at the boundary.

***2.130.** Using the results of Exercise 2.114, show that the Euler equation for the functional

$$\lambda = \frac{\int_a^b \left(pu'^2/2 + qu^2/2\right) dx}{\int_a^b (ru^2/2)\, dx}$$

with $u(a) = 0$ and $u(b) = 0$ is

$$(pu')' + (\lambda r - qu) = 0$$

The functional λ is known as the *Rayleigh quotient*. In vibrations, the numerator is usually a potential energy and the denominator is a kinetic energy. In elastic stability theory the numerator is an internal elastic energy, with the denominator a type of external potential energy.

***2.131.** Determine how the numerator of the Rayleigh quotient in Exercise 2.130 must be altered in order for the Euler equation and boundary conditions to be

$$(pu')' + (\lambda r - qu) = 0$$

$$-p(a)u'(a) + \alpha u(a) = 0$$

$$p(b)u'(b) + \beta u(b) = 0$$

***2.132.** In the Rayleigh quotient, represent the functionals in the numerator and denominator respectively as $N(u)$ and $D(u)$. Discretize the interval (a, b) in the usual way, and use linear interpolation to arrive at

$$N(u) = \frac{\mathbf{u}_G^T \mathbf{K}_G \mathbf{u}_G}{2}$$

$$D(u) = \frac{\mathbf{u}_G^T \mathbf{M}_G \mathbf{u}_G}{2}$$

where

$$\mathbf{K}_G = \sum (\mathbf{p}_e + \mathbf{q}_e)$$

$$\mathbf{M}_G = \sum \mathbf{r}_e$$

with

$$\mathbf{p}_e = \int_{x_i}^{x_{i+1}} \mathbf{N}' p \mathbf{N}'^T\, dx$$

$$\mathbf{q}_e = \int_{x_i}^{x_{i+1}} \mathbf{N} q \mathbf{N}^T\, dx$$

$$\mathbf{r}_e = \int_{x_i}^{x_{i+1}} \mathbf{N} r \mathbf{N}^T\, dx$$

Thus, the functional $\lambda = N(u)/D(u)$ has been converted to a function of the variables u_i, $i = 1, 2, \ldots, u_{i+1}$, that is,

$$\lambda = \frac{\mathbf{u}_G^T \mathbf{K}_G \mathbf{u}_G}{\mathbf{u}_G^T \mathbf{M}_G \mathbf{u}_G}$$

The stationary value of the function λ is now obtained by requiring that the partial derivatives with respect to each of the u_i vanish. Show that this leads to

$$\mathbf{K_G u_G} - \lambda \mathbf{M_G u_G} = \mathbf{0}$$

or

$$(\mathbf{K_G} - \lambda \mathbf{M_G})\mathbf{u_G} = \mathbf{0}$$

as was obtained by another approach in Section 2.5.

Section 2.8

2.133. Verify the special integration formulas developed in Section 2.8 for $\mathbf{p_e}$, $\mathbf{q_e}$, and $\mathbf{f_e}$.

2.134. Verify the special integration formulas developed in Section 2.8 for $\mathbf{p_e}$, $\mathbf{q_e}$, and $\mathbf{f_e}$ are exact when p, q, and f are constants, linear, or quadratic functions of x.

2.135. In each of the following, evaluate $\mathbf{p_e}$ for $p = x^3$ for a one-element quadratically interpolated model on the interval (1,3).
- **a.** By using an $N = 2$ Gauss quadrature.
- **b.** By using an $N = 3$ Gauss quadrature.
- **c.** By using an $N = 4$ Gauss quadrature.
- **d.** By evaluating the integrals exactly.
- **e.** By using the special integration formulas.

2.136. Repeat Exercise 2.135 for $\mathbf{q_e}$ with $q(x) = x^3$.

2.137. Repeat Exercise 2.135 for $\mathbf{f_e}$ with $f(x) = x^3$.

2.138–2.140. Repeat Exercises 2.135, 2.136, and 2.137 for two equal-length elements on the interval (1,3).

2.141. Set up and solve the finite element model using one quadratically interpolated element for

$$(x^3u')' + x^3u + x^3 = 0$$
$$u(1) = 1$$
$$u'(3) + u(3) = 0$$

corresponding to each of the methods a–c in Exercise 2.135 above for evaluating the elemental matrices $\mathbf{p_e}$, $\mathbf{q_e}$, and $\mathbf{f_e}$.

2.142. Repeat Exercise 2.141 for two equal-length elements.

2.143. In each of the following, evaluate $\mathbf{p_e}$ for $p = x^4$ for a one-element model on the interval (1,3).
- **a.** By using an $N = 2$ Gauss quadrature.
- **b.** By using an $N = 3$ Gauss quadrature.
- **c.** By using an $N = 4$ Gauss quadrature.
- **d.** By evaluating the integrals exactly.
- **e.** By using the special integration formulas.

2.144. Repeat Exercise 2.143 for $\mathbf{q_e}$ with $q(x) = x^4$.

2.145. Repeat Exercise 2.143 for $\mathbf{f_e}$ with $f(x) = x^4$.

2.146–2.148. Repeat Exercises 2.143, 2.144, and 2.145 for two equal-length elements on the interval (1,3).

2.149. Set up and solve the finite element model using one quadratically interpolated element for

$$(x^4u')' + x^4u + x^4 = 0$$
$$u(1) = 1$$
$$u'(3) + u(3) = 0$$

corresponding to each of the methods a–c above for evaluating the elemental matrices $\mathbf{p_e}$, $\mathbf{q_e}$, and $\mathbf{f_e}$.

2.150. Repeat Exercise 2.149 for two equal-length elements.

The point of these exercises is to provide some calibration for the applicability of the special integration formulas in Section 2.8 for evaluating the elemental matrices, and to indicate that as the mesh size is decreased, the errors of the special integration formulas decrease rapidly.

Section 2.9.1

In each of the following use quadratic interpolation and then compare the results to the corresponding linearly interpolated solutions from the exercises of Section 2.4.1.

2.151. Set up and solve a two-element model of the hanging bar problem in Section 2.9.1 and show that all the results are exact.

2.152. The uniform bar is loaded by a varying intensity distributed loading as shown. Set up and solve one-, two-, and four-element models for the displacements and internal forces. Make plots for both the displacements and internal forces. Connect the Gauss point values for the internal forces and compare the results with the exact solution.

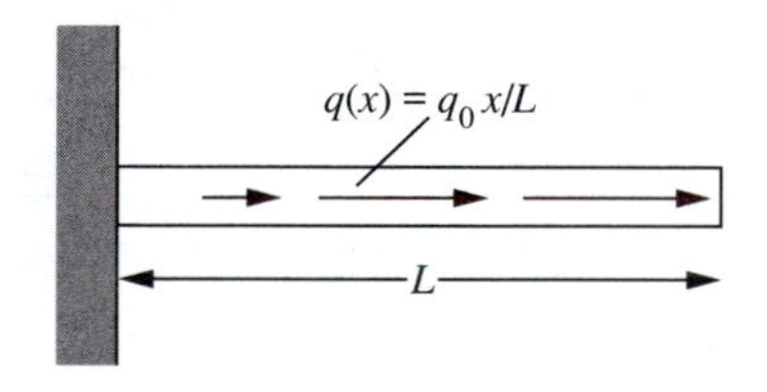

2.153. The uniform bar is loaded by a uniform intensity distributed loading over half of the bar as shown. Set up and solve one-, two-, and four-element models for the displacements and internal forces. Make plots for both the displacements and internal forces. Connect the Gauss point values for the internal forces and compare the results with the exact solution.

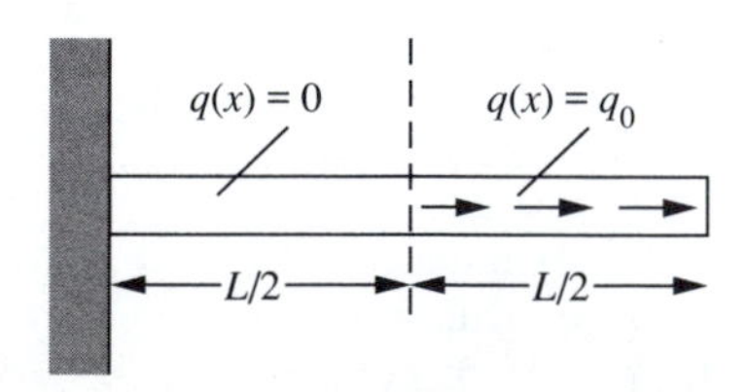

2.154. A variable-cross-section bar hangs under its own weight as shown. Set up and solve one-, two-, and four-element models for the displacements and internal forces. Make plots for both the displacements and internal forces. Connect the Gauss point values for the internal forces and compare the results with the exact solution.

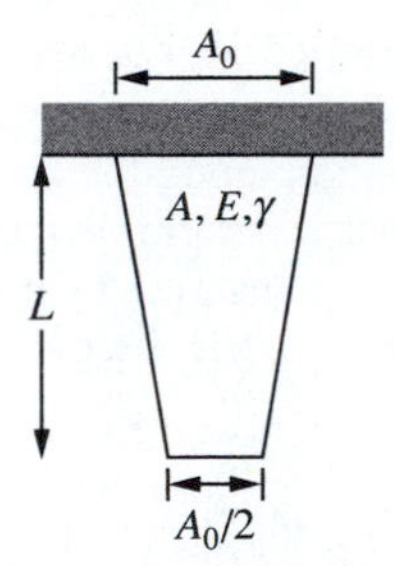

2.155. A variable-cross-section column is supported at the top and bottom and hangs under its own weight as shown. Set up and solve for the displacements and internal forces using models with one and two elements for each of the two sections. Make plots for both the displacements and internal forces. Take $\gamma = 150$ lbf/ft^2, $E = 3 \times 10^6$ psi, $D = 6$ ft, $d = 3$ ft, $L_1 = 10$ ft and $L_2 = 10$ ft.

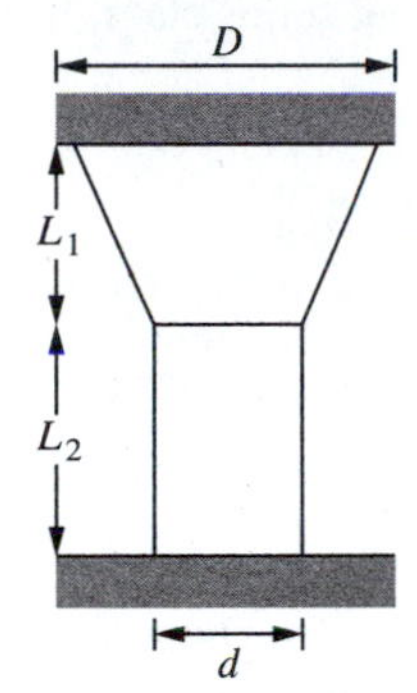

2.156. A uniform-cross-section bar rotates with angular velocity ω about one end as shown. Set up and solve one-, two-, and four-element models for the displacements and internal forces. Make plots for both the displacements and internal forces. Connect the Gauss point values for the internal forces and compare the results with the exact solution.

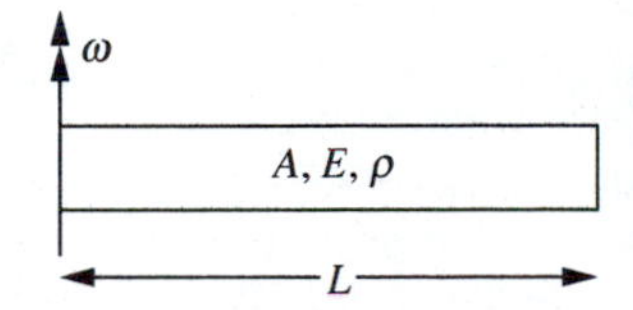

2.157. A uniform-cross-section bar is connected to a shaft of radius r_0 rotating with angular velocity ω as shown. Set up and solve one-, two-, and four-element models for the displacements and internal forces. Make plots for both the displacements and internal forces. Connect the Gauss point values for the internal forces and compare the results with the exact solution.

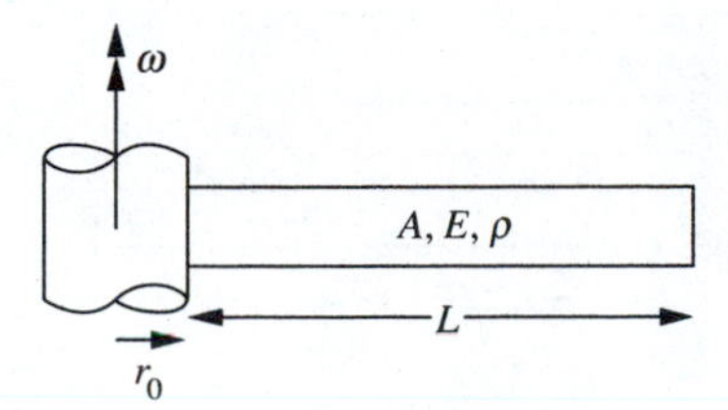

2.158. A variable-diameter, circular-cross-section bar is attached to a rigid shaft of radius R_0 which rotates at a constant angular rate ω as indicated. The radius $r(x)$ of the bar is a linear function of position along the axis of the bar. Set up and solve one-, two-, and four-element models to determine the axial displacement of the bar. Plot the displacement and the force transmitted for each of the models. At what point should the derived variable $P = AEu'$ be evaluated within an element?

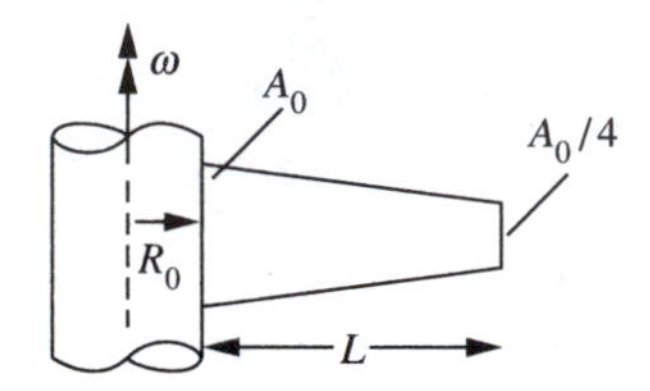

As developed in a mechanics of solids class, the governing equation of equilibrium for the torsion problem is

$$(JG\phi')' + t = 0$$

where $J = \pi a^4/2$ is the polar moment of the area, G is the shear modulus of the elastic material, t represents the distributed torques and ϕ is the rotation. The essential boundary condition is on the rotation ϕ and the natural boundary condition is usually on the torque $T = JG\phi'$. Thus the torsion problem is entirely analogous to the axial problem with AE replaced by JG and u replaced by ϕ. Use this analogy and the results developed in Section 2.9.1 to solve the following torsional problems.

2.159. The uniform circular-cross-section shaft is loaded by a varying intensity, distributed torsional loading as shown. Set up and solve one-, two-, and four-element models for the displacements and internal forces. Make plots for both the rotations and internal torques. Connect the Gauss point values for the internal torques and compare the results with the exact solution.

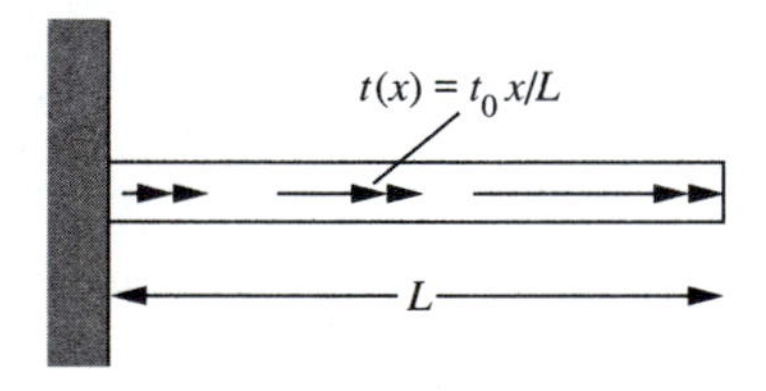

2.160. The uniform shaft is loaded by a uniform intensity, distributed torsional loading over half of the bar as shown. Set up and solve one-, two-, and four-element models for the displacements and internal forces. Make plots for both the displacements and internal torques. Connect the Gauss point values for the internal torques and compare the results with the exact solution.

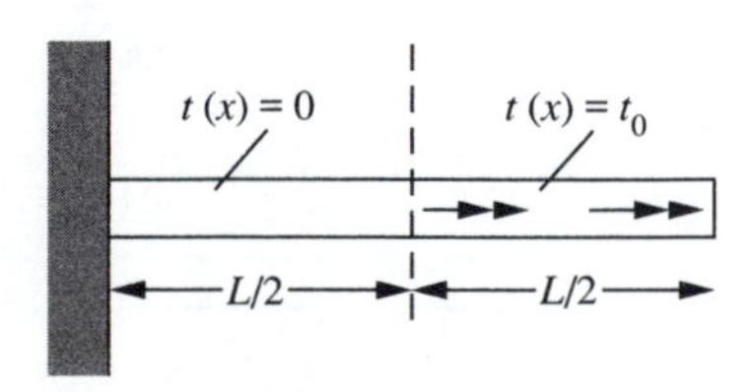

2.161. The variable-radius shaft transmits the torque T_0 as shown. Set up and solve one-, two-, and four-element models for determining the rotations and internal torques. Connect the Gauss point values for the internal torques and compare the results with the exact solution.

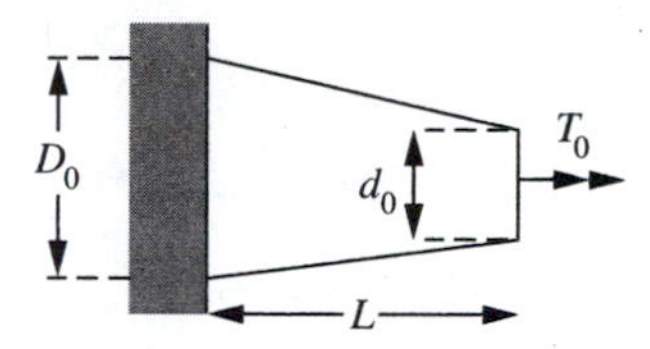

2.162. The composite circular-cross-section shaft is loaded by the torque T_0 as indicated. r is the linear function of x for the variable-cross-section portion of the bar. Using one, two, and four elements for the variable-section portion and one element for the constant-cross-section portion, set up and solve two-, three-, and five-element models for determining the rotations and internal torques. Connect the Gauss point values for the internal torques. Take $T_0 = 10^4$ in.-lbf, $G = 10^7$ lbf/in^2, $L_1 = L_2 = 15$ in., $D_0 = 4$ in., and $d_0 = 2$ in.

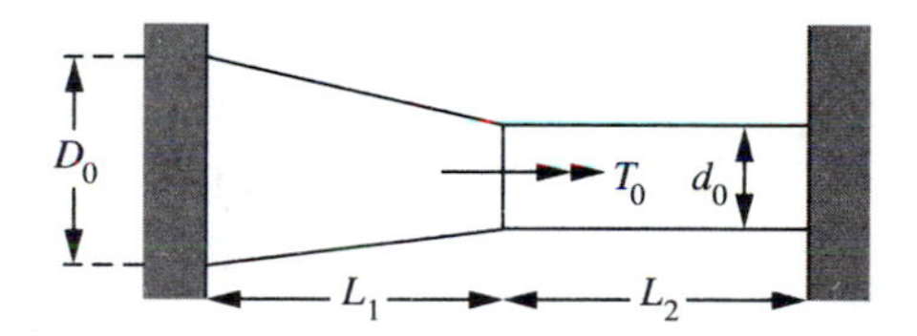

Section 2.9.2

2.163. For the simple slider bearing indicated, set up and solve one, two, and four equal-length quadratically interpolated finite element models for the pressure $p(x)$. For each of the models, plot $p(x)$ and determine what vertical load P can be sustained. Take $L = 12$ in., $h_0 = 10^{-3}$ in., $U = 10$ in./s, $p_{atm} = 14.7$ psi, and $\eta = 10^{-4}$ lbf-s/ft^2.

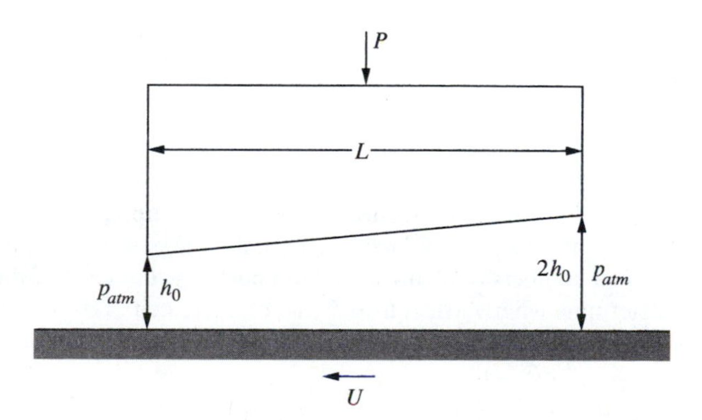

2.164. For the simple stepped wedge slider bearing indicated, set up and solve two, four, and eight equal-length quadratically interpolated finite element models for the pressure $p(x)$. For each of the models, plot $p(x)$ and determine what vertical load P can be sustained. Take $L = 6$ in., $h_0 = 10^{-3}$ in., $U = 10$ in./s, $p_{\text{atm}} = 14.7$ psi, and $\eta = 10^{-4}$ lbf-s/ft^2.

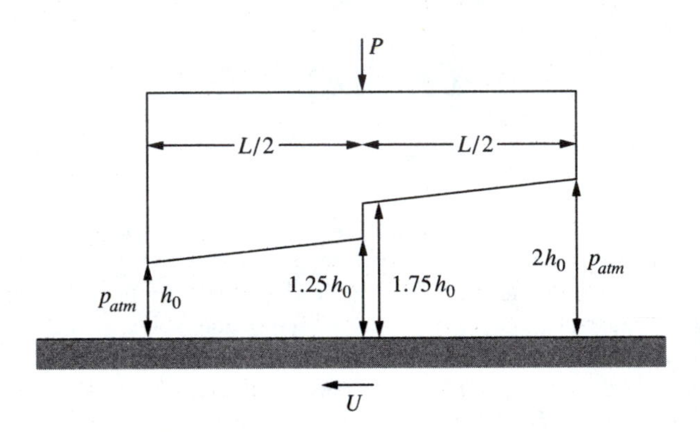

2.165. For the wedge slider bearing indicated, set up and solve 3, 6, and 12 equal-length quadratically interpolated finite element models for the pressure $p(x)$. For each of the models, plot $p(x)$ and determine what total vertical load P can be sustained. Take $L = 6$ in., $h_0 = 4 \times 10^{-3}$ in., $h_0/h_1 = 4$, $\delta/h_1 = 0.5$, $U = 20$ in./s, $p_{\text{atm}} = 14.7$ psi, and $\eta = 10^{-4}$ lbf-s/ft^2. What conclusion do you reach about the convergence of the solutions?

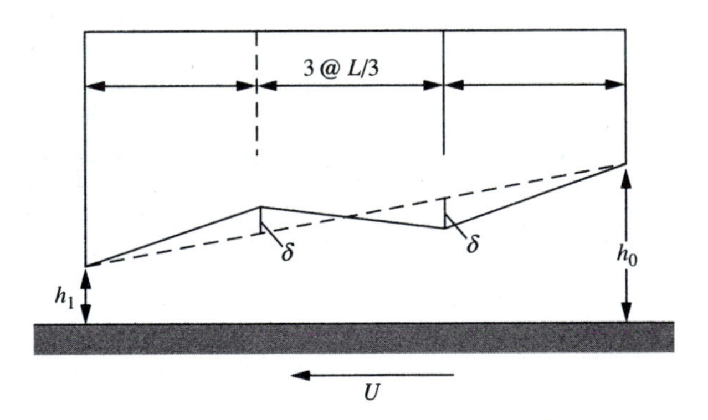

2.166. Repeat problem 2.165 for $\delta/h_1 = 1$.

2.167. For the compound wedge slider bearing indicated, set up and solve two, four, and eight equal-length quadratically interpolated finite element models for the pressure $p(x)$. Use equal numbers of elements in both portions. For each of the models, plot $p(x)$ and determine what vertical load P can be sustained. Take $L_1 = L_2 = 12$ in., $h_0 = 10^{-3}$ in., $h_1 = 2 \times 10^{-3}$ in., $U = 10$ in./s, $p_{\text{atm}} = 14.7$ psi, and $\eta = 10^{-4}$ lbf-s/ft^2.

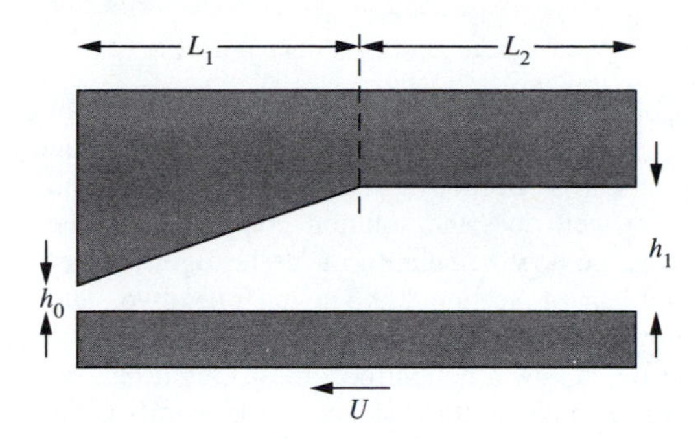

2.168. For the exponential wedge slider bearing indicated, set up and solve two, four, and eight equal-length quadratically interpolated finite element models for the pressure $p(x)$. With x as shown take $h(x) = h_0 \exp(-\alpha x)$. For each of the models, plot $p(x)$ and determine what vertical load P can be sustained. Take $L = 12$ in., $h_0 = 10^{-3}$ in., $h_1 = 0.5 \times 10^{-3}$ in., $U = 10$ in./s, $p_{\text{atm}} = 14.7$ psi, and $\eta = 10^{-4}$ lbf-s/ft^2.

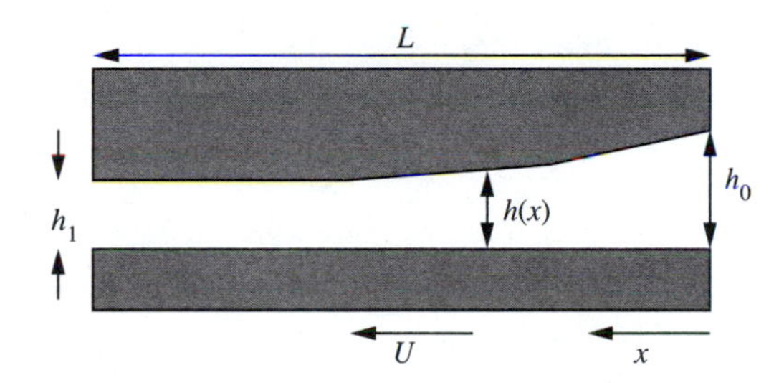

2.169. For the compound wedge slider bearing indicated, set up and solve two, four, and eight equal-length quadratically interpolated finite element models for the pressure $p(x)$. Use equal numbers of elements in both portions.

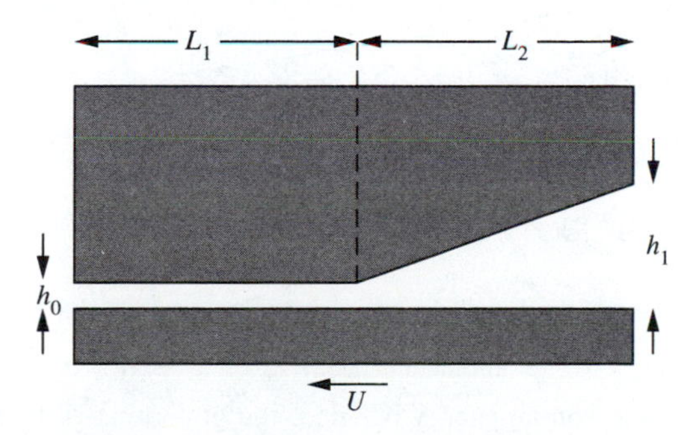

For each of the models, plot $p(x)$ and determine what vertical load P can be sustained. Take $L_1 = L_2 = 12$ in., $h_0 = 0.5 \times 10^{-3}$ in., $h_1 = 10^{-3}$ in., $U = 10$ in./s, $p_{\text{atm}} = 14.7$ psi, and $\eta = 10^{-4}$ lbf-s/ft^2.

Section 2.9.3

2.170. Repeat the problem of Section 2.9.3 using one, two, and four elements when $\phi^2 = 40$. How well does the solution compare with the corresponding exact solution?

2.171. Repeat the problem of Section 2.9.3 using one, two, and four elements when $\phi^2 = 100$. How well does the solution compare with the corresponding exact solution? Investigate how a nonuniform mesh might be used to advantage.

2.172. Repeat the problem of Section 2.9.3 using one, two, and four elements when $\phi^2 = 400$. How well does the solution compare with the corresponding exact solution? Investigate how a nonuniform mesh might be used to advantage.

2.173. A variable-area rectangular-cross-section fin transmits heat away from a mass as indicated in the figure. The thickness of the fin in the direction perpendicular to the paper is 10 times that shown in the plane of the paper. There is a convection on the entire lateral surface.

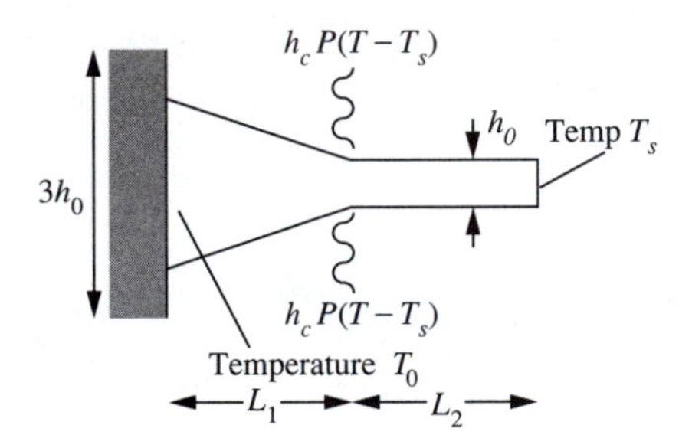

With $h_0 = 5$ cm, $L_1 = L_2 = 10$ cm, $T_0 = 400°C$, $T_s = 100°C$, $h_c = 10^{-3}$ W/mm$^2-°$C, and $k = 0.30$ W/mm$-°$C, set up one, two, and four equal length-linearly interpolated finite element models for the temperature T. Plot the results for T and for the derived variable $q_x = -kAT'$.

2.174. Heat is transferred through a thick-walled cylinder by conduction as shown in the figure.

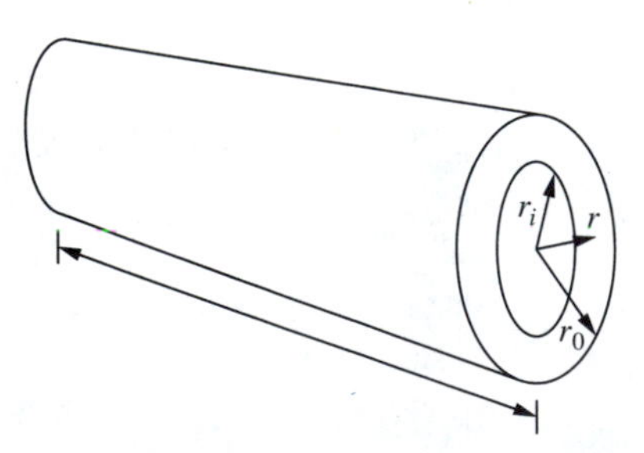

Show that based on an energy balance, the governing differential equation is

$$\frac{d}{dr}\left[(2\pi r L)k\frac{dT}{dr}\right] = 0$$

Appropriate boundary conditions are $T(r_i) = T_i$ and $T(r_0) = T_0$. Set up and solve one-, two-, and four-element models for the case $T_0/T_i = 3$. Plot the temperature and compare with the exact solution. Show that rT', properly interpreted, is a constant.

2.175. Repeat the previous problem when $T(r_i) = T_i$ and the boundary condition at $r = r_0$ is changed to $T' = 0$. Take $r_0 = 2r_i$.

2.176. Repeat the previous problem when $T(r_i) = T_i$ and the boundary condition at $r = r_0$ is changed to

$$-k\frac{dT}{dr} = h_c(T - T_e)$$

Take the dependent variable to be $u = (T - T_e)/(T_i - T_e)$ and the independent variable to be $s = r/r_i$. Take $r_0 = 2r_i$ and $hr_i/k = 10$.

2.177. When internal sources are present, steady-state radial heat conduction in the cylinder is governed by

$$-\frac{d}{dr}\left[(2\pi rL)k\frac{dT}{dr}\right] = 2\pi rLQ$$

where Q (power/unit volume) is the strength of the internal energy source. Appropriate boundary conditions are $T(r_i) = T_i$ and $T(r_0) = T_0$. Set up and solve one-, two-, and four-element models for the case $Q = Q_0, T_i = T_0 = 0$. For $r_0 = 2r_i$ plot $u = kT/Q_0r_i^2$ and the flux $q = -rT'$.

2.178. Repeat the previous problem when the boundary condition at r_0 is replaced by $T'(r_0) = 0$.

2.179. A radial fin whose thickness varies linearly with r transmits heat away from a region as indicated. There is also convection on the surface $r_i \le r \le r_0$.

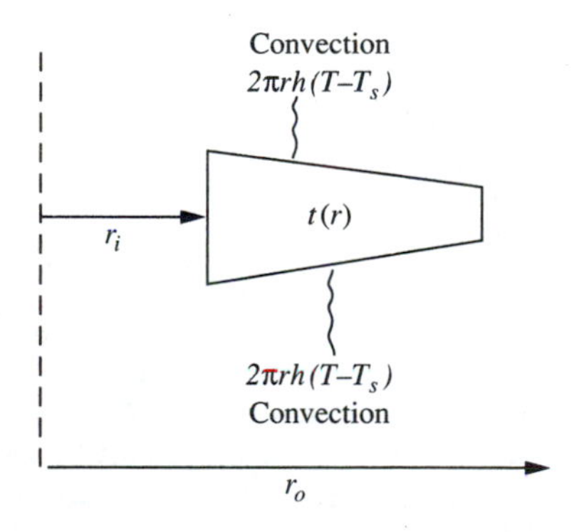

Show that based on an energy balance, the governing differential equation is

$$\frac{d[k(r)rt(r)dT/dr]}{dr} - 2rhT + 2rhT_s = 0$$

Take the boundary conditions to be $T(r_i) = T_i$ and $T(r_0) = T_0$. For $t_i = 5$ cm, $t_0 = 2$ cm, $r_i = 5$ cm, $r_0 = 20$ cm, $T_i = 300°C$, $T_0 = 80°C$ and $T_s = 80°C$. With $k = 0.2$ W/mm$-$°C, and $h = 10^{-3}$ W/mm$^2-$°C, set up and solve equal-length linearly interpolated finite element models with one, two, and four elements. For each of the elements, plot the temperature and the derived variable $q_r = -krtT$. Where within each of the elements should the derived variable be evaluated?

Quadratic interpolation exercises for eigenvalue problems. For each of the following, set up and solve the algebraic eigenvalue problem based on a finite element model using one, two, and then four equal-length quadratically interpolated elements.

2.180. $u'' + \lambda u = 0, u(0) = 0, u(L) = 0$
2.181. $u'' + \lambda x u = 0, u(0) = 0, u(L) = 0$
2.182. $x^2u'' + xu' + \lambda x^2 u = 0, u'(1) = 0, u(2) = 0$
2.183. $xu'' + 2u' + \lambda x^{-1} u = 0, u(1) = 0, u'(2) + u(2) = 0$

Section 2.9.4

2.184. Set up and solve a one-element model for the torsional vibrations of the uniform circular shaft fixed at both ends as shown.

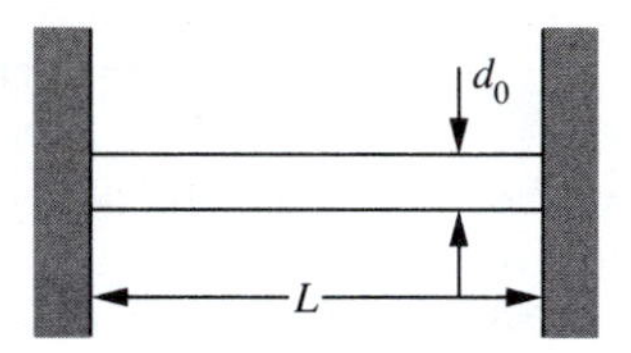

2.185. Repeat the previous problem using two elements.
2.186. Repeat the previous problem using three elements.
2.187. Set up and solve a one-element model for the torsional vibrations of the variable-radius shaft fixed at one end as shown. Take $d_0 = D_0/2$.

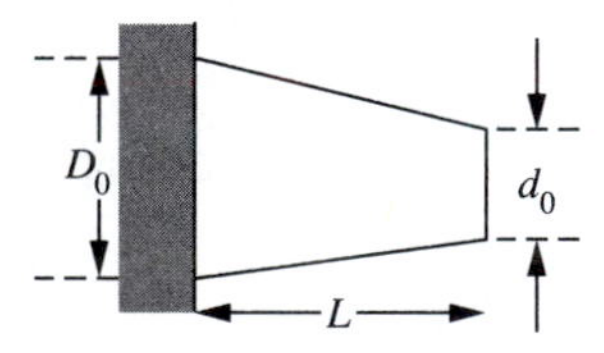

2.188. Repeat the previous problem using two elements.
2.189. Repeat the previous problem using four elements.
2.190. An important technical problem involves systems of shafts connected by gears. The study of such systems can begin by investigating the problem of a circular disk rigidly attached to a uniform shaft as shown.

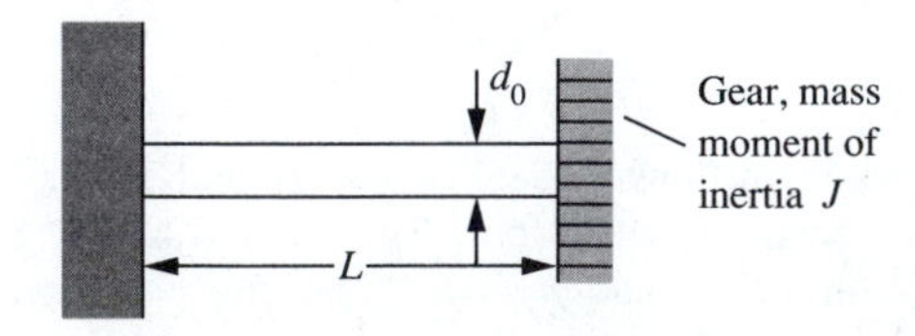

The corresponding finite element model is constructed by simply adding J to the corresponding main diagonal element of the global **M** matrix. The frequencies and mode shapes are quite accurate as long as J is large compared to the mass moment of inertia of the shaft. Set up and solve a 1 element model for the torsional vibrations of the circular shaft fixed at one end as shown. Take $J = 2J_{\text{shaft}} = 2(\pi d_0^4 \rho L/32)$.

2.191. Repeat the previous problem using two elements. Take $J = 2J_{\text{shaft}} = 2(\pi d_0^4 \rho L/32)$.

2.192. Repeat the previous problem using four elements. Take $J = 2J_{\text{shaft}} = 2(\pi d_0^4 \rho L/32)$.

2.193. Set up and solve a two-element model for the disk-shaft system shown. Take $J = 2J_{\text{shaft1}} = 2(\pi d_1^4 \rho L_1/32)$, $d_1 = d_2$ and $L_1 = L_2$.

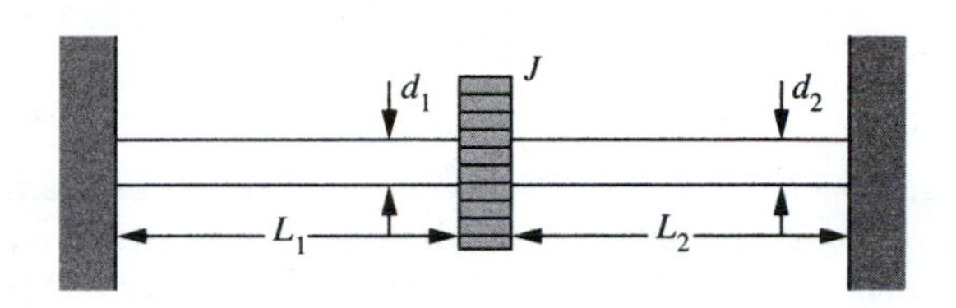

2.194. Repeat the previous problem using four equal length elements.

Axial vibrations. For axial vibrations the elemental stiffness and mass matrices are given by

$$\mathbf{k_e} = \int_{x_i}^{x_{i+1}} \mathbf{N}' AE \mathbf{N}'^{\mathrm{T}} dx \qquad \text{and} \qquad \mathbf{m_e} = \int_{x_i}^{x_{i+1}} \mathbf{N} A \rho \mathbf{N}^{\mathrm{T}} dx$$

Concentrated masses are handled by simply adding the mass M to the corresponding main diagonal element of the global $\mathbf{M}$ matrix. The rest of the basic steps in the finite element method are carried out in exactly the same manner as described in the previous section for the corresponding torsion problem.

2.195. Set up and solve a one-element model for the axial motion of the uniform bar shown. Compare your results with those from the corresponding exact solution.

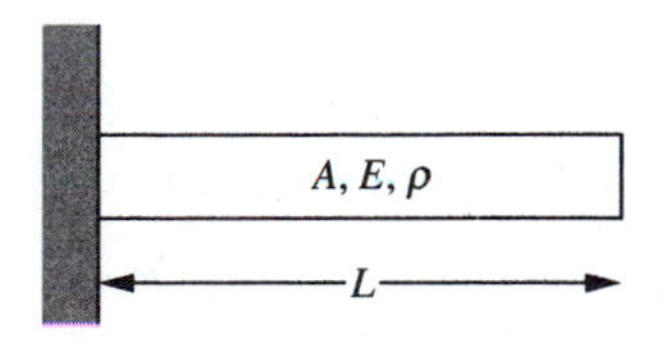

2.196. Repeat the previous problem using two equal-length elements.

2.197. Repeat the previous problem using three equal-length elements.

2.198. For the variable-radius, circular-cross-section bar shown set up a two-element model to determine the frequencies and mode shapes. Take $d_0 = D_0/2$.

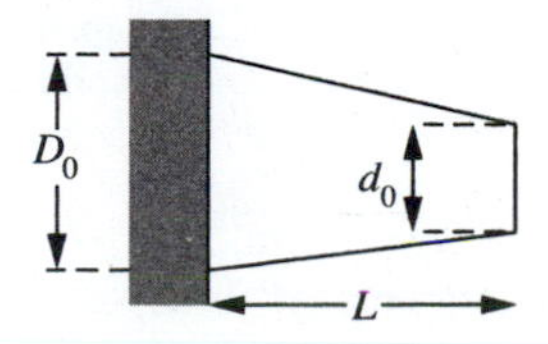

2.199. Repeat the previous problem using four elements.
2.200. Repeat the previous problem using eight elements.
2.201. Frequently the axial motion of a bar takes place in the presence of an additional mass M as shown.

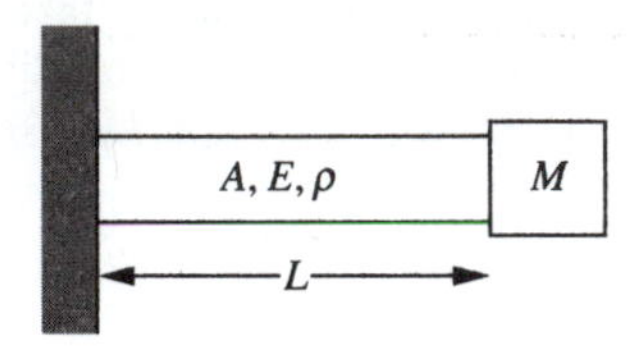

The corresponding finite element model is constructed by simply adding M to the corresponding main diagonal element of the global **M** matrix. The frequencies and mode shapes are quite accurate as long as M is large compared to the mass moment of inertia of the shaft. Set up and solve a one-element model for the axial vibrations of uniform bar fixed at one end as shown. Take $M = 2m_{\text{bar}} = 2\rho AL$.
2.202. Repeat the previous problem using two elements. Take $M = 2m_{\text{bar}}$.
2.203. Repeat the previous problem using four elements. Take $M = 2m_{\text{bar}}$.

Section 2.10

2.204. Verify Eq. (2.53) for $\mathbf{k_e}$.
2.205. Verify Eq. (2.54) for $\mathbf{q_e}$ for the linearly varying $q(x)$.
2.206. For the problem indicated below,

Differential equation: $EIw^{iv} = q(x)$

Boundary Conditions: $EIw''(0) = 0$

$$EIw'''(0) + k_L w(0) = Q_0$$

$$w(L) = 0$$

$$EIw''(L) + C_R w'(L) = -M_0$$

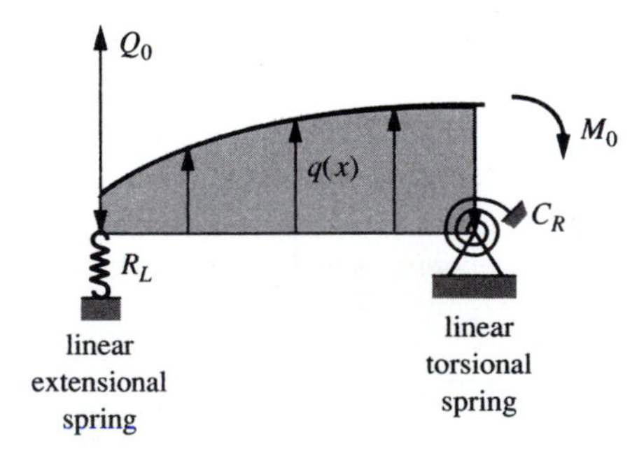

a. Develop the appropriate weak form.
b. What is the corresponding energy? Interpret each term.

For each of the following, set up and solve the indicated finite element models.

2.207. Take $EI = 3 \times 10^7$ lbf-in^2, $q_0 = 10$ lbf/in., and $L = 36$ in.

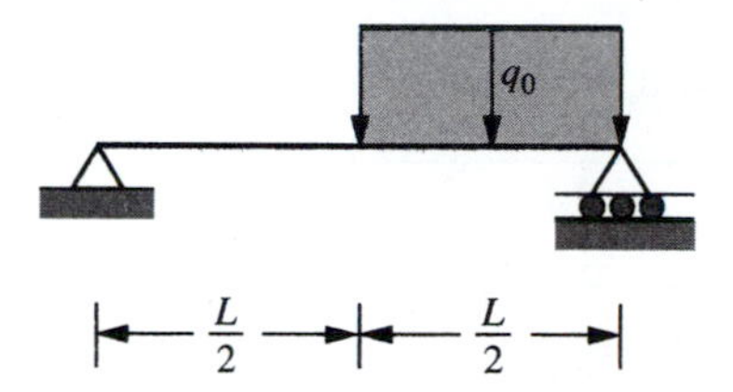

a. Use two elements, one for each half.
b. Three elements, one for $0 \le x \le L/2$, two for $L/2 \le x \le L$.

2.208. Take $EI = 3 \times 10^7$ lbf-in^2, $q_0 = 20$ lbf/in., and $L = 48$ in.

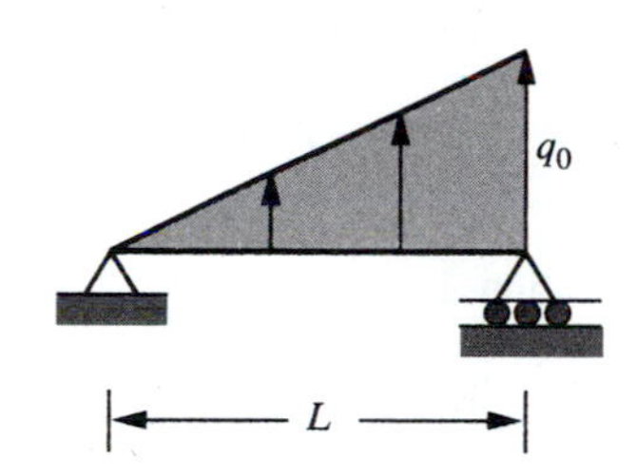

a. One element
b. Two elements
c. Four elements

2.209. Take $EI = 3 \times 10^7$ lbf-in^2, $q_0 = 30$ lbf/in., and $L = 24$ in.

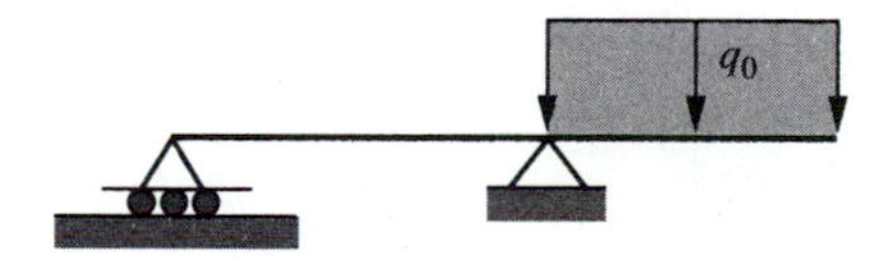

a. Two elements, one for each span.
b. Three elements, one for the first span, two for the second span.
c. Five elements, one for the first span, four for the second span.

2.210. Two-dimensional *framed* structures are generally composed of members that transmit all of axial and shear forces and bending moments. Within a given straight element, it is assumed that the axial and bending problems are uncoupled. The stiffness matrices for the axial and bending problems can then be superposed to yield the 6×6 stiffness matrix $\mathbf{k_e}$ indicated.

$$\begin{array}{cccccc} u_1 & v_1 & \theta_1 & u_2 & v_2 & \theta_2 \end{array}$$

$$\mathbf{k_e} = \begin{bmatrix} AE/l_e & 0 & 0 & -AE/l_e & 0 & 0 \\ 0 & 12EI/l_e^3 & 6EI/l_e^2 & 0 & -12EI/l_e^3 & 6EI/l_e^2 \\ 0 & & 4EI/l_e & 0 & -6EI/l_e^2 & 2EI/l_e \\ & & & AE/l_e & 0 & 0 \\ & \text{symm} & & & 12EI/l_e^3 & -6EI/l_e^2 \\ & & & & & 4EI/l_e \end{bmatrix}$$

The degrees of freedom are indicated in the figure.

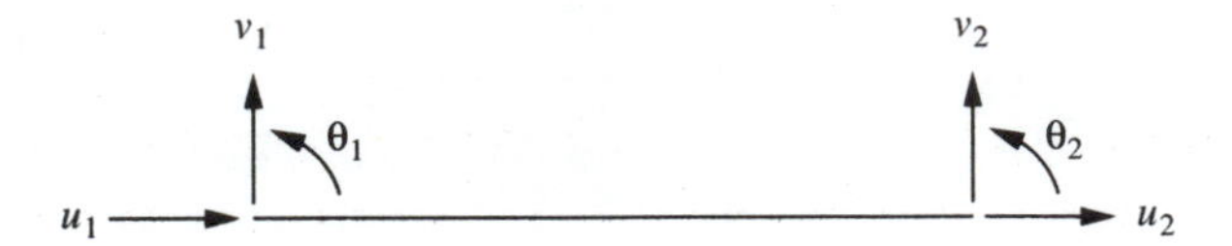

Each of these elemental stiffness matrices must be referred to the global coordinate system. After going through exactly the same transformation as was presented in Section 1.4.5, there results

$$\mathbf{k}_G = \begin{bmatrix} \mathbf{R} & \mathbf{O} \\ \mathbf{O} & \mathbf{R} \end{bmatrix} \mathbf{k}_e \begin{bmatrix} \mathbf{R}^T & \mathbf{O} \\ \mathbf{O} & \mathbf{R}^T \end{bmatrix}$$

where

$$\mathbf{R} = \begin{bmatrix} \cos\theta & -\sin\theta & 0 \\ \sin\theta & \cos\theta & 0 \\ 0 & 0 & 1 \end{bmatrix}$$

with θ defined as in Section 1.4.5. Assembly, constraints, and computation of derived variables are carried out in the usual fashion.

Apply these ideas to the solution of the problem indicated, and hence determine the axial force, shear force, and bending moment in each segment.

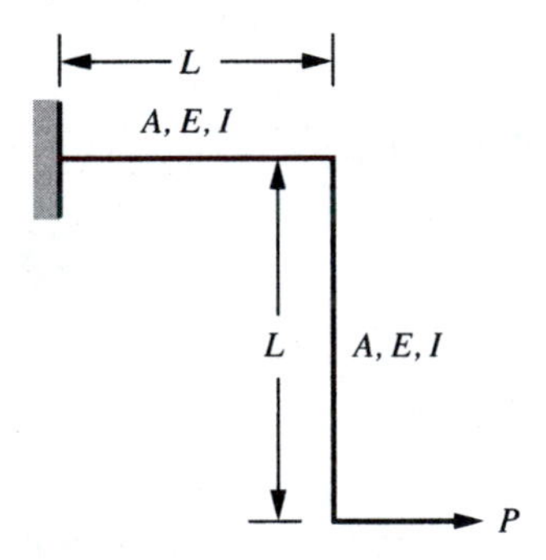

Determine also u, v, and θ at the point of application of the load P. Verify your answer by an independent calculation. Take $L = 20$ in., $AE = 3 \times 10^7$ lbf, $EI = 2.5 \times 10^6$ lbf-in^2, and $P = 100$ lbf.

CHAPTER 3

Two-Dimensional Boundary Value Problems

Chapter Contents

3.1 Introduction
 3.1.1 Problem statement
3.2 The Galerkin finite element model
3.3 Variational finite element models
 3.3.1 The weak formulation and variational principles
 3.3.2 The Ritz finite element model
3.4 Evaluation of matrices—linearly interpolated triangular elements
3.5 Example applications
 3.5.1 Torsion of a rectangular section
 3.5.2 Conduction heat transfer with boundary convection
3.6 Rectangular elements
 3.6.1 Evaluation of matrices
 3.6.2 Torsion of a rectangular section
 3.6.3 Conduction heat transfer with boundary convection
3.7 Eigenvalue problems
 3.7.1 Introduction
 3.7.2 Finite element models for the Helmholtz equation
 3.7.3 Examples for the Helmholtz equation
3.8 Isoparametric elements and numerical integration
 3.8.1 Four-node quadrilateral (Q4) elements
  3.8.1.1 Transformations and shape functions
  3.8.1.2 Evaluation of Q4 elemental matrices
  3.8.1.3 Application—stress concentration for the torsion of an angle section
  3.8.1.4 Application—steady-state heat transfer
 3.8.2 Eight-node quadrilateral (Q8) elements
  3.8.2.1 Transformations and shape functions
  3.8.2.2 Evaluation of Q8 elemental matrices
  3.8.2.3 Application—stress concentration for the torsion of an angle section
  3.8.2.4 Application—steady-state heat transfer
 3.8.3 Six-node triangular (T6) elements
  3.8.3.1 Transformations and shape functions
  3.8.3.2 Evaluation of T6 elemental matrices
  3.8.3.3 Application—stress concentration for the torsion of an angle section
  3.8.3.4 Application—steady-state heat transfer
3.9 Axisymmetric problems
3.10 Closure
References
Computer projects
Exercises

3.1 INTRODUCTION

Laplace's equation, written in rectangular Cartesian coordinates as

$$\nabla^2 u = \frac{\partial^2 u}{\partial x^2} + \frac{\partial^2 u}{\partial y^2} = 0$$

and Poisson's equation, written as

$$\nabla^2 u + f = \frac{\partial^2 u}{\partial x^2} + \frac{\partial^2 u}{\partial y^2} + f = 0$$

occur with remarkable frequency in all areas of engineering and applied mathematics. A few of the physical situations that can have models involving these equations include:

1. Steady-state heat conduction problems
2. Torsion problems in solid mechanics
3. Diffusion flow in porous media
4. Incompressible inviscid fluid flow
5. Electrostatic potentials
6. Gravitational or Newtonian potentials
7. Magnetostatics

All of these physical problems involve equilibrium or time-independent states. With suitable assumptions regarding the physics, the application of the appropriate physical law or principle results in what is termed an **elliptic boundary value problem.** That is, one governed by Laplace's equation, Poisson's equation, or several generalizations.

A corresponding two-dimensional eigenvalue problem is the **Helmholz equation** written in rectangular Cartesian coordinates as

$$\nabla^2 u + \lambda u = \frac{\partial^2 u}{\partial x^2} + \frac{\partial^2 u}{\partial y^2} + \lambda u = 0$$

Equations of this type occur in various situations involving vibrations and in the solution of parabolic and hyperbolic partial differential equations. In this section we will outline several important engineering application areas and will use these to demonstrate the finite element method throughout the remainder of the chapter.

SOLID MECHANICS APPLICATION. An important technical application from solid mechanics is the problem of the torsion of a homogeneous isotropic prismatic bar of arbitrary cross section as shown in Fig. 3–1. The three basic ideas of solid mechanics—kinematics, kinetics, and constitution—are used to develop the strain-displacement relations, the equilibrium equations, and the stress-strain relationships, after which the results are combined to formulate the governing equations.

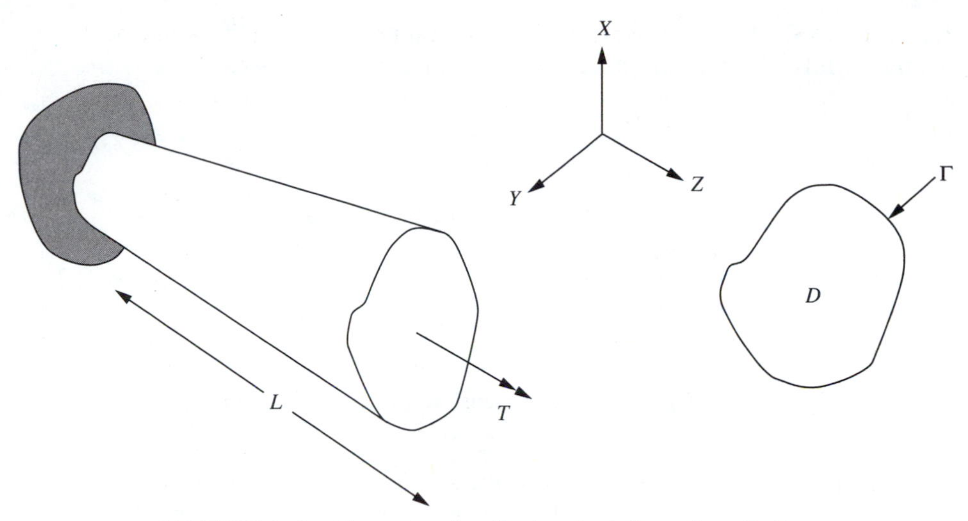

FIGURE 3–1 Geometry for the torsion of a prismatic bar

Using these basic ideas, the task of determining the deformations and stresses in a bar of arbitrary cross section [Boresi and Chong, 1] can be reduced to consideration of the two-dimensional boundary value problem

$$\nabla^2\phi(X, Y) + 2G\theta = 0 \qquad \text{in } D \tag{a}$$

$$\phi(X, Y) = 0 \qquad \text{on } \Gamma \tag{b}$$

where ϕ is the Prandtl stress function, G is the shear modulus, and θ is the constant rate of twist along the axis of the bar. D refers to the interior and Γ to the boundary of the cross section shown in Fig. 3–1. The nonzero stress components are given in terms of the stress function by

$$\tau_{XZ} = \frac{\partial\phi}{\partial Y} \qquad \text{and} \qquad \tau_{YZ} = -\frac{\partial\phi}{\partial X} \tag{c}$$

The total torque transmitted is

$$T = 2\int\int_D \phi \, dA \tag{d}$$

Generally the determination of the stress function ϕ and the stresses in terms of the applied torque T follows these steps:

1. Determine ϕ by solving the differential equation and boundary conditions, $\phi = \phi(G, \theta;\ X, Y)$.
2. Express $T = T(G\theta)$ using Eq. (d). $G\theta$ can then be eliminated so that, using Eq. (c), the stresses are expressed in terms of the transmitted torque T.

HEAT TRANSFER APPLICATION. The general problem of steady-state heat conduction [Mills, 2] in a thin plane body is depicted in Fig. 3–2.

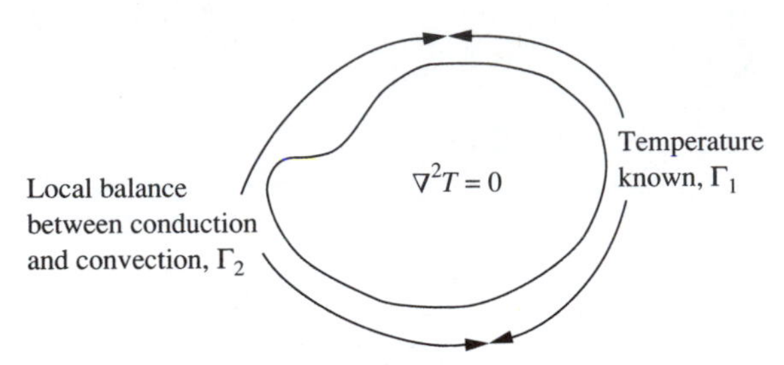

FIGURE 3–2 Two-dimensional heat conduction

With suitable simplifying assumptions, the application of the law of conservation of energy in the interior of the body leads to the differential equation

$$\nabla^2 T = 0$$

governing the temperature distribution. As indicated in Fig. 3–2, it is frequently the case that the temperature is known on a portion of the boundary and that there is a local balance between conduction $q = -kA(\partial T/\partial n)$ in the interior and exterior convection $q = hA(T - T_\infty)$ on the remaining portion of the boundary. Using a set of rectangular coordinates for reference, the physical problem is stated as

$$\nabla^2 T(x, y) = 0 \qquad \text{in } D$$

$$T(x, y) = T_0(s) \qquad \text{on } \Gamma_1$$

$$-k\frac{\partial T}{\partial n} + h(T - T_\infty) = 0 \qquad \text{on } \Gamma_2$$

where k is referred to as the thermal diffusivity, h is a convective heat transfer coefficient, T_∞ is a free stream temperature and $\partial/\partial n$ refers to the derivative in the outward direction normal to the boundary. Γ_1 and Γ_2 refer to portions of the boundary. The differential equation must be solved subject to the boundary conditions. Also of interest is the heat flux or energy transfer given by $\mathbf{q} = -kA\nabla T$, computed after the solution is obtained.

3.1.1 Problem Statement

Generally, the mathematical statement of a two-dimensional elliptic boundary value problem assumes the form

$$\nabla^2 u(x, y) + f(x, y) = 0 \qquad \text{in } D \tag{3.1a}$$

$$u = g(s) \qquad \text{on } \Gamma_1 \tag{3.1b}$$

$$\frac{\partial u}{\partial n} + \alpha(s)u = h(s) \qquad \text{on } \Gamma_2 \tag{3.1c}$$

where D is the interior of the domain, and Γ_1 and Γ_2 constitute the boundary as indicated in Fig. 3–3. The differential equation is generally a balance equation.

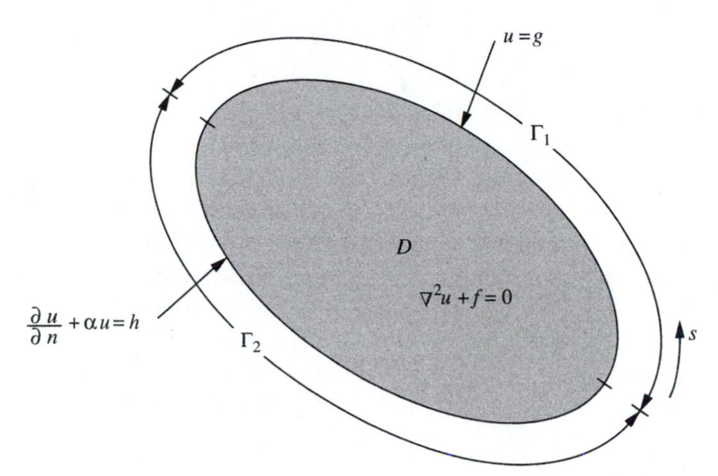

FIGURE 3–3 Typical region showing D, Γ_1, and Γ_2

A boundary condition that specifies the function u on Γ_1 will be referred to as a type one or **Dirichlet** boundary condition. A boundary condition that specifies a linear combination of the function and its normal derivative, that is, $\partial u/\partial n + \alpha u = h(s)$, on Γ_2 will be referred to as a type two boundary condition. If the α appearing in the type two boundary condition is zero (i.e., $\partial u/\partial n = h(s)$), the boundary condition is referred to as a **Neumann** boundary condition, whereas if α is not zero the boundary condition is known as a **Robins** boundary condition. The boundary condition of type two is generally a local balance equation which must be satisfied at the boundary. If the entire boundary is of type one, the boundary value problem is known as a Dirichlet problem. If the entire boundary is of type two with $\alpha = 0$, the boundary value problem is known as a Neumann problem.

The *classical* solution of the boundary value problem is a function $u(x, y)$ that possesses second partial derivatives everywhere in D, satisfies the partial differential equation, and satisfies the boundary conditions on Γ_1 and Γ_2. Except in situations where the domain D has a particular regular shape such as that of a rectangle or circle, and except where the boundary conditions are prescribed in certain relatively restricted ways, a classical or analytical solution is generally not feasible, and a numerical approach such as the finite element method is indicated. It is the purpose of this chapter to discuss in detail the numerical solution of the boundary value problems associated with Laplace's and Poisson's equations, and the eigenvalue problem in terms of Helmholz's equation, using the finite element method. The finite element models based on both the Ritz and Galerkin methods will be presented and discussed.

3.2 THE GALERKIN FINITE ELEMENT MODEL

In this section the finite element model is developed for the problem

$$\nabla^2 u + f = 0 \qquad \text{in } D$$

$$u = g(s) \qquad \text{on } \Gamma_1$$

$$\frac{\partial u}{\partial n} + \alpha(s)u = h(s) \qquad \text{on } \Gamma_2$$

The final set of linear algebraic equations is generated using the classical method of Galerkin applied directly to the differential equation. Integration by parts is then used to enter the natural boundary conditions $\partial u/\partial n + \alpha u = h$ into the formulation of the problem. The linearly interpolated triangular element is used in order for the student to be able to most easily detect the basic structure of the finite element method without the unnecessary details associated with more complicated elements and interpolations. The Galerkin finite element model is formulated by carrying out in a systematic manner the basic steps of discretization, interpolation, elemental formulation, assembly, constraints, solution, and computation of derived variables.

Discretization. The first step in developing a finite element model is discretization. A typical discretization using straight-sided triangular elements is indicated in Fig. 3–4. *Nodes, elements,* and *interelement boundaries* are indicated in the figure.

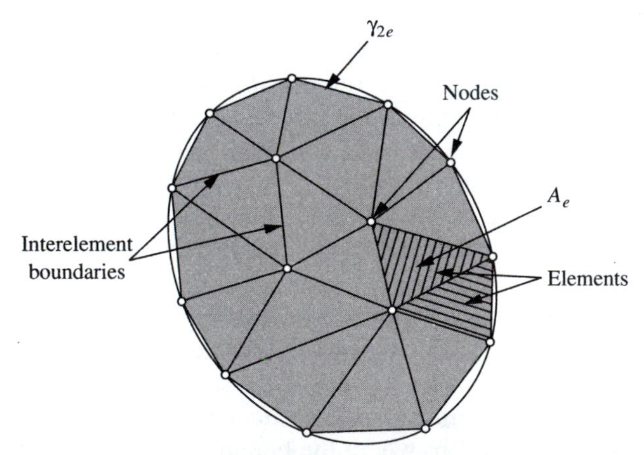

FIGURE 3–4 Typical mesh for the two-dimensional elliptic boundary value problem using linearly interpolated elements

It is immediately obvious that with straight-sided elements there is in general an inherent error in modeling the shape of the domain D. The effect of these errors is expected to diminish as the size of the elements is reduced. In addition to the problem associated with the errors in geometry, it will not be possible in general to enforce the boundary conditions exactly since the data that are prescribed on the actual boundary will not be correct for the replacement boundary. We will rely on the knowledge that elliptic boundary value problems of this type are well posed [Courant and Hilbert, 3], which means in part that the small changes in the input resulting from the new boundary and the new boundary values will result in small changes in the solution. Elements with curved sides can be expected to improve upon but not entirely eliminate such errors.

Generally the *mesh* (nodes, elements, and interelement boundaries) should be selected so as to be relatively fine in regions where large gradients or slopes are expected. Transitions from regions where the mesh is relatively fine to regions where it is relatively crude should be gradual. Extreme geometries such as triangles with "small" angles ($\theta \leq$ approximately $\pi/8 = 22.5°$ as a rule of thumb) are to be avoided.

Interpolation. Interpolation for the straight-sided three-node triangle corresponds to assuming that the solution surface $u(x, y)$ is represented by a plane defined above the element in question. This linear representation is indicated in Fig. 3–5. The solution surface is assumed to consist of segments of planes over each of the elements, that is, to be piecewise linear.

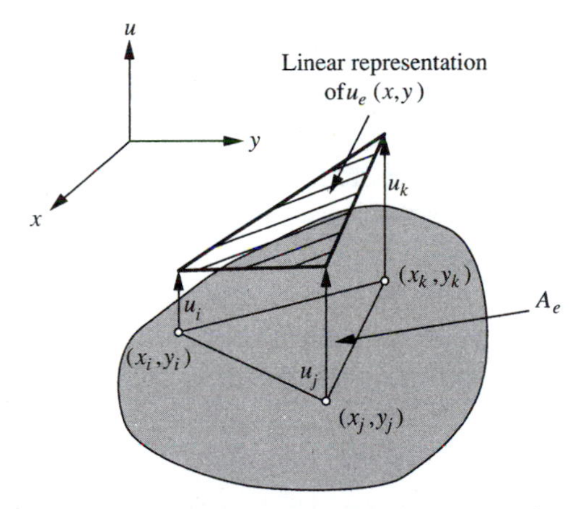

FIGURE 3–5 Linear interpolation on a triangle

Analytically we write

$$u_e(x, y) = \alpha + \beta x + \gamma y \tag{3.2}$$

where the three constants α, β, and γ are determined by the three equations requiring that u assume the nodal values at the vertices of the triangle, namely

$$\begin{aligned} u_e(x_i, y_i) &= \alpha + \beta x_i + \gamma y_i = u_i \\ u_e(x_j, y_j) &= \alpha + \beta x_j + \gamma y_j = u_j \\ u_e(x_k, y_k) &= \alpha + \beta x_k + \gamma y_k = u_k \end{aligned} \tag{3.3}$$

where i, j, and k denote the three vertices of the triangle. The x_i, y_i, x_j, y_j, x_k, y_k, u_i, u_j, and u_k are all indicated in Fig. 3–5. When the three equations (3.3) are solved for α, β, and γ and the results substituted back into Eq. (3.2), the elemental expression for $u_e(x, y)$ can be written as

$$u_e(x, y) = u_i N_i + u_j N_j + u_k N_k$$

where

$$N_i = \frac{a_i + b_i x + c_i y}{2A_e} \qquad i = 1, 2, 3$$

with

$$\begin{aligned} a_i &= x_j y_k - x_k y_j \\ b_i &= y_j - y_k \\ c_i &= x_k - x_j \end{aligned} \tag{3.4}$$

where i, j, and k are to be permuted cyclically. $2A_e$ is the determinant of the coefficients, namely

$$2A_e = \begin{vmatrix} 1 & x_i & y_i \\ 1 & x_j & y_j \\ 1 & x_k & y_k \end{vmatrix}$$

with A_e as the area of the element. Any numbering that proceeds counterclockwise around the element is acceptable: (i, j, k), (j, k, i), and (k, i, j). This counterclockwise direction is necessary in order to ensure that the area A_e be positive. N_i, N_j, and N_k, termed the **elemental linear interpolation functions**, are indicated in Fig. 3–6.

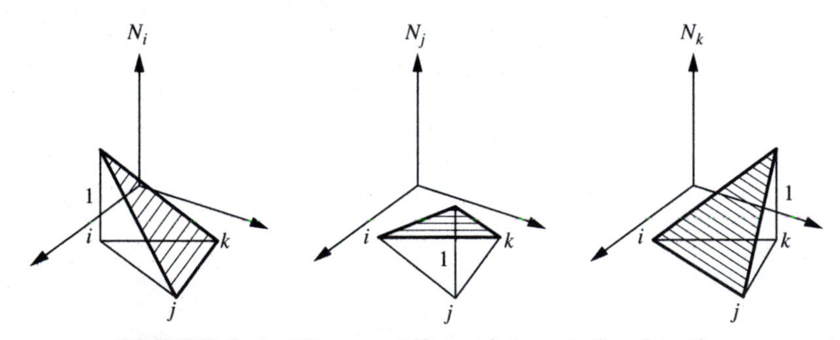

FIGURE 3–6 Elemental linear interpolation functions

With N as the number of nodes, an approximate Galerkin solution is taken to have the form

$$v(x, y) = \sum_{1}^{N} v_i n_i(x, y) \tag{3.5}$$

where the $n_i(x, y)$, shown in Fig. 3–7, are termed the **nodal linear interpolation functions.**

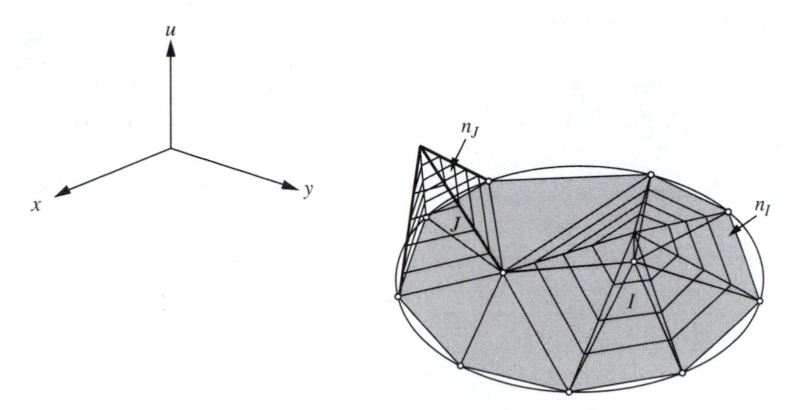

FIGURE 3–7 Nodal linear interpolation functions

It is clear from Fig. 3–7 that a given interior nodal interpolation function $n_I(x, y)$ can be considered to be composed of all the elemental $N_I(x, y)$ touching the interior node I. Similarly, a boundary nodal interpolation function $n_J(x, y)$ can be considered to be composed of all the elemental $N_J(x, y)$ touching the boundary node J. Note that as defined, the $n_I(x, y)$ have the property

$$n_I(x_J, y_J) = \delta_{IJ} = 1 \qquad \text{if } I \neq J$$

$$n_I(x_J, y_J) = 0 \qquad \text{if } I \neq J$$

In other words, $n_I(x, y)$ has a value of unity only at the node I and is zero at *all* other nodes. Thus using Eq. (3.5) it follows that

$$v(x_J, y_J) = \sum_1^N v_i n_i(x_J, y_j) = \sum_1^N v_i \delta_{iJ} = v_J$$

that is, that the coefficients v_i in Eq. (3.5) are actually the nodal values of the solution. The nodally based linear interpolation functions n_I pictured in Fig. 3–7 are sometimes called two-dimensional *hat* or *roof* functions for obvious reasons. It is also sometimes said that these functions have local support from the standpoint that they have nonzero values only in a small local neighborhood of the node in question. This small local neighborhood consists of only those elements whose definition involves node I.

In summary, the actual or exact solution $u(x, y)$ can be pictured as a surface above the domain D as indicated in Fig. 3.8*a*. The approximate solution $v(x, y) = \sum n_i v_i$ can be pictured as a piecewise linear or piecewise planar surface above the approximate domain D^- as shown in Fig. 3–8*b*. We expect that properly executed, the finite element method will generate values for the v_i such that $v(x, y) \approx u(x, y)$.

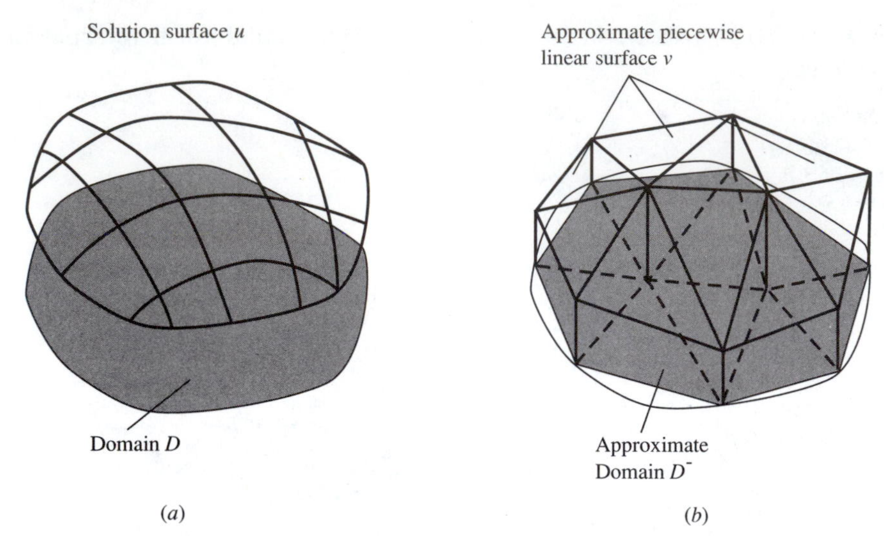

FIGURE 3–8 Piecewise planar character of the approximate solution

Elemental formulation. Substitution of the approximate solution $v(x, y)$ into the governing differential equation yields

$$\frac{\partial}{\partial x}\left(\frac{\partial v}{\partial x}\right)+\frac{\partial}{\partial y}\left(\frac{\partial v}{\partial y}\right)+f = E_N(x, y, v_1, v_2, \ldots, v_N)$$

where E_N is the **error** resulting from the fact that, in general, v is not a solution of the differential equation. The Galerkin method requires that the integral over D^- of the product of the error function E_N with *each* of the nodal interpolation functions $n_i(x, y)$ be zero; that is,

$$\int\int_{D^-} E_N n_k(x, y) dA = 0 \qquad k = 1, 2, \ldots, N$$

or

$$\int\int_{D^-}\left[\frac{\partial}{\partial x}\left(\frac{\partial v}{\partial x}\right)+\frac{\partial}{\partial y}\left(\frac{\partial v}{\partial y}\right)+f\right] n_k(x, y) dA = 0 \qquad k = 1, 2, \ldots, N$$

which we will refer to collectively as the Galerkin equations. D^- is the area occupied by the collection of triangular elements of the mesh.

Note that the term $\partial(\partial v/\partial x)/\partial x$ involves second derivatives of the nodal interpolation functions whereas the nodal interpolation functions n_k also appear without any derivatives in the equations. We can remedy this asymmetry and at the same time bring the boundary conditions into the formulation by integrating by parts. To prepare for the integration by parts, write

$$n_k \frac{\partial}{\partial x}\left(\frac{\partial v}{\partial x}\right) = \frac{\partial}{\partial x}\left(n_k \frac{\partial v}{\partial x}\right) - \frac{\partial n_k}{\partial x}\frac{\partial v}{\partial x}$$

and

$$n_k \frac{\partial}{\partial y}\left(\frac{\partial v}{\partial y}\right) = \frac{\partial}{\partial y}\left(n_k \frac{\partial v}{\partial y}\right) - \frac{\partial n_k}{\partial y}\frac{\partial v}{\partial y}$$

so that the Galerkin equations can be expressed as

$$\iint_{D^-} \left[\frac{\partial}{\partial x}\left(n_k \frac{\partial v}{\partial x}\right) + \frac{\partial}{\partial y}\left(n_k \frac{\partial v}{\partial y}\right)\right] dA$$

$$-\iint_{D^-} \left[\frac{\partial n_k}{\partial x}\frac{\partial v}{\partial x} + \frac{\partial n_k}{\partial y}\frac{\partial v}{\partial y} - f n_k(x,y)\right] dA = 0 \qquad k = 1, 2, \ldots, N$$

The divergence theorem [4] states that

$$\iint_D \frac{\partial Q}{\partial x} dA = \int_\Gamma n_x Q\, ds \qquad \text{and} \qquad \iint_D \frac{\partial P}{\partial y} dA = \int_\Gamma n_y P\, ds$$

where n_x and n_y are the components of the unit external normal around the bounding curve Γ. The unit normal **n** and the positive direction for integrating around Γ are shown in Fig. 3–9.

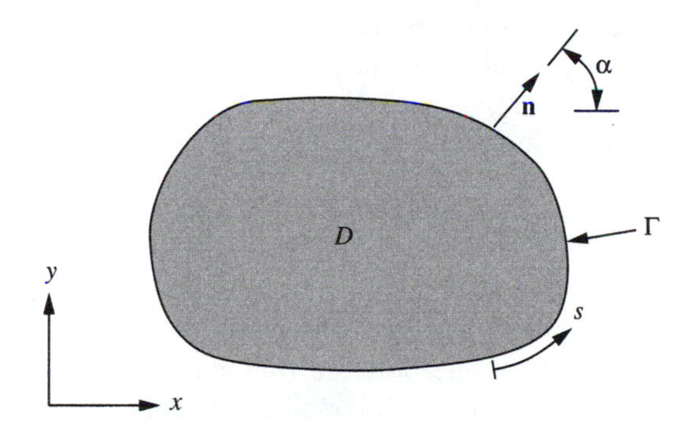

FIGURE 3–9 Positive directions for n and *s*

Applying the divergence theorem to the terms in the first integral, the Galerkin equations can be expressed as

$$\int_{\Gamma-} \left(n_x \frac{\partial v}{\partial x} + n_y \frac{\partial v}{\partial y}\right) n_k\, ds - \iint_{D-} \left[\frac{\partial n_k}{\partial x}\frac{\partial v}{\partial x} + \frac{\partial n_k}{\partial y}\frac{\partial v}{\partial y} - f n_k(x,y)\right] dA = 0$$

$$k = 1, 2 \ldots, N$$

where Γ^- is the collection of straight-line segments approximating the bounding curve Γ. Using the fact that

$$n_x \frac{\partial v}{\partial x} + n_y \frac{\partial v}{\partial y} = \frac{\partial v}{\partial n}$$

and that from the boundary condition we are taking $\partial v/\partial n \approx h(s) - \alpha v(s)$ on Γ_2, the Galerkin equations can now be written as

$$\int_{\Gamma_1-}\left[n_x\frac{\partial v}{\partial x} + n_y\frac{\partial v}{\partial y}\right]n_k ds + \int_{\Gamma_2-}[h(s) - \alpha v(s)]n_k ds$$
$$- \iint_{D-}\left[\frac{\partial n_k}{\partial x}\frac{\partial v}{\partial x} + \frac{\partial n_k}{\partial y}\frac{\partial v}{\partial y} - f n_k(x, y)\right]dA = 0 \qquad k = 1, 2 \ldots, N \tag{3.6}$$

where Γ_1^- indicates the collection of straight-line segments approximating the part of the boundary on which the *essential* boundary conditions are prescribed, and Γ_2^- the collection of straight-line segments approximating the part of the boundary on which the *natural* boundary conditions are prescribed.

The traditional Galerkin method is based upon an approximate solution that exactly satisfies both the essential boundary conditions on Γ_1 and the natural boundary conditions on Γ_2. When an integration by parts is carried out as above and the natural boundary conditions on Γ_2 become a part of the formulation of the problem, it turns out that the natural boundary conditions are satisfied approximately by the solution. As the mesh is refined, the satisfaction of the natural boundary conditions becomes more exact.

The essential boundary conditions must still be specifically enforced. This can be done in two different ways. First, the essential boundary conditions can be built into the approximate solution according to

$$v(x) = \sum_{i=1}^{M_1} n_i g_i + \sum_{i=1}^{N-M_1} n_i v_i$$

M_1 is the number of nodes on Γ_1 where u is prescribed to be $g_i = g(s_i)$ and the n_i are the corresponding nodal interpolation functions. The second term involves all the remaining unknowns on Γ_2 and in the interior of D. This approach is tedious to carry through algebraically and difficult to interpret.

The second approach assumes the solution to be of the form given by Eq. (3.5) (i.e., not including the constraints), then simply deletes the line integral on Γ_1 in Eq. (3.6) and imposes on the final set of equations the constraints $v_i = g_i, i = 1, 2, \ldots, M_1$, corresponding to the essential boundary conditions imposed at the nodes on Γ_1. That this second approach is exactly equivalent to the first is left for the student to show as an exercise.

Proceeding with the second approach, the Galerkin equations can be written as

$$\int_{\Gamma_2-}[h(s) - \alpha v(s)]n_k ds$$
$$- \iint_{D-}\left(\frac{\partial n_k}{\partial x}\frac{\partial v}{\partial x} + \frac{\partial n_k}{\partial y}\frac{\partial v}{\partial y} - f n_k\right)dA = 0 \qquad k = 1, 2 \ldots, N$$

or

$$\iint_{D-}\left(\frac{\partial n_k}{\partial x}\frac{\partial v}{\partial x}+\frac{\partial n_k}{\partial y}\frac{\partial v}{\partial y}\right)dA+\int_{\Gamma_{2-}}\alpha v(s)n_k ds$$

$$=\iint_{D-} f n_k dA+\int_{\Gamma_{2-}} h(s)n_k ds \qquad k=1,2\ldots,N$$

Then with $v = \Sigma n_i v_i$ the Galerkin equations can be written as

$$\iint_{D-}\left(\frac{\partial n_k}{\partial x}\sum v_i\frac{\partial n_i}{\partial x}+\frac{\partial n_k}{\partial y}\sum v_i\frac{\partial n_i}{\partial y}\right)dA+\int_{\Gamma_{2-}}\alpha\sum v_i n_i n_k ds$$

$$=\iint_{D-} f n_k dA+\int_{\Gamma_{2-}} h(s)n_k ds \qquad k=1,2\ldots,N$$

We write this $N \times N$ set of linear algebraic equations as

$$\sum_{i=1}^{N} K_{ki}v_i = F_k \qquad k=1,2,\ldots,N \tag{3.7}$$

$$K_{ki}=\iint_{D-}\left(\frac{\partial n_k}{\partial x}\frac{\partial n_i}{\partial x}+\frac{\partial n_k}{\partial y}\frac{\partial n_i}{\partial y}\right)dA+\int_{\Gamma_{2-}}\alpha n_k n_i\, ds$$

and

$$F_k=\iint_{D-} n_k f\, dA+\int_{\Gamma_2^-} n_k h(s)\, ds$$

or in matrix notation as

$$\mathbf{Kv}=\mathbf{F}$$

Assembly. In actuality, assembly is implicit in the process of developing the final set of algebraic equations $\mathbf{Kv} = \mathbf{F}$. In order to see the essential nature of this set of equations and how $\mathbf{K}$ and $\mathbf{F}$ can be considered to have been formed by contributions from the individual elements, consider the simplified domain D^- and corresponding mesh indicated in Fig. 3–10.

Consider first the portions that arise from the area elements,

$$K_{ki}=\iint_{D-}\left(\frac{\partial n_k}{\partial x}\frac{\partial n_i}{\partial x}+\frac{\partial n_k}{\partial y}\frac{\partial n_i}{\partial y}\right)dA$$

and

$$F_k=\iint_{D-} n_k f\, dA$$

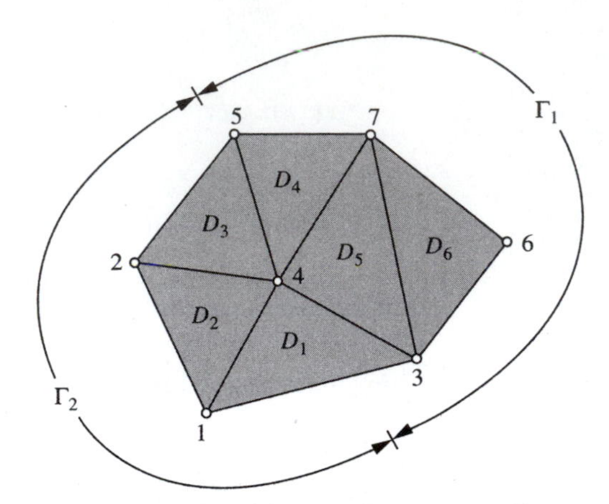

FIGURE 3–10 Domain *D* and corresponding mesh

where k and $i = 1, 2, \ldots, 7$. In order to avoid excessive writing in what follows, we define

$$a_{ki} = \frac{\partial n_k}{\partial x}\frac{\partial n_i}{\partial x} + \frac{\partial n_k}{\partial y}\frac{\partial n_i}{\partial y}$$

so that

$$K_{ki} = \iint_{D-} a_{ki}\, dA$$

The first equation ($k = 1$) is

$$\Sigma K_{1i} u_i = F_1$$

where

$$K_{1i} = \iint_{D-} a_{1i}\, dA$$

As n_1 and its derivatives are nonzero only in domains D_1 and D_2,

$$K_{1i} = \left[\iint_{D_1} + \iint_{D_2}\right] a_{1i}\, dA$$

In the integrals over D_1 and D_2 there will be contributions only from those nodes whose n_i are nonzero in either D_1 or D_2, that is, for n_1, n_2, n_3, and n_4. Thus v_1, v_2, v_3, and v_4 will appear in the first equation with the corresponding coefficients

$$K_{11} = \left[\iint_{D_1} + \iint_{D_2}^{*}\right] a_{11}\, dA$$

$$K_{12} = \iint_{D_2}^{*} a_{12}\, dA$$

$$K_{13} = \iint_{D_1} a_{13}\, dA$$

$$K_{14} = \left[\iint_{D_1} + \iint_{D_2}^{*}\right] a_{14}\, dA$$

The starred terms are those arising specifically from element 2.

Similarly for the second equation, only the integrals over D_2 and D_3 will contribute, with interaction between n_1, n_2, n_4, and n_5. It follows that v_1, v_2, v_4, and v_5 will appear in the equation with

$$K_{21} = \iint_{D_2}^{*} a_{21}\, dA$$

$$K_{22} = \left[\iint_{D_2}^{*} + \iint_{D_3}\right] a_{22}\, dA$$

$$K_{24} = \left[\iint_{D_2}^{*} + \iint_{D_3}\right] a_{24}\, dA$$

$$K_{25} = \iint_{D_3} a_{25}\, dA$$

with the starred terms again indicating the contribution from D_2.

For the third equation, integrals over D_1, D_2, and D_6 will arise with v_1, v_3, v_4, v_6, and v_7 appearing. The coefficients are

$$K_{31} = \iint_{D_1} a_{31}\, dA$$

$$K_{33} = \left[\iint_{D_1} + \iint_{D_5} + \iint_{D_6}\right] a_{33}\, dA$$

$$K_{34} = \left[\iint_{D_1} + \iint_{D_5}\right] a_{34}\, dA$$

$$K_{36} = \iint_{D_6} a_{36}\, dA$$

$$K_{37} = \left[\int\int_{D_5} + \int\int_{D_6}\right] a_{37}\, dA$$

For the fourth equation, domains D_1, D_2, D_3, D_4, and D_5 contribute with v_1, v_2, v_3, v_5, and v_7 occurring with

$$K_{41} = \left[\int\int_{D_1} + \int\int_{D_2}^{*}\right] a_{41}\, dA$$

$$K_{42} = \left[\int\int_{D_2}^{*} + \int\int_{D_3}\right] a_{42}\, dA$$

$$K_{43} = \left[\int\int_{D_1} + \int\int_{D_5}\right] a_{43}\, dA$$

$$K_{44} = \left[\int\int_{D_1} + \int\int_{D_2}^{*} + \int\int_{D_3} + \int\int_{D_4} + \int\int_{D_5}\right] a_{44}\, dA$$

$$K_{45} = \left[\int\int_{D_3} + \int\int_{D_4}\right] a_{45}\, dA$$

$$K_{47} = \left[\int\int_{D_4} + \int\int_{D_5}\right] a_{47}\, dA$$

Here again the contributions from D_2 are starred. The fifth, sixth, and seventh equations could be discussed in the same detail.

Consider specifically all the contributions from element 2 or domain D_2, displayed in a suggestive form as the 3×3 matrix

$$\mathbf{k_{e2}} = \begin{bmatrix} \int\int_{D_2} a_{11}\, dA & \int\int_{D_2} a_{12}\, dA & \int\int_{D_2} a_{14}\, dA \\ \int\int_{D_2} a_{21}\, dA & \int\int_{D_2} a_{22}\, dA & \int\int_{D_2} a_{42}\, dA \\ \int\int_{D_2} a_{41}\, dA & \int\int_{D_2} a_{42}\, dA & \int\int_{D_2} a_{44}\, dA \end{bmatrix}$$

Recalling that

$$a_{ki} = \frac{\partial n_k}{\partial x}\frac{\partial n_i}{\partial x} + \frac{\partial n_k}{\partial y}\frac{\partial n_i}{\partial y}$$

and recognizing that on D_2, the nodal interpolation functions and elemental interpolation functions are related according to $n_1 = N_1, n_2 = N_2$, and $n_4 = N_4$, the integrands of each of the above integrals can be written as

$$a_{ki} = \frac{\partial N_k}{\partial x}\frac{\partial N_i}{\partial x} + \frac{\partial N_k}{\partial y}\frac{\partial N_i}{\partial y}$$

In other words, the integrands can be expressed in terms of the elemental interpolation functions for the element corresponding to D_2. In terms of the elemental interpolation vector $\mathbf{N}^T = [N_1, N_4, N_2]^T$, the 3×3 matrix denoted above by $\mathbf{k_{e2}}$ can be written as

$$\mathbf{k_{e2}} = \iint_{D_2} \left[\frac{\partial \mathbf{N}}{\partial x} \frac{\partial \mathbf{N}^T}{\partial x} + \frac{\partial \mathbf{N}}{\partial y} \frac{\partial \mathbf{N}^T}{\partial y} \right] dA = \begin{bmatrix} c_{11} & c_{12} & c_{13} \\ c_{21} & c_{22} & c_{23} \\ c_{31} & c_{32} & c_{33} \end{bmatrix}$$

Recall that the elements of the elemental interpolation vector $\mathbf{N}$ must be ordered in a manner such that the element node numbers indicate a counterclockwise direction around the element. When $\mathbf{k_{e2}}$ is expanded to the global level there results

$$\mathbf{k_{g2}} = \begin{array}{c} \begin{matrix} 1 & 2 & 3 & 4 & 5 & 6 & 7 \end{matrix} \\ \begin{bmatrix} c_{11} & c_{13} & 0 & c_{12} & 0 & 0 & 0 \\ c_{31} & c_{33} & 0 & c_{32} & 0 & 0 & 0 \\ 0 & 0 & 0 & 0 & 0 & 0 & 0 \\ c_{21} & c_{23} & 0 & c_{22} & 0 & 0 & 0 \\ 0 & 0 & 0 & 0 & 0 & 0 & 0 \\ 0 & 0 & 0 & 0 & 0 & 0 & 0 \\ 0 & 0 & 0 & 0 & 0 & 0 & 0 \end{bmatrix} \end{array} \begin{matrix} 1 \\ 2 \\ 3 \\ 4 \\ 5 \\ 6 \\ 7 \end{matrix}$$

where column and row interchanges have been carried out in order that the coefficients of elemental stiffness matrix add to the correct positions in the global stiffness matrix. This process is further indicated in Fig. 3–11, where the node numbers at the elemental and global levels are clearly indicated. Each of the elemental $\mathbf{k_{ei}}$ is expanded to the global level, indicated by $\mathbf{k_{gi}}$, and assembled in the appropriate manner as part of the global stiffness matrix.

For the part of the load vector F arising from the function f on the domains D_i, it is necessary to evaluate and interpret

$$\iint_{D-} n_k f \, dA$$

For $k = 1$ the domains involved are D_1 and D_2, leading to

$$f_1 = \left[\iint_{D_1} + \iint_{D_2}^{*} \right] n_1 f \, dA$$

In a similar fashion the rest of the F_i that involve D_2 are

$$f_2 = \left[\iint_{D_2}^{*} + \iint_{D_3} \right] n_2 f \, dA$$

and

$$f_4 = \left[\iint_{D_1} + \iint_{D_2}^{*} + \iint_{D_3} + \iint_{D_4} + \iint_{D_5} \right] n_4 f \, dA$$

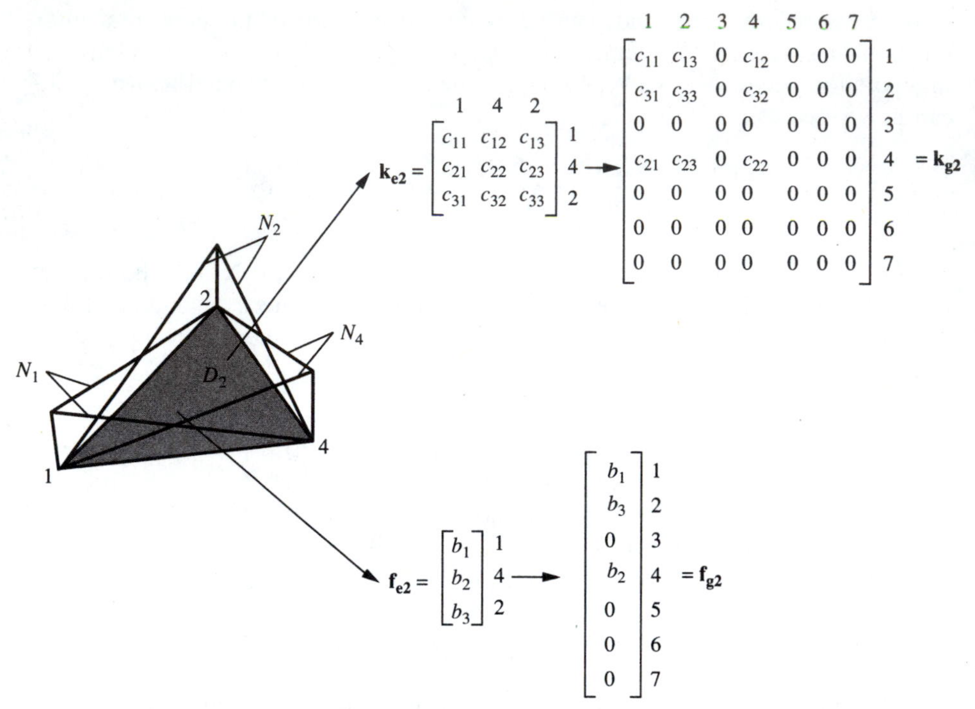

FIGURE 3–11 Elemental contributions and assembly for $\mathbf{k}_e$ and $\mathbf{f}_e$

Again, the collection of the starred terms is written in suggestive form as

$$\mathbf{f}_{e2} = \iint_{D_2} \begin{bmatrix} N_1 \\ N_4 \\ N_2 \end{bmatrix} f \, dA = \iint_{D_2} \mathbf{N} f \, dA = \begin{bmatrix} f_1 \\ f_2 \\ f_3 \end{bmatrix}$$

where, as mentioned previously, the elements of $\mathbf{N}$ must be arranged so as to traverse the element D_2 in the counterclockwise direction. When $\mathbf{f}_{e2}$ is expanded to the global level, there results

$$\mathbf{f}_{g2} = \begin{bmatrix} f_1 \\ f_3 \\ 0 \\ f_2 \\ 0 \\ 0 \\ 0 \end{bmatrix}$$

where row interchanges have been carried out in order that the coefficients of elemental load matrix add to the correct positions in the global load matrix. The process is further indicated in Fig. 3–11 where the node numbers at the elemental and global levels are clearly indicated.

Consider next the contributions to the global stiffness matrix **K** arising from the integrals on the Γ_2 portion of the boundary, $\int n_k \alpha n_i \, ds$. Only nodes 5, 2, 1, and 3 are involved; these will be considered in turn for their contributions by traversing the boundary in the positive direction.

$k = 5$: For node 5, n_5 is zero only on interval 5-2 of Γ^-. The contribution to equation 5 is

$$v_5 \int_5^2 n_5 \alpha n_5 \, ds + v_2 \int_5^2 n_5 \alpha n_2 \, ds$$

$k = 2$: For node 2, n_2 is nonzero on segments 5-2 and 2-1 and so interacts with n_5, n_2, and n_1 with the contribution

$$v_5 \int_5^2 n_2 \alpha n_5 \, ds + v_2 \left[\int_5^2 n_2 \alpha n_2 \, ds + \int_2^1 n_2 \alpha n_2 \, ds \right] + v_1 \int_2^1 n_2 \alpha n_1 \, ds$$

$k = 1$: For node 1, n_1 is nonzero on segments 2-1 and 1-3 and so interacts with n_2, n_1, and n_3 with the contribution

$$v_2 \int_2^1 n_1 \alpha n_2 \, ds + v_1 \left[\int_2^1 n_1 \alpha n_1 \, ds + \int_1^3 n_1 \alpha n_1 \, ds \right] + v_3 \int_1^3 n_1 \alpha n_3 \, ds$$

$k = 3$: For node 3, n_3 is nonzero on segment 1-3 of Γ^- and so interacts with n_1 and n_3 with the contribution

$$v_1 \int_1^3 n_3 \alpha n_1 \, ds + v_3 \int_1^3 n_3 \alpha n_3 \, ds$$

The total contribution arising from segment 5-2 is

$$\mathbf{a}_{e1} = \begin{bmatrix} \int_5^2 n_5 \alpha n_5 \, ds & \int_5^2 n_5 \alpha n_2 \, ds \\ \int_5^2 n_2 \alpha n_5 \, ds & \int_5^2 n_2 \alpha n_2 \, ds \end{bmatrix} = \int_5^2 \mathbf{N} \alpha \mathbf{N}^{\mathrm{T}} \, ds = \begin{bmatrix} \alpha_{11} & \alpha_{12} \\ \alpha_{21} & \alpha_{22} \end{bmatrix}$$

where $\mathbf{N}^{\mathrm{T}} = [N_5 \quad N_2]^{\mathrm{T}} = [1 - s/\ell_e \quad s\ell_e]$ are essentially the linear interpolation functions for the one-dimensional problem. ℓ_e is the length of segment 2-5.

The total contribution arising from segment 2-1 is

$$\mathbf{a}_{e2} = \begin{bmatrix} \int_2^1 n_2 \alpha n_2 \, ds & \int_2^1 n_2 \alpha n_1 \, ds \\ \int_2^1 n_1 \alpha n_2 \, ds & \int_2^1 n_1 \alpha n_1 \, ds \end{bmatrix} = \int_5^2 \mathbf{N} \alpha \mathbf{N}^{\mathrm{T}} \, ds = \begin{bmatrix} \beta_{11} & \beta_{12} \\ \beta_{21} & \beta_{22} \end{bmatrix}$$

where again $\mathbf{N}^{\mathrm{T}} = [N_2 \quad N_1]^{\mathrm{T}} = [1 - s/\ell_e \quad s/\ell_e]^{\mathrm{T}}$, ℓ_e being the length of segment 1-2. Last, the total contribution arising from segment 1-3 is

$$\mathbf{a}_{e3} = \begin{bmatrix} \int_1^3 n_1 \alpha n_1 \, ds & \int_1^3 n_1 \alpha n_3 \, ds \\ \int_1^3 n_3 \alpha n_1 \, ds & \int_1^3 n_1 \alpha n_1 \, ds \end{bmatrix} = \int_3^1 \mathbf{N} \alpha \mathbf{N}^{\mathrm{T}} \, ds = \begin{bmatrix} \gamma_{11} & \gamma_{12} \\ \gamma_{21} & \gamma_{22} \end{bmatrix}$$

where $\mathbf{N}^{\mathrm{T}} = [N_1 \; N_3]^{\mathrm{T}} = [1 - s/\ell_e \;\; s/\ell_e]^{\mathrm{T}}$, with ℓ_e as the length of segment 1-3. The main point here is that terms of this type that arise from the natural boundary conditions can also be considered on an element basis.

Expanding to the global level these can be expressed as

$$\mathbf{a}_{g1} = \begin{bmatrix} 0 & 0 & 0 & 0 & 0 & 0 & 0 \\ 0 & \alpha_{22} & 0 & 0 & \alpha_{21} & 0 & 0 \\ 0 & 0 & 0 & 0 & 0 & 0 & 0 \\ 0 & 0 & 0 & 0 & 0 & 0 & 0 \\ 0 & \alpha_{12} & 0 & 0 & \alpha_{11} & 0 & 0 \\ 0 & 0 & 0 & 0 & 0 & 0 & 0 \\ 0 & 0 & 0 & 0 & 0 & 0 & 0 \end{bmatrix}$$

$$\mathbf{a}_{g2} = \begin{bmatrix} \beta_{22} & \beta_{21} & 0 & 0 & 0 & 0 & 0 \\ \beta_{12} & \beta_{11} & 0 & 0 & 0 & 0 & 0 \\ 0 & 0 & 0 & 0 & 0 & 0 & 0 \\ 0 & 0 & 0 & 0 & 0 & 0 & 0 \\ 0 & 0 & 0 & 0 & 0 & 0 & 0 \\ 0 & 0 & 0 & 0 & 0 & 0 & 0 \\ 0 & 0 & 0 & 0 & 0 & 0 & 0 \end{bmatrix}$$

$$\mathbf{a}_{g3} = \begin{bmatrix} \gamma_{11} & 0 & \gamma_{12} & 0 & 0 & 0 & 0 \\ 0 & 0 & 0 & 0 & 0 & 0 & 0 \\ \gamma_{21} & 0 & \gamma_{22} & 0 & 0 & 0 & 0 \\ 0 & 0 & 0 & 0 & 0 & 0 & 0 \\ 0 & 0 & 0 & 0 & 0 & 0 & 0 \\ 0 & 0 & 0 & 0 & 0 & 0 & 0 \\ 0 & 0 & 0 & 0 & 0 & 0 & 0 \end{bmatrix}$$

where it is seen that in general the elements must be repositioned before assembling into the global stiffness matrix. These matrices are then added to $\Sigma \mathbf{k}_{gi}$, discussed previously, to complete the assembly of the **K** matrix. The ideas associated with the elemental contributions of $\int n_k \alpha n_i \, ds$ to the global matrices are further indicated in Fig. 3–12.

For the contribution to the load matrix **F** arising from the integrals on the Γ_2 portion of the boundary, we need to consider $\int n_k h \, ds$. Again only nodes 5, 2, 1, and 3 are involved and will be considered in turn for their contributions upon traversing the boundary in the positive direction.

$k = 5$**:** For node 5, n_5 is zero only on interval 5-2 of Γ^-. The contribution to equation 5 is

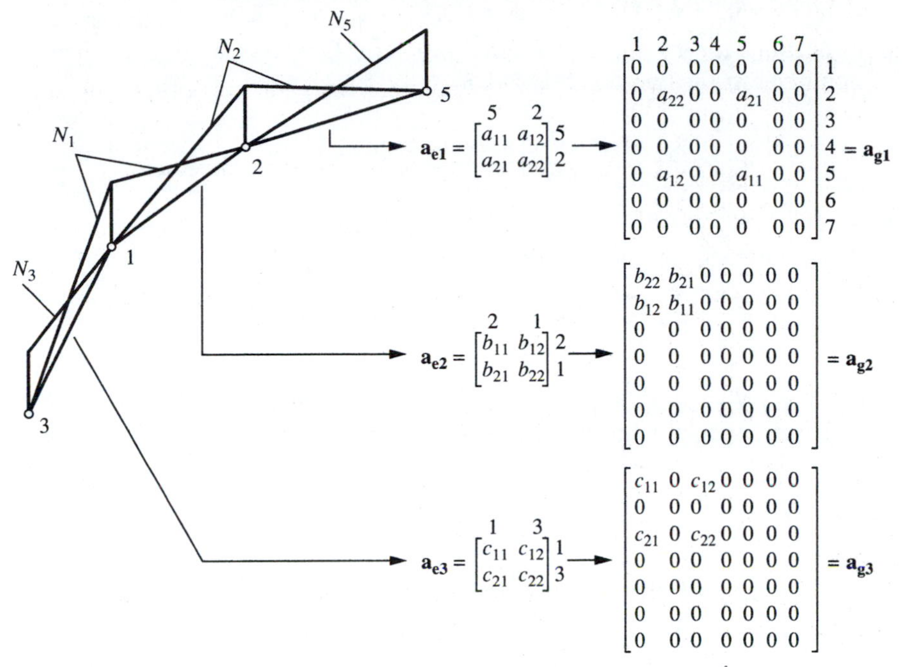

FIGURE 3–12 Elemental contributions and assembly for $\int n_k \alpha n_i \, ds$

$$\int_5^2 n_5 h \, ds$$

$k = 2$: For node 2, n_2 is nonzero on segments 5-2 and 2-1 with the contribution

$$\int_5^2 n_2 h \, ds + \int_2^1 n_2 h \, ds$$

$k = 1$: For node 1, n_1 is nonzero on segments 2-1 and 1-3 with the contribution

$$\int_2^1 n_1 h \, ds + \int_1^3 n_1 h \, ds$$

$k = 3$: For node 3, n_3 is nonzero on segment 1-3 of Γ^- with the contribution

$$\int_1^3 n_3 h \, ds$$

The contribution for line segment 5-2 is

$$\mathbf{h}_{e1} = \int_5^2 \begin{bmatrix} n_5 \\ n_2 \end{bmatrix} h \, ds = \int_5^2 \mathbf{N} h \, ds = \begin{bmatrix} a_1 \\ a_2 \end{bmatrix}$$

where $\mathbf{N}^{\mathrm{T}} = [N_5 \quad N_2]^{\mathrm{T}} = [1 - s/\ell_e \quad s/\ell_e]$ are essentially the linear interpolation functions for the one-dimensional problem. ℓ_e is the length of segment 5-2. The contribution for line segment 2-1 is

$$\mathbf{h}_{\mathbf{e2}} = \int_2^1 \begin{bmatrix} n_2 \\ n_1 \end{bmatrix} h\, ds = \int_2^1 \mathbf{N} h\, ds = \begin{bmatrix} b_1 \\ b_2 \end{bmatrix}$$

where $\mathbf{N}^{\mathrm{T}} = [N_2 \quad N_1]^{\mathrm{T}} = [1 - s/\ell_e \quad s/\ell_e]$, with ℓ_e as the length of segment 2-1. Last, the contribution for line segment 1-3 is

$$\mathbf{h}_{\mathbf{e3}} = \int_1^3 \begin{bmatrix} n_1 \\ n_3 \end{bmatrix} h\, ds = \int_1^3 N h\, ds = \begin{bmatrix} c_1 \\ c_2 \end{bmatrix}$$

where $\mathbf{N}^{\mathrm{T}} = [N_1 \quad N_3]^{\mathrm{T}} = [1 - s/\ell_e \quad s/\ell_e]$, ℓ_e being the length of segment 1-3.

The corresponding contributions to the load matrix at the global level are

$$\mathbf{h}_{\mathbf{g1}} = \begin{bmatrix} 0 \\ a_2 \\ 0 \\ 0 \\ a_1 \\ 0 \\ 0 \end{bmatrix} \qquad \mathbf{h}_{\mathbf{g2}} = \begin{bmatrix} b_2 \\ b_1 \\ 0 \\ 0 \\ 0 \\ 0 \\ 0 \end{bmatrix} \qquad \mathbf{h}_{\mathbf{g3}} = \begin{bmatrix} c_1 \\ 0 \\ c_2 \\ 0 \\ 0 \\ 0 \\ 0 \end{bmatrix}$$

The ideas associated with the elemental contributions to the global matrices for $\int n_k h\, ds$ are further indicated in Fig. 3–13.

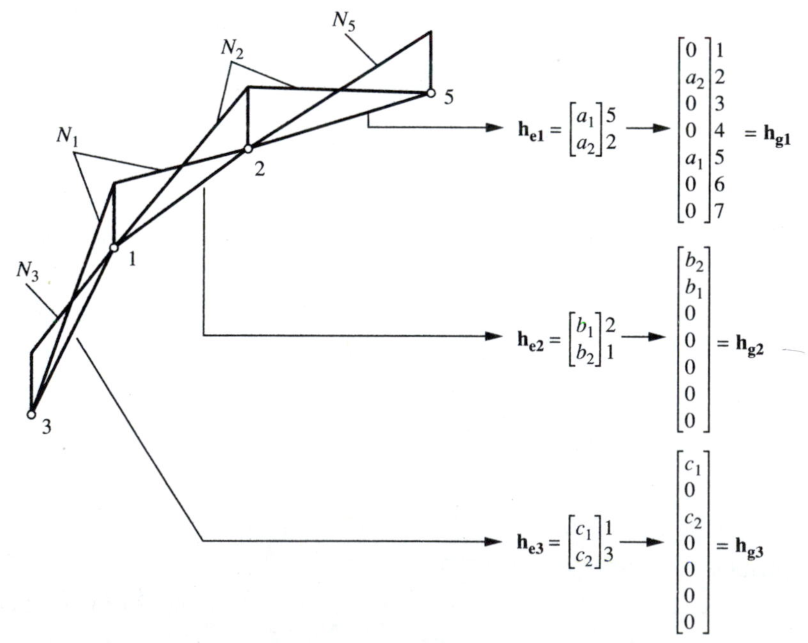

FIGURE 3–13 Elemental contributions and assembly for $\int n_k h\, ds$

Thus the global set of equations $\mathbf{Kv} = \mathbf{F}$ can be considered to have been assembled according to

$$\mathbf{K} = \sum \mathbf{k}_{gi} + \sum \mathbf{a}_{gi} \qquad \mathbf{F} = \sum \mathbf{f}_{gi} + \sum \mathbf{h}_{gi}$$

where all $\mathbf{k}_{gi}$, $\mathbf{a}_{gi}$, $\mathbf{f}_{gi}$, and $\mathbf{h}_{gi}$ have been expanded to the global level from the elemental matrices

$$\mathbf{k}_{ei} = \iint_{Di} \left[\frac{\partial \mathbf{N}}{\partial x} \frac{\partial \mathbf{N}^T}{\partial x} + \frac{\partial \mathbf{N}}{\partial y} \frac{\partial \mathbf{N}^T}{\partial y} \right] dA \tag{3.8}$$

$$\mathbf{a}_{ei} = \int_{\gamma_{2i}} \mathbf{N}\alpha\mathbf{N}^T \, ds \tag{3.9}$$

$$\mathbf{f}_{ei} = \iint_{D_i} \mathbf{N} f \, dA \tag{3.10}$$

$$\mathbf{h}_{ei} = \int_{\gamma_{2i}} \mathbf{N} h \, ds \tag{3.11}$$

respectively, with $\mathbf{N}$ the elemental interpolation vector for the element or line segment in question. In the above elemental matrices γ_{2i} indicates one of the line segments on the Γ_2 part of the boundary.

Constraints, solution, and derived variables. For a discussion of constraints, solution, and derived variables, see Section 3.3.2, pages 247–49.

3.3 VARIATIONAL FINITE ELEMENT MODELS

In this section, we seek to determine whether there is a functional from which we can proceed to develop the finite element model for the elliptic boundary value problem using the Ritz method. We will do this by establishing the weak formulation for the present class of elliptic boundary value problems and determining whether it is of the required form to permit the subsequent conversion to a variational principle. The development will parallel that used for the corresponding class of one-dimensional boundary value problems studied in Chapter 2. It will be seen that for the class of boundary value problems studied in this chapter, the Ritz finite element model coincides with the Galerkin finite element model developed in the previous section.

3.3.1 The Weak Formulation and Variational Principles

The first step in developing the weak formulation is to multiply the differential equation by an arbitrary test function $w(x, y)$ having the following properties.

1. It is a function possessing first partial derivatives on the domain D.
2. It vanishes on Γ_1, the portion of the boundary on which u is prescribed.

We then integrate the result over the domain D to obtain

$$\iint_D w[\nabla^2 u(x, y) + f(x, y)]dA = 0 \tag{3.12}$$

As in the previous chapter we argue that if w is truly arbitrary, the satisfaction of the above statement or equation is tantamount to saying that the differential equation is satisfied, at least on average, over every small subdomain in D. Also as before, there is a lack of symmetry as regards the requirements on w and u. Although the Galerkin method could be applied directly to the above statement, it would turn out that the resulting set of linear algebraic equations would be asymmetric, an undesirable situation. The problem is easily rectified by using integration by parts, which in the present context is accomplished by using Green's first identity, or the two-dimensional form of the divergence theorem. Recall that the two-dimensional form of the divergence theorem [Kaplan and Lewis, 4] states that

$$\iint_D \nabla \cdot \mathbf{u}\, dA = \int_\Gamma \mathbf{n} \cdot \mathbf{u}\, ds$$

where $\mathbf{u}$ is a sufficiently regular vector field. As indicated in Fig. 3–14, the line integral is evaluated in a counterclockwise sense around the bounding curve Γ with $\mathbf{n}$ the outward pointing normal.

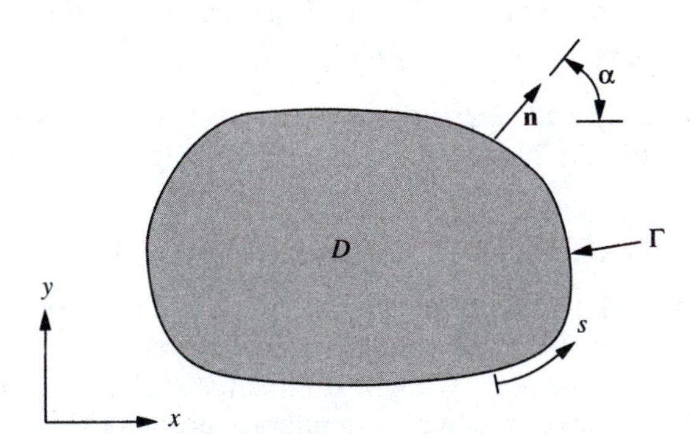

FIGURE 3–14 Positive n and integration direction

With $\mathbf{u} = \phi\mathbf{i} + \psi\mathbf{j}$, $n_x = \cos\alpha$, $n_y = \sin\alpha$, and α as indicated in Fig. 3–14, this can be expressed in component form as

$$\iint_D \left(\frac{\partial\phi}{\partial x} + \frac{\partial\psi}{\partial y}\right) dA = \int_\Gamma (n_x\phi + n_y\psi)\, ds$$

$$= \int_\Gamma (\phi\cos\alpha + \psi\sin\alpha)\, ds$$

By taking $\psi = 0$, there results the formula

$$\iint_D \frac{\partial\phi}{\partial x}\, dA = \int_\Gamma n_x\phi\, ds = \int_\Gamma \phi\cos\alpha\, ds$$

which converts a partial integration with respect to x of a function over the domain D into a line integral around the boundary Γ. Similarly, by taking $\phi = 0$ there results

$$\iint_D \frac{\partial \psi}{\partial y}\, dA = \int_\Gamma n_y \psi\, ds = \int_\Gamma \psi \sin\alpha\, ds$$

which we will view as partial integration with respect to y.

Returning to Eq. (3.12), consider the individual term

$$\iint_D w\nabla^2 u\, dA = \iint_D w\left(\frac{\partial^2 u}{\partial x^2} + \frac{\partial^2 u}{\partial y^2}\right) dA \tag{3.13}$$

We state the obvious identity

$$w\frac{\partial^2 u}{\partial x^2} = \frac{\partial(w\, \partial u/\partial x)}{\partial x} - \frac{\partial w}{\partial x}\frac{\partial u}{\partial x}$$

so that as a result of integrating both sides over the domain D, applying integration by parts to the first integral on the right, and employing the divergence theorem, it follows that

$$\iint_D w\frac{\partial^2 u}{\partial x^2}\, dA = \int_\Gamma n_x w\frac{\partial u}{\partial x}\, ds - \iint_D \frac{\partial w}{\partial x}\frac{\partial u}{\partial x}\, dA$$

In a completely similar fashion, the second term of (3.13) can be integrated to obtain

$$\iint_D w\frac{\partial^2 u}{\partial y^2}\, dA = \int_\Gamma n_y w\frac{\partial u}{\partial y}\, ds - \iint_D \frac{\partial w}{\partial y}\frac{\partial u}{\partial y}\, dA$$

Thus it follows that

$$\iint_D w\left(\frac{\partial^2 u}{\partial x^2} + \frac{\partial^2 u}{\partial y^2}\right) dA = \int_\Gamma w\left(n_x\frac{\partial u}{\partial x} + n_y\frac{\partial u}{\partial y}\right) ds - \iint_D \left(\frac{\partial w}{\partial x}\frac{\partial u}{\partial x} + \frac{\partial w}{\partial y}\frac{\partial u}{\partial y}\right) dA$$

which, when expressed completely in the notation of the ∇ operator, becomes

$$\begin{aligned}\iint_D w\nabla^2 u\, dA &= \int_\Gamma w\mathbf{n}\cdot\nabla u\, ds - \iint_D \nabla w\cdot\nabla u\, dA \\ &= \int_\Gamma w\frac{\partial u}{\partial n}\, ds - \iint_D \nabla w\cdot\nabla u\, dA\end{aligned} \tag{3.14}$$

The first integral on the right is a line integral taken in the counterclockwise direction around the boundary Γ. Using the identity (3.14) to replace the $\nabla^2 u$ term in Eq. (3.12) leads to

$$\iint_D \nabla w\cdot\nabla u\, dA = \int_\Gamma w\frac{\partial u}{\partial n}\, ds + \iint_D wf\, dA$$

The line integral can be decomposed into the sum of two line integrals, one on Γ_1 and one on Γ_2, to yield

$$\int\!\!\int_D \nabla w \cdot \nabla u \, dA = \left[\int_{\Gamma_1} + \int_{\Gamma_2}\right] w \frac{\partial u}{\partial n} \, ds + \int\!\!\int_D wf \, dA$$

Recalling that the test function w vanishes on Γ_1, and that on Γ_2

$$\frac{\partial u}{\partial n} = h(s) - \alpha(s)u$$

there results

$$\int\!\!\int_D \nabla w \cdot \nabla u \, dA + \int_{\Gamma_2} w\alpha u \, ds = \int\!\!\int_D wf \, dA + \int_{\Gamma_2} wh \, ds \tag{3.15}$$

The integration by parts has resulted in a statement in which the continuity requirements on w and u are the same, and has automatically entered the natural boundary condition on the Γ_2 portion of the boundary. Equation (3.15) is the required weak form of the original class of elliptic boundary value problems. The weak form can be written as

$$B(w, u) - l(w) = 0$$

where

$$B(w, u) = \int\!\!\int_D \nabla w \cdot \nabla u \, dA + \int_{\Gamma_2} w\alpha u \, ds$$

and

$$l(w) = \int\!\!\int_D wf \, dA + \int_{\Gamma_2} wh \, ds$$

B and l are bilinear and linear functionals respectively.

With the observation that B is symmetric, that is, $B(w, u) = B(u, w)$, it follows that the variational principle corresponding to the original boundary value problem is

$$\delta I = 0 \qquad \text{with } u = g \text{ on } \Gamma_1$$

where

$$I(u) = \frac{B(u, u)}{2} - l(u) \tag{3.16}$$

with

$$B(u, u) = \int\!\!\int_D \nabla u \cdot \nabla u \, dA + \int_{\Gamma_2} \alpha u^2 \, ds$$

and

$$l(u) = \int\!\!\int_D uf \, dA + \int_{\Gamma_2} uh \, ds$$

In the next section, the functional given by Eq. (3.16) will be used to develop the Ritz finite element model for the elliptic boundary value problem. As already mentioned, it will turn out to be identical to the corresponding Galerkin finite element model developed in Section 3.2.

3.3.2 Ritz Finite Element Model

The Ritz finite element model is based on the functional Eq. (3.16) developed in the previous section:

$$I(u) = \frac{1}{2}\left(\iint_D \left(\left(\frac{\partial u}{\partial x}\right)^2 + \left(\frac{\partial u}{\partial y}\right)^2\right) dA + \int_{\Gamma_2} \alpha u^2\, ds\right) - \iint_D uf\, dA - \int_{\Gamma_2} uh\, ds$$

With this functional as the starting point, we now discuss each of the basic steps of the finite element method.

Discretization. The first step in developing a finite element model is discretization. A typical discretization using straight-sided triangular elements is indicated in Fig. 3–15.

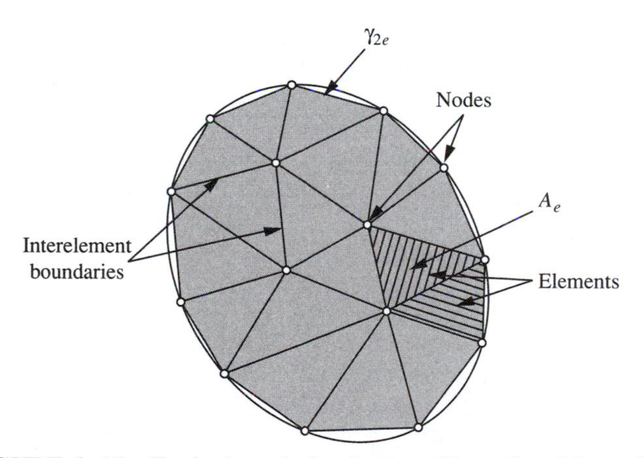

FIGURE 3–15 Typical mesh for the two-dimensional boundary value problem using linearly interpolated triangular elements

Nodes, elements, and interelement boundaries are indicated in the figure. Repeating the observations from Section 3.2 for this discretization, it is immediately obvious that with straight-sided elements there is in general an inherent error in modeling the curved shape of the domain D. The effect of these errors is expected to diminish as the size of the elements is reduced. In addition to the problem associated with the errors in geometry, it will not be possible in general to enforce the boundary conditions exactly since the data that are prescribed on the actual boundary will not be correct for the replacement boundary. We will rely on the knowledge that elliptic boundary value problems of this type are well posed [Courant and Hilbert, 3], which means that the small changes in the input resulting from a slightly different boundary and boundary values will result in small changes in the solution. Elements with curved sides can be expected to improve on but not entirely eliminate such errors.

Generally the mesh should be selected so as to be relatively fine in regions where large gradients or slopes are expected. Transitions from regions where the mesh is relatively fine to regions where it is relatively crude should be gradual. Extreme geometries such as triangles with small angles ($\leq$ approximately $\pi/8 = 22.5°$ as a rule of thumb) or with large angles (approaching π) are to be avoided.

In terms of the discretization, the functional I is now represented as the sum of integrals over the areas of the elements in D, and as the sum of integrals over the corresponding line segments on the Γ_2 portion(s) of the boundary according to

$$I(u) \approx \frac{1}{2}\left(\sum \iint_{A_e} \left(\left(\frac{\partial u}{\partial x}\right)^2 + \left(\frac{\partial u}{\partial y}\right)^2\right) dA + \sum{}' \int_{\gamma_{2e}} \alpha u^2 \, ds\right) - \sum \iint_{A_e} uf \, dA - \sum{}' \int_{\gamma_{2e}} uh \, ds \tag{3.17}$$

where the sum $\sum$ is over the elemental areas, and the $\sum{}'$ is over the elemental segments γ_{2e} of the Γ_2 portion(s) of the boundary.

Interpolation. The most elementary type of interpolation for the straight-sided three-node triangle corresponds to assuming that the solution surface $u(x, y)$ is represented by a plane defined on the element in question. This linear representation is indicated in Fig. 3–16. The solution surface is assumed to consist of segments of planes over each of the elements, that is, to be piecewise linear.

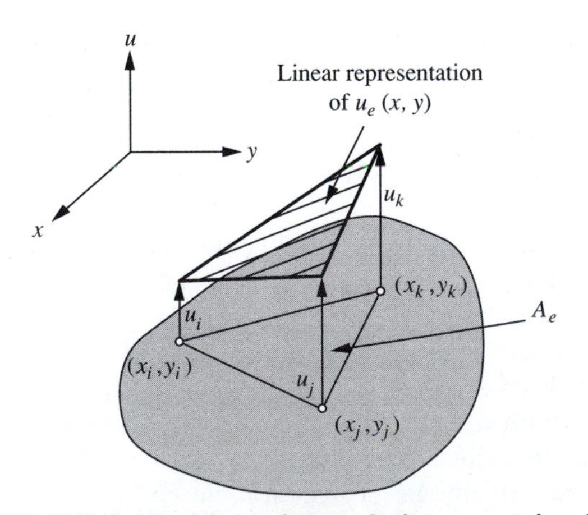

FIGURE 3–16 Linear interpolation on a triangle

Analytically we write

$$u_e(x, y) = \alpha + \beta x + \gamma y \tag{3.18}$$

where the three constants α, β, and γ are determined by the three equations requiring that u assume the nodal values at the vertices of the triangle,

$$\begin{aligned} u_e(x_i, y_i) &= \alpha + \beta x_i + \gamma y_i = u_i \\ u_e(x_j, y_j) &= \alpha + \beta x_j + \gamma y_j = u_j \\ u_e(x_k, y_k) &= \alpha + \beta x_k + \gamma y_k = u_k \end{aligned} \tag{3.19}$$

where i, j, and k denote the three vertices of the triangle. When the three Eqs. (3.19) are solved for α, β, and γ and the results substituted back into (3.18), the elemental expression for $u(x, y)$ can be written as

$$u_e(x, y) = u_i N_i + u_j N_j + u_k N_k$$

where

$$N_i = \frac{a_i + b_i x + c_i y}{2A_e} \qquad i = 1, 2, 3$$

with

$$\begin{aligned} a_i &= x_j y_k - x_k y_j \\ b_i &= y_j - y_k \\ c_i &= x_k - x_j \end{aligned}$$

where i, j, and k are to be permuted cyclically. The N's are called interpolation functions or shape functions. The determinant of the coefficients is

$$2A_e = \begin{vmatrix} 1 & x_i & y_i \\ 1 & x_j & y_j \\ 1 & x_k & y_k \end{vmatrix}$$

with A_e as the area of the element. Any numbering that proceeds counterclockwise around the element is acceptable, that is, (i, j, k), (j, k, i), and (k, i, j). This counterclockwise direction is necessary in order that the area A_e (as calculated according to the determinant above) be positive. The shape functions N_i, N_j, and N_k are indicated in Fig. 3–17.

In matrix notation, the elemental representation can be expressed as

$$u_e(x, y) = \mathbf{u}_\mathbf{e}^\mathrm{T}\mathbf{N} = \mathbf{N}^\mathrm{T}\mathbf{u}_\mathbf{e} \tag{3.20}$$

with both forms of the representation being necessary in different settings. Additionally, the derivatives of the elemental representations will be needed. These are easily computed to be

$$\frac{\partial u_e(x, y)}{\partial x} = \mathbf{u}_\mathbf{e}^\mathrm{T}\frac{\partial \mathbf{N}}{\partial x} = \frac{\partial \mathbf{N}^\mathrm{T}}{\partial x}\mathbf{u}_\mathbf{e}$$

and

$$\frac{\partial u_e(x, y)}{\partial y} = \mathbf{u}_\mathbf{e}^\mathrm{T}\frac{\partial \mathbf{N}}{\partial y} = \frac{\partial \mathbf{N}^\mathrm{T}}{\partial y}\mathbf{u}_\mathbf{e}$$

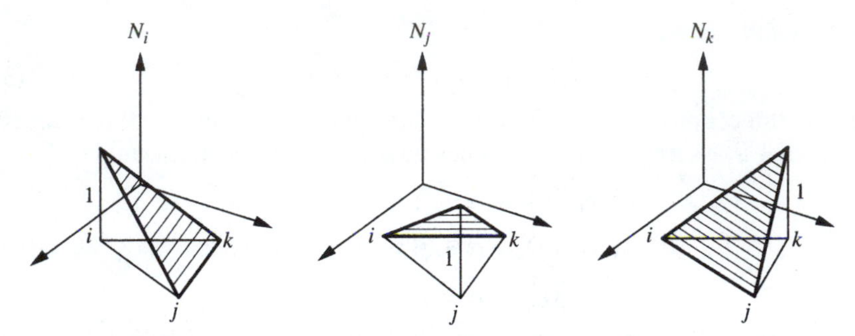

FIGURE 3–17 Linear interpolation functions for the triangle

Recalling the expressions for the N_i, it follows that

$$\frac{\partial \mathbf{N}}{\partial x} = \frac{\mathbf{b_e}}{2A_e} \tag{3.21a}$$

and

$$\frac{\partial \mathbf{N}}{\partial y} = \frac{\mathbf{c_e}}{2A_e} \tag{3.21b}$$

where $\mathbf{b}_\mathbf{e}^\mathrm{T} = [b_1\ b_2\ b_3]$ and $\mathbf{c}_\mathbf{e}^\mathrm{T} = [c_1\ c_2\ c_3]$. It is apparent that the partial derivatives of the solution within an element will be constant when linear interpolation is used. Practically, this so-called *constant strain element* turns out to have major deficiencies with regard to accuracy and convergence.

In view of the fact that the interpolation functions are linear on any edge, interelement continuity is assured. Additionally, the basic representation given by Eq. (3.18) contains the terms necessary for the solution and the partial derivatives $\partial u/\partial x$ and $\partial u/\partial y$ to assume constant values within the element, so that the completeness criterion is also satisfied.

Elemental formulation. We repeat Eq. (3.17) for the functional associated with the elliptic boundary value problem in connection with a typical discretization, namely

$$I(u) = \frac{1}{2}\sum\left(\iint_{A_e}\left(\left(\frac{\partial u}{\partial x}\right)^2 + \left(\frac{\partial u}{\partial y}\right)^2\right)dA + \sum{}'\int_{\gamma_{2e}} \alpha u^2\, ds\right)$$
$$-\sum\iint_{A_e} uf\, dA - \sum{}'\int_{\gamma_{2e}} uh\, ds$$

where again, the $\sum$ indicates a sum over all the triangular elements A_e, and the $\sum'$ indicates a sum over the straight-line segments γ_{2e} representing the boundary. Rewrite (3.17) as

$$I(u) = \sum_e \frac{I_{e1}}{2} + \sum_e{}' \frac{I_{e2}}{2} - \sum_e I_{e3} - \sum_e{}' I_{e4}$$

where

$$I_{e1} = \iint_{A_e} \left(\left(\frac{\partial u}{\partial x} \right)^2 + \left(\frac{\partial u}{\partial y} \right)^2 \right) dA$$

$$I_{e2} = \int_{\gamma_{2e}} \alpha u^2 \, ds$$

$$I_{e3} = \iint_{A_e} uf \, dA$$

$$I_{e4} = \int_{\gamma_{2e}} uh \, ds$$

Using the results for the linearly interpolated triangle, each of these four types of elemental contributions to the global stiffness and load matrices is evaluated in turn.

I_{e1}**:** The integral I_{e1} is given by

$$I_{e1} = \iint_{A_e} \left(\frac{\partial u}{\partial x} \frac{\partial u}{\partial x} + \frac{\partial u}{\partial y} \frac{\partial u}{\partial y} \right) dA$$

The first $\partial u / \partial x$ is replaced by its representation in terms of the derivatives of the interpolation functions as $\mathbf{u_e^T} \, \partial \mathbf{N} / \partial x$ and the second $\partial u / \partial x$ by $\partial \mathbf{N^T} / \partial x \; \mathbf{u_e}$. With an entirely similar treatment of the second term there results

$$I_{e1} = \iint_{A_e} \left(\mathbf{u_e^T} \frac{\partial \mathbf{N}}{\partial x} \frac{\partial \mathbf{N^T}}{\partial x} \mathbf{u_e} + \mathbf{u_e^T} \frac{\partial \mathbf{N}}{\partial y} \frac{\partial \mathbf{N^T}}{\partial y} \mathbf{u_e} \right) dA$$

$$= \mathbf{u_e^T} \left(\iint_{A_e} \left(\frac{\partial \mathbf{N}}{\partial x} \frac{\partial \mathbf{N^T}}{\partial x} + \frac{\partial \mathbf{N}}{\partial y} \frac{\partial \mathbf{N^T}}{\partial y} \right) dA \right) \mathbf{u_e} = \mathbf{u_e^T k_e u_e} \qquad (3.22)$$

where

$$\mathbf{k_e} = \iint_{A_e} \left(\frac{\partial \mathbf{N}}{\partial x} \frac{\partial \mathbf{N^T}}{\partial x} + \frac{\partial \mathbf{N}}{\partial y} \frac{\partial \mathbf{N^T}}{\partial y} \right) dA$$

is called an elemental stiffness matrix. Note that $\mathbf{k_e}$ is symmetric. Using the specific expressions developed previously for the linearly interpolated three-node triangle, $\mathbf{k_e}$ can be further simplified to

$$\mathbf{k_e} = \iint_{Ae} \left(\frac{\mathbf{b_e b_e^T} + \mathbf{c_e c_e^T}}{4A_e^2} \right) dA$$

In view of the fact that the integrand is constant, $\mathbf{k_e}$ reduces to

$$\mathbf{k_e} = \frac{\mathbf{b_e b_e^T} + \mathbf{c_e c_e^T}}{4A_e} \qquad (3.23)$$

During assembly, the 3×3 elemental stiffness matrix $\mathbf{k_e}$ will contribute or add to the global stiffness matrix at those locations corresponding to the nodal degrees of freedom associated with the particular A_e in question.

I_{e2}: The integral I_{e2} is given by

$$I_{e2} = \int_{\gamma_{2e}} \alpha u^2 \, ds$$

which, on using the interpolated representation for u on the boundary, becomes

$$I_{e2} = \int_{\gamma_{2e}} \mathbf{u}_\mathbf{e}^\mathrm{T} \mathbf{N} \alpha \mathbf{N}^\mathrm{T} \mathbf{u}_\mathbf{e} \, ds$$

$$= \mathbf{u}_\mathbf{e}^\mathrm{T} \left(\int_{\gamma_{2e}} \mathbf{N} \alpha \mathbf{N}^\mathrm{T} \, ds \right) \mathbf{u}_\mathbf{e} = \mathbf{u}_\mathbf{e}^\mathrm{T} \mathbf{a}_\mathbf{e} \mathbf{u}_\mathbf{e} \tag{3.24}$$

where

$$\mathbf{a}_\mathbf{e} = \int_{\gamma_{2e}} \mathbf{N} \alpha \mathbf{N}^\mathrm{T} \, ds \tag{3.25}$$

is a 2×2 elemental stiffness matrix which contributes or adds during assembly to the global stiffness matrix at the positions corresponding to the two boundary nodes associated with the particular γ_{2e} in question.

I_{e3}: The integral I_{e3} is given by

$$I_{e3} = \iint_{A_e} uf \, dA = \iint_{A_e} \mathbf{u}_\mathbf{e}^\mathrm{T} \mathbf{N} f \, dA = \mathbf{u}_\mathbf{e}^\mathrm{T} \mathbf{f}_\mathbf{e} \tag{3.26}$$

where

$$\mathbf{f}_\mathbf{e} = \iint_{A_e} \mathbf{N} f \, dA \tag{3.27}$$

is a 3×1 elemental load vector which contributes or adds during assembly to the global load vector at the positions corresponding to the nodal degrees of freedom associated with the particular A_e in question.

I_{e4}: The integral I_{e4} is given by

$$I_{e4} = \int_{\gamma_{2e}} uh \, ds = \int_{\gamma_{2e}} \mathbf{u}_\mathbf{e}^\mathrm{T} \mathbf{N} h \, ds = \mathbf{u}_\mathbf{e}^\mathrm{T} \mathbf{h}_\mathbf{e} \tag{3.28}$$

where

$$\mathbf{h}_\mathbf{e} = \int_{\gamma_{2e}} \mathbf{N} h(s) \, ds \tag{3.29}$$

is a 2×1 elemental load contributing to the global load vector at the positions corresponding to the boundary nodal degrees of freedom associated with the particular γ_{2e} in question.

Using Eqs. (3.22), (3.24), (3.26), and (3.28), the functional Eq. (3.17) can be expressed as

$$I(u_1, u_2, u_3, u_4, \ldots) = \sum \left[\frac{\mathbf{u}_\mathbf{e}^\mathrm{T} \mathbf{k}_\mathbf{e} \mathbf{u}_\mathbf{e}}{2} - \mathbf{u}_\mathbf{e}^\mathrm{T} \mathbf{f}_\mathbf{e} \right] + \sum{}' \left[\frac{\mathbf{u}_\mathbf{e}^\mathrm{T} \mathbf{a}_\mathbf{e} \mathbf{u}_\mathbf{e}}{2} - \mathbf{u}_\mathbf{e}^\mathrm{T} \mathbf{h}_\mathbf{e} \right] \tag{3.30}$$

where the first sum is over the area elements A_e of D and the second sum is over the line elements γ_{2e} of the Γ_2 portion of the boundary.

This completes the elemental formulation and indicates that there are two types of stiffnesses: (1) the $\mathbf{k_e}$ arising from the ∇^2 operator in the interior and on the boundary, and (2) the $\mathbf{a_e}$ arising from the type two boundary conditions which contain a nonzero α. The loads or the right-hand side of the final set of equations arise from (1) the f term in the differential equation giving rise to the $\mathbf{f_e}$, and (2) the nonhomogeneous term h in the type two boundary conditions that produce the $\mathbf{h_e}$.

It is important to note that, by virtue of the finite element methodology, the original functional Eq. (3.17) has been converted to a *function* Eq. (3.30) with the nodal unknowns u_i as the independent variables.

Assembly. Assembly is implied by the sums in Eq. (3.30). Each of the elemental matrices $\mathbf{k_e}$, $\mathbf{a_e}$, $\mathbf{f_e}$, and $\mathbf{h_e}$ is expanded to the global level, with the result

$$I = \frac{\mathbf{u_G^T K_G u_G}}{2} - \mathbf{u_G^T F_G} = I(\mathbf{u_G})$$

where

$$\mathbf{K_G} = \sum_e \mathbf{k_G} + \sum_e{}' \mathbf{a_G}$$

and

$$\mathbf{F_G} = \sum_e \mathbf{f_G} + \sum_e{}' \mathbf{h_G}$$

Note that the elemental matrices $\mathbf{k_G}$ and $\mathbf{f_G}$ each arise from an elemental area A_e, whereas the matrices $\mathbf{a_G}$ and $\mathbf{h_G}$ each arise from an elemental line segment γ_{2e} on the Γ_2 portion of the boundary.

As just mentioned above, the finite element methodology has converted the functional Eq. (3.17) into a function Eq. (3.30). The stationary value of this function is obtained by requiring the partial derivatives with respect to each of the u_i to vanish,

$$\frac{\partial I}{\partial u_i} = 0 \qquad i = 1, 2, \ldots, N$$

leading to

$$\frac{(\mathbf{K_G} + \mathbf{K_G^T})\mathbf{u_G}}{2} = \mathbf{F_G}$$

or since $\mathbf{K_G} = \mathbf{K_G^T}$,

$$\mathbf{K_G u_G} = \mathbf{F_G}$$

an $N \times N$ set of linear algebraic equations for $\mathbf{u_G}$ or equivalently, the u_i.

Constraints. The constraints arise from the essential boundary conditions

$$u = g(s) \text{ on } \Gamma_1$$

The values of the constraints are usually taken to be the function g evaluated at the node in question. For this reason it is convenient to select the nodes *on* the bounding curve Γ. Each of the constraints is imposed in the usual manner by suitably altering the coefficient matrix $\mathbf{K_G}$ and the load matrix $\mathbf{P_G}$ as discussed in Chapters 1 and 2.

Solution. It was pointed out that each of the elementary stiffness matrices $\mathbf{k_e}$ is singular and, hence, so also is the global stiffness matrix $\mathbf{K_G} = \sum \mathbf{k_G}$. As long as Dirichlet boundary conditions are specified on at least a part of the boundary, a subset of the dependent variables is constrained. The resulting constrained $\mathbf{K_G}$ is nonsingular, leading to a unique solution. If the entire boundary is of type two with $\alpha \neq 0$ on at least a portion of the boundary, the $\mathbf{a_e}$ added at assembly render $\mathbf{K_G}$ nonsingular, with a corresponding unique solution. If the entire boundary is of type two with $\alpha \equiv 0$ on Γ_2, $\mathbf{K_G}$ will be singular. The resulting set of unconstrained algebraic equations will have a solution if and only if the functions f and h satisfy certain auxiliary conditions. If these conditions are satisfied the existing solution is not unique. This nonunique solution can be determined by specifying one nodal value u_i. If the auxiliary conditions are not satisfied, there is no solution to the algebraic equations or the boundary value problem. See Appendix D for a general discussion of this issue.

Once the character of the equations has been established, the equations are then solved. For a set of any size this is accomplished by calling on a suitable linear equation solver, several of which are normally available. One should seek out and employ a robust code which exploits any special character of the set of equations such as symmetry and bandedness, both of which often occur in finite element analyses. Appendix B contains routines BDecomp and BSolve, which are suitable for the solution of sets of algebraic equations arising from the application of the finite element method to two-dimensional boundary value problems.

Computation of derived variables. In a usual displacement formulation such as discussed in this text, the variables associated with the essential boundary conditions are termed **primary variables**, whereas the derivatives that occur in the natural boundary conditions are termed **secondary** or **derived variables**. The name derived variables comes from the fact that they are computed as derivatives of the primary variables. In the present case u is the primary variable and $\partial u/\partial n = n_x\, \partial u/\partial x + n_y\, \partial u/\partial y$ can be considered as the secondary or derived variable. We recall from our treatment of the derivatives of the solution in terms of the interpolation functions, that the partial derivatives have the form

$$\frac{\partial u}{\partial x} = \mathbf{u_e^T}\frac{\partial \mathbf{N}}{\partial x} = \frac{\mathbf{u_e^T b_e}}{2A_e} = \frac{\mathbf{b_e^T u_e}}{2A_e}$$

and

$$\frac{\partial u}{\partial y} = \mathbf{u_e^T}\frac{\partial \mathbf{N}}{\partial y} = \frac{\mathbf{u_e^T c_e}}{2A_e} = \frac{\mathbf{c_e^T u_e}}{2A_e}$$

both of which, as mentioned previously, are constants within each element. Herein lies the main objection to the use of the linearly interpolated triangular element, namely the constant strain aspect of the solution. The dependent variable, as determined by the finite element solution, is continuous at all the interelement boundaries, but in general suffers a jump discontinuity in the normal derivative at an interelement boundary as was the case for the linearly interpolated one-dimensional problems. This results in reduced accuracy and relatively slow convergence of the solution. Higher-order interpolations, which will be discussed in detail later, can substantially improve the accuracy and convergence of the finite element solution. The linearly interpolated element is presented here because of its simplicity and because it enables us to see the basic structure of the finite element method without the complicating details of higher-order interpolations. These more appropriate types of interpolation will be used later.

3.4 EVALUATION OF MATRICES—LINEARLY INTERPOLATED TRIANGULAR ELEMENTS

As indicated in the previous section, the final sets of equations obtained using the method of Ritz in connection with the energy functional and Galerkin's method in connection with the weak formulation are exactly the same for the class of elliptic boundary value problems being considered. The variational approach makes it possible to easily think of the final set of equations as having been generated from the *elemental* contributions to the stiffnesses and loads, rather than in terms of the *nodal* contributions, as is the case when using Galerkin's method. We will generally find it preferable to think in terms of the elemental contributions rather than the nodal contributions.

Recall from Sections 3.2 and 3.3 that the elemental matrices for the elliptic boundary value problem are

$$\mathbf{k_e} = \iint_{A_e} \left(\frac{\partial \mathbf{N}}{\partial x} \frac{\partial \mathbf{N}^{\mathrm{T}}}{\partial x} + \frac{\partial \mathbf{N}}{\partial y} \frac{\partial \mathbf{N}^{\mathrm{T}}}{\partial y} \right) dA = \frac{(\mathbf{b_e b_e^{\mathrm{T}}} + \mathbf{c_e c_e^{\mathrm{T}}})}{4A_e}$$

$$\mathbf{f_e} = \iint_{A_e} \mathbf{N} f \, dA$$

$$\mathbf{a_e} = \int_{\gamma_{2e}} \mathbf{N} \alpha \mathbf{N}^{\mathrm{T}} \, ds$$

$$\mathbf{h_e} = \int_{\gamma_{2e}} \mathbf{N} h \, ds$$

The evaluation of these matrices involving integrals over the domains A_e can be tedious due to the general location and shape of the element A_e. In order to simplify the evaluation of integrals over the triangular domains, we introduce a local set of coordinates called area coordinates. In Fig. 3–18, consider a typical A_e.

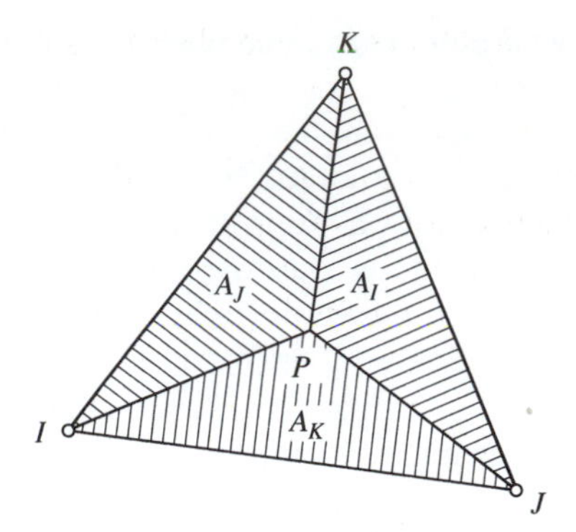

FIGURE 3–18 Typical triangular region and area coordinates

A point P in the interior is joined to the vertices and the areas A_I, A_J, and A_K are defined as shown. Obviously A_I, A_J, and A_K are related according to

$$A_I + A_J + A_K = A_e \tag{3.31}$$

where A_e is the area of the element. Dividing Eq. (3.31) by A_e yields

$$\frac{A_I}{A_e} + \frac{A_J}{A_e} + \frac{A_K}{A_e} = 1$$

or

$$L_I + L_J + L_K = 1 \tag{3.32}$$

where L_I, L_J, and L_K are referred to as the **area coordinates.** As P approaches any point on the line KJ, $A_I \to 0$ indicating that $L_I \to 0$, with $L_J + L_K = 1$. As $P \to I$, $A_I \to A_e$, and $L_I \to 1$, with L_J and L_K both approaching 0. Similar statements apply for P approaching other points and lines. It should be verified that the global coordinates x and y are related to the area or local coordinates according to

$$x = x_I L_I + x_J L_J + x_K L_K$$

$$y = y_I L_I + y_J L_J + y_K L_K$$

that is, as P varies within the element, x and y range over the x and y coordinates within A_e. As P approaches I (J or K) for instance, x approaches x_I (x_J or x_K) and y approaches y_I (y_J or y_K).

Collectively we have

$$\begin{aligned} x &= x_I L_I + x_J L_J + x_K L_K \\ y &= y_I L_I + y_J L_J + y_K L_K \\ 1 &= L_I + L_J + L_K \end{aligned} \tag{3.33}$$

If we consider these equations to be a set of linear equations that define L_I, L_J, and L_k, it can easily be shown that the solution produces exactly the same expressions as were developed earlier for N_i, N_j, and N_k. Thus L_I, L_J, and L_K are simply new names for and a new way of thinking about the N_i, N_j, and N_k.

It is also important in what follows to remember that since $L_I = 0$ on the line JK,

$$L_J + L_K = 1$$

At J, $L_J = 1$ and $L_K = 0$, and at K, $L_J = 0$ and $L_K = 1$. In view of the fact that the L's are linear functions of position within the element, it follows that

$$L_J = 1 - \frac{s}{l_{JK}}$$

$$L_K = \frac{s}{l_{JK}}$$

where s and l_{JK} are indicated in Fig. 3–19.

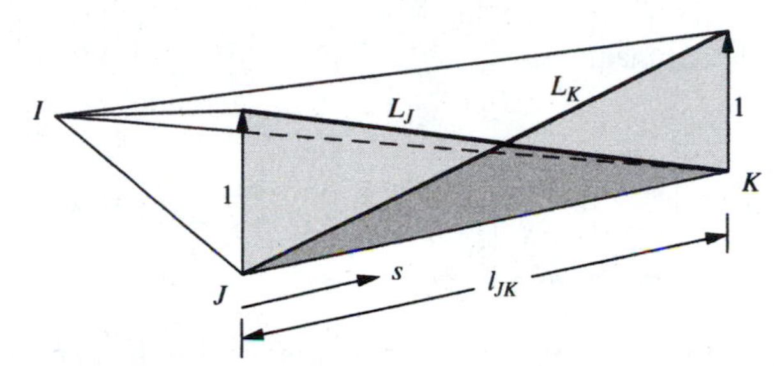

FIGURE 3–19 Interpolation functions on γ_{2e}

Thus on the boundary of the element the L's reduce to the linear one-dimensional interpolation functions discussed in Chapter 2. It is seen that the integrals on the segments of the boundary γ_{2e} will involve the N's or L's on that edge and hence will reduce to the evaluation of the same sorts of integrals already considered in detail in Chapter 2.

The equation $1 = L_I + L_J + L_K$ indicates that the area or local variables L_I, L_J, and L_K are not linearly independent. Generally we select two of the three as the independent variables and eliminate the third. Eliminating L_K from the last of Eqs. (3.33) according to $L_K = 1 - L_I - L_J$ and substituting into the first two of Eqs. (3.33) yields

$$\begin{aligned} x &= x_3 + (x_1 - x_3)L_I + (x_2 - x_3)L_J = x(L_I, L_J) \\ y &= y_3 + (y_1 - y_3)L_I + (y_2 - y_3)L_J = y(L_I, L_J) \end{aligned} \tag{3.34}$$

representing a transformation between the global variables (x,y) and the local variables (L_I, L_J). In this case it is worth pointing out that the transformation is linear in (x, y) and (L_I, L_J) and that, in principle, Eqs. (3.34) can be inverted to express the L's in terms of x and y. Generally, this is not practical for many other types of interpolations.

Evaluation of $\mathbf{k_e}$. When we wish to use the local or area coordinates for evaluating the intervals, the following development of the stiffness matrix $\mathbf{k_e}$ is typical of the analysis necessary to accomplish the task. Although it was shown in Section 3.3 that $\mathbf{k_e}$ has an easily determined simple form for the linearly interpolated triangular element, we will in what follows rederive those results using the ideas associated with the transformation between the global and local (or area) coordinates. Again,

$$\mathbf{k_e} = \int\int \left(\frac{\partial \mathbf{N}}{\partial x} \frac{\partial \mathbf{N}^{\mathrm{T}}}{\partial x} + \frac{\partial \mathbf{N}}{\partial y} \frac{\partial \mathbf{N}^{\mathrm{T}}}{\partial y} \right) dA = \int\int \mathbf{G}(x, y)\, dx\, dy$$

Using the transformation Eq. (3.34), we wish to transform this into an integral whose integrand depends only on L_I and L_J, that is,

$$\mathbf{k_e} = \int\int \mathbf{G}\big(x(L_I, L_J), y(L_I, L_J)\big)\, |\mathbf{J}|\, dL_I\, dL_J$$

where $|\mathbf{J}|$ is the determinant

$$|\mathbf{J}| = \begin{vmatrix} \dfrac{\partial x}{\partial L_1} & \dfrac{\partial y}{\partial L_1} \\ \dfrac{\partial x}{\partial L_2} & \dfrac{\partial y}{\partial L_2} \end{vmatrix} = \begin{vmatrix} x_1 - x_3 & y_1 - y_3 \\ x_2 - x_3 & y_2 - y_3 \end{vmatrix} = 2A_e$$

For the purpose of transforming the entire integrand of $\mathbf{k_e}$ into the variables L_I and L_J, it is necessary to convert the partial derivatives $\partial/\partial x$ and $\partial/\partial y$ appearing in the integrand of $\mathbf{k_e}$ to equivalent expressions in terms of $\partial/\partial L_I$ and $\partial/\partial L_J$. To this end we use the chain rule to write the operator equations

$$\frac{\partial}{\partial L_I} = \frac{\partial}{\partial x}\frac{\partial x}{\partial L_I} + \frac{\partial}{\partial y}\frac{\partial y}{\partial L_I}$$

and

$$\frac{\partial}{\partial L_J} = \frac{\partial}{\partial x}\frac{\partial x}{\partial L_J} + \frac{\partial}{\partial y}\frac{\partial y}{\partial L_J}$$

or in matrix notation as

$$\frac{\boldsymbol{\partial}}{\boldsymbol{\partial}\mathbf{L}} = \mathbf{J}\frac{\boldsymbol{\partial}}{\boldsymbol{\partial}\mathbf{x}} \tag{3.35}$$

where

$$\mathbf{J} = \begin{bmatrix} \dfrac{\partial x}{\partial L_I} & \dfrac{\partial y}{\partial L_I} \\ \dfrac{\partial x}{\partial L_J} & \dfrac{\partial y}{\partial L_J} \end{bmatrix}$$

with

$$|\mathbf{J}| = \begin{vmatrix} \dfrac{\partial x}{\partial L_I} & \dfrac{\partial y}{\partial L_I} \\ \dfrac{\partial x}{\partial L_J} & \dfrac{\partial y}{\partial L_J} \end{vmatrix}$$

and

$$\frac{\boldsymbol{\partial}}{\boldsymbol{\partial}\mathbf{L}} = \begin{bmatrix} \dfrac{\partial}{\partial L_I} \\ \dfrac{\partial}{\partial L_J} \end{bmatrix} \qquad \frac{\boldsymbol{\partial}}{\boldsymbol{\partial}\mathbf{x}} = \begin{bmatrix} \dfrac{\partial}{\partial x} \\ \dfrac{\partial}{\partial y} \end{bmatrix}$$

In Eq. (3.35), $\mathbf{J}$ is the *Jacobian matrix* of the transformation. The determinant of the Jacobian matrix $|\mathbf{J}|$ is the more usually encountered Jacobian determinant discussed in an advanced calculus setting. Equation (3.35) can be inverted to yield

$$\frac{\boldsymbol{\partial}}{\boldsymbol{\partial}\mathbf{x}} = \mathbf{J}^{-1}\frac{\boldsymbol{\partial}}{\boldsymbol{\partial}\mathbf{L}} \tag{3.36}$$

which is perfectly well defined as long as the Jacobian determinant $|\mathbf{J}|$ is nonzero. Generally such a determinant is a function of (x, y) or (L_I, L_J) although in this case it is a constant equal to twice the area of the element. The matrix Eq. (3.36) can be partitioned to yield

$$\frac{\partial}{\partial x} = \mathbf{J}_1\frac{\boldsymbol{\partial}}{\boldsymbol{\partial}\mathbf{L}} \qquad \text{and} \qquad \frac{\partial}{\partial y} = \mathbf{J}_2\frac{\boldsymbol{\partial}}{\boldsymbol{\partial}\mathbf{L}}$$

where $\mathbf{J}_1$ and $\mathbf{J}_2$ are the first and second rows of $\mathbf{J}^{-1}$. The partial derivatives $\partial/\partial x$ and $\partial/\partial y$ are now represented entirely in terms of the L's.

Returning to the evaluation of the elemental stiffness matrix $\mathbf{k_e}$, we compute

$$\frac{\partial \mathbf{N}^{\mathrm{T}}}{\partial x} = \mathbf{J}_1\frac{\partial \mathbf{N}^{\mathrm{T}}}{\partial \mathbf{L}} = \mathbf{J}_1\boldsymbol{\Delta}^{\mathrm{T}}$$

and

$$\frac{\partial \mathbf{N}^{\mathrm{T}}}{\partial y} = \mathbf{J}_2\frac{\partial \mathbf{N}^{\mathrm{T}}}{\partial \mathbf{L}} = \mathbf{J}_2\boldsymbol{\Delta}^{\mathrm{T}}$$

where

$$\boldsymbol{\Delta}^{\mathrm{T}} = \frac{\partial \mathbf{N}^{\mathrm{T}}}{\partial \mathbf{L}} = \begin{bmatrix} \dfrac{\partial L_I}{\partial L_I} & \dfrac{\partial L_J}{\partial L_I} & \dfrac{\partial L_K}{\partial L_I} \\ \dfrac{\partial L_I}{\partial L_J} & \dfrac{\partial L_J}{\partial L_J} & \dfrac{\partial L_K}{\partial L_J} \end{bmatrix} = \begin{bmatrix} 1 & 0 & -1 \\ 0 & 1 & -1 \end{bmatrix}$$

where we have used the fact that $L_K = 1 - L_I - L_J$. It follows that

$$\frac{\partial \mathbf{N}}{\partial x} = \mathbf{\Delta J}_1^{\mathrm{T}} \qquad \text{and} \qquad \frac{\partial \mathbf{N}}{\partial y} = \mathbf{\Delta J}_2^{\mathrm{T}}$$

and that $\mathbf{k_e}$ can be expressed as

$$\begin{aligned} \mathbf{k_e} &= \int\int [\mathbf{\Delta J}_1^{\mathrm{T}}\mathbf{J}_1\mathbf{\Delta}^{\mathrm{T}} + \mathbf{\Delta J}_2^{\mathrm{T}}\mathbf{J}_2\mathbf{\Delta}^{\mathrm{T}}]\,|\mathbf{J}|\,dL_I\,dL_J \\ &= \int\int \mathbf{\Delta}[\mathbf{J}_1^{\mathrm{T}}\mathbf{J}_1 + \mathbf{J}_2^{\mathrm{T}}\mathbf{J}_2]\mathbf{\Delta}^{\mathrm{T}}\,|\mathbf{J}|\,dL_I\,dL_J \\ &= \int\int \mathbf{\Delta JJ\Delta}^{\mathrm{T}}\,dL_I\,dL_J \end{aligned} \tag{3.37}$$

where $\mathbf{JJ} = (\mathbf{J}_1^{\mathrm{T}}\mathbf{J}_1 + \mathbf{J}_2^{\mathrm{T}}\mathbf{J}_2)|\mathbf{J}|$. In view of the fact that the integrand is constant, it follows that

$$\mathbf{k_e} = \frac{\mathbf{\Delta}^{\mathrm{T}}\mathbf{JJ\Delta}}{2}$$

In the Exercises, the student is asked to show that this coincides with the previous result

$$\mathbf{k_e} = \frac{(\mathbf{b_e b_e^{\mathrm{T}}} + \mathbf{c_e c_e^{\mathrm{T}}})}{4A_e}$$

In any event, $\mathbf{k_e}$ is a 3×3 stiffness matrix which is to be added at assembly to the global stiffness matrix $\mathbf{K_G}$ at the positions corresponding to the nodes of the elements.

Before continuing with the evaluation of $\mathbf{f_e}$, $\mathbf{a_e}$, and $\mathbf{h_e}$ we state without proof the following integration formulas:

$$\int\int_{A_e} L_I^a L_J^b L_K^c\,dA = a!b!c!\frac{2A_e}{(a+b+c+2)!}$$

or (3.38)

$$\int\int_{A_e} N_I^a N_J^b N_K^c\,dA = a!b!c!\frac{2A_e}{(a+b+c+2)!}$$

and

$$\int_{l_e} L_I^a L_J^b\,ds = a!b!\frac{l_e}{(a+b+1)!}$$

or (3.39)

$$\int_{l_e} N_I^a N_J^b\,ds = a!b!\frac{l_e}{(a+b+1)!}$$

The area integral formulas will obviously be used in general to evaluate integrals over interior area elements A_e, and the line integrals to evaluate integrals over segments γ_{2e} on the boundary.

Evaluation of $\mathbf{f_e}$. Consider a typical element as shown in Fig. 3–20.

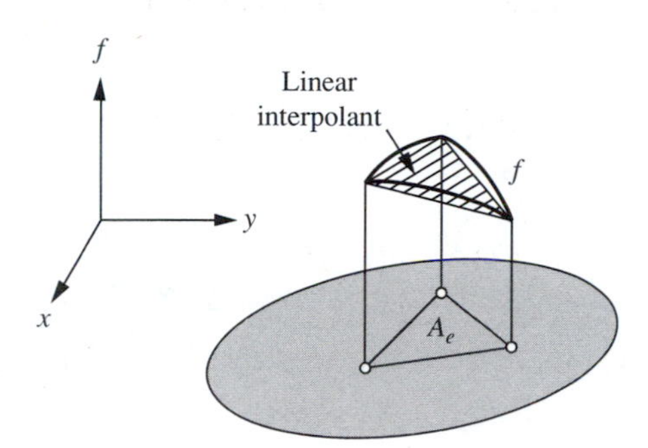

FIGURE 3–20 ***f(x, y)* and its linear interpolant on a typical triangular element**

In general, the integral for $\mathbf{f_e}$,

$$\mathbf{f_e} = \int\int_{A_e} \mathbf{N} f(x, y)\, dA$$

can be somewhat tedious to evaluate when the integrand is considered to be a function of x and y, with the integration to take place over the triangular region A_e. Alternately, the integrand can be expressed entirely in terms of L_I and L_J after which it might then be possible, depending on the form of f, to use the above quoted integration formulas Eqs. (3.38) to evaluate the integrals. From Fig. 3–20, however, one might suspect that a good approximation to the value of the integral would be obtained if f were replaced by its linear interpolant, that is, $f = \mathbf{N}^{\mathrm{T}}\mathbf{f_i}$, where $\mathbf{f_i} = [f_1\ f_2\ f_3]$, with the f_i usually chosen to be the function f evaluated at the node in question. With this assumption there results

$$\mathbf{f_e} \approx \int\int_{A_e} \mathbf{N}\mathbf{N}^{\mathrm{T}}\mathbf{f_i}\, dA = \left(\int\int_{A_e} \mathbf{N}\mathbf{N}^{\mathrm{T}}\, dA\right)\mathbf{f_i}$$

which, after using the integration formulas of Eqs. (3.38), yields

$$\mathbf{f_e} \approx \frac{A_e}{12}\begin{bmatrix} 2 & 1 & 1 \\ 1 & 2 & 1 \\ 1 & 1 & 2 \end{bmatrix}\mathbf{f_i}$$

or

$$\mathbf{f_e} \approx \frac{A_e}{12}\begin{bmatrix} 2f_i + f_j + f_k \\ f_i + 2f_j + f_k \\ f_i + f_j + 2f_k \end{bmatrix} \tag{3.40}$$

This elemental matrix is loaded during the assembly process into the correct locations, that is, into the positions corresponding to nodes i, j, and k, in the global load vector $\mathbf{F_G}$.

Evaluation of $\mathbf{h_e}$. For the evaluation of the $\mathbf{h_e}$ we consider a typical segment γ_{2e} on the boundary as shown in Fig. 3–21.

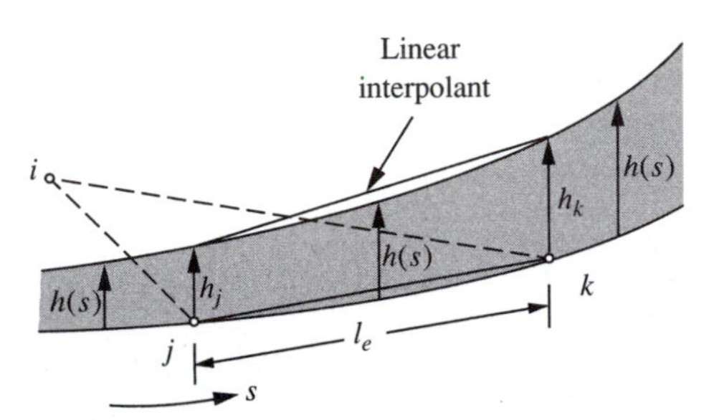

FIGURE 3–21 Typical γ_{2e} segment on the boundary showing $h(s)$ and its linear interpolant

With the integral to be evaluated *on* γ_{2e}, N_i, N_j, and N_k are also to be evaluated on the boundary, yielding, for the i, j, and k pictured, $N_i = 0$, $N_j = (1 - s/l_e)$, and $N_k = s/l_e$, where s and l_e are as indicated in Fig. 3–21. The expression for $\mathbf{h_e}$ then becomes

$$\mathbf{h_e} = \int_{\gamma_{2e}} \begin{bmatrix} 1 - \dfrac{s}{l_e} \\ \dfrac{s}{l_e} \end{bmatrix} h(s)\, ds$$

with contributions to the global load vector $\mathbf{F_G}$ at positions j and k. Depending on the form of $h(s)$ it may be possible to evaluate this integral exactly. Generally, however, an approximate approach is taken. To this end we assume that h is sufficiently smooth so that it can be approximated by its linear interpolant according to $h(s) = \mathbf{N}^\mathrm{T}\mathbf{h_i}$, where $\mathbf{h_i} = [h_j \;\; h_k]$. Using the integration formulas Eqs. (3.39) there results

$$\mathbf{h_e} \approx \int_{\gamma_{2e}} \mathbf{N}\mathbf{N}^\mathrm{T}\mathbf{h_i}\, ds = \left(\int_{\gamma_{2e}} \mathbf{N}\mathbf{N}^\mathrm{T}\, ds\right)\mathbf{h_i}$$

$$= \frac{l_e}{6}\begin{bmatrix} 2 & 1 \\ 1 & 2 \end{bmatrix}\mathbf{h_i}$$

or finally

$$\mathbf{h_e} \approx \frac{l_e}{6}\begin{bmatrix} 2h_j + h_k \\ h_j + 2h_k \end{bmatrix} \tag{3.41}$$

which again will be added in the correct locations, that is, at nodes j and k in the global load vector $\mathbf{F_G}$, during the assembly process. Note that $\mathbf{h}_e$ has precisely the same form as the $\mathbf{f_e}$ for the linearly interpolated finite element model in Chapter 2.

Evaluation of the $\mathbf{a_e}$. The evaluation of the $\mathbf{a_e}$ is essentially the same as for the $\mathbf{h_e}$ except for the extra $\mathbf{N}^T$ that will appear in the integrand. Specifically, when $\alpha(s)$ is approximated by its linear interpolant according to $\alpha(s) = \alpha_j N_j + \alpha_k N_k$ there results

$$
\begin{aligned}
\mathbf{a_e} &= \int_{\gamma_{2e}} \mathbf{N}\alpha\mathbf{N}^T\, ds \\
&\approx \begin{bmatrix} \int_{\gamma_{2e}} N_j(\alpha_j N_j + \alpha_k N_k)N_j\, ds & \int_{\gamma_{2e}} N_j(\alpha_j N_j + \alpha_k N_k)N_k\, ds \\ \int_{\gamma_{2e}} N_k(\alpha_j N_j + \alpha_k N_k)N_j\, ds & \int_{\gamma_{2e}} N_k(\alpha_j N_j + \alpha_k N_k)N_k\, ds \end{bmatrix}
\end{aligned}
\tag{3.42}
$$

which, after using the integration formulas Eqs. (3.39), becomes

$$
\mathbf{a_e} = \frac{l_e}{12}\begin{bmatrix} 3\alpha_j + \alpha_k & \alpha_j + \alpha_k \\ \alpha_j + \alpha_k & \alpha_j + 3\alpha_k \end{bmatrix}
\tag{3.43}
$$

a 2×2 stiffness matrix to be added to the appropriate locations in the global stiffness matrix $\mathbf{K_G}$ at assembly. Note that $\mathbf{a_e}$ has precisely the same form as the $\mathbf{q_e}$ for the linearly interpolated finite element model in Chapter 2.

To reiterate, these elemental matrices enter into the assembly process to produce the final unconstrained global equations

$$
\mathbf{K_G u_G} = \mathbf{F_G}
$$

according to

$$
\begin{aligned}
\mathbf{K_G} &= \sum_e \mathbf{k_G} + \sum_e{}' \mathbf{a_G} \\
\mathbf{F_G} &= \sum_e \mathbf{f_G} + \sum_e{}' \mathbf{h_G}
\end{aligned}
$$

Recall that the $\sum$ refers to contributions from the elemental areas A_e, and that the $\sum'$ indicates a sum over the elemental line segments γ_{2e}.

3.5 EXAMPLE APPLICATIONS

In this section we examine the application of the finite element method to several practical two-dimensional problems of general interest in engineering.

3.5.1 Torsion of a Rectangular Section

The problem of the torsion of a solid prismatic bar was introduced in Section 3.1. In this section we will investigate in detail the problem of the torsion of a bar whose rectangular cross section occupies the region $-a \leq X \leq a, \ -b \leq Y \leq b$ as shown in Fig. 3–22.

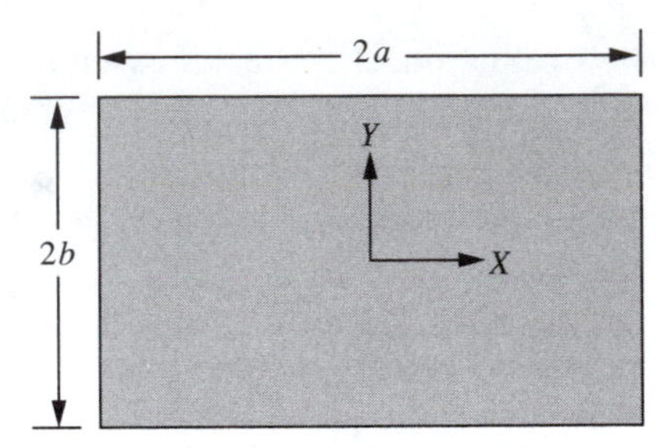

FIGURE 3–22 Rectangular cross section

Before generating the finite element model, it is desirable from a numerical standpoint to nondimensionalize the problem according to the transformations on the independent and dependent variables given by

$$\frac{X}{a} = x, \quad \frac{Y}{a} = y, \qquad \text{and} \qquad \frac{\phi}{2G\theta a^2} = \psi$$

so that the boundary value problem becomes

$$\frac{\partial^2 \psi}{\partial x^2} + \frac{\partial^2 \psi}{\partial y^2} + 1 = 0 \qquad \text{in } D$$

$$\psi = 0 \qquad \text{on } \Gamma$$

D now refers to the domain $-1 \leq x \leq 1, -\alpha \leq y \leq \alpha$, where $\alpha = b/a$. It also follows that

$$\tau_{xz} = 2G\theta a \frac{\partial \psi}{\partial y} \qquad \tau_{yz} = -2G\theta a \frac{\partial \psi}{\partial x}$$

and

$$T = 4G\theta a^4 \int\int \psi \, dx \, dy$$

Finite element model. In constructing a finite element model, any symmetries in the solution should be exploited. In the present case, there is symmetry with respect to both the x and y axes enabling us to use any of the four quadrants as the region to be modeled. (In order to assist with the visualization of the symmetries of the ψ (or ϕ) surface, the student may want to recall or review the so-called *membrane analogy* [Timoshenko and Goodier, 5].) In terms of the quadrant region $0 \leq x \leq 1, \ 0 \leq y \leq \alpha$, indicated in Fig. 3–23, the boundary value problem becomes

$$\nabla^2\psi + 1 = 0 \qquad 0 \le x \le 1,\ 0 \le y \le \alpha$$

$$\psi = 0 \qquad \text{on } AB \text{ and } BC$$

$$\frac{\partial \psi}{\partial n} = 0 \qquad \text{on } OA \text{ and } OC$$

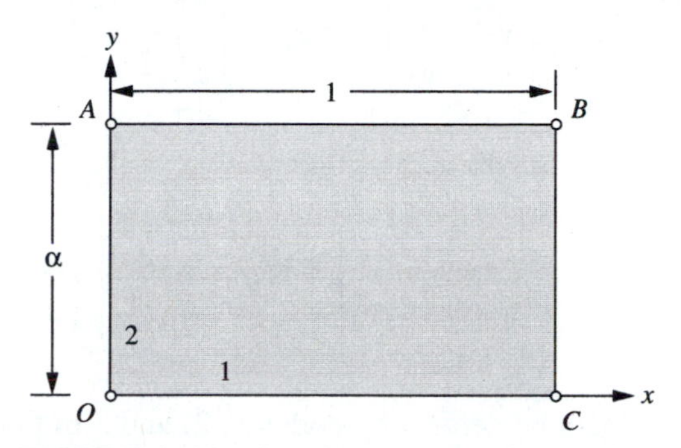

FIGURE 3–23 Quadrant region for the finite element model accounting for symmetries

The zero normal derivatives on the two exposed interior boundaries OA and OC reflect the symmetric character of the solution ψ, or equivalently, state that the appropriate shear stress component on that interior surface is zero. Note that if the cross section were a square, additional symmetries would enable us to further reduce the region to an octant.

Discretization and interpolation. The simplest possible mesh for the rectangular region using linearly interpolated triangular elements consists of the four nodes and two triangular elements indicated in Fig. 3–24.

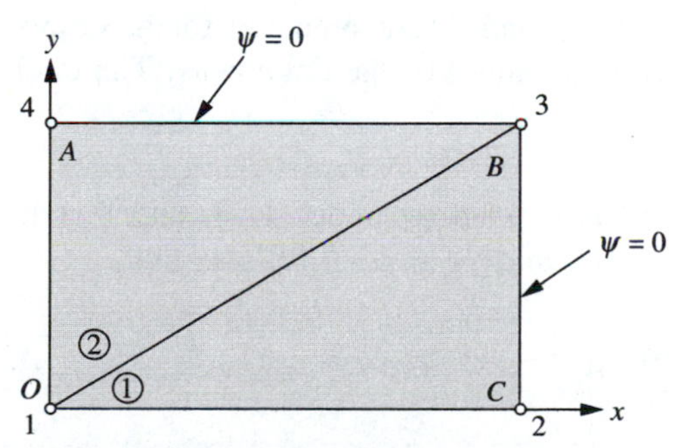

FIGURE 3–24 Simple mesh for rectangular region

Elemental formulation. For the linearly interpolated triangular element, the formulation was completed in Sections 3.2 and 3.3. We have only to evaluate the elemental matrices corresponding to that formulation. Recall that in general

$$\mathbf{k_e} = \frac{\mathbf{b_e b_e^T} + \mathbf{c_e c_e^T}}{4A_e}$$

where

$$\mathbf{b_e} = \begin{bmatrix} y_2 - y_3 \\ y_3 - y_1 \\ y_1 - y_2 \end{bmatrix} \qquad \mathbf{c_e} = \begin{bmatrix} x_3 - x_2 \\ x_1 - x_3 \\ x_2 - x_1 \end{bmatrix}$$

with A_e the area of the element. Also, since in this case $f = 1$,

$$\mathbf{f_e} = \iint f\mathbf{N}\,dA = \iint \mathbf{N}\,dA = \frac{A_e}{3}\begin{bmatrix} 1 \\ 1 \\ 1 \end{bmatrix}$$

Element 1. Element 1 can be defined by nodes 1, 2, and 3 in that order. It follows that $x_1 = 0$, $x_2 = 1$, $x_3 = 1$, $y_1 = 0$, $y_2 = 0$, and $y_3 = \alpha = b/a$, with

$$\mathbf{b_e^T} = [-\alpha \quad \alpha \quad 0] \qquad \mathbf{c_e^T} = [0 \quad -1 \quad 1] \qquad A_e = \frac{\alpha}{2}$$

from which

$$\mathbf{k_e} = \frac{1}{2\alpha}\begin{array}{c} \begin{matrix} 1 & 2 & 3 \end{matrix} \\ \begin{bmatrix} \alpha^2 & -\alpha^2 & 0 \\ -\alpha^2 & 1+\alpha^2 & -1 \\ 0 & -1 & 1 \end{bmatrix} \end{array} \begin{matrix} 1 \\ 2 \\ 3 \end{matrix} \qquad \mathbf{f_e} = \frac{\alpha}{6}\begin{bmatrix} 1 \\ 1 \\ 1 \end{bmatrix} \begin{matrix} 1 \\ 2 \\ 3 \end{matrix}$$

The elemental matrices $\mathbf{k_e}$ and $\mathbf{f_e}$ are prepared for assembly by identifying the rows and columns with the names of the unknowns. The circled numbers refer to the nodes in question.

Element 2. Element 2 is defined by nodes 1, 3, and 4, with $x_1 = 0$, $x_2 = 1$, $x_3 = 0$, $y_1 = 0$, $y_2 = \alpha$, and $y_3 = \alpha$. It follows that

$$\mathbf{b_e^T} = [0 \quad \alpha \quad -\alpha] \qquad \mathbf{c_e^T} = [-1 \quad 0 \quad 1] \qquad A_e = \frac{\alpha}{2}$$

$$\mathbf{k_e} = \frac{1}{2\alpha}\begin{array}{c} \begin{matrix} 1 & 3 & 4 \end{matrix} \\ \begin{bmatrix} 1 & 0 & -1 \\ 0 & \alpha^2 & -\alpha^2 \\ -1 & -\alpha^2 & 1+\alpha^2 \end{bmatrix} \end{array} \begin{matrix} 1 \\ 3 \\ 4 \end{matrix} \qquad \mathbf{f_e} = \frac{\alpha}{6}\begin{bmatrix} 1 \\ 1 \\ 1 \end{bmatrix} \begin{matrix} 1 \\ 3 \\ 4 \end{matrix}$$

Assembly. The global stiffness and load matrices are determined according to

$$\mathbf{K_G} = \sum \mathbf{k_G} + \sum{}' \mathbf{a_G} = \sum \mathbf{k_G}$$

$$\mathbf{F_G} = \sum \mathbf{f_G} + \sum{}' \mathbf{h_G} = \sum \mathbf{f_G}$$

leading to

$$\mathbf{K_G} = \begin{bmatrix} 1+\alpha^2 & -\alpha^2 & 0 & -1 \\ -\alpha^2 & 1+\alpha^2 & -1 & 0 \\ 0 & -1 & 1+\alpha^2 & -\alpha^2 \\ -1 & 0 & -\alpha^2 & 1+\alpha^2 \end{bmatrix} \frac{1}{2\alpha}$$

and

$$\mathbf{F_G^T} = \frac{\alpha}{6}[2 \quad 1 \quad 2 \quad 1]$$

In augmented form the unconstrained global equations are

$$\left[\begin{array}{cccc|c} 1+\alpha^2 & -\alpha^2 & 0 & -1 & 2\beta \\ -\alpha^2 & 1+\alpha^2 & -1 & 0 & \beta \\ 0 & -1 & 1+\alpha^2 & -\alpha^2 & 2\beta \\ -1 & 0 & -\alpha^2 & 1+\alpha^2 & \beta \end{array}\right]$$

where $\beta = \alpha^2/3$.

Constraints. For the current model, with $\psi = 0$ on AB and BC, it follows that $\psi_2 = \psi_3 = \psi_4 = 0$. The constrained equations appear as

$$\left[\begin{array}{cccc|c} 1+\alpha^2 & 0 & 0 & 0 & 2\beta \\ 0 & 1 & 0 & 0 & 0 \\ 0 & 0 & 1 & 0 & 0 \\ 0 & 0 & 0 & 1 & 0 \end{array}\right]$$

Solution. The solution is clearly

$$\psi_1 = \frac{2\beta}{1+\alpha^2}$$

with

$$\phi_1 = \frac{4G\theta a^2\alpha^2}{3(1+\alpha^2)}$$

On the quadrant region being considered, the finite element solution would appear as in Fig. 3–25.

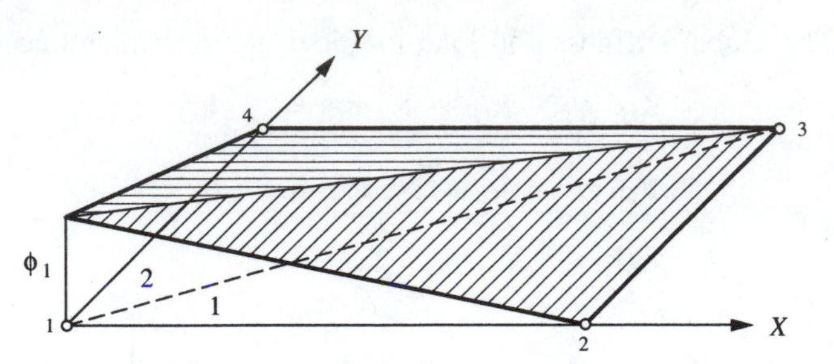

FIGURE 3–25 Finite element solution surface

Computation of derived variables. In this case the partial derivatives with respect to x and y within the elements are easily determined by inspection to be

Element 1.

$$\tau_{XZ} = \frac{\partial \phi}{\partial Y} = 0$$

$$\tau_{YZ} = -\frac{\partial \phi}{\partial X} = \frac{-(0-\phi_1)}{a} = \frac{4G\theta a\alpha^2}{3(1+\alpha^2)}$$

Element 2.

$$\tau_{XZ} = \frac{\partial \phi}{\partial Y} = \frac{0-\phi_1}{b} = \frac{-4G\theta a\alpha^2}{3(1+\alpha^2)}$$

$$\tau_{YZ} = -\frac{\partial \phi}{\partial X} = 0$$

In a general situation, it may not be possible to compute the derivatives on a visual basis, and the machinery developed previously will have to be employed. To this end we repeat the previous calculations using the expressions

$$\tau_{XZ} = \frac{\partial \phi}{\partial Y} = 2G\theta a\frac{\partial \psi}{\partial \eta} = 2G\theta a\frac{\mathbf{c_e^T \psi_e}}{2A_e}$$

$$\tau_{YZ} = -\frac{\partial \phi}{\partial X} = -2G\theta a\frac{\partial \psi}{\partial \xi} = -2G\theta a\frac{\mathbf{b_e^T \psi_e}}{2A_e}$$

Element 1. From the previous calculations for $\mathbf{k_e}$,

$$\mathbf{b_e^T} = [-\alpha \quad \alpha \quad 0] \qquad \mathbf{c_e^T} = [0 \quad -1 \quad 1]$$

which, with

$$\boldsymbol{\psi}_\mathbf{e} = \begin{bmatrix} \psi_1 \\ \psi_2 \\ \psi_3 \end{bmatrix} = \begin{bmatrix} \dfrac{2\alpha^2}{3(1+\alpha^2)} \\ 0 \\ 0 \end{bmatrix}$$

again yields

$$\tau_{XZ} = 2G\theta a[0 \quad -1 \quad 1]\begin{bmatrix} \dfrac{2\alpha^2}{3(1+\alpha^2)} \\ 0 \\ 0 \end{bmatrix}\frac{1}{\alpha} = 0$$

$$\tau_{YZ} = -2G\theta a[-\alpha \quad \alpha \quad 0]\begin{bmatrix} \dfrac{2\alpha^2}{3(1+\alpha^2)} \\ 0 \\ 0 \end{bmatrix}\frac{1}{\alpha} = \frac{4G\theta a\alpha^2}{3(1+\alpha^2)}$$

Element 2. Again, from the previous calculations for $\mathbf{k_e}$,

$$\mathbf{b}_\mathbf{e}^\mathrm{T} = [0 \quad \alpha \quad -\alpha] \qquad \mathbf{c}_\mathbf{e}^\mathrm{T} = [-1 \quad 0 \quad 1]$$

With

$$\boldsymbol{\psi}_\mathbf{e} = \begin{bmatrix} \psi_1 \\ \psi_3 \\ \psi_4 \end{bmatrix} = \begin{bmatrix} \dfrac{2\alpha^2}{3(1+\alpha^2)} \\ 0 \\ 0 \end{bmatrix}$$

there results

$$\tau_{XZ} = 2G\theta a[-1 \quad 0 \quad 1]\begin{bmatrix} \dfrac{2\alpha^2}{3(1+\alpha^2)} \\ 0 \\ 0 \end{bmatrix}\frac{1}{\alpha} = \frac{-4G\theta a\alpha^2}{3(1+\alpha^2)}$$

$$\tau_{YZ} = -2G\theta a[0 \quad \alpha \quad -\alpha]\begin{bmatrix} \dfrac{2\alpha^2}{3(1+\alpha^2)} \\ 0 \\ 0 \end{bmatrix}\frac{1}{\alpha} = 0$$

These calculations indicate an important practical detail: when the elemental stiffnesses or certain of the ingredients that go to making up the stiffnesses are formed, *they should be saved on an auxiliary storage device to be used later in the calculation of the derived variables*.

Recall that the total torque transmitted is given by

$$T = 2\int\!\!\int \phi \, dA = 4G\theta a^4 \int\!\!\int \psi \, dx \, dy$$

In terms of the quadrant currently being considered, we can write

$$\frac{T}{4} = 4G\theta a^4 \sum_e \iint_{A_e} \psi_e \, dx \, dy = 4G\theta a^4 \sum_e \iint_{A_e} \boldsymbol{\psi}_e^T \mathbf{N} \, dx \, dy$$

$$= 4G\theta a^4 \sum_e \frac{A_e}{3} (\psi_{ie} + \psi_{je} + \psi_{ke})$$

where the integration formulas $\iint N_1 \, dx \, dy = \iint N_2 \, dx \, dy = \iint N_3 \, dx \, dy = A_e/3$ have been used, and ψ_{ie}, ψ_{je}, and ψ_{ke} represent the nodal values for the element. For the present setting

$$\frac{T}{4} = 4G\theta a^4 \frac{A_e}{3} \big((\psi_1 + 0 + 0) + (\psi_1 + 0 + 0)\big)$$

leading to

$$T = \frac{32G\theta a^4 \alpha^3}{9(1 + \alpha^2)}$$

Specifically for $\alpha = b/a = 0.5$, $T = 0.355G\theta a^4$ as compared with the exact value of $0.458G\theta a^4$ [Timoshenko and Goodier, 5]. In view of the fact that generally $T = JG\theta$, where JG is the torsional stiffness, we see that this particular finite element model underestimates the torsional stiffness by approximately 22 percent.

The maximum shear stress is τ_{XZ} in element 2 with a value of

$$\tau_{XZ} = \frac{4G\theta \alpha a}{3(1 + \alpha^2)} = 0.533G\theta a = \frac{1.5T}{a^3}$$

to be compared with the exact value of $2.033T/a^3$. The maximum stress is underestimated by approximately 26 percent.

Consider next the mesh shown in Fig. 3–26 where the element size has been obtained by halving the dimensions of the elements of the previous model.

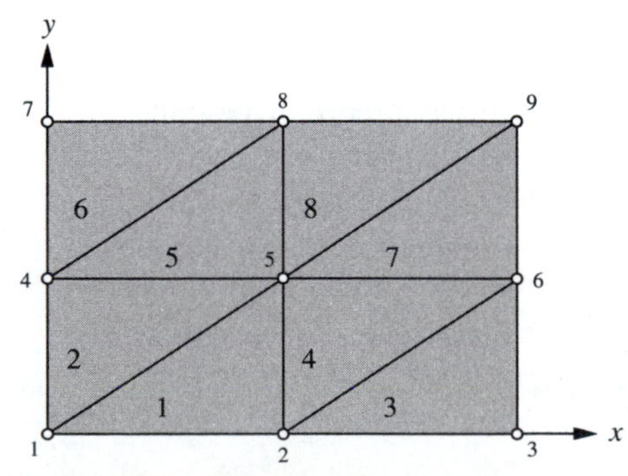

FIGURE 3–26 Finer mesh obtained by halving

The $\mathbf{k_e}$ for elements 2, 4, 6, and 8 coincide with the $\mathbf{k_e}$ of element 2 of the previous model, whereas the $\mathbf{k_e}$ for elements 1, 3, 5, and 7 coincide with the $\mathbf{k_e}$ for element 1 of the previous model. Upon assembly with $\beta = \alpha^2/3$, the augmented equations become

$$\left[\begin{array}{ccccccccc|c}
1+\alpha^2 & -\alpha^2 & 0 & -1 & 0 & 0 & 0 & 0 & 0 & 2\beta \\
-\alpha^2 & 2(1+\alpha^2) & -\alpha^2 & 0 & -2 & 0 & 0 & 0 & 0 & 3\beta \\
0 & -\alpha^2 & 1+\alpha^2 & 0 & 0 & -1 & 0 & 0 & 0 & \beta \\
-1 & 0 & 0 & 2(1+\alpha^2) & -2\alpha^2 & 0 & -1 & 0 & 0 & 3\beta \\
0 & -2 & 0 & -2\alpha^2 & 4(1+\alpha^2) & -2\alpha^2 & 0 & -2 & 0 & 6\beta \\
0 & 0 & -1 & 0 & -2\alpha^2 & 2(1+\alpha^2) & 0 & 0 & -1 & 3\beta \\
0 & 0 & 0 & -1 & 0 & 0 & 1+\alpha^2 & -\alpha^2 & 0 & \beta \\
0 & 0 & 0 & 0 & -2 & 0 & -\alpha^2 & 2(1+\alpha^2) & -\alpha^2 & 3\beta \\
0 & 0 & 0 & 0 & 0 & -1 & 0 & \alpha^2 & 1+\alpha^2 & 2\beta
\end{array}\right]$$

The constraints are $\psi_3 = \psi_6 = \psi_7 = \psi_8 = \psi_9 = 0$, resulting in

$$\left[\begin{array}{cccc|c}
1+\alpha^2 & -\alpha^2 & -1 & 0 & 2\beta \\
-\alpha^2 & 2(1+\alpha^2) & 0 & -2 & 3\beta \\
-1 & 0 & 2(1+\alpha^2) & -2\alpha^2 & 3\beta \\
0 & -2 & -2\alpha^2 & 4(1+\alpha^2) & 6\beta
\end{array}\right]$$

which for $\alpha = 0.5$ has the solution

$$\psi_1 = 0.1232 \qquad \psi_2 = 0.0947 \qquad \psi_4 = 0.0886 \qquad \psi_5 = 0.0717$$

The maximum shear stress is the component τ_{XZ} in element 6 and is

$$\tau_{XZ} = \frac{(0 - \psi_4)2G\theta a^2}{(b/2)} = 0.709G\theta a$$

The torque is computed to be

$$T = \frac{4G\theta a^4(2\psi_1 + 3\psi_2 + 3\psi_4 + 6\psi_5)}{24} = 0.409G\theta a^4$$

from which, upon eliminating $G\theta$, the maximum shear stress in terms of the torque T becomes

$$\tau_{\max} = \frac{1.733T}{a^3}$$

For this model the percent errors in the torsional stiffness and maximum stress are 10.7 and 14.8, respectively.

The results from the two- and eight-element models along with two additional models obtained by continued successive halving of the element size are presented in Table 3.1. As can be seen from the table, the convergence for the torsional constant is quite reasonable with somewhat poorer results for the maximum stress. This is to be expected in that the stress is a derived variable, whereas the torsional constant is based on the values of the dependent variable ϕ.

TABLE 3.1 Results from torque of a rectangular section

Mesh	# of nodes	# of elements	J/J_{ex}	τ_{max}/τ_{ex}
1	4	2	0.775	0.737
2	9	8	0.893	0.853
3	25	32	0.967	0.902
4	81	128	0.989	0.942

When the boundary value problem for the torsion of the rectangle is formulated with respect to the quadrant region of Fig. 3–23, the boundary conditions on lines *AB* and *BC* are essential, whereas on lines *OA* and *OC*, zero value natural boundary conditions are to be satisfied. In a finite element solution, these natural boundary conditions are not imposed directly but are generally satisfied only in the limit as the mesh size is decreased. Thus a measure of the convergence of the finite element model is available by looking at how well the zero normal derivative is satisfied. Specifically, for this example, the partial derivatives $\partial\phi/\partial Y$ along *OC* and $\partial\phi/\partial X$ along *OA* should both be zero.

In this vein, consider the results from the two-element model. In element 2, $\partial\phi/\partial X = 0$ along *OA*, and in element 1, $\partial\phi/\partial Y = 0$ along *OC*, so in a sense the natural boundary conditions are satisfied. A bit of reflection, however, indicates that for this mesh and constraints no other result is possible. For the eight-element model pictured in Fig. 3–27, $\partial\phi/\partial X$ is zero in element 6 and $\partial\phi/\partial Y$ is zero in element 3. In element 2, however, $\partial\phi/\partial X = (\phi_5 - \phi_4)/(a/2) = -0.0676G\theta a$, and in element 1, $\partial\phi/\partial Y = (\phi_5 - \phi_2)/(a/4) = -0.1840G\theta a$, the second of which is 26 percent of the maximum stress.

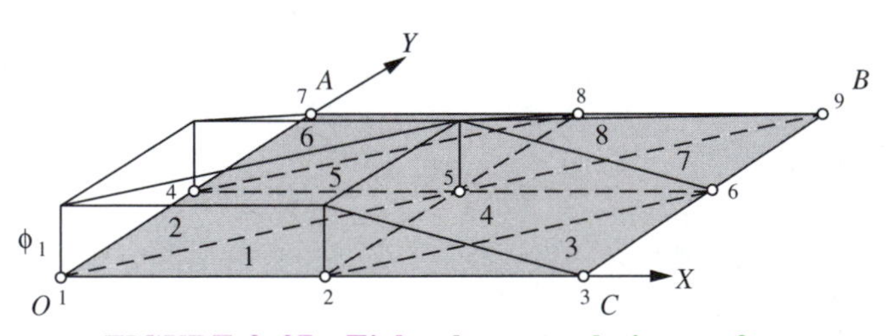

FIGURE 3–27 Eight-element solution surface

These are meaningful measures of the satisfaction of the natural boundary conditions, and also an indication that further mesh refinement is necessary to reduce the error in the satisfaction of the natural boundary conditions. Table 3.2 displays the results from the 8-, 32-, and 128-element models. For each of the three models, the absolute values of the normal derivatives in elements 1 and 2 are normalized with respect to the corresponding maximum stress predicted by the model.

TABLE 3.2 Normal derivatives — elements 1 and 2

# of elements	$(\partial\phi^*/\partial Y)_{ELT\ 1}$	$(\partial\phi^*/\partial X)_{ELT\ 2}$
8	0.259	0.095
32	0.135	0.048
128	0.065	0.023

These normalized normal derivatives, $(\partial\phi*/\partial Y)_{ELT\ 1}$ and $(\partial\phi*/\partial X)_{ELT\ 2}$, clearly seem to be approaching zero with the decreased mesh size. Note that the errors in the derivatives are approximately halved as a result of halving the dimensions of the element. This is consistent with the predictions of the theory for the linearly interpolated element. In an actual application, the exact solution is usually not known, so that calculations like these can be very valuable in assessing the validity and convergence of the results of a finite element model.

3.5.2 Conduction Heat Transfer with Boundary Convection

As an example of the type of heat conduction discussed in the introduction, consider the problem of two-dimensional heat conduction in an isotropic medium, with a convective-type boundary condition on one edge and prescribed temperatures on the other three edges, as indicated in Fig. 3–28. This example is presented primarily to indicate the manner in which the general derivative-type boundary condition is handled.

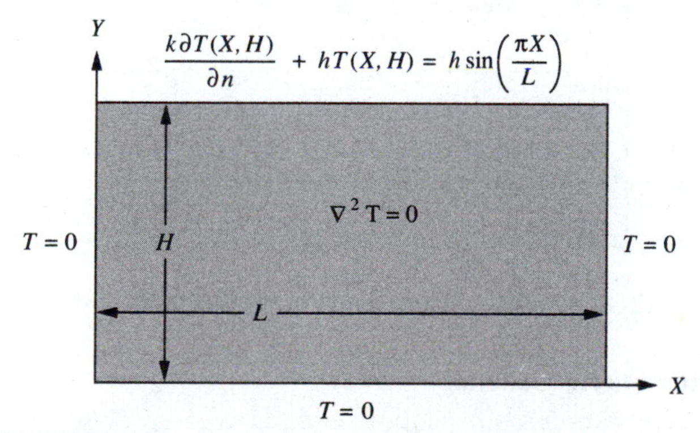

FIGURE 3–28 Heat conduction with boundary convection

In terms of the physical variables, the boundary value problem can be stated as

$$\nabla^2 T = 0 \qquad \text{in } D$$

$$T(0, Y) = 0$$

$$T(X, 0) = 0$$

$$T(L, Y) = 0$$

$$\frac{k\,\partial T(X,H)}{\partial n} + hT(X,H) = h\sin\left(\frac{\pi X}{L}\right)$$

or in terms of dimensionless independent variables $x = X/L, y = Y/L$, and $H/L = r$,

$$\nabla^2 T = 0 \qquad \text{in } D$$

$$T(0,y) = 0$$

$$T(x,0) = 0$$

$$T(1,y) = 0$$

$$\frac{\partial T(x,r)}{\partial n} + \beta T(x,r) = \beta \sin(\pi x)$$

with $\beta = hL/k$.

Discretization and interpolation. The first mesh chosen is a linearly interpolated nine-node, eight-element model as indicated in Fig. 3–29.

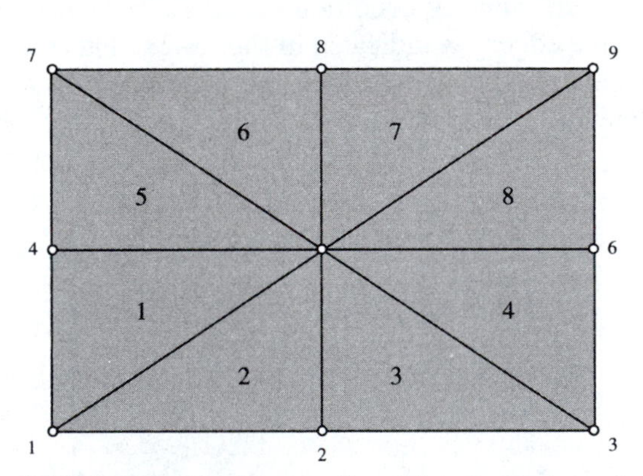

FIGURE 3–29 Nine-node, eight-element mesh

Elemental formulation. Generally,

$$\mathbf{K_G} = \sum \mathbf{k_G} + \sum{}' \mathbf{a_G}$$

$$\mathbf{F_G} = \sum \mathbf{f_G} + \sum{}' \mathbf{h_G}$$

where we will use the approximate expressions derived in Section 3.4,

$$\mathbf{a_e} = \frac{l_e}{12}\begin{bmatrix} 3\alpha_i + \alpha_j & \alpha_i + \alpha_j \\ \alpha_i + \alpha_j & \alpha_i + 3\alpha_j \end{bmatrix} \qquad \mathbf{h_e} = \frac{l_e}{6}\begin{bmatrix} 2h_i + h_j \\ h_i + 2h_j \end{bmatrix}$$

for the contributions from the convection on the boundary. Note that $\mathbf{k_e}$ and $\mathbf{f_e}$ are given by

$$\mathbf{k_e} = \frac{\mathbf{b_e b_e^T} + \mathbf{c_e c_e^T}}{4A_e} \qquad \text{and} \qquad \mathbf{f_e} = \mathbf{0}$$

In turn, the $\mathbf{k_e}$ appear as

$$2r\mathbf{k_{e1}} = \begin{array}{c} \begin{matrix} 1 & 5 & 4 \end{matrix} \\ \begin{bmatrix} 1 & 0 & -1 \\ 0 & r^2 & -r^2 \\ -1 & -r^2 & 1+r^2 \end{bmatrix} \end{array} \begin{matrix} 1 \\ 5 \\ 4 \end{matrix} \qquad 2r\mathbf{k_{e7}} = \begin{array}{c} \begin{matrix} 5 & 9 & 8 \end{matrix} \\ \begin{bmatrix} 1 & 0 & -1 \\ 0 & r^2 & -r^2 \\ -1 & -r^2 & 1+r^2 \end{bmatrix} \end{array} \begin{matrix} 5 \\ 9 \\ 8 \end{matrix}$$

$$2r\mathbf{k_{e2}} = \begin{array}{c} \begin{matrix} 1 & 2 & 5 \end{matrix} \\ \begin{bmatrix} r^2 & -r^2 & 0 \\ -r^2 & 1+r^2 & -1 \\ 0 & -1 & 1 \end{bmatrix} \end{array} \begin{matrix} 1 \\ 2 \\ 5 \end{matrix} \qquad 2r\mathbf{k_{e8}} = \begin{array}{c} \begin{matrix} 5 & 6 & 9 \end{matrix} \\ \begin{bmatrix} r^2 & -r^2 & 0 \\ -r^2 & 1+r^2 & -1 \\ 0 & -1 & 1 \end{bmatrix} \end{array} \begin{matrix} 5 \\ 6 \\ 9 \end{matrix}$$

$$2r\mathbf{k_{e3}} = \begin{array}{c} \begin{matrix} 2 & 3 & 5 \end{matrix} \\ \begin{bmatrix} 1+r^2 & -r^2 & -1 \\ -r^2 & r^2 & 0 \\ -1 & 0 & 1 \end{bmatrix} \end{array} \begin{matrix} 2 \\ 3 \\ 5 \end{matrix} \qquad 2r\mathbf{k_{e5}} = \begin{array}{c} \begin{matrix} 4 & 5 & 7 \end{matrix} \\ \begin{bmatrix} 1+r^2 & -r^2 & -1 \\ -r^2 & r^2 & 0 \\ -1 & 0 & 1 \end{bmatrix} \end{array} \begin{matrix} 4 \\ 5 \\ 7 \end{matrix}$$

$$2r\mathbf{k_{e4}} = \begin{array}{c} \begin{matrix} 3 & 6 & 5 \end{matrix} \\ \begin{bmatrix} 1 & -1 & 0 \\ -1 & 1+r^2 & -r^2 \\ 0 & -r^2 & r^2 \end{bmatrix} \end{array} \begin{matrix} 3 \\ 6 \\ 5 \end{matrix} \qquad 2r\mathbf{k_{e6}} = \begin{array}{c} \begin{matrix} 5 & 8 & 7 \end{matrix} \\ \begin{bmatrix} 1 & -1 & 0 \\ -1 & 1+r^2 & -r^2 \\ 0 & -r^2 & r^2 \end{bmatrix} \end{array} \begin{matrix} 5 \\ 8 \\ 7 \end{matrix}$$

The $\mathbf{a_e}$ and $\mathbf{h_e}$ are evaluated to be

$$\mathbf{a_{e1}} = \frac{\beta}{12} \begin{array}{c} \begin{matrix} 9 & 8 \end{matrix} \\ \begin{bmatrix} 2 & 1 \\ 1 & 2 \end{bmatrix} \end{array} \begin{matrix} 9 \\ 8 \end{matrix} \qquad \mathbf{a_{e2}} = \frac{\beta}{12} \begin{array}{c} \begin{matrix} 8 & 7 \end{matrix} \\ \begin{bmatrix} 2 & 1 \\ 1 & 2 \end{bmatrix} \end{array} \begin{matrix} 8 \\ 7 \end{matrix}$$

and

$$\mathbf{h_{e1}} = \frac{\beta}{12} \begin{bmatrix} 1 \\ 2 \end{bmatrix} \begin{matrix} 9 \\ 8 \end{matrix} \qquad \mathbf{h_{e2}} = \frac{\beta}{12} \begin{bmatrix} 2 \\ 1 \end{bmatrix} \begin{matrix} 8 \\ 7 \end{matrix}$$

respectively. For each of the elemental matrices above, the positions to be occupied in the assembled global stiffness and load matrices are indicated by the circled numbers.

Assembly. Taking $r = \beta = 0.5$, the assembled equations can be written in augmented form as

$$\left[\begin{array}{ccccccccc|c} 5 & -1 & 0 & -4 & 0 & 0 & 0 & 0 & 0 & 0 \\ -1 & 10 & -1 & 0 & -8 & 0 & 0 & 0 & 0 & 0 \\ 0 & -1 & 5 & 0 & 0 & -4 & 0 & 0 & 0 & 0 \\ -4 & 0 & 0 & 10 & -2 & 0 & -4 & 0 & 0 & 0 \\ 0 & -8 & 0 & -2 & 20 & -2 & 0 & -8 & 0 & 0 \\ 0 & 0 & -4 & 0 & -2 & 10 & 0 & 0 & -4 & 0 \\ 0 & 0 & 0 & -4 & 0 & 0 & 16/3 & -5/6 & 0 & 1/6 \\ 0 & 0 & 0 & 0 & -8 & 0 & -5/6 & 32/3 & -5/6 & 4/6 \\ 0 & 0 & 0 & 0 & 0 & -4 & 0 & -5/6 & 16/3 & 1/6 \end{array}\right]$$

Constraints. In general, points on the boundary at the intersection of Γ_1 and Γ_2 segments of the boundary may need special consideration. This is because in some situations, the idealization of the actual physical boundary conditions can result in a local discontinuity at such points of intersection. In the present example, the points $P : (0, H)$ and $Q : (L, H)$ are such points. However, since $h(s)$ is zero at these points, there is no discontinuity in view of the fact that in the limit as P is approached along $x = 0$, both T and its normal derivative are zero, as is h. A similar result is found as Q is approached along $x = L$. In terms of the finite element model, this translates into the fact that if we consider nodes 7 and 9 to lie on the Γ_1 portion of the boundary and hence constrain them to have zero values, the boundary condition on Γ_2 at P and Q will be identically satisfied since, as just mentioned, both T and its normal derivative will be zero as y approaches H along $x = 0$ and $x = L$. Thus, the constraints will be taken as

$$T_1 = T_2 = T_3 = T_4 = T_6 = T_7 = T_9 = 0$$

Solution. The resulting 2×2 set of equations

$$\left[\begin{array}{cc|c} 20 & -8 & 0 \\ -8 & 32/3 & 4/6 \end{array}\right]$$

is solved to yield $T_5 = 0.0357$ and $T_8 = 0.0893$. The corresponding exact solutions are $T_5 = 0.0481$ and $T_8 = 0.1274$, showing considerable error for this crude mesh.

Derived variables. We compute the normal derivative at point 8 according to

$$\frac{\partial T}{\partial n} = \frac{(T_8 - T_5)}{0.25} = 0.2144$$

compared to the exact value of 0.4363.

One can get an idea of the accuracy of the finite element solution by checking the satisfaction of the boundary condition at point 8, namely

$$\frac{\partial T}{\partial n} + \frac{1}{2}T = 0.2144 + 0.0446 = 0.2590$$

as compared to the prescribed value of 0.5. These results show that the solution is not sufficiently accurate and that refinement of the mesh is needed.

To this end we consider the two additional meshes shown in Fig. 3–30.

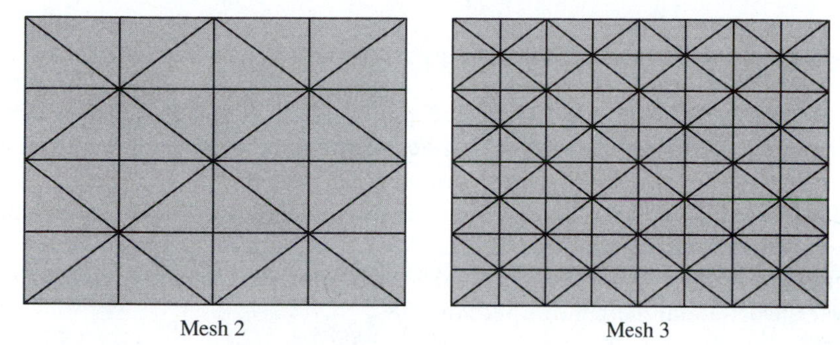

FIGURE 3–30 Two additional meshes obtained by successive halving

As can be seen, each of these meshes is obtained by quartering the element size of its predecessor. Results for the solution, the normal derivative, and the satisfaction of the boundary condition along $y = H$ are displayed in Tables 3.3, 3.4, and 3.5. The three successive meshes are indicated by $M1$, $M2$, and $M3$.

TABLE 3.3 Solution T at $y = H$

x/L	$M1$	$M2$	$M3$	T_{ex}
0.125			0.0476	0.0487
0.250		0.0822	0.0880	0.0901
0.375			0.1150	0.1177
0.500	0.0893	0.1163	0.1245	0.1274

TABLE 3.4 Normal derivative $\partial T/\partial n$ at $y = H$

x/L	$M1$	$M2$	$M3$	$\partial T/\partial n_{ex}$
0.125			0.1488	0.1670
0.250		0.2338	0.2749	0.3085
0.375			0.3592	0.4031
0.500	0.2144	0.3306	0.3887	0.4363

TABLE 3.5 Satisfaction of natural boundary condition $\partial T/\partial n + \beta T$ at $y = H$

x/L	$M1$	$M2$	$M3$	$(\partial T/\partial n + \beta T)_{ex}$
0.125			0.1726	0.1914
0.250		0.2749	0.3189	0.3535
0.375			0.4167	0.4619
0.500	0.2590	0.3887	0.4510	0.5000

The data clearly indicate the slow convergence. This is due primarily to the fact that the linearly interpolated element has been used. The improved accuracy and more rapid convergence available using higher-order interpolations will be illustrated later in this chapter.

3.6 RECTANGULAR ELEMENTS

In some applications it may be desirable to discretize the domain D such that the elements are four-sided, that is, either rectangular or quadrilateral in shape. We begin with a discussion of the rectangular element in view of its relative simplicity. Consider then the four-node rectangular element with sides parallel to the x-y axes as shown in Fig. 3–31. This restriction on the orientation of the rectangular element is unnecessary and will be removed in the treatment of the general quadrilateral element in Section 3.8.

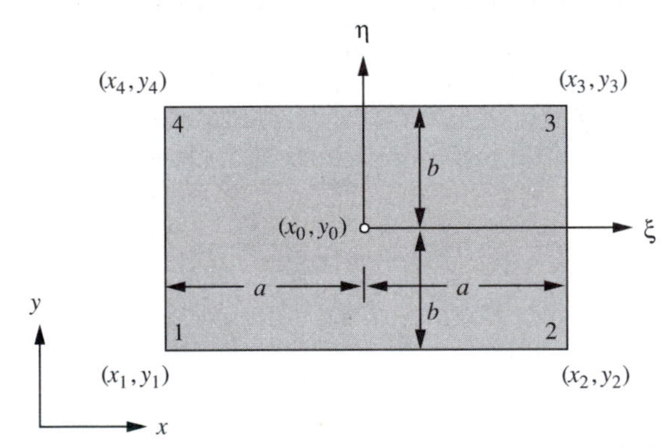

FIGURE 3–31 Rectangular element with *x*-*y* parallel sides

For a typical elliptic boundary value problem there will be four degrees of freedom, one for each node, which we will designate as u_1, u_2, u_3, and u_4. We introduce a bilinear form of interpolation as

$$u_e(x, y) = \alpha + \beta x + \gamma y + \delta x y \tag{3.44}$$

that is, linear in both x and y and containing the bilinear term xy. The xy term is selected from the second-order monomials x^2, xy, and y^2 because of its symmetry with respect to both x and y. The student is asked to consider the other two choices in the Exercises.

The four constants α, β, γ, and δ are determined by requiring

$$u_e(x_i, y_i) = \alpha + \beta x_i + \gamma y_i + \delta x_i y_i = u_i \qquad i = 1, 2, 3, 4$$

that is, four equations for the four unknowns α, β, γ, and δ. Determining α, β, γ, and δ and back-substituting into the general expression for $u_e(x, y)$, the results can be put in the form

$$u_e(x, y) = \sum u_i N_i(x, y) = \mathbf{u}_\mathbf{e}^\mathrm{T}\mathbf{N} = \mathbf{N}^\mathrm{T}\mathbf{u}_\mathbf{e} \tag{3.45}$$

where the interpolation functions N_i are given by

$$N_1(x, y) = \frac{(x_2 - x)(y_3 - y)}{(x_2 - x_1)(y_3 - y_1)}$$

$$N_2(x, y) = \frac{(x_1 - x)(y_3 - y)}{(x_1 - x_2)(y_3 - y_2)}$$

$$N_3(x, y) = \frac{(x_1 - x)(y_1 - y)}{(x_1 - x_3)(y_1 - y_3)}$$

$$N_4(x, y) = \frac{(x - x_2)(y_1 - y)}{(x_4 - x_2)(y_1 - y_4)}$$

and are indicated in Fig. 3–32.

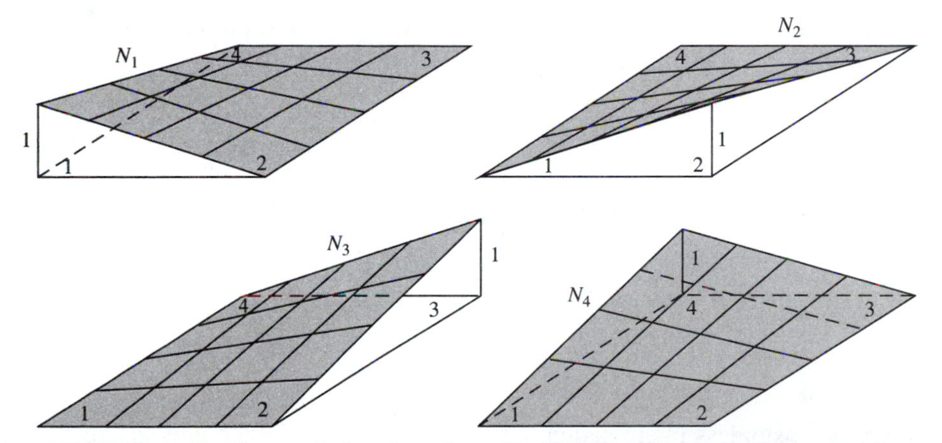

FIGURE 3–32 Interpolation functions N_1, N_2, N_3, and N_4 for the rectangle

The shapes of the N_i are such that there is no curvature in directions parallel to the sides of the element, but there is twist arising from the xy term. More importantly, the partial derivatives are no longer constant in an element but rather are linear functions according to

$$\begin{aligned}\frac{\partial u_e}{\partial x} &= \frac{(u_2 - u_1)(y_3 - y) + (u_3 - u_4)(y - y_1)}{(x_2 - x_1)(y_3 - y_1)} \\ &= \text{a linear function of } y \\ \frac{\partial u_e}{\partial y} &= \frac{(u_3 - u_2)(x - x_1) + (u_4 - u_1)(x_2 - x)}{(x_2 - x_1)(y_3 - y_1)} \\ &= \text{a linear function of } x\end{aligned}$$

Interelement continuity is clearly present in that the form of the solution is linear on any edge. This element also suffers from a discontinuity in the normal derivative at interelement boundaries, as was the case for the linearly interpolated triangular element.

If we choose to represent the interpolation functions in terms of the local coordinates (ξ, η) shown in Fig. 3-31, there results

$$\begin{aligned} N_1(\xi, \eta) &= \frac{(a - \xi)(b - \eta)}{4ab} \\ N_2(\xi, \eta) &= \frac{(a + \xi)(b - \eta)}{4ab} \\ N_3(\xi, \eta) &= \frac{(a + \xi)(b + \eta)}{4ab} \\ N_4(\xi, \eta) &= \frac{(a - \xi)(b + \eta)}{4ab} \end{aligned} \tag{3.46}$$

These in turn can be expressed in terms of the nondimensional coordinates $s = \xi/a$ and $t = \eta/b$ as

$$\begin{aligned} N_1(s, t) &= \frac{(1 - s)(1 - t)}{4} \\ N_2(s, t) &= \frac{(1 + s)(1 - t)}{4} \\ N_3(s, t) &= \frac{(1 + s)(1 + t)}{4} \\ N_4(s, t) &= \frac{(1 - s)(1 + t)}{4} \end{aligned} \tag{3.47}$$

These dimensionless interpolation functions are defined on the unit square $-1 \leq s \leq 1$, $-1 \leq t \leq 1$. The representation of the solution can thus be expressed as

$$u = \sum_1^4 u_i N_i$$

It is left for the student to show that in terms of this same set of four interpolation functions the geometry of the element can be represented as

$$x = \sum_1^4 x_i N_i \qquad \text{and} \qquad y = \sum_1^4 y_i N_i$$

that is, that the geometry can be represented in terms of the same set of interpolation or shape functions as are used for the interpolation of the solution.

3.6.1 Evaluation of Matrices

The general *forms* of the elemental matrices, which must be evaluated for generating the finite element model for the elliptic boundary value problem, are independent of the type of interpolation used. From Section 3.3 these matrices are

$$\mathbf{k_e} = \int\int_{A_e} \left(\frac{\partial \mathbf{N}}{\partial x} \frac{\partial \mathbf{N}^{\mathrm{T}}}{\partial x} + \frac{\partial \mathbf{N}}{\partial y} \frac{\partial \mathbf{N}^{\mathrm{T}}}{\partial y} \right) dA$$

$$\mathbf{f_e} = \int\int_{A_e} \mathbf{N} f \, dA$$

$$\mathbf{h_e} = \int_{\gamma_{2e}} \mathbf{N} h \, ds$$

$$\mathbf{a_e} = \int_{\gamma_{2e}} \mathbf{N} \alpha \mathbf{N}^{\mathrm{T}} \, ds$$

As we have seen, the N_i and their derivatives appear in different combinations in evaluating the elemental matrices. Care must be taken so that the integrations are carried out properly, depending on the coordinate system used for evaluating the integrals.

For integrals over a typical elemental area A_e, the integrals to be evaluated are of the form

$$\int\int_{A_e} F(x, y) \, dx \, dy = \int_{y_2}^{y_3} \int_{x_1}^{x_2} F(x, y) \, dx \, dy$$

Using the translational transformation $\xi = x - x_0$ and $\eta = y - y_0$ suggested in Fig. 3–31, the above integral can be expressed as

$$\int_{-b}^{b} \int_{-a}^{a} F(x_0 + \xi, y_0 + \eta) \frac{\partial(x, y)}{\partial(\xi, \eta)} \, d\xi \, d\eta$$

where $\partial(x, y)/\partial(\xi, \eta)$ is the Jacobian of the transformation, easily computed to be 1. Using the transformation $\xi = as$, $\eta = bt$, this can in turn be transformed into

$$\int_{-1}^{1} \int_{-1}^{1} F(x_0 + as, y_0 + bt) \frac{\partial(\xi, \eta)}{\partial(s, t)} \, ds \, dt$$

where the Jacobian $\partial(\xi, \eta)/\partial(s, t)$ has the value $ab = A_e/4$.

Ultimately it will turn out that the evaluation of the integrals will be most easily accomplished when using the dimensionless coordinates (s, t). Care must be taken to include the Jacobian determinant $A_e/4$ when transforming to the s-t coordinates. In what follows, we will discuss in detail the evaluation of the integrals for $\mathbf{k_e}$, $\mathbf{f_e}$, $\mathbf{h_e}$, and $\mathbf{a_e}$.

Evaluation of $\mathbf{f_e}$. The general expression for the elemental load vector $\mathbf{f_e}$ is

$$\mathbf{f_e} = \int\int_{A_e} \mathbf{N}(x, y) f(x, y) \, dA$$

which in principle can be evaluated knowing $f(x, y)$. Thinking in terms of using a computer code, it is usual to provide nodal values of such a function as input rather than to be able to actually specify the function over the entire domain. With this in mind consider Fig. 3–33, where a typical f is indicated.

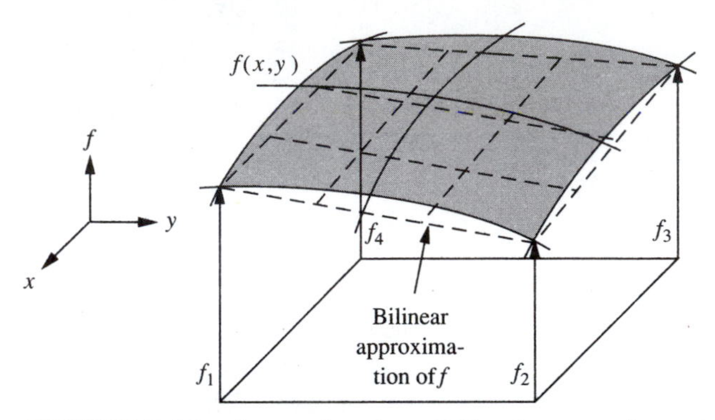

FIGURE 3–33 Approximation of $f(x, y)$ on an element

In order to obtain a specific representation for the integral, $f(x, y)$ will be *approximated* by $\mathbf{N}^T\mathbf{f_i}$, where

$$\mathbf{f_i}^T = [f_1 \ f_2 \ f_3 \ f_4]$$

This is equivalent to replacing the actual function f by the bilinear function defined by the interpolation functions and the f_i. The f_i are frequently taken to be the values $f_i = f(x_i, y_i)$, that is, the nodal values of the function f. There results

$$\mathbf{f_e} \approx \left[\iint_{A_e} \mathbf{N}\mathbf{N}^T \, dx \, dy\right] \mathbf{f_i}$$

or when expressed in terms of the dimensionless coordinates s and t,

$$\mathbf{f_e} = \left[\int_{-1}^{1}\int_{-1}^{1} \frac{\mathbf{N}\mathbf{N}^T A_e}{4} \, ds \, dt\right] \mathbf{f_i}$$

Using the integration formulas given in Eqs. (3.31), these integrals can be evaluated to yield

$$\mathbf{f_e} = \frac{A_e}{36}\begin{bmatrix} 4f_1 + 2f_2 + f_3 + 2f_4 \\ 2f_1 + 4f_2 + 2f_3 + f_4 \\ f_1 + 2f_2 + 4f_3 + 2f_4 \\ 2f_1 + f_2 + 2f_3 + 4f_4 \end{bmatrix} \tag{3.48}$$

Note that if $f = f_0$, a constant, the expression for $\mathbf{f_e}$ indicates that the total input $f_0 A_e$ from the element is split equally among the four nodes.

Evaluation of $\mathbf{h_e}$, $\mathbf{a_e}$. In Fig. 3–34, where part of the Γ_2 boundary occurs along one of the sides of a rectangular element, we indicate the function $h(s)$ as well as the two interpolation functions N_k and N_l which have the nonzero values shown.

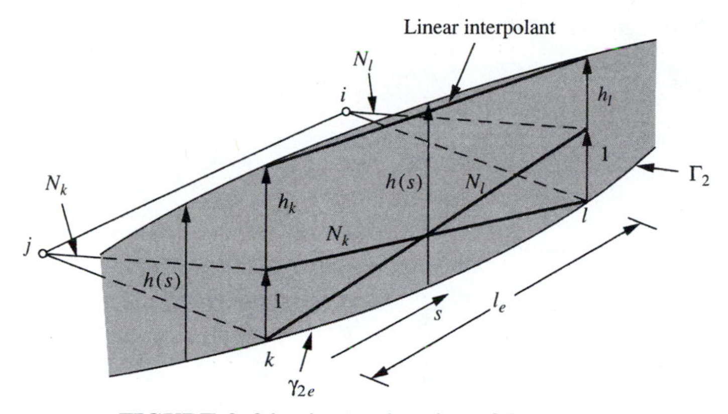

FIGURE 3–34 Approximation of $h(s)$ on γ_{2e}

If we choose to represent $h(s)$ *approximately* as $h_k N_k + h_l N_l$, there results

$$\mathbf{h_e} \approx \int_{\gamma_{2e}} \begin{bmatrix} N_k \\ N_l \end{bmatrix} (h_k N_k + h_l N_l)\, ds$$

which with $N_k = (1 - s/l_e)$ and $N_l = s/l_e$ becomes

$$\mathbf{h_e} = \frac{l_e}{6} \begin{bmatrix} 2h_k + h_l \\ h_k + 2h_l \end{bmatrix}$$

If h is constant, the total contribution $h_0 l_e$ is split equally between the two nodes. In a completely similar manner it can be shown that

$$\mathbf{a_e} \approx \frac{l_e}{12} \begin{bmatrix} 3\alpha_k + \alpha_l & \alpha_k + \alpha_l \\ \alpha_k + \alpha_l & \alpha_k + 3\alpha_l \end{bmatrix}$$

where α_k and α_l are appropriately chosen values of $\alpha(s)$. These are usually taken to be the function $\alpha(s)$ evaluated at the node in question. Note that the expressions for $\mathbf{a_e}$ and $\mathbf{h_e}$ are exactly the same as those developed for the linearly interpolated triangular element. This is to be expected in view of the fact that the interpolations for the linearly interpolated triangular element and the bilinearly interpolated rectangular element both reduce to the same one-dimensional interpolation functions on a boundary of an element.

Evaluation of $\mathbf{k_e}$. The expression developed for $\mathbf{k_e}$ contains partial derivatives of the interpolation functions both with respect to x and y. We again choose to develop relationships using the chain rule for converting to the dimensionless coordinates s and t. To this end, consider the transformation

$$x = x_0 + as \qquad y = y_0 + bt$$

or

$$x = x(s, t) \qquad y = y(s, t)$$

Using the chain rule we write

$$\frac{\partial}{\partial s} = \frac{\partial}{\partial x}\frac{\partial x}{\partial s} + \frac{\partial}{\partial y}\frac{\partial y}{\partial s} = a\frac{\partial}{\partial x}$$

$$\frac{\partial}{\partial t} = \frac{\partial}{\partial x}\frac{\partial x}{\partial t} + \frac{\partial}{\partial y}\frac{\partial y}{\partial t} = b\frac{\partial}{\partial y}$$

or written in matrix notation,

$$\frac{\boldsymbol{\partial}}{\boldsymbol{\partial}\mathbf{s}} = \mathbf{J}\frac{\boldsymbol{\partial}}{\boldsymbol{\partial}\mathbf{x}}$$

where $\mathbf{J}$ is the Jacobian matrix given by

$$\mathbf{J} = \begin{bmatrix} \dfrac{\partial x}{\partial s} & \dfrac{\partial y}{\partial s} \\ \dfrac{\partial x}{\partial t} & \dfrac{\partial y}{\partial t} \end{bmatrix} = \begin{bmatrix} a & 0 \\ 0 & b \end{bmatrix}$$

Inverting yields

$$\frac{\boldsymbol{\partial}}{\boldsymbol{\partial}\mathbf{x}} = \mathbf{J}^{-1}\frac{\boldsymbol{\partial}}{\boldsymbol{\partial}\mathbf{s}}$$

from which

$$\frac{\partial}{\partial x} = \mathbf{J}_1\frac{\boldsymbol{\partial}}{\boldsymbol{\partial}\mathbf{s}}$$

$$\frac{\partial}{\partial y} = \mathbf{J}_2\frac{\boldsymbol{\partial}}{\boldsymbol{\partial}\mathbf{s}}$$

where $\mathbf{J}_1$ and $\mathbf{J}_2$ are the first and second rows of $\mathbf{J}^{-1}$. We can now compute

$$\frac{\partial \mathbf{N}^{\mathrm{T}}}{\partial x} = \mathbf{J}_1\frac{\boldsymbol{\partial}\mathbf{N}^{\mathrm{T}}}{\boldsymbol{\partial}\mathbf{s}} = \mathbf{J}_1\boldsymbol{\Delta}^{\mathrm{T}}$$

and

$$\frac{\partial \mathbf{N}^{\mathrm{T}}}{\partial y} = \mathbf{J}_2\frac{\boldsymbol{\partial}\mathbf{N}^{\mathrm{T}}}{\boldsymbol{\partial}\mathbf{s}} = \mathbf{J}_2\boldsymbol{\Delta}^{\mathrm{T}}$$

so that

$$\mathbf{k_e} = \int_{-1}^{1}\int_{-1}^{1}\left[\boldsymbol{\Delta}\mathbf{J}_1^{\mathrm{T}}\mathbf{J}_1\boldsymbol{\Delta}^{\mathrm{T}} + \boldsymbol{\Delta}\mathbf{J}_2^{\mathrm{T}}\mathbf{J}_2\boldsymbol{\Delta}^{\mathrm{T}}\right]\frac{A_e}{4}\,ds\,dt$$

$$= \int_{-1}^{1}\int_{-1}^{1}[\boldsymbol{\Delta}\mathbf{J}\mathbf{J}\boldsymbol{\Delta}^{\mathrm{T}}]\,ds\,dt$$

where $\mathbf{JJ} = [\mathbf{J}_1^{\mathrm{T}}\mathbf{J}_1 + \mathbf{J}_2^{\mathrm{T}}\mathbf{J}_2]A_e/4$. Using Eqs. (3.46) the matrix $\boldsymbol{\Delta}^{\mathrm{T}}$ is easily computed to be

$$
\boldsymbol{\Delta}^{\mathrm{T}} = \begin{bmatrix} \dfrac{-(1-t)}{4} & \dfrac{(1-t)}{4} & \dfrac{(1+t)}{4} & \dfrac{-(1+t)}{4} \\ \dfrac{-(1-s)}{4} & \dfrac{-(1+s)}{4} & \dfrac{(1+s)}{4} & \dfrac{(1-s)}{4} \end{bmatrix}
$$

which with **JJ** as

$$
\mathbf{JJ} = \begin{bmatrix} \dfrac{b}{a} & 0 \\ 0 & \dfrac{a}{b} \end{bmatrix}
$$

permits $\mathbf{k_e}$ to easily be evaluated as

$$
\mathbf{k_e} = \frac{1}{6}\left[\frac{b}{a}\begin{bmatrix} 2 & -2 & -1 & 1 \\ -2 & 2 & 1 & -1 \\ -1 & 1 & 2 & -2 \\ 1 & -1 & -2 & 2 \end{bmatrix} + \frac{a}{b}\begin{bmatrix} 2 & 1 & -1 & -2 \\ 1 & 2 & -2 & -1 \\ -1 & -2 & 2 & 1 \\ -2 & -1 & 1 & 2 \end{bmatrix} \right]
$$

For the square element, $b = a$ and $\mathbf{k_e}$ becomes

$$
\mathbf{k_e} = \frac{1}{6}\begin{bmatrix} 4 & -1 & -2 & -1 \\ -1 & 4 & -1 & -2 \\ -2 & -1 & 4 & -1 \\ -1 & -2 & -1 & 4 \end{bmatrix}
$$

These results will be applied to specific examples in the following sections.

3.6.2 Torsion of a Rectangular Section

Consider again the nondimensionalized torsion problem of Section 3.5.1 given by

$$
\nabla^2\psi + 1 = 0 \qquad \text{in } D
$$

$$
\psi = 0 \qquad \text{on } \Gamma
$$

for the rectangular cross section. The symmetry is again exploited to use only the quadrant $1 \geq x \geq 0$ and $\alpha \geq y \geq 0$.

One-element model. The nodes are chosen as indicated in Fig. 3–35.
$\mathbf{K_e} = \mathbf{K_G}$ and $\mathbf{f_e} = \mathbf{f_G}$ are easily computed to be

$$
\mathbf{K_e} = \frac{1}{6\alpha}\begin{bmatrix} 2(1+\alpha^2) & 1-2\alpha^2 & -(1+\alpha^2) & -2+\alpha^2 \\ & 2(1+\alpha^2) & -2+\alpha^2 & -(1+\alpha^2) \\ & & 2(1+\alpha^2) & 1-2\alpha^2 \\ \text{symm} & & & 2(1+\alpha^2) \end{bmatrix}
$$

$$
\mathbf{f_e^T} = \frac{\alpha}{4}[1 \quad 1 \quad 1 \quad 1]
$$

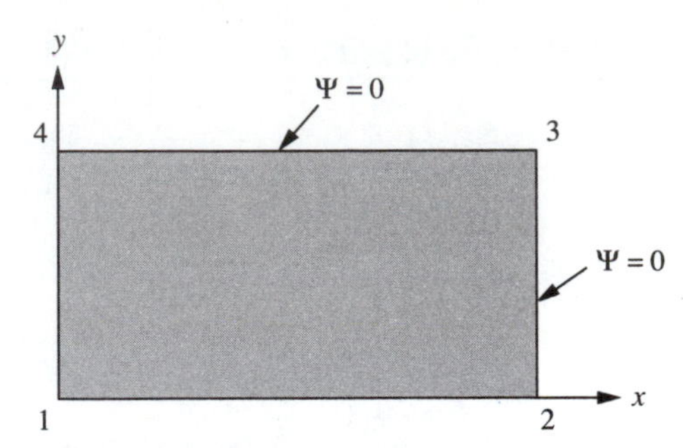

FIGURE 3–35 One-element model accounting for symmetries

The constraints are $\psi_2 = \psi_3 = \psi_4 = 0$, leading to the single equation

$$\frac{2(1+\alpha^2)}{6\alpha}\psi_1 = \frac{\alpha}{4}$$

or

$$\psi_1 = \frac{3\alpha^2}{4(1+\alpha^2)} = 0.15$$

and hence

$$\phi_1 = \frac{3G\theta a^2\alpha^2}{2(1+\alpha^2)} = 0.3G\theta a^2$$

where $\alpha = b/a$ has again been taken as 0.5. The torque is given by

$$T = 2\int\int \phi \, dA = 2\int\int \mathbf{N}^{\mathrm{T}}\boldsymbol{\phi}_{\mathbf{e}} \, dA$$

$$= \frac{2\sum_e[\phi_{1e} + \phi_{2e} + \phi_{3e} + \phi_{4e}]A_e}{4}$$

$$= \frac{3G\theta a^4\alpha^3}{(1+\alpha^2)} = 0.3G\theta a^4$$

where the result $\int\int \mathbf{N}^{\mathrm{T}} \, dA = A_e/4[1 \quad 1 \quad 1 \quad 1]$ has been used. The maximum shear stress is τ_{XZ} evaluated at $X = 0$ and is given by

$$\frac{\partial\phi}{\partial Y} = \frac{0-\phi_1}{b} = \frac{0-\phi_1}{a\alpha} = -0.6G\theta a$$

so that

$$|\tau_{XZ}| = (0.6a)\left(\frac{T}{0.3a^4}\right) = \frac{2T}{a^3}$$

The stress function and the stresses are indicated in Fig. 3–36.

$$\tau_{XZ} = \frac{-2T}{a^3}\left(1-\frac{X}{a}\right) \qquad \tau_{YZ} = \frac{T}{a^3}\left(1-\frac{Y}{b}\right)$$

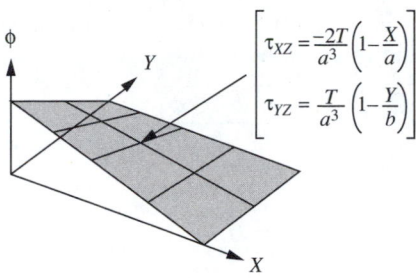

FIGURE 3–36 Stress function ϕ and stresses

Four-element model. The quadrant is modeled as four rectangular elements with the elements and nodes as indicated in Fig. 3–37.

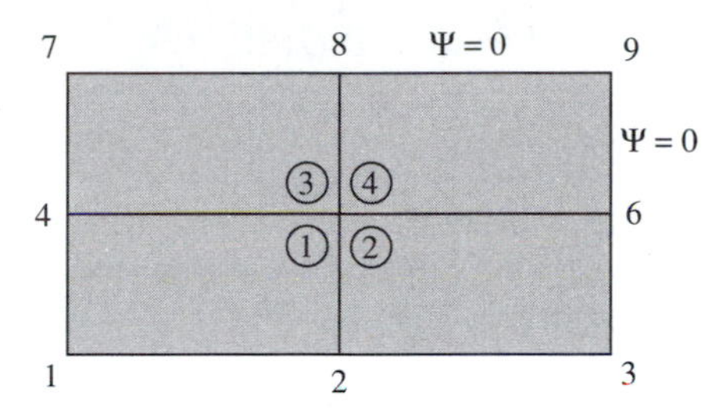

FIGURE 3–37 Four-element model

With $A = 1 + \alpha^2$, $B = 1 - 2\alpha^2$, and $C = -2 + \alpha^2$, the stiffness matrix is

$$\mathbf{K_G} = \frac{1}{6\alpha}\begin{bmatrix} 2A & B & 0 & C & -A & 0 & 0 & 0 & 0 \\ & 4A & B & -A & 2C & -A & 0 & 0 & 0 \\ & & 2A & 0 & -A & C & 0 & 0 & 0 \\ & & & 4A & 2B & 0 & C & -A & 0 \\ & & & & 8A & 2B & -A & 2C & -A \\ & & & & & 4A & 0 & -A & C \\ & & & & & & 2A & B & 0 \\ & \text{symm} & & & & & & 4A & B \\ & & & & & & & & 2A \end{bmatrix}$$

The load matrix is

$$\mathbf{f}_\mathbf{G}^\mathrm{T} = \frac{\alpha}{16}[1 \quad 2 \quad 1 \quad 2 \quad 4 \quad 2 \quad 1 \quad 2 \quad 1]$$

Constraining $\psi_3 = \psi_6 = \psi_7 = \psi_8 = \psi_9 = 0$ results in

$$\psi_1 = 0.119037$$

$$\psi_2 = 0.103719$$

$$\psi_4 = 0.089139$$

$$\psi_5 = 0.079767$$

where again α has been taken as 0.5. The maximum shear stress occurs at the left edge of element 3 and is given by

$$\tau_{XZ} = \frac{\partial \phi}{\partial Y} = \frac{\phi_7 - \phi_4}{b/2} = \frac{0 - \phi_4}{\alpha a/2}$$

or

$$|\tau_{XZ}| = 0.7131 G\theta a$$

The torque is given by

$$\frac{T}{4} = 2\frac{ab}{4}(\phi_1 + 2(\phi_2 + \phi_4) + 4\phi_3)$$

or

$$T = 0.4119 G\theta a^4$$

Eliminating $G\theta$ there results

$$|\tau_{XZ}| = \frac{1.731T}{a^3}$$

The ϕ function is again indicated in Fig. 3–38.

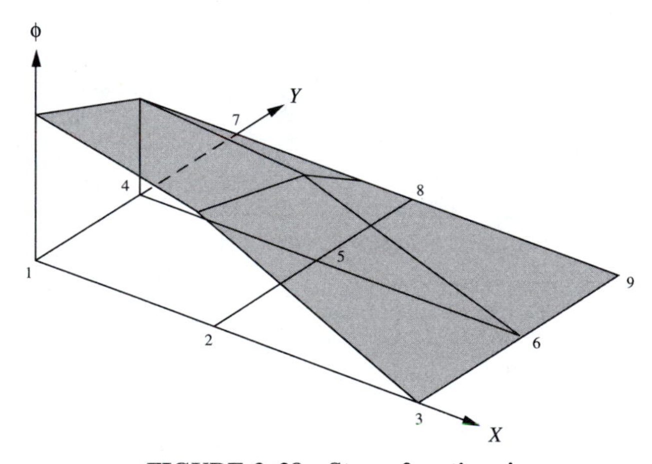

FIGURE 3–38 Stress function ϕ

It is clear that $\tau_{XZ} = \partial\phi/\partial Y$ is not continuous across the interelement boundaries between nodes 4 and 5 and between nodes 5 and 6. Similarly, $\tau_{YZ} = -\partial\phi/\partial X$ is not continuous across the interelement boundaries between nodes 2 and 5 and between nodes 5 and 8. Thus, as was the case for the linearly interpolated triangular element, the function ϕ and its tangential derivative are continuous across an interelement boundary whereas the normal derivative is in general discontinuous.

The results for two other models that successively quarter the element size of the previous two models are given in Table 3.6, where $J_{ex} = 0.458G\theta a^4$ and $\tau_{ex} = 2.033T/a^3$.

TABLE 3.6 J and τ comparisons

Mesh	# of nodes	# of elements	J/J_{ex}	τ/τ_{ex}
1	4	1	0.655	0.984
2	9	4	0.891	0.851
3	25	16	0.973	0.879
4	81	64	0.992	0.942

For this problem the convergence is clearly not monotonic, at least initially. Further refinement of the mesh would indicate a monotonic convergence for both J and τ. The stress, being a derived variable, does not converge as rapidly as the torsional constant J which is computed from an integral of the ϕ function. Comparison with the results of Table 3.1 for the triangular element models indicates that the results for the two element choices are comparable.

As in Section 3.5.1, it is of interest to investigate the satisfaction of the natural boundary conditions on the interior boundaries. To this end the terms $(\partial\phi*/\partial Y)$ and $(\partial\phi*/\partial X)$, as defined in Section 3.5.1 and evaluated on the interior boundaries of element 1, are displayed in Table 3.7. (Recall that the * indicates a comparison with the corresponding maximum stress in the region.)

TABLE 3.7 Normal derivatives—element 1

# of elements	$(\partial\phi^*/\partial Y)_{ELT\ 1}$	$(\partial\phi^*/\partial X)_{ELT\ 1}$
4	0.244	0.040
16	0.138	0.034
64	0.064	0.016

The normal derivatives clearly seem to be converging towards the desired zero values. The results are comparable to those given in Table 3.2 for the triangular element models.

3.6.3 Conduction Heat Transfer with Boundary Convection

Consider again the nondimensionalized heat transfer problem of Section 3.5.2, namely

$$\nabla^2 T = 0 \qquad \text{in } D$$

$$T(0, y) = 0$$

$$T(x, 0) = 0$$

$$T(1, y) = 0$$

$$\frac{\partial T(x,r)}{\partial n} + \beta T(x,r) = \beta \sin(\pi x)$$

The reader is referred to Section 3.5.2 for the details of the problem formulation.

The first mesh chosen is a nine-node, four-element model as indicated in Fig. 3–39.

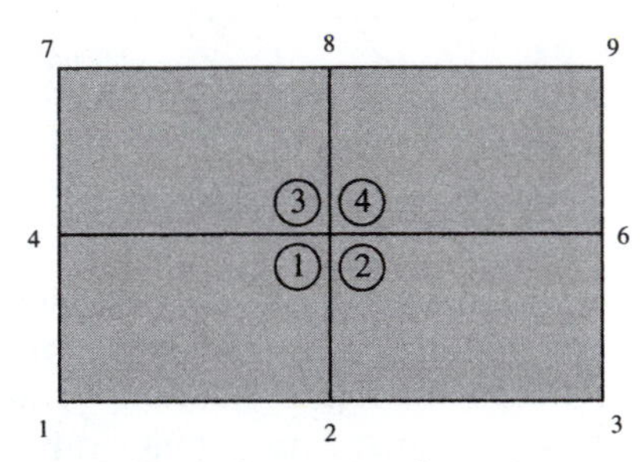

FIGURE 3–39 Nine-node, four-element model

The $\mathbf{a_e}$, $\mathbf{h_e}$, and $\mathbf{k_e}$ matrices are respectively

$$\mathbf{a_e} = \frac{l_e}{12}\begin{bmatrix} 3\alpha_i + \alpha_j & \alpha_i + \alpha_j \\ \alpha_i + \alpha_j & \alpha_i + 3\alpha_j \end{bmatrix} \qquad \mathbf{h_e} = \frac{l_e}{6}\begin{bmatrix} 2h_i + h_j \\ h_i + 2h_j \end{bmatrix}$$

$$\mathbf{k_e} = \frac{1}{6r}\begin{bmatrix} 2(1+r^2) & 1-2r^2 & -(1+r^2) & -2+r^2 \\ & 2(1+r^2) & -2+r^2 & -(1+r^2) \\ & & 2(1+r^2) & 1-2r^2 \\ \text{symm} & & & 2(1+r^2) \end{bmatrix}$$

Since $f = 0$, each $\mathbf{f_e} = 0$. Taking $\beta = r = 0.5$, the assembled global stiffness matrix is

$$\mathbf{K_G} = \begin{bmatrix} 2A & B & 0 & C & -A & 0 & 0 & 0 & 0 \\ & 4A & B & -A & 2C & -A & 0 & 0 & 0 \\ & & 2A & 0 & -A & C & 0 & 0 & 0 \\ & & & 4A & 2B & 0 & C & -A & 0 \\ & & & & 8A & 2B & -A & 2C & -A \\ & & & & & 4A & 0 & -A & C \\ & & & & & & D & E & 0 \\ \text{symm} & & & & & & & 2D & E \\ & & & & & & & & D \end{bmatrix}$$

where $A = 0.4167$, $B = 0.1667$, $C = -0.5833$, $D = 0.9167$, and $E = 0.2083$. Note the effect of the $\mathbf{a_e}$ in the bottom right 3×3 submatrix. The load matrix arising from the $\mathbf{h_e}$ contributions is

$$\mathbf{f_G^T} = [0 \quad 0 \quad 0 \quad 0 \quad 0 \quad 0 \quad 0.0417 \quad 0.1667 \quad 0.0417]$$

Constraining $\psi_1 = \psi_2 = \psi_3 = \psi_4 = \psi_6 = \psi_7 = \psi_9 = 0$ results in the constrained augmented equations

$$\left[\begin{array}{cc|c} 3.3333 & -1.1667 & 0.0000 \\ -1.1667 & 1.8333 & 0.1667 \end{array}\right]$$

with the solution as

$$\psi_5 = 0.040936$$

$$\psi_8 = 0.116959$$

where again, α has been taken as 0.5. The normal derivative at point 8 is computed according to

$$\frac{\partial T}{\partial n} = \frac{T_8 - T_5}{0.25} = 0.3041$$

and is to be compared to the exact value of 0.4363.

One can again get an idea of the accuracy of the finite element solution by checking the satisfaction of the boundary condition at point 8, namely

$$\frac{\partial T}{\partial n} + \frac{1}{2}T = 0.3041 + 0.0585 = 0.3626$$

as compared to the exact value of 0.5. These results show that the solution is not sufficiently accurate and that refinement of the mesh is needed.

To this end we consider the two additional meshes shown in Fig. 3–40.

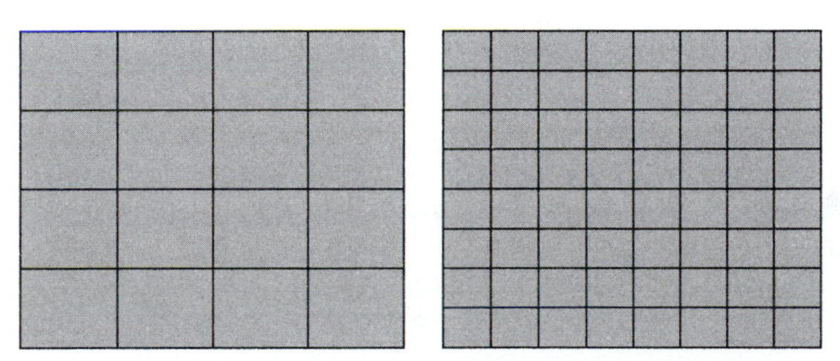

FIGURE 3–40 Two additional meshes obtained by quartering

As can be seen, each of these meshes is obtained by quartering the element size of its predecessor. Results for the solution, the normal derivative, and the satisfaction of the boundary condition along $y = H$ are displayed in Tables 3.8, 3.9, and 3.10. The three successive meshes are denoted by $M1$, $M2$, and $M3$.

TABLE 3.8 Solution T at $y = H$

x/L	$M1$	$M2$	$M3$	T_{ex}
0.125			0.0485	0.0487
0.250		0.0882	0.0896	0.0901
0.375			0.1171	0.1177
0.500	0.1170	0.1248	0.1267	0.1274

TABLE 3.9 Normal derivative $\partial T/\partial n$ at $y = H$

x/L	$M1$	$M2$	$M3$	$\partial T/\partial n_{ex}$
0.125			0.1529	0.1670
0.250		0.2592	0.2826	0.3085
0.375			0.3692	0.4031
0.500	0.3041	0.3665	0.3996	0.4363

TABLE 3.10 Satisfaction of natural boundary condition $\partial T/\partial n + \beta T$ at $y = H$

x/L	$M1$	$M2$	$M3$	$(\partial T/\partial n + \beta T)_{ex}$
0.125			0.1771	0.1914
0.250		0.3033	0.3274	0.3535
0.375			0.4278	0.4619
0.500	0.3626	0.4289	0.4630	0.5000

Comparison with Section 3.5.2, where the same problem was solved using the linearly interpolated triangular element, shows some improvement using the rectangular element. Again the data clearly indicate the slow convergence. The improved accuracy and more rapid convergence available using higher-order interpolations will be discussed later in the chapter.

3.7 EIGENVALUE PROBLEMS

Many of the techniques for solving two-dimensional time-dependent problems such as the diffusion and wave equations are very closely related to a corresponding boundary value problem. In this section we will investigate the relationship between the time-dependent problem and its corresponding boundary value problem and see how the finite element method can be used to extract information about the eigenvalue and eigenfunctions of the corresponding boundary value problems.

3.7.1 Introduction

Time-dependent diffusion problems can frequently be stated as

$$\nabla^2\phi - \gamma\frac{\partial\phi}{\partial t} + f(x, y, t) = 0 \qquad \text{in } D$$

$$\phi = g(s, t) \qquad \text{on } \Gamma_1$$

$$\frac{\partial\phi}{\partial n} + \alpha\phi = h(s, t) \qquad \text{on } \Gamma_2$$

$$\phi(x, y, 0) = c(x, y)$$

The first two auxiliary conditions are boundary conditions and the third an initial condition. Such problems are frequently referred to as initial-boundary value problems. The corresponding completely homogeneous boundary value problem, obtained by taking $f = g = h = 0$, is

$$\begin{aligned} \nabla^2\phi - \gamma\frac{\partial \phi}{\partial t} &= 0 \qquad \text{in } D \\ \phi &= 0 \qquad \text{on } \Gamma_1 \\ \frac{\partial \phi}{\partial n} + \alpha\phi &= 0 \qquad \text{on } \Gamma_2 \end{aligned} \tag{3.49}$$

Solutions of this completely homogeneous boundary value problem have great utility in solving the original nonhomogeneous initial-boundary value problem. For the diffusion problem, solutions of Eqs. (3.49) are known generally to behave according to $\phi(x, y, t) = \psi(x, y)\exp(-\beta t)$, leading to

$$\begin{aligned} \nabla^2\psi + \lambda\psi &= 0 \\ \psi &= 0 \qquad \text{on } \Gamma_1 \\ \frac{\partial \psi}{\partial n} + \alpha\psi &= 0 \qquad \text{on } \Gamma_2 \end{aligned} \tag{3.50}$$

where $\lambda = \beta\gamma$. The λ's and corresponding nontrivial ψ's that satisfy Eq. (3.50) are known as eigenvalues and eigenfunctions respectively. These are the two-dimensional counterparts of the λ_n and u_n discussed in Section 2.5.

In a completely similar fashion, the two-dimensional initial-boundary value problem associated with the scalar wave equation can be written as

$$\begin{aligned} c^2\nabla^2\phi - \frac{\partial^2 \phi}{\partial t^2} + f(x, y, t) &= 0 \qquad \text{in } D \\ \phi &= g(s, t) \qquad \text{on } \Gamma_1 \\ \frac{\partial \phi}{\partial n} + \alpha\phi &= h(s, t) \qquad \text{on } \Gamma_2 \\ \phi(x, y, 0) &= c(x, y) \qquad \text{in } D \\ \frac{\partial \phi(x, y, 0)}{\partial t} &= d(x, y) \qquad \text{in } D \end{aligned}$$

Here, the first two conditions are boundary conditions and the second two are initial conditions. Frequently, ϕ is a displacement or generalized displacement and hence $\partial\phi/\partial t$ is a velocity or generalized velocity. For a physical problem associated with the wave equation it is usually appropriate to investigate solutions to the homogeneous differential equation and boundary conditions of the form $\phi(x, y, t) = \exp(i\omega t)\psi(x, y)$, again leading to

$$\begin{aligned} \nabla^2\psi + \lambda\psi &= 0 \\ \psi &= 0 \qquad \text{on } \Gamma_1 \end{aligned}$$

$$\frac{\partial \psi}{\partial n} + \alpha\psi = 0 \qquad \text{on } \Gamma_2$$

where $\lambda = (\omega/c)^2$. This is precisely the same completely homogeneous boundary value problem given by Eq. (3.50) for the diffusion problem.

The differential equation (3.50) is known as the Helmholtz equation and occurs with remarkable frequency in mathematical models. The problem to be solved is that of finding the eigenvalues and eigenfunctions satisfying the differential equation and boundary conditions for the two-dimensional region D associated with the original initial-boundary value problem. Exact solutions are known only for a few very special regions D. Finite element solutions will be investigated in the next section. These solutions to the Helmholtz equation will be used to help construct solutions to time-dependent problems in Chapter 4.

3.7.2 Finite Element Models for the Helmholtz Equation

The finite element model for the Helmholtz equation will be discussed in terms of the typical region indicated in Fig. 3–41.

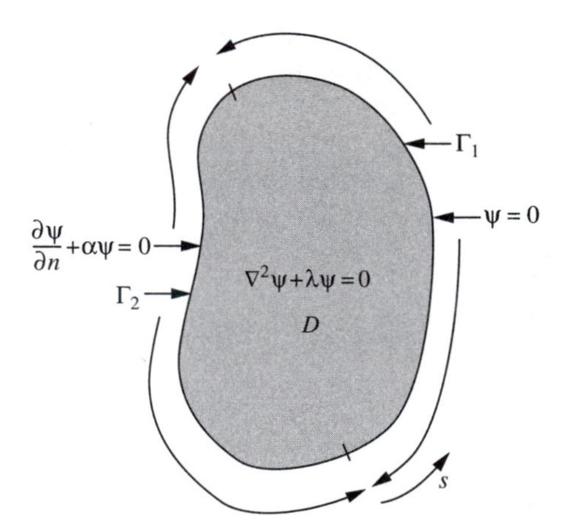

FIGURE 3–41 Typical region for the Helmholtz equation

Discretization and interpolation are carried out in the same fashion as in Sections 3.2 and 3.3. The Galerkin finite element method will be used to generate the desired algebraic equations for determining the approximate eigenvalues and eigenfunctions.

The form of the solution is taken as

$$\psi(x, y) = \sum_{1}^{N} \psi_i n_i(x, y)$$

where the n_i are the appropriate nodally based interpolation functions which were discussed in Section 3.2, and the ψ_i are the unknown nodal values of $\psi(x, y)$.

With the test function designated as ζ, the appropriate weak form (see Exercise 3.66) for this boundary value problem is

$$\iint_D \left(\frac{\partial \zeta}{\partial x} \frac{\partial \psi}{\partial x} + \frac{\partial \zeta}{\partial y} \frac{\partial \psi}{\partial y} \right) dA + \int_{\Gamma_2} \zeta \alpha \psi \, ds - \lambda \iint_D \zeta \psi \, dA = 0$$

Taking $\psi = \sum_1^N \psi_i n_i$ and $\zeta = n_k$, $k = 1, 2, \ldots, N$ yields

$$\sum_1^N \iint_D \left\{ \left(\frac{\partial n_k}{\partial x} \frac{\partial n_i}{\partial x} + \frac{\partial n_k}{\partial y} \frac{\partial n_i}{\partial y} \right) dA \right\} \psi_i + \sum_1^N \left(\int_{\Gamma_2} n_k \alpha n_i \, ds \right) \psi_i$$

$$- \lambda \sum_1^N \left(\iint_D n_k n_i \, dA \right) \psi_i = 0 \qquad k = 1, 2, \ldots, N$$

or

$$\sum_1^N (K_{ki} - \lambda M_{ki}) \psi_i = 0 \qquad k = 1, 2, \ldots, N$$

The student is asked to show in the Exercises that these algebraic equations are equivalent to

$$\mathbf{K_G} \boldsymbol{\psi}_\mathbf{G} - \lambda \mathbf{M_G} \boldsymbol{\psi}_\mathbf{G} = 0$$

where

$$\mathbf{K_G} = \sum_e \mathbf{k_G} + \sum_e{}' \mathbf{a_G}$$
$$\mathbf{M_G} = \sum_e \mathbf{m_G} \tag{3.51}$$

with

$$\mathbf{k_e} = \iint_{A_e} \left(\frac{\partial \mathbf{N}}{\partial x} \frac{\partial \mathbf{N}^\mathrm{T}}{\partial x} + \frac{\partial \mathbf{N}}{\partial y} \frac{\partial \mathbf{N}^\mathrm{T}}{\partial y} \right) dA$$

$$\mathbf{a_e} = \int_{\gamma_{2e}} \mathbf{N} \alpha \mathbf{N}^\mathrm{T} \, ds$$

$$\mathbf{m_e} = \iint_{A_e} \mathbf{N} \mathbf{N}^\mathrm{T} \, dA$$

where, as in Section 3.4, γ_{2e} is an appropriate segment of the boundary Γ_2. The *assembly* is implied by the $\sum$'s in Eq. (3.51). As mentioned in Section 2.5, Eq. (3.51) is an example of the generalized linear algebraic eigenvalue problem. In Eq. (3.51), $\mathbf{m_e}$ is referred to as an elemental consistent mass matrix. As indicated in the Exercises, it is possible to develop these same equations using the Rayleigh quotient as the starting point.

The boundary conditions for the Helmholtz equation are entirely homogeneous. In particular, the constraints associated with the essential boundary conditions on Γ_1 can be enforced by deleting the row and column corresponding to the degree of freedom in question. This results in

$$\mathbf{K}_G^* \boldsymbol{\psi}_G^* - \lambda \mathbf{M}_G^* \boldsymbol{\phi}_G^* = \mathbf{0} \tag{3.52}$$

where the rows and columns associated with the constraints have been removed. The solution is obtained by determining the eigenvalues and corresponding eigenvectors of Eq. (3.52). For a problem with a small number (two to four) of constrained degrees of freedom, it is perhaps feasible to extract the eigen information by hand. For larger problems, it is essential to have a robust computer code for this task. Appendix C contains Fortran source listings for several routines that may be used in this regard. In addition, Eispack [6,7] contains listings of many codes that are useful.

3.7.3 Examples for the Helmholtz Equation

Example 3.1.

Consider the problem of a classical square vibrating membrane with all edges fixed against transverse displacement. The differential equation of motion can be written as

$$T\nabla^2 w = \rho \frac{\partial^2 w}{\partial t^2}$$

where T is the initial tension in the membrane and ρ the area density. The boundary condition is that w vanish on all the edges of the membrane. Taking $w(x, y, t) = \psi(x, y)\exp(i\,\omega t)$ leads to

$$\nabla^2\psi + \lambda\psi = 0 \qquad \text{in } D$$

with

$$\psi = 0 \qquad \text{on } \Gamma_1$$

where $\lambda = \rho\omega^2/T$. This is the Helmholtz problem on the square, with the dependent variable ψ prescribed as zero everywhere on the boundary. The eigenvalues are related to the natural frequencies and the eigenfunctions to the mode shapes. The simplest possible model using linearly interpolated triangular elements is indicated in Fig. 3–42.

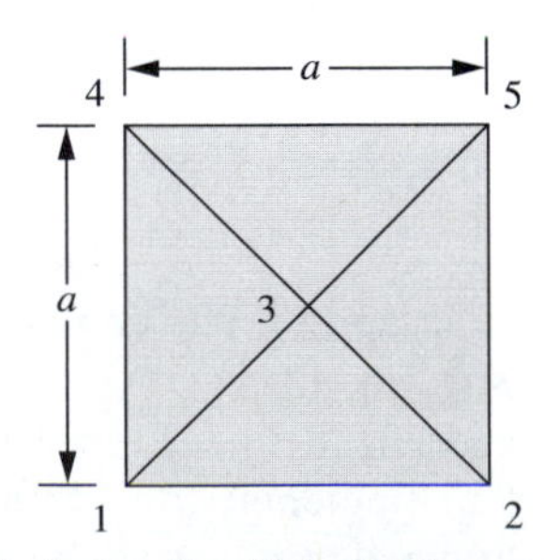

FIGURE 3–42 Mesh for the Helmholtz equation on the square

Using the results of Section 3.4, the assembled $\mathbf{K_G}$ and $\mathbf{M_G}$ matrices are determined to be

$$\mathbf{K_G} = \frac{1}{2}\begin{bmatrix} 2 & 0 & -2 & 0 & 0 \\ 0 & 2 & -2 & 0 & 0 \\ -2 & -2 & 8 & -2 & -2 \\ 0 & 0 & -2 & 2 & 0 \\ 0 & 0 & -2 & 0 & 2 \end{bmatrix}$$

and

$$\mathbf{M_G} = \frac{a^2}{4}\frac{1}{12}\begin{bmatrix} 4 & 1 & 2 & 1 & 0 \\ 1 & 4 & 2 & 0 & 1 \\ 2 & 2 & 8 & 2 & 2 \\ 1 & 0 & 2 & 4 & 1 \\ 0 & 1 & 2 & 1 & 4 \end{bmatrix}$$

Constraining ψ_1, ψ_2, ψ_4, and ψ_5 yields the single equation

$$(\mathbf{K_G})_{33} - \lambda(\mathbf{M_G})_{33} = \mathbf{0}$$

or

$$\lambda = \frac{24}{a^2}$$

This approximate value is to be compared to the exact value of $2\pi^2/a^2$, an error of approximately 22 percent. This is quite reasonable for the very crude mesh. When plotted, the corresponding eigenfunction is the pyramid-shaped function indicated in Fig. 3–43.

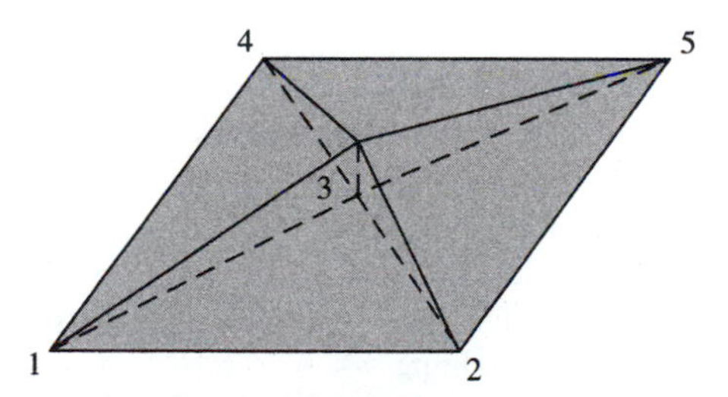

FIGURE 3–43 Approximate eigenfunction for the square

It is left for the student to show in the Exercises that a model using four square elements over the same region also yields $\lambda = 24/a^2$.

Example 3.2.

Consider the problem of the vibration of a circular membrane. From Example 3.1, it follows that solutions to the Helmholtz problem for the circle are related to the vibration of the circular membrane. Consider a circular region modeled with eight linearly interpolated triangular elements as indicated in Fig. 3–44.

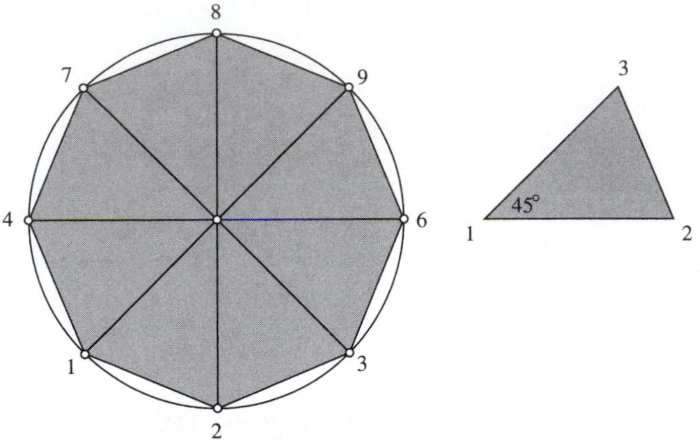

FIGURE 3–44 Domain and mesh for the Helmholtz equation on the circle

Note that the elemental stiffness matrix for the single triangle numbered as indicated in Fig. 3–44 is

$$\mathbf{k_e} = (0.7071)\begin{array}{c} \begin{array}{ccc} \quad 2 \quad & \quad 3 \quad & \quad 1 \quad \end{array} \\ \left[\begin{array}{ccc} 1.0000 & -0.7071 & -0.2929 \\ & 1.0000 & -0.2929 \\ \text{symm} & & 0.5858 \end{array}\right] \end{array} \begin{array}{c} 2 \\ 3 \\ 1 \end{array}$$

and the elemental mass matrix is

$$\mathbf{m_e} = \frac{0.3536a^2}{12}\begin{bmatrix} 2 & 1 & 1 \\ & 2 & 1 \\ \text{symm} & & 2 \end{bmatrix}$$

where a is the radius of the circle. With a little forward planning, it can be seen that all of nodes 1–4 and 6–9 will be constrained with only node 5 remaining. Thus only $(\mathbf{K_G})_{55}$ and $(\mathbf{M_G})_{55}$ need to be computed. The contribution to $(\mathbf{K_G})_{55}$ from each element will be the (1,1) element of the $\mathbf{k_e}$ above, yielding

$$(\mathbf{K_G})_{55} = 8(0.7071)(0.5858)$$

Similarly,

$$(\mathbf{M_G})_{55} = 8(0.3536a^2)/6$$

yielding

$$\lambda = \frac{(\mathbf{K_G})_{55}}{(\mathbf{M_G})_{55}} = \frac{7.03}{a^2}$$

compared to the exact value of $5.78/a^2$. The eigenfunction plots as the pyramidal cone indicated in Fig. 3–45.

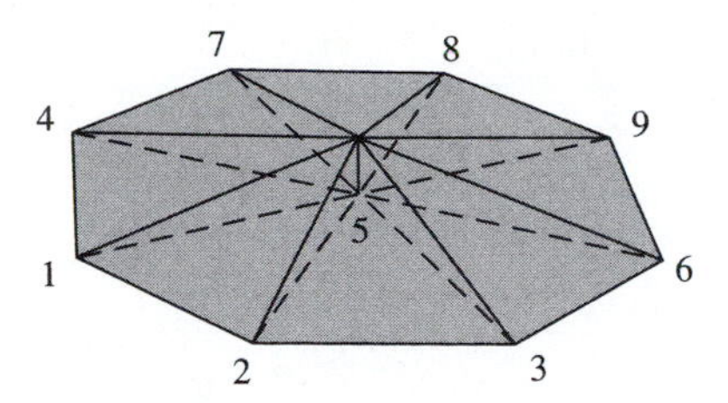

FIGURE 3–45 Approximate eigenfunction for the circle

For the circle, a better approximation to the fundamental eigenvalue can be obtained without having to solve an excessively large problem by considering the mesh indicated in Fig. 3–46. For this model, we will enforce constraints at nodes 1, 2, and 3 with nodes 4, 5, and 6 unconstrained. The portion of the boundary containing nodes 1–2–3 is Γ_1 and the two straight portions 1–4–6 and 3–5–6 must be considered as Γ_2. In the limit as the mesh is refined, the natural boundary conditions ($\partial\psi/\partial n = 0$) will be satisfied on Γ_2.

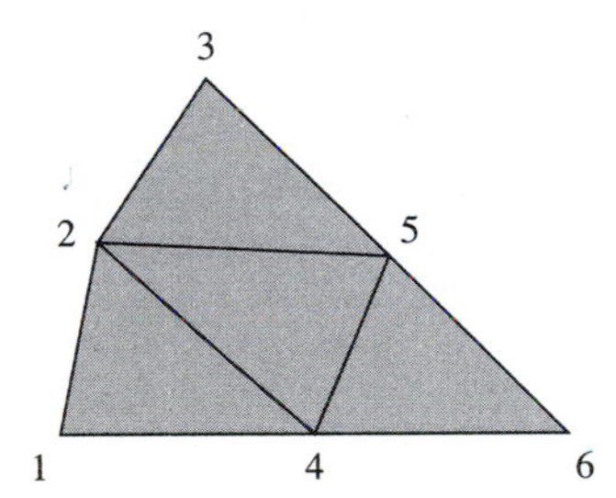

FIGURE 3–46 Symmetry mesh for the circle

As the results will indicate, the lowest eigenvalue is associated with an eigenfunction that is symmetric, that is, $\psi_4 = \psi_5$ with ψ_4, ψ_5, and ψ_6 all positive, so that an improved approximation for the lowest eigenvalue for the entire circle will be obtained. The other λ's and ψ's must be interpreted very carefully for a model consisting of such a subregion, in that correctly identifying conditions of symmetry or antisymmetry and the character of the corresponding modes can require considerable insight into the mathematics of the problem.

For the subregion of Fig. 3–46, the *constrained* $\mathbf{K_G}$ and $\mathbf{M_G}$ matrices respectively are

$$\begin{array}{ccc} \quad 6 & \quad 5 & \quad 4 \end{array}$$

$$\mathbf{K}_\mathbf{G}^* = \begin{bmatrix} 0.4142 & -0.2071 & -0.2071 \\ & 1.8968 & -1.1141 \\ \text{symm} & & 1.8968 \end{bmatrix} \begin{array}{c} 6 \\ 5 \\ 4 \end{array}$$

$$\begin{array}{ccc} \quad 6 & \quad 5 & \quad 4 \end{array}$$

$$\frac{12\mathbf{M}_\mathbf{G}^*}{a^2} = \begin{bmatrix} 0.1768 & 0.0884 & 0.0884 \\ & 0.5740 & 0.1913 \\ \text{symm} & & 0.5740 \end{bmatrix} \begin{array}{c} 6 \\ 5 \\ 4 \end{array}$$

The eigenvalues are determined to be

$$\lambda_1 = \frac{6.12}{a^2}$$

$$\lambda_2 = \frac{46.9}{a^2}$$

$$\lambda_3 = \frac{94.4}{a^2}$$

with the eigenvectors

$$\psi_1^* = [1.000 \quad 0.643 \quad 0.643]^T$$

$$\psi_2^* = [1.000 \quad -0.250 \quad -0.250]^T$$

$$\psi_3^* = [0.000 \quad -1.000 \quad 1.000]^T$$

The corresponding approximate eigenfunctions are indicated in Fig. 3–47.

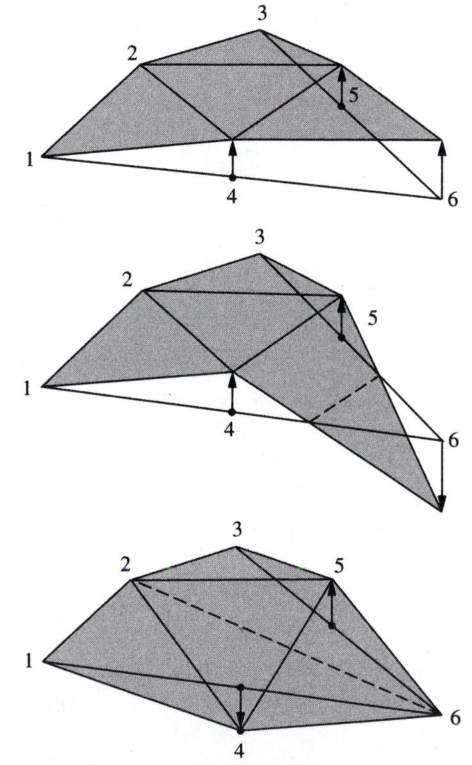

FIGURE 3–47 Approximate eigenfunctions

The first two eigenvectors clearly have a symmetric character in that $\psi_4 = \psi_5$. One would suspect that these eigenvalue-eigenfunction pairs correspond to radially symmetric modes for the circle, with the lowest eigenvalue for this model representing an improvement in the single eigenvalue obtained from the first model using eight elements. The third eigenvalue-eigenfunction demonstrates asymmetry with respect to the bisector of the angle subtended by the element. Finer meshes would need to be investigated to more clearly identify the character of this and higher eigenvalue-eigenfunction pairs.

3.8 ISOPARAMETRIC ELEMENTS AND NUMERICAL INTEGRATION

From a practical point of view, the linearly interpolated triangular element and the bilinearly interpolated rectangular element, which were discussed in Sections 3.2, 3.3, and 3.6, are deficient in the following ways:

1. Inability to represent curved boundaries.
2. Slow convergence and poor estimation of derived variables.

These deficiencies are addressed by the use of higher-order isoparametric elements, such as the quadratically interpolated triangular and quadrilateral elements indicated in the typical two-dimensional mesh shown in Fig. 3–48.

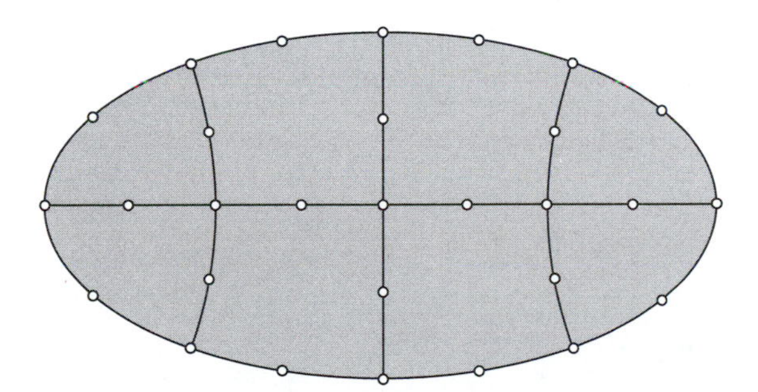

FIGURE 3–48 Mesh using higher-order T6 and Q8 elements

The term **isoparametric** means that *equal numbers of parameters* are used to represent the geometry and the dependent variable. Isoparametric also refers to the use of the same interpolation functions or shape functions for representing the dependent variable within the element and for representing the shape or geometry of the element. Thus with N_{gi} and N_{si} the interpolation functions for the geometry and the solution, respectively, and with n_g and n_s the number of N_{gi} and N_{si}, respectively, the geometry is represented as

$$x = \sum_{1}^{n_g} x_i N_{gi} \qquad y = \sum_{1}^{n_g} y_i N_{gi}$$

and the solution is represented as

$$u = \sum_{1}^{n_s} u_i N_{si}$$

with isoparametric representations $n_g = n_s$, and $N_{gi} = N_{si}$, for $i = 1, 2, \ldots, n_g$.

The linearly interpolated triangular element considered earlier in this chapter is an example of an isoparametric element. The element geometry can be defined by

$$x = \sum_{1}^{3} x_i L_i \qquad y = \sum_{1}^{3} y_i L_i$$

If the solution is also interpolated linearly according to

$$u = \sum_{1}^{3} u_i L_i$$

wherein each of x, y, and u the L_i are the linear interpolation functions for the triangle, the element is clearly isoparametric.

The rectangular element discussed in Section 3.6 is also an isoparametric element with $n_g = n_s = 4$ and with the N_i given by Eqs. (3.47).

A **subparametric** element is one for which the geometry can be defined by fewer parameters than are used to interpolate the solution. An example is the straight-sided quadratically interpolated triangular element indicated in Fig. 3–49

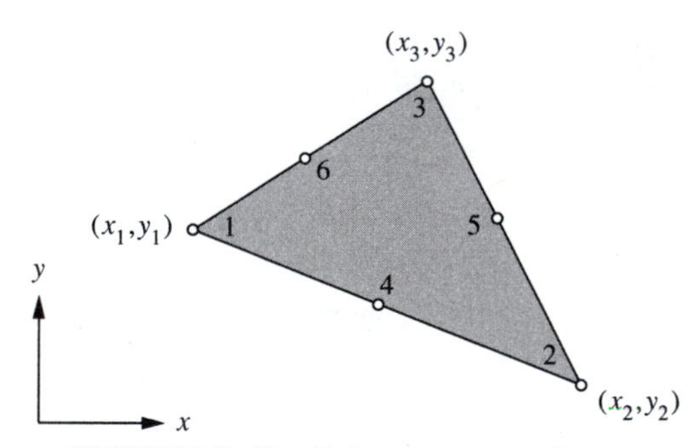

FIGURE 3–49 Subparametric element

where

$$x = \sum_{1}^{3} x_i L_i \qquad y = \sum_{1}^{3} y_i L_i$$

and

$$u = \sum_{1}^{6} u_i N_i$$

and where the L_i are the area coordinates for the triangle and the N_i are the appropriate quadratic interpolation functions for the six nodes shown. Here $n_g = 3$ (the geometry of the element can still be defined by the vertices of the triangle) and $n_s = 6$. The six interpolation functions N_i for the subparametric triangular element will be discussed later in this section. **Superparametric** elements employ more parameters for the geometry than for interpolation of the dependent variable.

In this section we consider several isoparametric elements and their general use in the finite element method as applied to the two-dimensional elliptic boundary value problem. The properties of each of these elements are based on so-called shape functions which transform or map the region associated with a so-called parent element into the shape occupied by the element in question. It follows that the transformation itself should be understood before embarking on the use of isoparametric elements in constructing finite element models.

3.8.1 Four-Node Quadrilateral (Q4) Elements

For the Q4 element the elemental solution is assumed to be

$$\begin{aligned} u_e(s,t) &= u_1N_1(s,t) + u_2N_2(s,t) + u_3N_3(s,t) + u_4N_4(s,t) \\ &= \mathbf{u_e^T N} = \mathbf{N^T u_e} \end{aligned}$$

where s and t are the coordinates indicated in Fig. 3–50.

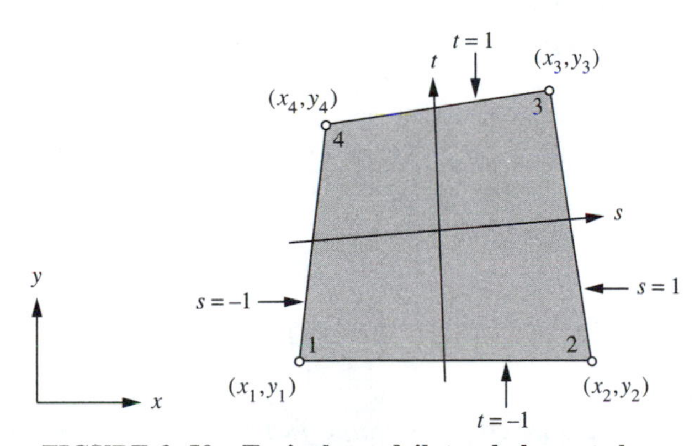

FIGURE 3–50 Typical quadrilateral element shape

The $N_i(s,t)$ have precisely the same form as those introduced in Section 3.6. For the proper choice of the nodal degrees of freedom, u can assume a constant value within the element and in addition is capable of assuming a constant slope (partial derivative) in either the x or y direction. Thus the convergence and completeness requirements are met. The properties and uses of the Q4 element will be investigated in this section.

3.8.1.1 TRANSFORMATIONS AND SHAPE FUNCTIONS. It is most convenient to think of the shape of the quadrilateral element as being defined by the transformation or mapping $x = x(s, t)$, $y = y(s, t)$ of a ***parent element*** indicated in Fig. 3–51a into the quadrilateral shape of Fig. 3–51b.

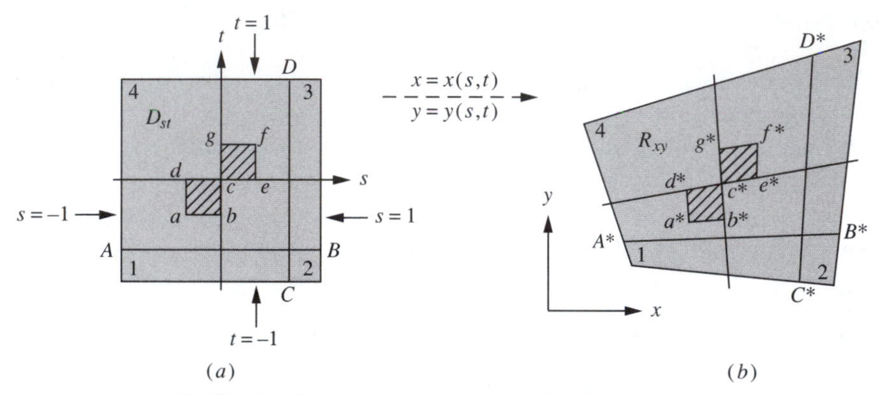

FIGURE 3–51 Parent and quadrilateral elements

The transformation that accomplishes this is

$$\begin{aligned} x &= x_1N_1 + x_2N_2 + x_3N_3 + x_4N_4 \\ y &= y_1N_1 + y_2N_2 + y_3N_3 + y_4N_4 \end{aligned} \tag{3.53}$$

where the N_i are given by

$$N_1 = \frac{(1-s)(1-t)}{4}$$

$$N_2 = \frac{(1+s)(1-t)}{4}$$

$$N_3 = \frac{(1+s)(1+t)}{4}$$

$$N_4 = \frac{(1-s)(1+t)}{4}$$

and in this setting are referred to as ***shape functions,*** in that when used in connection with Eqs. (3.53), they define the shape of the quadrilateral element. They are precisely the same as the interpolation functions for the rectangle discussed in Section 3.6.

As s and t vary between the limits $-1 \leq s \leq 1$, and $-1 \leq t \leq 1$, the parent element, that is, the region D_{st} of Fig. 3–51a, is mapped into the quadrilateral region R_{xy} of Fig. 3–51b. Straight lines in the s-t coordinate system are transformed or mapped into straight lines in the quadrilateral. As examples, lines AB and CD in the s-t system are mapped linearly into lines A^*B^* and C^*D^* in the

x-y plane. Consider for instance the straight line $t = -1$ connecting nodes 1 and 2 in the parent element. For $t = -1$, Eqs. (3.47) become

$$x = \frac{\big((x_1 + x_2) + (x_2 - x_1)s\big)}{2}$$

$$y = \frac{\big((y_1 + y_2) + (y_2 - y_1)s\big)}{2}$$

As s varies from -1 to $+1$, x varies linearly from x_1 to x_2, and y varies linearly from y_1 to y_2. Similarly, the line $s = 1/2$ is mapped into the straight line three quarters of the distance from line segment 1-4 to line segment 2-3.

As indicated in Fig. 3–51, two different differential areas *abcd* and *cefg* of the same size in the s-t plane are in general mapped into two different differential areas $a^*b^*c^*d^*$ and $c^*e^*f^*g^*$ of different sizes in the x-y plane. In order to determine the character of these changes, the chain rule is used to form the differential relations

$$\frac{\partial}{\partial s} = \frac{\partial}{\partial x}\frac{\partial x}{\partial s} + \frac{\partial}{\partial y}\frac{\partial y}{\partial s}$$

$$\frac{\partial}{\partial t} = \frac{\partial}{\partial x}\frac{\partial x}{\partial t} + \frac{\partial}{\partial y}\frac{\partial y}{\partial t}$$

relating the partial derivatives in the two systems. In matrix notation these can be written as

$$\frac{\boldsymbol{\partial}}{\boldsymbol{\partial}\mathbf{s}} = \mathbf{J}\frac{\boldsymbol{\partial}}{\boldsymbol{\partial}\mathbf{x}}$$

where

$$\frac{\boldsymbol{\partial}}{\boldsymbol{\partial}\mathbf{s}} = \begin{bmatrix} \dfrac{\partial}{\partial s} & \dfrac{\partial}{\partial t} \end{bmatrix}^{\mathrm{T}} \quad \text{and} \quad \frac{\boldsymbol{\partial}}{\boldsymbol{\partial}\mathbf{x}} = \begin{bmatrix} \dfrac{\partial}{\partial x} & \dfrac{\partial}{\partial y} \end{bmatrix}^{\mathrm{T}}$$

are matrix operators and

$$\mathbf{J} = \begin{bmatrix} \dfrac{\partial x}{\partial s} & \dfrac{\partial y}{\partial s} \\ \dfrac{\partial x}{\partial t} & \dfrac{\partial y}{\partial t} \end{bmatrix}$$

is the Jacobian matrix. The more familiar Jacobian of the transformation $|\mathbf{J}|$, is the determinant of $\mathbf{J}$ given by

$$|\mathbf{J}| = \frac{\partial x}{\partial s}\frac{\partial y}{\partial t} - \frac{\partial y}{\partial s}\frac{\partial x}{\partial t} = \begin{vmatrix} \dfrac{\partial x}{\partial s} & \dfrac{\partial y}{\partial s} \\ \dfrac{\partial x}{\partial t} & \dfrac{\partial y}{\partial t} \end{vmatrix}$$

The Jacobian $|\mathbf{J}|$ is used to test the invertibility of the transformation $x = x(s, t)$, $y = y(s, t)$. When $|\mathbf{J}|$ is everywhere positive in the D_{st} region, the transformation can in principle be inverted to determine $s = s(x, y)$ and $t = t(x, y)$, that is, given a point x, y in R_{xy}, there is a unique corresponding point s, t in D_{st}. For the quadrilateral element of Fig. 3–50, the elements of the Jacobian matrix are computed using Eqs. (3.53) as

$$\begin{aligned}
\frac{\partial x}{\partial s} &= \frac{(x_2 - x_1)(1 - t) + (x_3 - x_4)(1 + t)}{4} \\
\frac{\partial y}{\partial s} &= \frac{(y_2 - y_1)(1 - t) + (y_3 - y_4)(1 + t)}{4} \\
\frac{\partial x}{\partial t} &= \frac{(x_3 - x_2)(1 + s) + (x_4 - x_1)(1 - s)}{4} \\
\frac{\partial y}{\partial t} &= \frac{(y_3 - y_2)(1 + s) + (y_4 - y_1)(1 - s)}{4}
\end{aligned} \tag{3.54}$$

It follows that $|\mathbf{J}|$ is of the form

$$|\mathbf{J}| = j_1 + j_2 s + j_3 t + j_4 s t$$

where the constants j_i are functions of the coordinates of the nodes, that is, functions of the geometry of the element. It is left to the Exercises to show that j_4 is actually zero so that $|\mathbf{J}|$ is a linear function of s and t. Geometrically, $|\mathbf{J}|$ represents the relationship between a differential area $dA(s, t) = ds\,dt$ and the differential area $dA(x, y)$ into which it is mapped, according to

$$dA(x, y) = |\mathbf{J}|\, dA(s, t) = |\mathbf{J}|\, ds\, dt$$

Consider the special case where the element is rectangular and oriented with its sides parallel to the x-y axes, as shown in Fig. 3–52.

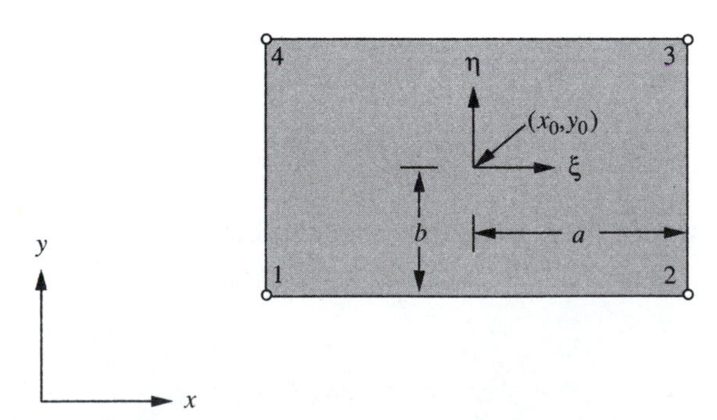

FIGURE 3–52 Rectangular element with sides parallel to the *x*-*y* axes

It is left to the Exercises to show that

$$|\mathbf{J}| = \begin{vmatrix} \dfrac{(x_2 - x_1)}{2} & 0 \\ 0 & \dfrac{(y_4 - y_1)}{2} \end{vmatrix} = ab = \frac{A_e}{4}$$

that is, the Jacobian of the transformation is everywhere positive and equal to the constant ab. Furthermore,

$$\frac{\partial}{\partial s} = \frac{x_2 - x_1}{2}\frac{\partial}{\partial x} \qquad \text{or} \qquad \frac{\partial}{\partial x} = \frac{1}{a}\frac{\partial}{\partial s}$$

$$\frac{\partial}{\partial t} = \frac{y_4 - y_1}{2}\frac{\partial}{\partial y} \qquad \text{or} \qquad \frac{\partial}{\partial y} = \frac{1}{b}\frac{\partial}{\partial t}$$

In other words, $a\,ds = dx$ and $b\,dt = dy$. These simple relations for the specially oriented rectangle also follow directly from the relations $(x - x_0)/a = s$ and $(y - y_0)/b = t$, used in Section 3.6 to convert the interpolation functions expressed in terms of x and y to their equivalent expressions in terms of s and t. For the special rectangle of Fig. 3–52, $|\mathbf{J}|$ is the constant ab and it follows that

$$dx\,dy = |\mathbf{J}|\,ds\,dt = ab\,ds\,dt$$

and

$$\int\!\!\int_{R_{xy}} dx\,dy = \int_{-1}^{1}\int_{-1}^{1} |\mathbf{J}|\,ds\,dt = \int_{-1}^{1}\int_{-1}^{1} ab\,ds\,dt = 4ab$$

For the more general case of the transformation appropriate for the quadrilateral element, $|\mathbf{J}|$ is a relative measure of the increase or decrease in the area of an infinitesimal element $dA(s, t) = ds\,dt$ when mapped into the corresponding $dA(x, y)$. These ideas are discussed further in the examples that follow.

Example 3.3.

Consider a quadrilateral element whose nodes are as indicated in Fig. 3–53.

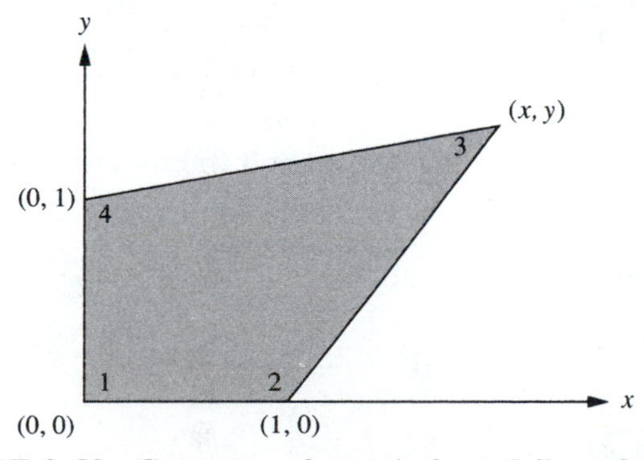

FIGURE 3–53 Geometry of a typical quadrilateral element

The x and y locations of node 3 will be considered as parameters that can be varied in order to demonstrate certain aspects of the transformation.

Using Eqs. (3.54) it follows easily that

$$\frac{\partial x}{\partial s} = \frac{(1 + x) + t(x - 1)}{4} \qquad \frac{\partial y}{\partial s} = \frac{(y - 1)(t + 1)}{4}$$

$$\frac{\partial x}{\partial t} = \frac{(x - 1)(s + 1)}{4} \qquad \frac{\partial y}{\partial t} = \frac{(1 + y) + s(y - 1)}{4}$$

and that

$$|\mathbf{J}| = \frac{(x + y) + (x - 1)t + (y - 1)s}{8}$$

Consider the special choice $x = y = 2$. The transformation is then given by

$$x = \frac{(1 + s)(3 + t)}{4}$$

$$y = \frac{(1 + t)(3 + s)}{4}$$

The parent element in the s-t system and the shape of the element are indicated in Fig. 3–54.

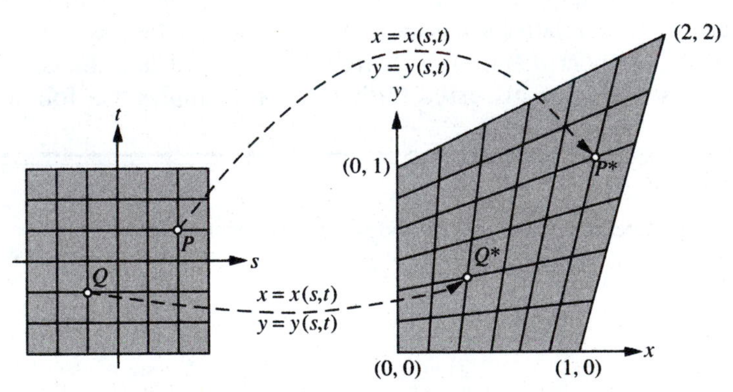

FIGURE 3–54 Parent element and transformed region R_{xy}

The Jacobian, given by

$$|\mathbf{J}| = \frac{4 + s + t}{8}$$

is indicated in Fig. 3–55.

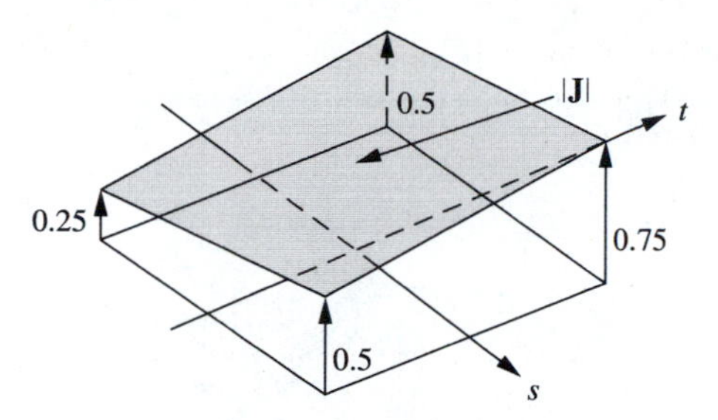

FIGURE 3–55 Jacobian as a function of position in D_{st}

$|\mathbf{J}|$ is clearly everywhere positive for $-1 \leq s \leq 1$ and $-1 \leq t \leq 1$, indicating that there are no problems with the transformation. In particular, we compute the area of the element in R_{xy} as

$$A = \iint_{R_{xy}} dx\, dy = \int_{-1}^{1}\int_{-1}^{1} |\mathbf{J}|\, ds\, dt$$

$$= \int_{-1}^{1}\int_{-1}^{1} \frac{4 + s + t}{8}\, ds\, dt = 2$$

Consider next the choice $x = y = 0.4$. The transformation is given by

$$x = \frac{(1 + s)(1.4 - 0.6t)}{4}$$

$$y = \frac{(1 + t)(1.4 - 0.6s)}{4}$$

The parent element in the s-t system and the shape of the element are indicated in Fig. 3–56.

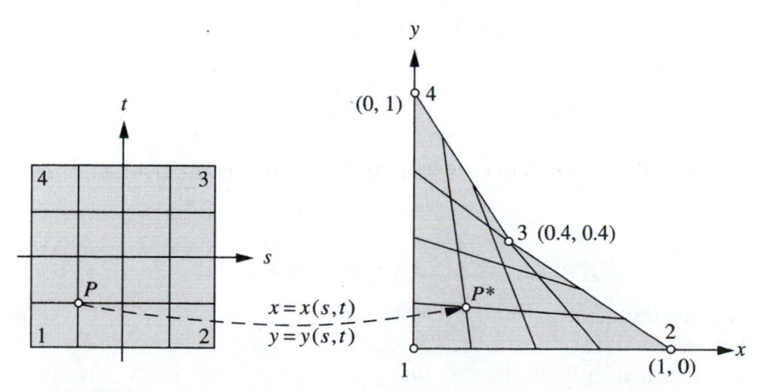

FIGURE 3–56 Parent element and transformed region R_{xy}

The Jacobian given by

$$|\mathbf{J}| = \frac{0.8 - 0.6(s + t)}{8}$$

is indicated in Fig. 3–57.

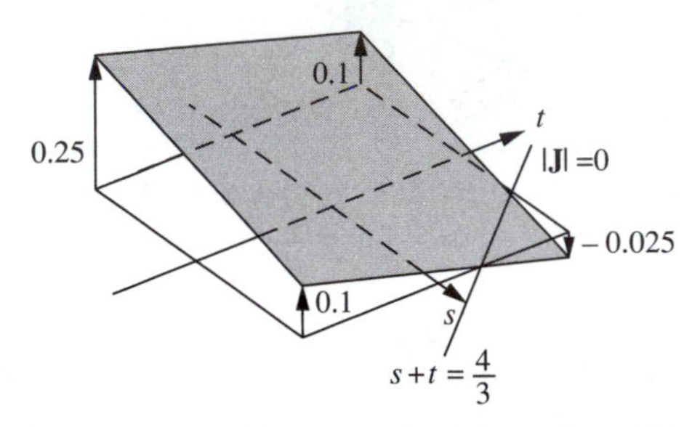

FIGURE 3–57 Jacobian as a function of position in D_{st}

$|\mathbf{J}|$ is clearly negative for $(s + t) \geq 4/3$, indicating that there are problems with the transformation for these values of s and t. In particular, consider the transformation

$$x = \frac{(1 + s)(1.4 - 0.6t)}{4}$$

$$y = \frac{(1 + t)(1.4 - 0.6s)}{4}$$

along the line $s = t$. Then both x and y are given by

$$x = y = (1 + s)(0.35 - 0.15s)$$

which is plotted in Fig. 3–58.

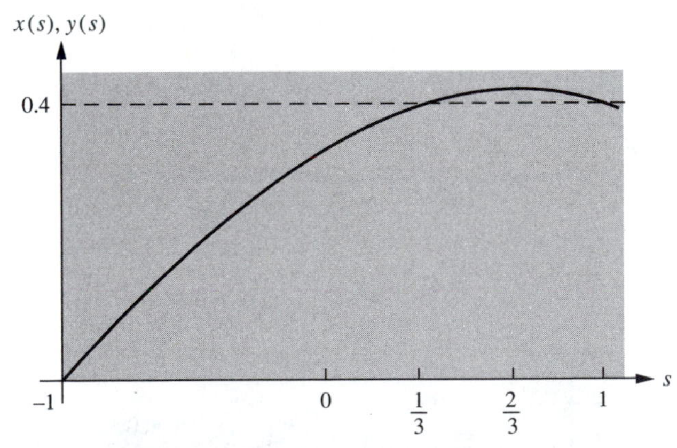

FIGURE 3–58 $x(s)$ and $y(s)$ along $s = t$

This plot indicates clearly that both the points $s = t = 1/3$ and $s = t = 1$ are mapped into $x = y = 0.4$, and that *the points* $1/3 < s < 1$ *are actually mapped outside the region.* The problem is associated with the fact that the region is not convex, meaning that there are sets of points in R_{xy} that cannot be connected by straight lines lying entirely in R_{xy}.

Generally, when the shape of a quadrilateral element is such that an interior angle approaches zero or π, the Jacobian approaches zero at the node in question, indicating that such geometries should be avoided. When choosing the mesh using quadrilateral elements, the shape of an individual quadrilateral element should be somewhat rectangular with an aspect ratio (ratio of the length of the longest to the shortest side of the quadrilateral) not exceeding approximately 2.5 to 3.

Derived variables. The derived variables for the Laplace or Poisson equation are the partial derivatives $\partial u/\partial x$ and $\partial u/\partial y$. These are computed as functions of position, per element, using the shape functions and the nodal values of the solution. If desired, these partial derivatives can be combined to yield the normal or directional derivative, given by

$$\frac{\partial u}{\partial n} = \mathbf{n} \cdot \nabla \mathbf{u} = n_x \frac{\partial u}{\partial x} + n_y \frac{\partial u}{\partial y}$$

The directional derivative measures the rate of change of u in the direction of the normal $\mathbf{n}$ given by $\mathbf{n} = \mathbf{i}n_x + \mathbf{j}n_y$ as indicated in Fig. 3–59.

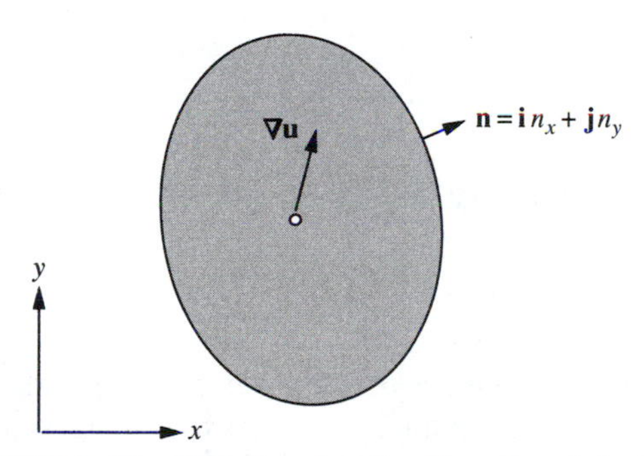

FIGURE 3–59 Notation for the directional derivative

Recall that the relationships between the partial derivatives in the s-t and x-y systems are given by

$$\begin{bmatrix} \dfrac{\partial u}{\partial s} \\ \dfrac{\partial u}{\partial t} \end{bmatrix} = \begin{bmatrix} \dfrac{\partial x}{\partial s} & \dfrac{\partial y}{\partial s} \\ \dfrac{\partial x}{\partial t} & \dfrac{\partial y}{\partial t} \end{bmatrix} \begin{bmatrix} \dfrac{\partial u}{\partial x} \\ \dfrac{\partial u}{\partial y} \end{bmatrix}$$

or

$$\frac{\partial \mathbf{u}}{\partial \mathbf{s}} = \mathbf{J} \frac{\partial \mathbf{u}}{\partial \mathbf{x}}$$

It follows that

$$\frac{\partial \mathbf{u}}{\partial \mathbf{x}} = \mathbf{J}^{-1} \frac{\partial \mathbf{u}}{\partial \mathbf{s}}$$

from which the scalars $\partial u/\partial x$ and $\partial u/\partial y$ are

$$\frac{\partial u}{\partial x} = \mathbf{J_1}\frac{\partial \mathbf{u}}{\partial \mathbf{s}}$$

$$\frac{\partial u}{\partial y} = \mathbf{J_2}\frac{\partial \mathbf{u}}{\partial \mathbf{s}}$$

where $\mathbf{J_1}$ and $\mathbf{J_2}$ are respectively the first and second rows of $\mathbf{J}^{-1}$. Futhermore, using

$$\frac{\partial \mathbf{u}}{\partial \mathbf{s}} = \begin{bmatrix} \dfrac{\partial u}{\partial s} \\ \dfrac{\partial u}{\partial t} \end{bmatrix} = \begin{bmatrix} \dfrac{\partial[\mathbf{N}^{\mathrm{T}}\mathbf{u_e}]}{\partial s} \\ \dfrac{\partial[\mathbf{N}^{\mathrm{T}}\mathbf{u_e}]}{\partial t} \end{bmatrix} = \Delta^{\mathrm{T}}\mathbf{u_e}$$

it follows that

$$\frac{\partial u}{\partial x} = \mathbf{J_1}\Delta^{\mathrm{T}}\mathbf{u_e}$$

$$\frac{\partial u}{\partial y} = \mathbf{J_2}\Delta^{\mathrm{T}}\mathbf{u_e}$$

The directional or normal derivative is then given by

$$\frac{\partial u}{\partial n} = (n_x\mathbf{J}_1 + n_y\mathbf{J}_2)\Delta^{\mathrm{T}}\mathbf{u_e}$$

All of $\partial u/\partial x$, $\partial u/\partial y$, and hence $\partial u/\partial n$ are functions of s and t within the element in question. Common points at which to evaluate the derived variables are the Gauss points, the centroid, and the nodes.

As was the case for the one-dimensional problems treated in Chapter 2, there are special points within the element [Moan, 8] at which the derivatives are more accurate. These special points are located very near the Gauss points associated with the degree of polynomial interpolation used for the dependent variable. For instance, when the solution is interpolated linearly as with the Q4 element, the special or Gauss point is located at the centroid of the element. Hence for the Q4 element, the partial derivatives $\partial u/\partial x$ and $\partial u/\partial y$ are often computed at the centroid as the best representation of the gradient within the element.

3.8.1.2 EVALUATION OF Q4 ELEMENTAL MATRICES. As stated several times previously in this chapter, the elemental matrices for the standard elliptic boundary value problem indicated in Fig. 3–60 are

$$\mathbf{k_e} = \int\int_{R_{xy}} \left(\frac{\partial \mathbf{N}}{\partial x}\frac{\partial \mathbf{N}^{\mathrm{T}}}{\partial x} + \frac{\partial \mathbf{N}}{\partial y}\frac{\partial \mathbf{N}^{\mathrm{T}}}{\partial y}\right) dA$$

$$\mathbf{f_e} = \iint_{R_{xy}} \mathbf{N} f \, dA$$

$$\mathbf{h_e} = \int_{\gamma_{2e}} \mathbf{N} h \, ds$$

$$\mathbf{a_e} = \int_{\gamma_{2e}} \mathbf{N} \alpha \mathbf{N} \, ds$$

where R_{xy} refers to the region occupied by a typical element and γ_{2e} to a typical line segment on the Γ_2 portion of the boundary. In what follows, we will indicate the evaluation of each of these integrals for the general Q4 element.

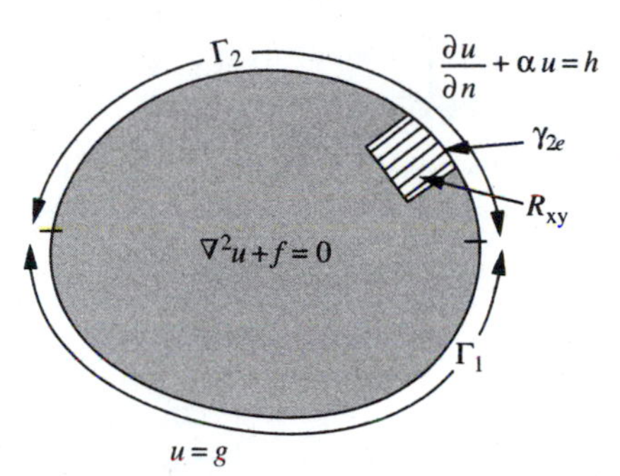

FIGURE 3–60 Standard elliptic boundary value problem

Evaluation of $\mathbf{a_e}$, $\mathbf{h_e}$. The evaluation of the elemental matrices $\mathbf{a_e}$ and $\mathbf{h_e}$ associated with the boundary conditions on Γ_2 will depend on the form of the interpolation functions $\mathbf{N}$ on the line segment γ_{2e}. For the Q4 element the interpolation functions reduce to linear functions on γ_{2e}, enabling us to use the results obtained for either the linearly interpolated triangular element (Sections 3.2 and 3.3) or the bilinearly interpolated rectangular element (Section 3.6) to which the reader is referred for details. Assuming that $h(s)$ and $\alpha(s)$ are sufficiently smooth and that the mesh is sufficiently refined, the results from either of Sections 3.2, 3.3, or 3.6 are

$$\mathbf{h_e} = \frac{l_e}{6} \begin{bmatrix} 2h_k + h_l \\ h_k + 2h_l \end{bmatrix}$$

$$\mathbf{a_e} = \frac{l_e}{12} \begin{bmatrix} 3\alpha_k + \alpha_l & \alpha_k + \alpha_l \\ \alpha_k + \alpha_l & \alpha_k + 3\alpha_l \end{bmatrix}$$

where, as usual, $h(s)$ and $\alpha(s)$ have been approximated by their linear interpolants on the segment γ_{2e} as indicated in Fig. 3–61.

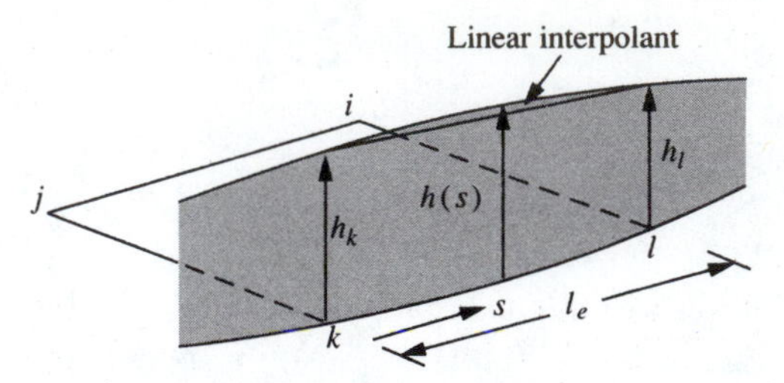

FIGURE 3–61 Representation of $h(s)$ and $\alpha(s)$ on γ_{2e}

Note that a code for solving the general elliptic boundary value problem would likely require that nodal values of h and α be specified.

Evaluation of $\mathbf{f_e}$. The general expression for $\mathbf{f_e}$ is

$$\mathbf{f_e} = \iint_{R_{xy}} \mathbf{N}(x, y) f(x, y)\, dA$$

We assume that it is generally possible to choose nodal values of f so as to be able to approximate f with reasonable accuracy according to

$$f(x, y) \approx f_1 N_1 + f_2 N_2 + f_3 N_3 + f_4 N_4 = \mathbf{N}^{\mathrm{T}}\mathbf{f}$$

as indicated in Fig. 3–62.

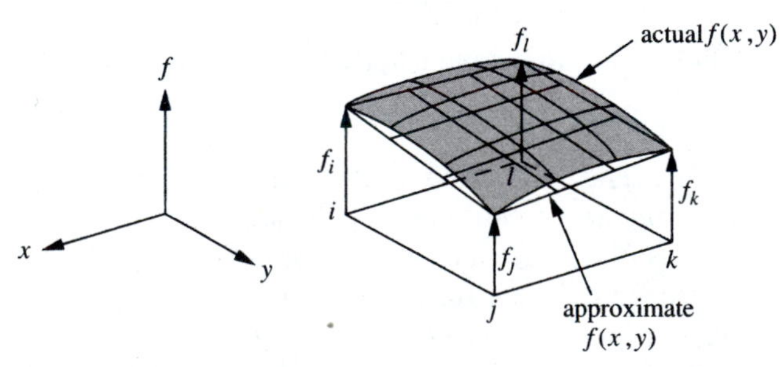

FIGURE 3–62 Shape function representation of $f(x, y)$

Then $\mathbf{f_e}$ becomes

$$\mathbf{f_e} = \left[\iint_{R_{xy}} \mathbf{N}(x, y)\mathbf{N}^{\mathrm{T}}(x, y)\, dA\right]\mathbf{f}$$

Transforming this to the s-t coordinates yields

$$\mathbf{f_e} = \left[\int_{-1}^{1}\int_{-1}^{1} \mathbf{N}(s, t)\mathbf{N}^{\mathrm{T}}(s, t)\, |\mathbf{J}|\, ds\, dt\right]\mathbf{f} \tag{3.55}$$

indicating that the problem of determining $\mathbf{f_e}$ has been reduced to evaluating the double integral

$$\int_{-1}^{1}\int_{-1}^{1} \mathbf{N}(s,t)\mathbf{N}^{\mathrm{T}}(s,t)\,|\mathbf{J}|\,ds\,dt$$

over D_{st}. We shall return to this task presently. Note that if this approximation is not felt to be sufficiently accurate one can always choose to evaluate

$$\int_{-1}^{1}\int_{-1}^{1} \mathbf{N}(s,t)\, f\big(x(s,t), y(s,t)\big)\,|\mathbf{J}|\,ds\,dt \tag{3.56}$$

without approximating f.

Evaluation of $\mathbf{k_e}$. The development of the elemental stiffness matrix $\mathbf{k_e}$ for the rectangular element was carried out in Section 3.6. The result is

$$\mathbf{k_e} = \int_{-1}^{1}\int_{-1}^{1} \mathbf{\Delta JJ\Delta}^{\mathrm{T}}\,ds\,dt$$

where

$$\mathbf{\Delta}^{\mathrm{T}} = \begin{bmatrix} -(1-t) & (1-t) & (1+t) & -(1+t) \\ -(1-s) & -(1+s) & (1+s) & (1-s) \end{bmatrix} \frac{1}{4}$$

and

$$\mathbf{JJ} = (\mathbf{J}_1^{\mathrm{T}}\mathbf{J}_1 + \mathbf{J}_2^{\mathrm{T}}\mathbf{J}_2)|\mathbf{J}|$$

with $\mathbf{J}_1$ and $\mathbf{J}_2$ the first and second rows of the inverse of the Jacobian matrix. For the general Q4 element, the 2×2 matrix $\mathbf{JJ}$ is a rational function of s and t with the Jacobian $|\mathbf{J}|$ in the denominator, making it essentially impossible to evaluate the integrals except by numerical means.

Thus it is seen that each of the evaluations of the elemental matrices $\mathbf{f_e}$ and $\mathbf{k_e}$ lead to integrals of the form

$$\mathbf{I} = \int_{-1}^{1}\int_{-1}^{1} \mathbf{G}(s,t)\,ds\,dt$$

where $\mathbf{G}(s,t)$ is a somewhat complicated function of the variables s and t. In principle it might be possible to evaluate $\mathbf{f_e}$ in general, although in practice it turns out that a numerical evaluation is entirely satisfactory. For $\mathbf{k_e}$ the appearance of the Jacobian $|\mathbf{J}|$ in the denominator of the integrand strongly suggests the use of a quadrature. In particular, the fact that s and t both range from -1 to 1 makes the Gauss-Legendre quadrature ideal for evaluating the integrals appearing in both $\mathbf{f_e}$ and $\mathbf{k_e}$.

As representative of the type of integrals for either $\mathbf{f_e}$ or $\mathbf{k_e}$, consider the general integral

$$\mathbf{I} = \int_{-1}^{1}\int_{-1}^{1} \mathbf{G}(s, t)\, ds\, dt$$

to be an iterated integral and first use an N-term Gauss-Legendre quadrature in the s-direction to obtain

$$\mathbf{I} = \int_{-1}^{1}\left[\sum_{i=1}^{N} W_i \mathbf{G}(s_i, t)\right] dt$$

Following this with an N-term Gauss-Legendre quadrature in the t-direction then yields

$$\mathbf{I} = \sum_{j=1}^{N}\sum_{i=1}^{N} W_j W_i \mathbf{G}(s_i, t_j) = \sum_{i=1}^{N}\sum_{j=1}^{N} W_i W_j \mathbf{G}(s_i, t_j)$$

Specifically for $\mathbf{f_e}$,

$$\mathbf{f_e} = \left[\sum_{i=1}^{N}\sum_{j=1}^{N} W_i W_j \mathbf{N}(s_i, t_j)\mathbf{N}^{\mathrm{T}}(s_i, t_j)\, |\mathbf{J}|_{ij}\right]\mathbf{f}$$

when using Eq. (3.55), or

$$\mathbf{f_e} = \sum_{i=1}^{N}\sum_{j=1}^{N} W_i W_j \mathbf{N}(s_i, t_j) f\big(x(s_i, t_j), y(s_i, t_j)\big)\, |\mathbf{J}|_{ij}$$

when using Eq. (3.56). In the above, $|\mathbf{J}|_{ij} \equiv |\mathbf{J}(s_i, t_j)|$.

In an entirely similar fashion,

$$\mathbf{k_e} = \sum_{i=1}^{N}\sum_{j=1}^{N} W_i W_j \boldsymbol{\Delta}_{\mathbf{ij}} \mathbf{JJ}_{\mathbf{ij}} \boldsymbol{\Delta}_{\mathbf{ij}}^{\mathrm{T}}$$

where the ij subscripts again refer to the evaluation of the quantity involved at s_i and t_j.

In the domain D_{st}, the integration points for the one-, two-, and three-point Gauss quadratures are located as indicated in Figs. 3–63*a*, 3–63*b*, and 3–63*c*, respectively.

For implementing these calculations in a code, each of the matrices $\mathbf{N}$, $\boldsymbol{\Delta}$, and $\mathbf{JJ}$, as well as the scalar $|\mathbf{J}|$ would be computed at the Gauss point in question, after which the appropriate matrix multiplications would be carried out with the weighted results over all the Gauss points then summed. As will be seen in the following examples, a two-point Gauss scheme in both directions is generally adequate for evaluating the Q4 elemental matrices with sufficient accuracy.

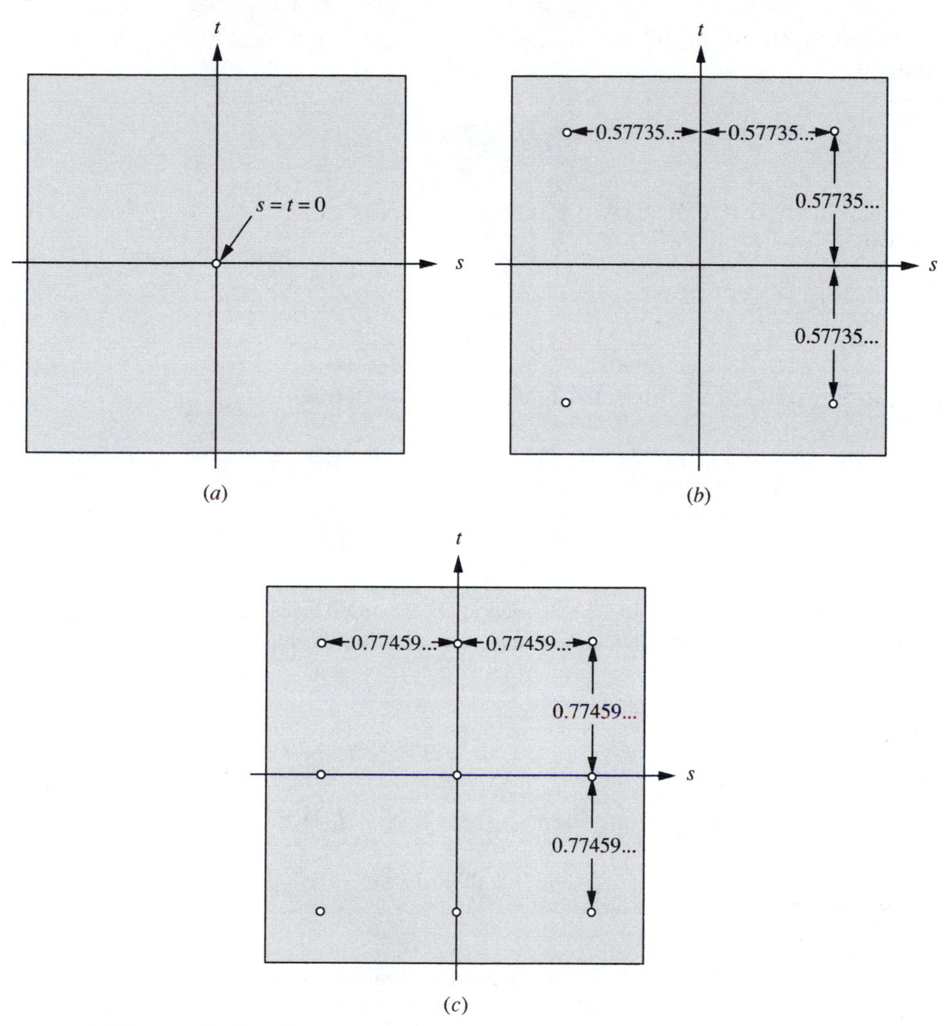

FIGURE 3–63 Gauss point locations for $N = 1$, $N = 2$, and $N = 3$

Example 3.4.

Consider the problem of the evaluation of

$$\mathbf{f_e} = \int\int_{R_{xy}} \mathbf{N}(x, y) f(x, y)\, dx\, dy$$

for $f = xy$ with R_{xy} the region shown in Fig. 3–64.

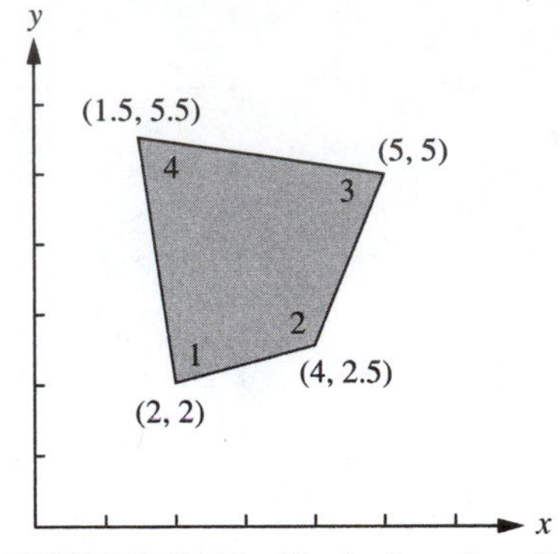

FIGURE 3–64 Typical region R_{xy}

We are interested in two aspects of the problem. First, what are the errors generally introduced by the approximation

$$f(x, y) = \sum_{1}^{4} f_i N_i = \mathbf{N}^{\mathrm{T}}\mathbf{f}$$

in the evaluation of the integrals, and second, what order Gauss-Legendre quadrature is generally necessary for reasonable accuracy in evaluating the integrals?

Consider first the evaluation of the exact expression for $\mathbf{f_e}$ given by

$$\mathbf{f_e} = \int\int_{R_{xy}} \mathbf{N}(x, y)xy \, dx \, dy$$

$$= \int_{-1}^{1}\int_{-1}^{1} \mathbf{N}(s, t)\mathbf{x_e^{\mathrm{T}}}\mathbf{N}(s, t)\mathbf{y_e^{\mathrm{T}}}\mathbf{N}(s, t) \, |\mathbf{J}| \, ds \, dt$$

using various orders of Gauss-Legendre integration. The results for each of the four components of $\mathbf{f_e}$ are displayed in Table 3.11.

TABLE 3.11 Numerical evaluation of exact $\mathbf{f_e}$

N	f_{e1}	f_{e2}	f_{e3}	f_{e4}
1	24.1699	24.1699	24.1699	24.1699
2	17.2509	20.6910	33.6276	27.0868
3	17.2545	20.6708	33.6497	27.0813
4	17.2545	20.6708	33.6497	27.0813

Consider also the evaluation of the approximate expression for $\mathbf{f_e}$ given by

$$\mathbf{f_e} = \left[\int_{-1}^{1}\int_{-1}^{1} \mathbf{N}(s, t)\mathbf{N}(s, t)^{\mathrm{T}} \, |\mathbf{J}| \, ds \, dt\right]\mathbf{f}$$

where $\mathbf{f}$ is taken to be [4.0 10.0 25.0 8.25], that is, the values of $f = xy$ at the nodes. The results appear in Table 3.12.

TABLE 3.12 Numerical evaluation of approximate $\mathbf{f_e}$

N	f_{e1}	f_{e2}	f_{e3}	f_{e4}
1	24.3633	24.3633	24.3633	24.3633
2	17.1936	21.0469	33.7491	26.6667
3	17.1936	21.0469	33.7491	26.6667

First, notice from Table 3.11 for the exact evaluation of $\mathbf{f_e}$, that the $N = 3$ and $N = 4$ quadrature results are identical. This is easily explained by the fact that the maximum power of either s (or t) appearing in the integrand is four and is integrated exactly by a third-order ($2 \times 3 - 1 = 5$) Gauss-Legendre quadrature. Similarly, since the maximum power of s (or t) in the integrand of the integrals for the approximate $\mathbf{f_e}$ is three, the second-order ($2 \times 2 - 1 = 3$) Gauss scheme is sufficient.

Second, from Table 3.12, it is clear that the approximate evaluation is very accurate for the typical element of Fig. 3–64. Assuming an $N = 2$ scheme is generally used in practice, the maximum error occurs in the second element of $\mathbf{f_e}$ and is approximately 1.8 percent.

Last, from Table 3.11, when evaluating the exact expression for $\mathbf{f_e}$, the 2×2 Gauss-Legendre integration produces results accurate to within approximately 0.1 percent, showing that for the present typical example, the 2×2 scheme is entirely adequate.

Example 3.5.

Consider also the evaluation of a typical elemental stiffness matrix given by

$$\mathbf{k_e} = \int_{-1}^{1}\int_{-1}^{1} \mathbf{\Delta J J \Delta}^{\mathrm{T}}\, ds\, dt$$

for the region indicated in Fig. 3–65.

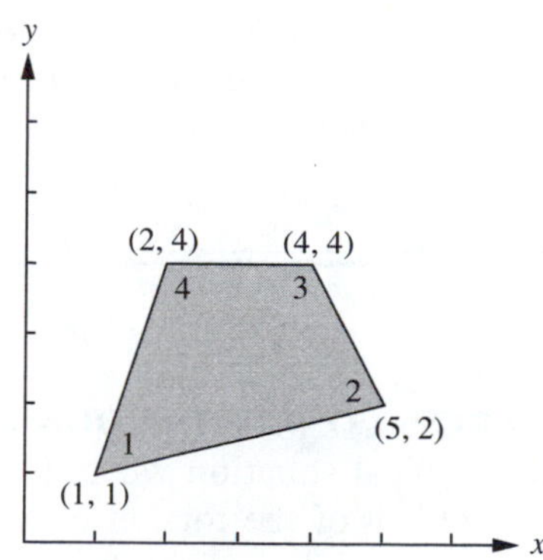

FIGURE 3–65 Typical geometry for a Q4 element

Each of the elements of Δ is linear in s or t whereas each of the elements of **JJ** is a rational function of s and t, that is, a function with polynomials in both the numerator and denominator. Thus it follows that there is no order of Gauss quadrature that will evaluate the integrals exactly. However, we will demonstrate in this example what effect the order of the Gauss quadrature generally has on the accuracy of the results.

As mentioned previously, the calculations are performed by evaluating the matrices Δ, **JJ**, and the Jacobian $|\mathbf{J}|$ at each Gauss point, performing the matrix multiplications and then summing over all the integration points. Indicated below are the resulting $\mathbf{k_e}$ for the typical region of Fig. 3–65, having used Gauss quadrature with $N = 1$, 2, 3, and 4.

$$N = 1 \qquad \mathbf{k_e} = \begin{bmatrix} 0.43333 & 0.10000 & -0.43333 & -0.10000 \\ & 0.60000 & -0.10000 & -0.60000 \\ & & 0.43333 & 0.10000 \\ \text{symm} & & & 0.60000 \end{bmatrix}$$

$$N = 2 \qquad \mathbf{k_e} = \begin{bmatrix} 0.48413 & 0.02353 & -0.29314 & -0.21471 \\ & 0.71471 & -0.31029 & -0.42794 \\ & & 0.81887 & -0.21544 \\ \text{symm} & & & 0.85809 \end{bmatrix}$$

$$N = 3 \qquad \mathbf{k_e} = \begin{bmatrix} 0.48519 & 0.02212 & -0.29072 & -0.21668 \\ & 0.71668 & -0.31392 & -0.42498 \\ & & 0.82552 & -0.22088 \\ \text{symm} & & & 0.86254 \end{bmatrix}$$

$$N = 4 \qquad \mathbf{k_e} = \begin{bmatrix} 0.48522 & 0.02218 & -0.29066 & -0.21674 \\ & 0.71674 & -0.31402 & -0.42490 \\ & & 0.82570 & -0.22102 \\ \text{symm} & & & 0.86266 \end{bmatrix}$$

The $N = 1$ results are clearly not acceptable, whereas the results using the 2×2 Gauss quadrature are generally within 5 percent of those using the 4×4 scheme. Although the 4×4 results are not exact, the general agreement to the fourth or fifth decimal between the $N = 3$ and $N = 4$ results is an indication of convergence to that number of significant decimals. The conclusion is that practically, the 2×2 Gauss scheme is adequate for evaluation of the $\mathbf{f_e}$ and $\mathbf{k_e}$ matrices for Q4 elements.

3.8.1.3 APPLICATION—STRESS CONCENTRATION FOR THE TORSION OF AN ANGLE SECTION. For a typical situation where the use of the Q4 elements is appropriate, consider the problem of the torsion of an angle section as shown in Fig. 3–66.

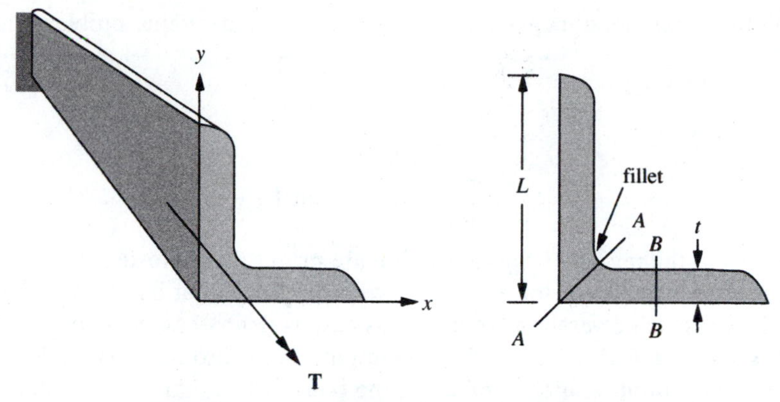

FIGURE 3–66 Torsion of an angle section

As outlined in Section 3.5, the boundary value problem

$$\nabla^2\phi + 2G\theta = 0 \qquad \text{in } D$$

with

$$\phi = 0 \qquad \text{on } \Gamma$$

is solved, after which the stresses are computed according to

$$\tau_{xz} = \frac{\partial\phi}{\partial y} \qquad \tau_{yz} = -\frac{\partial\phi}{\partial x}$$

Of particular interest for this application is the stress concentration due to the presence of the fillet contained in section *A-A*.

In selecting the region to be modeled, we will make the following assumptions:

1. The thickness t and the length L are such that the plane *A-A* is a plane of symmetry.
2. Section *B-B* is far enough removed from the fillet at section *A-A* that ϕ is independent of x, that is, $\partial\phi/\partial x = \partial\phi/\partial n = 0$ on *B-B*.

With these assumptions in mind, we will construct a model of the region between sections *A-A* and *B-B* as shown in Fig. 3–67.

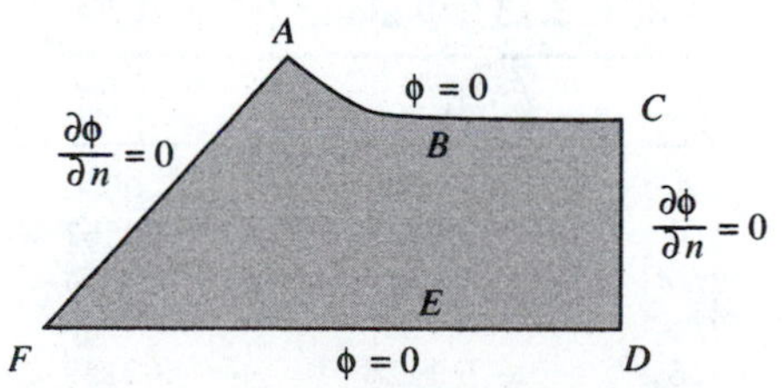

FIGURE 3–67 Modeled region for torsion of an angle

We will use the nondimensional form of the boundary value problem, namely

$$\nabla^2\psi + 1 = 0 \qquad \text{in } D$$

$$\psi = 0 \qquad \text{on } \Gamma_1$$

and

$$\frac{\partial\psi}{\partial n} = 0 \qquad \text{on } \Gamma_2$$

with respect to the region of Fig. 3–67. We are primarily interested in determining the stress concentration factor associated with the fillet, that is, at point *A* in Fig. 3–67. The stress concentration factor in this case is defined as the maximum stress through section *A-A* divided by the maximum stress through section *B-B*. Both maxima will occur at, and be parallel to, the boundaries, that is, at *A*, and at *D* or *C*. Assumption 2 is equivalent to stating that the nominal stress, computed from the elementary theory, will occur at section *B-B* or along line *CD*.

The mesh that will be used is shown in Fig. 3–68.

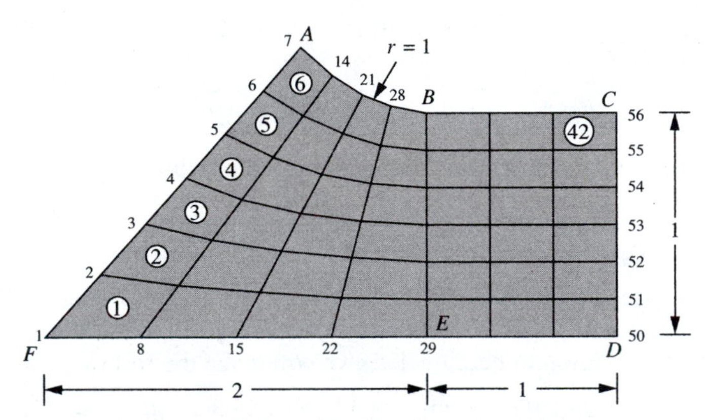

FIGURE 3–68 Mesh for torsion of angle section

The nodes along the portions *AB* and *EF* are chosen at equal intervals as are those along the portions *AF* and *BE*. The geometry modeled corresponds to an r/t ratio of 1, r being the radius of the fillet and t the nominal thickness of the angle.

Partial results for ψ are presented in Table 3.13 and displayed in Figs. 3–69*a* and *b*.

TABLE 3.13 ψ along *DC* and *FA*

x/l_e	ψ_{DC}	ψ_{FA}
0	0.0000	0.0000
1/6	0.0703	0.0567
2/6	0.1125	0.1364
3/6	0.1267	0.1988
4/6	0.1126	0.2159
5/6	0.0702	0.1623
1	0.0000	0.0000

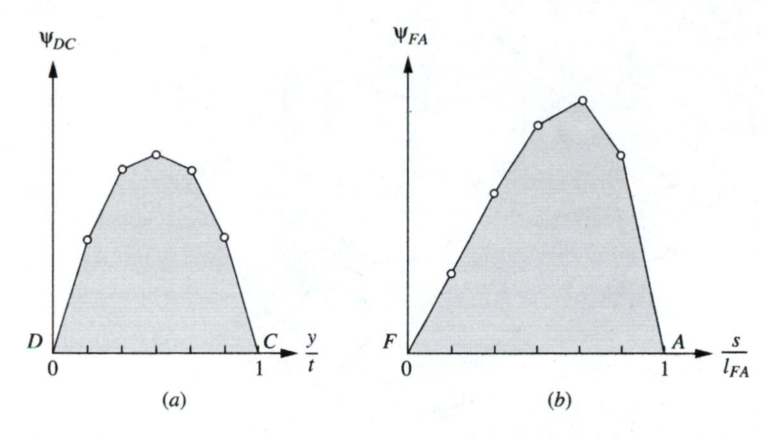

FIGURE 3–69 ψ along *DC* and *FA*

The component of stress at a point in the interior or on the boundary of the region is proportional to the derivative of the ψ function in the perpendicular direction at the point. Thus at C, the component of stress tangent to the boundary is proportional to the derivative in the direction normal to the boundary, that is,

$$\tau_C \propto \frac{\psi_{56} - \psi_{55}}{1/6} = -0.4142$$

and in a completely similar vein,

$$\tau_A \propto \frac{\psi_7 - \psi_6}{l_{FA}/6} = -0.5327$$

On the basis of assumptions 1 and 2, the stress at C is the nominal stress computed on the basis of the theory of thin-walled open sections so that the stress concentration factor is the ratio of the stress at A to the nominal stress at C,

$$k_T = \frac{0.5327}{0.4142} = 1.29$$

This is to be compared with tabulated experimental value of approximately $k_T = 1.4$ [Griffith and Taylor, 9].

The somewhat disappointing predicted value of the stress concentration factor is due primarily to the relatively crude mesh and the linear interpolation used in the Q4 element. These results could be improved somewhat by refining the mesh. As will be seen in the next sections, substantial improvement can be realized with the use of elements that employ higher-order interpolations with essentially the same nodes.

3.8.1.4 APPLICATION—STEADY-STATE HEAT TRANSFER. An important technical problem in heat transfer concerns a thin fin intended to transmit energy away from a region as indicated in Fig. 3–70.

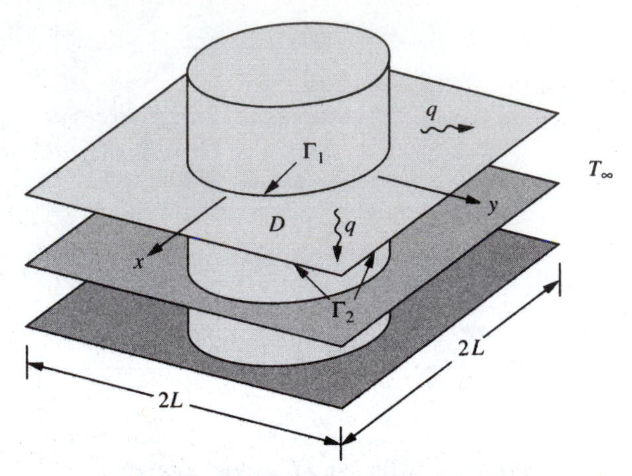

FIGURE 3–70 Typical fin geometry

A simple model considers that energy transfer takes place only in the plane of an individual fin resulting in the two-dimensional boundary value problem

$$\begin{aligned} \nabla^2 T(x, y) &= 0 && \text{in } D \\ T &= T_0 && \text{on } \Gamma_1 \\ -k\frac{\partial T}{\partial n} &= h(T - T_\infty) && \text{on } \Gamma_2 \end{aligned}$$

where D, Γ_1, and Γ_2 are indicated in Fig. 3–70, and k and h are physical constants associated with the conduction in D and the convection at the boundary Γ_2, respectively. The free stream convection temperature is T_∞.

The symmetries of the problem permit solution in the region indicated in Fig. 3–71.

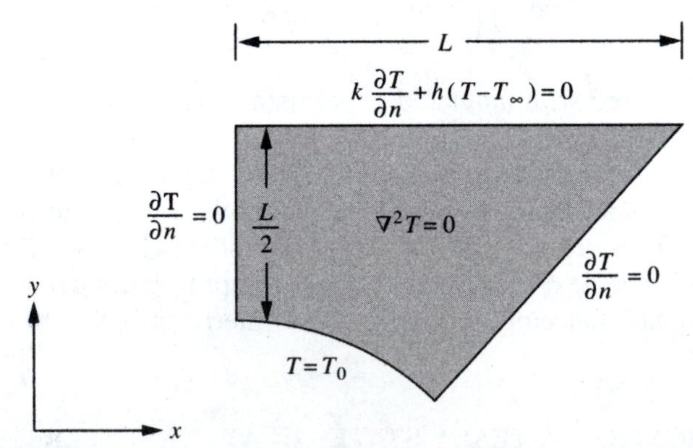

FIGURE 3–71 Replacement region considering symmetries

Define new independent variables $x/L = \xi$, $y/L = \eta$, and a new dependent variable $t = (T - T_\infty)/(T_0 - T_\infty)$, in terms of which the boundary value problem can be stated as indicated in Fig. 3–72.

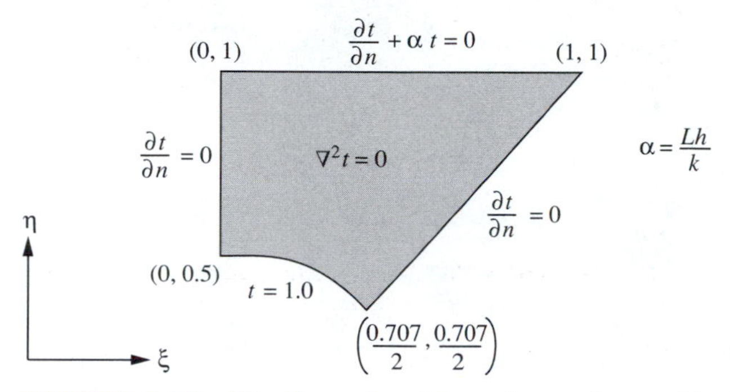

FIGURE 3–72 Nondimensional boundary value problem

The mesh to be used is indicated in Fig. 3–73.

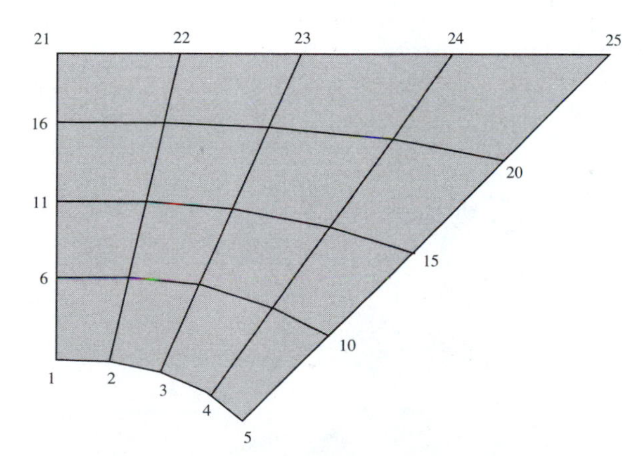

FIGURE 3–73 Mesh for Q4 elements

With $\alpha = 1$, the nodal values for t are presented in Fig. 3–74.

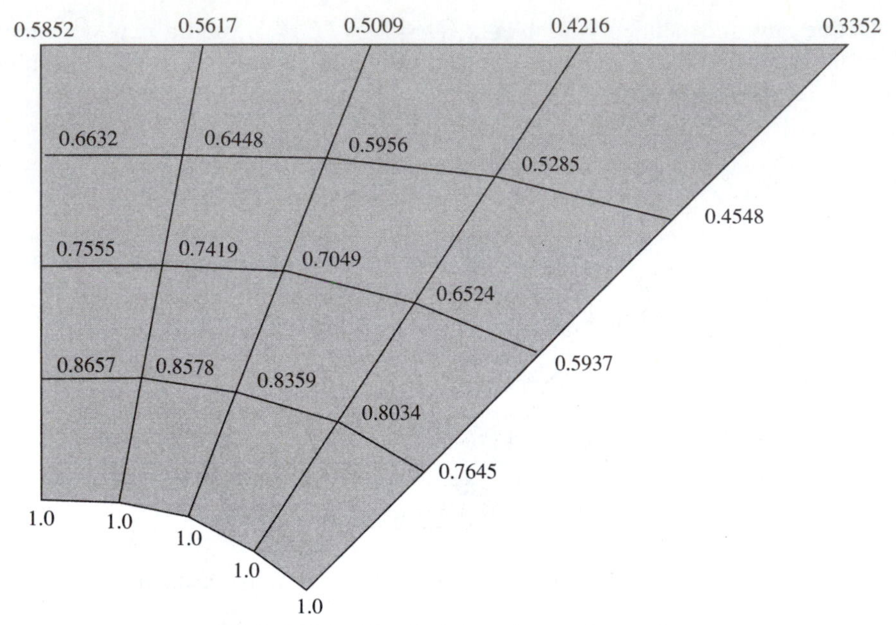

FIGURE 3–74 Nodal *t* values—Q4 model

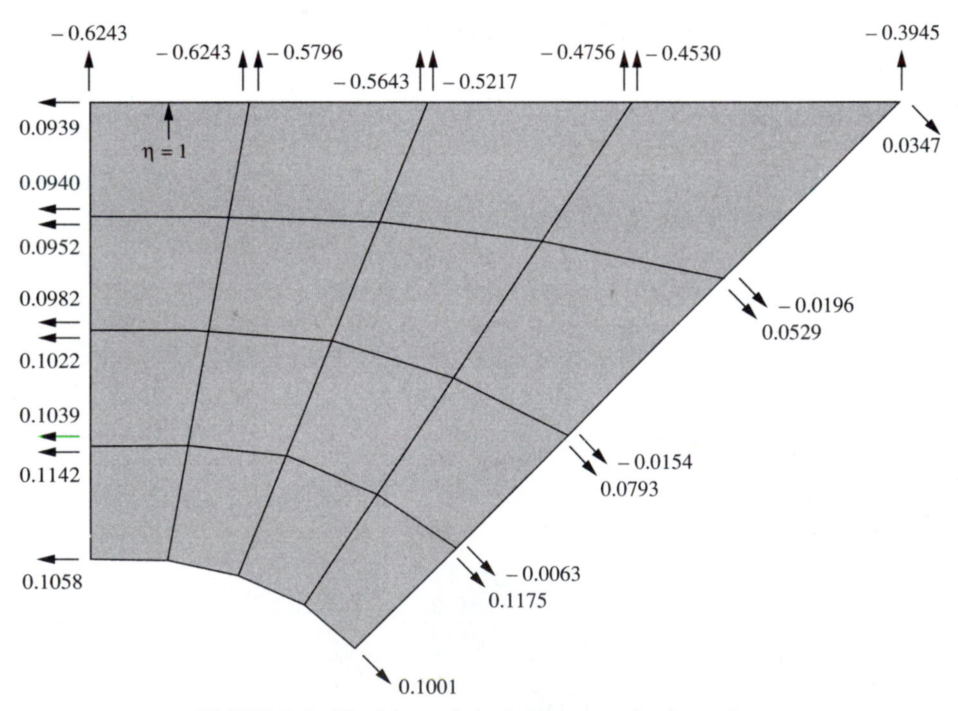

FIGURE 3–75 Normal derivatives on the boundary

Figure 3–75 shows the values of the normal derivatives on the Γ_2 portion of the boundary. Each of these nodal values is calculated according to

$$\frac{\partial \phi}{\partial n} = (n_x \mathbf{J}_1 + n_y \mathbf{J}_2)\boldsymbol{\Delta}^{\mathrm{T}}\mathbf{t}_{\mathbf{e}}$$

Note that at each boundary node where there are two adjacent elements, the normal derivative in each of the two adjacent elements is computed. Equality of these derivatives is an indication of convergence.

The normal derivatives on the portions of the boundary defined by nodes 1, 6, 11, 16, and 21, and by nodes 5, 10, 15, 20, and 25 are relatively large compared to the exact values of zero, indicating that further refinement of the mesh is in order for this application when using Q4 elements.

Averaging the values of the normal derivatives computed in adjacent elements at nodes 22–25, the satisfaction of the natural boundary condition $\partial t/\partial n + t = 0$ on $\eta = 1$ can be checked as follows.

Node 21: $\quad \dfrac{\partial t}{\partial n} + t = -0.6243 + 0.5852 = -0.0391$

Node 22: $\quad \dfrac{\partial t}{\partial n} + t = -0.6020 + 0.5617 = -0.0403$

Node 23: $\quad \dfrac{\partial t}{\partial n} + t = -0.5430 + 0.5009 = -0.0421$

Node 24: $\quad \dfrac{\partial t}{\partial n} + t = -0.4643 + 0.4216 = -0.0427$

Node 25: $\quad \dfrac{\partial t}{\partial n} + t = -0.3945 + 0.3352 = -0.0593$

These relatively small values are an indication that the satisfaction of the natural boundary condition on this portion of Γ_2 is reasonable.

The results from this mesh give a reasonable idea as to the correct solution. If more accurate results are desired, as judged for instance on better satisfaction of the natural boundary conditions, a finer mesh or elements using higher-order interpolation should be investigated.

3.8.2 Eight-Node Quadrilateral (Q8) Elements

A logical next step in the hierarchy of quadrilateral elements is to place midside nodes along the sides of the quadrilateral. Additionally, any side of the quadrilateral can be curved, helping to more accurately model curved boundaries. Finally, interpolation along any side will be quadratic, resulting generally in improved accuracy relative to elements employing linear interpolation.

3.8.2.1 TRANSFORMATIONS AND SHAPE FUNCTIONS. The eight-node quadrilateral element is a generalization of the eight-node rectangular element indicated in Fig. 3–76.

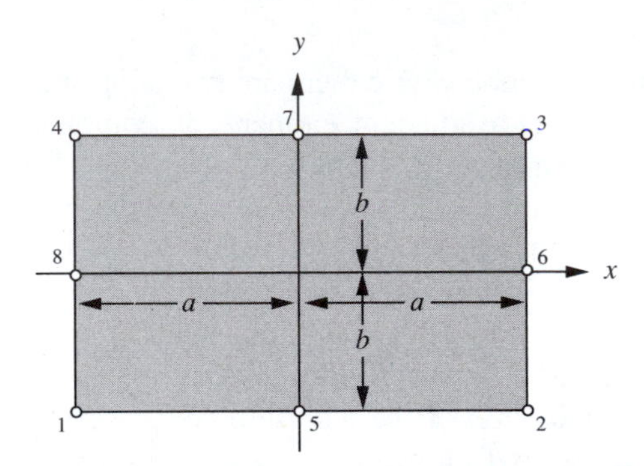

FIGURE 3–76 Eight-node rectangular element

For the eight-node rectangular element, the interpolation functions for the solution $u(x, y)$ are based on the assumption

$$u_e(x, y) = a + bx + cy + dx^2 + exy + fy^2 + gx^2y + hxy^2 \tag{3.57}$$

For x = constant, u_e is quadratic in y and similarly for y = constant, u_e is quadratic in x. For this reason this type of interpolation is frequently referred to as quadratic interpolation for the rectangle. Interelement continuity and the completeness conditions are satisfied by representation given by Eq. (3.57). Requiring

$$u_e(x_i, y_i) = u_i \qquad i = 1, \ldots, 8$$

yields a set of eight algebraic equations to determine the constants a through h. These equations can be solved in principle with the results being back-substituted into Eq. (3.57) to yield

$$u_e(x, y) = \sum_{1}^{8} u_i N_i(x, y) = \mathbf{u}_\mathbf{e}^\mathrm{T}\mathbf{N} = \mathbf{N}^\mathrm{T}\mathbf{u}_\mathbf{e} \tag{3.58}$$

where the $N_i(x, y)$ are the eight quadratic interpolation functions. Each $N_i(x, y)$ has the property that

$$N_i(x_j, y_j) = \delta_{ij} \qquad i, j = 1, \ldots, 8$$

that is, $N_i(x_j, y_j) = 1$ if $i = j$, and is zero otherwise.

It turns out, however, that these interpolation functions can be developed in a much more instructive fashion as follows. Consider Fig. 3–77, where the equations of each of the sides as well as those of each of four diagonals are indicated.

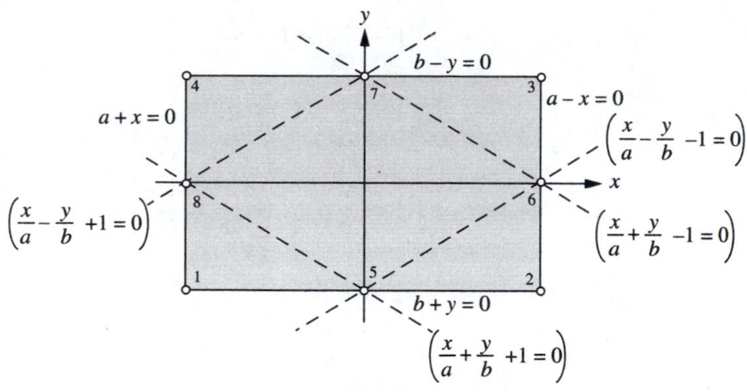

FIGURE 3–77 Rectangular Q8 elements

If we form the product of the equations of the straight lines passing through nodes 4-7-3, 3-6-2, and 8-5 respectively, there results the function

$$n_1(x, y) = (b - y)(a - x)(bx + ay + ab)$$

which vanishes at all nodes other than node 1. If we set $x = -a$ and $y = -b$, that is, the coordinates of node 1, there results

$$n_1(-a, -b) = -4a^2b^2$$

We then form

$$N_1(x, y) = \frac{n_1(x, y)}{n_1(-a, -b)} = \left(1 - \frac{x}{a}\right)\left(1 - \frac{y}{b}\right)\left(-1 - \frac{x}{a} - \frac{y}{b}\right)\frac{1}{4}$$

a function that vanishes at all nodes but node 1 where it has the value $+1$. $N_1(x, y)$ is clearly of the form given in Eq. (3.57) and is thus the same as the N_1 function that appears in Eq. (3.58). The interpolation functions for the other corner nodes are developed in an entirely analogous fashion.

For the midside nodes, node 5 for example, we take the product of the lines containing nodes 1-8-4, 4-7-3, and 3-6-2 to obtain

$$n_5(x, y) = (a + x)(b - y)(a - x)$$

and then form

$$N_5(x, y) = \frac{n_5(x, y)}{n_5(0, -b)} = \left(1 - \frac{x}{a}\right)\left(1 + \frac{x}{a}\right)\left(1 - \frac{y}{b}\right)$$

again of the form as given by Eq. (3.57). Note that N_5 is quadratic in x and linear in y.

The eight different quadratic interpolation functions, each determined in this way, are

$$N_1(x,y) = \left(1 - \frac{x}{a}\right)\left(1 - \frac{y}{b}\right)\left(-1 - \frac{x}{a} - \frac{y}{b}\right)\frac{1}{4}$$

$$N_2(x,y) = \left(1 + \frac{x}{a}\right)\left(1 - \frac{y}{b}\right)\left(-1 + \frac{x}{a} - \frac{y}{b}\right)\frac{1}{4}$$

$$N_3(x,y) = \left(1 + \frac{x}{a}\right)\left(1 + \frac{y}{b}\right)\left(-1 + \frac{x}{a} + \frac{y}{b}\right)\frac{1}{4}$$

$$N_4(x,y) = \left(1 - \frac{x}{a}\right)\left(1 + \frac{y}{b}\right)\left(-1 - \frac{x}{a} + \frac{y}{b}\right)\frac{1}{4}$$

$$N_5(x,y) = \left(1 - \frac{x^2}{a^2}\right)\left(1 - \frac{y}{b}\right)\frac{1}{2}$$

$$N_6(x,y) = \left(1 + \frac{x}{a}\right)\left(1 - \frac{y^2}{b^2}\right)\frac{1}{2}$$

$$N_7(x,y) = \left(1 - \frac{x^2}{a^2}\right)\left(1 + \frac{y}{b}\right)\frac{1}{2}$$

$$N_8(x,y) = \left(1 - \frac{x}{a}\right)\left(1 - \frac{y^2}{b^2}\right)\frac{1}{2}$$

and appear typically as the shape functions N_1 and N_5 shown in Fig. 3–78.

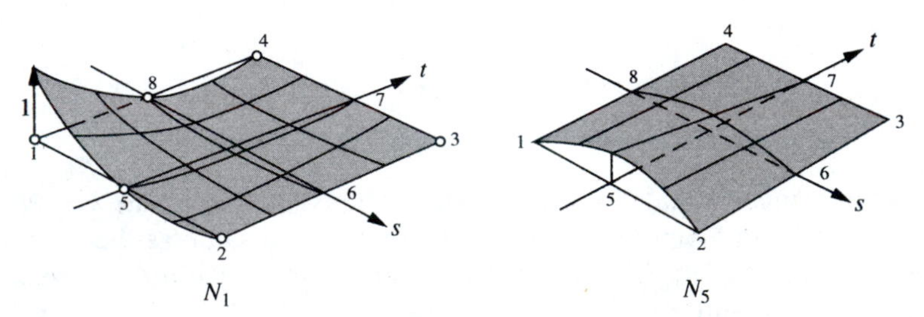

FIGURE 3–78 Q8 interpolation functions N_1 and N_5

Expressed in terms of the dimensionless variables $s = x/a$ and $t = y/b$, the interpolation functions become

$$N_1(s,t) = (1-s)(1-t)(-1-s-t)\frac{1}{4}$$

$$N_2(s,t) = (1+s)(1-t)(-1+s-t)\frac{1}{4}$$

$$N_3(s,t) = (1+s)(1+t)(-1+s+t)\frac{1}{4}$$

$$N_4(s,t) = (1-s)(1+t)(-1-s+t)\frac{1}{4}$$

$$
\begin{aligned}
N_5(s,t) &= \left(1-s^2\right)(1-t)\frac{1}{2} \\
N_6(s,t) &= (1+s)\left(1-t^2\right)\frac{1}{2} \\
N_7(s,t) &= \left(1-s^2\right)(1+t)\frac{1}{2} \\
N_8(s,t) &= (1-s)\left(1-t^2\right)\frac{1}{2}
\end{aligned} \tag{3.59}
$$

In what follows, we will use the dimensionless forms of the interpolation functions.

Consider next a generally shaped Q8 element as indicated in Fig. 3–79*b*.

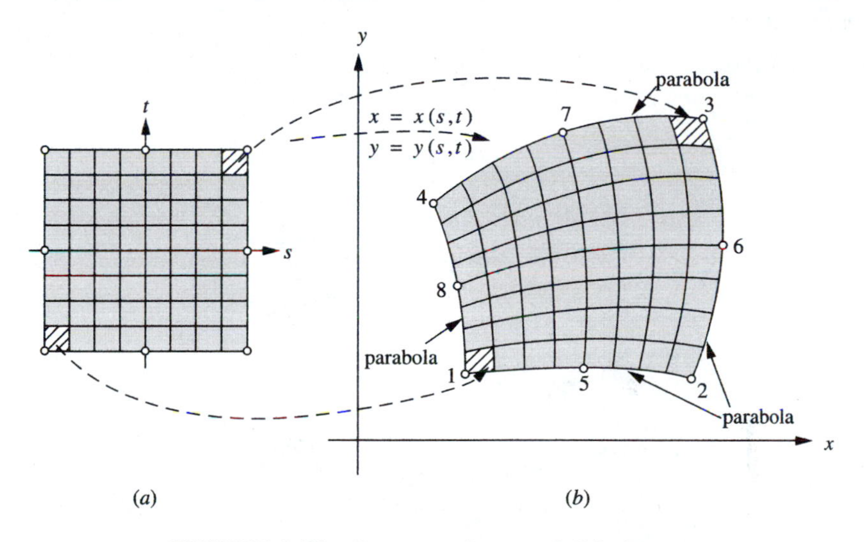

FIGURE 3–79 Parent and general Q8 elements

The shape of the element, which is determined by the location of the nodes, is defined by the transformation

$$
\begin{aligned}
x(s,t) &= \sum_{1}^{8} x_i N_i(s,t) = \mathbf{x}_\mathbf{e}^\mathrm{T}\mathbf{N} \\
y(s,t) &= \sum_{1}^{8} y_i N_i(s,t) = \mathbf{y}_\mathbf{e}^\mathrm{T}\mathbf{N}
\end{aligned} \tag{3.60}
$$

where

$$\mathbf{x}_\mathbf{e}^\mathrm{T} = [x_1 \quad x_2 \quad x_3 \quad x_4 \quad x_5 \quad x_6 \quad x_7 \quad x_8]$$

and

$$\mathbf{y}_\mathbf{e}^\mathrm{T} = [y_1 \quad y_2 \quad y_3 \quad y_4 \quad y_5 \quad y_6 \quad y_7 \quad y_8]$$

are vectors containing the nodal coordinates. If we also represent the solution as

$$u_e(x, y) = \mathbf{u}_\mathbf{e}^\mathrm{T}\mathbf{N}$$

the element is clearly isoparametric in view of the fact that the shape and interpolation functions are identical.

The transformation maps the parent element of Fig. 3–79*a* with domain D_{st} given by $-1 \leq s \leq 1$, $-1 \leq t \leq 1$ into the Q8 element of Fig. 3–79*b*. In practice the positions of the eight nodes given by $\mathbf{x_e}^\mathrm{T}$ and $\mathbf{y_e}^\mathrm{T}$ are specified, with the geometry of the element then fixed on the basis of the transformation of Eqs. (3.60). Differential area elements $dA = ds\, dt$ in D_{st} of Fig. 3–79*a* are mapped into curvilinear differential area elements as indicated in Fig. 3–79*b*.

Example 3.6.

Consider for instance the specific transformation indicated in Fig. 3–80.

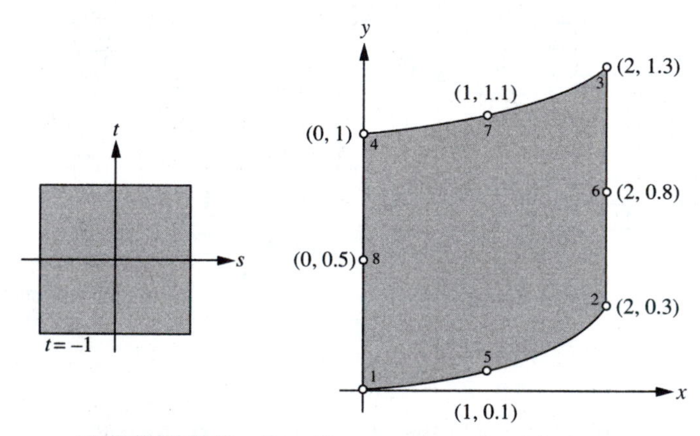

FIGURE 3–80 Specific geometry of a Q8 element

The two vertical sides are straight with the two horizontal sides curved. It is easily verified that the line $t = -1$ of the parent element is mapped into

$$x(s, -1) = 1 + s$$

$$y(s, -1) = 0.1 + 0.15s + 0.05s^2$$

which is clearly the parabola

$$y(x) = 0.1 + 0.15(x - 1) + 0.05(x - 1)^2$$

passing through the points 1-5-2 in the mapped region R_{xy}. In other words, the straight line $t = -1$ containing nodes 1-5-2 in D_{st} is mapped into the parabola defined by nodes 1-5-2. A more detailed study would indicate that, in general, straight lines parallel to the s axis in the parent element are mapped into parabolas in R_{xy} and that straight lines parallel to the t axis become straight lines parallel to the y axis.

Any side of a Q8 element is a parabola passing through the three nodes that define the side. It follows that the boundary of a region D which is being discretized using Q8 elements is represented by a sequence of parabolas as indicated in Fig. 3–81.

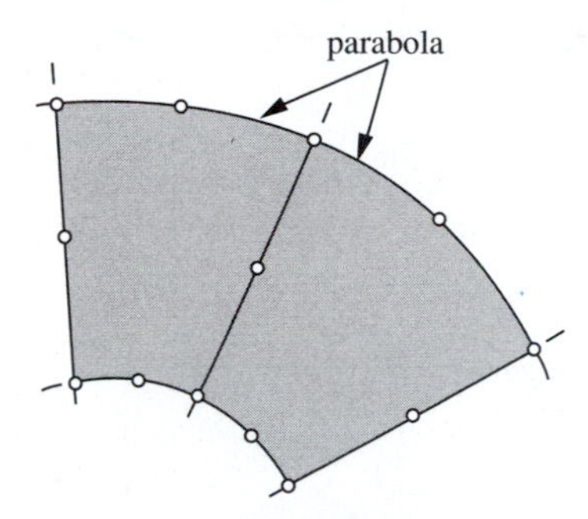

FIGURE 3–81 Character of the boundary of a region discretized using Q8 elements

Care must be taken in the modeling process to avoid defining nodes that result in an angle α approaching 0 or π as indicated in Fig. 3–82. This idea will be investigated further in the following examples.

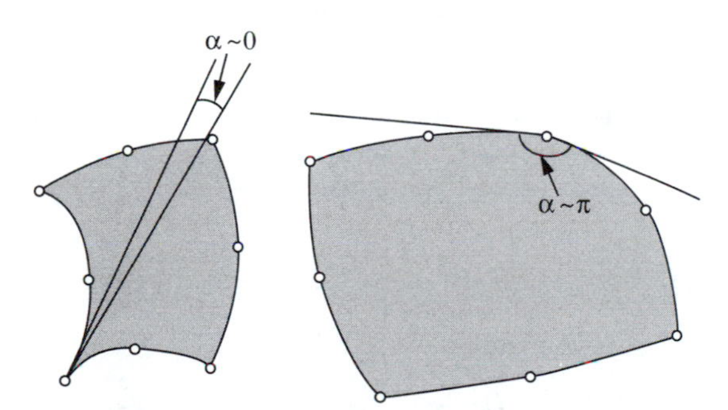

FIGURE 3–82 Undesirable Q8 geometry types

The character of the transformation in the interior of the mapped region R_{xy} is determined by the determinant of the Jacobian matrix

$$|\mathbf{J}(s,t)| = \begin{vmatrix} \dfrac{\partial x}{\partial s} & \dfrac{\partial y}{\partial s} \\ \dfrac{\partial x}{\partial t} & \dfrac{\partial y}{\partial t} \end{vmatrix} = \frac{\partial x}{\partial s}\frac{\partial y}{\partial t} - \frac{\partial y}{\partial s}\frac{\partial x}{\partial t}$$

$$= \left(\mathbf{x}_\mathbf{e}^\mathrm{T}\frac{\partial \mathbf{N}}{\partial s}\right)\left(\mathbf{y}_\mathbf{e}^\mathrm{T}\frac{\partial \mathbf{N}}{\partial t}\right) - \left(\mathbf{y}_\mathbf{e}^\mathrm{T}\frac{\partial \mathbf{N}}{\partial s}\right)\left(\mathbf{x}_\mathbf{e}^\mathrm{T}\frac{\partial \mathbf{N}}{\partial t}\right)$$

which in general is a rather messy function of s and t. The Jacobian $|\mathbf{J}(s,t)|$ is again an indication of the expansion or contraction of an area element of D_{st} according to

$$dx\,dy = |\mathbf{J}(s,t)|\,ds\,dt$$

Example 3.7.

Consider the Q8 element indicated in Fig. 3–83.

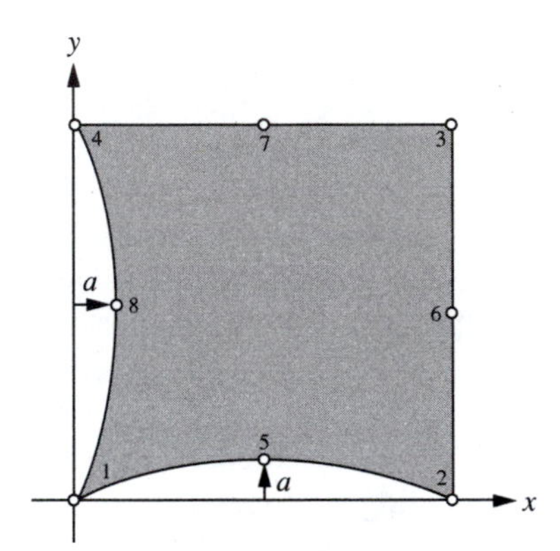

FIGURE 3–83 Specific Q8 geometry

The purpose of this example is to investigate the range of values of the parameter a for which the Jacobian is everywhere positive in D_{st}. The nodal coordinate vectors are given by

$$\mathbf{x}_\mathbf{e}^\mathrm{T} = [0 \quad 1 \quad 1 \quad 0 \quad 0.5 \quad 1 \quad 0.5 \quad a]$$

and

$$\mathbf{y}_\mathbf{e}^\mathrm{T} = [0 \quad 0 \quad 1 \quad 1 \quad a \quad 0.5 \quad 1 \quad 0.5]$$

It is left to the Exercises to show that

$$|\mathbf{J}(s,t)| = \begin{vmatrix} \dfrac{(1-a)}{2} + \dfrac{at^2}{2} & -as(1-t) \\ -at(1-s) & \dfrac{(1-a)}{2} + \dfrac{as^2}{2} \end{vmatrix}$$

Evaluating the Jacobian at $s = t = -1$, the point at which $|\mathbf{J}|$ is expected to first become negative for increasing a, results in

$$|\mathbf{J}(-1,-1)| = 0.25 - 4a^2$$

It follows that $|\mathbf{J}(-1,-1)|$ first vanishes when $a = \pm 0.25$. These values correspond to the geometries shown in Fig. 3–84. Each can easily be shown to represent a situation where there is a common tangent to the two parabolas in question. The first case corresponds to an interior angle of 0 and the second to an interior angle of π, precisely the two geometries indicated previously.

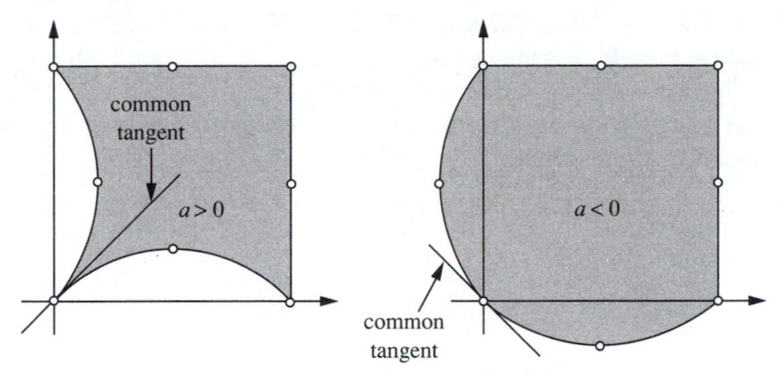

FIGURE 3–84 Geometries for limiting values of *a*

The area of the element is computed according to

$$A = \iint_{R_{xy}} dx\, dy = \int_{-1}^{1}\int_{-1}^{1} |\mathbf{J}|\, ds\, dt$$

$$= \int_{-1}^{1}\int_{-1}^{1}\left[\left(\frac{(1-a)}{2} + \frac{at^2}{2}\right)\left(\frac{(1-a)}{2} + \frac{as^2}{2}\right) - a^2 st(1-s)(1-t)\right] ds\, dt$$

$$= 1 - \frac{4a}{3}$$

which can be verified to be correct by the elementary means of using the integral calculus.

Derived variables. As stated several times previously, the derived variables are $\partial\phi/\partial x$ and $\partial\phi/\partial y$. These are frequently combined to yield the normal derivative given by

$$\frac{\partial\phi}{\partial n} = n_x\frac{\partial\phi}{\partial x} + n_y\frac{\partial\phi}{\partial y} = (n_x\mathbf{J}_1 + n_y\mathbf{J}_2)\mathbf{\Delta}^{\mathrm{T}}\boldsymbol{\phi}_{\mathbf{e}}$$

which is a function of position within the element. The derived variables are frequently computed directly at the nodes. However, as mentioned previously Moan [8] has shown that the accuracy of the derived variable is best when computed at certain special locations within the element. For the Q8 element these points are the $N = 2$ Gauss points, $s = \pm\sqrt{1/3}$ and $t = \pm\sqrt{1/3}$.

Based on the values of the derived variables computed at the Gauss points, a technique referred to as smoothing [Hinton and Campbell, 10] can be used to extrapolate the derived variables to the nodes, generally resulting in better nodal estimates than if the derived variables are computed directly at the nodes. The reader is referred to Section 5.8 and to the Exercises for further details.

3.8.2.2 EVALUATION OF Q8 ELEMENTAL MATRICES. The matrices that need to be evaluated are $\mathbf{k_e}$, $\mathbf{f_e}$, $\mathbf{h_e}$, and $\mathbf{a_e}$. $\mathbf{k_e}$ and $\mathbf{f_e}$ are defined as area integrals over the region R_{xy} with $\mathbf{h_e}$ and $\mathbf{a_e}$ as line integrals over a typical segment γ_{2e}. Essentially, $\mathbf{h_e}$ and $\mathbf{a_e}$ can be evaluated by comparison with results of Chapter 2, whereas for $\mathbf{k_e}$ and $\mathbf{f_e}$, a Gaussian quadrature is required.

Evaluation of $\mathbf{h_e}$, $\mathbf{a_e}$. Consider a typical γ_{2e} segment of the Γ_2 portion of the boundary as indicated in Fig. 3–85.

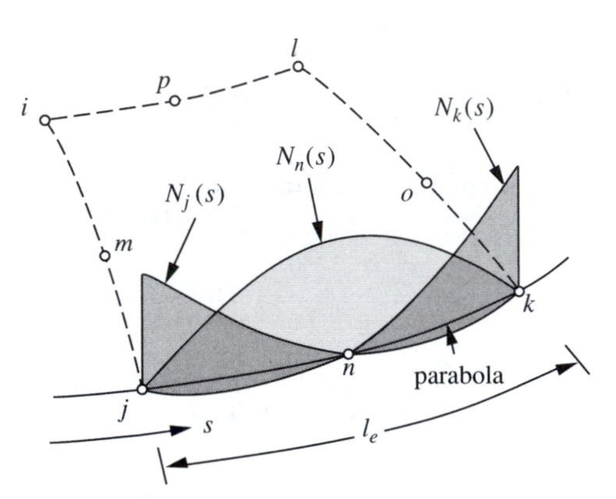

FIGURE 3–85 Q8 interpolation functions on a typical γ_{2e}

On γ_{2e}, the three nonzero interpolation functions are

$$N_j(s) = \left(\frac{2s}{l_e} - 1\right)\left(\frac{s}{l_e} - 1\right)$$

$$N_n(s) = 4\left(\frac{s}{l_e}\right)\left(1 - \frac{s}{l_e}\right)$$

$$N_k(s) = \left(\frac{s}{l_e}\right)\left(\frac{2s}{l_e} - 1\right)$$

Note that s is a measure of the *arc length* and l_e is the length of the approximating parabola passing through nodes j, n, and k. This representation for the N's on γ_{2e} assumes that the node n is positioned at the midpoint of the arc length l_e in question. When defining the mesh, care should be taken so as to position the midside nodes as accurately as possible in this regard.

It is assumed that on Γ_2, $h(s)$ is sufficiently smooth so as to allow the approximation

$$h(s) = N_j h_j + N_n h_n + N_k h_k = \mathbf{N}^{\mathrm{T}}\mathbf{h}$$

where $\mathbf{h} = [h_j \quad h_n \quad h_k]$ is a vector of suitably chosen nodal values for the function $h(s)$. On the basis of this approximation $\mathbf{h_e}$ can be represented as

$$\mathbf{h_e} = \left[\int_0^{l_e} \mathbf{N}\mathbf{N}^{\mathrm{T}}\, ds\right]\mathbf{h} = \left\{\int_0^{l_e} \begin{bmatrix} N_jN_j & N_jN_n & N_jN_k \\ N_nN_j & N_nN_n & N_nN_k \\ N_kN_j & N_kN_n & N_kN_k \end{bmatrix} ds\right\}\mathbf{h}$$

These integrals were encountered in Section 2.8 in the course of evaluating the elemental matrices $\mathbf{f_e}$. From those results, it follows that

$$\mathbf{h_e} = \begin{bmatrix} 4h_j + 2h_n - \ \ h_k \\ 2h_j + 16h_n + 2h_k \\ -h_j + 2h_n + 4h_k \end{bmatrix} \frac{l_e}{30}$$

In a completely similar fashion, representing $\alpha(s)$ as

$$\alpha(s) = \alpha_j N_j + \alpha_n N_n + \alpha_k N_k = \mathbf{N}^{\mathrm{T}}\boldsymbol{\alpha}$$

leads to

$$\mathbf{a_e} = \left[\int_0^{l_e} \mathbf{N}(\alpha_j N_j + \alpha_n N_n + \alpha_k N_k)\mathbf{N}^{\mathrm{T}}\, ds\right]$$

Again referring to the results of Section 2.8, specifically to the evaluation of $\mathbf{q_e}$, it follows that

$$[\mathbf{a_e}]_{11} = \frac{l_e}{420}(39\alpha_j + 20\alpha_n - 3\alpha_k)$$

$$[\mathbf{a_e}]_{12} = \frac{l_e}{420}(20\alpha_j + 16\alpha_n - 8\alpha_k)$$

$$[\mathbf{a_e}]_{13} = \frac{l_e}{420}(-3\alpha_j - 8\alpha_n - 3\alpha_k)$$

$$[\mathbf{a_e}]_{22} = \frac{l_e}{420}(16\alpha_j + 192\alpha_n + 16\alpha_k)$$

$$[\mathbf{a_e}]_{23} = \frac{l_e}{420}(-8\alpha_j + 16\alpha_n + 20\alpha_k)$$

$$[\mathbf{a_e}]_{33} = \frac{l_e}{420}(-3\alpha_j + 20\alpha_n + 39\alpha_k)$$

with $[\mathbf{a_e}]_{ij} = [\mathbf{a_e}]_{ji}$.

For both of $\mathbf{h_e}$ and $\mathbf{a_e}$, l_e is the *arc length* of the parabola passing through nodes j, n, and k and should be so computed. To this end consider the expression for the differential arc length on γ_{2e},

$$(dl_e)^2 = (dx)^2 + (dy)^2$$

where x and y are expressed in terms of the natural coordinate $-1 \leq u \leq 1$ as indicated in Fig. 3–86.

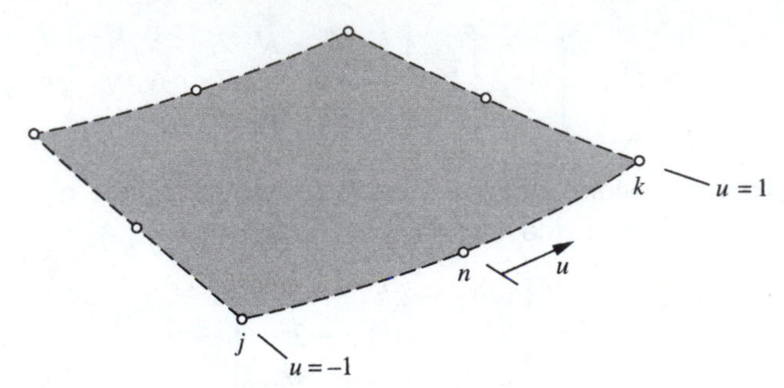

FIGURE 3–86 Natural coordinates on γ_{2e}

Specifically, $x(u)$ and $y(u)$ are given by

$$x = x(u) = x_j N_j + x_n N_n + x_k N_k = \mathbf{x}_\mathbf{e}^\mathrm{T}\mathbf{N}(u)$$

$$y = y(u) = y_j N_j + y_n N_n + y_k N_k = \mathbf{y}_\mathbf{e}^\mathrm{T}\mathbf{N}(u)$$

where

$$N_j(u) = \frac{u(u-1)}{2}$$

$$N_n(u) = 1 - u^2$$

$$N_k(u) = \frac{u(u+1)}{2}$$

It follows that

$$dl_e = \left((x'(u))^2 + (y'(u))^2\right)^{1/2} du$$

where

$$x'(u) = \frac{(x_k - x_j)}{2} + u(x_j - 2x_n + x_k)$$

and

$$y'(u) = \frac{(y_k - y_j)}{2} + u(y_j - 2y_n + y_k)$$

so that the arc length is given by

$$l_e = \int_{-1}^{1} \left((x'(u))^2 + (y'(u))^2\right)^{1/2} du$$

For element geometries fitting within the framework of those discussed in the previous section, l_e can be evaluated with sufficient accuracy using a two-term Gaussian quadrature. This yields

$$l_e = \left(\left(x'\left(\frac{-1}{\sqrt{3}}\right)\right)^2 + \left(y'\left(\frac{-1}{\sqrt{3}}\right)\right)^2\right)^{1/2} + \left(\left(x'\left(\frac{1}{\sqrt{3}}\right)\right)^2 + \left(y'\left(\frac{1}{\sqrt{3}}\right)\right)^2\right)^{1/2}$$

which can easily be computed given the coordinates of nodes j, n, and k.

Evaluation of $\mathbf{f_e}$. The general expression for $\mathbf{f_e}$ is

$$\mathbf{f_e} = \int\int_{R_{xy}} \mathbf{N} f \, dx \, dy$$

which can be represented in terms of the variables s and t and the domain D_{st} of the parent element as

$$\mathbf{f_e} = \int_{-1}^{1}\int_{-1}^{1} \mathbf{N}(s,t)\, f\big(x(s,t), y(s,t)\big)\, |\mathbf{J}| \, ds \, dt$$

We again assume f is sufficiently regular so as to be able to accurately approximate f according to

$$f \approx \sum_{1}^{8} f_i N_i = \mathbf{N}^{\mathrm{T}}\mathbf{f}$$

leading to

$$\mathbf{f_e} \approx \left[\int_{-1}^{1}\int_{-1}^{1} \mathbf{N}(s,t)\mathbf{N}^{\mathrm{T}}(s,t)\, |\mathbf{J}| \, ds \, dt\right]\mathbf{f}$$

The vector $\mathbf{f} = [f_1 \quad f_2 \quad f_3 \quad f_4 \quad f_5 \quad f_6 \quad f_7 \quad f_8]^{\mathrm{T}}$ is usually taken to be the values of $f(x,y)$ at the nodes. Determining $\mathbf{f_e}$ is thus reduced to evaluating the integral

$$\left[\int_{-1}^{1}\int_{-1}^{1} \mathbf{N}(s,t)\mathbf{N}^{\mathrm{T}}(s,t)\, |\mathbf{J}| \, ds \, dt\right]$$

We will investigate this issue further in the examples that follow.

Evaluation of $\mathbf{k_e}$. The general expression for $\mathbf{k_e}$ is

$$\mathbf{k_e} = \int\int_{R_{xy}} \left[\frac{\partial \mathbf{N}}{\partial x}\frac{\partial \mathbf{N}^{\mathrm{T}}}{\partial x} + \frac{\partial \mathbf{N}}{\partial y}\frac{\partial \mathbf{N}^{\mathrm{T}}}{\partial y}\right] dx \, dy$$

Using the chain rule in the usual way, this can be converted into

$$\mathbf{k_e} = \int_{-1}^{1}\int_{-1}^{1} \mathbf{\Delta}(s,t)\mathbf{JJ}(s,t)\mathbf{\Delta}^{\mathrm{T}}(s,t) \, ds \, dt$$

where

$$\mathbf{\Delta}(s,t) = \left[\frac{\partial \mathbf{N}}{\partial s} \quad \frac{\partial \mathbf{N}}{\partial t} \right] = \begin{bmatrix} \dfrac{(1-t)(2s+t)}{4} & \dfrac{(1-s)(s+2t)}{4} \\ \dfrac{(1-t)(2s-t)}{4} & \dfrac{(1+s)(-s+2t)}{4} \\ \dfrac{(1+t)(2s+t)}{4} & \dfrac{(1+s)(s+2t)}{4} \\ \dfrac{(1+t)(2s-t)}{4} & \dfrac{(1-s)(-s+2t)}{4} \\ -s(1-t) & \dfrac{-(1-s^2)}{2} \\ \dfrac{(1-t^2)}{2} & -t(1+s) \\ -s(1+t) & \dfrac{(1-s^2)}{2} \\ \dfrac{-(1-t^2)}{2} & -t(1-s) \end{bmatrix}$$

and $\mathbf{JJ} = (\mathbf{J}_1^T\mathbf{J}_1 + \mathbf{J}_2^T\mathbf{J}_2)|\mathbf{J}|$, with $\mathbf{J}_1$ and $\mathbf{J}_2$ the first and second rows, respectively, of the inverse of the Jacobian matrix. The Jacobian matrix is given by

$$\mathbf{J} = \begin{bmatrix} \mathbf{x}_\mathbf{e}^T \dfrac{\partial \mathbf{N}}{\partial s} & \mathbf{y}_\mathbf{e}^T \dfrac{\partial \mathbf{N}}{\partial s} \\ \mathbf{x}_\mathbf{e}^T \dfrac{\partial \mathbf{N}}{\partial t} & \mathbf{y}_\mathbf{e}^T \dfrac{\partial \mathbf{N}}{\partial t} \end{bmatrix}$$

with $|\mathbf{J}|$ the determinant of the Jacobian matrix.

We are thus again confronted with the two integrals

$$\int_{-1}^{1}\int_{-1}^{1} \mathbf{N}(s,t)\mathbf{N}^T(s,t)\,|\mathbf{J}|\,ds\,dt$$

and

$$\int_{-1}^{1}\int_{-1}^{1} \mathbf{\Delta}(s,t)\mathbf{JJ}(s,t)\mathbf{\Delta}^T(s,t)\,ds\,dt$$

which are ideally suited for Gaussian quadrature. To this end we write

$$\mathbf{f_e} \approx \left[\sum_1^N \sum_1^N W_i W_j \mathbf{N}_{ij} \mathbf{N}_{ij}^T |\mathbf{J}|_{ij} \right] \mathbf{f}$$

and

$$\mathbf{k_e} \approx \sum_{1}^{N} \sum_{1}^{N} W_i W_j \mathbf{\Delta_{ij} J J_{ij} \Delta_{ij}^T}$$

where the subscripts refer to evaluation at the Gauss points $s = s_i$, $t = t_j$, the W_i to the weights, and N to the order. These will be discussed in the following examples.

Example 3.8.

Consider the evaluation of

$$\mathbf{f_e} = \int_{-1}^{1} \int_{-1}^{1} \mathbf{N}(s, t) f\left(x(s, t), y(s, t)\right) |\mathbf{J}|\, ds\, dt$$

and

$$\mathbf{f_e} \approx \left[\int_{-1}^{1} \int_{-1}^{1} \mathbf{N}(s, t) \mathbf{N}^{\mathrm{T}}(s, t) |\mathbf{J}|\, ds\, dt \right] \mathbf{f}$$

that is, the exact and approximate expressions for $\mathbf{f_e}$. Specifically, we will take $f(x, y) = x^2 y$ and R_{xy} as indicated in Fig. 3–87. The nodal coordinates in R_{xy} are given by

$$\mathbf{x_e^T} = [0.866 \quad 1.732 \quad 1.000 \quad 0.500 \quad 1.400 \quad 1.414 \quad 0.850 \quad 0.707]$$

and

$$\mathbf{y_e^T} = [0.500 \quad 1.000 \quad 1.732 \quad 0.866 \quad 0.577 \quad 1.414 \quad 1.200 \quad 0.707]$$

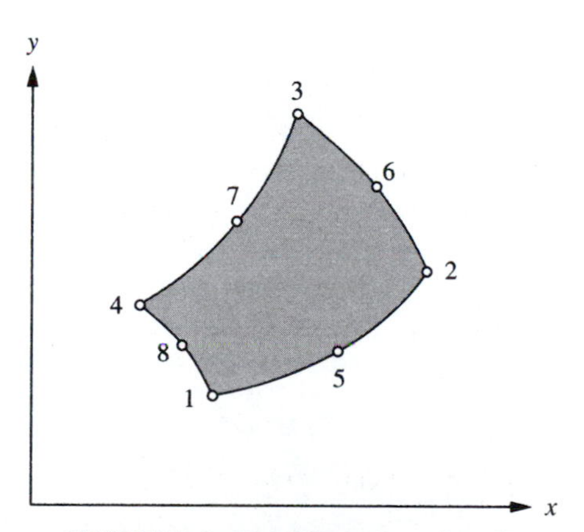

FIGURE 3–87 Typical region R_{xy}

Results using various orders of Gauss quadrature and the first or exact expression above for the evaluation of $\mathbf{f_e}$ are presented in Table 3.14. The results using Gauss quadrature and the approximate expression for $\mathbf{f_e}$ are presented in Table 3.15.

TABLE 3.14 Numerical evaluation of exact $\mathbf{f_e}$

$\mathbf{f_e}$	$N = 1$	$N = 2$	$N = 3$	$N = 4$	$N = 5$
f_{e1}	−0.25836	−0.13624	−0.12554	−0.12550	−0.12550
f_{e2}	−0.25836	−0.04925	−0.03981	−0.03976	−0.03976
f_{e3}	−0.25836	−0.07204	−0.06569	−0.06569	−0.06569
f_{e4}	−0.25836	−0.13606	−0.12649	−0.12651	−0.12651
f_{e5}	0.51672	0.41621	0.38530	0.38526	0.38525
f_{e6}	0.51672	0.54460	0.56176	0.56165	0.56165
f_{e7}	0.51672	0.37098	0.34683	0.34689	0.34689
f_{e8}	0.51672	0.24259	0.24471	0.24475	0.24475

TABLE 3.15 Numerical evaluation of approximate $\mathbf{f_e}$

$\mathbf{f_e}$	N = 1	$N = 2$	$N = 3$	$N = 4$	$N = 5$
f_{e1}	−0.26090	−0.13769	−0.12696	−0.12691	−0.12691
f_{e2}	−0.26090	−0.04926	−0.04036	−0.04025	−0.04025
f_{e3}	−0.26090	−0.07189	−0.06603	−0.06591	−0.06591
f_{e4}	−0.26090	−0.13525	−0.12662	−0.12650	−0.12650
f_{e5}	0.52180	0.41428	0.38541	0.38519	0.38519
f_{e6}	0.52180	0.54588	0.56291	0.56304	0.56304
f_{e7}	0.52180	0.37390	0.34984	0.34955	0.34955
f_{e8}	0.52180	0.24230	0.24561	0.24559	0.24559

These results indicate that for the element of Fig. 3–87, Gauss quadrature of at least order two should be used in the evaluation of the $\mathbf{f_e}$ integrals. For applications where the elements are substantially distorted from the parent element, it may be necessary to use $N = 3$.

Results from the two tables also indicate that for any N, the accuracy resulting from the use of the approximate expression is adequate, with errors on the order of 1 percent.

Example 3.9.

Consider also the evaluation of

$$\mathbf{k_e} = \int_{-1}^{1}\int_{-1}^{1} \boldsymbol{\Delta}\mathbf{J}\mathbf{J}\boldsymbol{\Delta}^{\mathrm{T}}\, ds\, dt$$

for several regions. The first region considered is the square indicated in Fig. 3–88.

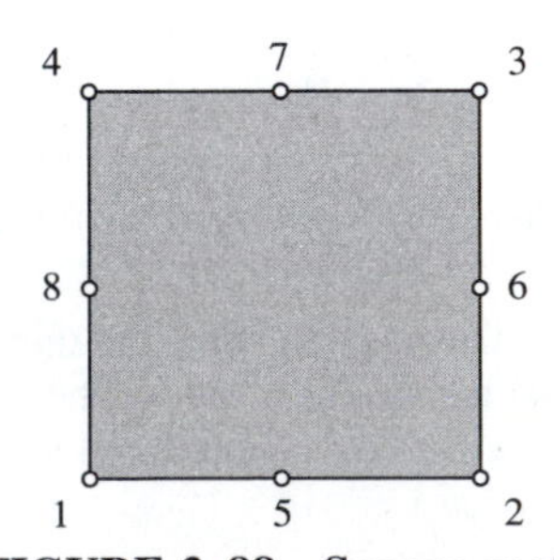

FIGURE 3–88 Square region

The resulting $\mathbf{k_e}$ using Gauss quadrature of order $N = 1$, 2, and 3, respectively, are indicated as follows.

$\mathbf{k_e}, N = 1$

$$
\begin{matrix}
0.0000 & 0.0000 & 0.0000 & 0.0000 & 0.0000 & 0.0000 & 0.0000 & 0.0000 \\
0.0000 & 0.0000 & 0.0000 & 0.0000 & 0.0000 & 0.0000 & 0.0000 & 0.0000 \\
0.0000 & 0.0000 & 0.0000 & 0.0000 & 0.0000 & 0.0000 & 0.0000 & 0.0000 \\
0.0000 & 0.0000 & 0.0000 & 0.0000 & 0.0000 & 0.0000 & 0.0000 & 0.0000 \\
0.0000 & 0.0000 & 0.0000 & 0.0000 & 1.0000 & 0.0000 & -1.0000 & 0.0000 \\
0.0000 & 0.0000 & 0.0000 & 0.0000 & 0.0000 & 1.0000 & 0.0000 & -1.0000 \\
0.0000 & 0.0000 & 0.0000 & 0.0000 & -1.0000 & 0.0000 & 1.0000 & 0.0000 \\
0.0000 & 0.0000 & 0.0000 & 0.0000 & 0.0000 & -1.0000 & 0.0000 & 1.0000
\end{matrix}
$$

$\mathbf{k_e}, N = 2$

$$
\begin{matrix}
1.1111 & 0.5000 & 0.5556 & 0.5000 & -0.7778 & -0.5556 & -0.5556 & -0.7778 \\
0.5000 & 1.1111 & 0.5000 & 0.5556 & -0.7778 & -0.7778 & -0.5556 & -0.5556 \\
0.5556 & 0.5000 & 1.1111 & 0.5000 & -0.5556 & -0.7778 & -0.7778 & -0.5556 \\
0.5000 & 0.5556 & 0.5000 & 1.1111 & -0.5556 & -0.5556 & -0.7778 & -0.7778 \\
-0.7778 & -0.7778 & -0.5556 & -0.5556 & 2.2222 & 0.0000 & 0.4444 & 0.0000 \\
-0.5556 & -0.7778 & -0.7778 & -0.5556 & 0.0000 & 2.2222 & 0.0000 & 0.4444 \\
-0.5556 & -0.5556 & -0.7778 & -0.7778 & 0.4444 & 0.0000 & 2.2222 & 0.0000 \\
-0.7778 & -0.5556 & -0.5556 & -0.7778 & 0.0000 & 0.4444 & 0.0000 & 2.2222
\end{matrix}
$$

$\mathbf{k_e}, N = 3$

$$
\begin{matrix}
1.1556 & 0.5000 & 0.5111 & 0.5000 & -0.8222 & -0.5111 & -0.5111 & -0.8222 \\
0.5000 & 1.1556 & 0.5000 & 0.5111 & -0.8222 & -0.8222 & -0.5111 & -0.5111 \\
0.5111 & 0.5000 & 1.1556 & 0.5000 & -0.5111 & -0.8222 & -0.8222 & -0.5111 \\
0.5000 & 0.5111 & 0.5000 & 1.1556 & -0.5111 & -0.5111 & -0.8222 & -0.8222 \\
-0.8222 & -0.8222 & -0.5111 & -0.5111 & 2.3111 & 0.0000 & 0.3556 & 0.0000 \\
-0.5111 & -0.8222 & -0.8222 & -0.5111 & 0.0000 & 2.3111 & 0.0000 & 0.3556 \\
-0.5111 & -0.5111 & -0.8222 & -0.8222 & 0.3556 & 0.0000 & 2.3111 & 0.0000 \\
-0.8222 & -0.5111 & -0.5111 & -0.8222 & 0.0000 & 0.3556 & 0.0000 & 2.3111
\end{matrix}
$$

For a square or rectangular element, it turns out that **JJ** is a constant and the integrals can be evaluated exactly. The exact results coincide with the ones given above for $N = 3$. This fact can be deduced by considering the order of a typical integrand of $\mathbf{k_e}$. Each integrand is generally the sum of products of quadratic functions and is hence quartic. $N = 2(2 \times 2 - 1 = 3)$ is not sufficient whereas $N = 3(2 \times 3 - 1 = 5)$ integrates all such terms exactly. Note that the $N = 2$ results agree reasonably well with the exact results, with errors on the order of 5 percent.

The results for this simplest geometry are presented to indicate that even when there is no distortion of the parent element as a result of the transformation (the element is square or rectangular), Gauss quadrature of order three is required for exact evaluation of the integrals. Generally, for any other Q8 element, $\mathbf{k_e}$ is evaluated only approximately using Gauss quadrature of order three.

Consider in this regard the evaluation of $\mathbf{k_e}$ for the distorted Q8 element shown in Figure 3–89.

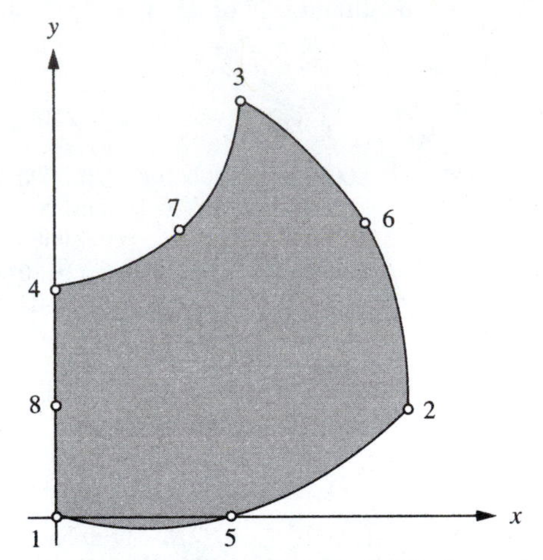

FIGURE 3–89 Distorted Q8 element

The nodal coordinates in R_{xy} are given by

$$\mathbf{x}_e^T = [0.000 \quad 0.400 \quad 0.200 \quad 0.000 \quad 0.200 \quad 0.350 \quad 0.125 \quad 0.000]$$

and

$$\mathbf{y}_e^T = [0.000 \quad 0.100 \quad 0.400 \quad 0.200 \quad 0.000 \quad 0.300 \quad 0.275 \quad 0.100]$$

The results for $N = 1$ through 5 are presented as follows.

$\mathbf{k}_e, N = 1$

0.0000	0.0000	0.0000	0.0000	0.0000	0.0000	0.0000	0.0000
0.0000	0.0000	0.0000	0.0000	0.0000	0.0000	0.0000	0.0000
0.0000	0.0000	0.0000	0.0000	0.0000	0.0000	0.0000	0.0000
0.0000	0.0000	0.0000	0.0000	0.0000	0.0000	0.0000	0.0000
0.0000	0.0000	0.0000	0.0000	1.4607	0.2584	−1.4607	−0.2584
0.0000	0.0000	0.0000	0.0000	0.2584	0.7303	−0.2584	−0.7303
0.0000	0.0000	0.0000	0.0000	−1.4607	−0.2584	1.4607	0.2584
0.0000	0.0000	0.0000	0.0000	−0.2584	−0.7303	0.2584	0.7303

$\mathbf{k}_e, N = 2$

1.1436	0.5866	0.5915	0.6910	−0.4144	−0.7166	−0.5541	−1.3276
0.5866	1.7198	0.6482	0.7910	−0.8453	−1.2246	−0.6924	−0.9833
0.5915	0.6482	1.1066	0.5360	−0.5282	−0.9507	−0.6045	−0.7990
0.6910	0.7910	0.5360	1.4709	−0.6462	−0.7886	−0.5468	−1.5072
−0.4144	−0.8453	−0.5282	−0.6462	1.8416	0.3441	0.1733	0.0750
−0.7166	−1.2246	−0.9507	−0.7886	0.3441	2.3293	−0.0326	1.0397
−0.5541	−0.6924	−0.6045	−0.5468	0.1733	−0.0326	1.9302	0.3269
−1.3276	−0.9833	−0.7990	−1.5072	0.0750	1.0397	0.3269	3.1754

$\mathbf{k_e}, N = 3$

1.2600	0.5661	0.5000	0.6128	−0.5111	−0.6236	−0.4477	−1.3566
0.5661	1.7593	0.5989	0.6741	−0.9203	−1.2392	−0.5790	−0.8599
0.5000	0.5989	1.1412	0.5169	−0.4436	−0.9390	−0.6702	−0.7043
0.6128	0.6741	0.5169	1.5076	−0.5297	−0.6772	−0.6405	−1.4640
−0.5111	−0.9203	−0.4436	−0.5297	2.0027	0.3573	−0.0167	0.0614
−0.6236	−1.2392	−0.9390	−0.6772	0.3573	2.3474	−0.0763	0.8506
−0.4477	−0.5790	−0.6702	−0.6405	−0.0167	−0.0763	2.0986	0.3316
−1.3566	−0.8599	−0.7043	−1.4640	0.0614	0.8506	0.3316	3.1410

$\mathbf{k_e}, N = 4$

1.2706	0.5673	0.4966	0.6137	−0.5173	−0.6215	−0.4421	−1.3674
0.5673	1.7594	0.5987	0.6738	−0.9209	−1.2391	−0.5782	−0.8610
0.4966	0.5987	1.1428	0.5165	−0.4417	−0.9399	−0.6725	−0.7005
0.6137	0.6738	0.5165	1.5127	−0.5284	−0.6767	−0.6421	−1.4693
−0.5173	−0.9209	−0.4417	−0.5286	2.0065	0.3561	−0.0205	0.0664
−0.6215	−1.2391	−0.9399	−0.6767	0.3561	2.3481	−0.0753	0.8485
−0.4421	−0.5782	−0.6725	−0.6421	−0.0205	−0.0753	2.1035	0.3274
−1.3674	−0.8610	−0.7005	−1.4693	0.0664	0.8485	0.3274	3.1559

$\mathbf{k_e}, N = 5$

1.2713	0.5674	0.4964	0.6136	−0.5177	−0.6214	−0.4418	−1.3679
0.5674	1.7594	0.5987	0.6738	−0.9210	−1.2392	−0.5781	−0.8610
0.4964	0.5987	1.1429	0.5165	−0.4416	−0.9400	−0.6726	−0.7003
0.6136	0.6738	0.5165	1.5130	−0.5285	−0.6767	−0.6423	−1.4695
−0.5177	−0.9210	−0.4416	−0.5285	2.0068	0.3560	−0.0208	0.0667
−0.6214	−1.2392	−0.9400	−0.6767	0.3560	2.3482	−0.0753	0.8483
−0.4418	−0.5781	−0.6726	−0.6423	−0.0208	−0.0753	2.1037	0.3272
−1.3679	−0.8610	−0.7003	−1.4695	0.0667	0.8483	0.3272	3.1566

Here it is seen that even Gauss quadrature of order $N = 5$ is not sufficient to evaluate the integrals exactly. Comparison of the $N = 4$ and $N = 5$ results indicate that either of the $N = 4$ or the $N = 5$ results are approximately correct. Based on this observation, the $N = 2$ and $N = 3$ results exhibit errors of the order of 20 percent and up to 1 percent, respectively. For substantially distorted elements $N = 3$ is usually appropriate, whereas for geometries mildly distorted from rectangular, $N = 2$ may suffice. In practice, experience will dictate whether Gauss quadrature of order $N = 2$ or $N = 3$ is required.

We note in passing that when the Q8 element is rectangular or square, it is essentially a subparametric element in that the geometry can be determined on the basis of only the corner nodes, that is, only the four shape functions for the Q4 element are required. The Jacobian matrix is a diagonal matrix with constant entries, such that the integrand of $\mathbf{k_e}$ is substantially simplified.

3.8.2.3 APPLICATION—STRESS CONCENTRATION FOR THE TORSION OF AN ANGLE SECTION. Consider again the problem of the torsion of the angle section, presented in Section 3.8.1.3. The boundary value problem is indicated in Fig. 3–90.

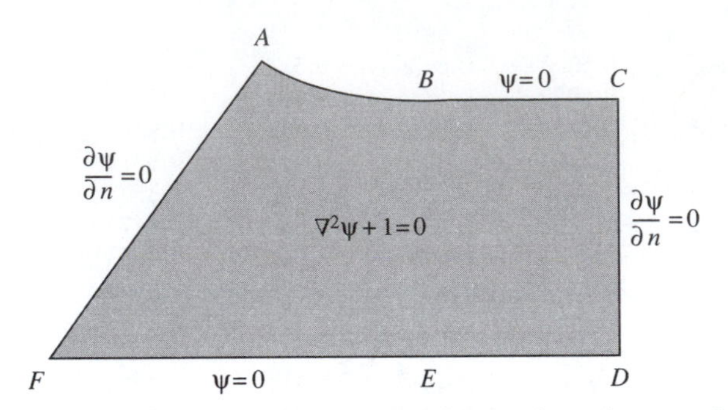

FIGURE 3–90 Modeled region for torsion of an angle

The mesh that will be used in connection with the Q8 elements is shown in Fig. 3–91.

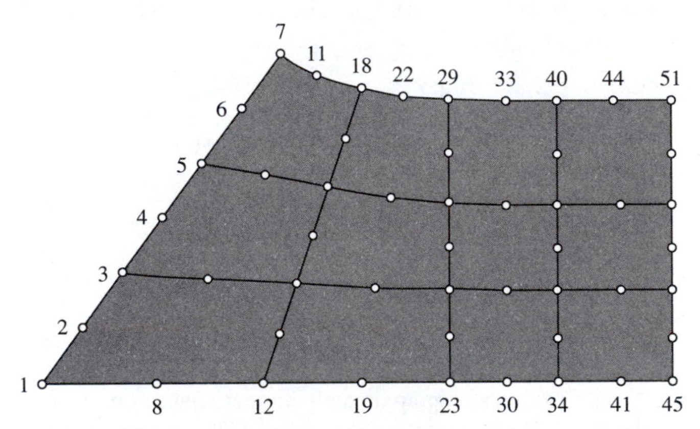

FIGURE 3–91 Q8 mesh for torsion of angle section

Partial results for ψ are presented in Table 3.16 and displayed in Figs. 3–92*a* and *b*.

TABLE 3.16 ψ along *DC* and *FA*

x/l_e	ψ_{DC}	ψ_{FA}
0	0.0000	0.0000
1/6	0.0696	0.0586
2/6	0.1147	0.1299
3/6	0.1254	0.1982
4/6	0.1114	0.2162
5/6	0.0698	0.1587
1	0.0000	0.0000

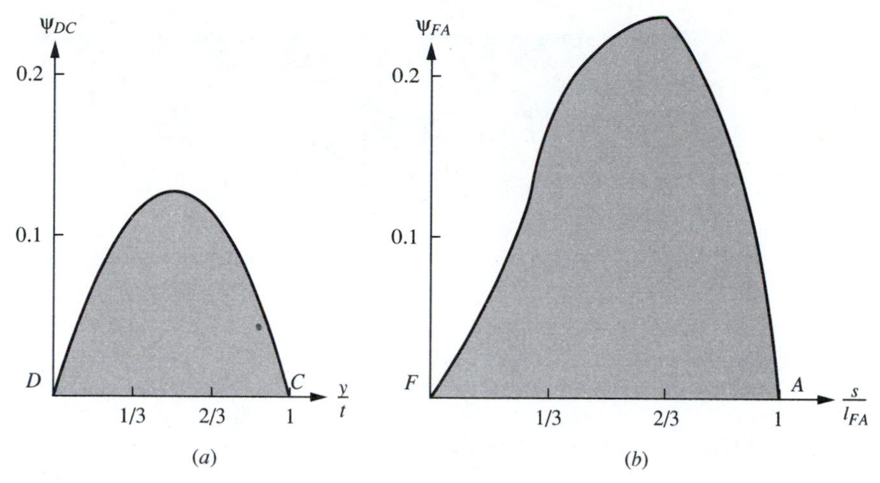

FIGURE 3–92 ψ along *DC* and *FA*

The maximum stress along FA is clearly at A and is proportional to the slope of the ψ surface along line FA at A. Similarly, the maximum stress along DC is proportional to the slope of the ψ surface along line DC at either D or C. Computing the slopes at points C and A respectively yields the values

$$\frac{\partial \psi}{\partial n} \text{ at } C = -0.5029$$

and

$$\frac{\partial \psi}{\partial n} \text{ at } A = -0.6872$$

with the resulting approximate stress concentration factor of

$$k_T = \frac{0.6872}{0.5029} = 1.37$$

As indicated in Section 3.8.1.3, this is to be compared with the tabulated experimental value of $k_T = 1.4$ [Griffith and Taylor, 9].

3.8.2.4 APPLICATION—STEADY-STATE HEAT TRANSFER. Consider again the steady-state heat transfer application discussed previously in Section 3.8.1.4. Using the same mesh as in that example, Q8 elements are as defined in Fig. 3–93.

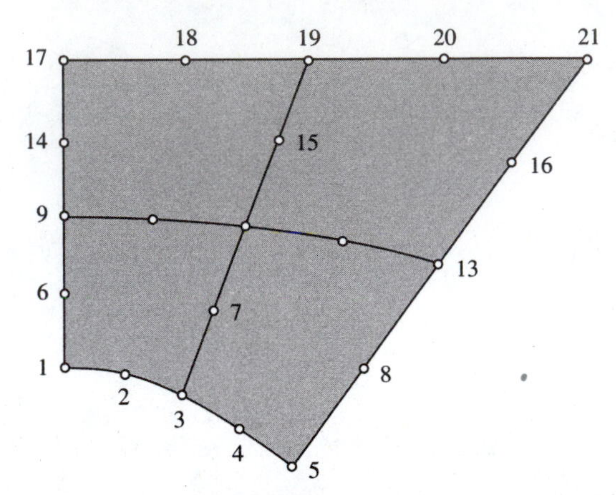

FIGURE 3–93 Q8 elements for heat transfer application

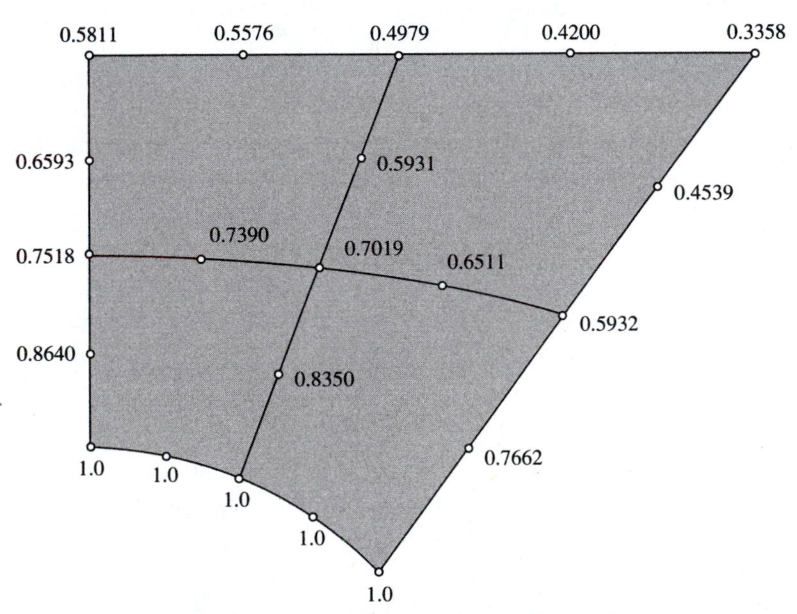

FIGURE 3–94 Nodal t values—Q8 model

The resulting nodal values for t are presented in Fig. 3–94. Figure 3–95 shows the values of the elemental normal derivatives at the nodes on the Γ_2 portion of the boundary, which are calculated, as outlined in Section 3.8.2.1, according to

$$\frac{\partial t}{\partial n} = (n_x \mathbf{J_1} + n_y \mathbf{J_2}) \mathbf{\Delta}^{\mathbf{T}} \mathbf{t_e}$$

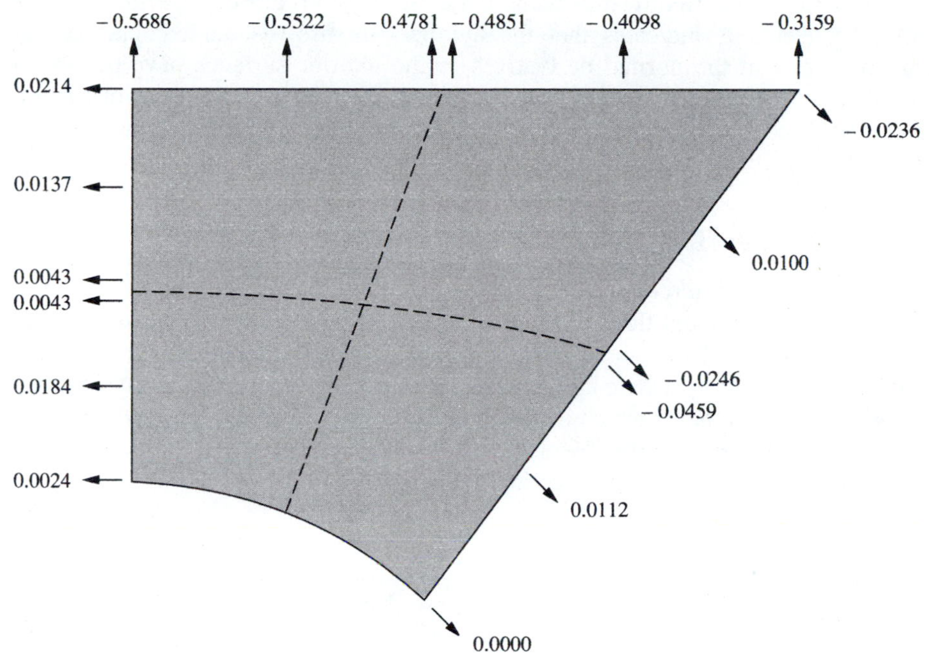

FIGURE 3–95 $\partial t/\partial n$ **on the** Γ_2 **portion of the boundary**

For this model using Q8 elements, the values of the normal derivatives on the sides defined by nodes 1, 6, 9, 14, and 17 and by nodes 5, 8, 13, 16, and 21 are quite small, comparing favorably with the correct values of zero.

Averaging the values of the normal derivatives computed in adjacent elements at node 19, the satisfaction of the natural boundary conditions along the line defined by nodes 17 through 21 can be checked as follows.

Node 17: $$\frac{\partial t}{\partial n} + t = -0.5686 + 0.5811 = 0.0125$$

Node 18: $$\frac{\partial t}{\partial n} + t = -0.5522 + 0.5576 = 0.0054$$

Node 19: $$\frac{\partial t}{\partial n} + t = -0.4816 + 0.4979 = 0.0163$$

Node 20: $$\frac{\partial t}{\partial n} + t = -0.4098 + 0.4200 = 0.0102$$

Node 21: $$\frac{\partial t}{\partial n} + t = -0.3159 + 0.3358 = 0.0199$$

These checks are a good indication that the solution is quite accurate.

Comparing the current results using Q8 elements with those of Section 3.8.1.4 using Q4 elements indicates that the numbers themselves are comparable but that the errors in the normal derivatives in the interior surfaces have decreased significantly. The errors in satisfying the natural boundary on the exterior surface have decreased also. We conclude for this example that a model using four Q8 elements is superior to a model using 16 Q4 elements.

3.8.3 Six-Node Triangular (T6) Elements

The quadratically interpolated triangular element is an element with six nodes, three at the vertices and three at the middle of each side. The sides can be either straight or curved. This can be useful in modeling particular geometries such as indicated in Fig. 3–96, where it is desirable to use a T6 element rather than a group of Q8 elements, some of which may be unduly deformed from rectangular. In this section we will develop the T6 element and indicate its use in the applications considered in the previous two sections.

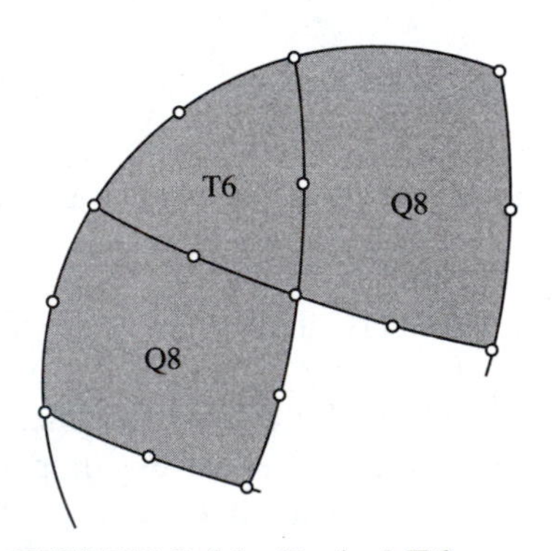

FIGURE 3–96 Typical T6 usage

3.8.3.1 TRANSFORMATIONS AND SHAPE FUNCTIONS. There are several approaches to developing the interpolation or shape functions for the quadratically interpolated triangular element. The first approach to be considered here is based on representing the geometry and dependent variable as functions of the appropriate monomials in terms of the global coordinates x and y, whereas the second approach begins with a parent element where the interpolation and shape functions are expressed in terms of the area coordinates introduced in Section 3.4. Both approaches are instructive. For the first approach, consider a typical straight-sided triangular element as shown in Fig. 3–97.

The six nodes consist of those at the vertices and those at the midpoints of the sides. The form of the solution is taken to be

$$u_e(x, y) = a + bx + cy + dx^2 + exy + fy^2 \tag{3.61}$$

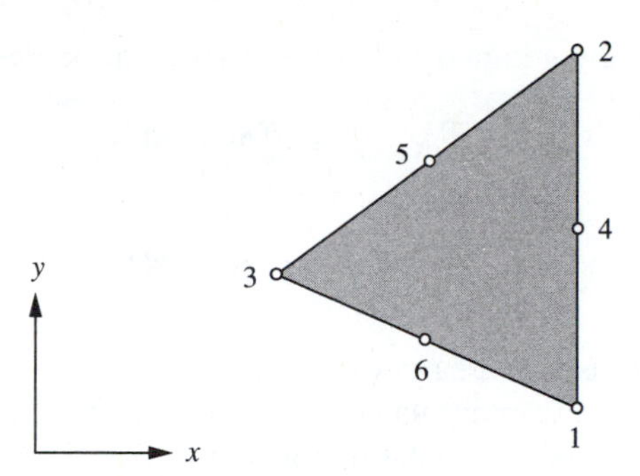

FIGURE 3–97 Typical straight-sided T6 triangular element

that is, all the monomials up to and including those of second degree. These six monomials match the required number of degrees of freedom for the element. In the usual way we require that

$$u(x_i, y_i) = u_i = a + bx_i + cy_i + dx_i^2 + ex_iy_i + fy_i^2 \qquad i = 1, \ldots, 6$$

be satisfied. This leads to a 6×6 set of linear algebraic equations from which the six constants c_i can be determined. Back-substituting the results into Eq. (3.61), we have

$$u_e(x, y) = \mathbf{u}_\mathbf{e}^\mathrm{T}\mathbf{N} = \mathbf{N}^\mathrm{T}u_\mathbf{e}$$

where $\mathbf{u}_\mathbf{e}^\mathrm{T} = [u_1 \; u_2 \; u_3 \; u_4 \; u_5 \; u_6]$ and

$$\begin{aligned}
N_1 &= \frac{\big(x_{23}(y - y_3) - y_{23}(x - x_3)\big)\big(x_{46}(y - y_6) - y_{46}(x - x_6)\big)}{(x_{23}y_{13} - y_{23}x_{13})(x_{46}y_{16} - y_{46}x_{16})} \\
N_2 &= \frac{\big(x_{31}(y - y_1) - y_{31}(x - x_1)\big)\big(x_{54}(y - y_4) - y_{54}(x - x_4)\big)}{(x_{31}y_{21} - y_{31}x_{21})(x_{54}y_{24} - y_{54}x_{24}))} \\
N_3 &= \frac{\big(x_{21}(y - y_1) - y_{21}(x - x_1)\big)\big(x_{56}(y - y_6) - y_{56}(x - x_6)\big)}{(x_{21}y_{31} - y_{21}x_{31})(x_{56}y_{36} - y_{56}x_{36})} \\
N_4 &= \frac{\big(x_{31}(y - y_1) - y_{31}(x - x_1)\big)\big(x_{23}(y - y_3) - y_{23}(x - x_3)\big)}{(x_{31}y_{41} - y_{31}x_{41})(x_{23}y_{43} - y_{23}x_{43})} \\
N_5 &= \frac{\big(x_{31}(y - y_1) - y_{31}(x - x_1)\big)\big(x_{21}(y - y_1) - y_{21}(x - x_1)\big)}{(x_{31}y_{51} - y_{31}x_{51})(x_{21}y_{51} - y_{21}x_{51})} \\
N_6 &= \frac{\big(x_{21}(y - y_1) - y_{21}(x - x_1)\big)\big(x_{23}(y - y_3) - y_{23}(x - x_3)\big)}{(x_{21}y_{61} - y_{21}x_{61})(x_{23}y_{63} - y_{23}x_{63})}
\end{aligned} \tag{3.62}$$

where $x_{ij} = x_i - x_j$ and $y_{ij} = y_i - y_j$.

It also follows that the geometry of the element can be described according to

$$x = \sum_{1}^{3} x_i N_i = \mathbf{x}_\mathbf{e}^\mathrm{T}\mathbf{N} = \mathbf{N}^\mathrm{T}\mathbf{x}_\mathbf{e}$$

$$y = \sum_{1}^{3} y_i N_i = \mathbf{y}_\mathbf{e}^\mathrm{T}\mathbf{N} = \mathbf{N}^\mathrm{T}\mathbf{y}_\mathbf{e}$$

with the N_i the same as the area coordinates given in Section 3.4, so that the straight-sided quadratically interpolated triangular element can be considered to be subparametric. The six interpolation or shape functions expressed in terms of the global coordinates x and y and given by Eqs. (3.62) are rarely used in practice. Rather, the equivalent functions expressed in terms of the area coordinates introduced in Section 3.4 are preferred because of their simplicity of form and relative ease in evaluating the corresponding elemental matrices. The interpolation functions expressed in terms of the area coordinates will be developed in what follows.

The second approach for generating the interpolation or shape functions for the quadratically interpolated triangle involves the use of the local or area coordinates indicated in Fig. 3–98. The reader is referred to Section 3.4 for a review of area coordinates.

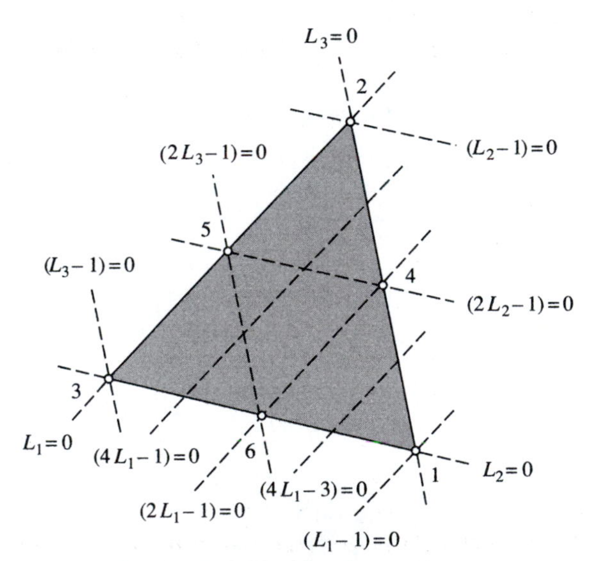

FIGURE 3–98 Area coordinates

Indicated in Fig. 3–98 are several sets of lines representing constant values of the area coordinates. In particular, $L_1 = 0$ containing nodes 2, 5, and 3, $L_1 = 1/2$ containing nodes 4 and 6, and $L_1 = 1$ containing node 1 are indicated. All other constant values of L_1 are proportionally spaced along the L_1 direction as indicated. Similar $L_2 =$ constant and $L_3 =$ constant lines are also shown.

The interpolation functions are formed as follows. For N_1, a function n_1 is formed first according to

$$n_1 = L_1(2L_1 - 1)$$

that is, the product of the function L_1 such that n_1 vanishes at nodes 2, 5, and 3, and the function $(2L_1 - 1)$ such that n_1 vanishes at nodes 4 and 6. Evaluating n_1 at node 1 where $L_1 = 1$ yields

$$n_1 = 1\big(2(1) - 1\big) = 1$$

It follows that the interpolation function N_1 is given by

$$N_1 = L_1(2L_1 - 1)$$

Thus as required, N_1 vanishes at nodes 2 through 6, and has the value one at node 1.

In a completely similar fashion it can be shown that

$$N_2 = L_2(2L_2 - 1)$$

and

$$N_3 = L_3(2L_3 - 1)$$

are the appropriate interpolation functions for nodes 2 and 3, respectively. N_1 is indicated in Fig. 3–99, with N_2 and N_3 similar in appearance with respect to nodes 2 and 3, respectively.

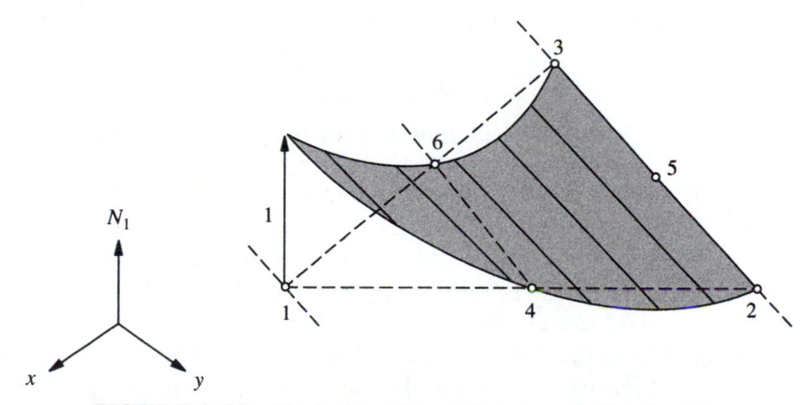

FIGURE 3–99 Typical vertex node interpolation function

As seen, N_1 is linear along L_1 = constant lines and quadratic along the direction perpendicular to the side defined by nodes 2, 5, and 3, that is, in the L_1 direction.

For the midside nodes, node 4 for instance, we first form

$$n_4 = L_1 L_2$$

The presence of the L_1 factor insures that n_4 vanishes at nodes 2, 5, and 3, with the presence of the L_2 factor such that n_4 vanishes at nodes 3, 6, and 1. Dividing n_4 by its value at point 4, which with $L_1 = L_2 = 1/2$, is $n_4(1/2, 1/2) = 1/4$, yields

$$N_4 = 4L_1L_2$$

which has all the desired properties for the interpolation function at node 4. In a completely similar fashion,

$$N_5 = 4L_2L_3$$

and

$$N_6 = 4L_3L_1$$

N_4 is indicated in Fig. 3–100, with N_5 and N_6 having similar shapes with respect to nodes 5 and 6.

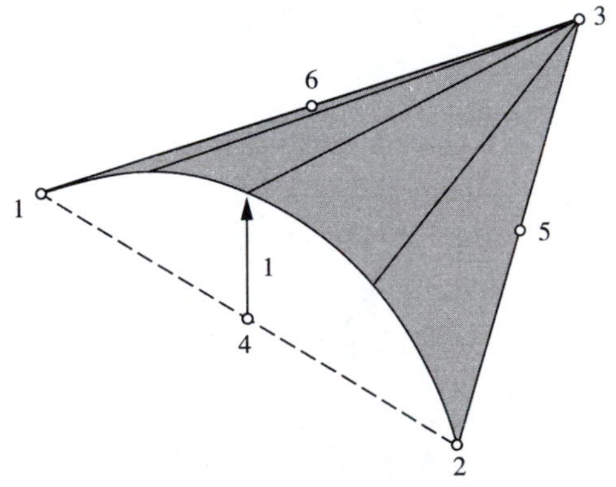

FIGURE 3–100 Typical midsize node interpolation function

N_1 through N_6 constitute the quadratic interpolation functions for the straight-sided triangle expressed in terms of the area coordinates.

Consider next the generation of the general curvilinear six-noded triangular element. The parent element is specially located as indicated in Fig. 3–101*a*, with the general curvilinear element shown in Fig. 3–101*b*.

Any edge or side of the curvilinear element can be either straight or curved. Generally, the positions of the nodes of the curvilinear element are specified, with the shape of the element then determined on the basis of the transformation, which is given by

$$x(L_1, L_2, L_3) = \sum_1^6 x_i N_i(L_1, L_2, L_3)$$

$$y(L_1, L_2, L_3) = \sum_1^6 y_i N_i(L_1, L_2, L_3)$$

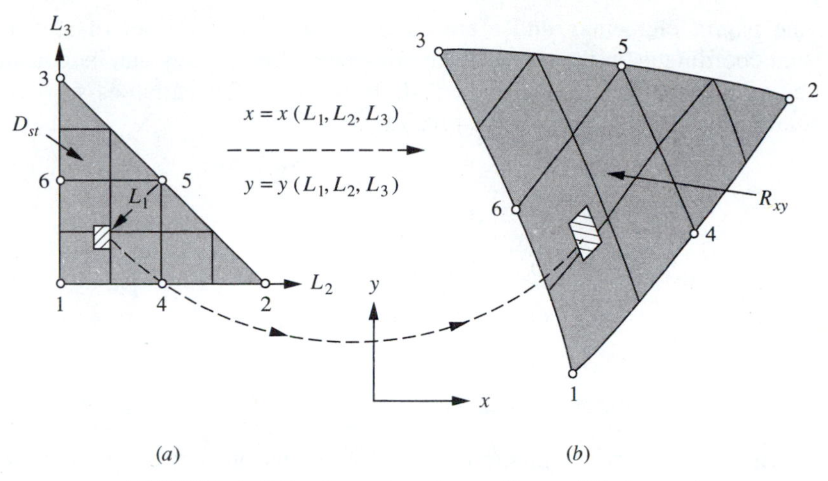

FIGURE 3–101 Parent and curvilinear T6 elements

Keeping in mind that there are only two independent L_i, we eliminate L_1 by using the relation $L_1 + L_2 + L_3 = 1$ to rewrite these transformations in terms of the independent variables L_2 and L_3 as

$$x(L_2, L_3) = \sum_1^6 x_i N_i(L_2, L_3)$$

$$y(L_2, L_3) = \sum_1^6 y_i N_i(L_2, L_3)$$

where L_1 has been replaced by $1 - L_2 - L_3$.

At this junction it is convenient to replace L_2 by s, L_3 by t, and L_1 by $1 - s - t$. The parent and mapped elements then appear as in Fig. 3–102.

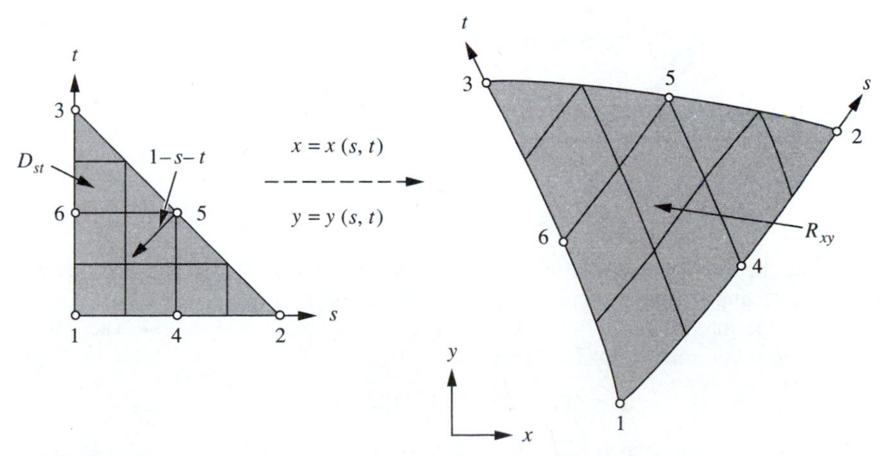

FIGURE 3–102 Parent and mapped element in terms of $s = L_2$ and $t = L_3$

In the parent element s and t are clearly equivalent to a set of rectangular Cartesian coordinates in the plane. In the mapped element they can be considered to be curvilinear coordinates as indicated. In terms of the variables s and t, the interpolation or shape functions are given by

$$
\begin{aligned}
N_1(s,t) &= (1-s-t)(1-2s-2t) \\
N_2(s,t) &= s(2s-1) \\
N_3(s,t) &= t(2t-1) \\
N_4(s,t) &= 4s(1-s-t) \\
N_5(s,t) &= 4st \\
N_6(s,t) &= 4t(1-s-t)
\end{aligned}
\tag{3.63}
$$

The reader should verify that these shape functions are correctly expressed in terms of s and t.

When the dependent variable is also represented in terms of the six interpolation functions above, the T6 element is isoparametric.

Example 3.10.

In order to better appreciate the essential character of the transformation from the parent element to R_{xy}, consider the specific transformation indicated in Fig. 3–103.

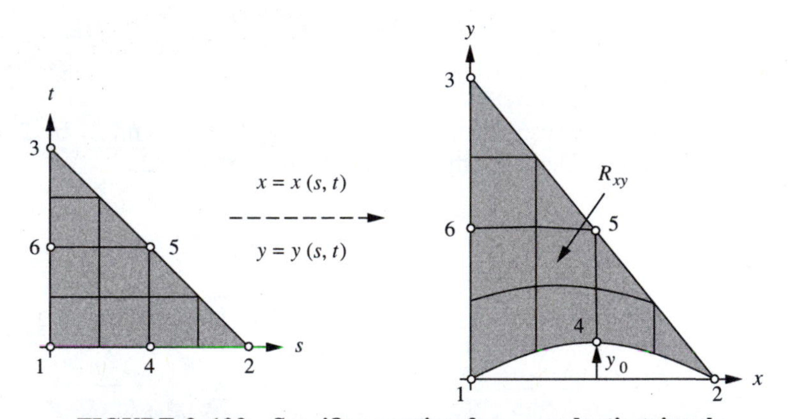

FIGURE 3–103 Specific mapping for a quadratic triangle

The straight line in the parent element containing nodes 1, 4, and 2 is the line $t = 0$. This line is mapped into the corresponding curve defined by nodes 1, 4, and 2 in the region R_{xy}. Of interest is the nature of the curve connecting nodes 1, 4, and 2 in R_{xy}. With the nodal coordinates in R_{xy} given by

$$\mathbf{x}_{\mathbf{e}}^{\mathrm{T}} = [0.0 \quad 1.0 \quad 0.0 \quad 0.5 \quad 0.5 \quad 0.0]$$

and

$$\mathbf{y}_{\mathbf{e}}^{\mathrm{T}} = [0.0 \quad 0.0 \quad 1.0 \quad y_0 \quad 0.5 \quad 0.5]$$

the transformation mapping the parent element into R_{xy} becomes

$$x = s(2s - 1) + 2s(1 - s - t) + 2st$$

$$y = t(2t - 1) + 4y_0 s(1 - s - t) + 2st + 2t(1 - s - t)$$

Specializing this transformation to the line $t = 0$ in the parent element, it follows that $x(s, 0)$ and $y(s, 0)$ can be written as

$$x(s, 0) = s$$

$$y(s, 0) = y_0 4s(1 - s)$$

which describes in parametric form the parabola passing through points 1, 4, and 2 in R_{xy}. It is left to the Exercises to show that generally, straight lines in the parent element are mapped into parabolas in R_{xy}.

Having adopted $L_2 = s$ and $L_3 = t$ as the two independent variables, we proceed to develop the Jacobian matrix and transformations necessary to express the elemental matrices in terms of the area coordinates. The chain rule is used to write

$$\begin{bmatrix} \dfrac{\partial}{\partial s} \\ \dfrac{\partial}{\partial t} \end{bmatrix} = \begin{bmatrix} \dfrac{\partial x}{\partial s} & \dfrac{\partial y}{\partial s} \\ \dfrac{\partial x}{\partial t} & \dfrac{\partial y}{\partial t} \end{bmatrix} \begin{bmatrix} \dfrac{\partial}{\partial x} \\ \dfrac{\partial}{\partial y} \end{bmatrix}$$

or

$$\frac{\boldsymbol{\partial}}{\boldsymbol{\partial}\mathbf{s}} = \mathbf{J}\frac{\boldsymbol{\partial}}{\boldsymbol{\partial}\mathbf{x}}$$

Inverting these relationships yields

$$\frac{\boldsymbol{\partial}}{\boldsymbol{\partial}\mathbf{x}} = \mathbf{J}^{-1}\frac{\boldsymbol{\partial}}{\boldsymbol{\partial}\mathbf{s}}$$

from which

$$\frac{\partial}{\partial x} = \mathbf{J}_1\frac{\boldsymbol{\partial}}{\boldsymbol{\partial}\mathbf{s}}$$

$$\frac{\partial}{\partial y} = \mathbf{J}_2\frac{\boldsymbol{\partial}}{\boldsymbol{\partial}\mathbf{s}}$$

These will be used later in the development of $\mathbf{k_e}$.

The Jacobian of the transformation is of importance in assessing the character of the mapping. In terms of the variables s and t, the Jacobian is

$$|\mathbf{J}(s,t)| = \begin{vmatrix} \dfrac{\partial x}{\partial s} & \dfrac{\partial y}{\partial s} \\ \dfrac{\partial x}{\partial t} & \dfrac{\partial y}{\partial t} \end{vmatrix}$$

$$= \left(\sum_1^6 x_i \frac{\partial N_i}{\partial s}\right)\left(\sum_1^6 y_i \frac{\partial N_i}{\partial t}\right) - \left(\sum_1^6 y_i \frac{\partial N_i}{\partial s}\right)\left(\sum_1^6 x_i \frac{\partial N_i}{\partial t}\right)$$

The mapping is invertible if $|\mathbf{J}(s,t)|$ is positive everywhere in the domain of the parent element. The Jacobian $|\mathbf{J}(s,t)|$ is a measure of the expansion or contraction of a differential area element resulting from the transformation according to

$$dx\ dy = |\mathbf{J}(s,t)|\, ds\ dt$$

Example 3.11.

In view of the fact that for a specific element, the transformation and thus the Jacobian of the transformation are determined by the locations of the nodes, it follows that some care must be exercised in order to avoid undesirable geometries. In this regard consider the geometry defined by the nodes shown in Fig. 3–104.

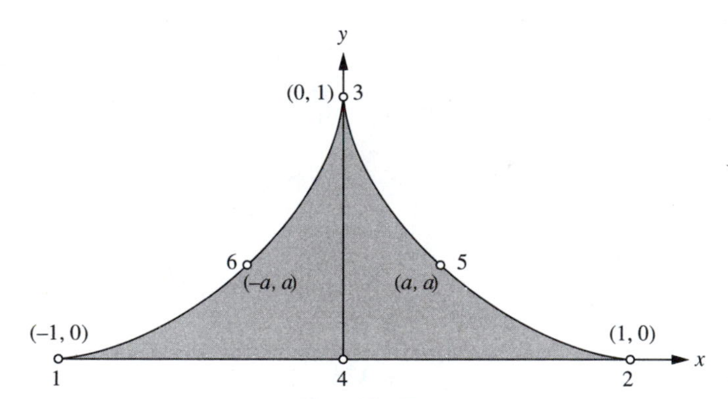

FIGURE 3–104 Specific T6 element geometry

The nodal coordinates are given by

$$\mathbf{x}_\mathbf{e}^\mathrm{T} = [-1.0\ \ 1.0\ \ 0.0\ \ 0.0\ \ a\ \ -a]$$

$$\mathbf{y}_\mathbf{e}^\mathrm{T} = [0.0\ \ 0.0\ \ 1.0\ \ 0.0\ \ a\ \ a]$$

with the corresponding transformation given by

$$x = -(1-s-t)(1-2s-2t) + s(2s-1) + 4at(2s+t-1)$$

$$y = t(2t-1) + 4at(1-t)$$

The resulting Jacobian is given by

$$|\mathbf{J}(s,t)| = \big(2 - 4t(1-2a)\big)\big(-1 + 4a + 4t(1-2a)\big)$$

The potential problem with the geometry will most likely arise at node 3, at which $t = 1$. The corresponding value of the Jacobian is

$$|\mathbf{J}(s,1)| = (-2 + 8a)(3 - 4a)$$

which is easily determined to be positive for $1/4 < a < 3/4$. The geometries for these limiting values of a are indicated in Fig. 3–105.

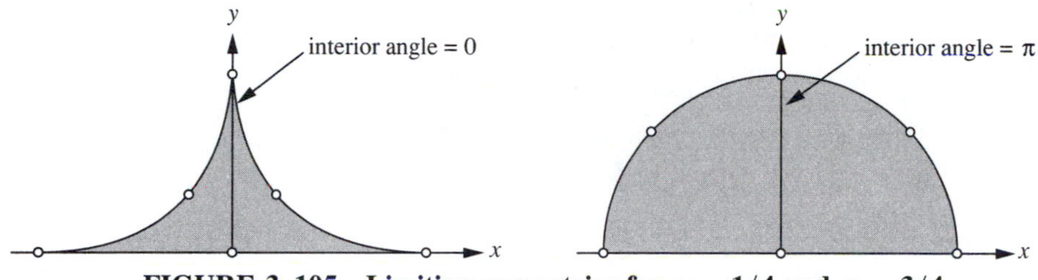

FIGURE 3–105 Limiting geometries for $a = 1/4$ and $a = 3/4$

Note that $a = 1/4$ corresponds to an interior angle of zero at node 3 and that $a = 3/4$ corresponds to an interior angle of π at node 3. One can also verify that when $t = 0$, $|\mathbf{J}|$ is zero for $a = 1/4$, indicating that the interior angles at nodes 1 and 2 are zero.

For the T6 element the special points [Moan, 8] within the element at which the derived variables exhibit improved accuracy are the $N = 4$ Gauss points (see Section 3.8.3.2) for the triangle, and are located at the centroid (1/3,1/3) and the symmetrically placed points (0.2,0.2), (0.2,0.6), and (0.6,0.2). These values can be employed with smoothing [Hinton and Campbell, 10] to extrapolate the derived variables to the nodes with accuracy that is generally improved relative to the values computed directly at the nodes. The reader is referred to the Exercises for further details.

3.8.3.2 EVALUATION OF T6 ELEMENTAL MATRICES. As stated several times in preceding sections, the general elemental matrices for the classical elliptic boundary value problem are

$$\mathbf{f_e} = \int\!\!\int_{R_{xy}} \mathbf{f}(x,y)\mathbf{N}\,dx\,dy$$

$$\mathbf{h_e} = \int_{\gamma_{2e}} h(s)\mathbf{N}\,ds$$

$$\mathbf{a_e} = \int_{\gamma_{2e}} \mathbf{N}\alpha(s)\mathbf{N}^{\mathrm{T}}\,ds$$

$$\mathbf{k_e} = \iint_{R_{xy}} \left(\frac{\partial \mathbf{N}}{\partial x} \frac{\partial \mathbf{N}^{\mathrm{T}}}{\partial x} + \frac{\partial \mathbf{N}}{\partial y} \frac{\partial \mathbf{N}^{\mathrm{T}}}{\partial y} \right) dx\, dy$$

In what follows we will indicate the techniques commonly used for evaluating these matrices for models using quadratically interpolated triangles.

Evaluation of $\mathbf{h_e}$, $\mathbf{a_e}$. Consider a typical γ_{2e} segment on the Γ_2 portion of the boundary as indicated in Fig. 3–106.

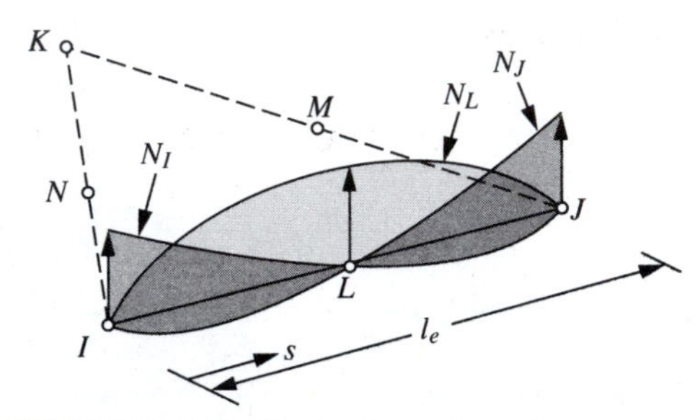

FIGURE 3–106 Typical γ_{2e} boundary segment

The nonvanishing interpolation functions N_I, N_J, and N_L are indicated. The coordinate s measures arc length along γ_{2e} and should not be confused with the s used in connection with the s-t coordinate system for the parent element. Approximating the functions $h(s)$ and $\alpha(s)$ by their quadratic interpolants according to

$$h(s) = h_I N_I + h_J N_J + h_L N_L$$

and

$$\alpha(s) = \alpha_I N_I + \alpha_J N_J + \alpha_L N_L$$

yields

$$\mathbf{h_e} = \left(\int_0^{l_e} \mathbf{N}\mathbf{N}^{\mathrm{T}} ds \right) \mathbf{h}$$

$$\mathbf{a_e} = \int_0^{l_e} \mathbf{N}(\alpha_I N_I + \alpha_J N_J + \alpha_L N_L)\mathbf{N}^{\mathrm{T}}\, ds$$

As is clear from Fig. 3–106, the elemental matrices $\mathbf{h_e}$ and $\mathbf{a_e}$ essentially depend on the character of the one-dimensional quadratic interpolation functions N_I, N_J, and N_L along the segment γ_{2e}. The results of evaluating these integrals have already been presented in Section 3.8.2.2, to which the reader is referred for the details. The problem of the evaluation of the arc length of a typical segment γ_{2e} was also treated in Section 3.8.2.2.

Evaluation of $\mathbf{f_e}$, $\mathbf{k_e}$. The evaluations of $\mathbf{f_e}$ and $\mathbf{k_e}$ are most easily carried out by first transforming the integrals evaluated over R_{xy} into integrals over the region of the corresponding parent element. For $\mathbf{f_e}$ this results in

$$\mathbf{f_e} = \iint_{R_{xy}} f(x, y)\mathbf{N}\, dx\, dy$$

$$= \iint_{D_{st}} f\big(x(s, t), y(s, t)\big)\mathbf{N}(s, t)\, |\mathbf{J}|\, ds\, dt$$

If we represent f on R_{xy} approximately according to

$$f\big(x(s, t), y(s, t)\big) = \mathbf{N}^{\mathrm{T}}(s, t)\mathbf{f}$$

the elemental vector $\mathbf{f_e}$ becomes

$$\mathbf{f_e} = \left[\iint_{D_{st}} \mathbf{N}(s, t)\mathbf{N}^{\mathrm{T}}(s, t)\, |\mathbf{J}|\, ds\, dt\right]\mathbf{f}$$

with $\mathbf{f}$ a vector usually consisting of the nodal values of $f(x, y)$.

For $\mathbf{k_e}$ the chain rule results of the previous section are used to convert the general expression

$$\mathbf{k_e} = \iint_{R_{xy}} \left(\frac{\partial \mathbf{N}}{\partial x}\frac{\partial \mathbf{N}^{\mathrm{T}}}{\partial x} + \frac{\partial \mathbf{N}}{\partial y}\frac{\partial \mathbf{N}^{\mathrm{T}}}{\partial y}\right) dx\, dy$$

into

$$= \iint_{D_{st}} \mathbf{\Delta JJ\Delta}^{\mathrm{T}}\, ds\, dt$$

where

$$\mathbf{\Delta}(s, t) = \begin{bmatrix} \dfrac{\partial \mathbf{N}}{\partial s} & \dfrac{\partial \mathbf{N}}{\partial t} \end{bmatrix} = \begin{bmatrix} -3 + 4s + 4t & -3 + 4s + 4t \\ 4s - 1 & 0 \\ 0 & 4t - 1 \\ 4 - 8s - 4t & -4s \\ 4t & 4s \\ -4t & 4 - 4s - 8t \end{bmatrix}$$

and $\mathbf{JJ} = (\mathbf{J}_1^{\mathrm{T}}\mathbf{J}_1 + \mathbf{J}_2^{\mathrm{T}}\mathbf{J}_2)\,|\mathbf{J}|$, with $\mathbf{J}_1$ and $\mathbf{J}_2$ the first and second rows, respectively, of the inverse of the Jacobian matrix. The Jacobian matrix is given by

$$\mathbf{J} = \begin{bmatrix} \mathbf{x_e^{\mathrm{T}}}\dfrac{\partial \mathbf{N}}{\partial s} & \mathbf{y_e^{\mathrm{T}}}\dfrac{\partial \mathbf{N}}{\partial s} \\ \mathbf{x_e^{\mathrm{T}}}\dfrac{\partial \mathbf{N}}{\partial t} & \mathbf{y_e^{\mathrm{T}}}\dfrac{\partial \mathbf{N}}{\partial t} \end{bmatrix}$$

with $|\mathbf{J}|$ the determinant of the Jacobian matrix.

We are thus confronted with the two integrals

$$\int_0^1\int_0^{1-t} \mathbf{N}(s, t)\mathbf{N}^{\mathrm{T}}(s, t)\, |\mathbf{J}|\, ds\, dt \qquad \text{and} \qquad \int_0^1\int_0^{1-t} \mathbf{\Delta}(s, t)\mathbf{JJ}(s, t)\mathbf{\Delta}^{\mathrm{T}}(s, t)\, ds\, dt$$

In what follows, the appropriate Gauss quadrature for evaluating these integrals over the parent region D_{st} will be discussed.

The development of the Gauss quadrature rules for triangles uses precisely the same reasoning as was presented in Section 2.6 for the evaluation of single integrals. The reader is referred to that section and to the References [Cowper, 11] for details. The results are of the form

$$\iint_{D_{st}} f(s,t)\,ds\,dt = \int_0^1 \int_0^{1-t} f(s,t)\,ds\,dt = \frac{1}{2}\sum_1^N f(s_{iN}, t_{iN}) w_{iN}$$

For integrals over R_{xy}

$$\iint_{R_{xy}} f(x,y)\,dx\,dy = \iint_{D_{st}} f\big(x(s,t), y(s,t)\big)|\mathbf{J}|\,ds\,dt = \iint_{D_{st}} F(s,t)\,|\mathbf{J}|\,ds\,dt$$

$$= \frac{1}{2}\sum_1^N F(s_{iN}, t_{iN})\,|\mathbf{J}|_{iN} w_{iN} \tag{3.64}$$

The 1/2 factor accounts for the area of the parent triangle. The weights and integration points are given in Table 3.17. Accuracy refers to the maximum degree of the polynomial that is integrated exactly. For example, accuracy = 3 means that the rule will integrate exactly any third-degree polynomial function. Note that all the integration points are either at the centroid or consist of sets of three points symmetrically located within the triangle.

TABLE 3.17 Gauss points and weights for triangles

N	s_{iN}	t_{iN}	W_{iN}	Accuracy
1	0.33333 33333	0.33333 33333	1.00000 00000	1
3	0.66666 66667	0.16666 66667	0.33333 33333	2
	0.16666 66667	0.66666 66667	0.33333 33333	
	0.16666 66667	0.16666 66667	0.33333 33333	
4	0.33333 33333	0.33333 33333	−0.56250 00000	3
	0.60000 00000	0.20000 00000	0.52083 33333	
	0.20000 00000	0.20000 00000	0.52083 33333	
	0.20000 00000	0.60000 00000	0.52083 33333	
6	0.81684 75730	0.09157 62135	0.10995 17437	4
	0.09157 62135	0.09157 62135	0.10995 17437	
	0.09157 62135	0.81684 75730	0.10995 17437	
	0.10810 30182	0.44594 84909	0.22338 15897	
	0.44594 84909	0.44594 84909	0.22338 15897	
	0.44594 84909	0.10810 30182	0.22338 15897	
7	0.33333 33333	0.33333 33333	0.22500 00000	5
	0.79742 69854	0.10128 65073	0.12593 91805	
	0.10128 65073	0.10128 65073	0.12593 91805	
	0.10128 65073	0.79742 69854	0.12593 91805	
	0.47014 20641	0.05971 58718	0.13239 41528	
	0.47014 20641	0.47014 20641	0.13239 41528	
	0.05971 58718	0.47014 20641	0.13239 41528	

Example 3.12.

A typical application of T6 elements arises in connection with modeling a curved boundary. One of the sides is curved and the other two often straight as indicated in Fig. 3–107. We will consider the evaluation of several integrals over this typical region.

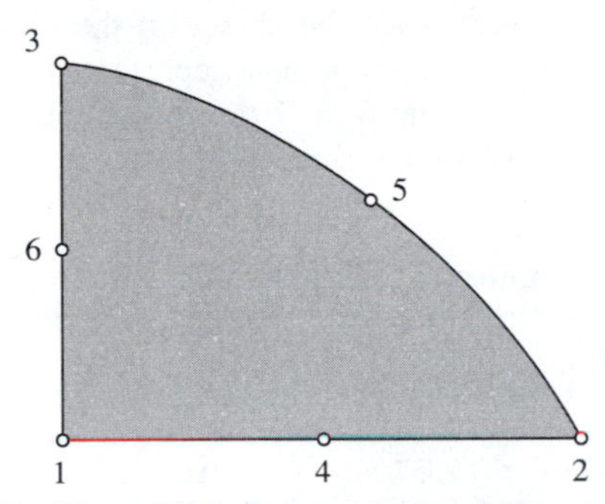

FIGURE 3–107 Use of T6 element in modeling curved boundary

Consider first the evaluation of the area of the element of Fig. 3–107. The area is expressed as

$$A = \int\int_{R_{xy}} dx\ dy = \int\int_{D_{st}} |\mathbf{J}|\, ds\ dt$$

With the nodal coordinates of the element given as

$$\mathbf{x}_\mathbf{e}^\mathrm{T} = [5.00 \quad 5.80 \quad 5.00 \quad 5.40 \quad 5.49 \quad 5.00]$$

$$\mathbf{y}_\mathbf{e}^\mathrm{T} = [7.00 \quad 7.00 \quad 7.60 \quad 7.00 \quad 7.42 \quad 7.30]$$

it follows that the Jacobian determinant is given by

$$|\mathbf{J}| = \frac{\partial x}{\partial s}\frac{\partial y}{\partial t} - \frac{\partial y}{\partial s}\frac{\partial x}{\partial t} = 0.480 + 0.384s + 0.216t$$

and hence

$$A = \int_0^1 \int_0^{1-t} [0.480 + 0.384s + 0.216t]\, ds\, dt = 0.340$$

by direct integration.

Consider also the use of the quadrature rules of Eq. (3.64) and Table 3.17.

$$N = 1 \qquad A \approx \left(\frac{1}{2}\right)(1)\left(0.480 + \frac{0.384}{3} + \frac{0.216}{3}\right) = 0.340$$

which for a linear function of s and t is exact. It is left to the Exercises to show that any higher-order scheme also produces an exact evaluation.

Consider also the evaluation of

$$\mathbf{f_e} = \int\int_{D_{st}} f\big(x(s,t), y(s,t)\big)\mathbf{N}(s,t)\,|\mathbf{J}|\, ds\ dt$$

and the corresponding approximation

$$\mathbf{f_e} = \left[\iint_{D_{st}} \mathbf{N}(s,t)\mathbf{N}^T(s,t)\,|\mathbf{J}|\,ds\,dt\right]\mathbf{f}$$

over the same element as in Fig. 3–107. Take $f(x, y) = xy$. The results of using the quadrature rules of Eq. (3.64) and Table 3.17 are presented in Tables 3.18 and 3.19 for several N. These results indicate that for the choice of the element and the function f, there is negligible error in using the approximate representation for $\mathbf{f_e}$. Furthermore, the close agreement between the $N = 6$ and $N = 7$ results for both cases is an indication that both sets of results are reasonably accurate.

TABLE 3.18 Numerical integration results, exact $\mathbf{f_e}$

$\mathbf{f_e}$	$N = 1$	$N = 3$	$N = 4$	$N = 6$	$N = 7$
f_{e1}	−1.4541	−0.2753	−0.2509	−0.2433	−0.2431
f_{e2}	−1.4541	0.2557	0.2310	0.2276	0.2279
f_{e3}	−1.4541	0.0196	0.0196	0.0140	0.0141
f_{e4}	5.8164	4.3430	4.3371	4.3250	4.3247
f_{e5}	5.8164	4.6379	4.6976	4.7143	4.7142
f_{e6}	5.8164	4.1069	4.0551	4.0520	4.0520

TABLE 3.19 Numerical integration results, approximate $\mathbf{f_e}$

$\mathbf{f_e}$	$N = 1$	$N = 3$	$N = 4$	$N = 6$	$N = 7$
f_{e1}	−1.4546	−0.2754	−0.2509	−0.2434	−0.2433
f_{e2}	−1.4546	0.2556	0.2310	0.2274	0.2277
f_{e3}	−1.4546	0.0198	0.0198	0.0141	0.0142
f_{e4}	5.8185	4.3436	4.3375	4.3257	4.3254
f_{e5}	5.8185	4.6388	4.6985	4.7152	4.7150
f_{e6}	5.8185	4.1079	4.0560	4.0529	4.0529

Last, consider the evaluation of the elemental stiffness $\mathbf{k_e}$ given by

$$\mathbf{k_e} = \iint_{D_{st}} \mathbf{\Delta J J \Delta}^T\,ds\,dt$$

over the region of Fig. 3–107. The results are given in Table 3.20 for a number of different N.

Table 3.20 clearly shows that the results using the $N = 6$ and $N = 7$ Gauss quadrature rules differ only slightly. Collectively, the results from Tables 3.18, 3.19, and 3.20 indicate that for T6 elements, an $N = 6$ Gauss rule is generally adequate for the evaluation of the integrals representing $\mathbf{f_e}$ and $\mathbf{k_e}$.

TABLE 3.20 Numerical integration results for $\mathbf{k_e}$

		$N = 1$			
0.081699	−0.029412	−0.052288	0.209150	−0.326797	0.117647
−0.029412	0.048366	−0.018954	0.075817	0.117647	−0.193464
−0.052288	−0.018954	0.071242	−0.284967	0.209150	0.075817
0.209150	0.075817	−0.284967	1.139869	−0.836601	−0.303268
−0.326797	0.117647	0.209150	−0.836601	1.307190	−0.470588
0.117647	−0.193464	0.075817	−0.303268	−0.470588	0.773856
		$N = 3$			
0.853698	0.127306	0.212655	−0.368386	−0.111205	−0.714067
0.127306	0.569544	0.067881	−0.492999	−0.160073	−0.111660
0.212655	0.067881	0.838440	−0.147888	−0.153368	−0.817722
−0.368386	−0.492999	−0.147888	2.120745	−0.926325	−0.185148
−0.111205	−0.160073	−0.153368	−0.926325	1.833854	−0.482883
−0.714067	−0.111660	−0.817722	−0.185148	−0.482883	2.311480
		$N = 4$			
0.835407	0.111355	0.189737	−0.351997	−0.098541	−0.685961
0.111355	0.554409	0.063949	−0.465077	−0.158156	−0.106481
0.189737	0.063949	0.813204	−0.144736	−0.149521	−0.772634
−0.351997	−0.465077	−0.144736	2.075972	−0.935362	−0.178800
−0.098541	−0.158156	−0.149521	−0.935362	1.838100	−0.496521
−0.685961	−0.106481	−0.772634	−0.178800	−0.496521	2.240397
		$N = 6$			
0.868232	0.117017	0.198403	−0.377265	−0.087159	−0.719228
0.117017	0.560502	0.055789	−0.483091	−0.153399	−0.096818
0.198403	0.055789	0.833711	−0.125640	−0.151786	−0.810476
−0.377265	−0.483091	−0.125640	2.131863	−0.951577	−0.194290
−0.087159	−0.153399	−0.151786	−0.951577	1.843572	−0.499651
−0.719228	−0.096818	−0.810476	−0.194290	−0.499651	2.320465
		$N = 7$			
0.868186	0.117150	0.198125	−0.377606	−0.087098	−0.718759
0.117150	0.560627	0.055632	−0.483464	−0.153298	−0.096647
0.198125	0.055632	0.833843	−0.125136	−0.151941	−0.810523
−0.377606	−0.483464	−0.125136	2.132965	−0.951867	−0.194893
−0.087098	−0.153298	−0.151941	−0.951867	1.843643	−0.499439
−0.718759	−0.096647	−0.810523	−0.194893	−0.499439	2.320260

3.8.3.3 APPLICATION—STRESS CONCENTRATION FOR THE TORSION OF AN ANGLE SECTION. Consider again the problem of the torsion of an angle section as shown in Fig. 3–108. The reader should consult Section 3.8.1.3 for discussion of the corresponding boundary value problem. The mesh to be used in connection with the Q8 elements is shown in Fig. 3–109.

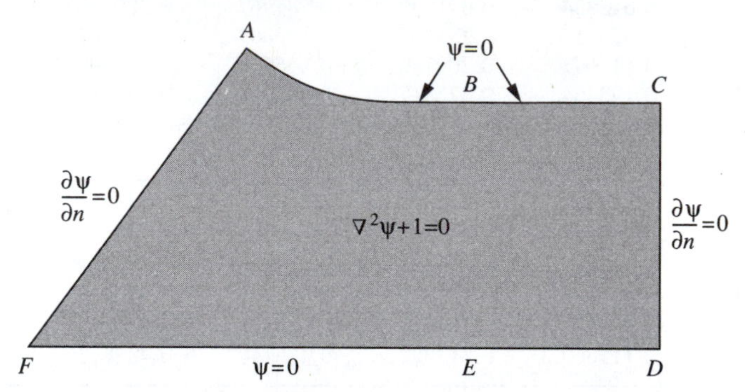

FIGURE 3–108 Modeled region for torsion of an angle

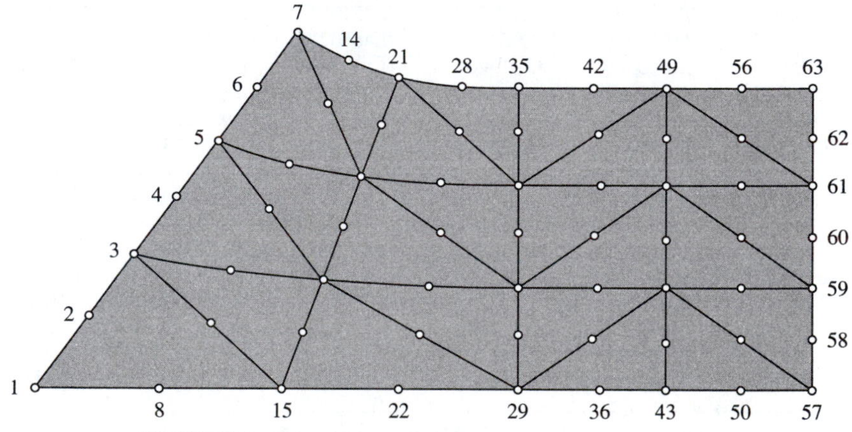

FIGURE 3–109 T6 mesh for torsion of angle section

Partial results for ψ are presented in Table 3.21 and displayed in Figs. 3–110a and 3–110b. An $N = 6$ Gauss rule was used to evaluate the integrals for $\mathbf{f_e}$ and $\mathbf{k_e}$.

TABLE 3.21 ψ along *DC* and *FA*

x/l_m	ψ_{DC}	ψ_{FA}
0	0.0000	0.0000
1/6	0.0696	0.0585
2/6	0.1114	0.1337
3/6	0.1254	0.1972
4/6	0.1114	0.2170
5/6	0.0696	0.1546
1	0.0000	0.0000

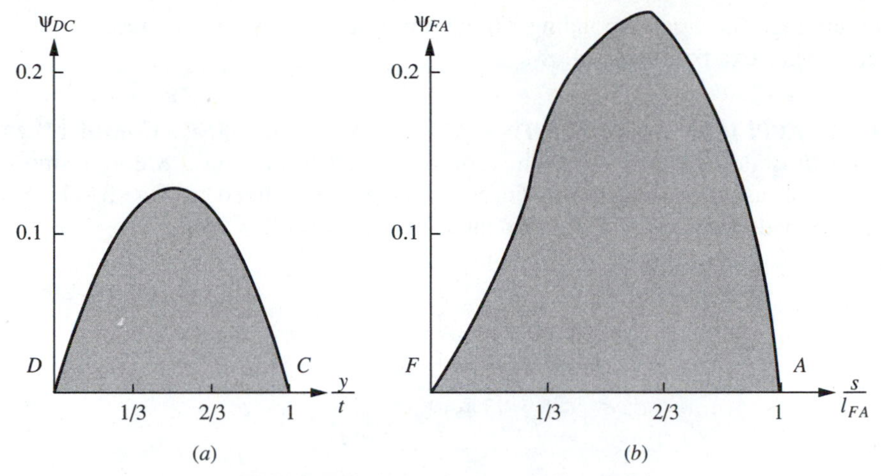

FIGURE 3–110 ψ **along** ***DC*** **and** ***FA***

The maximum stress along FA is clearly at A and is proportional to the slope of the ψ surface along line FA at A. Similarly, the maximum stress along DC is proportional to the slope of the surface along line DC at either D or C. Computing the slopes at points C and A, respectively, yields the values

$$\frac{\partial \psi}{\partial n} \text{ at } C = -0.5010$$

and

$$\frac{\partial \psi}{\partial n} \text{ at } A = -0.6587$$

with the resulting approximate stress concentration factor of

$$k_T = \frac{0.6587}{0.5010} = 1.31$$

As indicated in Section 3.8.1.3, this is to be compared with the tabulated experimental value of $k_T = 1.4$ [Griffith and Taylor, 9]. The predicted value of $k_T = 1.31$ for the stress concentration factor using T6 elements is to be compared to the values of $k_T = 1.26$ using Q4 elements, and $k_T = 1.37$ using Q8 elements.

For this example, the prediction of the stress concentration using Q8 elements is closer to the exact value than the results using T6 elements. This is due in part to the fact that although the dependent variable and geometry behave quadratically along interelement boundaries for each of the Q8 and T6 elements, the interpolation functions for the Q8 elements involve third-order terms in the local coordinates s and t, whereas the T6 interpolation functions are truly quadratic in both s and t. Results using Q8 elements are generally, but not always, more accurate than for a corresponding T6 element mesh. This will be observed in the heat transfer example that follows.

3.8.3.4 APPLICATION—STEADY-STATE HEAT TRANSFER. Consider again the steady-state heat transfer application discussed in Section 3.8.2.4. Using the same mesh as in that example, T6 elements are as defined in Fig. 3–111. The resulting nodal values for t are presented in Fig. 3–112.

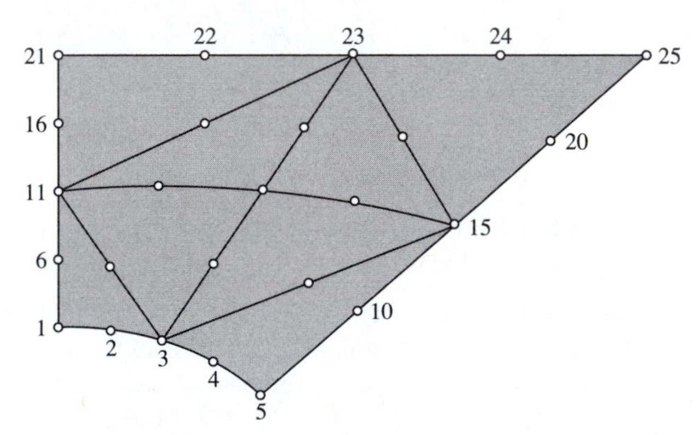

FIGURE 3–111 T6 elements for heat transfer application

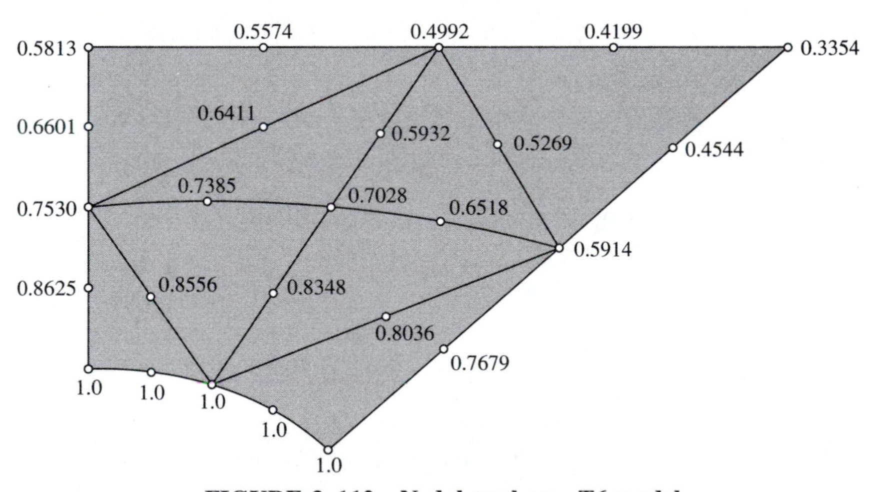

FIGURE 3–112 Nodal t values—T6 model

Fig. 3–113 shows the values of the elemental normal derivatives at the nodes on the Γ_2 portion of the boundary, calculated according to

$$\frac{\partial t}{\partial n} = (n_x\mathbf{J_1} + n_y\mathbf{J_2})\mathbf{\Delta}^{\mathrm{T}}\mathbf{t_e}$$

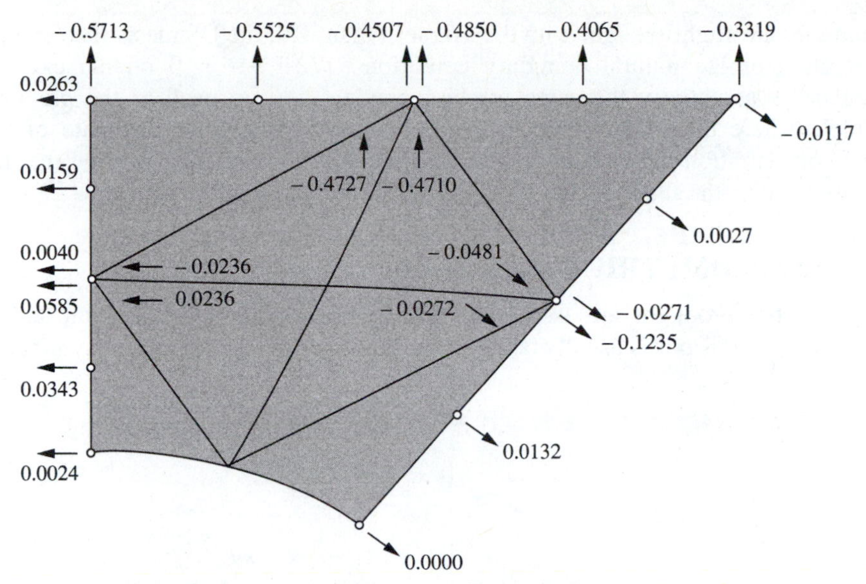

FIGURE 3–113 $\partial t/\partial n$ **on the** Γ_2 **portion of the boundary**

For this model using T6 elements, the values of the normal derivatives on the sides defined by nodes 1, 6, 11, 16, and 21 and by nodes 5, 10, 15, 20, and 25 are quite small and compare favorably with the correct value of zero.

By averaging the values of the normal derivatives computed in adjacent elements at node 23, the satisfaction of the natural boundary conditions along the line defined by nodes 21 through 25 can be checked as follows.

Node 21: $$\frac{\partial t}{\partial n} + t = -0.5713 + 0.5813 = 0.0100$$

Node 22: $$\frac{\partial t}{\partial n} + t = -0.5525 + 0.5574 = 0.0049$$

Node 23: $$\frac{\partial t}{\partial n} + t = -0.4679 + 0.4992 = 0.0313$$

Node 24: $$\frac{\partial t}{\partial n} + t = -0.4065 + 0.4199 = 0.0134$$

Node 25: $$\frac{\partial t}{\partial n} + t = -0.3319 + 0.3354 = 0.0035$$

These relatively small values, compared to zero, are a good check on the accuracy and convergence of the solution.

Comparing these results with those of Section 3.8.2.4, where Q8 elements were used, it is seen that the errors in the normal derivatives on the two interior boundaries are slightly larger with the T6 mesh than with the Q8 mesh. The overall satisfaction of the natural boundary condition $\partial t/\partial n + t = 0$ on the exterior boundary is essentially the same for both models. As opposed to the previous example where the model using a Q8 mesh provided a better estimate of the stress concentration factor than did the model using T6 elements, we find that for this application the results from the Q8 and T6 are essentially the same.

3.9 AXISYMMETRIC PROBLEMS

Axisymmetric problems are associated with bodies of revolution as indicated in Fig. 3–114. The z-axis is usually taken as the axis of revolution or axis of symmetry.

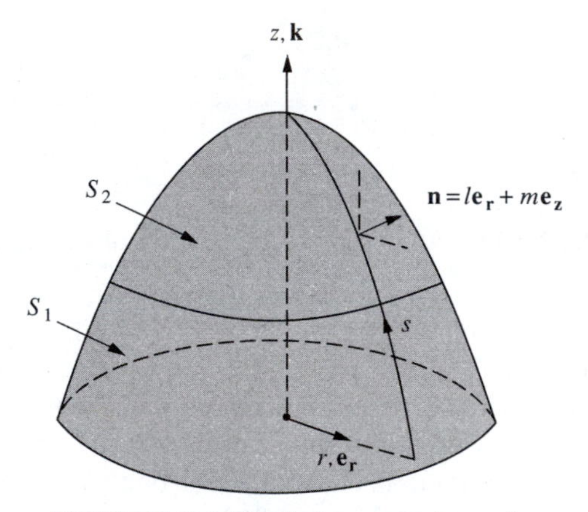

FIGURE 3–114 Axisymmetric region

Axisymmetric problems are sometimes referred to as radially symmetric problems. They are geometrically three dimensional but mathematically only two dimensional in that the physics of the problem is such that the dependent variable u is assumed to be independent of the angle θ; that is, $u = u(r, z)$. For these situations the Laplacian operator reduces to

$$\nabla^2 u = \frac{1}{r}\frac{\partial(r\partial u/\partial r)}{\partial r} + \frac{\partial^2 u}{\partial z^2}$$

which can be rewritten as

$$\nabla^2 u = \frac{1}{r}\left(\frac{\partial(r\partial u/\partial r)}{\partial r} + \frac{\partial(r\partial u/\partial z)}{\partial z}\right)$$

The boundary value problem of interest is

$$\begin{aligned} \nabla^2 u + f &= 0 && \text{in } D \\ u &= g(S) && \text{on } S_1 \\ \frac{\partial u}{\partial n} + \alpha u &= h(S) && \text{on } S_2 \end{aligned} \tag{3.65}$$

where the surface S is considered to consist of a portion S_1 on which Dirichlet type boundary conditions are prescribed and a portion S_2 on which Neumann or Robins boundary conditions are prescribed. In view of the assumed axisymmetric character of the problem, it follows that the boundary conditions must also be independent of θ so that the boundary conditions can be restated as

$$u = g(s) \text{ on } S_1$$

$$\frac{\partial u}{\partial n} + \alpha u = h(s) \text{ on } S_2$$

where s is a coordinate measuring arc length along a meridian as indicated in Fig. 3–114.

Weak formulation. When developing the weak form the integration is taken over the volume of the region. The appropriate differential volume for the solid of revolution is

$$d\text{Vol} = 2\pi r\ dr\ dz$$

The factor of 2π is common and can be canceled. With $w(r, z)$ a test function vanishing on Γ_1, multiply the differential equation by $w d\text{Vol}/2\pi$ and integrate over the volume to obtain

$$\iint_D w\left(\frac{\partial(r\partial u/\partial r)}{\partial r} + \frac{\partial(r\partial u/\partial z)}{\partial z} + rf\right) dr\ dz = 0$$

which can be rewritten as

$$\iint_D \left(\frac{\partial(rw\,\partial u/\partial r)}{\partial r} + \frac{\partial(rw\,\partial u/\partial z)}{\partial z} + wrf\right) dr\ dz - \iint_D \left(\frac{\partial w}{\partial r} r \frac{\partial u}{\partial r} + \frac{\partial w}{\partial z} r \frac{\partial u}{\partial z}\right) dr\ dz = 0 \tag{3.66}$$

The appropriate form of the divergence theorem for the axisymmetric situation is

$$\iint_D \left(\frac{\partial(r\phi)}{\partial r} + \frac{\partial(r\psi)}{\partial z}\right) dr\ dz = \int_S (l\phi + m\psi)\ r\ ds$$

where l and m are the direction cosines of the exterior normal as indicated in Fig. 3–114. Applying the divergence theorem to the second volume integral in Eq. (3.66) yields the result

$$\int_S \left(lw\frac{\partial u}{\partial r} + mw\frac{\partial u}{\partial z}\right) r\ ds - \iint_D \left(\frac{\partial w}{\partial r}\ r\ \frac{\partial u}{\partial r} + \frac{\partial w}{\partial z} r \frac{\partial u}{\partial z} - wrf\right) dr\ dz = 0 \tag{3.67}$$

Write the integral over S as the sum of the integrals over S_1 and S_2. Recall that $w = 0$ on S_1 and that on S_2

$$\frac{\partial u}{\partial n} = l\ \frac{\partial u}{\partial r} + m\ \frac{\partial u}{\partial z} = h(s) - \alpha(s)u$$

It follows that Eq. (3.67) can be written as

$$\int\int_D \left(\frac{\partial w}{\partial r} r \frac{\partial u}{\partial r} + \frac{\partial w}{\partial z} r \frac{\partial u}{\partial z}\right) dr\ dz + \int_{S_2} w\alpha u r\ ds$$
$$= \int\int_D wrf\ dr\ dz + \int_{S_2} whr\ ds \qquad (3.68)$$

which is the required weak form. The bilinear functional B and the linear functional l are

$$B(w, u) = \int\int_D \left(\frac{\partial w}{\partial r} r \frac{\partial u}{\partial r} + \frac{\partial w}{\partial z} r \frac{\partial u}{\partial z}\right) dr\ dz + \int_{S_2} w\alpha u r\ ds$$

$$l(w) = \int\int_D wrf\ dr\ dz + \int_{S_2} whr\ ds$$

The corresponding functional can be written as

$$I(u) = \frac{1}{2}\int\int_D \left(r\left(\frac{\partial u}{\partial r}\right)^2 + r\left(\frac{\partial u}{\partial z}\right)^2\right) dr\ dz + \frac{1}{2}\int_{S_2} \alpha r u^2\ ds$$
$$- \int\int_D urf\ dr\ dz - \int_{S_2} hru\ ds \qquad (3.69)$$

and will serve as the basis for the development of the Ritz finite element model.

Due to the mathematical nature of the problem, the analysis can be carried out within the two-dimensional region in the rz-plane which is revolved about the z-axis to form the corresponding three-dimensional region. A typical case is indicated in Fig. 3–115, where D, S_1, S_2, s, l, and m are shown. Also indicated is the situation regarding any body of revolution: that an area revolved around the z-axis produces a toroid.

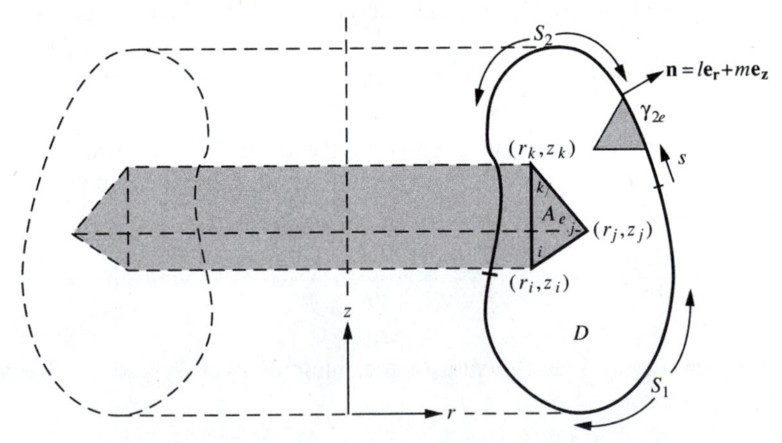

FIGURE 3–115 Typical two-dimensional region and element for the axisymmetric boundary value problem

The symbols S_1 and S_2 are used to remind us that we are actually dealing with a surface, and that the coordinate s corresponds to an arc, analogous to a line of constant latitude on the surface of the earth.

Discretization. The first step in developing the finite element model is discretization. Discretization is carried out in precisely the same fashion as in Section 3.3. A typical element in the rz-plane is indicated in Fig. 3–115. A_e denotes the area and γ_{2e} a typical line segment on the boundary. It should be apparent that although the element appears as a triangular area in Fig. 3–115, the form of the actual three-dimensional element is a ring whose cross section is that of the triangular element. All of the comments made in Section 3.3 regarding the shape of the triangular element continue to be true for the application to the axisymmetric boundary value problem.

In terms of the discretization, the functional I is now represented as the sum of integrals over the areas of the elements in D, and as the sum of integrals over the corresponding line segments on the Γ_2 portion(s) of the boundary according to

$$I(u) \approx \frac{1}{2}\left(\sum \int\int_{A_e} \left\{ r\left(\frac{\partial u}{\partial r}\right)^2 + r\left(\frac{\partial u}{\partial z}\right)^2 \right\} dr\, dz + \sum{}' \int_{\gamma_{2e}} \alpha r u^2\, ds \right)$$

$$- \sum \int\int_{A_e} urf\, dA - \sum{}' \int_{\gamma_{2e}} urh\, ds \tag{3.70}$$

where the sum $\sum$ is over the elemental areas, and the $\sum'$ is over the elemental segments γ_{2e} of the Γ_2 portion(s) of the boundary.

Interpolation. Linearly interpolated triangular elements will be used for developing the Ritz finite element model. The dependent variable is represented as

$$u_e(r, z) = \alpha + \beta r + \gamma z \tag{3.71}$$

where the three constants α, β, and γ are determined by the three equations requiring that u assume the nodal values at the vertices of the triangle,

$$\begin{aligned} u_e(r_i, z_i) &= \alpha + \beta r_i + \gamma z_i = u_i \\ u_e(r_j, z_j) &= \alpha + \beta r_j + \gamma z_j = u_j \\ u_e(r_k, z_k) &= \alpha + \beta r_k + \gamma z_k = u_k \end{aligned} \tag{3.72}$$

where i, j, and k denote the three vertices of the triangle. The r_i, z_i, r_j, z_j, r_k, z_k, u_i, u_j, and u_k are all indicated in Fig. 3–115. When the three Eqs. (3.72) are solved for α, β, and γ and the results back-substituted into Eq. (3.71), the elemental expression for $u(r, z)$ can be written as

$$u_e(r, z) = u_i N_i + u_j N_j + u_k N_k$$

where

$$N_i = \frac{(a_i + b_i r + c_i z)}{2A_e} \qquad i = 1, 2, 3$$

with

$$a_i = r_j z_k - r_k z_j$$

$$b_i = z_j - z_k$$

$$c_i = r_k - r_j$$

The i, j, and k are to be permuted cyclically. The determinant of the coefficients is

$$2A_e = \begin{vmatrix} 1 & r_i & z_i \\ 1 & r_j & z_j \\ 1 & r_k & z_k \end{vmatrix}$$

with A_e as the area of the element. Any numbering that proceeds counterclockwise around the element is acceptable, that is, (i, j, k), (j, k, i), and (k, i, j). This counterclockwise direction is necessary in order that the area A_e be positive. In matrix notation, the elemental representation can be expressed as

$$u_e(r, z) = \mathbf{u_e^T N} = \mathbf{N^T u_e} \tag{3.73}$$

with both forms of the representation being necessary in different settings. Additionally, the derivatives of the elemental representations will be needed. These are easily computed to be

$$\frac{\partial u_e(r, z)}{\partial r} = \mathbf{u_e^T} \frac{\partial \mathbf{N}}{\partial r} = \frac{\partial \mathbf{N^T}}{\partial r} \mathbf{u_e}$$

and

$$\frac{\partial u_e(r, z)}{\partial z} = \mathbf{u_e^T} \frac{\partial \mathbf{N}}{\partial z} = \frac{\partial \mathbf{N^T}}{\partial z} \mathbf{u_e}$$

Recalling the expressions for the N_i it is easily seen that

$$\frac{\partial \mathbf{N}}{\partial r} = \frac{\mathbf{b_e}}{2\mathrm{A_e}} \tag{3.74a}$$

and

$$\frac{\partial \mathbf{N}}{\partial z} = \frac{\mathbf{c_e}}{2\mathrm{A_e}} \tag{3.74b}$$

where $\mathbf{b_e^T} = [b_1 \; b_2 \; b_3]$ and $\mathbf{c_e^T} = [c_1 \; c_2 \; c_3]$. It is apparent that the partial derivatives of the solution within an element will be constant when linear interpolation is used.

In view of the fact that the interpolation functions are linear on any edge, interelement continuity is assured. Additionally, the basic representation given by Eq. (3.73) contains the terms necessary for the solution and the partial derivatives $\partial u/\partial r$ and $\partial u/\partial z$ to be constant within the element, so that the completeness criterion is also satisfied.

Elemental formulation. We repeat Eq. (3.70) for the functional associated with the elliptic boundary value problem in connection with a typical discretization:

$$I(u) = \frac{1}{2}\sum\left(\iint_{A_e}\left\{r\left(\frac{\partial u}{\partial r}\right)^2 + r\left(\frac{\partial u}{\partial z}\right)^2\right\} dr\,dz + \sum{}'\int_{\gamma_{2e}} \alpha r u^2\,ds\right)$$

$$-\sum\iint_{A_e} urf\,dA - \sum{}'\int_{\gamma_{2e}} urh\,ds$$

where again the $\sum$ indicates a sum over all the triangular elements A_e, and the $\sum'$ indicates a sum over the straight line segments γ_{2e} representing the boundary. Rewrite Eq. (3.70) as

$$I(u) = \sum\frac{I_{e1}}{2} + \sum{}'\frac{I_{e2}}{2} - \sum I_{e3} - \sum{}' I_{e4}$$

where

$$I_{e1} = \iint_{A_e}\left\{r\left(\frac{\partial u}{\partial r}\right)^2 + r\left(\frac{\partial u}{\partial z}\right)^2\right\} dr\,dz$$

$$I_{e2} = \int_{\gamma_{2e}} \alpha r u^2\,ds$$

$$I_{e3} = \iint_{A_e} urf\,dr\,dz$$

$$I_{e4} = \int_{\gamma_{2e}} urh\,ds$$

Using the results for the linearly interpolated triangle, the form of the elemental matrices arising for each of the above terms is indicated as follows.

I_{e1}: The integral I_{e1} is given by

$$I_{e1} = \iint_{A_e}\left(\frac{\partial u}{\partial r}\,r\,\frac{\partial u}{\partial r} + \frac{\partial u}{\partial z}\,r\,\frac{\partial u}{\partial z}\right) dr\,dz$$

The first $\partial u/\partial r$ is replaced by its representation in terms of the derivatives of the interpolation functions as $\mathbf{u}_\mathbf{e}^\mathrm{T}\partial\mathbf{N}/\partial r$ and the second $\partial u/\partial r$ by $\partial\mathbf{N}^\mathrm{T}/\partial r\,\mathbf{u}_\mathbf{e}$. With an entirely similar treatment of the second term there results

$$I_{e1} = \iint_{A_e}\left[\mathbf{u}_\mathbf{e}^\mathrm{T}\frac{\partial\mathbf{N}}{\partial r}\,r\,\frac{\partial\mathbf{N}^\mathrm{T}}{\partial r}\mathbf{u}_\mathbf{e} + \mathbf{u}_\mathbf{e}^\mathrm{T}\frac{\partial\mathbf{N}}{\partial z}\,r\,\frac{\partial\mathbf{N}^\mathrm{T}}{\partial z}\mathbf{u}_\mathbf{e}\right] dr\,dz$$

$$= \mathbf{u}_\mathbf{e}^\mathrm{T}\left\{\iint_{A_e}\left[\frac{\partial\mathbf{N}}{\partial r}\,r\,\frac{\partial\mathbf{N}^\mathrm{T}}{\partial r} + \frac{\partial\mathbf{N}}{\partial z}\,r\,\frac{\partial\mathbf{N}^\mathrm{T}}{\partial z}\right] dr\,dz\right\}\mathbf{u}_\mathbf{e} = \mathbf{u}_\mathbf{e}^\mathrm{T}\mathbf{k}_\mathbf{e}\mathbf{u}_\mathbf{e}$$

where

$$\mathbf{k}_\mathbf{e} = \iint_{A_e}\left[\frac{\partial\mathbf{N}}{\partial r}\,r\,\frac{\partial\mathbf{N}^\mathrm{T}}{\partial r} + \frac{\partial\mathbf{N}}{\partial z}\,r\,\frac{\partial\mathbf{N}^\mathrm{T}}{\partial z}\right] dr\,dz \tag{3.75}$$

is an elemental stiffness matrix. Using the specific expressions developed previously, $\mathbf{k_e}$ can be further simplified to

$$\mathbf{k_e} = \iint_{A_e} \frac{[\mathbf{b_e} r \mathbf{b_e^T} + \mathbf{c_e} r \mathbf{c_e^T}]}{4A_e^2} \, dA$$

where $\mathbf{k_e}$ is a 3×3 elemental stiffness matrix that will contribute or add during assembly to the global stiffness matrix at those locations corresponding to the degrees of freedom associated with the particular element in question.

I_{e2}: The integral I_{e2} is given by

$$I_{e2} = \int_{\gamma_{2e}} \alpha r u^2 \, ds$$

where γ_{2e} is a typical straight-line segment of the Γ_2 portion of the boundary of the area in the rz-plane. Using the interpolated representation for u on the boundary results in

$$\begin{aligned} I_{e2} &= \int_{\gamma_{2e}} \mathbf{u_e^T N} \alpha r \mathbf{N^T u_e} \, ds = \mathbf{u_e^T} \int_{\gamma_{2e}} \mathbf{N} \alpha r \mathbf{N^T} \, ds \, \mathbf{u_e} \\ &= \mathbf{u_e^T a_e u_e} \end{aligned}$$

where

$$\mathbf{a_e} = \int_{\gamma_{2e}} \mathbf{N} \alpha r \mathbf{N^T} \, ds \tag{3.76}$$

is a 2×2 elemental stiffness matrix that contributes or adds during assembly to the global stiffness matrix at the positions corresponding to the two boundary nodes associated with the particular γ_{2e} in question.

I_{e3}: The integral I_{e3} is given by

$$I_{e3} = \iint_{A_e} urf \, dr \, dz = \iint_{A_e} \mathbf{u_e^T N} rf \, dr \, dz = \mathbf{u_e^T f_e}$$

where

$$\mathbf{f_e} = \iint_{A_e} \mathbf{N} fr \, dr \, dz \tag{3.77}$$

is a 3×1 elemental load vector that contributes or adds during assembly to the global load vector at the positions corresponding to the degrees of freedom associated with the particular A_e in question.

I_{e4}: The integral I_{e4} is given by

$$I_{e4} = \int_{\gamma_{2e}} urh \, ds = \int \mathbf{u_e^T N} rh \, ds = \mathbf{u_e^T h_e}$$

where

$$\mathbf{h_e} = \int_{\gamma_{2e}} \mathbf{N} r\, h(s)\, ds \tag{3.78}$$

is a 2×1 elemental load that contributes to the global load vector at the positions corresponding to the boundary degrees of freedom associated with the particular γ_{2e} in question.

Using Eqs. (3.75), (3.76), (3.77), and (3.78), the functional given by Eq. (3.69) can be expressed as

$$I(u_1, u_2, u_3, u_4, \ldots) = \sum \left[\frac{\mathbf{u_e^T k_e u_e}}{2} - \mathbf{u_e^T f_e} \right] + \sum{}' \left[\frac{\mathbf{u_e^T a_e u_e}}{2} - \mathbf{u_e^T h_e} \right] \tag{3.79}$$

where the first sum is over the area elements A_e of D and the second sum is over the line elements γ_{2e} of the Γ_2 portion of the boundary.

Assembly. Assembly is implied by the sums in Eq. (3.79). Each of the elemental matrices $\mathbf{k_e}$, $\mathbf{a_e}$, $\mathbf{f_e}$, and $\mathbf{h_e}$ is expanded to the global level with the result

$$I = \frac{\mathbf{u_G^T K_G u_G}}{2} - \mathbf{u_G^T F_G} = \mathbf{I(u_G)}$$

where

$$\mathbf{K_G} = \sum_e \mathbf{k_G} + \sum_e{}' \mathbf{a_G}$$

and

$$\mathbf{F_G} = \sum_e \mathbf{f_G} + \sum_e{}' \mathbf{h_G}$$

Note that the elemental matrices $\mathbf{k_G}$ and $\mathbf{f_G}$ each arise from an elemental area A_e, whereas the elemental matrices $\mathbf{a_G}$ and $\mathbf{h_G}$ each arise from an elemental line segment γ_{2e} on the Γ_2 portion of the boundary.

As mentioned several times previously, the finite element methodology has converted the functional given by Eq. (3.69) into the function given by Eq. (3.79). The stationary value of this function is obtained by requiring the partial derivatives with respect to each of the u_i to vanish; that is,

$$\frac{\partial I}{\partial u_i} = 0 \qquad i = 1, 2, \ldots, N$$

leading to

$$\frac{(\mathbf{K_G} + \mathbf{K_G^T})\mathbf{u_G}}{2} = \mathbf{F_G}$$

or since $\mathbf{K_G} = \mathbf{K_G^T}$,

$$\mathbf{K_G u_G} = \mathbf{F_G}$$

Thus we have an N × N set of linear algebraic equations for $\mathbf{u_G}$, or equivalently, the $u_i, i = 1, 2, \ldots, N$.

Constraints. The constraints arise from the essential boundary conditions

$$u = g(s) \qquad \text{on } \Gamma_1$$

The nodal values of the constraints are usually taken to be the function g evaluated at the nodal point in question. For this reason it is convenient to select the nodes *on* the bounding curve Γ. Each of the constraints is imposed in the usual manner by suitably altering the coefficient matrix $\mathbf{K_G}$ and the load matrix $\mathbf{F_G}$ as discussed in Chapters 1 and 2.

Solution. The reader is referred to Section 3.1 for a discussion of the character of the solution to the two-dimensional elliptic boundary value problem. Appendix B contains routines BSolve and BDecomp, which can be used for extracting the solution from the final set of algebraic equations.

Computation of derived variables. Recall from our treatment of the derivatives of the solution in terms of the interpolation functions that the partial derivatives have the form

$$\frac{\partial u}{\partial r} = \mathbf{u_e^T}\frac{\partial \mathbf{N}}{\partial r} = \frac{\mathbf{u_e^T b_e}}{2A_e} = \frac{\mathbf{b_e^T u_e}}{2A_e}$$

and

$$\frac{\partial u}{\partial z} = \mathbf{u_e^T}\frac{\partial \mathbf{N}}{\partial z} = \frac{\mathbf{u_e^T c_e}}{2A_e} = \frac{\mathbf{c_e^T u_e}}{2A_e}$$

both of which, as mentioned previously, are constants within each element.

Evaluation of matrices. The elemental matrices are given by

$$\mathbf{k_e} = \int\!\!\int_{A_e} \frac{[\mathbf{b_e} r \mathbf{b_e^T} + \mathbf{c_e} r \mathbf{c_e^T}]}{4A_e^2}\, dr\, dz$$

$$\mathbf{a_e} = \int_{\gamma_{2e}} \mathbf{N}\alpha r \mathbf{N}^T\, ds$$

$$\mathbf{f_e} = \int\!\!\int_{A_e} \mathbf{N} f r\, dr\, dz$$

$$\mathbf{h_e} = \int_{\gamma_{2e}} \mathbf{N} r h(s)\, ds$$

and are essentially the same as their counterparts of Section 3.3, with the obvious difference of the r appearing in each of the integrals. We will evaluate each of these in turn.

Evaluation of $\mathbf{k_e}$. Write $\mathbf{k_e}$ as

$$\mathbf{k_e} = \frac{[\mathbf{b_e b_e^T} + \mathbf{c_e c_e^T}]}{4A_e^2} \iint_{A_e} r\,dr\,dz$$

which can be evaluated exactly as

$$\mathbf{k_e} = \frac{[\mathbf{b_e b_e^T} + \mathbf{c_e c_e^T}]R}{4A_e}$$

where $R = (r_i + r_j + r_k)/3$ is the r-coordinate of the centroid of the element.

Evaluation of $\mathbf{f_e}$. Replacing r by its exact representation $\mathbf{r_e^T N}$, and assuming that f can be approximated by its linear interpolant as $\mathbf{N^T f}$, the elemental matrix $\mathbf{f_e}$ becomes

$$\mathbf{f_e} = \iint_{A_e} \mathbf{N r_e^T N N^T f}\,dr\,dz$$

Using the integration formulas introduced in Section 3.4,

$$\iint_{A_e} N_I^a N_J^b N_K^c\,dA = \frac{a!b!c!2A_e}{(a+b+c+2)!}$$

$\mathbf{f_e}$ becomes

$$\mathbf{f_e} = \frac{A_e}{60}\begin{bmatrix} 6r_I + 2r_J + 2r_K & 2r_I + 2r_J + r_K & 2r_I + r_J + 2r_K \\ & 2r_I + 6r_J + 2r_K & r_I + 2r_J + 2r_K \\ \text{symm} & & 2r_I + 2r_J + 6r_K \end{bmatrix}\begin{bmatrix} f_I \\ f_J \\ f_K \end{bmatrix}$$

If f is a constant, say f_0, this reduces to

$$\mathbf{f_e} = \frac{f_0 A_e}{12}\begin{bmatrix} 2r_I + r_J + r_K \\ r_I + 2r_J + r_K \\ r_I + r_J + 2r_K \end{bmatrix}$$

Evaluation of $\mathbf{a_e}$. Since $\mathbf{a_e}$ is evaluated on a line segment γ_{2e}, the interpolation functions reduce to their one-dimensional counterparts $(1-\mathbf{s}/l_e)$ and $\mathbf{s}/l_e$. Denote the two boundary nodes by J and K as shown in Fig. 3–116.

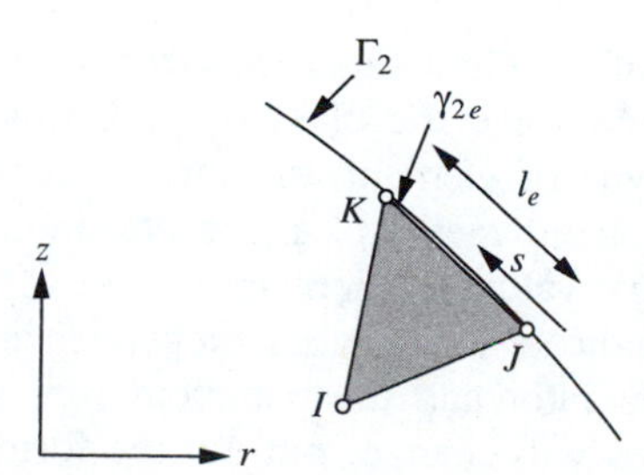

FIGURE 3–116 Typical γ_{2e} segment for the axisymmetric case

Again replacing r by its exact representation $\mathbf{r}_\mathbf{e}^T\mathbf{N}$ and assuming that α can be approximated by its linear interpolant as $\mathbf{N}^T\boldsymbol{\alpha}$, $\mathbf{a}_\mathbf{e}$ becomes

$$\mathbf{a}_\mathbf{e} = \int_{\gamma_{2e}} \mathbf{N}\mathbf{r}_\mathbf{e}^T\mathbf{N}\mathbf{N}^T\alpha\mathbf{N}^T ds$$

Using the integration formulas presented previously,

$$\int_{\gamma_{2e}} N_1^a N_2^b\, ds = a!b!\frac{l_e}{(a+b+1)!}$$

it follows that the components of $\mathbf{a}_\mathbf{e}$ are

$$[\mathbf{a}_\mathbf{e}]_{11} = \frac{l_e}{60}(12\alpha_J r_J + 3(\alpha_J r_K + \alpha_K r_J) + 2\alpha_k r_K)$$

$$[\mathbf{a}_\mathbf{e}]_{12} = \frac{l_e}{60}(3\alpha_J r_J + 2(\alpha_J r_K + \alpha_K r_J) + 3\alpha_k r_K)$$

$$[\mathbf{a}_\mathbf{e}]_{21} = [a_e]_{12}$$

$$[\mathbf{a}_\mathbf{e}]_{22} = \frac{l_e}{60}(2\alpha_J r_J + 3(\alpha_J r_K + \alpha_K r_J) + 12\alpha_k r_K)$$

Evaluation of $\mathbf{h}_\mathbf{e}$. In a completely similar fashion with h replaced by its linear interpolant $\mathbf{N}^T\mathbf{h}$ and r represented by $\mathbf{r}_\mathbf{e}^T\mathbf{N}$, $\mathbf{h}_\mathbf{e}$ becomes

$$\mathbf{h}_\mathbf{e} = \frac{l_e}{12}\begin{bmatrix} 3r_J + r_K & r_J + r_K \\ r_J + r_K & r_J + 3r_K \end{bmatrix}\begin{bmatrix} h_J \\ h_K \end{bmatrix}$$

completing the evaluation of the elemental matrices. It should be quite clear that any of the elements introduced in Section 3.8 can be used in connection with the functional given by Eq. (3.69) to develop the corresponding finite element model.

3.10 CLOSURE

Several points should be made regarding the general two-dimensional elliptic boundary value problem discussed in this chapter.

1. The general forms of the elementary matrices $\mathbf{k}_\mathbf{e}$, $\mathbf{a}_\mathbf{e}$, $\mathbf{f}_\mathbf{e}$, and $\mathbf{h}_\mathbf{e}$, which are necessary to formulate and solve the class of problems treated in this chapter, are independent of the type of element and interpolation used. As indicated in Fig. 3–117, the finite element method can be considered as a sort of operator that converts the boundary value problem into a set of corresponding algebraic equations. The arrows indicate the explicit dependence of the global matrices $\mathbf{K}_\mathrm{G}$ and $\mathbf{F}_\mathrm{G}$ on the ∇^2 operator and the parameters f, α, and h. The size and complexity of the matrices will change, but not the form.

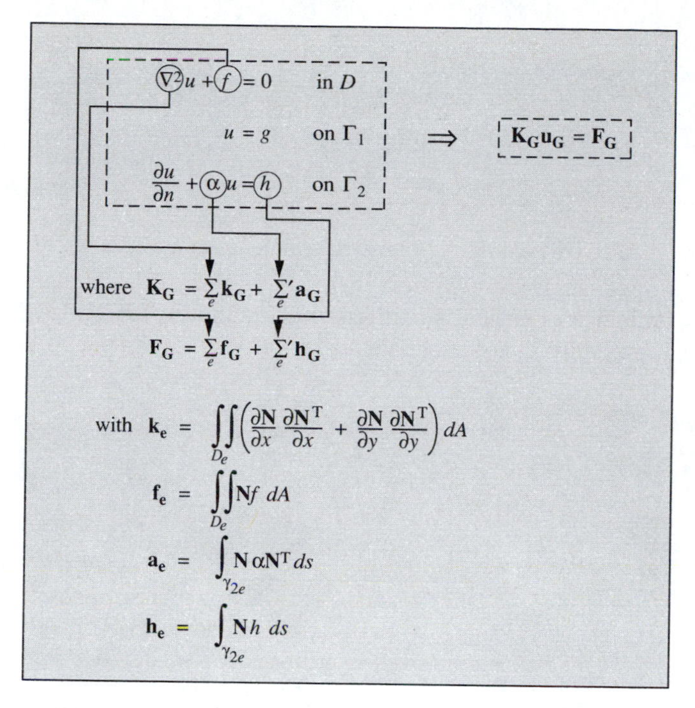

FIGURE 3–117 Conversion of the elliptic boundary value problem into the corresponding algebraic equations

2. Other classes of problems, such as ones governed by differential equations of the form

$$\frac{\partial(c_1 \partial u/\partial x)}{\partial x} + \frac{\partial(c_2 \partial u/\partial y)}{\partial y} + c_3 u + f = 0$$

with appropriate boundary conditions, can easily be handled using straightforward extensions of the material already presented. These extensions are left to the Exercises at the end of this chapter.

3. It must always be kept in mind that in choosing a mesh, extreme element shapes should be avoided. As mentioned previously, the equilateral triangle is ideal with substantial deviations likely to result in numerical difficulties or inaccuracies. Similarly, for the quadrilateral element, an aspect ratio (the ratio of the larger to the smaller side) much larger than 2.5 to 3 should be avoided. The student is encouraged to experiment with such parameters and find out what geometrical limits are permissible.

4. Interelement continuity should be observed. Shown in Fig. 3–118 are several permissible combinations of elements.

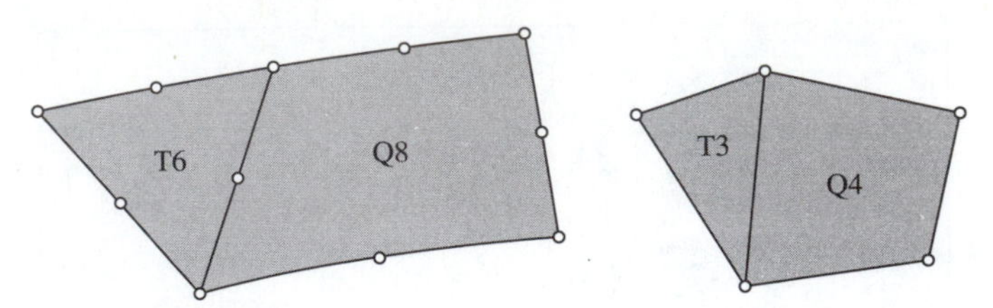

FIGURE 3–118 Permissible element combinations

Fig. 3–119 indicates typical situations that should be avoided, that is, where interelement continuity is not satisfied.

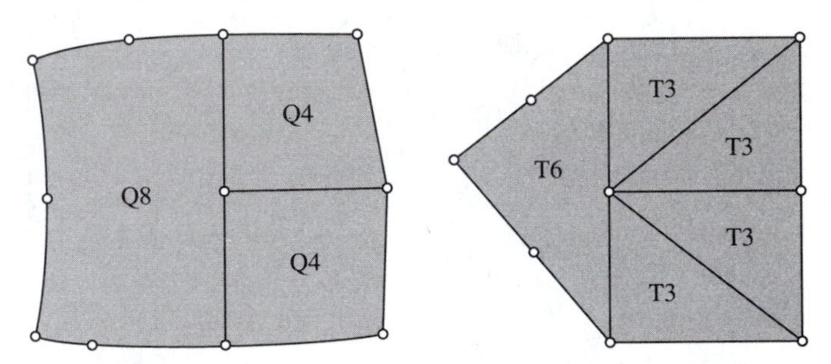

FIGURE 3–119 Absence of interelement continuity

It is also possible to develop elements that are useful in transition between variable mesh regions such as indicated in Fig. 3–120.

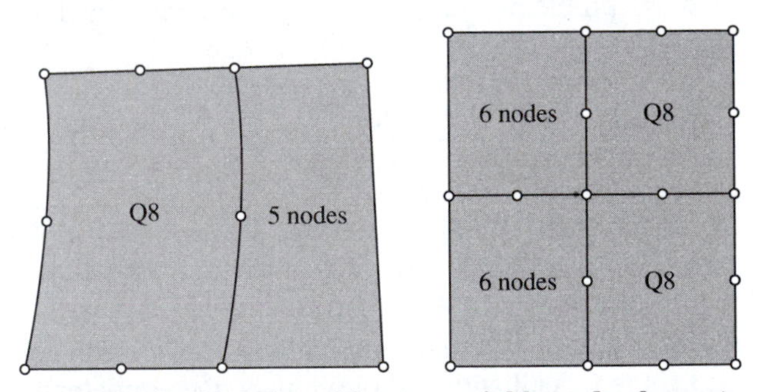

FIGURE 3–120 Transition or variable node elements

This type of element is often called a variable node element. The reader is referred to the References for a discussion of these elements.

5. Another important practical aspect of choosing the mesh has to do with the bandwidth of the global stiffness matrix. Generally a matrix is said to be banded if all the nonzero elements are positioned near the main diagonal. The two following matrices indicate bandedness.

$$\begin{bmatrix} 2 & -1 & & & & 0\text{'s} \\ -1 & 3 & -1 & & & \\ & -1 & 4 & -2 & & \\ & & -2 & 4 & -2 & \\ & & & -2 & 5 & -1 \\ 0\text{'s} & & & & -2 & 4 \end{bmatrix} \begin{bmatrix} 3 & -2 & -1 & & & & 0\text{'s} \\ -2 & 4 & -2 & -1 & & & \\ -1 & -2 & 5 & -2 & -1 & & \\ & -1 & -2 & 5 & -2 & -1 & \\ & & -1 & -2 & 5 & -2 & -1 \\ & & & -1 & -2 & 4 & -1 \\ 0\text{'s} & & & & -1 & -1 & 3 \end{bmatrix}$$

For a symmetric matrix **K**, we state that if

$$[K]_{ij} = 0 \qquad \text{for } j > i + \text{NB} - 1$$

K has a half bandwidth of NB and a bandwidth of 2NB − 1. For the matrices above, the bandwidths are 3 and 5 with NB = 2 and NB = 3, respectively.

When a symmetric coefficient matrix is known to be banded it is only necessary to store the elements on and below (or above) the main diagonal, often resulting in a significant savings in core storage required. The storage required for a symmetric matrix is approximately $n^2/2$, while for a symmetric banded matrix the required storage is approximately $n \times \text{NB}$. Equation solvers are available for banded matrices that can result in substantial reduction in the cost (time) of solving the final equations. The ratio of the times to solve a symmetric banded set of equations compared to a symmetric unbanded set of equations is, for large n, approximately $3(\text{NB}/n)^2$, which can be significant.

These ideas are illustrated by the following example. Consider the two numbering choices for the region shown in Fig.3–121.

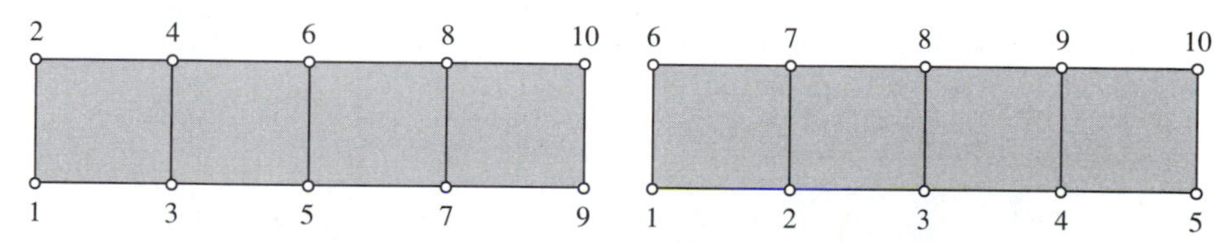

FIGURE 3–121 Two numbering schemes for a region

We assume that there is one degree of freedom per node. For the choice on the left, the maximum difference in node numbers that define any element is three, whereas for the choice on the right, it is six. With x's indicating the potential nonzero elements, the two resulting stiffness matrices are:

$$\begin{bmatrix} x & x & x & x & & & & & & \\ x & x & x & x & & & & & & \\ x & x & x & x & x & x & & & & \\ x & x & x & x & x & x & & & & \\ & & x & x & x & x & x & x & & \\ & & x & x & x & x & x & x & & \\ & & & & x & x & x & x & x & x \\ & & & & x & x & x & x & x & x \\ & & & & & & x & x & x & x \\ & & & & & & x & x & x & x \end{bmatrix} \begin{bmatrix} x & x & & & & x & x & & & \\ x & x & x & & & x & x & x & & \\ & x & x & x & & & x & x & x & \\ & & x & x & x & & & x & x & x \\ & & & x & x & & & & x & x \\ x & x & & & & x & x & & & \\ x & x & x & & & x & x & x & & \\ & x & x & x & & & x & x & x & \\ & & x & x & x & & & x & x & x \\ & & & x & x & & & & x & x \end{bmatrix}$$

For the two choices, NB is clearly 4 and 7, respectively. Generally, a smaller bandwidth results when numbering takes place across the smaller dimension first.

Finally, it is worth mentioning again that when setting up and solving a finite element model of a physical problem, the analyst must be careful to make sure that all dimensional aspects are considered and that a consistent set of units is used.

REFERENCES

1. Boresi, A. P., and K. P. Chong: *Elasticity in Engineering Mechanics,* Elsevier, New York, 1987.
2. Mills, A. F.: *Heat Transfer,* Richard D. Irwin, Homewood, Illinois, 1992.
3. Courant, R., and D. Hilbert: *Methods of Mathematical Physics,* vol.II, Wiley Interscience, New York, 1962.
4. Kaplan, W., and D. J. Lewis: *Calculus and Linear Algebra,* John Wiley, New York, 1971.
5. Timoshenko, S. P., and J. N. Goodier: *Theory of Elasticity,* McGraw-Hill, New York, 1970.
6. Smith, B. T., et al.: *Lecture Notes in Computer Science, Matrix Eigensystem Routines—EISPACK Guide,* Springer-Verlag, Berlin, 1976.
7. Garbow, G. S., et al.: *Lecture Notes in Computer Science, Matrix Eigensystem Routines—EISPACK Guide Extension,* Springer-Verlag, Berlin, 1977.
8. Moan, T.: "On the Local Distribution of Errors by Finite Element Approximations," in *Theory and Practice in Finite Element Structural Analysis*, Y. Yamada and R. H. Gallagher, eds., University of Tokyo Press, Tokyo, 1973.
9. Griffith, A. A., and G. I. Taylor: "The Use of Soap Films in Solving Torsional Problems," *Proceedings of the Institution of Mechanical Engineers*, Oct.–Dec., 1917, p. 755.
10. Hinton, E., and J. S. Campbell: "Local and Global Smoothing of Discontinuous Finite Element Functions Using a Least Squares Method," *International Journal for Numerical Methods in Engineering,* vol. 8, pp. 461–480, 1974.
11. Cowper, G. R.: "Gaussian Quadrature Formulas for Triangles," *International Journal for Numerical Methods in Engineering,* vol. 7, pp. 405–408, 1973.

GENERAL REFERENCES

Akin, J. E.: *Finite Element Analysis for Undergraduates,* Academic Press, London, 1986.

Allaire, P. E.: *Basics of the Finite Element Method—Solid Mechanics, Heat Transfer and Fluid Mechanics,* Wm. C. Brown, Dubuque, Iowa, 1985.

Axelsson, O., and V. A. Barker: *Finite Element Solution of Boundary Value Problems—Theory and Computation,* Academic Press, Orlando, 1984.

Bathe, K. J.: *Finite Element Procedures in Engineering Analysis,* Prentice Hall, Englewood Cliffs, New Jersey, 1982.

Bathe, K. J., and E. L. Wilson: *Numerical Methods in Finite Element Analysis,* Prentice Hall, Englewood Cliffs, New Jersey, 1976.

Becker, E. B., et al.: *Finite Elements—An Introduction,* vol. I, Prentice Hall, Englewood Cliffs, New Jersey, 1981.

Burnett, D. S.: *Finite Element Analysis, From Concepts to Applications,* Addison-Wesley, Reading, Massachusetts, 1987.

Carey, G. F., and J. T. Oden: *Finite Elements—A Second Course,* vol. II, Prentice Hall, Englewood Cliffs, New Jersey, 1983.

——: *Finite Elements—Computational Aspects,* vol. III, Prentice Hall, Englewood Cliffs, New Jersey, 1984.

Chung, T. J.: *Finite Element Analysis in Fluid Dynamics,* McGraw-Hill, New York, 1978.
Ciarlet, P. G.: *The Finite Element Method for Elliptic Problems,* North-Holland, Amsterdam, 1978.
Connor, J. C., and C. A. Brebbia: *Finite Element Techniques for Fluid Flow,* Butterworths, London, 1976.
Cook, R. D.: *Concepts and Applications of Finite Element Analysis,* John Wiley, New York, 1981.
Desai, C. S., and J. F. Abel: *Introduction to the Finite Element Method,* Van Nostrand Reinhold, New York, 1972.
Fletcher, C. A. J.: *Computational Galerkin Methods,* Springer-Verlag, New York, 1984.
Gallagher, R. H.: *Finite Element Analysis, Fundamentals,* Prentice Hall, Englewood Cliffs, New Jersey, 1975.
Gelfand, I. M., and S. V. Fomin: *Calculus of Variations,* Prentice Hall, Englewood Cliffs, New Jersey, 1963.
Gould, S. H.: *Variational Methods for Eigenvalue Problems,* Toronto Press, Toronto, 1966.
Huebner, K. H., and E. A. Thornton: *The Finite Element Method for Engineers,* John Wiley, New York, 1982.
Hughes, T. J. R.: *The Finite Element Method, Linear Static and Dynamic Finite Element Analysis,* Prentice Hall, Englewood Cliffs, New Jersey, 1987.
Livesley, R. K.: *Finite Elements: An Introduction for Engineers,* Cambridge University Press, Cambridge, UK, 1983.
Martin, H. C., and G. F. Carey: *Introduction to Finite Element Analysis—Theory and Application,* McGraw-Hill, New York, 1973.
Noble, B., and J. W. Daniel: *Applied Linear Algebra,* Prentice Hall, Englewood Cliffs, New Jersey, 1977.
Norrie, D. H., and G. de Vries: *The Finite Element Method: Fundamentals and Applications,* Academic Press, Orlando, 1973.
——: *An Introduction to Finite Element Analysis,* Academic Press, Orlando, 1978.
Rao, S. S.: *The Finite Element Method in Engineering,* Pergamon, Oxford, 1982.
Reddy, J. N.: *Energy and Variational Methods in Applied Mechanics with an Introduction to the Finite Element Method,* Wiley-Interscience, New York, 1984.
——: *An Introduction to the Finite Element Method,* McGraw-Hill, New York, 1984.
Strang, G.: *Linear Algebra and its Applications,* Academic Press, New York, 1980.
Strang, G., and G. Fix: *An Analysis of the Finite Element Method,* Prentice Hall, Englewood Cliffs, New Jersey, 1973.
Tong, P., and J. N. Rossettos: *Finite Element Method: Basic Technique and Implementation,* MIT Press, Cambridge, Massachusetts, 1977.
Wait, R., and A. R. Mitchell: *Finite Element Analysis and Applications,* John Wiley, New York, 1985.
Weinstock, R.: *Calculus of Variations with Applications to Physics and Engineering,* McGraw-Hill, New York, 1952.
White, R. E.: *An Introduction to the Finite Element Method with Applications to Non-linear Problems,* Wiley-Interscience, New York, 1985.
Zienkiewicz, O. C.: *The Finite Element Method,* Third Edition, McGraw-Hill, New York, 1977.

COMPUTER PROJECTS

Develop a computer code for the general two-dimensional boundary value problem

$$\begin{aligned} \nabla^2 u + f &= 0 \qquad \text{in } D \\ u &= g \qquad \text{on } \Gamma_1 \\ \frac{\partial u}{\partial n} + \alpha u &= h \qquad \text{on } \Gamma_2 \end{aligned}$$

Structure the program in roughly the following fashion.

INPUT.

1. NN — Number of nodes.
2. NE — Number of elements.
3. Nodal and f coordinates
   - X(I) I = 1,NN
   - Y(I) I = 1,NN
   - F(I) I = 1,NN
4. Element connectivity, NI,NJ,NK, . . .
   - NI(I) I = 1,NE
   - NJ(I) I = 1,NE
   - NK(I) I = 1,NE

   ⋮

The number of sets entered depends upon the type of element being used, that is, three sets for the linearly interpolated triangular element, eight sets for the Q8 element.

5. N2, the number of segments on which type two boundary conditions are prescribed.

N2S(I)	I = 1 to N2	Each of these integers represents the number of nodes in the Ith segment.

Then for each segment, input:

M2(I,J)	J = 1 to N2S(I)	Representing the node numbers on the Ith segment.
H(I,J)	J = 1 to N2S(I)	Each of these real numbers represents the value of the function h at the Jth node on the Ith segment.
A(I,J)	J = 1 to N2S(I)	Each of these real numbers represents the value of the function α at the Jth node on the Ith segment.

6. NC, the number of constrained nodes.

ND(I)	I = 1, NC	The constrained node numbers.
C(I)	I = 1, NC	The constrained values.

DATA REFLECTION. It is a good idea to print out all of the above data for a check.

ELEMENT FORMULATION AND ASSEMBLY. $\mathbf{k_e}$ and $\mathbf{f_e}$ need to be formed and assembled for each of the NE elements. These steps are probably best handled in separate subroutines or procedures. The X, Y, F, NI, NJ, . . . data are needed here. Also, $\mathbf{a_e}$ and $\mathbf{h_e}$ need to be formed and assembled for each of the N2 segments on the boundary. The X, Y, N2, N2S, H, and A data are required.

CONSTRAINTS. All of the constraint data must be imposed on the assembled equations. The NC, ND, and C data are required. The equations should also be put into symmetric form before solving.

SOLUTION. An appropriate equation solver is employed to determine the solution for the nodal u's. Routines BDecomp and BSolve are provided in Appendix B.

COMPUTATION OF DERIVED VARIABLES. The derived variable $\partial\phi/\partial n$ is computed for each element. The point(s) at which pu' is evaluated should be the best points mentioned in the text. Information generated during the course of determining the elemental stiffnesses should have been saved for reuse at this stage.

Specifically, the following codes are easily generated, based on the material presented in the chapter.

1. Linearly interpolated triangular elements.
2. Bilinearly interpolated quadrilateral (Q4) elements.
3. A combination of 1 and 2.
4. Quadratically interpolated (T6) triangular elements.
5. Quadratically interpolated (Q8) quadrilateral elements.
6. A combination of 4 and 5.

Each of these projects can also be carried out for the Helmholtz eigenvalue problem. The equation solver becomes a suitable routine for extracting the eigenvalues and eigenvectors. There are several routines in Appendix C for this purpose.

Codes using T3 and T6 elements to solve the elliptic boundary value problem are contained on the diskette available from your instructor.

EXERCISES

Section 3.2

3.1. Go through the details of showing that for the generalized problem

$$\frac{\partial}{\partial x}\left(a(x,y)\frac{\partial u}{\partial x}\right)+\frac{\partial}{\partial y}\left(b(x,y)\frac{\partial u}{\partial y}\right)-c(x,y)u+f=0 \qquad \text{in } D$$

$$u=g(s) \quad \text{on } \Gamma_1$$

$$n_x a\frac{\partial u}{\partial x}+n_x b\frac{\partial u}{\partial y}+\alpha u=h(s) \quad \text{on } \Gamma_2$$

with the approximate solution taken as $v(x,y)=\sum n_i v_i$, the final set of equations is given by $\mathbf{Kv}=\mathbf{F}$, where

$$K_{ki}=\iint_{D^-}\left[\frac{\partial n_k}{\partial x}a(x,y)\frac{\partial n_i}{\partial x}+\frac{\partial n_k}{\partial y}b(x,y)\frac{\partial n_i}{\partial y}+n_k c(x,y)n_i\right]dA+\int_{\Gamma_{2-}}\alpha n_k n_i\,ds$$

and

$$F_k=\iint_{D^-}n_k f\,dA+\int_{\Gamma_{2-}}n_k h(s)\,ds$$

3.2. For the generalization of the previous problem show that the elemental matrices can be written as

$$\mathbf{k_{ei}}=\iint_{D_i}\left[\frac{\partial\mathbf{N}}{\partial x}a(x,y)\frac{\partial\mathbf{N}^{\mathrm{T}}}{\partial x}+\frac{\partial\mathbf{N}}{\partial y}b(x,y)\frac{\partial\mathbf{N}^{\mathrm{T}}}{\partial y}+\mathbf{N}c(x,y)\,\mathbf{N}^{\mathrm{T}}\right]dA$$

$$\mathbf{a_{ei}}=\int_{\alpha_{2i}}\mathbf{N}\alpha\mathbf{N}^{\mathrm{T}}\,ds \qquad \text{and} \qquad \mathbf{f_{ei}}=\iint_{D_i}\mathbf{N}f\,dA \qquad \text{and} \qquad \mathbf{h_{ei}}=\int_{\alpha_{2i}}\mathbf{N}h\,ds$$

In solving the next group of problems the reader can refer to the integration formulas for triangles stated in Section 3.4.

3.3. Verify that the solution to Eqs. (3.2) can be expressed in terms of the a_i, b_i, and c_i given by Eqs. (3.3).

3.4. Show that the derivatives of the elemental interpolation vector **N** can be expressed as

$$\frac{\partial \mathbf{N}}{\partial x} = \frac{\mathbf{b}_\mathbf{e}^\mathrm{T}}{2A_e} \quad \text{and} \quad \frac{\partial \mathbf{N}}{\partial y} = \frac{\mathbf{c}_\mathbf{e}^\mathrm{T}}{2A_e}$$

where

$$\mathbf{b_e} = [x_j - x_k \quad x_k - x_i \quad x_i - x_j]^\mathrm{T} \quad \text{and} \quad \mathbf{c_e} = [y_k - y_j \quad y_i - y_k \quad y_j - y_i]^\mathrm{T}$$

3.5. Show that the elemental stiffness matrix

$$\mathbf{k_{ei}} = \iint_{D_i} \left[\frac{\partial \mathbf{N}}{\partial x} \frac{\partial \mathbf{N}^\mathrm{T}}{\partial x} + \frac{\partial \mathbf{N}}{\partial y} \frac{\partial \mathbf{N}^\mathrm{T}}{\partial y} \right] dA$$

can be evaluated explicitly to yield

$$\mathbf{k_{ei}} = \frac{\mathbf{b_e b_e^T} + \mathbf{c_e c_e^T}}{4A_\mathbf{e}}$$

3.6. Show that with the nodes ordered as shown, the stiffness matrix for a right isosceles triangle becomes

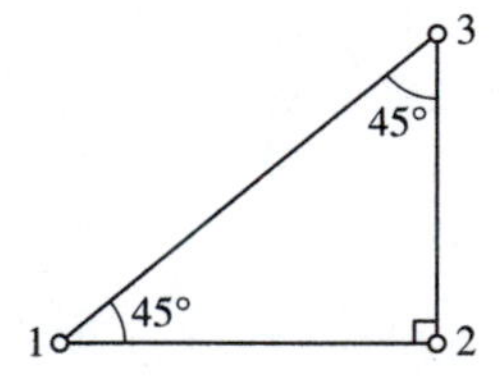

$$k_e = \frac{1}{2} \begin{bmatrix} 1 & -1 & 0 \\ -1 & 2 & -1 \\ 0 & -1 & 1 \end{bmatrix}$$

3.7. Show that for $f = f_0$, a constant,

$$\mathbf{f_e} = \frac{f_0 A_e}{3} \begin{bmatrix} 1 \\ 1 \\ 1 \end{bmatrix}$$

3.8. Show that if f is a linear function of position on an element, i.e., $f = f_i N_i + f_j N_j + f_k N_k$,

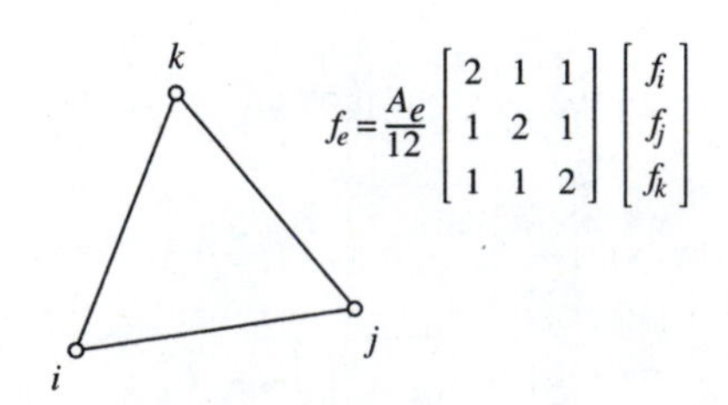

$$f_e = \frac{A_e}{12} \begin{bmatrix} 2 & 1 & 1 \\ 1 & 2 & 1 \\ 1 & 1 & 2 \end{bmatrix} \begin{bmatrix} f_i \\ f_j \\ f_k \end{bmatrix}$$

3.9. Show that if α is a constant, say α_0

$$\mathbf{a_e} = \left(\frac{\alpha_0 \ell_e}{6}\right)\begin{bmatrix} 2 & 1 \\ 1 & 2 \end{bmatrix}$$

3.10. Show that if α is a linear function of position on a segment of the boundary δ_{2i} defined by nodes I and J (i.e., $\alpha = \alpha_I N_I + \alpha_J N_J$),

$$\boldsymbol{\alpha}_\mathbf{e} = \frac{\ell_e}{12}\begin{bmatrix} 3\alpha_I + \alpha_J & \alpha_I + \alpha_J \\ \alpha_I + \alpha_J & \alpha_I + 3\alpha_J \end{bmatrix}$$

Section 3.3

***3.11.** Determine the functional $I(u)$ corresponding to the boundary value problem described in 3.1.

***3.12.** Verify that for the functional

$$2I(u) = \int\int_D \left(\left(\frac{\partial u}{\partial x}\right)^2 + \left(\frac{\partial u}{\partial y}\right)^2 - 2fu\right) dA + \int_{\Gamma_2} (\alpha u^2 - 2hu)\, ds$$

with $u = g$ on Γ_1, requiring $\delta I = 0$ leads to

$$\begin{aligned} \nabla^2 u + f &= 0 \qquad \text{in } D \\ u &= g \qquad \text{on } \Gamma_1 \\ \frac{\partial u}{\partial n} + \alpha u &= h \qquad \text{on } \Gamma_2 \end{aligned}$$

Proceed by forming

$$\begin{aligned} \Delta I &= I(u + \delta u) - I(u) = I(u + \epsilon\eta) - I(u) \\ &= \epsilon\, \delta I + \frac{\epsilon^2\, \delta^2 I}{2!} + \cdots \end{aligned}$$

and requiring

$$\frac{dI(0)}{d\epsilon} = \lim_{\epsilon \to 0} \frac{\Delta I}{\epsilon} = \delta I = 0$$

as was described in the Exercises in Chapter 2. You will need to use the two-dimensional form of the divergence theorem introduced in Section 3.2.

***3.13.** Repeat 3.12 for the functional developed in 3.11.

Section 3.4

3.14. For linear interpolation, verify that the two expressions for the elemental stiffness $\mathbf{k_e}$, namely,

$$\mathbf{k_e} = \int\int (\Delta\{\mathbf{J}_1^T\mathbf{J}_1 + \mathbf{J}_2^T\mathbf{J}_2\}\, \Delta^T)|\mathbf{J}|\, dA(L_1, L_2)$$

and

$$\mathbf{k_e} = \frac{\mathbf{b_e b_e^T} + \mathbf{c_e c_e^T}}{4A_e}$$

are exactly the same.

3.15. The nodal values of u for the linearly interpolated triangular element shown are $u_1 = 1.6548$, $u_2 = 2.6529$, and $u_3 = 3.0864$.

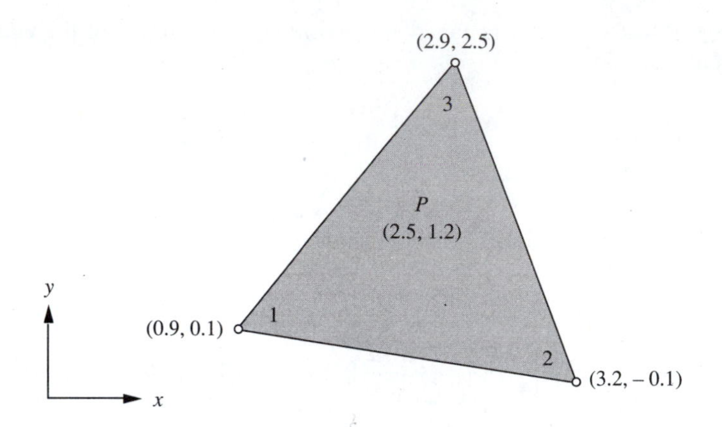

Determine $u(P)$. Determine also $\partial u/\partial n$ along sides 1-2, 2-3, and 3-1 using

$$\frac{\partial u}{\partial n} = \frac{[n_x \mathbf{b}_\mathbf{e}^\mathrm{T} + n_y \mathbf{c}_\mathbf{e}^\mathrm{T}]\mathbf{u}_\mathbf{e}}{2A_e}$$

3.16. Show that

$$\frac{\partial u}{\partial x} = \frac{\mathbf{b}_\mathbf{e}^\mathrm{T}\mathbf{u}_\mathbf{e}}{2A_e} \quad \text{and} \quad \frac{\partial u}{\partial x} = \mathbf{J}_1 \Delta^\mathrm{T} \mathbf{u}_\mathbf{e}$$

are precisely the same. Similarly show that

$$\frac{\partial u}{\partial y} = \frac{\mathbf{c}_\mathbf{e}^\mathrm{T}\mathbf{u}_\mathbf{e}}{2A_e} \quad \text{and} \quad \frac{\partial u}{\partial y} = \mathbf{J}_2 \Delta^\mathrm{T} \mathbf{u}_\mathbf{e}$$

are precisely the same.

3.17. Repeat the second part of Exercise 3.15 using

$$\frac{\partial u}{\partial n} = [n_x \mathbf{J}_1 + n_y \mathbf{J}_2]\mathbf{\Delta}^\mathrm{T}\mathbf{u}_\mathbf{e}$$

3.18. Verify the approximate evaluation for $\mathbf{f}_\mathbf{e}$, namely

$$\mathbf{f}_\mathbf{e} = \frac{A_e}{12}\begin{bmatrix} 2 & 1 & 1 \\ 1 & 2 & 1 \\ 1 & 1 & 2 \end{bmatrix}\begin{bmatrix} f_i \\ f_j \\ f_k \end{bmatrix}$$

3.19. Consider the specific case where $f = xy$ and the element is as follows.

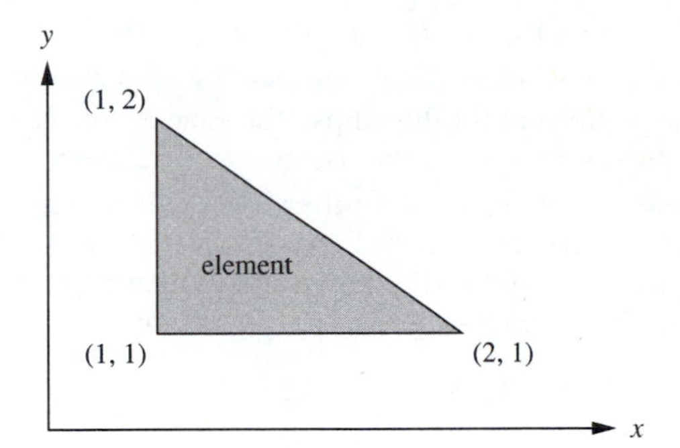

Evaluate $\mathbf{f_e}$ exactly by constructing the interpolation functions for the element shown, and compare the results with those obtained using the approximate evaluation scheme of Exercise 3.18.

3.20. Repeat Exercise 3.19 for the region follows:

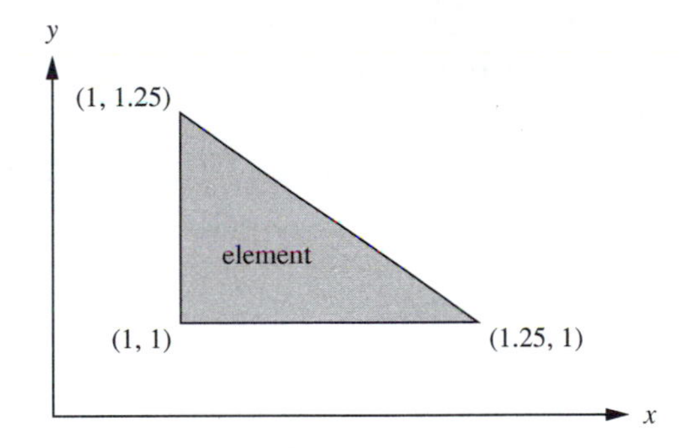

The point of Exercises 3.19 and 3.20 is that the results from the approximate formula for $\mathbf{f_e}$ approach the exact results as the mesh is refined.

Section 3.5

3.21. Set up and solve the problem of the torsion of an ellipse for the mesh shown. Determine the maximum shear stress and the torsional constant. State the boundary conditions that must be satisfied on each part of the boundary.

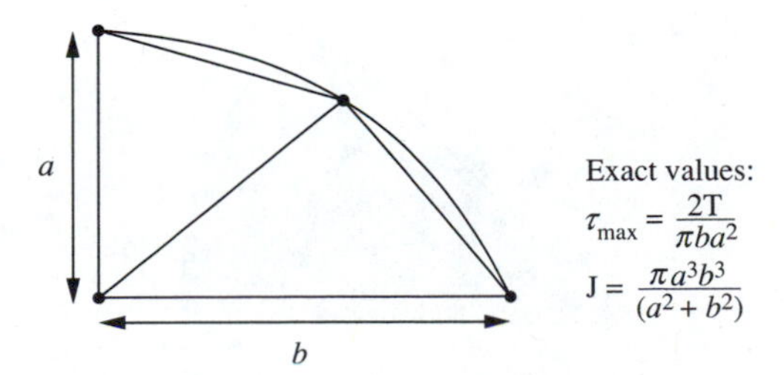

3.22. Repeat the torsion problem for the ellipse for a mesh obtained by "halving" the one shown for the previous problem, that is, one with eight elements.

3.23. Repeat the torsion problem for the ellipse for a mesh obtained by halving the one shown for the previous problem, that is, one with 32 elements.

3.24. For the mesh shown, set up and solve the torsion problem for the circle. Compare your results for the maximum shear stress expressed in terms of the torque T with the exact expression $2T/\pi R^3$. Also compare the expression for the torsional constant with the exact expression $\pi R^4/2$. Recall that $T = 2\int\int \phi\, dA$.

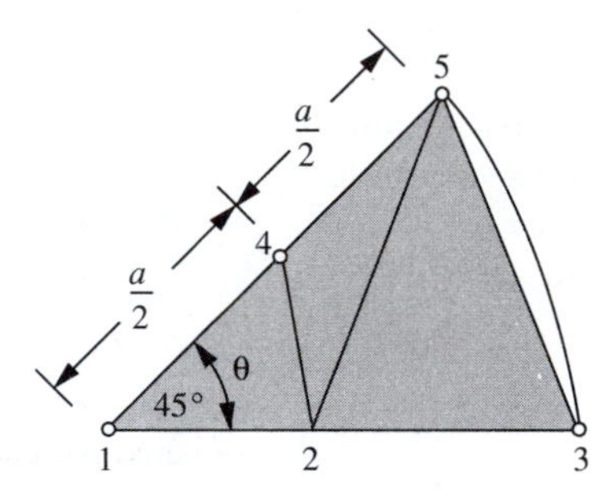

Should the answers depend upon the angle θ? What boundary conditions should be used on the radial lines of the model? Check to see how well these boundary conditions are satisfied.

3.25. Repeat Exercise 3.24 for the mesh indicated.

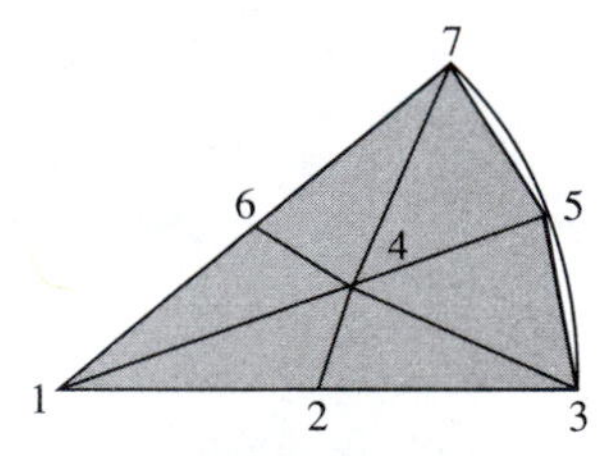

3.26. Repeat Exercise 3.24 for the mesh indicated.

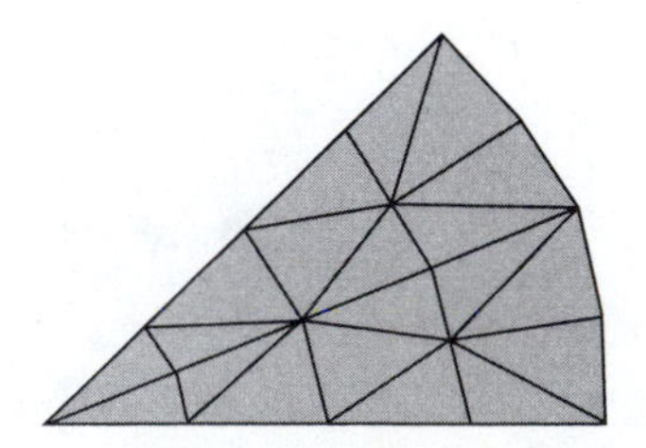

3.27. Using eightfold symmetry for torsion of the square cross section, set up and solve the model indicated. Determine the maximum shear stress in terms of the torque T, and also the torsional constant. Compare your results with the exact expressions $\tau_{max} = 4.80T/a^3$ and $J = 0.141a^4$ respectively.

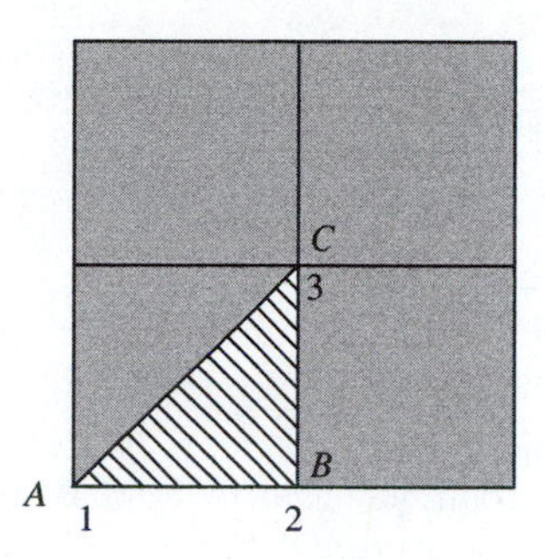

What boundary conditions should be used on lines AC and BC? Check to see how well these boundary conditions are satisfied for each of the models.

3.28. Repeat Exercise 3.27 for the mesh indicated.

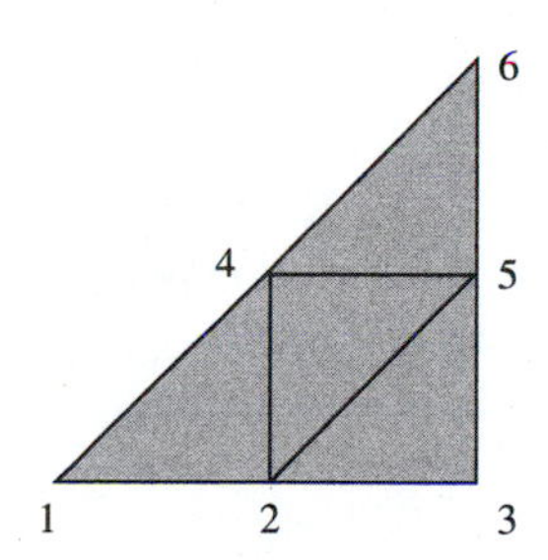

3.29. Repeat Exercise 3.27 for the mesh indicated.

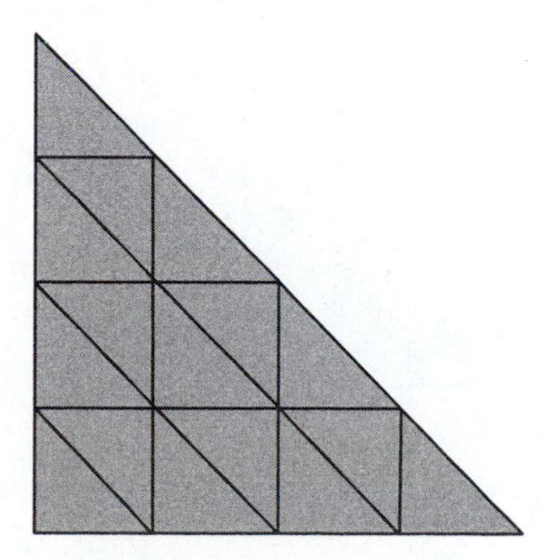

3.30. One of the situations that results when using the triangular element of Section 3.4 is that there is a bias depending on the manner in which the diagonals are chosen. For example, for the meshes of Exercises 3.28 and 3.29 the diagonals could be chosen in either of the two ways indicated.

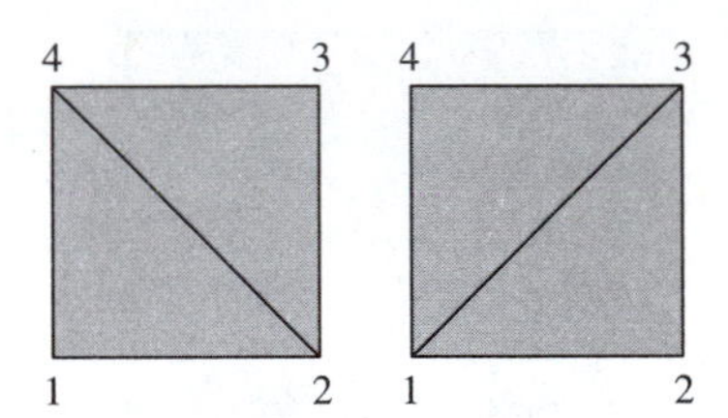

What is the 4×4 contribution to the global stiffness matrix in each case? Compute a third case as the average of the two. Do the same for the contributions to the global load matrix for a constant $f(x, y) = f_0$. Does the bias in the case of the triangular elements result from the $\mathbf{k_e}$ or from the $\mathbf{f_e}$? Note that this is the special case of the diagonals being chosen in the two possible ways for a rectangle. The more general case of the quadrilateral is posed in the next Exercise.

3.31. Repeat Exercise 3.30 for a general quadrilateral shape as indicated.

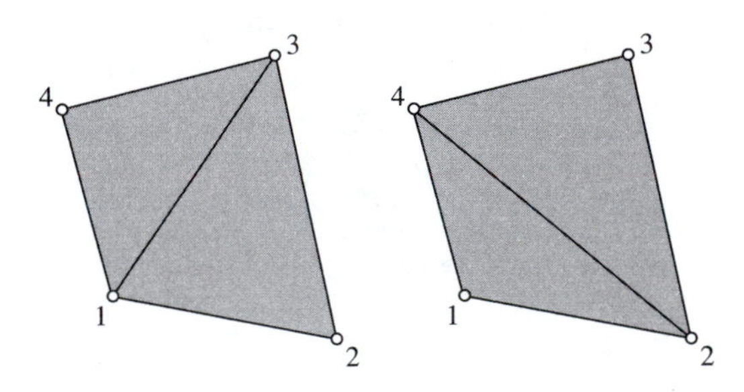

3.32. Return to Exercise 3.24 and resolve the torsion problem using the mesh indicated. Compare the ϕ values with those obtained in Exercise 3.24.

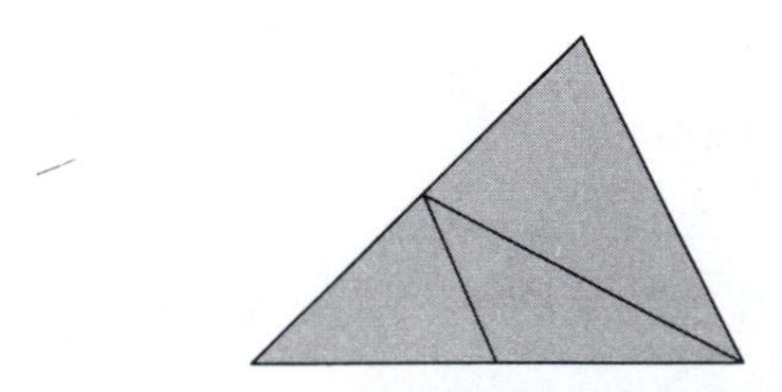

3.33. Repeat Exercise 3.24 for the mesh indicated. Compare the ϕ values with those obtained in Exercise 3.25.

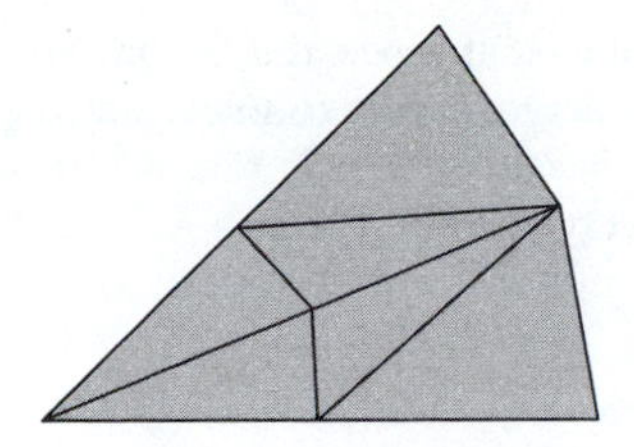

3.34. Repeat Exercise 3.24 for the mesh indicated. Compare the ϕ values with those obtained in Exercise 3.26.

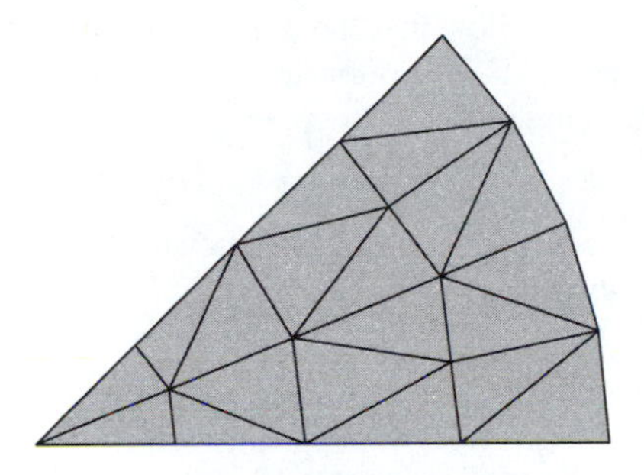

Using each of the following meshes, set up and resolve Exercise 3.27.

3.35. Triangular mesh 1. Compare the results for the ϕ's with those for Exercise 3.28.

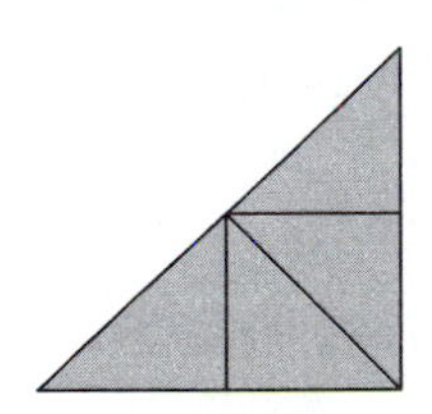

3.36. Triangular mesh 2. Compare also the results for the ϕ's with those for Exercise 3.29.

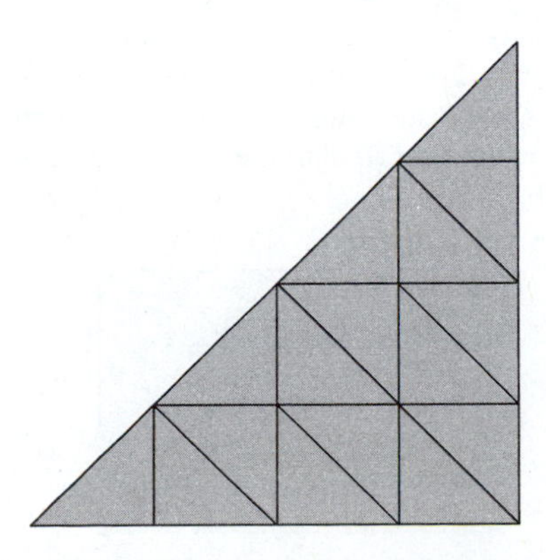

In Exercises 3.35 and 3.36, note that since the global stiffness matrices are the same regardless of the choice of the diagonals, it follows that the solution for the average of the two inputs $\mathbf{F}_1$ and $\mathbf{F}_2$ is $\boldsymbol{\phi}_3 = (\boldsymbol{\phi}_1 + \boldsymbol{\phi}_2)/2$, that is, since the two problems using the two triangular meshes respectively produce

$$\mathbf{K}_G\boldsymbol{\phi}_1 = \mathbf{F}_1$$

and

$$\mathbf{K}_G\boldsymbol{\phi}_2 = \mathbf{F}_2$$

the solution to $(\mathbf{F}_1 + \mathbf{F}_2)/2$ is $(\boldsymbol{\phi}_1 + \boldsymbol{\phi}_2)/2$. For the quadrilateral cases of Exercises 3.32, 3.33, and 3.34, the $\mathbf{K}_G$'s and $\mathbf{F}_G$'s are both different for the two different choices of the diagonals and hence it does not follow that the average of the two outputs corresponds to the average of the two inputs. It is still often useful to consider the average of the two solutions for the quadrilateral case.

****

An electric current is generated in the copper plate shown. Assume that the energy created is constant q_0 throughout the volume of the plate.

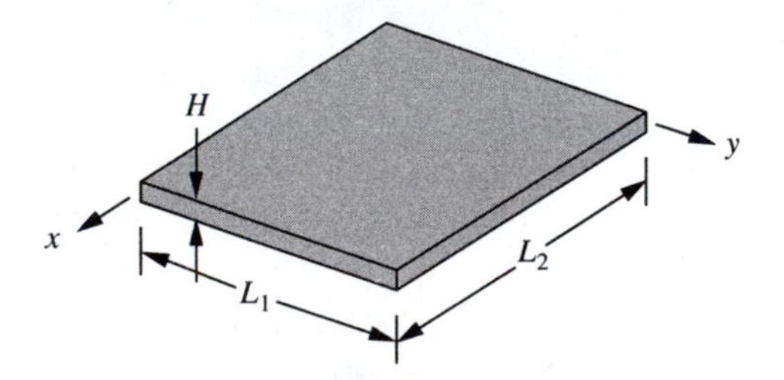

Take $k = 40$ BTU/hr-ft-°F, $H = 1.0$ in., $L_1 = 15$ in., $L_2 = 10$ in., $h = 200.0$ BTU/hr-ft^2-°F, $q_0 = 10^5$ BTU/hr-ft^3, and $T_0 = 100°$ F. Set up and solve finite element models for two-dimensional steady-state heat conduction for each of the following physical models and meshes.

3.37. The surfaces $z = \pm H/2$ are insulated (no convection) and the edges of the plate are held at a fixed temperature T_0.

3.38. The surfaces $z = \pm H/2$ are insulated (no convection) with the temperature at the edges satisfying a convective boundary condition $k\, \partial T/\partial n = -h(T - T_0)$.

3.39. Both surfaces $z = \pm H/2$ convect energy according to $h(T - T_0)$ with the temperature at the edges also satisfying a convective boundary condition $k\, \partial T/\partial n = -h(T - T_0)$. The governing differential equation for this case is of the form given in Exercise 3.1 with $c \neq 0$.

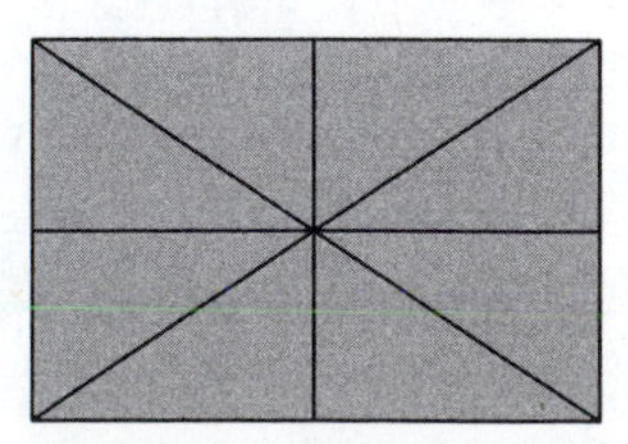

Mesh for Exercises 3.37–3.39

3.40–3.42. Repeat Exercises 3.37–3.39 for the following mesh.

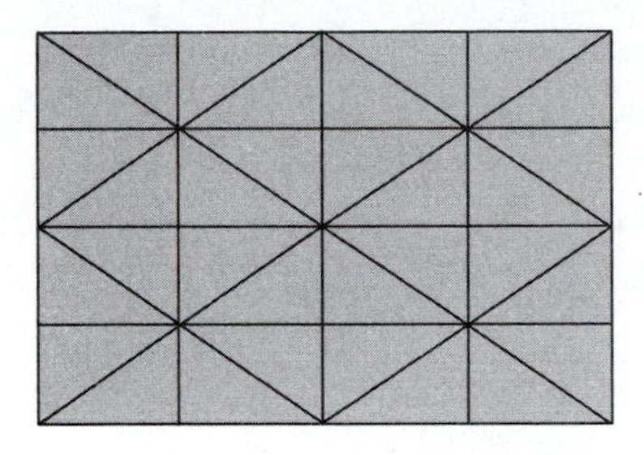

****

The two-dimensional irrotational inviscid flow of a fluid can be described in terms of either the stream function ψ or in terms of the velocity potential ϕ. In either case the governing differential equation is Laplace's equation,

$$\nabla^2\psi = 0 \qquad \text{or} \qquad \nabla^2\phi = 0$$

In terms of the stream function the velocity components are given by $u_x = \partial\psi/\partial y$ and $u_y = -\partial\psi/\partial x$. For the velocity potential $u_x = \partial\phi/\partial x$ and $u_y = \partial\phi/\partial y$. For the stream function formulation the boundary condition on a fixed surface such as indicated below is ψ = constant.

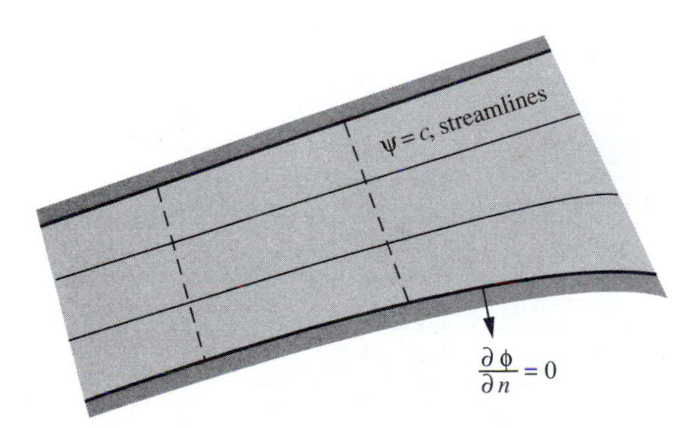

For the velocity potential, the appropriate boundary condition is $\partial\phi/\partial n = 0$ since the velocity must be tangent to the boundary. Note that the functions ϕ and ψ are conjugate harmonic functions, which means that each ϕ = constant curve is orthogonal to all ψ = constant curves that intersect it. In either case, only the derivatives of ϕ or ψ are of importance and ϕ or ψ can be assigned arbitrarily at some point or curve. As a classical example, consider the problem of the flow in the region indicated next. The fluid is assumed to enter with a constant velocity at the left end of the region and to exit with the same constant velocity at the right end. A classical problem of this type is that of the effect that a circular solid obstruction, as indicated, has on the flow.

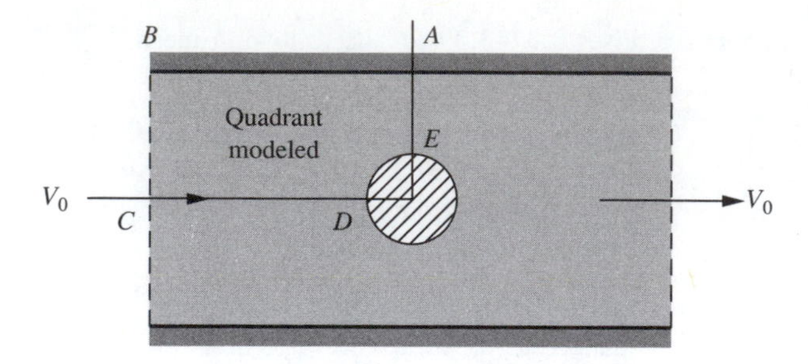

By using the symmetry associated with the region, it is possible to model only the quadrant indicated.

3.43. Stream function formulation. The portion of the boundary CDE is a streamline. Recalling that only the derivatives are of importance in establishing the flow, we arbitrarily set $\psi = 0$ on the portion of the boundary CDE. On the portion of the boundary BC the velocity $u_x = \partial\psi/\partial y = V_0$. This can be integrated to yield

$$\psi = V_0 y + \text{constant}$$

which, when coupled with the fact that we have taken $\psi = 0$ at $y = 0$, gives

$$\psi = V_0 y$$

Thus on the portion of the boundary AB, $\psi = V_0 H$. ψ depends linearly on y between C and B. On EA, the y-component of velocity is zero by symmetry resulting in $u_y = -\partial\psi/\partial x = 0$. Set up and solve for the stream function ψ for the crude mesh indicated below. By computing $\mathbf{v} = \mathbf{i}\,\partial\psi/\partial y - \mathbf{j}\,\partial\psi/\partial x$, determine the velocity within each element. Show these on a sketch.

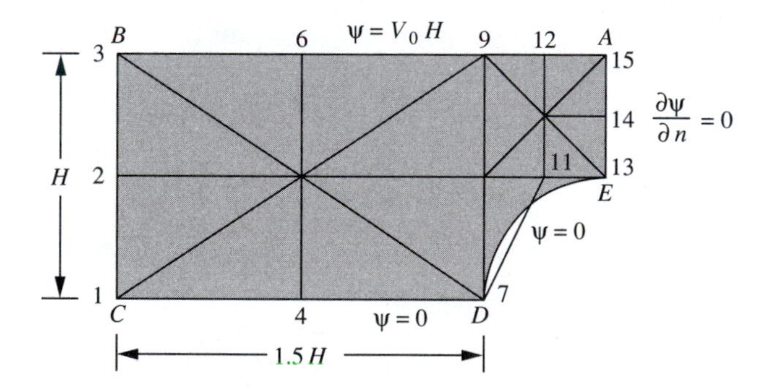

3.44. Repeat Exercise 3.43 for the mesh indicated.

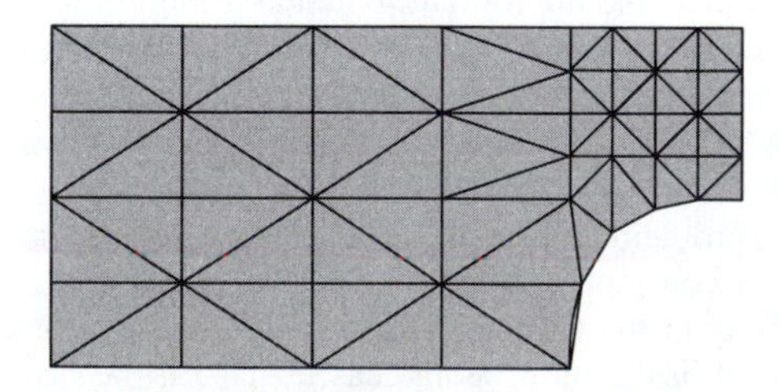

3.45. Velocity potential formulation. For the velocity potential formulation there is no flow perpendicular to lines CDE and AB, that is, $\mathbf{n} \cdot \nabla\phi = \partial\phi/\partial n = 0$. On the line EA, by taking $\phi = 0$ it follows that $u_y = \partial\phi/\partial y = 0$, consistent with the stream function formulation. On line BC, $u_x = \partial\phi/\partial x = V_0$ is known. Set up and solve for the velocity potential ϕ for the crude mesh indicated below. By computing $\mathbf{v} = \mathbf{i}\,\partial\phi/\partial x + \mathbf{j}\,\partial\phi/\partial y$, determine the velocity within each element. Show these on a sketch.

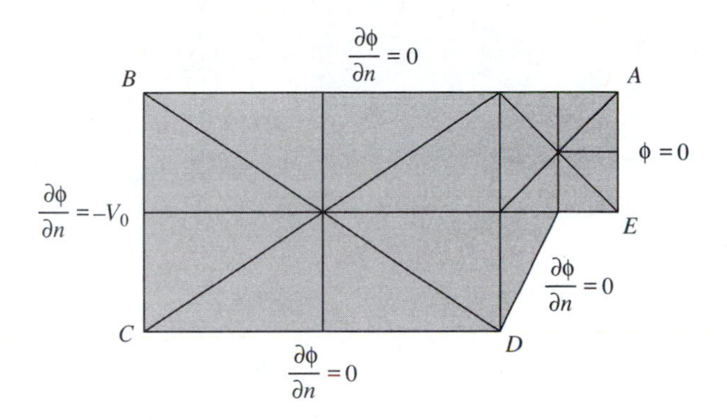

3.46. Repeat Exercise 3.45 for the mesh of Exercise 3.44.

3.47. Set up and solve the flow problem for the rectangular obstruction indicated in the figure below. Use the stream function formulation and the mesh given.

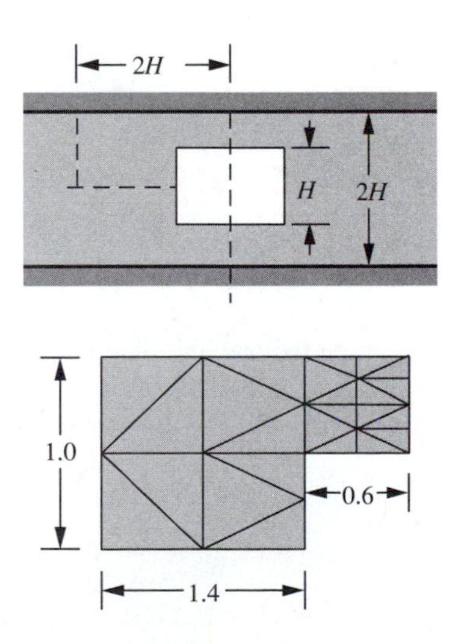

3.48. Repeat Exercise 3.47 using the velocity potential formulation.

3.49. Repeat Exercise 3.47 using the mesh indicated.

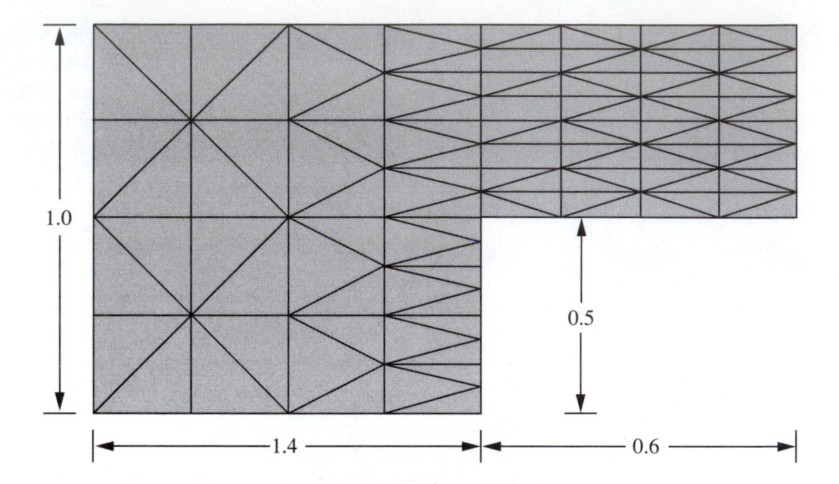

3.50. Repeat Exercise 3.48 using the mesh of Exercise 3.49.

Section 3.6

3.51. Set up the equations necessary to determine the interpolation functions for the rectangular element arbitrarily oriented with respect to the global axes as indicated.

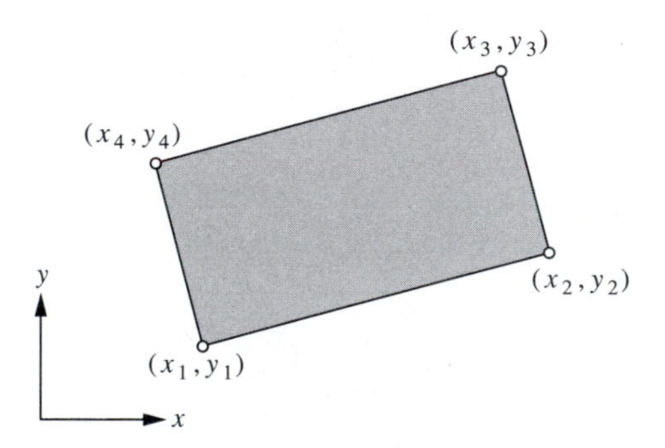

Solving these in general can be tedious and time consuming! Specialize the equations for the case where the sides of the element are parallel to the axes and then solve the equations. Show that when these interpolation functions are transformed according to

$$x = \frac{\left(x_1 + x_2 + s(x_2 - x_1)\right)}{2}$$

$$y = \frac{\left(y_1 + y_4 + t(y_4 - y_1)\right)}{2}$$

the interpolation functions of Section 3.6, expressed in terms of s and t, result.

3.52. The nodal values for u are as indicated in the figure.

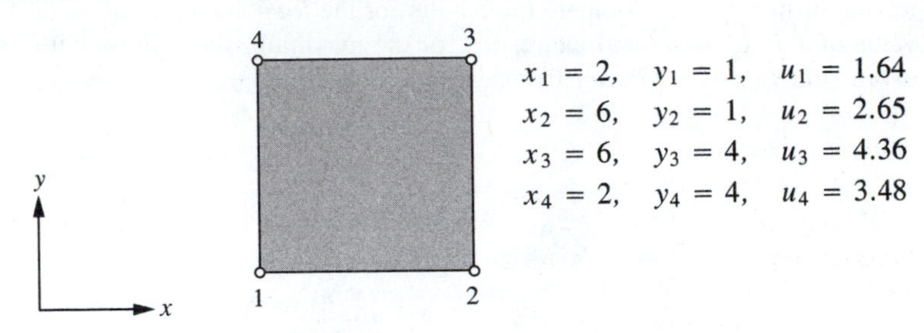

Determine

a. $u(5, 3.5)$.

b. $\partial u/\partial n$ along each of the lines 1-2, 2-3, 3-4, 4-1.

Compare the values along line 1-2 with those along line 3-4, and those along lines 2-3 and 4-1, respectively.

3.53. Verify the expression given for $\mathbf{f_e}$ in Section 3.6.

3.54. Verify the expression given for $\mathbf{k_e}$ in Section 3.6.

3.55. For the element shown, evaluate $\mathbf{f_e}$ exactly for $f = xy$. Compare the results with those using the approximate evaluation scheme of Exercise 3.53.

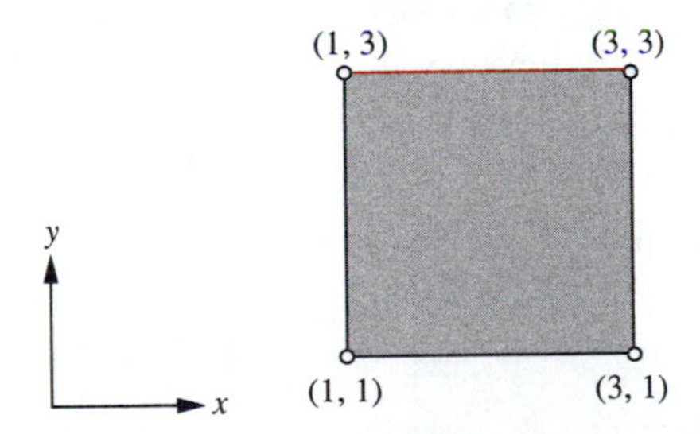

3.56. Repeat Exercise 3.55 for the mesh indicated below.

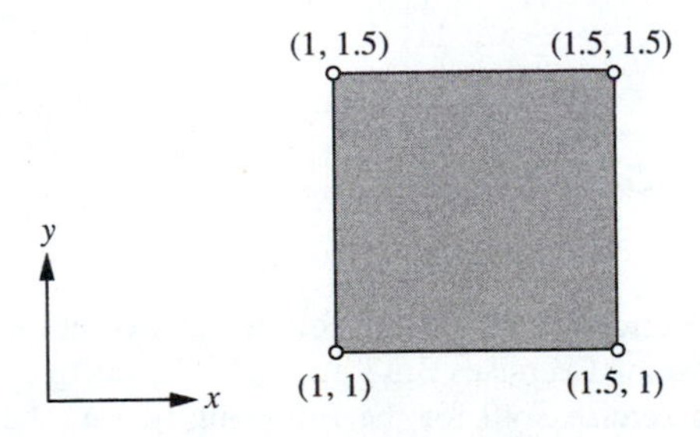

The point of these two exercises is that as a mesh is refined, the approximations become more exact.

3.57. Using quarter symmetry and the mesh shown, set up and solve the problem of the torsion of the square. Compare the results for the torsional constant with the exact value of $J = 0.141a^4$ and the results for the maximum shear stress with the exact expression $\tau_{\max} = 4.8T/a^3$. Compare the results with those of Exercise 3.27.

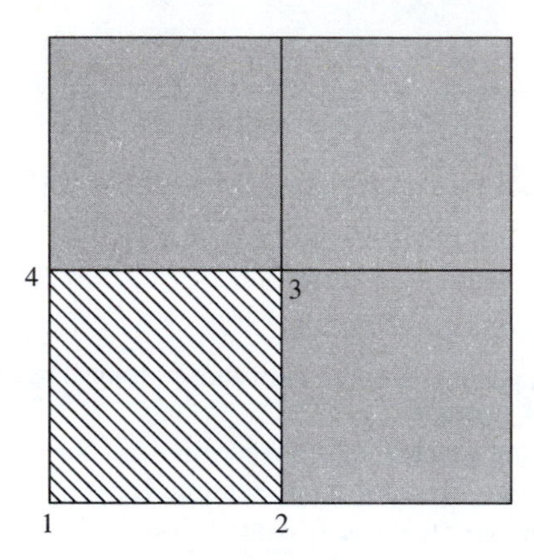

3.58 and 3.59. Repeat Exercise 3.57 for each of the following meshes. Compare the results with those of Exercises 3.28 and 3.29, respectively.

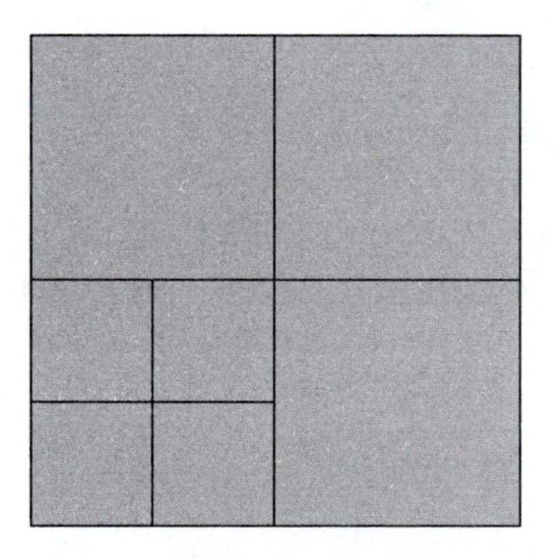

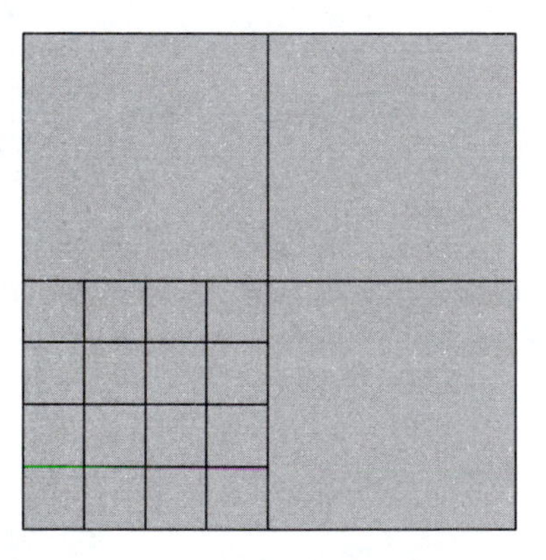

3.60 and 3.61. Repeat Exercise 3.39 for the following two meshes. Compare the results with those of Exercises 3.37 and 3.40, respectively.

3.62 and 3.63. Repeat Exercise 3.40 for the following two meshes. Compare the results with those of Exercises 3.38 and 3.41, respectively.

3.64 and 3.65. Repeat Exercise 3.41 for the following two meshes. Compare the results with those of Exercises 3.39 and 3.42, respectively.

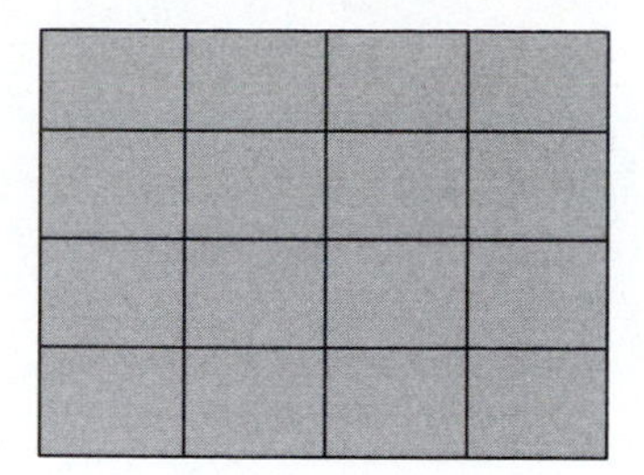
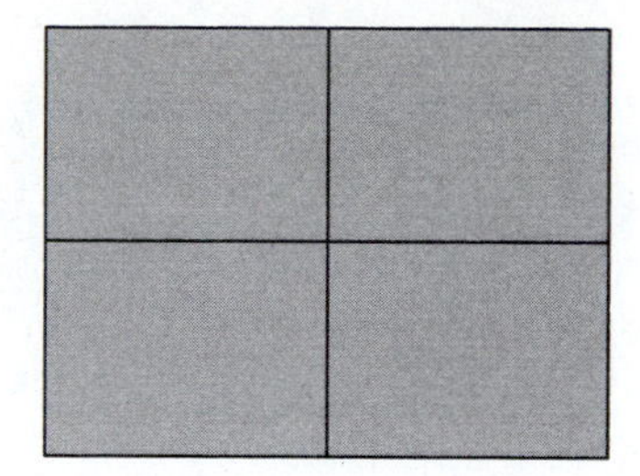

The second mesh should be considered as having been obtained from the first by halving. For some of the previous exercises, the student may first want to complete certain of the computer projects.

Section 3.7

***3.66.** Consider the Helmholtz differential equation and boundary conditions

$$\nabla^2\psi + \lambda\psi = 0 \qquad \text{in } D$$

$$\psi = 0 \text{ on } \Gamma_1 \qquad \text{on } \Gamma_1$$

$$\frac{\partial\psi}{\partial n} + \alpha\psi = 0 \text{ on } \Gamma_2 \qquad \text{on } \Gamma_2$$

Show that the weak form for this boundary value problem is

$$\int\int_D \left(\frac{\partial\zeta}{\partial x}\frac{\partial\psi}{\partial x} + \frac{\partial\zeta}{\partial y}\frac{\partial\psi}{\partial y}\right) dA + \int_{\Gamma_2} \zeta\alpha\psi \, ds - \lambda\int\int_D \zeta\psi \, dA = 0$$

***3.67.** The so-called Rayleigh quotient for the Helmholtz problem is the functional

$$\lambda = \frac{\int\int_D \left((\partial\psi/\partial x)^2 + (\partial\psi/\partial y)^2\right) dA + \int_{\Gamma_2} \alpha\psi^2 \, ds}{\int\int_D \psi^2 \, dA}$$

Show that $\delta\lambda = 0$ leads to the boundary value problem

$$\nabla^2\psi + \lambda\psi = 0 \qquad \text{in } D$$

$$\psi = 0 \text{ on } \Gamma_1 \qquad \text{on } \Gamma_1$$

$$\frac{\partial\psi}{\partial n} = 0 \text{ on } \Gamma_2 \qquad \text{on } \Gamma_2$$

***3.68.** Show that for the Rayleigh quotient, discretization and interpolation of D leads to

$$\lambda(\psi) = \frac{\{\sum \boldsymbol{\psi}_e^T \mathbf{k}_e \boldsymbol{\psi}_e + \sum' \boldsymbol{\psi}_e^T \mathbf{a}_e \boldsymbol{\psi}_e\}}{\sum \boldsymbol{\psi}_e^T \mathbf{m}_e \boldsymbol{\psi}_e} = \frac{\boldsymbol{\psi}_G^T \mathbf{K}_G \boldsymbol{\psi}_G}{\boldsymbol{\psi}_G^T \mathbf{M}_G \boldsymbol{\psi}_G}$$

where

$$\mathbf{k_e} = \iint_{A_e} \left(\frac{\partial \mathbf{N}}{\partial x} \frac{\partial \mathbf{N}^{\mathrm{T}}}{\partial x} + \frac{\partial \mathbf{N}}{\partial y} \frac{\partial \mathbf{N}^{\mathrm{T}}}{\partial y} \right) dA$$

$$\mathbf{a_e} = \int_{\Gamma_{2e}} \mathbf{N} \alpha \mathbf{N}^{\mathrm{T}} \, ds$$

$$\mathbf{m_e} = \iint_{A_e} \mathbf{N}\mathbf{N}^{\mathrm{T}} \, dA$$

and

$$\mathbf{K_G} = \sum_e \mathbf{k_G} + \sum_e{}' \mathbf{a_G}$$

$$\mathbf{M_G} = \sum_e \mathbf{m_G}$$

***3.69.** The stationary value of the function λ is now calculated by setting the appropriate partial derivatives equal to zero according to

$$\frac{\partial \lambda}{\partial \boldsymbol{\psi}_{\mathbf{G}}} = \mathbf{0}$$

Show that this leads to the generalized linear algebraic eigenvalue problem

$$(\mathbf{K_G} - \lambda \mathbf{M_G})\boldsymbol{\psi}_{\mathbf{G}} = \mathbf{0}$$

Imposition of constraints and solution then follow in the usual manner.

3.70. Set up and solve the Helmholtz problem for the square using four square elements as indicated and hence show that $\lambda = 24/a^2$.

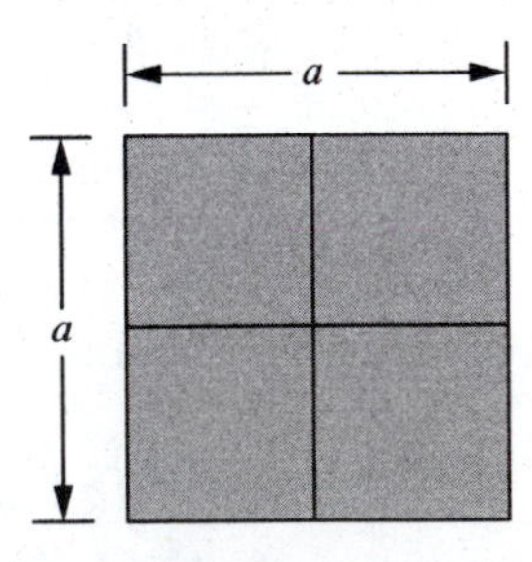

3.71. For a better estimate of the fundamental (lowest) frequency for the square, set up the quarter symmetry model as indicated. Determine the smallest frequency.

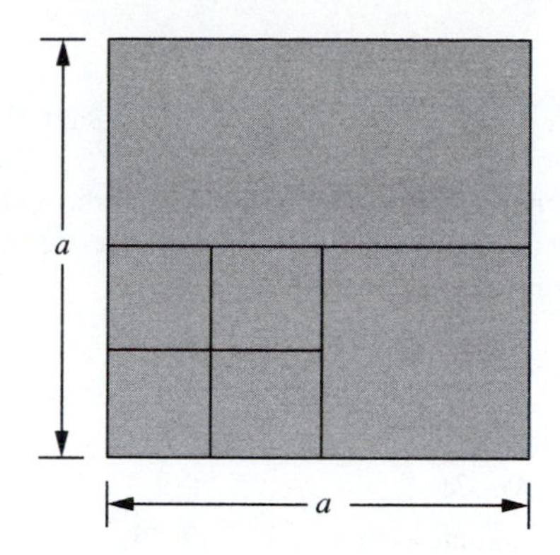

3.72. Set up the Helmholtz problem for the circle using the mesh indicated. Determine the lowest two λ's and ψ's. Compare with the exact results $\lambda_1 = 5.78/a^2$ and $\lambda_2 = 30.5/a^2$.

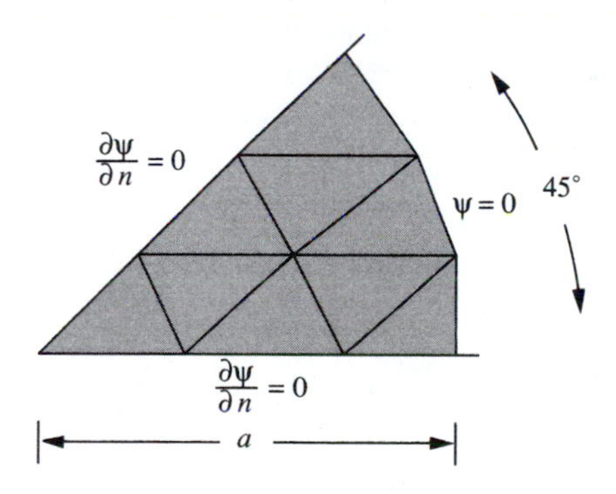

3.73. Set up the Helmholtz problem for the L-shaped region and mesh indicated. Determine the lowest eigenvalue-eigenvector pair. Take $\phi = 0$ everywhere on the boundary.

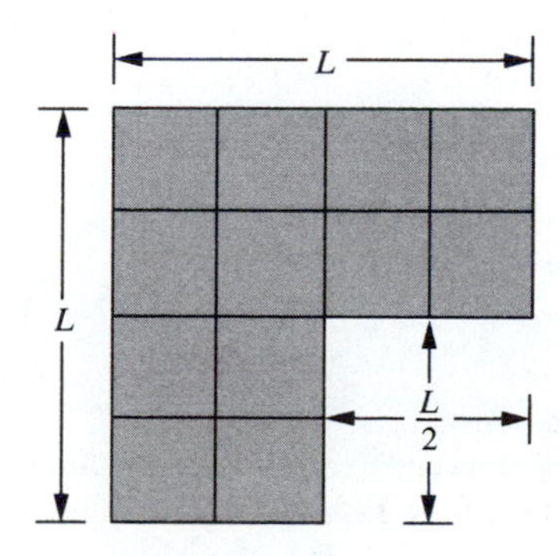

Section 3.8.1

3.74. For the Q4 element, show that any line s = constant (in the parent element) is transformed into a straight line in R_{xy}. Repeat for t = constant.

3.75. Show that the Jacobian of the transformation for the Q4 element is of the form $j_1 + j_2 s + j_3 t$, that is, that the coefficient of the st term is zero.

3.76. Show that the area of the Q4 element is given by

$$R_{xy} = \int_{-1}^{1}\int_{-1}^{1} |\mathbf{J}|\, ds\, dt = 4j_1$$

and hence that j_1 = Area/4.

3.77. Verify that for a rectangular element, Eq. (3.55) for $\mathbf{f_e}$ reduces to Eq. (3.48) of Section 3.6.

3.78. Verify that if the Q4 element is rectangular, $\mathbf{k_e}$ given by

$$\mathbf{k_e} = \int_{-1}^{1}\int_{-1}^{1} \mathbf{\Delta J J \Delta}\, ds\, dt$$

reduces to Eq. (3.49) given in Section 3.6.

Section 3.8.2

3.79. Verify the expression given for the Jacobian determinant of Section 3.8.2.2.

3.80. For the curve indicated, evaluate the arc length exactly and compare it with the approximate 2-point Gauss evaluation outlined in Section 3.8.2.2,

$$l = \left(\left(x'\left(\frac{-1}{\sqrt{3}}\right)\right)^2 + \left(y'\left(\frac{-1}{\sqrt{3}}\right)\right)^2\right)^{1/2} + \left(\left(x'\left(\frac{1}{\sqrt{3}}\right)\right)^2 + \left(y'\left(\frac{1}{\sqrt{3}}\right)\right)^2\right)^{1/2}$$

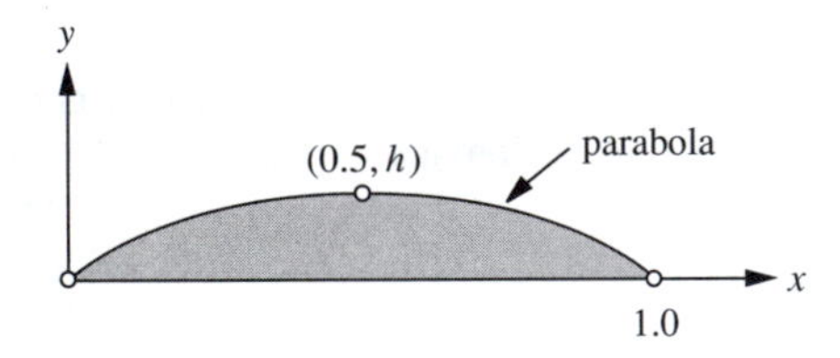

Take h = 0.05, 0.10, 0.15, 0.20, and 0.25.

3.81. Evaluate $|\mathbf{J}|$ for a Q8 rectangular element.

3.82. For a rectangular Q8 element evaluate

$$\mathbf{f_e} = \iint_{R_{xy}} \mathbf{N} f\, dA = \int_{-1}^{1}\int_{-1}^{1} \mathbf{N} f\, |\mathbf{J}|\, ds\, dt$$

for the case $f = f_0$, a constant.

3.83. Verify at least one of the main diagonal terms and one off diagonal term of $\mathbf{k_e}$ for the Q8 square element given in Example 3.9.

3.84. For the application of Section 3.8.2.3, the normal derivative at point 7 was computed by passing a parabola through nodes 5, 6, and 7 and then computing the slope at point 7. Verify that applying the general result

$$\frac{\partial \phi}{\partial n} = (n_x \mathbf{J}_1 + n_y \mathbf{J}_2)\mathbf{\Delta}^{\mathrm{T}}\boldsymbol{\phi}_{\mathbf{e}}$$

evaluated at node 7 produces the same result. The required nodal values are

$$\psi_5 = 0.2162,\ \psi_6 = 0.1587,\ \psi_7 = 0.0000,\ \psi_{10} = 0.1982,$$

$$\psi_{11} = 0.0000,\ \psi_{16} = 0.1693,\ \psi_{17} = 0.1982,\ \psi_{18} = 0.0000,$$

$$x_5 = y_5 = 0.8619,\ x_6 = y_6 = 1.0774,\ x_7 = y_7 = 1.2929$$

$$x_{10} = 1.1296,\ y_{10} = 0.7790,\ x_{11} = 1.4444,\ y_{11} = 1.1685$$

$$x_{16} = 1.4115,\ y_{16} = 0.7717,\ x_{17} = 1.5144,\ y_{17} = 0.8968$$

$$x_{18} = 1.6173,\ y_{18} = 1.0761$$

***3.85.** For the Q8 element the process of extrapolation or smoothing [10] mentioned in Section 3.8.2.1 can be carried out for $\partial\psi/\partial x$ (or $\partial\psi/\partial y$) as follows:

Step 1. Determine the derived variable $\partial\phi/\partial x$ at the $N = 2$ Gauss points I, II, III, and IV as indicated in the figure.

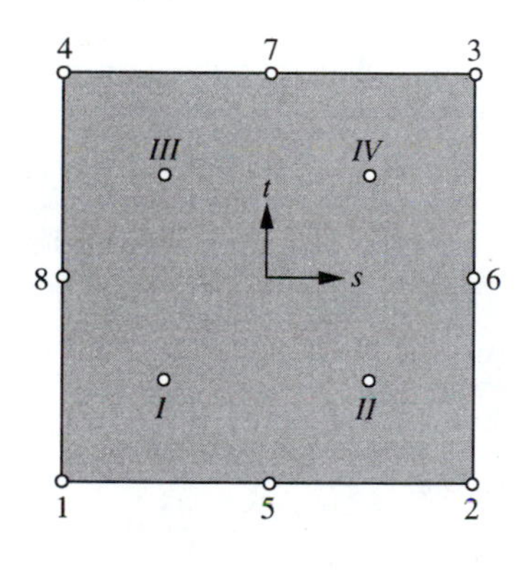

Node	s	t
I	-1	-1
II	1	-1
III	-1	1
IV	1	1
1	$-\sqrt{3}$	$-\sqrt{3}$
5	0	$-\sqrt{3}$
2	$+\sqrt{3}$	$-\sqrt{3}$
8	$-\sqrt{3}$	0
6	$+\sqrt{3}$	0
4	$-\sqrt{3}$	$+\sqrt{3}$
7	0	$+\sqrt{3}$
3	$+\sqrt{3}$	$+\sqrt{3}$

Step 2. Define a smoothed $\partial\phi/\partial x$ on the element according to

$$\left(\frac{\partial\phi}{\partial x}\right)_S = \left(\frac{\partial\phi}{\partial x}\right)_I N_I + \left(\frac{\partial\phi}{\partial x}\right)_{II} N_{II} + \left(\frac{\partial\phi}{\partial x}\right)_{III} N_{III} + \left(\frac{\partial\phi}{\partial x}\right)_{IV} N_{IV}$$

where

$$N_I = \frac{(1-s)(1-t)}{4}$$

$$N_{II} = \frac{(1+s)(1-t)}{4}$$

$$N_{III} = \frac{(1-s)(1+t)}{4}$$

$$N_{IV} = \frac{(1+s)(1+t)}{4}$$

that is, a bilinear function passing through the values of $\partial\phi/\partial x$ at the Gauss points. The basic underlying idea is that taking derivatives of a quadratic function produces a linear function, and that the best linear function is constructed in terms of the values which have the best accuracy within the element, the $N = 2$ Gauss points.

Step 3. Compute the extrapolated or smoothed variable at the nodes using the table shown next to the previous figure.

Apply this idea to the example of Section 3.8.2.3 as follows. Using the nodal coordinates and the nodal values for ψ from Exercise 3.84, compute the partial derivatives $\partial\phi/\partial x$ and $\partial\phi/\partial y$ at the Gauss points and then extrapolate them to the nodes. The shear stress at node 7 can then be computed as being proportional to the normal derivative at point 7. Repeat these steps for element 12 to arrive at the same quantity at point 51 and hence compute a ratio for the stress concentration factor.

How does this compare with the values computed on the basis of evaluating the derived variables at the nodes? The ψ values for element 12 are

$$\psi_{38} = 0.1127,\ \psi_{39} = 0.0701,\ \psi_{40} = 0.0000,\ \psi_{43} = 0.1116,$$

$$\psi_{44} = 0.0000,\ \psi_{49} = 0.1114,\ \psi_{50} = 0.0698,\ \psi_{51} = 0.0000$$

$$x_{38} = x_{39} = x_{40} = 2.6667,\ x_{43} = x_{44} = 2.6667$$

$$x_{49} = x_{50} = x_{51} = 3.0000$$

$$y_{38} = y_{43} = y_{49} = 0.6667,\ y_{39} = y_{50} = 0.8333$$

$$y_{40} = y_{44} = y_{51} = 1.0000$$

Section 3.8.3

3.86. Verify that in terms of s and t the expressions

$$N_1(s,t) = (1-s-t)(1-2s-2t) \qquad N_4(s,t) = 4s(1-s-t)$$

$$N_2(s,t) = s(2s-1) \qquad N_5(s,t) = 4st$$

$$N_3(s,t) = t(2t-1) \qquad N_6(s,t) = 4t(1-s-t)$$

are correct.

3.87. Verify the transformation of Example 3.10.

3.88. Verify the results of Example 3.10.

3.89. Show that any integral of the form

$$\int_0^1 \int_0^{1-t} (\alpha + \beta s + \gamma t)\, ds\, dt$$

is exactly integrated by the $N = 1$ Gauss quadrature for triangles. Show also that the $N = 3$ and $N = 4$ schemes are exact.

3.90. When $f = f_0$, show that for the straight-sided triangle

$$\mathbf{f}_\mathbf{e}^\mathrm{T} = \frac{f_0 A_e}{3}[0\ 0\ 0\ 1\ 1\ 1]$$

that is, the total load of $f_0 A_e$ is divided equally between the three midside nodes.

3.91. Show that for the straight-sided isoparametric triangle,

$$\mathbf{k}_\mathbf{e} = \left(\frac{1}{3}\right)\begin{bmatrix} 3a & -b & a+b & 4b & 0 & -4(a+b) \\ & 3d & b+d & 4b & -4(b+d) & 0 \\ & & 3(a+2b+d) & 0 & -4(b+d) & -4(a+b) \\ & & & 8(a+b+d) & -8(a+b) & -8(b+d) \\ & & & & 8(a+b+d) & 8b \\ & & & & & 8(a+b+d) \end{bmatrix}$$

where

$$\mathbf{JJ} = \begin{bmatrix} a & b \\ b & d \end{bmatrix}$$

3.92. For the application of Section 3.8.3.3, the normal derivative at point 7 was computed by passing a parabola through nodes 5, 6, and 7 (nodes in element 5) and then computing the slope at point 7. Verify that applying the general result

$$\frac{\partial \phi}{\partial n} = (n_x \mathbf{J}_1 + n_y \mathbf{J}_2)\boldsymbol{\Delta\phi}_\mathbf{e}$$

evaluated at the node in question produces the same result. The required nodal values for the computation within element 5 are

$$\psi_5 = 0.2170,\ \psi_6 = 0.1546,\ \psi_7 = 0.0000,$$
$$\psi_{12} = 0.1978,\ \psi_{13} = 0.1347,\ \psi_{19} = 0.1676$$

Repeat the calculations for element 6, evaluating the result at node 7 with

$$\psi_7 = 0.0000,\ \psi_{13} = 0.1347,\ \psi_{14} = 0.0000,$$
$$\psi_{19} = 0.1676,\ \psi_{20} = 0.1105,\ \psi_{21} = 0.0000$$

If one now computes an average of these two values, what is the value of the stress concentration factor based on the results from Section 3.8.3.3 for element 24?

***3.93.** Using the smoothing technique described in Exercise 3.85 compute the smoothed normal derivatives at node 7 in elements 5 and 6, and also the smoothed normal derivative at node 61 in element 24. The situation as regards the smoothing procedure for the triangle is indicated in the figure.

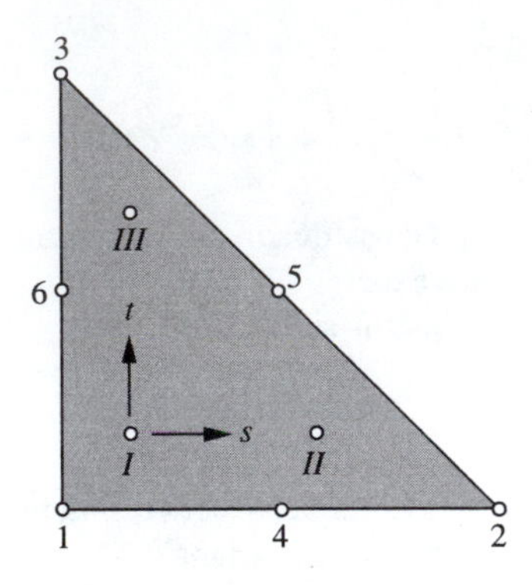

Node	s	t
I	0	0
II	1	0
III	0	1
1	−1/3	−1/3
2	5/3	−1/3
3	−1/3	5/3
4	2/3	−1/3
5	2/3	2/3
6	−1/3	2/3

A derived variable Φ to be smoothed is represented according to

$$\Phi_S = \Phi_I N_I + \Phi_{II} N_{II} + \Phi_{III} N_{III}$$

where $N_I = 1 - s - t$, $N_{II} = s$, and $N_{III} = t$.

Compute a stress concentration factor based on the nodal averaging at node 7. The nodal ψ for element 24 are

$$\psi_{49} = 0.0000,\ \psi_{55} = 0.0698,\ \psi_{56} = 0.0000,\ \psi_{61} = 0.1114,$$
$$\psi_{62} = 0.0696,\ \psi_{63} = 0.0000$$

for calculating the required smoothed normal derivative at node 63. The nodal coordinates are

$$\begin{array}{lll} x_{49} = 2.3333 & x_{56} = 2.6667 & x_{63} = 3.0000 \\ y_{49} = 1.0000 & y_{56} = 1.0000 & y_{63} = 1.0000 \\ & x_{55} = 2.6667 & x_{62} = 3.0000 \\ & x_{55} = 0.8333 & x_{62} = 0.8333 \\ & & x_{61} = 3.0000 \\ & & x_{61} = 0.6667 \end{array}$$

***3.94.** For a cylindrical region with the general boundary conditions as indicated, carry through the details of showing that the weak form is

$$\int_{-H/2}^{H/2}\int_0^a \left(\frac{\partial \psi}{\partial r}\, r\, \frac{\partial u}{\partial r} + \frac{\partial \psi}{\partial z}\, r\, \frac{\partial u}{\partial z}\right) dr\ dz + \int_{-H/2}^{H/2} a\psi(a,z)\alpha_1(z)u(a,z)\ dz$$
$$+ \int_0^a \psi\left(r, \frac{H}{2}\right)\alpha_2(r)u\left(r, \frac{H}{2}\right) r\ dr + \int_0^a \psi\left(r, \frac{-H}{2}\right)\alpha_3(r)u\left(r, \frac{-H}{2}\right) r\ dr$$
$$= \int_{-H/2}^{H/2}\int_o^a rfw\ dr\ dz + \int_{-H/2}^{H/2} aw(a,z)h_1(z)\ dz$$
$$+ \int_0^a w\left(r, \frac{-H}{2}\right) h_3(r) r\ dr + \int_0^a w\left(r, \frac{H}{2}\right) h_2(r) r\ dr$$

with $B(w, u)$ as the left side of the equation and $l(w)$ as the right side. The corresponding functional is

$$I(u) = \frac{1}{2}\left\{ \int_{-H/2}^{H/2}\int_0^a \left(\left(\frac{\partial u}{\partial r}\right)^2 + \left(\frac{\partial u}{\partial z}\right)^2\right) r\, dr\, dz + \int_{-H/2}^{H/2} a\alpha_1(z)u(a,z)^2\, dz \right.$$

$$\left. + \int_0^a \left[\alpha_2(r)u\left(r, \frac{H}{2}\right)^2 + \alpha_3(r)u\left(r, -\frac{H}{2}\right)^2\right] r\, dr \right\} - \int_{-H/2}^{H/2}\int_0^a rfu\, dr\, dz$$

$$- \int_{-H/2}^{H/2} au(a,z)h_1(z)\, dz - \int_o^a \left(u\left(r, \frac{H}{2}\right)h_2(r) + u\left(r, \frac{-H}{2}\right)h_3(r)\right) r\, dr$$

With a suitable choice for the discretization and interpolation,

$$I(u) = \frac{1}{2}\left\{\sum \mathbf{u}_e^T\mathbf{k}_e\mathbf{u}_e + \sum{}' \mathbf{u}_e^T\mathbf{a}_e\mathbf{u}_e\right\} - \sum \mathbf{u}_e^T\mathbf{f}_e - \sum{}' \mathbf{u}_e^T\mathbf{h}_e$$

Determine the form of each of $\mathbf{k}_e$, $\mathbf{a}_e$, $\mathbf{f}_e$, and $\mathbf{h}_e$.

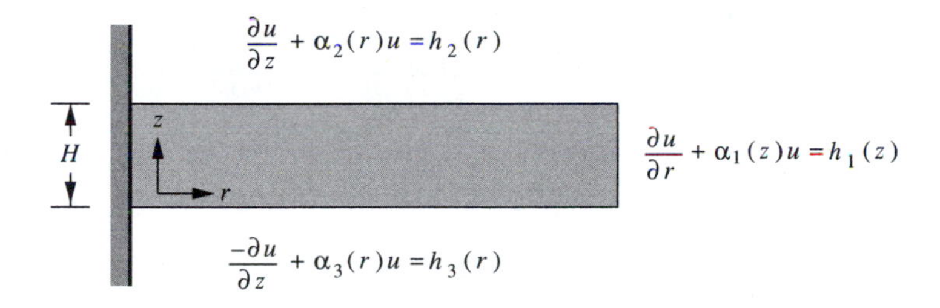

$\frac{\partial u}{\partial z} + \alpha_2(r)u = h_2(r)$
H
z
r
$\frac{\partial u}{\partial r} + \alpha_1(z)u = h_1(z)$
$\frac{-\partial u}{\partial z} + \alpha_3(r)u = h_3(r)$

3.95. For the problem and mesh indicated set up and solve the finite element model. Take

$$u_1 = u_2 = u_3 = u_6 = u_7 = u_8 = u_9 = 0$$

The appropriate boundary condition at $r = 0$ is $\partial u/\partial r = 0$ so that u_4 is unspecified. Evaluate the area integrals by evaluating the r that appears at the centroid of the element in question.

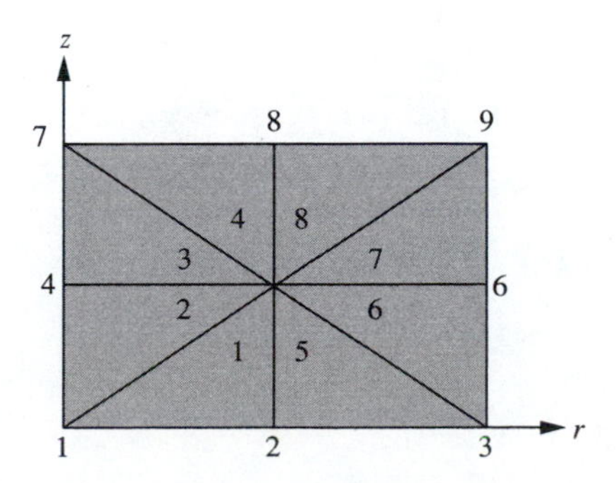

3.96. Repeat Exercise 3.95 for the mesh indicated.

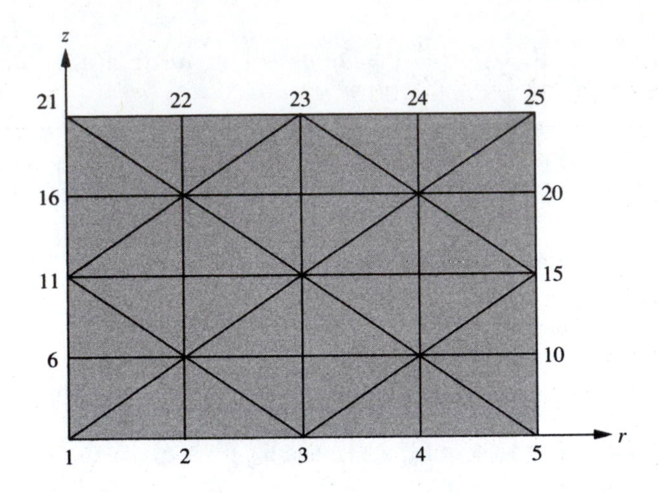

Require $u_1 \rightarrow u_5$, u_{10}, u_{15}, u_{20}, and $u_{21} \rightarrow u_{25} = 0$.

CHAPTER 4

Time-Dependent Problems

Chapter Contents

4.1 Introduction
4.2 One-dimensional diffusion or parabolic equations
    4.2.1 The Galerkin finite element model
    4.2.2 Example of one-dimensional diffusion
    4.2.3 Analytical integration techniques
    4.2.4 Time domain integration techniques—first-order systems
        4.2.4.1 The Euler method
        4.2.4.2 Improved Euler or Crank-Nicolson method
        4.2.4.3 Analysis of algorithms
4.3 One-dimensional wave or hyperbolic equations
    4.3.1 The Galerkin finite element model
    4.3.2 One-dimensional wave example
    4.3.3 Analytical integration techniques
    4.3.4 Time domain integration techniques—second-order systems
        4.3.4.1 Central difference method
        4.3.4.2 Newmark's method
        4.3.4.3 Analysis of algorithms
4.4 Two-dimensional diffusion
4.5 Two-dimensional wave equations
4.6 Closure

References
Computer projects
Exercises

4.1 INTRODUCTION

The previous three chapters dealt exclusively with steady-state problems, that is, problems where time did not enter explicitly into the formulation or solution of the problem. The types of problems considered in Chapters 2 and 3, respectively, were one- and two-dimensional elliptic boundary value problems. In this chapter, finite element models for parabolic and hyperbolic equations, such as the one-dimensional transient heat conduction and the one-dimensional scalar wave equation, respectively, will be developed. The finite element models for these two types of initial-boundary value problems will turn out to be, respectively, first- and second-order systems of ordinary differential equations with time as the independent variable. Analytical and numerical algorithms for the solution of these systems of equations will be presented and discussed.

4.2 ONE-DIMENSIONAL DIFFUSION OR PARABOLIC EQUATIONS

The example to be used to develop a model for one-dimensional diffusion processes is the classical heat conduction problem indicated in Fig. 4–1.

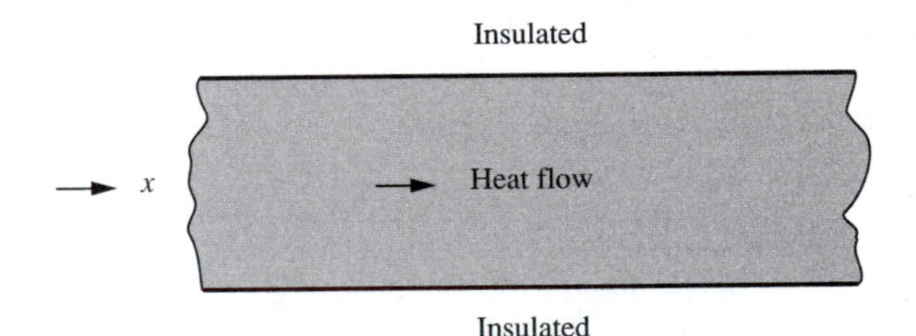

FIGURE 4–1 Typical geometry for one-dimensional diffusion

We will assume that energy in the form of heat flows only in the x-direction, that is, that there is no flux perpendicular to the x-axis. The basic physical principle for this type of problem is balance of energy. A differential element of length Δx as shown in Fig. 4–2 is isolated and an energy balance performed as

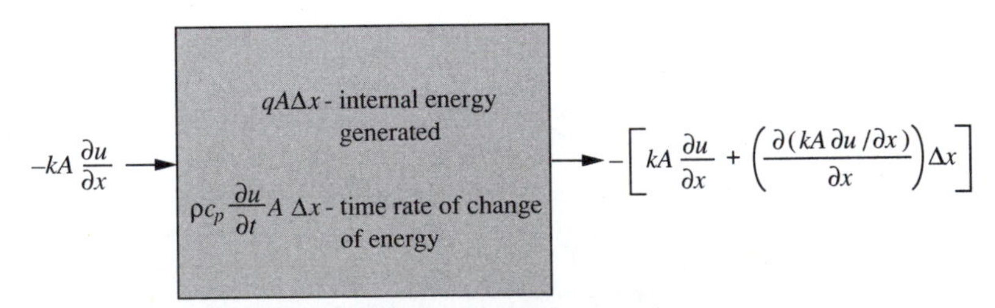

FIGURE 4–2 Differential element for energy balance

energy in − energy out + internal energy generated =

time rate of change of energy within the element

With the energy terms as indicated in Fig. 4–2, the balance of energy statement becomes

$$+\left(-kA\frac{\partial u}{\partial x}\right)+-\left(-\left(kA\frac{\partial u}{\partial x}+\left(\frac{\partial\,(kA\,\partial u/\partial x)}{\partial x}\right)\Delta x\right)\right)+qA\Delta x = \rho c_p A\Delta x\frac{\partial u}{\partial t}$$

or in the limit as $\Delta x \longrightarrow 0$,

$$\frac{\partial(kA\,\partial u/\partial x)}{\partial x}+qA = \rho c_p A\frac{\partial u}{\partial t}$$

a second-order, linear partial differential equation. The auxiliary conditions consist of two boundary conditions and one initial condition. An appropriate boundary condition prescribes either

1. the dependent variable u
2. the flux $-kA(\partial u/\partial x)$
3. a linear combination of the flux and the dependent variable, $\pm kA(\partial u/\partial x)+hu$

This third type of boundary condition is called a **convective boundary condition** and is a local energy balance between the convection externally and the conduction internally. At the left boundary $x = a$, for instance, the external convective and internal conductive terms appear as in Fig. 4–3.

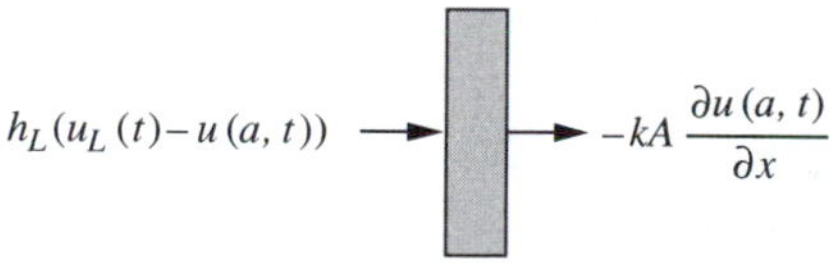

$h_L(u_L(t)-u(a,t))$ → → $-kA\dfrac{\partial u(a,t)}{\partial x}$

FIGURE 4–3 Local energy balance at $x = a$

The local energy balance produces

$$-kA\frac{\partial u(a,t)}{\partial x}+h_L u(a,t) = h_L u_L(t)$$

A similar energy balance at the right end $x = b$ yields

$$kA\frac{\partial u(b,t)}{\partial x}+h_R u(b,t) = h_R u_R(t)$$

For a time-dependent diffusion problem it is also necessary to specify an initial value for the dependent variable of the form

$$u(x,0) = u_0(x)$$

The complete statement of the initial-boundary value problem consists of the differential equation, two boundary conditions, and an initial condition,

$$\frac{\partial(kA\,\partial u/\partial x)}{\partial x}+qA = \rho c_p A\frac{\partial u}{\partial t} \qquad a \le x \le b,\ 0 \le t \qquad (4.1a)$$

$$-kA\frac{\partial u(a,t)}{\partial x} + h_L u(a,t) = h_L u_L(t) \qquad 0 \le t \tag{4.1b}$$

$$kA\frac{\partial u(b,t)}{\partial x} + h_R u(b,t) = h_R u_R(t) \qquad 0 \le t \tag{4.1c}$$

$$u(x,0) = u_0(x) \quad a \le x \le b \tag{4.1d}$$

and represents a well-posed problem [Courant and Hilbert, 1] in partial differential equations. When there is no convection at a boundary the other two types of boundary conditions appropriate at $x = a$ are

$$u(a,t) = u_a(t)$$

where the temperature is specified, or

$$-kA\frac{\partial u(a,t)}{\partial x} = Q(t)$$

where the energy flux is prescribed. The general development of the finite element model will assume type 3 conditions at both boundaries.

4.2.1 The Galerkin Finite Element Model

We will consider the general, one-dimensional diffusion problem developed in the previous section and stated in Eqs.(4.1).

Discretization. The first step in developing a finite element model is discretization. Nodes for the spatial domain $a \le x \le b$ are chosen as indicated in Fig. 4–4, with $a = x_1$ and $b = x_{N+1}$. As was the case for steady-state problems considered in Chapter 2, the nodes are usually selected at equally spaced intervals, keeping in mind that it may be desirable in some problems to concentrate the nodes in regions of high gradients.

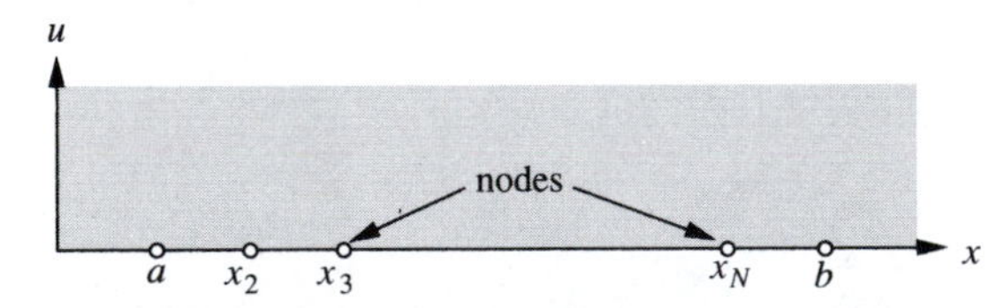

FIGURE 4–4 Nodes for the spatial domain

Interpolation. The interpolation functions are selected in exactly the same fashion as for the time-independent problem *except* that the nodal values are now taken to be functions of time rather than constants; that is,

$$u(x,t) = \sum_{1}^{N+1} u_i(t)\, n_i(x) \tag{4.2}$$

The $n_i(x)$ are nodally based interpolation functions such as discussed in Section 2.5.2, and can be linear, quadratic, or as otherwise desired. The representation in Eq. (4.2) is referred to as **semidiscretization** in that the spatial variable x is discretized whereas the temporal variable t is not. A *finite difference* model of a time-dependent partial differential equation typically involves discretization of both the spatial and temporal variables.

Elemental formulation. The elemental formulation for the diffusion problem of Eqs. (4.1) is based on a corresponding weak statement. The weak form is developed by multiplying the differential equation (4.1*a*) by a test function $v(x)$ satisfying any homogeneous essential boundary conditions, and integrating over the spatial region according to

$$\int_a^b v\left(\frac{\partial\,(kA\,\partial u/\partial x)}{\partial x} + qA - \rho c_p A\frac{\partial u}{\partial t}\right) dx = 0$$

Integrating by parts and eliminating the derivative terms from the boundary conditions (4.1*b*) and (4.1*c*) yields

$$\int_a^b \left(v'kA\frac{\partial u}{\partial x} + \rho c_p A v\frac{\partial u}{\partial t}\right) dx + h_L v(a)u(a,t) + h_R v(b)u(b,t)$$
$$= \int_a^b vAq\,dx + v(a)h_L u_L(t) + v(b)h_R u_R(t) \qquad (4.3)$$

which is the required weak statement for the class of one-dimensional diffusion problems of Eqs. (4.1). The finite element model is obtained by substituting the approximate solution given by Eq. (4.2), and $v = n_k, k = 1, 2, \ldots, N+1$, successively, into Eq. (4.3) to obtain

$$\sum_1^{N+1}\int_a^b \left(n_k' kAn_i' u_i + n_k\rho c_p An_i\dot{u}_i\right) dx + h_L\delta_{k1}u_1(t) + h_R\delta_{kN+1}u_{N+1}(t)$$
$$= \int_a^b n_k qA\,dx + \delta_{k1}h_L u_L(t) + \delta_{kN+1}h_R u_R(t) \qquad k = 1, 2, \ldots, N+1 \quad (4.4)$$

which can be written as

$$\sum \left(A_{ki}u_i(t) + B_{ki}\dot{u}_i(t)\right) = q_k(t) \qquad k = 1, 2, \ldots, N+1 \qquad (4.5)$$

where

$$A_{ki} = \int_a^b n_k' kAn_i'\,dx + h_L\delta_{k1}\delta_{ik} + h_R\delta_{kN+1}\delta_{ik}$$

$$B_{ki} = \int_a^b n_k\rho c_p An_i\,dx$$

$$q_k = \int_a^b qAn_k\,dx + h_L\delta_{k1}u_L(t) + h_R\delta_{kN+1}u_R(t)$$

The student is asked to show in the Exercises that in matrix notation these can be expressed as

$$\mathbf{Au} + \mathbf{B\dot{u}} = \mathbf{q} \tag{4.6}$$

where

$$\mathbf{A} = \sum_e \mathbf{k_G} + \mathbf{BT} \tag{4.7a}$$

$$\mathbf{B} = \sum_e \mathbf{m_G} \tag{4.7b}$$

$$\mathbf{q} = \sum_e \mathbf{q_G} + \mathbf{bt} \tag{4.7c}$$

with

$$\mathbf{k_e} = \int_{x_i}^{x_j} \mathbf{N}' k A \mathbf{N}'^{\mathrm{T}} \, dx \tag{4.8a}$$

$$\mathbf{m_e} = \int_{x_i}^{x_j} \mathbf{N} \rho c_p A \mathbf{N}^{\mathrm{T}} \, dx \tag{4.8b}$$

$$\mathbf{q_e} = \int_{x_i}^{x_j} q A \mathbf{N} \, dx \tag{4.8c}$$

$$\mathbf{BT} = \begin{bmatrix} h_L & & & & & & \\ & 0 & & & & & \\ & & 0 & & & & \\ & & & 0 & & & \\ & & & & 0 & & \\ & & & & & 0 & \\ & & & & & & h_R \end{bmatrix} \tag{4.8d}$$

$$\mathbf{bt}^{\mathrm{T}} = [h_L u_L(t) \quad 0 \quad 0 \quad \ldots \quad 0 \quad 0 \quad h_R u_R(t)] \tag{4.8e}$$

Thus the original initial-boundary value problem has been converted into the initial value problem

$$\mathbf{Au} + \mathbf{B\dot{u}} = \mathbf{q} \tag{4.9a}$$

with

$$\mathbf{u}(0) = \mathbf{u_0} \tag{4.9b}$$

The initial vector $\mathbf{u_0}$ is usually taken to be a vector consisting of the values of $u_0(x)$ at the nodes, or

$$\mathbf{u}(0) = \mathbf{u_0} = [\, u_0(a) \quad u_0(x_2) \quad u_0(x_3) \quad \ldots \quad u_0(x_N) \quad u_0(b)\,]$$

Note that the assembly process has taken place implicitly during the process of carrying out the details of obtaining the governing equations using the Galerkin method.

It is instructive to note that when time is not involved, Eqs. (4.6) through (4.9) are exactly what would result from the finite element model developed in Chapter 2 for the corresponding boundary value problem.

Enforcement of constraints is necessary if either of the boundary conditions is essential, that is, if the dependent variable is prescribed at either boundary point. The system of Eqs. (4.9) must be altered to reflect these constraints. Consider for example the case where the boundary condition at $x = a$ is $u(a, t) = u_a(t)$. The h_L terms in both **BT** and **bt** would be taken as zero and the first equation of (4.9*a*) would be replaced by the constraint resulting in

$$\begin{aligned} u_1 \qquad\qquad\qquad\qquad\qquad\qquad\qquad\qquad\qquad &= u_a(t) \\ a_{21}u_1 + a_{22}u_2 + a_{23}u_3 + \cdots + b_{21}\dot{u}_1 + b_{22}\dot{u}_2 + b_{23}\dot{u}_3 &= q_2(t) \\ a_{31}u_1 + a_{32}u_2 + a_{33}u_3 + \cdots + b_{31}\dot{u}_1 + b_{32}\dot{u}_2 + b_{33}\dot{u}_3 &= q_3(t) \\ &\;\vdots \end{aligned}$$

The u_1 and $\dot{u}_1$ terms in the remaining equations are transferred to the right-hand side to yield

$$\begin{aligned} u_1 \qquad\qquad\qquad\qquad\qquad\qquad &= u_a(t) \\ a_{22}u_2 + a_{23}u_3 + \cdots + b_{22}\dot{u}_2 + b_{23}\dot{u}_3 &= q_2(t) - a_{21}u_1 - b_{21}\dot{u}_1 \\ a_{32}u_2 + a_{33}u_3 + \cdots + b_{32}\dot{u}_2 + b_{33}\dot{u}_3 &= q_3(t) - a_{31}u_1 - b_{31}\dot{u}_1 \\ &\;\vdots \end{aligned}$$

showing the effect of the constraint on the remaining equations. For a linearly interpolated model the half bandwidth is two and only the terms involving u_1 and $\dot{u}_1$ in the second equation need to be transferred to the right-hand side. For a quadratically interpolated model the half bandwidth is three and terms from the first two equations need to be transferred. If the constraint is at the right end, the Nth, $(N - 1)$st, ... equations would be similarly altered.

Write the constrained set of equations as

$$\mathbf{M}\dot{\mathbf{u}} + \mathbf{K}\mathbf{u} = \mathbf{f} \tag{4.10}$$

subject to the initial condition

$$\mathbf{u}(0) = \mathbf{u}_0$$

Algorithms for integrating these equations are studied in the following sections.

4.2.2 Example of One-Dimensional Diffusion

As a typical example consider the specific heat conduction problem

$$\begin{aligned} \frac{\partial(kA \quad \partial u/\partial x)}{\partial x} &= \rho c_p A \frac{\partial u}{\partial t} \qquad 0 \le x \le L, 0 \le t \\ u(0, t) &= u_0 \qquad\qquad\qquad\qquad 0 \le t \end{aligned} \tag{4.11}$$

$$u(L, t) = 0 \qquad 0 \le t$$

$$u(x, 0) = 0 \qquad 0 \le x \le L$$

This corresponds to the idealized situation of a region initially at zero temperature and whose left end $x = 0$ is instantaneously forced to assume the value u_0 for all time greater than zero. With A a constant and $k/\rho c_p = \alpha$, the initial boundary value problem can be written as

$$
\begin{aligned}
\alpha \frac{\partial^2 u}{\partial x^2} &= \frac{\partial u}{\partial t} && 0 \le x \le L, 0 \le t \\
u(0, t) &= u_0 && 0 \le t \\
u(L, t) &= 0 && 0 \le t \\
u(x, 0) &= 0 && 0 \le x \le L
\end{aligned}
\tag{4.12}
$$

Discretization. For purposes of illustration, a four-element model will be investigated. The mesh is indicated in Fig. 4–5.

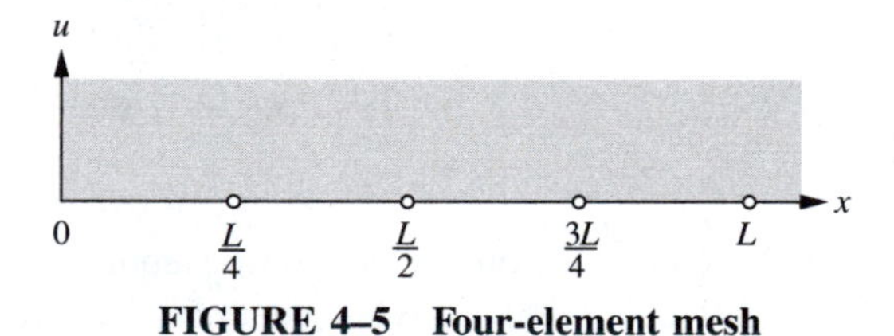

FIGURE 4–5 Four-element mesh

Interpolation. Linear interpolation will be used for the four elements.

Elemental formulation. The elemental matrices are easily seen from Eqs. (4.7) and (4.8) to be

$$\mathbf{k_e} = \int_{x_i}^{x_{i+1}} \mathbf{N}'\alpha\mathbf{N}'^{\mathrm{T}}\, dx = \frac{4\alpha}{L}\begin{bmatrix} 1 & -1 \\ -1 & 1 \end{bmatrix}$$

$$\mathbf{m_e} = \int_{x_i}^{x_{i+1}} \mathbf{N}\mathbf{N}^{\mathrm{T}}\, dx = \frac{L}{24}\begin{bmatrix} 2 & 1 \\ 1 & 2 \end{bmatrix}$$

$$\mathbf{q_e} = \mathbf{0}$$

Assembly. With both the boundary conditions essential, $\mathbf{BT} = \mathbf{0}$ and $\mathbf{bt} = \mathbf{0}$. It follows that the assembled equations are

$$\mathbf{Au} + \mathbf{B\dot{u}} = \mathbf{0}$$

with

$$\mathbf{A} = \sum_e \mathbf{k_G} = \frac{4\alpha}{L}\begin{bmatrix} 1 & -1 & 0 & 0 & 0 \\ -1 & 2 & -1 & 0 & 0 \\ 0 & -1 & 2 & -1 & 0 \\ 0 & 0 & -1 & 2 & -1 \\ 0 & 0 & 0 & -1 & 1 \end{bmatrix}$$

$$\mathbf{B} = \sum_e \mathbf{m_G} = \frac{L}{24}\begin{bmatrix} 2 & 1 & 0 & 0 & 0 \\ 1 & 4 & 1 & 0 & 0 \\ 0 & 1 & 4 & 1 & 0 \\ 0 & 0 & 1 & 4 & 1 \\ 0 & 0 & 0 & 1 & 2 \end{bmatrix}$$

$$\mathbf{u} = [u_1 \quad u_2 \quad u_3 \quad u_4 \quad u_5]^{\mathrm{T}}$$

The initial condition is homogeneous so that $\mathbf{u}(0) = \mathbf{0}$.

Constraints. The constraints follow from the boundary conditions as

$$u_1 = u_0 \qquad \text{and} \qquad u_5 = 0$$

As outlined in Section 4.2.1 the constrained equations become

$$\psi\begin{bmatrix} 2 & -1 & 0 \\ -1 & 2 & -1 \\ 0 & -1 & 2 \end{bmatrix}\mathbf{u} + \begin{bmatrix} 4 & 1 & 0 \\ 1 & 4 & 1 \\ 0 & 1 & 4 \end{bmatrix}\dot{\mathbf{u}} = \begin{bmatrix} \psi u_0 \\ 0 \\ 0 \end{bmatrix} \tag{4.13}$$

subject to the initial condition $\mathbf{u}(0) = \mathbf{u_0}$; $\psi = 96\alpha/L^2$. These approximate equations must now be integrated for an estimate of the time-dependent solution. Appropriate analytical and numerical methods of integration are presented and discussed in the following sections.

The elemental mass matrices $\mathbf{m_e}$ of Eq. (4.8*b*) are referred to as **consistent mass matrices** in that they are determined on the basis of the same interpolation functions as were used for the corresponding stiffnesses $\mathbf{k_e}$. Another approach to generating mass matrices is referred to as *lumping* with the results referred to as **lumped mass matrices**. The idea is simply that the total mass associated with the consistent mass matrix

$$\mathbf{m_e} = \rho c_p A \begin{bmatrix} \dfrac{2l_e}{6} & \dfrac{l_e}{6} \\ \dfrac{l_e}{6} & \dfrac{2l_e}{6} \end{bmatrix}$$

namely, $\rho c_p A l_e$, is split between the two nodes to form a diagonal mass matrix called a lumped mass matrix $\mathbf{m_{le}}$

$$\mathbf{m_{le}} = \rho c_p A \begin{bmatrix} \frac{l_e}{2} & 0 \\ 0 & \frac{l_e}{2} \end{bmatrix}$$

This lumped mass matrix has advantages in certain of the time integration algorithms to be discussed in later sections, as well as the interesting property that the resulting eigenvalues are generally smaller than the exact values, as opposed to the general overestimation of the eigenvalues when using the consistent mass matrices.

This suggests that a third possibility for treating the mass is to consider a weighted average of the consistent mass matrix $\mathbf{m_{ce}}$ and lumped mass matrix $\mathbf{m_{le}}$ according to

$$\mathbf{m_w} = \alpha \mathbf{m_{ce}} + (1 - \alpha)\mathbf{m_{le}}$$

The choice $\alpha = 0.5$ results in

$$\mathbf{m_w} = \frac{l_e}{12}\begin{bmatrix} 5 & 1 \\ 1 & 5 \end{bmatrix}$$

due to Goudreau [2]; it is one of the so-called **higher-order accurate mass matrices**. Its use results in improved estimates for the eigenvalues as compared with estimates using either of the consistent or lumped mass matrix formulations. The corresponding result for the quadratically interpolated element turns out to be

$$\mathbf{m_w} = \frac{l_e}{60}\begin{bmatrix} 9 & 2 & -1 \\ 2 & 36 & 2 \\ -1 & 2 & 9 \end{bmatrix}$$

The student is asked to investigate these ideas further in several Exercises at the end of the chapter.

4.2.3 Analytical Integration Techniques

An analytical approach to the integration of the set of equations (4.10) involves using the techniques discussed in Section 3.7. To use this approach it is convenient to decompose the solution of

$$\mathbf{Ku} + \mathbf{M\dot{u}} = \mathbf{f}$$

into

$$\mathbf{u} = \mathbf{u_h} + \mathbf{u_p}$$

where $\mathbf{u_h}$ is the homogeneous solution satisfying

$$\mathbf{Ku_h} + \mathbf{M\dot{u}_h} = \mathbf{0}$$

and $\mathbf{u_p}$ is any particular solution satisfying

$$\mathbf{K}\mathbf{u_p} + \mathbf{M}\dot{\mathbf{u}}_\mathbf{p} = \mathbf{f}$$

We will discuss each of these in turn.

Homogeneous solution. For the case where **K** and **M** are matrices of constants,

$$\mathbf{K}\mathbf{u_h}(t) + \mathbf{M}\dot{\mathbf{u}}_\mathbf{h} = \mathbf{0} \tag{4.14}$$

is a set of linear constant-coefficient, ordinary differential equations. When **K** and **M** are not constant matrices, it is necessary to use techniques such as discussed in the next section.

The standard approach to the solution of such a constant coefficient system is to assume

$$\mathbf{u_h}(t) = \mathbf{v}\exp(-\lambda t) \tag{4.15}$$

where **v** is a vector of constants. The negative sign in the exponential function is a matter of anticipating the decaying character of the solution of diffusion problems. Substitution of Eq. (4.15) into Eq. (4.14) yields

$$(\mathbf{K} - \lambda\mathbf{M})\mathbf{v}\exp(-\lambda t) = \mathbf{0}$$

from which

$$(\mathbf{K} - \lambda\mathbf{M})\mathbf{v} = \mathbf{0} \tag{4.16}$$

Thus the homogeneous solutions are obtained by solving the generalized linear algebraic eigenvalue problem given by Eq. (4.16).

Nontrivial solutions of Eq. (4.16) result from requiring

$$\det(\mathbf{K} - \lambda\mathbf{M}) = 0 \tag{4.17}$$

from which the eigenvalues are obtained. Denote these by $\lambda_1, \lambda_2, \ldots$. The corresponding eigenvectors $\mathbf{v_1}, \mathbf{v_2}, \ldots$ are then determined by back-substituting the eigenvalues one at a time into Eqs. (4.17). Several examples are given in Sections 2.5.1, 2.9.4, and 3.7.3. The analytical approach is quite valuable from the standpoint that the results can be immediately interpreted in terms of the decay rates [$\exp(-\lambda_i t)$ as determined by the eigenvalues] that will be present in the solution regardless of whether an analytical or a numerical approach is being used. These eigenvalues are an important part of the discussions of convergence and stability covered in later sections.

For the particular example developed in Section 4.2.2, the equation $\det(\mathbf{K} - \lambda\mathbf{M}) = 0$ can be written as

$$\begin{vmatrix} 2-4\phi & -(1+\phi) & 0 \\ -(1+\phi) & 2-4\phi & -(1+\phi) \\ & -(1+\phi) & 2-4\phi \end{vmatrix} = 0$$

where $\phi = \lambda L^2/96\alpha$. The roots of the corresponding characteristic equation

$$(2 - 4\phi)\left[(2 - 4\phi)^2 - 2(1 + \phi)^2\right] = 0$$

are $\phi_1 = 0.1082$, $\phi_2 = 0.5000$, and $\phi_3 = 1.3204$, from which

$$\lambda_1 = \frac{10.387\alpha}{L^2} \qquad \lambda_2 = \frac{48.000\alpha}{L^2} \qquad \lambda_3 = \frac{126.76\alpha}{L^2}$$

compared with the exact eigenvalues of $\lambda_{1,2,3} = \pi^2\alpha/L^2$, $4\pi^2\alpha/L^2$, and $9\pi^2\alpha/L^2$, respectively. It can easily be seen that, roughly, the solutions will decay to steady state too rapidly in view of the fact that the λ's predicted by the finite element solution are larger than the corresponding exact values. The corresponding eigenvectors obtained by back-substitution into Eqs. (4.16) are

$$\mathbf{v}_1^T = \begin{bmatrix} 1 & \sqrt{2} & 1 \end{bmatrix}, \quad \mathbf{v}_2^T = [1 \quad 0 \quad -1], \quad \mathbf{v}_3^T = \begin{bmatrix} 1 & -\sqrt{2} & 1 \end{bmatrix}$$

and the homogeneous solution can then be written as

$$\mathbf{u_h} = c_1\mathbf{v_1}\exp(-\lambda_1 t) + c_2\mathbf{v_2}\exp(-\lambda_2 t) + c_3\mathbf{v_3}\exp(-\lambda_3 t)$$

Particular solution. For the particular solution, rewrite Eq. (4.13) as

$$\begin{bmatrix} 2 & -1 & 0 \\ -1 & 2 & -1 \\ 0 & -1 & 2 \end{bmatrix}\mathbf{u_p} + \psi^{-1}\begin{bmatrix} 4 & 1 & 0 \\ 1 & 4 & 1 \\ 0 & 1 & 4 \end{bmatrix}\dot{\mathbf{u}}_\mathbf{p} = \begin{bmatrix} u_0 \\ 0 \\ 0 \end{bmatrix}$$

By inspection (Method of Intelligent Guessing!) it can be seen that by taking $\mathbf{u_p} = \mathbf{d}$, a constant, there results

$$\begin{bmatrix} 2 & -1 & 0 \\ -1 & 2 & -1 \\ 0 & -1 & 2 \end{bmatrix}\mathbf{d} = \begin{bmatrix} u_0 \\ 0 \\ 0 \end{bmatrix}$$

from which $\mathbf{d} = \mathbf{u_p} = (u_0/4)[3 \;\; 2 \;\; 1]$. The general solution can then be written as

$$\mathbf{u} = c_1\mathbf{v_1}\exp(-\lambda_1 t) + c_2\mathbf{v_2}\exp(-\lambda_2 t) + c_3\mathbf{v_3}\exp(-\lambda_3 t) + \mathbf{u_p}$$

with $\mathbf{u_p}$ given above.

Satisfying the initial condition $\mathbf{u}(0) = \mathbf{0}$ leads to the set of linear algebraic equations

$$\begin{bmatrix} 1 & 1 & 1 \\ \sqrt{2} & 0 & -\sqrt{2} \\ 1 & -1 & 1 \end{bmatrix}\begin{bmatrix} c_1 \\ c_2 \\ c_3 \end{bmatrix} = \begin{bmatrix} -\dfrac{3u_0}{4} \\ -\dfrac{2u_0}{4} \\ -\dfrac{u_0}{4} \end{bmatrix}$$

from which $c_1 = -0.4268u_0$, $c_2 = -0.2500u_0$, and $c_3 = -0.0732u_0$. The solution can finally be expressed as

$$\frac{u_2(t)}{u_0} = 0.7500 - 0.4268\exp(-\lambda_1 t) - 0.2500\exp(-\lambda_2 t) - 0.0732\exp(-\lambda_3 t)$$

$$\frac{u_3(t)}{u_0} = 0.5000 - 0.6036\exp(-\lambda_1 t) + 0.1036\exp(-\lambda_3 t) \tag{4.18}$$

$$\frac{u_4(t)}{u_0} = 0.2500 - 0.4268\exp(-\lambda_1 t) + 0.2500\exp(-\lambda_2 t) - 0.0732\exp(-\lambda_3 t)$$

Note that as $t \longrightarrow \infty$, $u_2(t) \longrightarrow 0.7500u_0$, $u_3(t) \longrightarrow 0.5000u_0$, and $u_4(t) \longrightarrow 0.2500u_0$, the correct steady-state values. As mentioned previously, the finite element model predicts that these steady-state values are reached too quickly. This will be discussed further in Section 4.2.4.3.

For a specific case assume that the bar is 0.2 m in length and is composed of an aluminum alloy for which $\alpha = 8.4 \times 10^{-4}$ m^2/s, from which $\alpha/L^2 = 0.021$ s^{-1}. The $u_2(t)$, $u_3(t)$, and $u_4(t)$ computed from Eqs. (4.18) are presented in Table 4.1 along with exact solutions for a range of values of t up to near steady state. A graphical comparison of the finite element and exact solutions at $x = L/3$ is shown in Fig. 4–6.

TABLE 4.1 Four-element analytical versus exact solution

t (sec)	(u_2/u_0)	$(u_2/u_0)_{ex}$	(u_3/u_0)	$(u_3/u_0)_{ex}$	(u_4/u_0)	$(u_4/u_0)_{ex}$
1.0	0.3105	0.2225	0.0219	0.0147	−0.0070	0.0003
2.0	0.4404	0.3884	0.1103	0.0845	0.0070	0.0096
3.0	0.5106	0.4812	0.1863	0.1589	0.0403	0.0342
4.0	0.5672	0.5419	0.2478	0.2223	0.0761	0.0650
5.0	0.6050	0.5852	0.2972	0.2742	0.1082	0.0953
6.0	0.6341	0.6180	0.3369	0.3164	0.1353	0.1224
7.0	0.6571	0.6435	0.3689	0.3508	0.1575	0.1455
8.0	0.6754	0.6638	0.3946	0.3787	0.1755	0.1647
9.0	0.6900	0.6801	0.4152	0.4014	0.1901	0.1805
10.0	0.7018	0.6933	0.4319	0.4199	0.2018	0.1934
15.0	0.7338	0.7299	0.4771	0.4716	0.2338	0.2299
20.0	0.7446	0.7429	0.4923	0.4899	0.2446	0.2429

As mentioned previously and as can easily be seen from these data, the finite element solution tends towards steady state too rapidly. The trend is of course helped by taking more elements, in which case the approximate eigenvalues arising from the finite element model approach the exact eigenvalues and the results are correspondingly closer to those given by the exact solution. In the Exercises, the student is asked to investigate other meshes and interpolations in connection with this example.

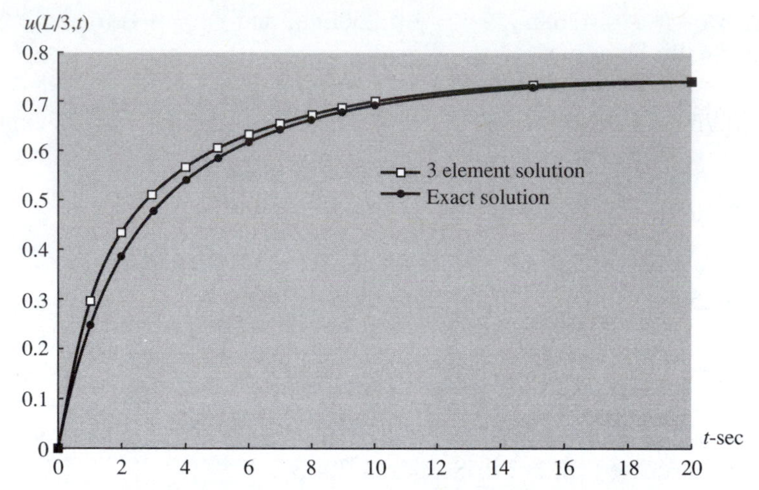

FIGURE 4–6 Comparison of exact and three-element solutions at $x = L/3$

Consider the following alternate analytical approach to integrating

$$\mathbf{M}\dot{\mathbf{u}} + \mathbf{K}\mathbf{u} = \mathbf{f}$$

$$\mathbf{u}(0) = \mathbf{u_0}$$

The first step is to determine the eigenvalues and eigenvectors as was indicated earlier in this section, that is, to solve

$$\mathbf{K}\mathbf{v_i} = \lambda_i \mathbf{M}\mathbf{v_i}$$

for the λ_i and the corresponding $\mathbf{v_i}$. In order to develop the necessary relationships between the eigenvalues and eigenvectors, write the same equation for a distinct λ_j ($\lambda_j \neq \lambda_i$), that is,

$$\mathbf{K}\mathbf{v_j} = \lambda_j \mathbf{M}\mathbf{v_j}$$

Multiply the i equation by $\mathbf{v_j^T}$, the j equation by $\mathbf{v_i^T}$, and subtract to obtain

$$\mathbf{v_j^T}\mathbf{K}\mathbf{v_i} - \mathbf{v_i^T}\mathbf{K}\mathbf{v_j} = \lambda_i \mathbf{v_j^T}\mathbf{M}\mathbf{v_i} - \lambda_j \mathbf{v_i^T}\mathbf{M}\mathbf{v_j} \tag{4.19}$$

Since **K** and **M** are symmetric, it follows that

$$(\mathbf{v_j^T}\mathbf{K}\mathbf{v_i})^{\mathrm{T}} = \mathbf{v_i^T}\mathbf{K}^{\mathrm{T}}\mathbf{v_j} = \mathbf{v_i^T}\mathbf{K}\mathbf{v_j}$$

and

$$(\mathbf{v_j^T}\mathbf{M}\mathbf{v_i})^{\mathrm{T}} = \mathbf{v_i^T}\mathbf{M}^{\mathrm{T}}\mathbf{v_j} = \mathbf{v_i^T}\mathbf{M}\mathbf{v_j}$$

so that Eq. (4.19) can be written as

$$0 = (\lambda_i - \lambda_j)\mathbf{v_j^T}\mathbf{M}\mathbf{v_i} \tag{4.20}$$

showing that eigenvectors corresponding to distinct eigenvalues are orthogonal with respect to the mass matrix **M**, that is,

$$\mathbf{v_j^T M v_i} = \mathbf{0} \qquad i \neq j$$

Note that premultiplying $\mathbf{Kv_i} = \lambda_i \mathbf{Mv_i}$ by $\mathbf{v_j^T}$ it also follows that $\mathbf{v_j^T K v_i} = \mathbf{0}$. Collecting all the eigenequations results in

$$\mathbf{K}[\mathbf{v}_1 \quad \mathbf{v}_2 \quad \ldots \quad \mathbf{v}_N] = \mathbf{M}[\mathbf{v}_1\lambda_1 \quad \mathbf{v}_2\lambda_2 \quad \ldots \quad \mathbf{v}_N\lambda_N]$$

which is written as

$$\mathbf{KV} = \mathbf{MV\Lambda} \tag{4.21}$$

where $\mathbf{V}$ is the *modal* matrix and $\mathbf{\Lambda}$ is a diagonal matrix consisting of the eigenvalues in the same order as the eigenvectors in $\mathbf{V}$. The order is usually that of increasing λ.

The mutually orthogonal character of the eigenvectors can similarly be collected to read

$$\begin{bmatrix} \mathbf{v}_1^T \\ \mathbf{v}_2^T \\ \vdots \\ \mathbf{v}_N^T \end{bmatrix} \mathbf{M}[\mathbf{v}_1 \quad \mathbf{v}_2 \quad \ldots \quad \mathbf{v}_N] = \mathbf{I}$$

or

$$\mathbf{V^T M V} = \mathbf{I}$$

Premultiplying Eq. (4.21) by $\mathbf{V^T}$ yields

$$\mathbf{V^T K V} = \mathbf{V^T M V \Lambda} = \mathbf{I\Lambda} = \mathbf{\Lambda}$$

The two relationships

$$\mathbf{V^T M V} = \mathbf{I} \qquad \mathbf{V^T K V} = \mathbf{\Lambda} \tag{4.22}$$

are often associated with what is referred to as the *simultaneous diagonalization of two quadratic forms*, that is, multiplication by $\mathbf{V^T}$ on the left and $\mathbf{V}$ on the right has simultaneously converted $\mathbf{M}$ and $\mathbf{K}$ to diagonal matrices. These relationships between the matrices $\mathbf{K}$, $\mathbf{M}$, $\mathbf{V}$, and $\mathbf{\Lambda}$ are the basis for the alternate analytical approach which we now complete. They are also fundamental in the consideration of stability of numerical integration algorithms discussed in Section 4.2.4.3.

Consider again the object of the present deliberations, namely

$$\mathbf{M\dot{u}} + \mathbf{Ku} = \mathbf{f}$$

$$\mathbf{u}(0) = \mathbf{u}_0$$

Take $\mathbf{u} = \mathbf{Vz}$ where $\mathbf{V}$ is the modal matrix defined above and $\mathbf{z}$ is a new set of dependent variables often referred to as normal coordinates or normal modes. Substitution and subsequent premultiplication by $\mathbf{V^T}$ yields

$$\mathbf{V^T M V \dot{z}} + \mathbf{V^T K V z} = \mathbf{V^T f}$$

which, upon using Eqs. (4.22), can be written as

$$\dot{\mathbf{z}} + \mathbf{\Lambda z} = \mathbf{g} \tag{4.23}$$

where $\mathbf{g} = \mathbf{V}^{\mathrm{T}}\mathbf{f}$. Each component of $\mathbf{z}$ is a linear combination of the components of $\mathbf{u}$ and each component of $\mathbf{g}$ is a linear combination of the components of $\mathbf{f}$.

Equation (4.24) is a set of uncoupled equations in the normal coordinates z_i which can be written as

$$\dot{z}_i + \lambda z_i = g_i(t) \qquad i = 1, 2, \ldots, N$$

A typical equation is solved to yield

$$z_i = c_i \exp(-\lambda_i t) + \int_0^t \exp\big(-\lambda_i(t-\tau)\big) g_i(\tau)\, d\tau \tag{4.24}$$

The particular solution can be thought of as having been generated by either of the methods of variation of parameters or Laplace transforms.

The initial condition specified originally on the variable $\mathbf{u}$ is expressed in terms of the variable $\mathbf{z}$ as

$$\mathbf{u}(0) = \mathbf{u}_0 = \mathbf{Vz}(0)$$

or, after premultiplying by $\mathbf{V}^{\mathrm{T}}\mathbf{M}$,

$$\mathbf{z}_0 = \mathbf{V}^{\mathrm{T}}\mathbf{M}\mathbf{u}_0 = \mathbf{V}^{\mathrm{T}}\mathbf{MVz}(0) = \mathbf{Iz}(0) = \mathbf{z}(0)$$

Enforcing the initial condition on the solution given in Eq. (4.24) yields

$$(\mathbf{z}_0)_i = c_i$$

so that the solution can be expressed as

$$z_i = (\mathbf{z}_0)_i \exp(-\lambda_i t) + \int_0^t \exp\big(-\lambda_i(t-\tau)\big) g_i(\tau)\, d\tau$$

The original variable $\mathbf{u}$ is recovered by the transformation $\mathbf{u} = \mathbf{Vz}$, giving back each of the original components as a linear combination of *each* of the normal coordinates. The student is asked to investigate the details of this approach in the Exercises.

4.2.4 Time Domain Integration Techniques—First-Order Systems

In situations where $\mathbf{M}$ and/or $\mathbf{K}$ in

$$\mathbf{M}\dot{\mathbf{u}} + \mathbf{Ku} = \mathbf{f}$$

are functions of t or where $\mathbf{f}$ is such that a particular solution by analytic means is difficult or impossible, the analytical technique discussed in Section 4.2.3 may be practically impossible to carry through. Depending on the particulars, numerical techniques are attractive or even necessary to carry out the integration. Two such schemes are presented and illustrated in the next sections.

4.2.4.1 THE EULER METHOD. Recall that for a first-order initial value problem of the form

$$y' = f(x, y)$$
$$y(x_0) = y_0$$

it is possible to develop numerical integration schemes for the approximate integration of the initial value problem. The problem is to determine a function $y(x)$ passing through the initial point (x_0, y_0) and satisfying the differential equation $y' = f$, as indicated in Fig. 4–7.

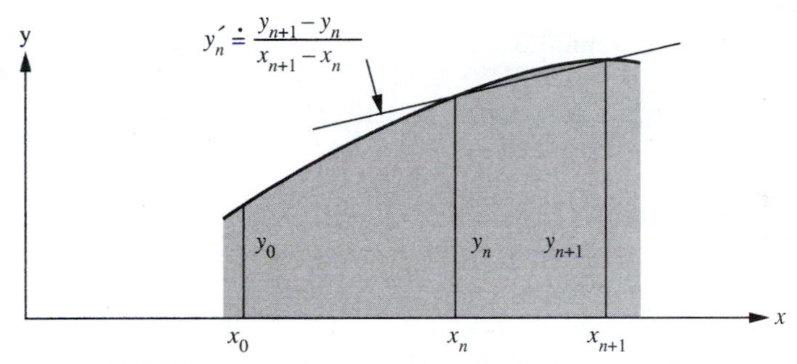

FIGURE 4–7 Geometry for the Euler algorithm

In the Euler method the derivative y' is represented as a **forward** difference according to

$$y_n' = \frac{(y_{n+1} - y_n)}{h}$$

giving a linear approximation to the derivative at x_n as indicated in Fig. 4–7, with $h = x_{n+1} - x_n$. The differential equation is evaluated at x_n to yield

$$y_n' = \frac{(y_{n+1} - y_n)}{h} = f(x_n, y_n)$$

or solving for y_{n+1}, there results

$$y_{n+1} = y_n + hf(x_n, y_n)$$

Starting with $y(x_0) = y_0$, this algorithm can be used to step ahead in the independent variable x to determine an approximate solution.

Note that the Euler method can also be viewed in the following manner. Integrate the differential equation between the limits of x_n and x_{n+1} to obtain

$$\int_{x_n}^{x_{n+1}} y'dx = y_{n+1} - y_n = \int_{x_n}^{x_{n+1}} f(x, y)\,dx$$

Approximate the remaining integral by $hf(x_n, y_n)$ to obtain

$$y_{n+1} = y_n + hf(x_n, y_n)$$

The integral has been approximated by evaluating f at the left end of the interval over which the integral is evaluated and multiplying by the interval h. This is again clearly the Euler algorithm.

The analogue of the Euler method for the system of equations

$$\mathbf{M}\dot{\mathbf{u}} + \mathbf{K}\mathbf{u} = \mathbf{f} \tag{4.25}$$

is obtained by again representing the derivative term $\dot{\mathbf{u}}$ at a particular value of the time t_n as a forward difference according to

$$\dot{\mathbf{u}}_\mathbf{n} = \frac{(\mathbf{u}_{\mathbf{n+1}} - \mathbf{u}_\mathbf{n})}{h}$$

and evaluating the differential equation (4.25) at $t = t_n$ to obtain

$$\mathbf{M}\dot{\mathbf{u}}_\mathbf{n} + \mathbf{K}\mathbf{u}_\mathbf{n} = \mathbf{f}_\mathbf{n} = \frac{\mathbf{M}(\mathbf{u}_{\mathbf{n+1}} - \mathbf{u}_\mathbf{n})}{h} + \mathbf{K}\mathbf{u}_\mathbf{n}$$

from which the Euler algorithm

$$\begin{aligned} \mathbf{M}\mathbf{u}_{\mathbf{n+1}} &= (\mathbf{M} - h\mathbf{K})\mathbf{u}_\mathbf{n} + h\mathbf{f}_\mathbf{n} \\ \mathbf{u}(0) &= \mathbf{u}_0 \end{aligned} \tag{4.26}$$

is obtained. With the initial condition this algorithm is used according to

$$\begin{aligned} \mathbf{M}\mathbf{u}_1 &= (\mathbf{M} - h\mathbf{K})\mathbf{u}_0 + h\mathbf{f}_0 \longrightarrow \mathbf{u}_1 \\ \mathbf{M}\mathbf{u}_2 &= (\mathbf{M} - h\mathbf{K})\mathbf{u}_1 + h\mathbf{f}_1 \longrightarrow \mathbf{u}_2 \\ \mathbf{M}\mathbf{u}_3 &= (\mathbf{M} - h\mathbf{K})\mathbf{u}_2 + h\mathbf{f}_2 \longrightarrow \mathbf{u}_3 \\ &\vdots \end{aligned}$$

to step ahead in time for the solution.

The utility and effectiveness of the algorithm of Eq. (4.26) is affected by its *stability*, that is, by whether for large time the solution predicted by the algorithm remains finite, independent of the step size h. As will be indicated in Section 4.2.4.3, the Euler algorithm is conditionally stable: there is a critical step size h_{cr} such that when $h > h_{cr}$, the solution oscillates with ever increasing amplitude, obviously negating the results of the algorithm.

The use of the Euler algorithm results in a local discretization error e, which depends upon the step size h according to

$$e = O(h^2)$$

indicating that when the time step h is halved the local discretization error is reduced approximately by 1/4. The accumulated discretization error E is given by

$$E = O(h)$$

that is, the accumulated discretization error is approximately halved by halving the integration time step. One might assume on the basis of these results that by decreasing the step size sufficiently the error could be decreased indefinitely. This is not the case in that the roundoff error begins to dominate the process for h too small. For a discussion of discretization and roundoff errors, the reader should consult Boyce and DiPrima [3] and Henrici [4].

Example 4.1.

Consider the application of this algorithm to the set of equations (4.13) developed in Section 4.2.3, namely

$$\begin{bmatrix} 4 & 1 & 0 \\ 1 & 4 & 1 \\ 0 & 1 & 4 \end{bmatrix} \dot{\mathbf{w}} + \phi \begin{bmatrix} 2 & -1 & 0 \\ -1 & 2 & -1 \\ 0 & -1 & 2 \end{bmatrix} \mathbf{w} = \phi \begin{bmatrix} 1 \\ 0 \\ 0 \end{bmatrix}$$

where $\phi = 96\alpha/L^2$ and $\mathbf{w} = \mathbf{u}/u_0$. Equations (4.26) become

$$\begin{bmatrix} 4 & 1 & 0 \\ 1 & 4 & 1 \\ 0 & 1 & 4 \end{bmatrix} \mathbf{w_{n+1}} = \left\{ \begin{bmatrix} 4 & 1 & 0 \\ 1 & 4 & 1 \\ 0 & 1 & 4 \end{bmatrix} - \phi h \begin{bmatrix} 2 & -1 & 0 \\ -1 & 2 & -1 \\ 0 & -1 & 2 \end{bmatrix} \right\} \mathbf{w_n} + \phi h \begin{bmatrix} 1 \\ 0 \\ 0 \end{bmatrix}$$

For this 3×3 example, the inverse of $\mathbf{M}$ is easily determined so that the equations can be written as

$$\mathbf{w_{n+1}} = \left\{ \begin{bmatrix} 1 & 0 & 0 \\ 0 & 1 & 0 \\ 0 & 0 & 1 \end{bmatrix} - \frac{\phi h}{56} \begin{bmatrix} 34 & -24 & 6 \\ -24 & 40 & -24 \\ 6 & -24 & 34 \end{bmatrix} \right\} \mathbf{w_n} + \frac{\phi h}{56} \begin{bmatrix} 15 \\ -4 \\ 1 \end{bmatrix} \tag{4.27}$$

With $\alpha/L^2 = 0.021\ \text{s}^{-1}$ as in the example in Section 4.2.3, and with $h = 0.1$ s, there results specifically

$$\mathbf{w_{n+1}} = \begin{bmatrix} 0.8776 & 0.0864 & -0.0216 \\ 0.0864 & 0.8560 & 0.0864 \\ -0.0216 & 0.0864 & 0.8776 \end{bmatrix} \mathbf{w_n} + \begin{bmatrix} 0.0540 \\ -0.0144 \\ 0.0036 \end{bmatrix}$$

or

$$\mathbf{w_{n+1}} = \mathbf{S}\mathbf{w_n} + \mathbf{b}$$

with

$$\mathbf{w}(0) = \mathbf{0}$$

Successively there results

$$\mathbf{w_1} = \mathbf{S}\,\mathbf{0} + \mathbf{b} = [0.0540 \quad -0.0144 \quad 0.0036]^T$$

$$\mathbf{w_2} = \mathbf{S}\mathbf{w_1} + \mathbf{b} = [0.1001 \quad -0.0217 \quad 0.0043]^T$$

The algorithm is repeatedly applied until the range of times of interest is covered.

In connection with this same example, the algorithm of Eq. (4.27) is used with several values of the step size $h = \Delta t$. The results for $u_3(t)$ are given in Table 4.2

along with the values as given by the exact solution computed from the infinite series representation for the solution.

TABLE 4.2 Approximate solution $w_3(t)$ for different h using the Euler algorithm

t (sec)	h = 0.083	h = 0.167	h = 0.333	h = 1.0	u_{ex}
1	0.0208	0.0198	0.0189	−0.1440	0.0147
2	0.1116	0.1131	0.1163	0.4170	0.0845
3	0.1881	0.1901	0.1940	−0.2683	0.1589
4	0.2497	0.2518	0.2560	1.0643	0.2223
5	0.2991	0.3013	0.3055	−0.9891	0.2742
6	0.3388	0.3409	0.3449	2.5436	0.3164
7	0.3706	0.3726	0.3763	−3.2333	0.3508
8	0.3962	0.3980	0.4014	6.4408	0.3787
9	0.4167	0.4183	0.4214	−9.5789	0.4014
10	0.4331	0.4346	0.4373	17.0888	0.4199
20	0.4926	0.4929	0.4935	2674.49	0.4899

These results show clearly several aspects of the numerical solution:

1. The approximate solution tends toward steady state more rapidly than the exact solution. As mentioned previously, this property is primarily attributable to the eigenvalues of $\mathbf{M} - \lambda\mathbf{K}$ being larger than the exact eigenvalues.
2. There is clearly a step size h above which the approximate solution is unstable as indicated by the $h = 1.0$ results.
3. When the step size is not exceeded, the approximate solution approaches the correct steady-state values for large t.
4. If lumping is used, the assembled mass matrix $\mathbf{M}$ is diagonal and easily inverted. The integration procedure is reduced at each step to a simple matrix multiplication and vector addition

$$\mathbf{u_{n+1}} = \mathbf{Su_n} + h\mathbf{M}^{-1}\mathbf{f_n}$$

where if $\mathbf{M}$ and $\mathbf{K}$ are constant matrices, $\mathbf{S} = \mathbf{I} - h\mathbf{M}^{-1}\mathbf{K}$ can be computed at the first time step and used thereafter for the subsequent applications of the basic algorithm.

4.2.4.2 IMPROVED EULER OR CRANK-NICOLSON METHOD. The *improved* Euler or Crank-Nicolson algorithm can be thought of in the following way. Again, thinking of the scalar first-order differential equation

$$y' = f(x, y)$$

$$y(x_0) = y_0$$

the improved Euler algorithm is developed according to the relation

$$y_{n+1} = y_n + \frac{h\big(f(x_n, y_n) + f(x_{n+1}, y_{n+1})\big)}{2}$$

that is, the integral

$$I = \int_{x_n}^{x_{n+1}} f(x, y)\, dx$$

has been evaluated by taking the average of f at the ends of the interval (x_n, x_{n+1}). For the vector equation in question, the corresponding expression is

$$\mathbf{u_{n+1}} = \mathbf{u_n} + \frac{h(\dot{\mathbf{u}}_\mathbf{n} + \dot{\mathbf{u}}_\mathbf{n+1})}{2}$$

which is equivalent to a central difference representation for the derivative $\dot{\mathbf{u}}$.

Multiplying through by $\mathbf{M}$ yields

$$\mathbf{Mu_{n+1}} = \mathbf{Mu_n} + \frac{h(\mathbf{M}\dot{\mathbf{u}}_\mathbf{n} + \mathbf{M}\dot{\mathbf{u}}_\mathbf{n+1})}{2}$$

which, on using the differential equation $\mathbf{M}\dot{\mathbf{u}} + \mathbf{Ku} = \mathbf{f}$, can be written as

$$\left(\mathbf{M} + \frac{h\mathbf{K}}{2}\right)\mathbf{u_{n+1}} = \left(\mathbf{M} - \frac{h\mathbf{K}}{2}\right)\mathbf{u_n} + \frac{h(\mathbf{f_n} + \mathbf{f_{n+1}})}{2} \tag{4.28}$$

which is the improved Euler or Crank-Nicolson algorithm. As opposed to the algorithm studied in the last section, this algorithm is unconditionally stable, that is, although the accuracy may suffer considerably and oscillations occur for a large step size h, the oscillations never become unbounded. In addition to being unconditionally stable, the improved Euler or Crank-Nicolson algorithm is one order more accurate than the previously developed Euler algorithm in that the accumulated discretization error E is given approximately by $E = O(h^2)$.

To implement the method, write Eq. (4.28) as

$$\mathbf{Au_{n+1}} = \mathbf{b_{n+1}} \tag{4.29}$$

where

$$\mathbf{A} = \mathbf{M} + \frac{h\mathbf{K}}{2}$$

$$\mathbf{b_{n+1}} = \frac{h(\mathbf{f_n} + \mathbf{f_{n+1}})}{2} + \left(\mathbf{M} - \frac{h\mathbf{K}}{2}\right)\mathbf{u_n}$$

This set of linear algebraic equations (4.29) must be solved at each time step. Given $\mathbf{u}(0) = \mathbf{u_0}$,

$$\mathbf{Au_1} = \mathbf{b_1} = \frac{h\big(\mathbf{f}(0) + \mathbf{f}(h)\big)}{2} + \left(\mathbf{M} - \frac{h\mathbf{K}}{2}\right)\mathbf{u_0}$$

is to be solved for $\mathbf{u_1}$, after which

$$\mathbf{Au_2} = \mathbf{b_2} = \frac{h\big(\mathbf{f}(h) + \mathbf{f}(2h)\big)}{2} + \left(\mathbf{M} - \frac{h\mathbf{K}}{2}\right)\mathbf{u_1}$$

is to be solved for $\mathbf{u_2}$, and so forth.

As long as h = constant, the same coefficient matrix $\mathbf{M} + h\mathbf{K}/2$ is involved at each step and it is economical to use an equation solver that decomposes $\mathbf{A}$ at the first step according to $\mathbf{A} = \mathbf{LU}$. This decomposition is then saved so that at each succeeding step, two triangular systems can be solved. This is substantially more economical than to solve $\mathbf{Ax} = \mathbf{b}$ at each step. In the event that $\mathbf{M}$ and $\mathbf{K}$ are functions of t, $\mathbf{M}$, $\mathbf{K}$, $\mathbf{M} + h\mathbf{K}/2$, and $\mathbf{M} - h\mathbf{K}/2$ must potentially be recomputed at each step. If the variation of these matrices with time is small, recalculation of the necessary matrices and decomposition of $\mathbf{M} + h\mathbf{K}/2$ can be done at suitable regular intervals rather than at each time step. Stability of this algorithm is discussed in Section 4.2.4.3

Example 4.2.

Consider the example of Sections 4.2.3 and 4.2.4.1,

$$\begin{bmatrix} 4 & 1 & 0 \\ 1 & 4 & 1 \\ 0 & 1 & 4 \end{bmatrix} \dot{\mathbf{w}} + \phi \begin{bmatrix} 2 & -1 & 0 \\ -1 & 2 & -1 \\ 0 & -1 & 2 \end{bmatrix} \mathbf{w} = \phi \begin{bmatrix} 1 \\ 0 \\ 0 \end{bmatrix}$$

where $\phi = 96\alpha/L^2$ and $\mathbf{w} = \mathbf{u}/u_0$. Generally,

$$\mathbf{M} \pm \frac{h\mathbf{K}}{2} = \begin{bmatrix} 4 & 1 & 0 \\ 1 & 4 & 1 \\ 0 & 1 & 4 \end{bmatrix} \pm \frac{48\alpha h}{L^2} \begin{bmatrix} 2 & -1 & 0 \\ -1 & 2 & -1 \\ 0 & -1 & 2 \end{bmatrix}$$

Taking $\alpha = 8.4 \times 10^{-4}$ m^2/s, $L = 0.2$ m, and $h = 0.1$ s results in

$$\mathbf{M} + \frac{h\mathbf{K}}{2} = \begin{bmatrix} 4.2016 & 0.8992 & 0.0000 \\ 0.8992 & 4.2016 & 0.8992 \\ 0.0000 & 0.8992 & 4.2016 \end{bmatrix}$$

and

$$\mathbf{M} - \frac{h\mathbf{K}}{2} = \begin{bmatrix} 3.7984 & 1.1008 & 0.0000 \\ 1.1008 & 3.7984 & 1.1008 \\ 0.0000 & 1.1008 & 3.7984 \end{bmatrix}$$

The 3×3 matrix $\mathbf{M} + h\mathbf{K}/2$ can be inverted so that there results specifically

$$\mathbf{w_{n+1}} = \begin{bmatrix} 0.8879 & 0.0754 & -0.0161 \\ 0.0754 & 0.8718 & 0.0754 \\ -0.0161 & 0.0754 & 0.8879 \end{bmatrix} \mathbf{w_n} + \begin{bmatrix} 0.0504 \\ -0.0113 \\ 0.0024 \end{bmatrix}$$

or

$$\mathbf{w_{n+1}} = \mathbf{S w_n} + \mathbf{b}$$

with

$$\mathbf{w}(0) = \mathbf{0}$$

The first couple of iterations yield

$$\mathbf{w_1} = \mathbf{S\,0} + \mathbf{b} = [0.0504 \quad -0.0113 \quad 0.0024]^{\mathrm{T}}$$

$$\mathbf{w}_2 = \mathbf{S}\mathbf{w}_1 + \mathbf{b} = [0.0943 \quad -0.0082 \quad 0.0029]^T$$

with iteration being continued until the time interval of interest is covered.

For this example, results for several values of h are presented in Table 4.3 for $u_3(t)$ along with those as given by the exact solution.

TABLE 4.3 Approximate solution $u_3(t)$ for different h using the improved Euler algorithm

t (sec)	$h = 0.083$	$h = 0.167$	$h = 0.333$	$h = 1.0$	u_{ex}
1	0.0219	0.0216	0.0207	0.0004	0.0147
2	0.1103	0.1103	0.1102	0.1126	0.0845
3	0.1863	0.1863	0.1864	0.1868	0.1589
4	0.2478	0.2478	0.2479	0.2487	0.2223
5	0.2972	0.2972	0.2973	0.2981	0.2742
6	0.3369	0.3370	0.3370	0.3378	0.3164
7	0.3689	0.3689	0.3690	0.3697	0.3508
8	0.3946	0.3946	0.3947	0.3953	0.3787
9	0.4153	0.4153	0.4153	0.4159	0.4014
10	0.4319	0.4319	0.4319	0.4324	0.4199
20	0.4923	0.4923	0.4923	0.4924	0.4899

Comparing these results with the results presented in Table 4.2 obtained using the Euler method, the following observations can be made:

1. The results using the improved Euler algorithm are more accurate than those from the Euler algorithm, as would be expected, with steady state not being approached as rapidly using the improved Euler algorithm.
2. For the values of h investigated, the improved Euler algorithm is stable as the results of Section 4.2.4.3 will indicate.
3. Lumping does not result in any computational advantage for the improved Euler algorithm since the coefficient matrix is $\mathbf{M} + h\mathbf{K}/2$ rather than $\mathbf{M}$ as for the Euler algorithm.

4.2.4.3 ANALYSIS OF ALGORITHMS. In this section, a bit of the theory behind the analysis of the stability of the time integration methods presented in Sections 4.2.4.1 and 4.2.4.2 for systems of first-order differential equations will be touched upon.

Both the Euler and improved Euler or Crank-Nicolson algorithms presented in the preceding sections can be considered as special cases of the so-called θ algorithm that assumes

$$\mathbf{u}_{n+1} = \mathbf{u}_n + h\,(\theta\dot{\mathbf{u}}_{n+1} + (1-\theta)\,\dot{\mathbf{u}}_n) \tag{4.30}$$

that is, a weighted sum of the derivatives at the beginning and end of the interval (t_n, t_{n+1}) is used to evaluate the integral

$$\int_{t_n}^{t_{n+1}} \dot{\mathbf{u}}_n\, dt$$

Multiplying Eq. (4.30) by **M** and subsequently using the differential equation to eliminate the **Mu** terms yields

$$(\mathbf{M} + h\theta\mathbf{K})\mathbf{u_{n+1}} = \big(\mathbf{M} - h(1-\theta)\mathbf{K}\big)\,\mathbf{u_n} + h\left(\theta\,\mathbf{f_{n+1}} + (1-\theta)\mathbf{f_n}\right) \qquad (4.31)$$

It is easily seen that

$$\theta = 0 \qquad \text{is the Euler method}$$

and

$$\theta = 1/2 \qquad \text{is the improved Euler or Crank-Nicolson method}$$

The value $\theta = 1$ corresponds to what is referred to as the *modified* Euler method and corresponds to using a **backward** difference scheme obtained by evaluating the differential equation at t_{n+1} and taking

$$\dot{\mathbf{u}}_{\mathbf{n+1}} = \frac{\mathbf{u_{n+1}} - \mathbf{u_n}}{h}$$

The stability of integrations performed using Eq. (4.31) will be discussed in what follows with specific attention paid to the three values $\theta = 0.0,\ 0.5$, and 1.0.

Write Eqns. (4.31) as

$$\mathbf{u_{n+1}} = \mathbf{Cu_n} + \mathbf{A}^{-1}\mathbf{b_n} = \mathbf{Cu_n} + \mathbf{F_n} \qquad (4.32)$$

where

$$\mathbf{C} = \mathbf{A}^{-1}\mathbf{B} = (\mathbf{M} + h\theta\mathbf{K})^{-1}\big(\mathbf{M} - h(1-\theta)\mathbf{K}\big)$$

Successive applications of the algorithm result in

$$\begin{aligned}
\mathbf{u_1} &= \mathbf{Cu_0} + \mathbf{F_0}\\
\mathbf{u_2} &= \mathbf{Cu_1} + \mathbf{F_1} = \mathbf{C}^2\mathbf{u_0} + \mathbf{CF_0} + \mathbf{F_1}\\
\mathbf{u_3} &= \mathbf{Cu_2} + \mathbf{F_2} = \mathbf{C}^3\mathbf{u_0} + \mathbf{C}^2\mathbf{F_0} + \mathbf{CF_1} + \mathbf{F_2}
\end{aligned}$$

The stability of the algorithm depends only on the behavior of the $\mathbf{C}^n\mathbf{u_0} = (\mathbf{A}^{-1}\mathbf{B})^n\mathbf{u_0}$ term. This algorithm is stable if, for any $\mathbf{u_0}$, the solution remains bounded. This in turn implies that $\mathbf{C}^n$ remains bounded as n becomes large. The investigation of the stability is best carried out in terms of the transformed coordinates associated with the modal matrix, which were introduced in Section 4.2.3.

To this end, let **V** be the modal matrix associated with

$$(\mathbf{K} - \lambda\mathbf{M})\mathbf{v} = \mathbf{0}$$

with the matrices **K**, **V**, **M**, and $\mathbf{\Lambda}$ satisfying

$$\mathbf{V}^{\mathrm{T}}\mathbf{MV} = \mathbf{I} \qquad \mathbf{V}^{\mathrm{T}}\mathbf{KV} = \mathbf{\Lambda}$$

Consider the term $(\mathbf{M} + h\theta\mathbf{K})$. Premultiply by $(\mathbf{V}^{\mathrm{T}})^{-1}\mathbf{V}^{\mathrm{T}}$ $(= \mathbf{I})$ and postmultiply by $\mathbf{VV}^{-1}$ $(= \mathbf{I})$ to obtain

$$\begin{aligned}
(\mathbf{M} + h\theta\mathbf{K}) &= (\mathbf{V}^{\mathrm{T}})^{-1}\mathbf{V}^{\mathrm{T}}(\mathbf{M} + h\theta\mathbf{K})\mathbf{VV}^{-1}\\
&= (\mathbf{V}^{\mathrm{T}})^{-1}(\mathbf{V}^{\mathrm{T}}\mathbf{MV} + h\theta\mathbf{V}^{\mathrm{T}}\mathbf{KV})\mathbf{V}^{-1}\\
&= (\mathbf{V}^{\mathrm{T}})^{-1}(\mathbf{I} + h\theta\mathbf{\Lambda})\mathbf{V}^{-1}
\end{aligned}$$

from which

$$(\mathbf{M} + h\theta\mathbf{K})^{-1} = \mathbf{V}(\mathbf{I} + h\theta\mathbf{\Lambda})^{-1}\mathbf{V}^{\mathrm{T}}$$

Similarly,

$$\big(\mathbf{M} - h(1-\theta)\mathbf{K}\big) = (\mathbf{V}^{\mathrm{T}})^{-1}\big(\mathbf{I} - (1-\theta)h\mathbf{\Lambda}\big)\mathbf{V}^{-1}$$

so that $\mathbf{C} = \big(\mathbf{M} + h\theta\mathbf{K})^{-1}(\mathbf{M} - h(1-\theta)\mathbf{K}\big)$ can be expressed as

$$\begin{aligned}\mathbf{C} &= \mathbf{V}(\mathbf{I} + h\theta\mathbf{\Lambda})^{-1}\mathbf{V}^{\mathrm{T}}(\mathbf{V}^{\mathrm{T}})^{-1}\big(\mathbf{I} - (1-\theta)h\mathbf{\Lambda}\big)\mathbf{V}^{-1}\\ &= \mathbf{V}(\mathbf{I} + h\theta\mathbf{\Lambda})^{-1}\big(\mathbf{I} - (1-\theta)h\mathbf{\Lambda}\big)\mathbf{V}^{-1} = \mathbf{V}\mathbf{D}\mathbf{V}^{-1}\end{aligned}$$

where $\mathbf{D}$ is the diagonal matrix

$$\mathbf{D} = (\mathbf{I} + h\theta\mathbf{\Lambda})^{-1}\big(\mathbf{I} - (1-\theta)h\mathbf{\Lambda}\big)$$

depending on the step size h and the eigenvalues λ_i. It follows that

$$\mathbf{C}^n = \mathbf{V}\mathbf{D}^n\mathbf{V}^{-1}$$

and that the growth depends on the behavior of $\mathbf{D}$. The elements of the diagonal matrix $\mathbf{D}$ are seen to be

$$d_{ii} = \frac{1 - h(1-\theta)\lambda_i}{1 + h\theta\lambda_i}$$

where λ_i are the eigenvalues of $\det(\mathbf{K} - \lambda\mathbf{M}) = 0$. If each of these ratios satisfies

$$\left|\frac{1 - h(1-\theta)\lambda_i}{1 + h\theta\lambda_i}\right| < 1 \tag{4.33}$$

the marching ahead from t_n to t_{n+1} takes place in a stable fashion in that the magnitude of the ith component in the transformed coordinates, and hence each of the original components, remains bounded for all t. The resulting inequality given in Eq. (4.33) can be written as

$$-1 < \frac{1 - h(1-\theta)\lambda_i}{1 + h\theta\lambda_i} < 1$$

eventually resulting in

$$-h\lambda_i < 0 \qquad \text{and} \qquad -2 < (2\theta - 1)h\lambda_i$$

When $\mathbf{A}$ and $\mathbf{B}$ are symmetric and positive definite (as is the case for the system of first-order equations being considered), the λ_i are positive so that the first inequality on $h\lambda_i$ is always satisfied. The second inequality is always satisfied as long as $\theta \geq 0.5$, showing that both the Crank-Nicolson and the modified Euler algorithms are **unconditionally stable**, that is, the solution will remain bounded for any h. For $\theta < 0.5$, the inequality is rewritten as

$$h\lambda_i < \frac{2}{1 - 2\theta}$$

so that in particular for $\theta = 0$, the Euler method,

$$h < \frac{2}{\lambda_i}$$

This inequality is governed by the largest eigenvalue of $\mathbf{K} - \lambda\mathbf{M}$ so that

$$h < \frac{2}{\lambda_{\max}}$$

As mentioned previously, the Euler method is **conditionally stable** in that h must be less than the critical value $2/\lambda_{\max}$. For matrices $\mathbf{M}$ and $\mathbf{K}$ of the type encountered in diffusion processes, $\lambda_{\max} = \text{cnst}/l_e^2$, where l_e is the length of an element in a uniform mesh. Hence

$$h < \frac{2l_e^2}{\text{cnst}}$$

showing that if the mesh size is halved, the critical time step size for the Euler method is reduced by a factor of four.

Example 4.3.

As a very simple demonstration, consider a two-element model for the problem of Section 4.2.2. The student is asked to show in the Exercises that the constrained system of equations reduces to the scalar equation

$$w + \phi w = \frac{\phi}{2} \tag{4.34}$$

with

$$w(0) = 0$$

where $w = u_2/u_0$ and $\phi = 12\alpha/L^2$. The eigenvalue is determined by taking $w = c\exp(-\lambda t)$ in the homogeneous equation leading to $\lambda = \phi = 12\alpha/L^2$. The θ method applied to Eq. (4.34) yields

$$(1 + \theta h\lambda)w_{n+1} = \Big(1 - (1 - \theta)h\lambda\Big)w_n + \frac{h\lambda}{2}$$

which can be written as

$$w_{n+1} = \frac{\Big(1 - (1 - \theta)h\lambda\Big)w_n + h\lambda/2}{(1 + \theta h\lambda)}$$

For the Euler method, $\theta = 0$ and

$$w_{n+1} = (1 - h\lambda)w_n + \frac{h\lambda}{2}$$

The critical condition is $h\lambda = 2$ for which $\mathbf{w}$ oscillates between 0 and 1. Note that if $h\lambda = 1$, steady state is reached in one step, obviously much too rapidly. Tables 4.4 through 4.7 show a comparison of the results from the Euler, the Crank-Nicolson, and the modified Euler algorithms, along with the analytical solution, for $h\lambda = 0.1, 1.0, 2.0$, and 2.2, respectively.

TABLE 4.4 Comparison of numerical and analytical solutions for $p = h\lambda = 0.1$

Step #	Euler	C-N	M-Euler	Analytical
1	0.0500	0.0476	0.0455	0.0476
2	0.0950	0.0907	0.0868	0.0906
3	0.1355	0.1297	0.1243	0.1296
4	0.1720	0.1650	0.1585	0.1648
5	0.2048	0.1969	0.1895	0.1967
6	0.2343	0.2257	0.2178	0.2256
7	0.2609	0.2519	0.2434	0.2517
8	0.2848	0.2755	0.2667	0.2753
9	0.3063	0.2969	0.2880	0.2967
10	0.3257	0.3162	0.3072	0.3161

TABLE 4.5 Comparison of numerical and analytical solutions for $p = h\lambda = 1.0$

Step #	Euler	C-N	M-Euler	Analytical
1	0.5000	0.3333	0.2500	0.3161
2	0.5000	0.4444	0.3750	0.4323
3	0.5000	0.4815	0.4375	0.4751
4	0.5000	0.4938	0.4688	0.4908
5	0.5000	0.4979	0.4844	0.4966
6	0.5000	0.4993	0.4922	0.4988
7	0.5000	0.4998	0.4961	0.4995
8	0.5000	0.4999	0.4980	0.4998
9	0.5000	0.5000	0.4990	0.4999
10	0.5000	0.5000	0.4995	0.5000

TABLE 4.6 Comparison of numerical and analytical solutions for $p = h\lambda = 2.0$

Step #	Euler	C-N	M-Euler	Analytical
1	1.0000	0.5000	0.3333	0.4323
2	0.0000	0.5000	0.4444	0.4908
3	1.0000	0.5000	0.4815	0.4988
4	0.0000	0.5000	0.4938	0.4998
5	1.0000	0.5000	0.4979	0.5000
6	0.0000	0.5000	0.4993	0.5000
7	1.0000	0.5000	0.4998	0.5000
8	0.0000	0.5000	0.4999	0.5000
9	1.0000	0.5000	0.5000	0.5000
10	0.0000	0.5000	0.5000	0.5000

TABLE 4.7 Comparison of numerical and analytical solutions for $p = h\lambda = 2.2$

Step #	Euler	C-N	M-Euler	Analytical
1	1.1000	0.5238	0.3438	0.4446
2	−0.2200	0.4989	0.4512	0.4939
3	1.3640	0.5001	0.4847	0.4993
4	−0.5368	0.5000	0.4952	0.4999
5	1.7442	0.5000	0.4985	0.5000
6	−0.9930	0.5000	0.4995	0.5000
7	2.2916	0.5000	0.4999	0.5000
8	−1.6499	0.5000	0.5000	0.5000
9	3.0799	0.5000	0.5000	0.5000
10	−2.5959	0.5000	0.5000	0.5000

The Euler and Crank-Nicolson algorithms both tend toward steady state too rapidly, with the modified Euler lagging consistently behind steady state. For small h ($h\lambda = 0.1$), the Crank-Nicolson algorithm essentially reproduces the analytical solution with the Euler and modified Euler algorithms above and below the analytical solutions, respectively.

For the step size h equal to the critical value for the Euler algorithm, the Euler algorithm diverges by oscillation between the values of 0 and 1, whereas the Crank-Nicolson and modified Euler solutions tend toward steady state for large t. For $h > h_{cr}$ the oscillations of the Euler algorithm become unbounded, whereas the Crank-Nicolson oscillates about and converges to the steady-state solution. The corresponding modified Euler solution converges from below to the steady-state solution. These tendencies, which are well known for large values of h ($h > h_{cr}$), are indicated symbolically in Fig. 4–8.

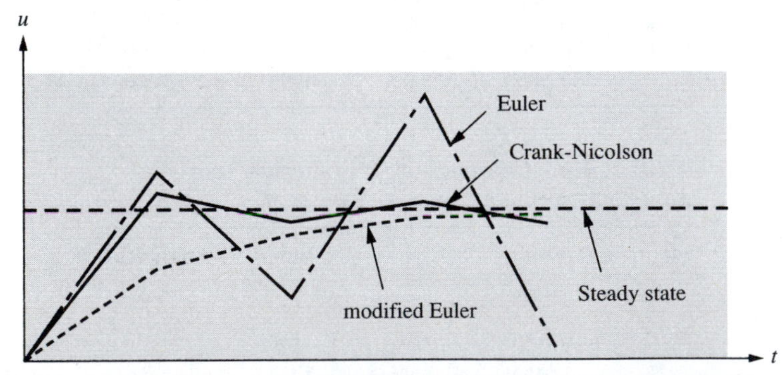

FIGURE 4–8 Character of solutions for various algorithms for large h

4.3 ONE-DIMENSIONAL WAVE OR HYPERBOLIC EQUATIONS

An example of a physical problem whose behavior is described by the classical one-dimensional wave equation is the problem of the longitudinal or axial motion of a straight prismatic elastic bar as indicated in Fig. 4–9.

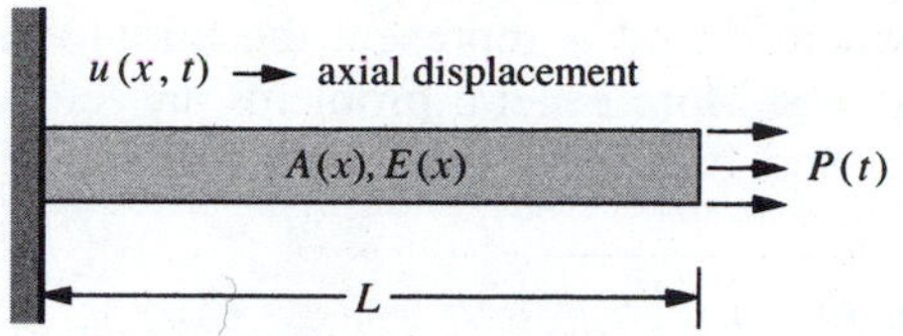

FIGURE 4–9 Longitudinal motion of an elastic bar

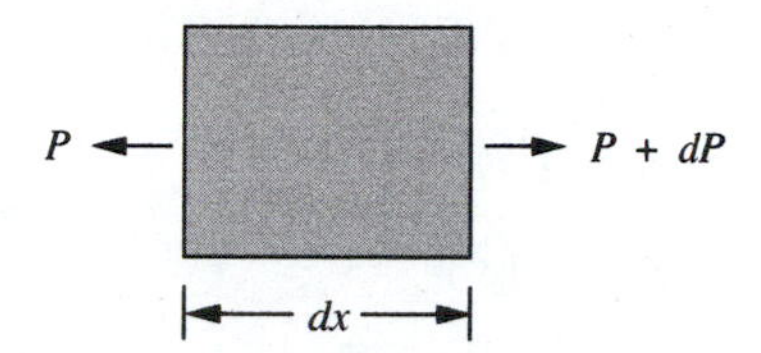

FIGURE 4–10 Free-body diagram for balance of momentum

The basic physical principle governing the motion is Newton's second law which, when applied to a typical differential element as shown in Fig. 4–10, yields

$$\sum F_x = -P + P + dP = \rho A \, dx \frac{\partial^2 u}{\partial t^2}$$

or with $P = A\sigma = AE\epsilon = AE \, \partial u/\partial x$,

$$\frac{\partial (AE \, \partial u/\partial x)}{\partial x} = A\rho \frac{\partial^2 u}{\partial t^2}$$

where A is the area, E is Young's modulus, and ρ is the mass density. This equation of motion is often referred to as the one-dimensional **wave equation** in that it is an example of the standard hyperbolic equation that predicts wave propagation in a one-dimensional setting. When A and E are constants, the equation is often written as

$$c^2 \frac{\partial^2 u}{\partial x^2} = \frac{\partial^2 u}{\partial t^2}$$

where $c = \sqrt{E/\rho}$ is the speed at which longitudinal waves propagate along the x-axis.

Appropriate boundary conditions for the situation pictured in Fig. 4–9 are

$$u(0, t) = 0 \qquad AE \frac{\partial u(L, t)}{\partial x} = P(t)$$

stating that the displacement is zero for all time at $x = 0$ and that there is a force $P(t)$ applied at $x = L$. Two initial conditions of the form

$$u(x, 0) = f(x) \qquad \frac{\partial u(x, 0)}{\partial t} = g(x)$$

are also prescribed, where f and g represent the initial axial displacement and axial velocity, respectively. More general problems are left to the Exercises.

Thus a typical initial-boundary value problem associated with the wave equation can be stated as

$$
\begin{aligned}
\frac{\partial(AE\,\partial u/\partial x)}{\partial x} &= A\rho\frac{\partial^2 u}{\partial t^2} && 0 \le x \le L, 0 \le t \\
u(0,t) &= 0 && \\
& && 0 \le t \\
AE\frac{\partial u(L,t)}{\partial x} &= P(t) && \\
u(x,0) &= f(x) && \\
& && 0 \le x \le L \\
\frac{\partial u(x,0)}{\partial t} &= g(x) &&
\end{aligned}
\tag{4.35}
$$

Many other physical situations such as the transverse motions of strings and membranes, propagation of sound, and dynamic disturbances in fluids and solids are governed by the wave equation.

4.3.1 The Galerkin Finite Element Model

Discretization and interpolation. As has been indicated numerous times in the preceding material, the first steps in developing a finite element model are discretization and interpolation. These are carried out exactly as in Section 4.2, to which the reader is referred for all the details. The discretization would appear as in Fig. 4–11, where $x_1 = 0$ and $x_{N+1} = L$.

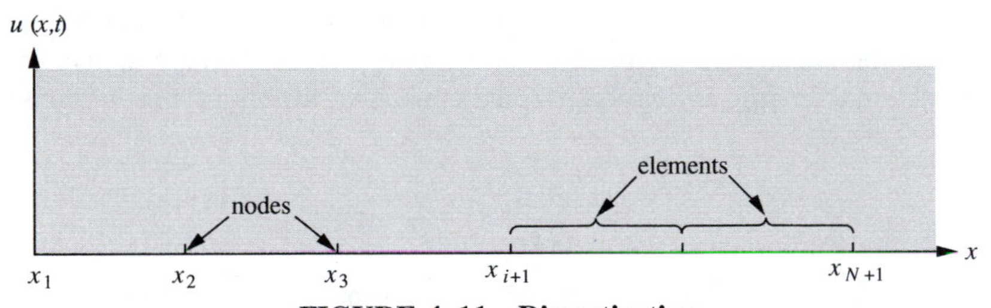

FIGURE 4–11 Discretization

Interpolation would again be semidiscrete, of the form

$$u(x,t) = \sum_{1}^{N+1} u_i(t)n_i(x) \tag{4.36}$$

where the $n_i(x)$ are nodally based interpolation functions and can be linear, quadratic, or higher-order if desired. The elements indicated in Fig. 4–11 are specifically for linear interpolation.

Elemental formulation. Consider again the initial-boundary value problem developed in the previous section:

$$\frac{\partial(AE\,\partial u/\partial x)}{\partial x} = A\rho\frac{\partial^2 u}{\partial t^2} \qquad 0 \le x \le L, 0 \le t \tag{4.37a}$$

$$u(0,t) = 0 \tag{4.37b}$$

$$0 \le t$$

$$AE\,\frac{\partial u(L,t)}{\partial x} = P(t) \tag{4.37c}$$

$$u(x,0) = f(x) \tag{4.37d}$$

$$0 \le x \le L$$

$$\frac{\partial u(x,0)}{\partial t} = g(x) \tag{4.37e}$$

The elemental formulation for the wave equation is based on a corresponding weak statement. The weak form is developed by multiplying the differential equation (4.37a) by a test function $v(x)$ satisfying any essential boundary conditions, with the result then integrated over the spatial region according to

$$\int_0^L v\left(\frac{\partial(AE\,\partial u/\partial x)}{\partial x} - A\rho\frac{\partial^2 u}{\partial t^2}\right)dx = 0$$

Integrating by parts and subsequently eliminating the integrated terms on the basis of the boundary conditions given by Eqs. (4.37b–c) yields

$$\int_0^L \left(v'AE\,\frac{\partial u}{\partial x} + vA\rho\,\frac{\partial^2 u}{\partial t^2}\right)dx - v(L)P(t) = 0 \tag{4.38}$$

which is the required weak statement for the initial-boundary value problem associated with the one-dimensional wave equation (4.37). The finite element model is obtained by substituting the approximate solution given by Eq. (4.36), and $v = n_k$, $k = 1, 2, \ldots, N+1$, successively, into Eq. (4.38) to obtain

$$\sum_1^{N+1}\int_0^L \left(n_k' AE n_i' u_i + n_k A\rho n_i \ddot{u}_i\right) dx - \delta_{kN+1}P(t) = 0$$

which can be written as

$$\sum_1^{N+1}\left(A_{ki}u_i(t) + B_{ki}\ddot{u}_i(t)\right) = F_k(t) \qquad k = 1, \ldots, N+1 \tag{4.39}$$

where

$$A_{ki} = \int_0^L n_k' AE n_i'\,dx$$

$$B_{ki} = \int_0^L n_k A\rho n_i \, dx$$

$$F_k = \delta_{kN+1} P(t)$$

The student is asked to show in the Exercises that in matrix notation, these can be expressed as

$$\mathbf{B\ddot{u}} + \mathbf{Au} = \mathbf{F} \tag{4.40}$$

where

$$\mathbf{A} = \sum_e \mathbf{k_G} \tag{4.41a}$$

$$\mathbf{B} = \sum_e \mathbf{m_G} \tag{4.41b}$$

$$\mathbf{F} = [0 \; 0 \; \ldots \; 0 \; P]^{\mathrm{T}}$$

with

$$\mathbf{k_e} = \int_{x_i}^{x_{i+1}} \mathbf{N}' AE \mathbf{N}'^{\mathrm{T}} \, dx \tag{4.42a}$$

$$\mathbf{m_e} = \int_{x_i}^{x_{i+1}} \mathbf{N} A\rho \mathbf{N}^{\mathrm{T}} \, dx \tag{4.42b}$$

Thus the original initial-boundary value problem has been converted into the initial value problem

$$\mathbf{B\ddot{u}} + \mathbf{Au} = \mathbf{F} \tag{4.43a}$$

with the initial conditions

$$\begin{aligned} \mathbf{u}(0) &= \mathbf{u_0} \\ \mathbf{\dot{u}}(0) &= \mathbf{\dot{u}_0} \end{aligned} \tag{4.43b}$$

The vectors $\mathbf{u_0}$ and $\mathbf{\dot{u}_0}$, representing the discretized version of the initial conditions f and g, are usually taken to be respectively the vectors consisting of the values of $f(x)$ and $g(x)$ at the nodes, that is,

$$\mathbf{u}(0) = \mathbf{u_0} = [\, f(0) \quad f(x_2) \quad f(x_3) \quad \ldots \quad f(x_N) \quad f(L)\,]$$

and

$$\mathbf{\dot{u}}(0) = \mathbf{\dot{u}_0} = [\, g(0) \quad g(x_2) \quad g(x_3) \quad \ldots \quad g(x_N) \quad g(L)\,]$$

Note that the assembly process has taken place implicitly, while carrying out the details of obtaining the governing equations, using the Galerkin method in connection with the weak formulation.

Enforcement of constraints is necessary if either of the boundary conditions is essential, that is, if the dependent variable is prescribed at either boundary point. The system of equations (4.43*a*) must be altered to reflect these constraints. Consider for example the case where the boundary condition at $x = 0$ is $u(0, t) = u_0(t)$. The first scalar equation of the set of equations (4.43*a*) would be replaced by the constraint so that there would result

$$\begin{aligned} u_1 \qquad\qquad\qquad\qquad\qquad\qquad\qquad\qquad &= u_0(t) \\ a_{21}u_1 + a_{22}u_2 + a_{23}u_3 + \cdots + b_{21}\ddot{u}_1 + b_{22}\ddot{u}_2 + \cdots &= 0 \\ a_{31}u_1 + a_{32}u_2 + a_{33}u_3 + \cdots + b_{31}\ddot{u}_1 + b_{32}\ddot{u}_2 + \cdots &= 0 \\ &\vdots \end{aligned}$$

The u_1 and $\ddot{u}_1$ terms in the remaining equations, which are known, are transferred to the right-hand side to yield

$$\begin{aligned} u_1 \qquad\qquad\qquad\qquad\qquad\qquad\qquad &= u_0(t) \\ a_{22}u_2 + a_{23}u_3 + \cdots + b_{22}\ddot{u}_2 + b_{23}\ddot{u}_3 + \cdots &= 0 - a_{21}u_0 - b_{21}\ddot{u}_0 \\ a_{32}u_2 + a_{33}u_3 + \cdots + b_{32}\ddot{u}_2 + b_{33}\ddot{u}_3 + \cdots &= 0 - a_{31}u_0 - b_{31}\ddot{u}_0 \\ &\vdots \end{aligned}$$

showing the effect of the constraint on the remaining equations. For a linearly interpolated model the half bandwidth is two, and only the u_1 and $\ddot{u}_1$ terms in the second equation need be transferred to the right-hand side. For a quadratically interpolated model the half bandwidth is three, and terms from the second and third equations need to be transferred. If the constraint is at the right end, the Nth, $(N - 1)$st, ... equations would be similarly altered.

Write the constrained set of equations as

$$\mathbf{M\ddot{u}} + \mathbf{Ku} = \mathbf{F} \tag{4.44}$$

subject to the initial conditions

$$\mathbf{u}(0) = \mathbf{u_0} \qquad \dot{\mathbf{u}}(0) = \dot{\mathbf{u}}_\mathbf{0}$$

Note that if there were distributed inputs resulting in a more general nodal distribution of forces

$$\mathbf{F} = [F_1(t) \quad F_2(t) \quad F_3(t) \quad \ldots]$$

the final set of equations would appear as

$$\begin{aligned} u_1 \qquad\qquad\qquad\qquad\qquad\qquad\qquad &= u_0(t) \\ a_{22}u_2 + a_{23}u_3 + \cdots + b_{22}\ddot{u}_2 + b_{23}\ddot{u}_3 + \cdots &= F_2(t) - a_{21}u_0 - b_{21}\ddot{u}_0 \\ a_{32}u_2 + a_{33}u_3 + \cdots + b_{32}\ddot{u}_2 + b_{33}\ddot{u}_3 + \cdots &= F_3(t) - a_{31}u_0 - b_{31}\ddot{u}_0 \\ &\vdots \end{aligned}$$

In any case, algorithms for integrating these equations (the solution step) are studied in the following sections. The derived variables, which are now functions of time, are computed per element in exactly the same fashion as outlined for the one-dimensional problems in Chapter 2.

4.3.2 One-Dimensional Wave Example

Consider again the problem outlined in Section 4.3.1,

$$
\begin{aligned}
AE\frac{\partial^2 u}{\partial x^2} - A\rho\frac{\partial^2 u}{\partial t^2} &= 0 \qquad 0 \le x \le L, 0 \le t \\
u(0,t) &= 0 \\
AE\frac{\partial u(L,t)}{\partial x} &= P_0 \qquad\qquad 0 \le t \\
u(x,0) &= 0 \\
\frac{\partial u(x,0)}{\partial t} &= 0 \qquad 0 \le x \le L
\end{aligned}
$$

corresponding to a uniform bar initially at rest and undeformed, acted on suddenly by a constant force P_0 at the unsupported end.

Discretization and interpolation. A mesh for three equal-length, linearly interpolated elements is indicated in Fig. 4–12.

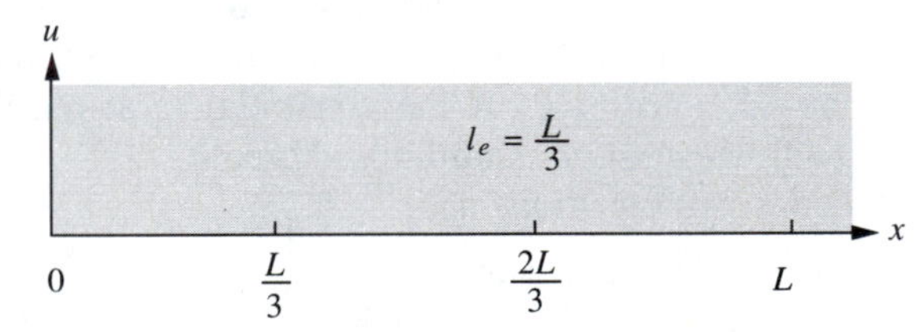

FIGURE 4–12 Mesh for three equal-length, linearly interpolated elements

Elemental formulation. Using Eqs. (4.36) developed in Section 4.3.1, each of the elemental matrices becomes

$$
\mathbf{k_e} = \frac{3AE}{L}\begin{bmatrix} 1 & -1 \\ -1 & 1 \end{bmatrix} \qquad \mathbf{m_e} = \frac{A\rho L}{18}\begin{bmatrix} 2 & 1 \\ 1 & 2 \end{bmatrix}
$$

with the assembled equations as

$$
\phi\begin{bmatrix} 2 & 1 & 0 & 0 \\ 1 & 4 & 1 & 0 \\ 0 & 1 & 4 & 1 \\ 0 & 0 & 1 & 2 \end{bmatrix}\ddot{\mathbf{u}} + \begin{bmatrix} 1 & -1 & 0 & 0 \\ -1 & 2 & -1 & 0 \\ 0 & -1 & 2 & -1 \\ 0 & 0 & -1 & 1 \end{bmatrix}\mathbf{u} = \begin{bmatrix} 0 \\ 0 \\ 0 \\ \Lambda/3 \end{bmatrix}
$$

where $\phi = \rho L^2/54E$ and $\Lambda = P_0 L/AE$.

The constraint from the boundary condition $u(0, t) = 0$ is $u_1(t) = 0$, resulting in

$$\phi \begin{bmatrix} 4 & 1 & 0 \\ 1 & 4 & 1 \\ 0 & 1 & 2 \end{bmatrix} \ddot{\mathbf{u}} + \begin{bmatrix} 2 & -1 & 0 \\ -1 & 2 & -1 \\ 0 & -1 & 1 \end{bmatrix} \mathbf{u} = \begin{bmatrix} 0 \\ 0 \\ \Lambda/3 \end{bmatrix} \tag{4.45}$$

with

$$\mathbf{u}(0) = \mathbf{u_0} = 0 \qquad \dot{\mathbf{u}}(0) = \dot{\mathbf{u}}_0 = 0$$

Several approaches to integrating equations of this type will be presented and discussed in the following sections.

The comments made in Section 4.2.2 regarding the different approaches available for handling the mass matrices in connection with one-dimensional diffusion equations are equally applicable for the wave equation. The forms of the mass matrices are identical, so that

$$\mathbf{m_{ce}} = \frac{A\rho l_e}{6} \begin{bmatrix} 2 & 1 \\ 1 & 2 \end{bmatrix} \qquad \mathbf{m_{le}} = \frac{A\rho l_e}{2} \begin{bmatrix} 1 & 0 \\ 0 & 1 \end{bmatrix}$$

for the consistent and lumped mass matrices respectively, and

$$\mathbf{m_w} = \frac{A\rho l_e}{12} \begin{bmatrix} 5 & 1 \\ 1 & 5 \end{bmatrix}$$

for an average of the consistent and lumped mass matrices.

4.3.3 Analytical Integration Techniques

Generally, for a one-dimensional wave equation the constrained system of ordinary differential equations resulting from the application of the finite element method is of the form

$$\begin{aligned} \mathbf{M}\ddot{\mathbf{u}} + \mathbf{K}\mathbf{u} &= \mathbf{F}(t) \\ \mathbf{u}(0) &= \mathbf{u_0} \\ \dot{\mathbf{u}}(0) &= \dot{\mathbf{u}}_0 \end{aligned} \tag{4.46}$$

that is, a coupled system of linear second-order ordinary differential equations. This system of differential equations will be treated analytically by decomposing the general solution $\mathbf{u}$ into a homogeneous solution $\mathbf{u_h}$ and a particular solution $\mathbf{u_p}$ according to

$$\mathbf{u} = \mathbf{u_h} + \mathbf{u_p}$$

The homogeneous equations are satisfied by $\mathbf{u_h}$,

$$\mathbf{M}\ddot{\mathbf{u}}_\mathbf{h} + \mathbf{K}\mathbf{u_h} = \mathbf{0} \tag{4.47}$$

and $\mathbf{u_p}$ is any solution of the nonhomogeneous equations

$$\mathbf{M}\ddot{\mathbf{u}}_\mathbf{p} + \mathbf{K}\mathbf{u_p} = \mathbf{F}(t)$$

This procedure is essentially the well-known **superposition principle**, valid for linear systems.

Homogeneous solution. For a system of second-order ordinary differential equations representing an undamped physical model, the homogeneous solution is taken to be of the form

$$\mathbf{u_h}(t) = \mathbf{v} \exp(i\,\omega t)$$

a solution that is harmonic or periodic in time. Substituting into Eq. (4.47) yields

$$(\mathbf{K} - \omega^2 \mathbf{M})\mathbf{v} = 0 \tag{4.48}$$

Equation (4.48) is the generalized linear algebraic eigenvalue problem discussed several times in previous sections. When **K** and **M** are symmetric and positive definite, as is the case for the one-dimensional problems currently being considered, all the eigenvalues ω_j^2 are positive and real with the eigenvectors $\mathbf{v_j}$ also real and M-orthogonal. The corresponding homogeneous solution is written as

$$\mathbf{u_h} = \sum c_j \mathbf{v_j} \exp(i\,\omega_j t)$$

where the c_j are complex constants. Expressed in real form,

$$\mathbf{u_h} = \sum \mathbf{v_j}\big(a_j \cos(\omega_j t) + b_j \sin(\omega_j t)\big)$$

Particular solution. The particular solution is any solution of

$$\mathbf{M\ddot{u}_p} + \mathbf{Ku_p} = \mathbf{F}(t)$$

and, depending on the specific form of **F**, can be determined by using

1. Undetermined coefficients (intelligent guessing)
2. Variation of parameters
3. Laplace transform methods

After determining the particular solution using one of these approaches, the general solution of Eq. (4.46) can be written as

$$\mathbf{u} = \sum \mathbf{v_j}\big(a_j \cos(\omega_j t) + b_j \sin(\omega_j t)\big) + \mathbf{u_p}(t) \tag{4.49}$$

The initial conditions are used to determine the $2N$ constants a_j and b_j, $j = 1, 2, \dots, N$, according to

$$\mathbf{u}(0) = \mathbf{u_0} = \sum \mathbf{v_j} a_j + \mathbf{u_p}(0)$$

or

$$\mathbf{Va} = \mathbf{u_0} - \mathbf{up}(0) \tag{4.50}$$

where **V** is the $N \times N$ matrix consisting of the eigenvectors as columns. N represents the number of constrained variables.

Similarly

$$\dot{\mathbf{u}}(0) = \dot{\mathbf{u}}_0 = \sum \mathbf{v_j} \omega_j b_j + \dot{\mathbf{u}}_\mathbf{p}(0)$$

so that

$$\mathbf{V\omega b} = \dot{\mathbf{u}}_0 - \dot{\mathbf{u}}_\mathbf{p}(0) \tag{4.51}$$

with $\boldsymbol{\omega}\mathbf{b} = [\omega_1 b_1 \quad \omega_2 b_2 \quad \omega_3 b_3 \ldots \omega_N b_N]^T$. A unique solution to each of the sets of equations (4.50) and (4.51) is guaranteed on the basis of the linearly independent character of the $\mathbf{v_j}$ for the case where $\mathbf{M}$ and $\mathbf{K}$ are symmetric and positive definite.

For the particular example developed in Section 4.3.2 and given by Eqs. (4.45), the eigenvalues and eigenvectors are determined from

$$\begin{bmatrix} 2-4\lambda & -(1+\lambda) & 0 \\ -(1+\lambda) & 2-4\lambda & -(1+\lambda) \\ 0 & -(1+\lambda) & 1-2\lambda \end{bmatrix} \mathbf{v} = \mathbf{0}$$

where $\lambda = \rho\omega^2 L^2/54E$. The roots of the corresponding characteristic equation

$$(1-2\lambda)\left[4(1-2\lambda)^2 - 3(1+\lambda)^2\right] = 0$$

are $\lambda_1 = 0.0467$, $\lambda_2 = 0.5000$, and $\lambda_3 = 1.6456$, from which

$$\omega_1 = \left(\frac{1.5887}{L}\right)c \qquad \omega_2 = \left(\frac{5.1962}{L}\right)c \qquad \omega_3 = \left(\frac{9.4266}{L}\right)c$$

where $c = \sqrt{E/\rho}$ is the speed of waves propagating along the bar. The corresponding exact values are

$$\beta_1 = \left(\frac{1.5708}{L}\right)c \qquad \beta_2 = \left(\frac{4.7124}{L}\right)c \qquad \beta_3 = \left(\frac{7.8540}{L}\right)c$$

with the corresponding eigenvectors as

$$\mathbf{v_1} = [0.5000 \quad 0.8660 \quad 1.0000]^T$$

$$\mathbf{v_2} = [1.0000 \quad 0.0000 \quad -1.0000]^T$$

$$\mathbf{v_3} = [0.5000 \quad -0.8660 \quad 1.0000]^T$$

so that the homogeneous solution is

$$\begin{aligned} \mathbf{u_h} = {} & \mathbf{v_1}\big(a_1 \cos(\omega_1 t) + b_1 \sin(\omega_1 t)\big) \\ & + \mathbf{v_2}\big(a_2 \cos(\omega_2 t) + b_2 \sin(\omega_2 t)\big) \\ & + \mathbf{v_3}\big(a_3 \cos(\omega_3 t) + b_3 \sin(\omega_3 t)\big) \end{aligned}$$

Note that there are six arbitrary constants to be determined from the six scalar equations represented by

$$\mathbf{u}(0) = \mathbf{u_0} \qquad \text{and} \qquad \dot{\mathbf{u}}(0) = \dot{\mathbf{u}}_\mathbf{0}$$

Observe that in

$$\mathbf{M}\ddot{\mathbf{u}}_\mathbf{p} + \mathbf{K}\mathbf{u}_\mathbf{p} = [0 \quad 0 \quad \Lambda/3]^{\mathrm{T}}$$

a particular solution is easily obtained by taking $\mathbf{u_p} = \mathbf{d}$, a constant, resulting in

$$\mathbf{K}\mathbf{d} = [0 \quad 0 \quad \Lambda/3]^{\mathrm{T}}$$

or

$$\left[\begin{array}{ccc|c} 2 & -1 & 0 & 0 \\ -1 & 2 & -1 & 0 \\ 0 & -1 & 1 & \Lambda/3 \end{array}\right]$$

yielding

$$\mathbf{d} = \frac{\Lambda}{3}[1 \quad 2 \quad 3]^{\mathrm{T}}$$

Applying the initial condition $\mathbf{u}(0) = 0$ yields

$$a_1\mathbf{v_1} + a_2\mathbf{v_2} + a_3\mathbf{v_3} = -\mathbf{d}$$

or

$$\left[\begin{array}{ccc|c} 0.5000 & 1.0000 & 0.5000 & -\Lambda/3 \\ 0.8660 & 0.0000 & -0.8660 & -2\Lambda/3 \\ 1.0000 & -1.0000 & 1.0000 & -3\Lambda/3 \end{array}\right]$$

from which $\mathbf{a} = [a_1 \ a_2 \ a_3]^{\mathrm{T}} = [-0.8294\Lambda \ 0.1111\Lambda \ -0.0595\Lambda]^{\mathrm{T}}$. With $\mathbf{u}(0) = \dot{\mathbf{u}}_0 = \mathbf{u_p}(0) = \mathbf{0}$, it follows that $\mathbf{b} = [b_1 \ b_2 \ b_3]^{\mathrm{T}} = \mathbf{0}$.

The solution can then be written as

$$\begin{aligned}\frac{\mathbf{u}}{\Lambda} &= -0.8294\mathbf{v_1}\cos(\omega_1 t) + 0.1111\mathbf{v_2}\cos(\omega_2 t) \\ &\quad -0.0595\mathbf{v}_3\cos(\omega_3 t) + [1 \quad 2 \quad 3]^{\mathrm{T}}\frac{1}{3}\end{aligned}$$

or

$$\frac{u_2}{\Lambda} = 0.3333 - 0.4147\cos(\omega_1 t) + 0.1111\cos(\omega_2 t) - 0.0298\cos(\omega_3 t)$$

$$\frac{u_3}{\Lambda} = 0.6667 - 0.7183\cos(\omega_1 t) + 0.0000\cos(\omega_2 t) + 0.0516\cos(\omega_3 t)$$

$$\frac{u_4}{\Lambda} = 1.0000 - 0.8294\cos(\omega_1 t) - 0.1111\cos(\omega_2 t) - 0.0595\cos(\omega_3 t) \quad (4.52)$$

The constant term represents, in a sense, the steady-state or static solution from the standpoint that if damping were included in the physical model, the terms in the homogeneous solution corresponding to the present cosine terms would eventually damp out, leaving only the constant term

$$\mathbf{u} = [1 \quad 2 \quad 3]^{\mathrm{T}} \frac{\Lambda}{3}$$

which is the corresponding static solution.

The corresponding exact solution can be represented in terms of the infinite series

$$\frac{u(x,t)}{\Lambda} = 2\sum \frac{\phi_n(\alpha_n L)\phi_n(\alpha_n x)(1 - \cos(\alpha_n ct))}{(\alpha_n L)^2}$$

where $(\alpha_n L) = (2n-1)\pi/2$ and $\phi(x) = \sin(\alpha_n x)$. Retaining the first three terms of the series solution at $x = L/3$, $2L/3$, and L yields

$$\frac{u(L/3,t)}{\Lambda} = 0.3314 - 0.4053\cos(\beta_1 t) + 0.0901\cos(\beta_2 t) - 0.0162\cos(\beta_3 t)$$

$$\frac{u(2L/3,t)}{\Lambda} = 0.7639 - 0.7020\cos(\beta_1 t) + 0.0901\cos(\beta_2 t) + 0.0281\cos(\beta_3 t)$$

$$\frac{u(L,t)}{\Lambda} = 0.9330 - 0.8106\cos(\beta_1 t) - 0.0901\cos(\beta_2 t) - 0.0324\cos(\beta_3 t) \tag{4.53}$$

where $\beta_n = \alpha_n c$. Note the general similarity between the three-term expansion of the exact solution and the approximate solution from the three-element finite element model.

The approximate lowest frequency ω_1 is quite close to the exact lowest frequency β_1, with

$$\frac{\omega_1}{\beta_1} = 1.0114$$

The other two ratios,

$$\frac{\omega_2}{\beta_2} = 1.1027 \qquad \text{and} \qquad \frac{\omega_3}{\beta_3} = 1.2002$$

are not quite as accurate. Recall from Section 3.7, the rule of thumb stating that approximately $2N$ unconstrained degrees of freedom are necessary in order that the first N frequencies be determined accurately. In this instance the first frequency should be quite accurate, which is certainly the case.

The exact solutions $u(L,t)$ and $u_4(t)$ are indicated for the first few oscillations in Fig. 4–13.

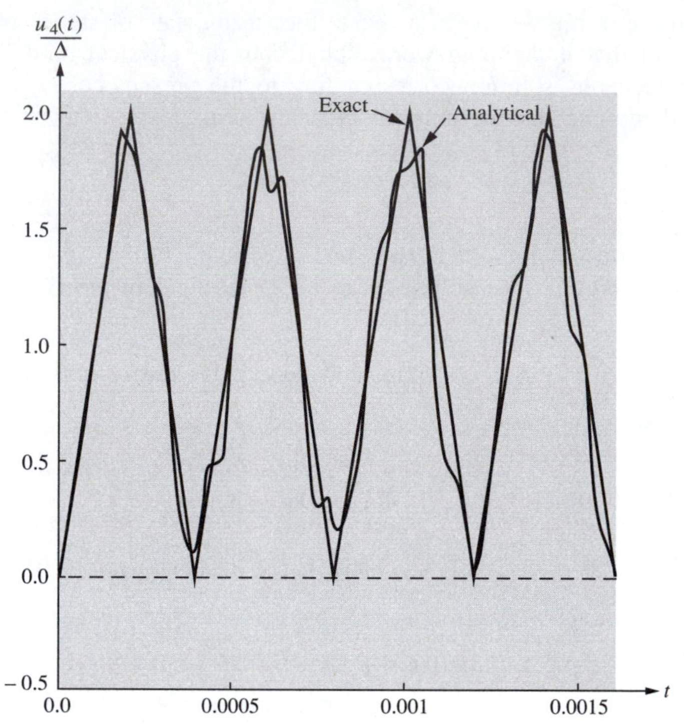

FIGURE 4–13 Comparison of exact and approximate solutions at $x = L$

The results in Fig. 4–13 are for $E = 3 \times 10^7$ psi, $\rho = 7.5 \times 10^{-4}$ lbf-s^2/in^4, and $L = 20$ in. The agreement is quite reasonable with the approximate solution beginning to peak early due to the fact that all the approximate frequencies exceed the exact values. A finer mesh would result in better agreement.

Consider the following alternate analytical approach to integrating

$$\mathbf{M\ddot{u}} + \mathbf{Ku} = \mathbf{f}$$

$$\mathbf{u}(0) = \mathbf{u_0}$$

$$\mathbf{\dot{u}}(0) = \mathbf{\dot{u}_0}$$

The first step is to determine the eigenvalues and eigenvectors as was indicated earlier in this section, that is, to solve

$$\mathbf{Kv_i} = \lambda_i \mathbf{Mv_i}$$

for the λ_i and the corresponding $\mathbf{v_i}$. The eigenvectors and eigenvalues are used collectively to form the matrices

$$\mathbf{V} = [\mathbf{v_1} \quad \mathbf{v_2} \quad \ldots \quad \mathbf{v_N}]$$

and

$$\mathbf{\Lambda} = \begin{bmatrix} \lambda_1 & & & & 0\text{'}s \\ & \lambda_2 & & & \\ & & \cdot & & \\ & & & \cdot & \\ & & & & \cdot \\ 0\text{'}s & & & & \lambda_N \end{bmatrix}$$

satisfying

$$\mathbf{V}^{\mathrm{T}}\mathbf{M}\mathbf{V} = \mathbf{I} \qquad \text{and} \qquad \mathbf{V}^{\mathrm{T}}\mathbf{K}\mathbf{V} = \mathbf{\Lambda}$$

which were discussed in detail in Section 4.2.3.

In the equations of motion, take $\mathbf{u} = \mathbf{Vz}$ and premultiply by $\mathbf{V}^{\mathrm{T}}$ to obtain

$$\ddot{\mathbf{z}} + \mathbf{\Lambda z} = \mathbf{V}^{\mathrm{T}}\mathbf{f} = \mathbf{g}$$

as the uncoupled system of equations in terms of the normal coordinates $\mathbf{z}$. The initial conditions are transformed into the z-coordinates according to

$$\mathbf{z}(0) = \mathbf{V}^{\mathrm{T}}\mathbf{M}\mathbf{u}(0) = \mathbf{V}^{\mathrm{T}}\mathbf{M}\mathbf{u}_0 = \mathbf{z}_0$$

$$\dot{\mathbf{z}}(0) = \mathbf{V}^{\mathrm{T}}\mathbf{M}\dot{\mathbf{u}}(0) = \mathbf{V}^{\mathrm{T}}\mathbf{M}\dot{\mathbf{u}}_0 = \dot{\mathbf{z}}_0$$

The uncoupled second-order system of equations above is equivalent to the scalar equations

$$\ddot{z}_i + \lambda_i z_i = g_i(t) \qquad i = 1, 2, \ldots, N$$

with solutions

$$z_i = a_i \cos(\omega_i t) + b_i \sin(\omega_i t) + \int_0^t \frac{\sin\big(\omega_i(t-\tau)\big) g_i(\tau)}{\omega_i}\, d\tau$$

where $(\omega_i)^2 = \lambda_i$. The initial conditions are evaluated to yield

$$z_i = (\mathbf{z}_0)_i \cos(\omega_i t) + \frac{(\dot{\mathbf{z}}_0)_i \sin(\omega_i t)}{\omega_i} + \int_0^t \frac{\sin\big(\omega_i(t-\tau)\big) g_i(\tau)}{\omega_i}\, d\tau$$

The transformation back to the original u-coordinates completes the solution. Specifics are left to the Exercises at the end of the chapter.

4.3.4 Time Domain Integration Techniques — Second-Order Systems

As was the case for the systems of first-order equations studied in Section 4.2.4, there may be situations where $\mathbf{M}$ and $\mathbf{K}$ are time-dependent or where $\mathbf{F}(t)$ is such that an analytical approach is not an intelligent way to proceed. Numerical integration techniques, which are appropriate in such situations, are presented and discussed in the next sections.

4.3.4.1 CENTRAL DIFFERENCE METHOD. The system of second-order linear ordinary differential equations in question is restated as

$$\begin{aligned} \mathbf{M}\ddot{\mathbf{u}} + \mathbf{K}\mathbf{u} &= \mathbf{F} \\ \mathbf{u}(0) &= \mathbf{u}_0 \\ \dot{\mathbf{u}}(0) &= \dot{\mathbf{u}}_0 \end{aligned} \tag{4.54}$$

Figure 4–14 indicates a discretization of the time variable with $t_n - t_{n-1} = t_{n+1} - t_n = h$, the time step.

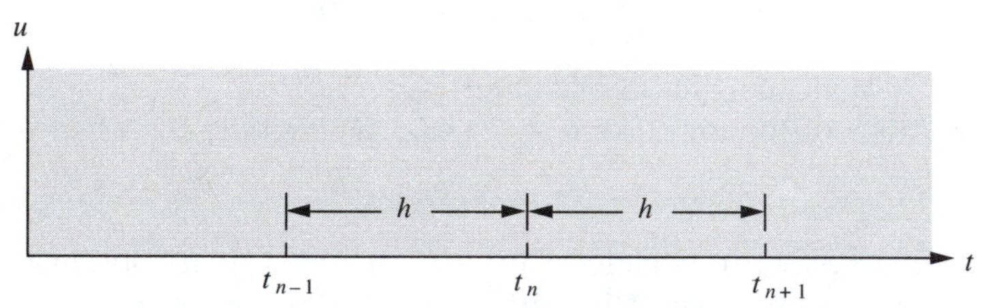

FIGURE 4–14 Discretization of the time variable

The differential equation (4.54) is evaluated at $t = t_n$ to yield

$$\mathbf{M}\ddot{\mathbf{u}}_\mathbf{n} + \mathbf{K}\mathbf{u}_\mathbf{n} = \mathbf{F}_\mathbf{n}$$

where $\mathbf{u}_\mathbf{n} \approx \mathbf{u}(t_n) = \mathbf{u}(nh)$, and $\mathbf{F}_\mathbf{n} = \mathbf{F}(t_n) = \mathbf{F}(nh)$. Central difference representations are used for the velocity and acceleration vectors, namely,

$$\dot{\mathbf{u}}_\mathbf{n} = \frac{\mathbf{u}_{\mathbf{n+1}} - \mathbf{u}_{\mathbf{n-1}}}{2h} \qquad \ddot{\mathbf{u}}_\mathbf{n} = \frac{\mathbf{u}_{\mathbf{n+1}} - 2\mathbf{u}_\mathbf{n} + \mathbf{u}_{\mathbf{n-1}}}{h^2} \tag{4.55}$$

Each is accurate to order h^2. Substituting the second of Eqs. (4.55) into Eq. (4.54) yields, after multiplying through by h^2 and rearranging,

$$\mathbf{M}\mathbf{u}_{\mathbf{n+1}} = (2\mathbf{M} - h^2\mathbf{K})\mathbf{u}_\mathbf{n} - \mathbf{M}\mathbf{u}_{\mathbf{n-1}} + h^2\mathbf{F}_\mathbf{n} \tag{4.56}$$

a three-term recurrence relation to be used for stepping ahead in time. A special starting procedure is necessary in that two successive **u**'s are required in order to accomplish the solution of Eq. (4.56). The procedure used is as follows: the vector function **u** is expanded in a Taylor's series about $t = 0$ according to

$$\mathbf{u}(-h) = \mathbf{u}(0) - h\dot{\mathbf{u}}(0) + \frac{h^2\ddot{\mathbf{u}}(0)}{2} + \cdots$$

with $\ddot{\mathbf{u}}(0)$ computed from the differential equation evaluated at $t = 0$,

$$\ddot{\mathbf{u}}(0) = \mathbf{M}^{-1}\big(\mathbf{F}(0) - \mathbf{K}\mathbf{u}(0)\big) \tag{4.57}$$

Usually $\mathbf{M}^{-1}$ is not computed; rather, the system of equations

$$\mathbf{M}\ddot{\mathbf{u}}(0) = \mathbf{F}(0) - \mathbf{K}\mathbf{u}(0)$$

is solved for $\ddot{\mathbf{u}}(0)$ using an **LU** decomposition. The special starting value $\mathbf{u}(-h)$ is then given formally by

$$\mathbf{u}(-h) = \mathbf{u}(0) - h\dot{\mathbf{u}}(0) + \frac{h^2\big(\mathbf{M}^{-1}(\mathbf{F}(0) - \mathbf{K}\mathbf{u}(0))\big)}{2}$$

or

$$\mathbf{u}_{-1} = \mathbf{u}_0 - h\dot{\mathbf{u}}_0 + \frac{h^2\big(\mathbf{M}^{-1}(\mathbf{F}_0 - \mathbf{K}\mathbf{u}_0)\big)}{2} \tag{4.58}$$

The recurrence relation of Eq. (4.56) is then evaluated for $n = 0$ to yield

$$\mathbf{M}\mathbf{u}_1 = (2\mathbf{M} - h^2\mathbf{K})\mathbf{u}_0 - \mathbf{M}\mathbf{u}_{-1} + h^2\mathbf{F}_0 \tag{4.59}$$

from which $\mathbf{u}_1$ is determined using $\mathbf{u}_{-1}$ from Eq. (4.58) and $\mathbf{u}_0$ from the initial conditions. The recurrence relation is then used successively for $n = 1, 2, \ldots$ until the desired time range is included. After determining $\mathbf{u}_{n+1}$, the velocity $\dot{\mathbf{u}}_n$ and the acceleration $\ddot{\mathbf{u}}_n$ at t_n are computed at each time step using Eqs. (4.55).

The entire process proceeds according to:

Given. The initial conditions $\mathbf{u}(0)$ and $\dot{\mathbf{u}}(0)$,

Compute.

$\ddot{\mathbf{u}}_0$ using Eq. (4.57)

$\mathbf{u}_{-1}$ using Eq. (4.58)

$\mathbf{u}_1$ using Eq. (4.59)

Then for $n = 1, 2, \ldots$ compute $\mathbf{u}_{n+1}$ using

$$\mathbf{M}\mathbf{u}_{n+1} = (2\mathbf{M} - h^2\mathbf{K})\mathbf{u}_n - \mathbf{M}\mathbf{u}_{n-1} + h^2\mathbf{F}_n$$

$$\dot{\mathbf{u}}_n = \frac{(\mathbf{u}_{n+1} - \mathbf{u}_{n-1})}{2h}$$

$$\ddot{\mathbf{u}}_n = \frac{(\mathbf{u}_{n+1} - 2\mathbf{u}_n + \mathbf{u}_{n-1})}{h^2}$$

As will be indicated in Section 4.3.4.3, this method is conditionally stable with the critical step size given by

$$h_{cr} = \frac{2}{\omega_{\max}}$$

where $(\omega_{\max})^2$ is the largest eigenvalue of the algebraic eigenvalue problem

$$(\mathbf{K} - \omega^2\mathbf{M})\mathbf{x} = \mathbf{0}$$

Just as for the first-order system discussed in Section 4.2.4, values of $h > h_{cr}$ result in an unbounded oscillation of the numerical solution.

Example 4.4.

Consider the one-dimensional problem

$$m\ddot{x} + kx = f_0$$
$$x(0) = 0$$
$$\dot{x}(0) = 0$$

Define the dimensionless displacement $z = kx/f_0$ and rewrite the differential equation as

$$\ddot{z} + \omega^2 z = \omega^2$$
$$z(0) = 0$$
$$\dot{z}(0) = 0$$

where $\omega^2 = k/m$. The student should fill in the details to show that the recurrence relation for this one-dimensional problem is

$$z_{n+1} = \left(2 - (\omega h)^2\right) z_n - z_{n-1} + (\omega h)^2$$

with $\ddot{z}(0)$ given by ω^2. It then follows that the starting procedure yields $z_{-1} = (h\omega)^2/2$ and also that for $n = 0$, the recurrence relation yields $z_1 = (h\omega)^2/2$. Then as outlined previously for $n = 1, 2, \ldots,$

$$z_{n+1} = \left(2 - (\omega h)^2\right) z_n - z_{n-1} + (\omega h)^2$$

with

$$\dot{z}_n = \frac{z_{n+1} - z_{n-1}}{2h} \qquad \ddot{z}_n = \frac{z_{n+1} - 2z_n + z_{n-1}}{h^2}$$

For $\omega = 1$, Figs. 4–15 through 4–17 present the results for the displacement, velocity, and acceleration, respectively, plotted as a function of time.

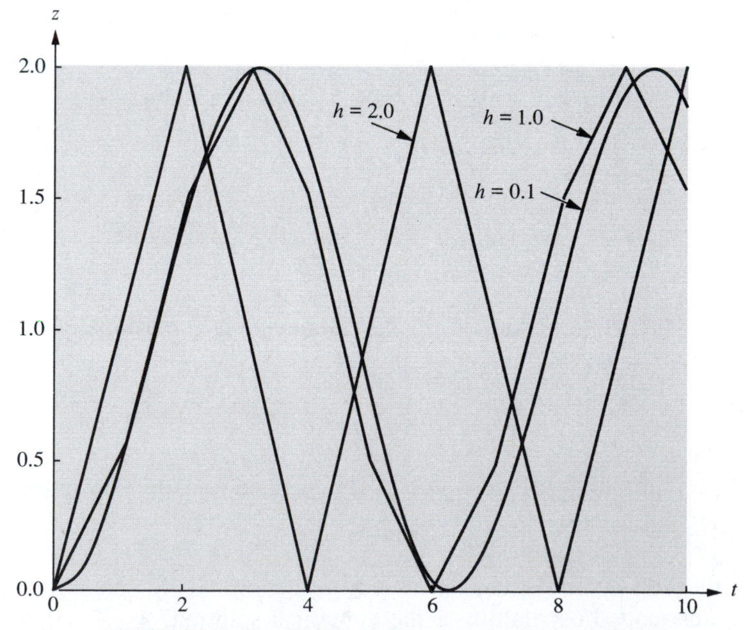

FIGURE 4–15 Displacement results for various h

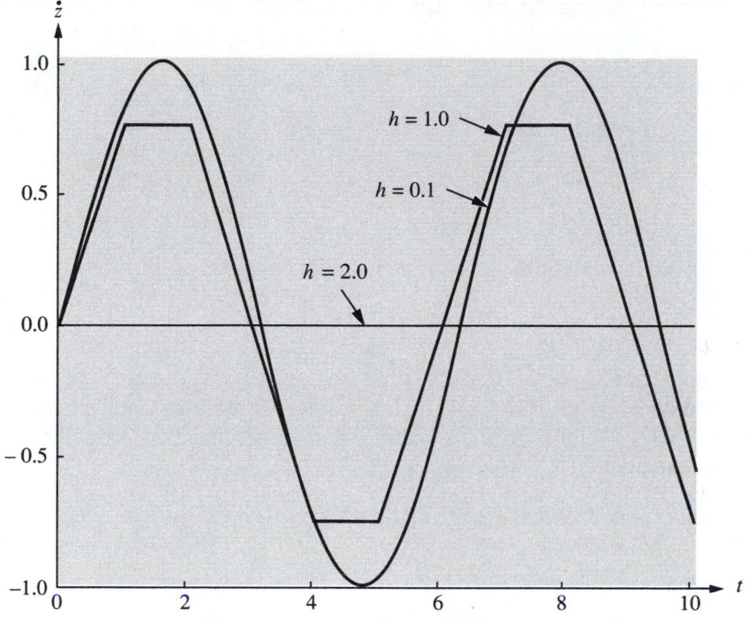

FIGURE 4–16 Velocity results for various h

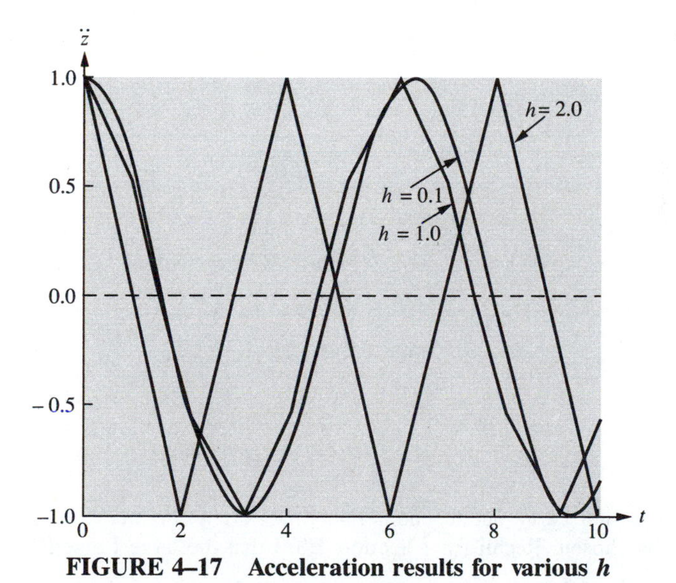

FIGURE 4–17 Acceleration results for various h

Figures 4–15 through 4–17 present results for $h = 0.1$, $h = 1.0$, and $h = 2.0$. The results for $h = 0.1$ are essentially the same as the exact results. The critical step size is represented by $h = 2.0$ and is thus the upper limit for stability. Values above $h = 2.0$ would result in unbounded oscillations. Results for other values of h could also be plotted but would not serve to demonstrate any other inherent properties of the numerical solution.

Example 4.5.

Consider again the example given by Eqs. (4.45) in Section 4.3.2:

$$\phi\begin{bmatrix}4 & 1 & 0\\ 1 & 4 & 1\\ 0 & 1 & 2\end{bmatrix}\ddot{\mathbf{v}} + \begin{bmatrix}2 & -1 & 0\\ -1 & 2 & -1\\ 0 & -1 & 1\end{bmatrix}\mathbf{v} = \begin{bmatrix}0\\ 0\\ 1/3\end{bmatrix}$$

or

$$\phi\mathbf{m}\ddot{\mathbf{v}} + \mathbf{k}\mathbf{v} = \mathbf{F}$$
$$\mathbf{v}(0) = \mathbf{0}$$
$$\dot{\mathbf{v}}(0) = \mathbf{0}$$

where $\mathbf{v} = \mathbf{u}/\Delta$ and $\phi = \rho L^2/54E$. Numerical results will be based on the values $E = 3\times 10^7$ psi, $\rho = 7.5\times 10^{-4}$ lbf-s²/in⁴, $L = 20$ in., $A = 1$ in², and $P = 1000$ lbf. Evaluating the differential equation at $t = 0$ yields

$$\phi\begin{bmatrix}4 & 1 & 0\\ 1 & 4 & 1\\ 0 & 1 & 2\end{bmatrix}\ddot{\mathbf{v}}_0 + \begin{bmatrix}2 & -1 & 0\\ -1 & 2 & -1\\ 0 & -1 & 1\end{bmatrix}\mathbf{0} = \begin{bmatrix}0\\ 0\\ 1/3\end{bmatrix}$$

from which

$$\ddot{\mathbf{v}}_0 = \begin{bmatrix}1\\ -4\\ 15\end{bmatrix}\frac{1}{78\phi}$$

Then $\mathbf{v}_{-1}$ is determined from

$$\mathbf{v}_{-1} = \mathbf{v}_0 - h\dot{\mathbf{v}}_0 + \frac{h^2\ddot{\mathbf{v}}_0}{2} = \frac{h^2}{156\phi}[1 \quad -4 \quad 15]^{\mathrm{T}}$$

The basic algorithm can be expressed as

$$\mathbf{m}\mathbf{v}_{\mathbf{n+1}} = (2\mathbf{m} - \psi\mathbf{k})\mathbf{v}_{\mathbf{n}} - \mathbf{m}\mathbf{v}_{\mathbf{n-1}} + \psi\mathbf{F}_{\mathbf{n}}$$

where $\psi = h^2/\phi = 54Eh^2/\rho L^2$. The first iteration yields

$$\mathbf{m}\mathbf{v}_1 = (2\mathbf{m} - \psi\mathbf{k})\mathbf{0} - \mathbf{m}\mathbf{v}_{-1} + \psi\mathbf{F}_0$$

or

$$\mathbf{v}_1^{\mathrm{T}} = \frac{\psi}{156}[1 \quad -4 \quad 15]$$

In order to further carry out the numerical integration for this example, a step size $\Delta t = h$ must be chosen. Recall from Section 4.3.3 that the largest eigenvalue is

$$(\omega_{\max})^2 = 1.6456\left(\frac{54E}{\rho L^2}\right)$$

so that the critical step size is given by

$$h_{cr} = \frac{2}{\omega_{\max}} \qquad \text{or} \qquad h_{cr} = 0.2212\left(\frac{\rho L^2}{E}\right)^{1/2}$$

or in terms of the parameter ψ appearing in the differential equation

$$\psi_{cr} = 2.4307$$

For values of ψ below ψ_{cr} the solution will remain bounded for large t, whereas for $\psi > \psi_{cr}$ the solution as given by the numerical procedure will oscillate with ever-increasing amplitude; that is, the algorithm is not stable when $\psi > \psi_{cr}$.

As was seen from the analytical solution presented in Section 4.3.3, all of the frequencies determined from $\det(\mathbf{k} - \omega^2\mathbf{M}) = 0$ are contained in the solution. In order to obtain numerical results that accurately contain the effects of all the frequency components, it is necessary to choose a step size that is relatively small compared with the period of the largest frequency. A rule of thumb is to break half the period of the largest frequency into 10 equal intervals; that is, take

$$h^* = \frac{\pi}{10\omega_{\max}}$$

For the present example

$$h^* = \frac{\pi L}{94.267c}$$

with the parameter ψ given by

$$\psi^* = \frac{(\pi L/94.267c)^2}{\phi} = 0.0600$$

Results for this example are presented in Figs. 4–18 through 4–20 for $h_1 = 4.2433 \times 10^{-6}$ sec and $h_2 = 2.1216 \times 10^{-5}$ sec. The critical step size is h_2, and $h_1 = h_2/5$ is

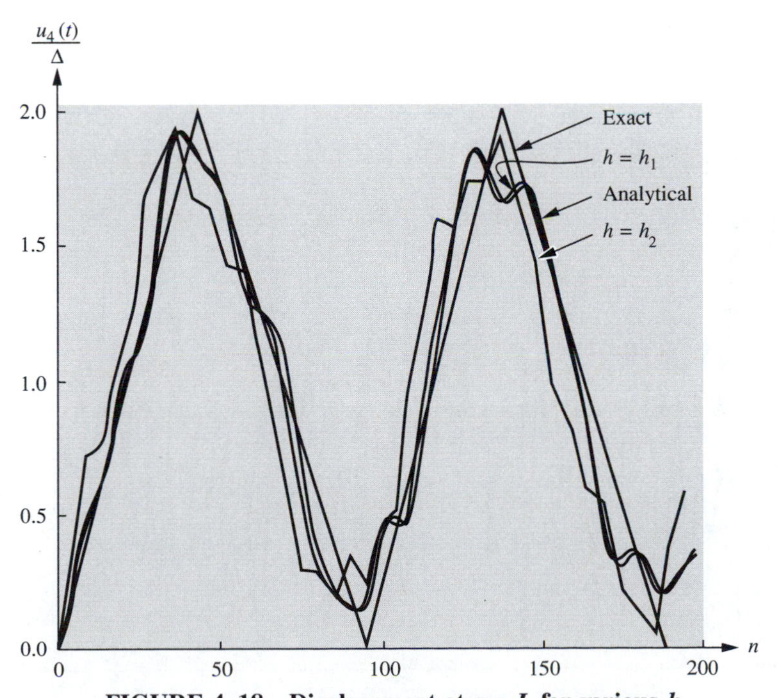

FIGURE 4–18 Displacement at $x = L$ for various h

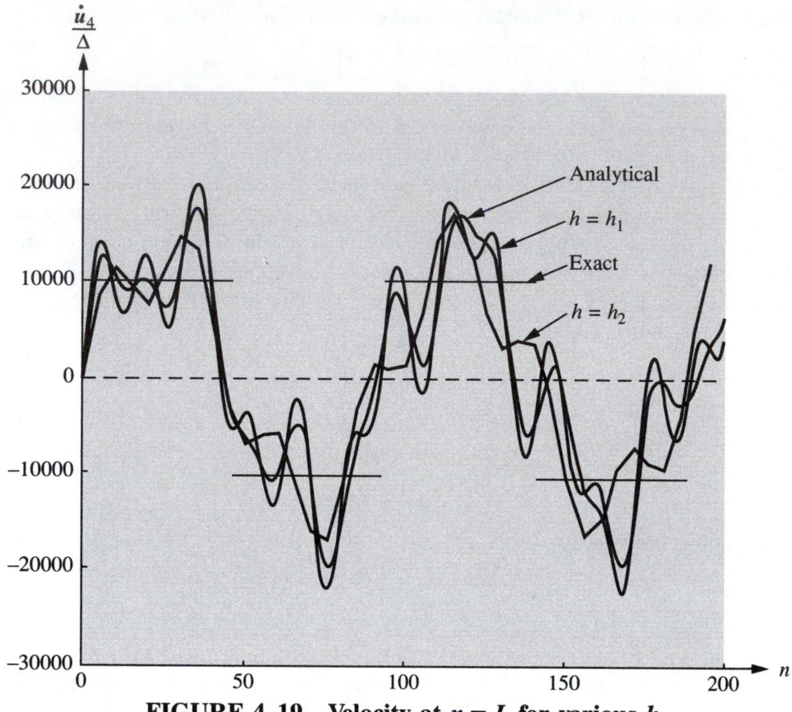

FIGURE 4–19 Velocity at $x = L$ for various h

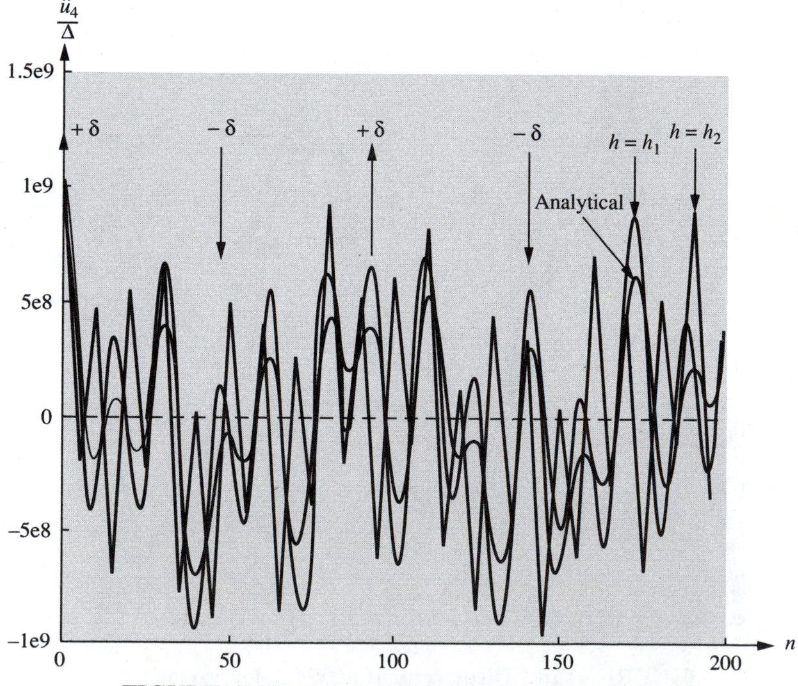

FIGURE 4–20 Acceleration at $x = L$ for various h

a value somewhat larger than the one corresponding to dividing the half period of the maximum frequency into 10 equal segments.

The results displayed in Fig. 4–18 for the displacement at $x = L$, that is, $u_4(t)$, show that the analytical solution and the central difference numerical solution for $\Delta t = h = 4.2433 \times 10^{-6}$ sec. agree well and constitute a reasonable approximation to the exact triangular wave solution also shown.

Figure 4–19 for the velocity $u_4(t)$ indicates that the analytical results and those from the central difference algorithm for both h_1 and h_2 are similar, both deviating substantially from the exact square wave solution shown. Results for the acceleration $u_4(t)$, shown in Fig. 4–20, are essentially meaningless in terms of being able to detect that the exact solution is the set of delta or impulse functions indicated.

Generally, the accuracy of the results improves with an increase in the number of elements used. This can be traced to the fact that more of the approximate eigenvalues corresponding to the exact solution are more accurately determined using more elements. The use of higher-order interpolations may also result in some improvement in accuracy, although not to the same extent as increasing the number of linearly interpolated elements. These issues are left to the reader to investigate in the Exercises.

As is apparent from the results of the example of Section 4.3.3, all three of the frequencies contribute to the solution. This means that the combined requirements of not exceeding the critical time step and integrating the effects of the higher modes accurately can lead to a very small h, and hence an expensive algorithm. Fortunately for large systems the higher modes do not contribute significantly to the solution so that an unconditionally stable algorithm with a larger time step can be used satisfactorily. Such an algorithm is discussed in the next section.

Finally, from Eq. (4.56) it is easily seen that if lumped mass matrices are used, **M** is a diagonal matrix and the computations involved in the central difference algorithm reduce at each step to a matrix multiplication and vector additions, that is, no solution of a set of algebraic equations is required at each step.

4.3.4.2 NEWMARK'S METHOD. Newmark's method [5] is based on an extension of the average acceleration method, which is conditionally stable. Newmark was able to generalize the algorithm so as to retain its simple form, yet produce an unconditionally stable algorithm.

The average acceleration method is based on the assumption that over a small time increment any nodal acceleration can be considered to be a linear function of time as indicated in Fig. 4–21. For the interval $0 < \tau < h$, the interval corresponding to the time step, the acceleration is expressed as

$$\ddot{\mathbf{u}}_{\mathbf{t}+\tau} = \ddot{\mathbf{u}}_{\mathbf{t}}\left(1 - \frac{\tau}{h}\right) + \ddot{\mathbf{u}}_{\mathbf{t+h}}\left(\frac{\tau}{h}\right)$$

Integrating yields

$$\dot{\mathbf{u}}_{\mathbf{t}+\tau} = \dot{\mathbf{u}}_{\mathbf{t}} + \ddot{\mathbf{u}}_{\mathbf{t}}\left(\tau - \frac{\tau^2}{2h}\right) + \ddot{\mathbf{u}}_{\mathbf{t+h}}\left(\frac{\tau^2}{2h}\right) \tag{4.60}$$

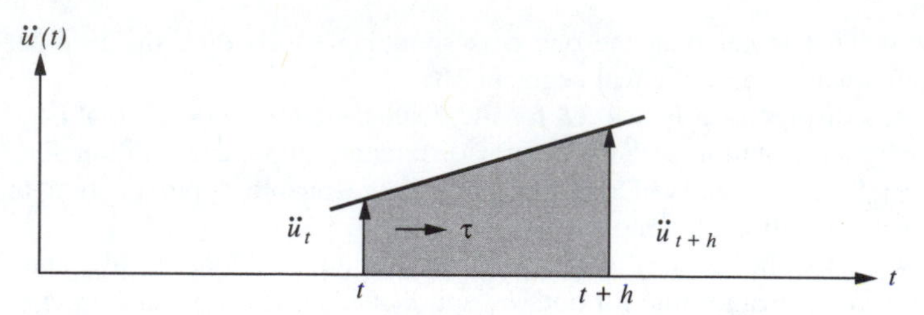

FIGURE 4–21 Linear acceleration on the interval $0 < \tau < h$

and specifically

$$\dot{\mathbf{u}}_{\mathbf{t+h}} = \dot{\mathbf{u}}_{\mathbf{t}} + \frac{(\ddot{\mathbf{u}}_{\mathbf{t}} + \ddot{\mathbf{u}}_{\mathbf{t+h}})h}{2} = \dot{\mathbf{u}}_{\mathbf{t}} + h\ddot{\mathbf{u}}_{\mathbf{AVG}} \tag{4.61}$$

that is, the increment in the velocity is based on the approximate average acceleration on the interval $(0, h)$. Integrating Eq. (4.60) yields

$$\mathbf{u}_{\mathbf{t}+\tau} = \mathbf{u}_{\mathbf{t}} + \tau\dot{\mathbf{u}}_{\mathbf{t}} + \ddot{\mathbf{u}}_{\mathbf{t}}\left(\frac{\tau^2}{2} - \frac{\tau^3}{6h}\right) + \ddot{\mathbf{u}}_{\mathbf{t+h}}\left(\frac{\tau^3}{6h}\right) \tag{4.62}$$

with

$$\mathbf{u}_{\mathbf{t+h}} = \mathbf{u}_{\mathbf{t}} + h\dot{\mathbf{u}}_{\mathbf{t}} + \frac{(2\ddot{\mathbf{u}}_{\mathbf{t}} + \ddot{\mathbf{u}}_{\mathbf{t+h}})h^2}{6} \tag{4.63}$$

These expressions are employed with the differential equations

$$\mathbf{M}\ddot{\mathbf{u}} + \mathbf{K}\mathbf{u} = \mathbf{F}$$

to yield the *conditionally stable* average acceleration algorithm.

Newmark generalized Equations (4.61) and (4.63) to read

$$\dot{\mathbf{u}}_{\mathbf{t+h}} = \dot{\mathbf{u}}_{\mathbf{t}} + \left((1-\delta)\ddot{\mathbf{u}}_{\mathbf{t}} + \delta\ddot{\mathbf{u}}_{\mathbf{t+h}}\right)h \tag{4.64}$$

and

$$\mathbf{u}_{\mathbf{t+h}} = \mathbf{u}_{\mathbf{t}} + h\dot{\mathbf{u}}_{\mathbf{t}} + \left(\left(\frac{1}{2} - \alpha\right)\ddot{\mathbf{u}}_{\mathbf{t}} + \alpha\ddot{\mathbf{u}}_{\mathbf{t+h}}\right)h^2 \tag{4.65}$$

where δ and α are parameters to be chosen for accuracy and stability. The method is *unconditionally stable* as long as the parameters δ and α are chosen to satisfy $\delta \geq 0.5$ and $\alpha \geq 0.25(\delta + 0.5)^2$. Note that $\delta = 1/2$ and $\alpha = 1/6$ corresponds to the average acceleration method. Eq. (4.65) is solved for $\ddot{\mathbf{u}}_{\mathbf{t+h}}$ and substituted into Eq. (4.64) to yield

$$\dot{\mathbf{u}}_{\mathbf{t+h}} = \dot{\mathbf{u}}_{\mathbf{t}} + \frac{\delta(\mathbf{u}_{\mathbf{t+h}} - \mathbf{u}_{\mathbf{t}} - h\dot{\mathbf{u}}_{\mathbf{t}})}{\alpha h} + c_2 h\ddot{\mathbf{u}}_{\mathbf{t}}$$

where $c_2 = 1 - \delta/2\alpha$, and then into the differential equation evaluated at $t + h$ to yield

$$(\mathbf{M} + \alpha h^2\mathbf{K})\mathbf{u}_{\mathbf{t+h}} = \mathbf{M}(\mathbf{u}_{\mathbf{t}} + h\dot{\mathbf{u}}_{\mathbf{t}} + c_1 h^2\ddot{\mathbf{u}}_{\mathbf{t}}) + \alpha h^2\mathbf{F}_{\mathbf{t+h}} \tag{4.66}$$

where $c_1 = 1/2 - \alpha$. This equation, together with the two equations for the velocity and acceleration at $t + h$, namely,

$$\begin{aligned} \dot{\mathbf{u}}_{\mathbf{t+h}} &= \dot{\mathbf{u}}_{\mathbf{t}} + \frac{\delta(\mathbf{u}_{\mathbf{t+h}} - \mathbf{u}_{\mathbf{t}} - h\dot{\mathbf{u}}_{\mathbf{t}})}{\alpha h} + c_2 h \ddot{\mathbf{u}}_{\mathbf{t}} \\ \ddot{\mathbf{u}}_{\mathbf{t+h}} &= \frac{\mathbf{u}_{\mathbf{t+h}} - \mathbf{u}_{\mathbf{t}} - h\dot{\mathbf{u}}_{\mathbf{t}}}{\alpha h^2} - \frac{c_1 \ddot{\mathbf{u}}_{\mathbf{t}}}{\alpha} \end{aligned} \tag{4.67}$$

can be used to step ahead in time to determine the solution. In order to start the process, the acceleration at $t = 0$ is needed and is determined by solving the governing equations evaluated at $t = 0$,

$$\mathbf{M}\ddot{\mathbf{u}}(0) = \mathbf{F}(0) - \mathbf{K}\mathbf{u}(0) \tag{4.68}$$

for the acceleration $\ddot{\mathbf{u}}(0)$. Equations (4.66) and (4.67) are then used to step ahead using the unconditionally stable Newmark algorithm.

The algorithm consists of:

Given. The initial conditions $\mathbf{u}_0$ and $\dot{\mathbf{u}}_0$.

Compute. $\ddot{\mathbf{u}}_0$ using Eq. (4.68), $\mathbf{u}_n$, $\dot{\mathbf{u}}_n$, and $\ddot{\mathbf{u}}_n$, $n = 1, 2, \ldots$ using Eqs. (4.66) and (4.67),

$$\begin{aligned} (\mathbf{M} + \alpha h^2 \mathbf{K})\mathbf{u}_{n+1} &= \mathbf{M}(\mathbf{u}_n + h\dot{\mathbf{u}}_n + c_1 h^2 \ddot{\mathbf{u}}_n) + \alpha h^2 \mathbf{F}_{n+1} \\ \dot{\mathbf{u}}_{n+1} &= \dot{\mathbf{u}}_n + \frac{\delta(\mathbf{u}_{n+1} - \mathbf{u}_n - h\dot{\mathbf{u}}_n)}{\alpha h} + hc_2 \ddot{\mathbf{u}}_n \\ \ddot{\mathbf{u}}_{n+1} &= \frac{\mathbf{u}_{n+1} - \mathbf{u}_n - h\dot{\mathbf{u}}_n}{\alpha h^2} - \frac{c_1 \ddot{\mathbf{u}}_n}{\alpha} \end{aligned}$$

Specifically, with $\mathbf{u}_0, \dot{\mathbf{u}}_0$, and $\ddot{\mathbf{u}}_0$ known,

$$\begin{aligned} (\mathbf{M} + \alpha h^2 \mathbf{K})\mathbf{u}_1 &= \mathbf{M}(\mathbf{u}_0 + h\dot{\mathbf{u}}_0 + c_1 h^2 \ddot{\mathbf{u}}_0) + \alpha h^2 \mathbf{F}_1 \\ \dot{\mathbf{u}}_1 &= \dot{\mathbf{u}}_0 + \frac{\delta(\mathbf{u}_1 - \mathbf{u}_0 - h\dot{\mathbf{u}}_0)}{\alpha h} + hc_2 \ddot{\mathbf{u}}_0 \\ \ddot{\mathbf{u}}_1 &= \frac{\mathbf{u}_1 - \mathbf{u}_0 - h\dot{\mathbf{u}}_0}{\alpha h^2} - \frac{c_1 \ddot{\mathbf{u}}_0}{\alpha} \end{aligned}$$

Then with $\mathbf{u}_1, \dot{\mathbf{u}}_1$, and $\ddot{\mathbf{u}}_1$ known,

$$\begin{aligned} (\mathbf{M} + \alpha h^2 \mathbf{K})\mathbf{u}_2 &= \mathbf{M}(\mathbf{u}_1 + h\dot{\mathbf{u}}_1 + c_1 h^2 \ddot{\mathbf{u}}_1) + \alpha h^2 \mathbf{F}_2 \\ \dot{\mathbf{u}}_2 &= \dot{\mathbf{u}}_1 + \frac{\delta(\mathbf{u}_2 - \mathbf{u}_1 - h\dot{\mathbf{u}}_1)}{\alpha h} + hc_2 \ddot{\mathbf{u}}_1 \\ \ddot{\mathbf{u}}_2 &= \frac{\mathbf{u}_2 - \mathbf{u}_1 - h\dot{\mathbf{u}}_1}{\alpha h^2} - \frac{c_1 \ddot{\mathbf{u}}_1}{\alpha} \end{aligned}$$

The algorithm is continued until the time range of interest is covered. Note that for the Newmark algorithm, lumping of the mass matrix results in no computational advantage.

Example 4.6.

Consider again the problem given by Eqs. (4.45) in Section 4.3.2:

$$\phi\begin{bmatrix}2/3 & 1/6 & 0\\ 1/6 & 2/3 & 1/6\\ 0 & 1/6 & 1/3\end{bmatrix}\ddot{\mathbf{v}}+\begin{bmatrix}2 & -1 & 0\\ -1 & 2 & -1\\ 0 & -1 & 1\end{bmatrix}\mathbf{v}=\begin{bmatrix}0\\ 0\\ 1/3\end{bmatrix}$$

or

$$\phi\mathbf{m}\ddot{\mathbf{v}}+\mathbf{kv}=\mathbf{f}$$
$$\mathbf{v}(0)=\mathbf{v}_0=\mathbf{0}$$
$$\dot{\mathbf{v}}(0)=\dot{\mathbf{v}}_0=\mathbf{0}$$

where $\phi = \rho L^2/9E$ and $\mathbf{v} = \mathbf{u}/\Delta$. The equations to be solved at the first step can be written as

$$(\mathbf{m}+\alpha\psi\mathbf{k})\mathbf{v}_1 = \mathbf{m}(\mathbf{v}_0+h\dot{\mathbf{v}}_0+c_1h^2\ddot{\mathbf{v}}_0)+\alpha\psi\mathbf{f}_1$$
$$\dot{\mathbf{v}}_1 = \dot{\mathbf{v}}_0+\frac{\delta(\mathbf{v}_1-\mathbf{v}_0-h\dot{\mathbf{v}}_0)}{\alpha h}+hc_2\ddot{\mathbf{v}}_0$$
$$\ddot{\mathbf{v}}_1 = \frac{\mathbf{v}_1-\mathbf{v}_0-h\dot{\mathbf{v}}_0}{\alpha h^2}-\frac{c_1\ddot{\mathbf{v}}_0}{\alpha}$$

where $\psi = h^2/\phi$.

Evaluating the differential equation at $t = 0$ yields

$$\phi\mathbf{m}\ddot{\mathbf{v}}_0+\mathbf{0}=\mathbf{f}$$

from which

$$\ddot{\mathbf{v}}_0 = 10^8\,[0.6923 \quad -2.7692 \quad 10.3846]^{\mathrm{T}}$$

Taking $\alpha = 0.25$, $\delta = 0.5$, and $h = 4.2433\times10^{-6}$ sec results in

$$\mathbf{m}+\alpha\psi\mathbf{k}=\begin{bmatrix}0.6748 & 0.1626 & 0.0000\\ 0.1626 & 0.6748 & 0.1626\\ 0.0000 & 0.1626 & 0.3374\end{bmatrix}$$

and

$$\alpha\psi\mathbf{f} = [0.0000 \quad 0.0000 \quad 0.0081]^{\mathrm{T}}$$

At step 1,

$$\begin{bmatrix}0.6748 & 0.1626 & 0.0000\\ 0.1626 & 0.6748 & 0.1626\\ 0.0000 & 0.1626 & 0.3374\end{bmatrix}\mathbf{v}_1 = \mathbf{m}[\mathbf{0}+\mathbf{0}+0.25h^2\ddot{\mathbf{v}}_0]+\alpha\psi\mathbf{f}$$

from which

$$\mathbf{v}_1 = [0.0006 \quad -0.0023 \quad 0.0091]^{\mathrm{T}}$$

with

$$\dot{\mathbf{v}}_1 = \dot{\mathbf{v}}_0+\frac{2}{h}[\mathbf{v}_1-\mathbf{0}-\mathbf{0}]+\mathbf{0}$$
$$= 10^4[0.0265 \quad -0.1101 \quad 0.4304]^{\mathrm{T}}$$

$$\ddot{\mathbf{v}}_1 = 4[\mathbf{v}_1 - \mathbf{0} - \mathbf{0}] - \ddot{\mathbf{v}}_0$$
$$= 10^8[0.5585 \quad -2.4209 \quad 9.9008]^T$$

After step 2,

$$\mathbf{v}_2 = [0.0020 \quad -0.0087 \quad 0.0357]^T$$
$$\dot{\mathbf{v}}_2 = 10^4[0.0423 \quad -0.1918 \quad 0.8211]^T$$
$$\ddot{\mathbf{v}}_2 = 10^8[0.1848 \quad -1.4300 \quad 8.5155]^T$$

The results for further integration are presented in Figs. 4–22 through 4–24. The step size $h_1 = 4.2433 \times 10^{-6}$ sec indicated above is the same as the smaller of the two values used for the central difference algorithm in the previous section. Integrations are also carried out for $h_2 = 4.2433 \times 10^{-5}$ sec $= 10h_1$, a value twice that of the critical value for the central difference algorithm of the previous section. In Figs. 4–22 through 4–24, the abscissa n represents the number of time steps of length h_1.

The results displayed in Fig. 4–22 indicate that for $h = h_1$, there is very good agreement between the numerical solution and the corresponding analytical solution, both comparing favorably with the exact solution. For $h = h_2$, the unconditionally stable Newmark algorithm is unable to predict the part of the response arising from the higher frequencies, but is able to predict the essential character of the displacement at the end $x = L$.

The results displayed in Fig. 4–23 for the velocity at $x = L$ indicate a rough similarity between the analytical solution and the Newmark solution for $h = h_1$. Similarly, the

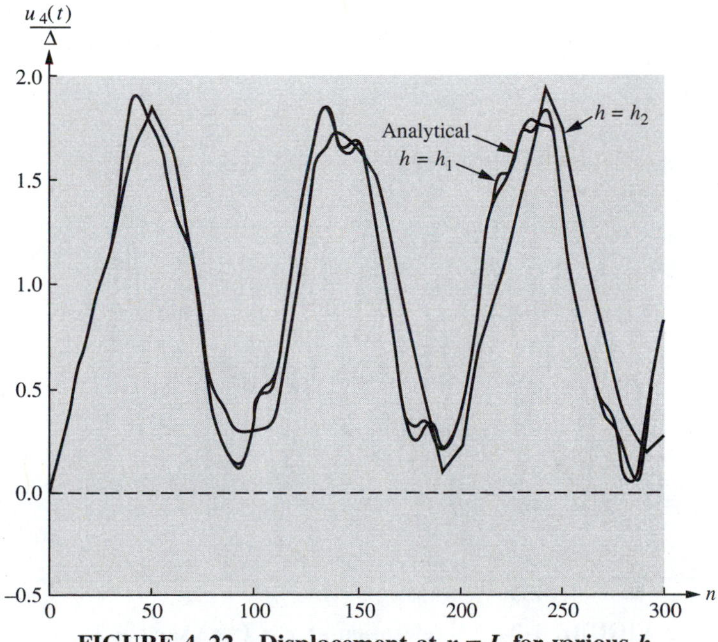

FIGURE 4–22 Displacement at $x = L$ for various h

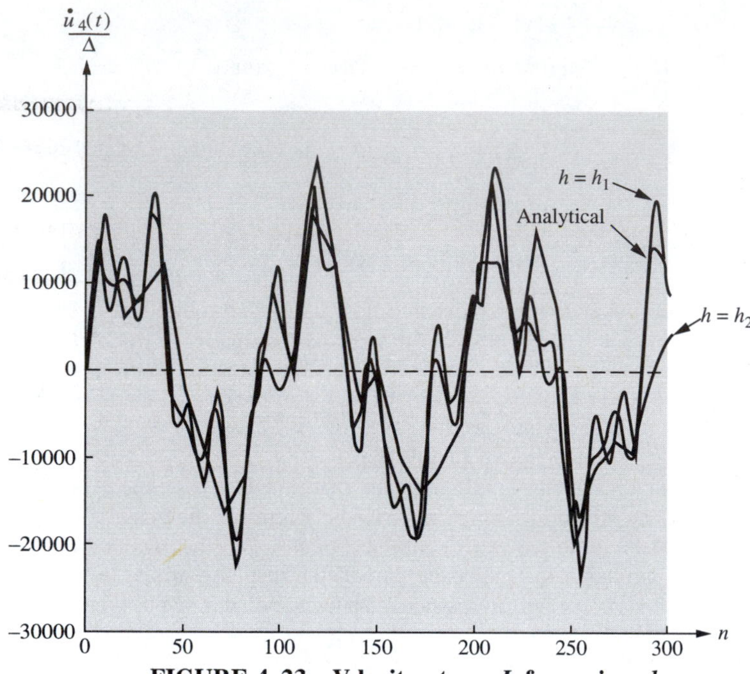

FIGURE 4–23 Velocity at $x = L$ for various h

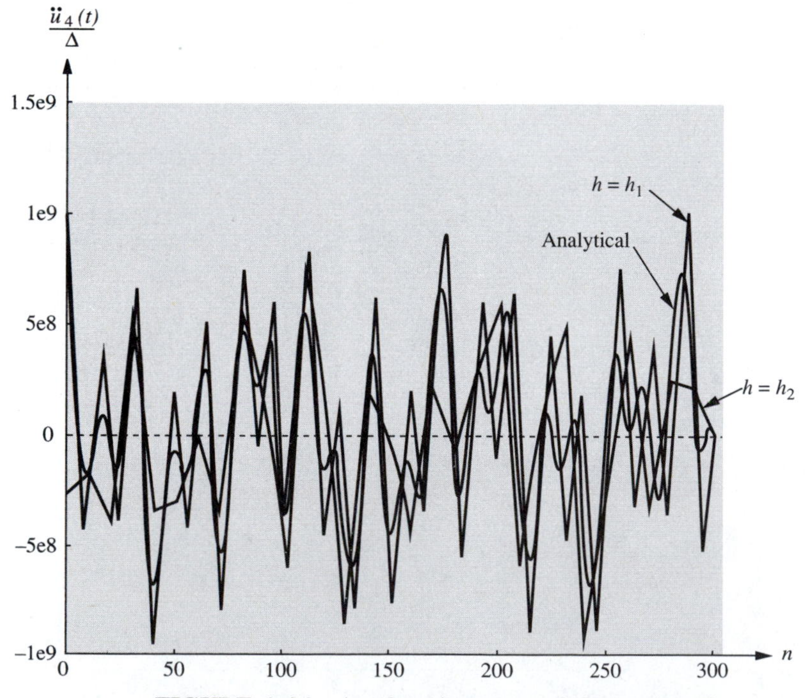

FIGURE 4–24 Acceleration at $x = L$ for various h

numerical results for $h = h_2$ bear some resemblance to the analytical and exact solutions, but are neither qualitatively nor quantitatively satisfactory. The results for the accelerations given in Fig. 4–24 are, as was the case for the central difference algorithm, completely unsatisfactory.

4.3.4.3 ANALYSIS OF ALGORITHMS. The purpose of this section is to provide an introduction to the analysis of algorithms for second-order systems resulting from the application of the finite element method to wave propagation or vibration problems. An elementary treatment of stability for the central difference and Newmark methods is presented. For further details the reader is referred to Hughes [6], Bathe [7], Bathe and Wilson [8], Strang and Fix [9], and Carey and Oden [10].

As was the case with the θ method as applied to first-order systems of ordinary differential equations as discussed in Section 4.2.4.3, there is a similar generalization for the second-order systems that can be used to analyze the stability of both the central difference and Newmark algorithms. The generalization consists of the three-term recurrence

$$[\mathbf{M} + \alpha h^2\mathbf{K}]\mathbf{u_{n+1}} + [-2\mathbf{M} + (0.5 - 2\alpha + \delta)h^2\mathbf{K}]\mathbf{u_n} + [\mathbf{M} + (0.5 + \alpha - \delta)h^2\mathbf{K}]\mathbf{u_{n-1}} - \mathbf{F_n}h^2 = 0 \qquad (4.69)$$

where α and δ are parameters that are to be selected for stability and accuracy. If $\alpha = 0$ and $\delta = 0.5$, there results

$$\mathbf{Mu_{n+1}} = [2\mathbf{M} - h^2\mathbf{K}]\mathbf{u_n} - \mathbf{Mu_{n-1}} + \mathbf{F_n}h^2$$

which is the central difference algorithm. For arbitrary α and δ the three-term recurrence relation given in Eq. (4.69) is equivalent [Zienkiewicz, 11] to the Newmark algorithm of Section 4.3.4.2.

The stability of the algorithm is best investigated in terms of the transformed coordinates $\mathbf{z}$, where $\mathbf{z}$ is related to $\mathbf{u}$ according to $\mathbf{u} = \mathbf{Vz}$, where $\mathbf{V}$ is the modal matrix introduced in Section 4.2.3, satisfying

$$\mathbf{V}^T\mathbf{MV} = \mathbf{I} \qquad \mathbf{V}^T\mathbf{KV} = \mathbf{\Lambda}$$

After setting $\mathbf{u} = \mathbf{Vz}$ in Eq. (4.69), premultiply by $\mathbf{V}^T$ to obtain

$$[\mathbf{V}^T\mathbf{MV} + \alpha h^2\mathbf{V}^T\mathbf{KV}]\mathbf{z_{n+1}} + [-2\mathbf{V}^T\mathbf{MV} + (0.5 - 2\alpha + \delta)h^2\mathbf{V}^T\mathbf{KV}]\mathbf{z_n} + [\mathbf{V}^T\mathbf{MV} + (0.5 + \alpha - \delta)h^2\mathbf{V}^T\mathbf{KV}]\mathbf{z_{n-1}} - h^2\mathbf{V}^T\mathbf{F_n} = \mathbf{0}$$

or

$$[\mathbf{I} + \alpha h^2\mathbf{\Lambda}]\mathbf{z_{n+1}} + [-2\mathbf{I} + (0.5 - 2\alpha + \delta)h^2\mathbf{\Lambda}]\mathbf{z_n} + [\mathbf{I} + (0.5 + \alpha - \delta)h^2\mathbf{\Lambda}]\mathbf{z_{n-1}} - \mathbf{V}^T\mathbf{F_n}h^2 = \mathbf{0}$$

This is equivalent to the uncoupled set of equations

$$(1 + \alpha h^2\omega_i^2)(z_i)_{n+1} + \left(-2 + (0.5 - 2\alpha + \delta)h^2\omega_i^2\right)(z_i)_n + \left(1 + (0.5 + \alpha - \delta)h^2\omega_i^2\right)(z_i)_{n-1} = \mathbf{0}$$

where only the homogeneous part of the equation has been retained for the purpose of assessing stability. Defining $\sigma = (\omega h)^2$, $\phi = (0.5 + \delta)\sigma/(1 + \alpha\sigma)$, and $\psi = (0.5 - \delta)\sigma/(1 + \alpha\sigma)$ results in

$$(z_i)_{n+1} - (2 - \phi)(z_i)_n + (1 + \psi)(z_i)_{n-1} = 0 \tag{4.70}$$

a simple difference equation for which a solution of the form $z_n = c\rho^n$ is assumed. This results in

$$\left(\rho^2 - (2 - \phi)\rho + 1 + \psi\right)c\rho^{n-1} = 0$$

from which the characteristic equation

$$\rho^2 - (2 - \phi)\rho + 1 + \psi = 0 \tag{4.71}$$

is obtained. The basic idea is that the solution should not become unbounded as $n \longrightarrow \infty$, which requires that both roots of the indicial equation satisfy

$$\left|\rho_i\right| \leq 1 \qquad i = 1, 2$$

Rewriting the characteristic equation in the form

$$\rho^2 - (2 - \phi)\rho + 1 + \psi = (\rho - \rho_1)(\rho - \rho_2)$$
$$= \rho^2 - (\rho_1 + \rho_2)\rho + \rho_1\rho_2 = 0$$

it is clear that

$$\rho_1\rho_2 = 1 + \psi$$

and that unless $\psi \leq 0$, at least one of the roots ρ_1, ρ_2 must have a modulus greater than unity. Since

$$\psi = \frac{(0.5 - \delta)\sigma}{(1 + \alpha\sigma)} \leq 0$$

it follows that a necessary condition for the absolute values of the roots to be less than one is

$$\delta \geq 0.5$$

Other conditions must be determined on the basis of looking at the actual roots of the characteristic equation (4.70). To this end the quadratic equation is solved to yield

$$\rho_{1,2} = \frac{2 - \phi \pm \left((2 - \phi)^2 - 4(1 + \psi)\right)^{1/2}}{2}$$

A careful consideration of the possibilities for the roots ρ_1 and ρ_2 leads to the conclusion that the discriminant

$$(2 - \phi)^2 - 4(1 + \psi) \tag{4.72}$$

must be negative; that is, Eq. (4.71) has complex conjugate roots. It follows after some algebra that Eq. (4.70) can be expressed as

$$\sigma\left(4\alpha - (0.5 + \delta)^2\right) > -4$$

It is possible to satisfy this inequality *for all* σ by taking

$$\alpha \geq 0.25(0.5 + \delta)^2$$

Thus by requiring that $\delta \geq 0.5$ and $\alpha \geq 0.25(0.5 + \delta)^2$, the roots of the difference equation (4.70) are less than one in absolute value and hence the Newmark algorithm is stable. More importantly, the results are independent of h and thus the Newmark algorithm is *unconditionally stable* for $\delta \geq 0.5$ and $\alpha \geq 0.25(0.5 + \delta)^2$.

For the central difference algorithm, $\alpha = 0$ and $\delta = 0.5$ in Eq. (4.69), and it follows that the corresponding difference equation is

$$u_{n+1} - (2 - \phi)u_n + u_{n-1} = 0$$

so that the characteristic equation becomes

$$\rho^2 - (2 - \sigma)\rho + 1 = 0$$

with

$$\rho = \frac{2 - \sigma \pm \left(\sigma(\sigma - 4)\right)^{1/2}}{2}$$

It again follows that the discriminant must be negative for both roots to have moduli less than one so that

$$\sigma = (\omega h)^2 \leq 4$$

or

$$h_{cr} = \frac{2}{\omega_{\max}} = \frac{T_{\min}}{\pi}$$

The maximum eigenvalue of $(\mathbf{K} - \omega^2\mathbf{K})\mathbf{x} = 0$ is $\omega_{\max}^2$, and $T_{\min}$ is the associated period. The central difference algorithm is conditionally stable with $h_{cr} = 2/\omega_{\max}$.

Generally the accuracy of the Newmark algorithm is comparable to that of the central difference algorithm for the same time step. The fact that the Newmark algorithm is unconditionally stable as long as the parameters α and δ are properly chosen usually makes it the algorithm of choice between the two. For a comparison of the Newmark algorithm with other unconditionally stable algorithms, see Bathe and Wilson [8].

There is still activity toward developing algorithms for the efficient and accurate integration of second-order systems of equations. Much of this is beyond the scope of this text and the reader is referred to the open literature for further study.

4.4 TWO-DIMENSIONAL DIFFUSION

The governing balance equations that describe diffusion processes in situations involving two independent variables appear typically [Dennemeyer, 12] as

$$\begin{aligned} \nabla \cdot (k \nabla u(x, y, t)) - \rho \frac{\partial u}{\partial t} + f(x, y, t) &= 0 && \text{in } D \\ u &= g(s, t) && \text{on } \Gamma_1 \\ k \frac{\partial u}{\partial n} + \alpha(s, t) u &= q(s, t) && \text{on } \Gamma_2 \\ u(x, y, 0) &= c(x, y) && \text{in } D \end{aligned} \tag{4.73}$$

where D is the interior of the domain, and Γ_1 and Γ_2 constitute the boundary as indicated in Fig. 4–25. The physical constants are k and ρ, and s is a linear measure of the position on the boundary. The type of boundary condition specified on Γ_2 results from a local balance between conduction in the interior and convection into the exterior.

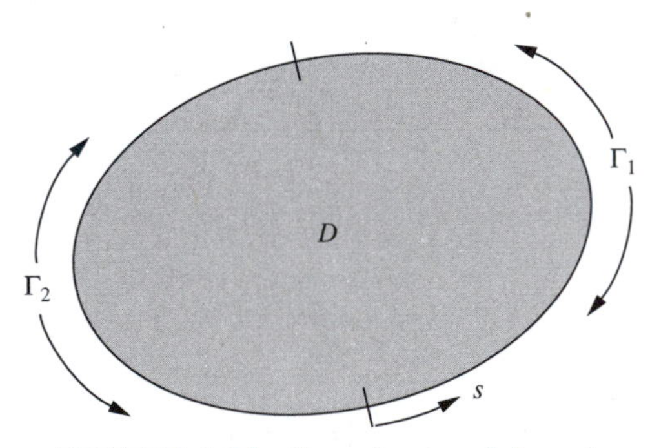

FIGURE 4–25 Boundary and domain

This equation is similar in form to the elliptic boundary value problem studied in Chapter 3, with the very important addition of the time derivative term in the differential equation and the corresponding initial condition. With these additions, the problem changes from an elliptic boundary value problem to a parabolic initial-boundary value problem.

The basic steps of discretization, interpolation, elemental formulation, assembly, constraints, solution, and computation of derived variables are presented in this section as they relate to the two-dimensional parabolic initial-boundary value problem stated in Eqs. (4.73). The Galerkin method, in connection with the corresponding weak formulation to be developed, will be used to generate the finite element model.

For discretization, the reader is asked to review the material in Section 3.3.2.

Interpolation. The solution is assumed to be expressible in terms of the nodally based interpolation functions $n_i(x, y)$ introduced and discussed in Section 3.2. In the present setting, these interpolation functions are used with the semidiscretization

$$u(x, y, t) = \sum_1^N u_i(t)n_i(x, y) \tag{4.74}$$

Elemental formulation. The starting point for the elemental formulation is the weak formulation of the initial-boundary value problem. The first step in developing the weak formulation is to multiply the differential equation (4.73) by an arbitrary test function $v(x, y)$ vanishing on Γ_1. The result is then integrated over the domain D to obtain

$$\iint v\left(\nabla \cdot (k\,\nabla u) - \rho \frac{\partial u}{\partial t} + f\right) dA = 0$$

Using the two-dimensional form of the divergence theorem to integrate the first term by parts, there results after rearranging

$$\iint_D \nabla v \cdot k\,\nabla u\, dA + \iint v\rho \frac{\partial u}{\partial t}\, dA = \int_\Gamma v\,\mathbf{n} \cdot k\,\nabla u\, ds + \iint vf\, dA$$

Writing $\Gamma = \Gamma_1 + \Gamma_2$ and recalling that v vanishes on Γ_1 and that $k\,\partial u/\partial n = k\mathbf{n} \cdot \nabla u = q - hu$ on Γ_2, it follows that

$$\iint_D \nabla v \cdot k\,\nabla u\, dA + \iint_D v\rho \frac{\partial u}{\partial t}\, dA + \int_{\Gamma_2} vhu\, ds = \iint_D vf\, dA + \int_{\Gamma_2} vq\, ds \tag{4.75}$$

Equation (4.75) is the required weak formulation for the two-dimensional diffusion problem stated in Eqs. (4.73).

Substituting the representation of Eq. (4.74) into the weak formulation of Eq. (4.75) and taking $v = n_k$, $k = 1, 2, \ldots$ yields

$$\iint_{D^-} \left(\frac{\partial n_k}{\partial x} k \sum u_i \frac{\partial n_i}{\partial x} + \frac{\partial n_k}{\partial y} k \sum u_i \frac{\partial n_i}{\partial y}\right) dA$$

$$+ \iint_{D^-} n_k \rho \sum \dot{u}_i n_i + \int_{\Gamma_2^-} n_k h \sum u_i n_i\, ds$$

$$= \iint_{D^-} n_k f\, dA + \int_{\Gamma_2^-} n_k q\, ds \qquad k = 1, 2, \ldots$$

D^- and Γ_2^- represent the collection of the elemental areas approximating D, and the collection of the elemental straight line segments approximating Γ_2, respectively.

This $N \times N$ set of linear algebraic equations can be written as

$$\sum B_{ki}\dot{u}_i + \sum A_{ki}u_i = F_k \qquad k = 1, 2, \ldots, N \tag{4.76}$$

where

$$A_{ki} = \iint_{D^-} \left(\frac{\partial n_k}{\partial x} k \frac{\partial n_i}{\partial x} + \frac{\partial n_k}{\partial y} k \frac{\partial n_i}{\partial y} \right) dA + \int_{\Gamma_2^-} n_k h n_i \, ds$$

$$B_{ki} = \iint_{D^-} n_k \rho n_i \, dA$$

$$F_k = \iint_{D^-} n_k f \, dA + \int_{\Gamma_2^-} n_k \, q \, ds$$

Note that assembly is contained implicitly within the formulation.

In terms of the corresponding elementally based interpolations

$$u_e(x, y) = \mathbf{N}^{\mathrm{T}}\mathbf{u_e} = \mathbf{u_e^{\mathrm{T}}N}$$

the finite element model can be expressed as

$$\begin{aligned} \mathbf{B\dot{u}} + \mathbf{Au} &= \mathbf{F} \\ \mathbf{u}(0) &= \mathbf{u_0} \end{aligned} \tag{4.77}$$

where

$$\mathbf{A} = \sum_e \mathbf{k_G} + \sum_e{}' \mathbf{a_G}$$

$$\mathbf{B} = \sum_e \mathbf{r_G}$$

$$\mathbf{F} = \sum_e \mathbf{f_G} + \sum_e{}' \mathbf{q_G}$$

with

$$\mathbf{k_e} = \iint_{A_e} \left[\frac{\partial \mathbf{N}}{\partial x} k \frac{\partial \mathbf{N}^{\mathrm{T}}}{\partial x} + \frac{\partial \mathbf{N}}{\partial y} k \frac{\partial \mathbf{N}^{\mathrm{T}}}{\partial y} \right] dA$$

$$\mathbf{r_e} = \iint_{A_e} \mathbf{N}\rho\mathbf{N}^{\mathrm{T}} \, dA \qquad \mathbf{a_e} = \int_{\gamma_{2e}} \mathbf{N}\alpha\mathbf{N}^{\mathrm{T}} \, ds$$

$$\mathbf{f_e} = \iint_{A_e} \mathbf{N}f \, dA \qquad \mathbf{q_e} = \int_{\gamma_{2e}} \mathbf{N}q(s) \, ds$$

The initial conditions for the system of first-order differential equations (4.77) are obtained from the initial conditions prescribed for the original initial-boundary value problem. Generally $\mathbf{u}(0)$ is determined by evaluating the function $c(x, y)$ at the nodes to obtain

$$\mathbf{u}(0) = \mathbf{u}_0 = [c_1 \quad c_2 \quad c_3 \quad \ldots \quad c_N]^{\mathrm{T}}$$

where $c_i = c(x_i, y_i)$, with (x_i, y_i) the coordinates of the ith node.

Constraints. The constraints arise from the boundary conditions specified on Γ_1. Generally the values of the constraints are determined from the q function with the constrained value of u at a node on Γ_1 being taken as the value of q at that point. These constraints are then enforced on the assembled equations as outlined in Section 4.2.1, resulting in the final global constrained set of linear first-order differential equations

$$\begin{aligned} \mathbf{M}\dot{\mathbf{u}} + \mathbf{K}\mathbf{u} &= \mathbf{F} \\ \mathbf{u}(0) &= \mathbf{u}_0 \end{aligned} \tag{4.78}$$

Solution. The system of equations (4.78) is precisely the same in form and character as the corresponding equations developed in Section 4.2.1 for the one-dimensional diffusion problem. An analytical method as well as the numerical methods of Euler and improved Euler or Crank-Nicolson can be used for integrating the above set of equations. The Euler method will be conditionally stable with the critical time step depending on the maximum eigenvalue of the associated problem $(\mathbf{K} - \lambda\mathbf{M})\mathbf{v} = \mathbf{0}$. The improved Euler or Crank-Nicolson algorithm will be unconditionally stable. Specific examples are left to the Exercises at the end of the chapter.

Derived variables. The derived variables will be time dependent, and depending on the particular problem being considered, may need to be computed at each time step. The computations would be per element and would be carried out using the techniques described in Chapter 3.

4.5 TWO-DIMENSIONAL WAVE EQUATIONS

The governing equations of motion that describe the propagation of waves in situations involving two independent variables appear typically [Dennemeyer, 12] as

$$\begin{aligned} \nabla \cdot \big(k \nabla u(x, y, t)\big) - \rho \frac{\partial^2 u}{\partial t^2} + f(x, y, t) &= 0 && \text{in } D \\ u &= g(s, t) && \text{on } \Gamma_1 \\ k \frac{\partial u}{\partial n} + \alpha(s, t) u &= q(s, t) && \text{on } \Gamma_2 \\ u(x, y, 0) &= c(x, y) && \text{in } D \\ \frac{\partial u(x, y, 0)}{\partial t} &= d(x, y) && \text{in } D \end{aligned} \tag{4.79}$$

where k and ρ are physical constants. The interior of the domain is D, and Γ_1 and Γ_2 constitute the boundary as indicated in Fig. 4–26, with s a linear measure of the position on the boundary. The type of boundary condition specified on Γ_2 results from a local balance between internal and external forces.

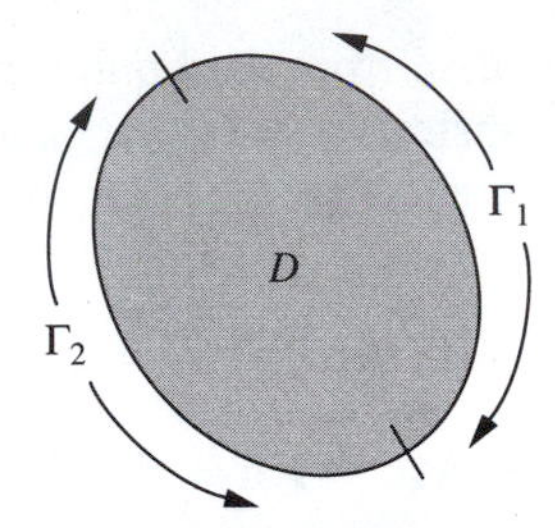

FIGURE 4–26 Boundary and domain

This equation is similar in form to the parabolic initial-boundary value problem presented in the previous section with the very important change in the time derivative term from a first to a second derivative, and with the addition of a second initial condition on the velocity. With these changes, the problem changes from a parabolic initial-boundary value problem to a hyperbolic initial-boundary value problem.

The basic steps of discretization, interpolation, elemental formulation, assembly, constraints, solution, and computation of derived variables are presented in this section as they relate to the two-dimensional hyperbolic initial-boundary value problem stated in Eqs. (4.79). The Galerkin method, in connection with the corresponding weak formulation to be developed, will be used to generate the finite element model.

For discretization, the reader is again referred to the material in Section 3.3.2.

Interpolation. The assumption for the solution is again in terms of the nodally based interpolation functions $n_i(x, y)$ introduced and discussed in Section 3.2. In the present setting, these interpolation functions are used with the semidiscretization

$$u(x, y, t) = \sum_{1}^{N} u_i(t) n_i(x, y) \tag{4.80}$$

Elemental formulation. The starting point for the elemental formulation is the weak formulation of the initial-boundary value problem. The first step in developing the weak formulation is to multiply the differential equation (4.79) by an arbitrary test function $v(x, y)$ vanishing on Γ_1. The result is then integrated over the domain D to obtain

$$\iint v \left(\nabla \cdot (k \nabla u) - \rho \frac{\partial^2 u}{\partial t^2} + f \right) dA = 0$$

Using the two-dimensional form of the divergence theorem to integrate the first term by parts, there results after rearranging

$$\iint_D \left(\nabla v \cdot k\,\nabla u + \rho v \frac{\partial^2 u}{\partial t^2}\right) dA = \int_\Gamma v\mathbf{n} \cdot k\,\nabla u\, ds + \iint vf\, dA$$

Writing $\Gamma = \Gamma_1 + \Gamma_2$ and recalling that v vanishes on Γ_1 and that $k\,\partial u/\partial n = k\mathbf{n} \cdot \nabla u = q - hu$ on Γ_2, it follows that

$$\iint_D \left(\nabla v \cdot k\,\nabla u + \rho v \frac{\partial^2 u}{\partial t^2}\right) dA + \int_{\Gamma_2} vhu\, ds = \iint_D vf\, dA + \int_{\Gamma_2} vq\, ds \qquad (4.81)$$

Equation (4.81) is the required weak formulation for the two-dimensional wave equation of Eq. (4.69).

Substituting the representation (4.80) into the weak formulation (4.81) and taking $v = n_k$, $k = 1, 2, \ldots$ yields

$$\iint_{D^-} \left(\frac{\partial n_k}{\partial x} k \sum u_i \frac{\partial n_i}{\partial x} + \frac{\partial n_k}{\partial y} k \sum u_i \frac{\partial n_i}{\partial y}\right) dA$$

$$+ \iint_{D^-} \rho n_k \sum \ddot{u}_i n_i + \int_{\Gamma_2^-} n_k h \sum u_i n_i\, ds$$

$$= \iint_{D^-} n_k f\, dA + \int_{\Gamma_2^-} n_k q\, ds \qquad k = 1, 2, \ldots$$

where D^- and Γ_2^- represent the collection of the elemental areas approximating D, and the collection of the elemental straight line segments approximating Γ_2, respectively.

This $N \times N$ set of linear algebraic equations can be written as

$$\sum B_{ki} \ddot{u}_i + \sum A_{ki} u_i = F_k \qquad k = 1, 2, \ldots, N \qquad (4.82)$$

where

$$A_{ki} = \iint_{D^-} \left(\frac{\partial n_k}{\partial x} k \frac{\partial n_i}{\partial x} + \frac{\partial n_k}{\partial y} k \frac{\partial n_i}{\partial y}\right) dA + \int_{\Gamma_2^-} n_k h n_i\, ds$$

$$B_{ki} = \iint_{D^-} n_k \rho n_i\, dA$$

$$F_k = \iint_{D^-} n_k f\, dA + \int_{\Gamma_2^-} n_k q\, ds$$

Note that assembly is contained implicitly within the formulation.

In terms of the corresponding elementally based interpolations

$$u_e(x, y) = \mathbf{N}^{\mathrm{T}}\mathbf{u_e} = \mathbf{u}_\mathbf{e}^{\mathrm{T}}\mathbf{N}$$

the finite element model can be expressed as

$$\begin{aligned}\mathbf{B\ddot{u}} + \mathbf{Au} &= \mathbf{F} \\ \mathbf{u}(0) &= \mathbf{u_0} \\ \dot{\mathbf{u}}(0) &= \dot{\mathbf{u}}_\mathbf{0}\end{aligned} \tag{4.83}$$

where

$$\mathbf{A} = \sum_e \mathbf{k_G} + \sum_e{}' \mathbf{a_G}$$

$$\mathbf{B} = \sum_e \mathbf{r_G}$$

$$\mathbf{F} = \sum_e \mathbf{f_G} + \sum_e{}' \mathbf{q_G}$$

with

$$\mathbf{k_e} = \iint_{A_e} \left[\frac{\partial \mathbf{N}}{\partial x} k \frac{\partial \mathbf{N}^\mathrm{T}}{\partial x} + \frac{\partial \mathbf{N}}{\partial y} k \frac{\partial \mathbf{N}^\mathrm{T}}{\partial y}\right] dA$$

$$\mathbf{r_e} = \iint_{A_e} \mathbf{N}\rho\mathbf{N}^\mathrm{T}\, dA \qquad \mathbf{a_e} = \int_{\gamma_{2e}} \mathbf{N}\alpha\mathbf{N}^\mathrm{T}\, ds$$

$$\mathbf{f_e} = \iint_{A_e} \mathbf{N} f\, dA \qquad \mathbf{q_e} = \int_{\gamma_{2e}} \mathbf{N} q(s)\, ds$$

The initial conditions for the system of first-order differential equations (4.83) are obtained from the initial conditions prescribed for the original initial-boundary value problem. Generally $\mathbf{u}(0)$ is determined by evaluating the function $c(x, y)$ at the nodes to obtain

$$\mathbf{u}(0) = \mathbf{u_0} = [c_1 \quad c_2 \quad c_3 \quad \dots \quad c_N]^\mathrm{T}$$

where $c_i = c(x_i, y_i)$, with (x_i, y_i) the coordinates of the ith node. In a completely similar fashion the initial condition on the velocity becomes

$$\dot{\mathbf{u}}(0) = \dot{\mathbf{u}}_\mathbf{0} = [d_1 \quad d_2 \quad d_3 \quad \dots \quad d_N]^\mathrm{T}$$

Constraints. The constraints arise from the boundary conditions specified on Γ_1. Generally the values of the constraints are determined from the q function with the constrained value of u at a node on Γ_1 being taken as the value of q at that point. These constraints are then enforced on the assembled equations as outlined in Section 4.3.1, resulting in the final global constrained set of linear second-order differential equations

$$\begin{aligned}\mathbf{M\ddot{u}} + \mathbf{Ku} &= \mathbf{F} \\ \mathbf{u}(0) &= \mathbf{u_0} \\ \dot{\mathbf{u}}(0) &= \dot{\mathbf{u}}_\mathbf{0}\end{aligned} \tag{4.84}$$

Solution. The system of equations (4.84) is precisely the same in form and character as the corresponding equations developed in Section 4.2.2 for the one-dimensional wave problem. The analytical method as well as the numerical methods using the central difference and Newmark algorithms can be used for integrating the above set of equations. The central difference algorithm will be conditionally stable with the critical time step depending on the maximum eigenvalue of the associated problem $(\mathbf{K} - \omega^2\mathbf{M})\mathbf{v} = 0$. The Newmark algorithm will be unconditionally stable for $\delta \geq 0.5$ and $\alpha \geq 0.25(\delta + 0.5)^2$. Specific examples are left to the exercises at the end of the chapter.

Derived variables. In a physical situation governed by a wave equation, the derived variables are usually the internal forces computed according to $\mathbf{F_e} = \mathbf{k_e u_e}$ per element for each time step.

4.6 CLOSURE

Time-dependent problems are inherently more difficult and expensive to solve than their corresponding steady-state counterparts. The expense of generating the global matrices is higher for the time-dependent problems because of the necessity of computing the mass matrices. The main extra expense, however, is in solving the resulting time-dependent global equations.

For an analytical approach to the solution, additional expense is incurred in terms of having to determine eigenvalues and eigenvectors. The actual amount of expense depends on the specific form of the stiffness and mass matrices and the algorithm used, but in any case it is significantly in excess of the expense of solving the single set of linear algebraic equations associated with the steady-state problem.

For a time domain integration technique, the additional expense is clearly related to the number of time steps necessary to trace out the desired time history. In addition to several matrix multiplications and additions, *each step* can involve the solution of a set of linear algebraic equations. In some instances this expense can be minimized by using a decomposition that can be reused for the computation of the solution at each new time.

In this regard recall that the Euler and central difference algorithms require that the size of the time step not exceed a value proportional to the inverse of the largest eigenvalue. For large systems this critical step size can be very small resulting in many applications of the algorithm to trace out the time history. The unconditionally stable Crank-Nicolson and Newmark algorithms, on the other hand, can be used with arbitrary step size that has been chosen so as to accurately integrate the lower modes, with significant improvement in the expense relative to the conditionally stable Euler and central difference algorithms. There are of course other algorithms available [Hughes, 6; Bathe, 7; Bathe and Wilson, 8] that are specifically tailored to address other numerical issues.

REFERENCES

1. Courant, R., and D. Hilbert: *Methods of Mathematical Physics,* Vol.II, Wiley Interscience, New York, 1962.

2. Goudreau, G. L.: "Evaluation of Discrete Methods for the Linear Dynamic Response of Elastic and Viscoelastic Solids," U C SESM Report 69–15, University of California, Berkeley, 1970.
3. Boyce, W. E., and R. C. DiPrima: *Elementary Differential Equations and Boundary Value Problems,* John Wiley, New York, 1986.
4. Henrici, P.: *Discrete Variable Methods in Ordinary Differential Equations,* John Wiley, New York, 1962.
5. Newmark, N. M.: "A Method of Computation for Structural Dynamics," *Journal of the Engineering Mechanics Division, ASCE,* vol. 85, pp. 69–74, 1959.
6. Hughes, T. J. R.: *The Finite Element Method, Linear Static and Dynamic Finite Element Analysis,* Prentice Hall, Englewood Cliffs, New Jersey, 1987.
7. Bathe, K. J.: *Finite Element Procedures in Engineering Analysis,* Prentice Hall, Englewood Cliffs, New Jersey, 1982.
8. Bathe, K. J., and E. L. Wilson: *Numerical Methods in Finite Element Analysis,* Prentice Hall, Englewood Cliffs, New Jersey, 1976.
9. Strang, G., and G. Fix: *An Analysis of the Finite Element Method,* Prentice Hall, Englewood Cliffs, New Jersey, 1973.
10. Carey, G. F., and J. T. Oden: *Finite Elements—Computational Aspects,* vol. III, Prentice Hall, Englewood Cliffs, New Jersey, 1984.
11. Zienkiewicz, O. C.: *The Finite Element Method, 3rd ed.,* McGraw-Hill, New York, 1977.
12. Dennemeyer, R.: *Introduction to Partial Differential Equations and Boundary Value Problems,* McGraw-Hill, New York, 1968.

GENERAL REFERENCES

Becker, E. B., et al.: *Finite Elements—An Introduction,* Vol. I, Prentice Hall, Englewood Cliffs, New Jersey, 1981.

Burnett, D. S.: *Finite Element Analysis, From Concepts to Applications,* Addison-Wesley, Reading, Massachussetts, 1987.

Fletcher, C. A. J.: *Computational Galerkin Methods,* Springer-Verlag, New York, 1984.

Huebner, K. H., and E. A. Thornton: *The Finite Element Method for Engineers,* John Wiley, New York, 1982.

Kikuchi, N.: *Finite Element Methods in Mechanics,* Cambridge University Press, Cambridge, UK, 1986.

Rao, S. S.: *The Finite Element Method in Engineering,* Pergamon, Oxford, 1982.

Reddy, J. N.: *An Introduction to the Finite Element Method,* McGraw-Hill, New York, 1984.

Tong, P., and J. N. Rossettos: *Finite Element Method: Basic Technique and Implementation,* MIT Press, Cambridge, Massachussetts, 1977.

Wait, R., and A. R. Mitchell: *Finite Element Analysis and Applications,* John Wiley, New York, 1985.

White, R. E.: *An Introduction to the Finite Element Method with Applications to Non-linear Problems,* Wiley-Interscience, New York, 1985.

COMPUTER PROJECTS

Develop a computer code for the one-dimensional initial-boundary value problem

$$\frac{\partial(p\,\partial u/\partial x)}{\partial x} - qu + r\frac{\partial u}{\partial t} + f = 0$$

$$-p(a)\frac{\partial u(a)}{\partial x} + \alpha u(a) = g(t)$$

$$p(b)\frac{\partial u(b)}{\partial x} + \beta u(b) = h(t)$$

$$u(x, 0) = u_0(x)$$

Note that when the general initial-boundary value problem is written in this way, the program can be closely patterned after the corresponding boundary value problem code suggested in Chapter 2. There are, of course, additional inputs and the *solution* phase must be altered! Assume that p, q, r, α, and β are not time dependent, and that f is separable; that is, $f(x, t) = F(x)G(t)$. Structure the program in roughly the following fashion.

INPUT. Input the information about p, q, r, f, a, b, α, β, g, h, and u_0.

1. NE		Number of elements, (number of nodes = NE+1 = NN.)
2. a, b		Boundary points.
3. p(I)	I=1,NN	Nodal p values.
4. q(I)	I=1,NN	Nodal q values.
5. r(I)	I=1,NN	Nodal r values.
6. F(I)	I=1,NN	Nodal F values.

Alternately, p, q, r, and F can be defined internally by subroutines, procedures, functions, and so forth, but would have to be changed for each new application.

7. $G(t)$, $g(t)$, and $h(t)$ need to be specified. These can be input as a "table," or defined internally as suggested for p, q, r and F.

8. NA	Boundary code at $x = a$. NA = 1 for an essential boundary condition and NA = 2 for a natural boundary condition.
9. NB	Boundary code at $x = b$. NB = 1 for an essential boundary condition and NB = 2 for a natural boundary condition.
10. $u_a(t)$ or α and $g(t)$	$u_a(t)$ as the prescribed value of u at a if NA = 1, or α and $a(t)$ if NA = 2.
11. $u_b(t)$ or β and $h(t)$	$u_b(t)$ as the prescribed value of u at b if NB = 1, or β and $h(t)$ if NB = 2.
12. Step size	The step size to be used in the numerical integration. Alternately, a subroutine or procedure for computing the largest eigenvalue of the constrained matrices can be included.

DATA REFLECTION. Output the input data for checking purposes.

NODE GENERATION. Generate the additional positions of all the internal nodes using NE, a, and b. Usually element lengths are taken to be equal but this is not necessary. An additional input could be used to specify a denser concentration of nodes near a or near b if desired.

ELEMENT FORMULATION AND ASSEMBLY. $\mathbf{p}_e$, $\mathbf{q}_e$, $\mathbf{r}_e$, and $\mathbf{F}_e$ need to be formed and assembled for each of the NE elements. Any of the methods discussed in Sections 2.6 and 2.8 can be used for evaluating the integrals. The information supplied in input items 1–6 is needed here. If either of the boundary conditions is natural, the appropriate α, β

must be added to the global stiffness matrix and the function $G(t)$, $g(t)$, or $h(t)$ must be available for computing the time-dependent, right-hand side of the system of equations.

CONSTRAINTS. If either boundary condition is essential, the first or last equation, respectively, must be altered to reflect the constraints $u_1 = u_a(t)$ and/or $u_{NN} = u_b(t)$. The equations should also be put into symmetric form, resulting, in general, in a change in the time-dependent character of the appropriate terms on the right-hand side of the equations. The information in items 7–11 is needed for these steps.

SOLUTION. There is no essential difficulty in writing the code to employ the general θ method discussed in Section 4.2.4.3. It is advisable for the student to have gone through several of the exercises that follow *by hand* in order to understand all that is involved. Appropriate coding, which avoids solving a set of equations at each step when lumping is used in connection with the Euler algorithm ($\theta = 0$), should be undertaken.

COMPUTATION OF DERIVED VARIABLES. The derived variable $-p\,\partial u/\partial x$ can be computed for each element at the end of each time step. The point(s) at which $p\,\partial u/\partial x$ is evaluated should be the best points mentioned in the text.

Specifically, the following codes are easily generated based on the material presented in Chapters 2 and 4:

1. Linearly interpolated elements using the special integration formulas indicated in Section 2.6 for evaluating the $\mathbf{p_e}$, $\mathbf{q_e}$, $\mathbf{r_e}$, and $\mathbf{F_e}$.
2. Linearly interpolated elements using a Gauss-Legendre quadrature for evaluating the $\mathbf{p_e}$, $\mathbf{q_e}$, $\mathbf{r_e}$, and $\mathbf{F_e}$. The order of the Gauss quadrature can be taken large enough so the $\mathbf{p_e}$, $\mathbf{q_e}$, $\mathbf{r_e}$, and $\mathbf{F_e}$ are essentially exact for most any of the p, q, r, and F functions encountered.
3. Quadratically interpolated elements using the special integration formulas indicated in Section 2.8 for evaluating the $\mathbf{p_e}$, $\mathbf{q_e}$, $\mathbf{r_e}$, and $\mathbf{F_e}$.
4. Quadratically interpolated elements using a Gauss-Legendre quadrature for evaluating the $\mathbf{p_e}$, $\mathbf{q_e}$, $\mathbf{r_e}$, and $\mathbf{F_e}$. Again, the order of the Gauss quadrature can be taken large enough so the $\mathbf{p_e}$, $\mathbf{q_e}$, $\mathbf{r_e}$, and $\mathbf{F_e}$ are essentially exact for most any of the p, q, r, and F functions encountered.

Repeat the previous project for the one-dimensional initial-boundary value problem

$$\frac{\partial(p\,\partial u/\partial x)}{\partial x} - qu + r\frac{\partial^2 u}{\partial t^2} + f = 0$$

$$-p(a)\frac{\partial u(a)}{\partial x} + \alpha u(a) = g(t)$$

$$p(b)\frac{\partial u(b)}{\partial x} + \beta u(b) = h(t)$$

$$u(x,0) = u_0(x)$$

Much of the code developed in the previous project for the diffusion type problem can be used again for this wave type problem. The major changes will be in the initial conditions and in the solution phase. The solution part of the program can be set up to handle either of the central difference or Newmark algorithms. Again it would be to the student's advantage to have gone through, by hand, the integration procedure(s) to the point where all of the basic steps are thoroughly understood.

Repeat for the two-dimensional diffusion type problem. Draw on the material covered in Chapter 3 for the elemental matrices and general form of the problem, and on the first computer project for the integration procedure(s).

Repeat for the two-dimensional wave type problem. Draw on the material covered in Chapter 3 for the elemental matrices and general form of the problem, and on the second computer project for the integration procedure(s).

EXERCISES

Section 4.2

4.1. Show that the unconstrained system of equations (4.5) can also be stated in matrix notation as expressed by Eqs. (4.6) through (4.8).

4.2. Show that when four linearly interpolated elements are used with lumped mass matrices for the example problem of Section 4.2.2, there results

$$\psi \begin{bmatrix} 2 & -1 & 0 \\ -1 & 2 & -1 \\ 0 & -1 & 2 \end{bmatrix} \mathbf{u} + \begin{bmatrix} 1 & 0 & 0 \\ 0 & 1 & 0 \\ 0 & 0 & 1 \end{bmatrix} \dot{\mathbf{u}} = \begin{bmatrix} \psi u_0 \\ 0 \\ 0 \end{bmatrix}$$

where $\psi = 16\alpha/L^2$. Determine the analytical solution and compare it with the results obtained by using four linearly interpolated consistent mass elements and with the exact results, both of which are presented in Table 4.1.

4.3. Show that when four linearly interpolated elements are used with the weighted mass matrices for the example problem of Section 4.2.2, there results

$$\psi \begin{bmatrix} 2 & -1 & 0 \\ -1 & 2 & -1 \\ 0 & -1 & 2 \end{bmatrix} \mathbf{u} + \begin{bmatrix} 10 & 1 & 0 \\ 1 & 10 & 1 \\ 0 & 1 & 10 \end{bmatrix} \dot{\mathbf{u}} = \begin{bmatrix} \psi u_0 \\ 0 \\ 0 \end{bmatrix}$$

where $\psi = 192\alpha/L^2$. Determine the analytical solution and compare it with the analytical results of Section 4.2.3, the analytical results of Exercise 4.2, and with the exact results which are presented in Table 4.1.

4.4. Show that when the consistent mass matrices are used with quadratic interpolation (two elements) for the example problem of Section 4.2.2, there results

$$\psi \begin{bmatrix} 16 & -8 & 0 \\ -8 & 14 & -8 \\ 0 & -8 & 16 \end{bmatrix} \mathbf{u} + \begin{bmatrix} 16 & 2 & 0 \\ 2 & 8 & 2 \\ 0 & 2 & 16 \end{bmatrix} \dot{\mathbf{u}} = \begin{bmatrix} 8\psi u_0 \\ -\psi u_0 \\ 0 \end{bmatrix}$$

where $\psi = 40\alpha/L^2$. Determine the analytical solution and compare it with the sets of results obtained by using four linearly interpolated elements, and with the exact results.

4.5. Show that when the lumped mass matrices are used with two quadratically interpolated elements for the example problem of Section 4.2.2, there results

$$\psi \begin{bmatrix} 16 & -8 & 0 \\ -8 & 14 & -8 \\ 0 & -8 & 16 \end{bmatrix} \mathbf{u} + \begin{bmatrix} 4 & 0 & 0 \\ 0 & 2 & 0 \\ 0 & 0 & 4 \end{bmatrix} \dot{\mathbf{u}} = \begin{bmatrix} 8\psi u_0 \\ -\psi u_0 \\ 0 \end{bmatrix}$$

where $\psi = 8\alpha/L^2$. Compare the results with the other analytical solutions.

4.6. Show that when the weighted mass matrices are used with two quadratically interpolated elements for the example problem of Section 4.2.2, there results

$$\psi \begin{bmatrix} 16 & -8 & 0 \\ -8 & 14 & -8 \\ 0 & -8 & 16 \end{bmatrix} \mathbf{u} + \begin{bmatrix} 36 & 2 & 0 \\ 2 & 18 & 2 \\ 0 & 2 & 36 \end{bmatrix} \dot{\mathbf{u}} = \begin{bmatrix} 8\psi u_0 \\ -\psi u_0 \\ 0 \end{bmatrix}$$

where $\psi = 80\alpha/L^2$. Compare the results with the other analytical solutions.

4.7. Set up and solve the problem in Section 4.2.2 for each of the following:

Case	Method	Interpolation	Elements	Mass	T_{max}	h
a	Euler	Linear	4	C	20	0.1
b	Euler	Linear	4	L	20	0.1
c	Euler	Linear	4	W	20	0.1
d	Cr-Nic	Linear	4	C	20	1.0
e	Cr-Nic	Linear	4	L	20	1.0
f	Cr-Nic	Linear	4	W	20	1.0
g	Euler	Quadratic	2	C	20	0.1
h	Euler	Quadratic	2	L	20	0.1
i	Euler	Quadratic	2	W	20	0.1
j	Cr-Nic	Quadratic	2	C	20	1.0
k	Cr-Nic	Quadratic	2	L	20	1.0
l	Cr-Nic	Quadratic	2	W	20	1.0

4.8. Consider the initial-boundary value problem with the time-dependent boundary condition given by

$$\frac{\partial(kA\,\partial u/\partial x)}{\partial x} - \rho c_p A \frac{\partial u}{\partial t} = 0 \qquad 0 \le t,\ 0 \le x \le L$$

$$u(0,t) = u_0\Big(1 - \exp(-\beta t)\Big)$$

$$u(L,t) = 0$$

$$u(x,0) = 0$$

where $A = A_0$, a constant. Treat β as a parameter that can be varied to adjust the rate at which $u(0,t) \longrightarrow u_0$. Specifically, take $\beta = 100\alpha/L^2$, where $\alpha = k/\rho c_p$. Take $\alpha/L^2 = 0.021\ \text{s}^{-1}$.

Set up and solve the finite element model with two linearly interpolated elements (consistent mass matrices) using:

a. An analytic approach.
b. The Euler method with $h = 1.0$ s (What is h_{cr}?)
c. The Crank-Nicolson method with $h = 1$ s.
d. The Crank-Nicolson method with $h = 4$ s.

In parts b, c, and d, integrate out to steady state.

4.9. Repeat Exercise 4.8 for each of the following:

Interpolation	Elements	Mass
Linear	2	L
Linear	2	W
Quadratic	1	C
Quadratic	1	L
Quadratic	1	W

4.10. Repeat Exercise 4.8 for each of the following. Take $h = 0.2$ for the Euler method and $h = 0.2$ and 0.4 for the Crank-Nicolson method.

Interpolation	Elements	Mass
Linear	4	C
Linear	4	L
Linear	4	W
Quadratic	2	C
Quadratic	2	L
Quadratic	2	W

4.11. For the initial-boundary value problem

$$\frac{\partial(kA\,\partial u/\partial x)}{\partial x} - \beta u - \rho c_p A\frac{\partial u}{\partial t} + f = 0$$

$$-kA\frac{\partial u(a,t)}{\partial x} + h_L u(a,t) = h_L u_a(t)$$

$$u(b,t) = u_b(t)$$

$$u(x,0) = c(x)$$

a. Develop the appropriate weak form for the initial-boundary value problem.

b. Determine the specific form of the elemental matrices that make up the final unconstrained equations

$$\mathbf{B}\dot{\mathbf{u}} + \mathbf{A}\mathbf{u} = \mathbf{f}$$

$$\mathbf{u}(0) = \mathbf{u}_0$$

4.12. For the problem of Exercise 4.11, take $\beta/kA = 40/L^2$, $u(0,t) = u(L,t) = 0$, $u(x,0) = 0$, $f = f_0 = 8kAu_0/L^2$, $\alpha/L^2 = 0.021\ \text{s}^{-1}$, and solve the following:

Case	Method	Interpolation	Elements	Mass	h
a	Analytic	Linear	2	C	
b	Analytic	Linear	2	L	
c	Analytic	Linear	2	W	
d	Euler	Linear	2	C	0.2
e	Euler	Linear	2	L	0.2
f	Euler	Linear	2	W	0.2
g	Cr-Nic	Linear	2	C	0.8
h	Cr-Nic	Linear	2	L	0.8
i	Cr-Nic	Linear	2	W	0.8

Integrate out to steady state in each case.

4.13. Repeat Exercise 4.12 using four elements with $h = 0.2$ for the Euler method and $h = 0.8$ for the Crank-Nicolson method.

Section 4.3

4.14. Show that the unconstrained system of equations (4.39) can also be stated in matrix notation as expressed by Eqs. (4.40)–(4.42).

4.15. Show that when three linearly interpolated elements are used with lumped mass matrices for the example problem of Section 4.3.2, there results

$$\begin{bmatrix} 2 & -1 & 0 \\ -1 & 2 & -1 \\ 0 & -1 & 1 \end{bmatrix} \mathbf{u} + \psi \begin{bmatrix} 2 & 0 & 0 \\ 0 & 2 & 0 \\ 0 & 0 & 1 \end{bmatrix} \ddot{\mathbf{u}} = \begin{bmatrix} 0 \\ 0 \\ \Delta/3 \end{bmatrix}$$

where $\psi = \rho L^2/18E$. Determine the analytical solution and compare it with the results obtained by using three linearly interpolated consistent mass elements, and with the exact results.

4.16. Show that when three linearly interpolated elements are used with the weighted mass matrices for the example problem of Section 4.3.2, there results

$$\begin{bmatrix} 2 & -1 & 0 \\ -1 & 2 & -1 \\ 0 & -1 & 2 \end{bmatrix} \mathbf{u} + \psi \begin{bmatrix} 10 & 1 & 0 \\ 1 & 10 & 1 \\ 0 & 1 & 10 \end{bmatrix} \ddot{\mathbf{u}} = \begin{bmatrix} 0 \\ 0 \\ \Delta/3 \end{bmatrix}$$

where $\psi = \rho L^2/108E$. Determine the analytical solution and compare it with the analytical results of Section 4.3.2, the analytical results of Exercise 4.2, and with the exact results presented in Table 4.1.

4.17. Show that when the consistent mass matrices are used with quadratic interpolation (one element) for the sample problem of Section 4.3.2, there results

$$\begin{bmatrix} 16 & -8 \\ -8 & 7 \end{bmatrix} \mathbf{u} + \psi \begin{bmatrix} 16 & 2 \\ 2 & 4 \end{bmatrix} \ddot{\mathbf{u}} = \begin{bmatrix} 0 \\ 3\Delta \end{bmatrix}$$

where $\psi = \rho L^2/10E$. Determine the analytical solution and compare it with the sets of results obtained by using four linearly interpolated elements, and with exact results.

4.18. Show that when the lumped mass matrices are used with quadratically interpolated elements for the example problem of Section 4.3.2, there results

$$\begin{bmatrix} 16 & -8 \\ -8 & 7 \end{bmatrix} \mathbf{u} + \psi \begin{bmatrix} 4 & 0 \\ 0 & 1 \end{bmatrix} \ddot{\mathbf{u}} = \begin{bmatrix} 0 \\ 3\Delta \end{bmatrix}$$

where $\psi = \rho L^2/2E$. Compare the results with the other analytical solutions.

4.19. Show that when the weighted mass matrices are used with one quadratically interpolated element for the example problem of Section 4.3.2, there results

$$\begin{bmatrix} 16 & -8 \\ -8 & 7 \end{bmatrix} \mathbf{u} + \psi \begin{bmatrix} 36 & 2 \\ 2 & 9 \end{bmatrix} \ddot{\mathbf{u}} = \begin{bmatrix} 0 \\ 3\Delta \end{bmatrix}$$

where $\psi = \rho L^2/20E$. Compare the results with the other analytical solutions.

4.20. Take $\sqrt{(E/\rho)}t/L = \tau$ and $\mathbf{u}/(PL/AE) = \mathbf{v}$ to recast the constrained equations for the example of Section 4.3.2 in the form

$$\begin{bmatrix} 6 & -3 & 0 \\ -3 & 6 & -3 \\ 0 & -3 & 3 \end{bmatrix} \mathbf{v} + \begin{bmatrix} \frac{2}{9} & \frac{1}{18} & 0 \\ \frac{1}{18} & \frac{2}{9} & \frac{1}{18} \\ 0 & \frac{1}{18} & \frac{1}{9} \end{bmatrix} \frac{d^2\mathbf{v}}{d\tau^2} = \begin{bmatrix} 0 \\ 0 \\ 1 \end{bmatrix}$$

Determine the analytical solution. Also set up and solve the example in Section 4.3.2 for each of the following:

Case	Method	Interpolation	Elements	Mass	h
a	Analytical	Linear	3	L	
b	Analytical	Linear	3	W	
c	CD	Linear	3	C	0.02
d	CD	Linear	3	L	0.02
e	CD	Linear	3	W	0.02
f	NM	Linear	3	C	0.02
g	NM	Linear	3	L	0.02
h	NM	Linear	3	W	0.02
i	CD	Linear	3	C	0.20
j	CD	Linear	3	L	0.20
k	CD	Linear	3	W	0.20
l	NM	Linear	3	C	0.20
m	NM	Linear	3	L	0.20
n	NM	Linear	3	W	0.20

4.21. The transverse motion of a tightly stretched string supported on an elastic medium is given by the equation

$$T\frac{\partial^2 w}{\partial x^2} - kw - \mu\frac{\partial^2 w}{\partial t^2} + p(x,t) = 0$$

Boundary conditions appropriate for the figure are

$$w(0,t) = w_0(t)$$
$$w(L,t) = 0$$

with initial conditions given by

$$w(x,0) = f(x)$$
$$\frac{\partial w(0,t)}{\partial t} = g(x)$$

a. Develop the corresponding weak formulation for the initial-boundary value problem.

b. Determine the form of the elemental matrices that make up the assembled unconstrained equations

$$\mathbf{B}\ddot{\mathbf{w}} + \mathbf{A}\mathbf{w} = \mathbf{F}$$
$$\mathbf{w}(0) = \mathbf{w_0}$$
$$\mathbf{w}(0) = \mathbf{v_0}$$

4.22. With $k = 0$, no springs at the boundaries, and three linearly interpolated elements, consider the problem of the string held motionless with no initial displacement and then released. Take $w(L, t) = 0$ and $w(0, t) = w_0 \sin \omega t$ where $\omega = \sqrt{(T/\mu)}/L$. Take $\sqrt{(T/\mu)}\, t/L = \tau$ and $\mathbf{w}/w_0 = \mathbf{v}$, and show that the result is

$$\begin{bmatrix} 6 & -3 \\ -3 & 6 \end{bmatrix} \mathbf{v} + \begin{bmatrix} \frac{2}{9} & \frac{1}{18} \\ \frac{1}{18} & \frac{2}{9} \end{bmatrix} \frac{d^2\mathbf{v}}{d\tau^2} = \begin{bmatrix} f(\tau) \\ 0 \end{bmatrix}$$

where $f(\tau) = (55/18) \sin \tau$. Solve this set of equations analytically. Also determine the solutions for each of the following combinations:

Case	Method	Interpolation	Elements	Mass	h
a	Analytical	Linear	3	L	
b	Analytical	Linear	3	W	
c	CD	Linear	3	C	0.02
d	CD	Linear	3	L	0.02
e	CD	Linear	3	W	0.02
f	NM	Linear	3	C	0.02
g	NM	Linear	3	L	0.02
h	NM	Linear	3	W	0.02
i	CD	Linear	3	C	0.20
j	CD	Linear	3	L	0.20
k	CD	Linear	3	W	0.20
l	NM	Linear	3	C	0.20
m	NM	Linear	3	L	0.20
n	NM	Linear	3	W	0.20

Section 4.4

4.23. For the two-dimensional diffusion problem indicated

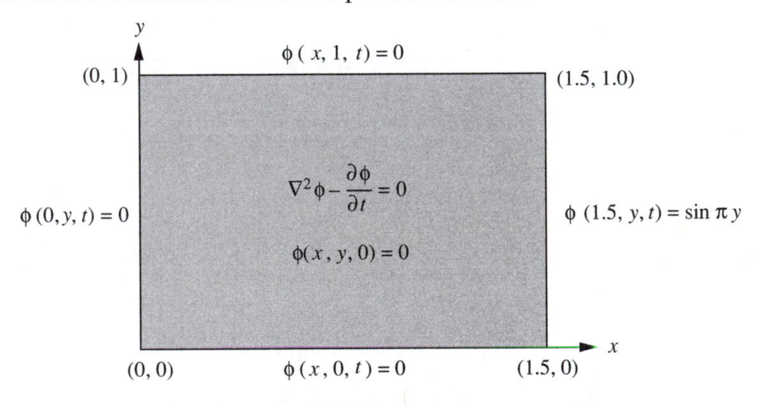

a. Set up the finite element model using six equal-size square elements as indicated. Use the results developed in Chapter 3 for the rectangular element.

b. Enforce the constraints, solve the equation analytically, and compare the results to the exact solution

$$\phi(x, y, t) = \left(\frac{8}{9\pi}\right)\sin \pi y \sum (-1)^{m+1}\left(\frac{1 - \exp\left(-\pi^2(1 + M^2)t\right)}{(1 + M^2)\pi^2}\right) \sin M\pi x$$

where $M = 2m/3$.

4.24. Repeat Exercise 4.23 using the lumped mass matrix

$$\mathbf{m}_{\mathrm{le}} = \frac{\rho A_e}{4}\begin{bmatrix} 1 & & & \\ & 1 & & \\ & & 1 & \\ & & & 1 \end{bmatrix}$$

4.25. Repeat Exercise 4.23 using the weighted mass matrix

$$\mathbf{m}_{\mathrm{we}} = \frac{\mathbf{m}_{\mathrm{ce}} + \mathbf{m}_{\mathrm{le}}}{2} = \frac{\rho A_e}{72}\begin{bmatrix} 13 & 2 & 1 & 2 \\ 2 & 13 & 2 & 1 \\ 1 & 2 & 13 & 2 \\ 2 & 1 & 2 & 13 \end{bmatrix}$$

4.26–4.28. Repeat the integration of each of 4.23–4.25 using the Euler algorithm with $h = h_{cr}/10$. What is h_{cr} in each case?

4.29–4.31. Repeat the integration of each of 4.23–4.25 using the Crank-Nicolson algorithm with $h = h_{cr}$. h_{cr} refers to the Euler algorithm.

Section 4.5

4.32. For the two-dimensional wave problem indicated

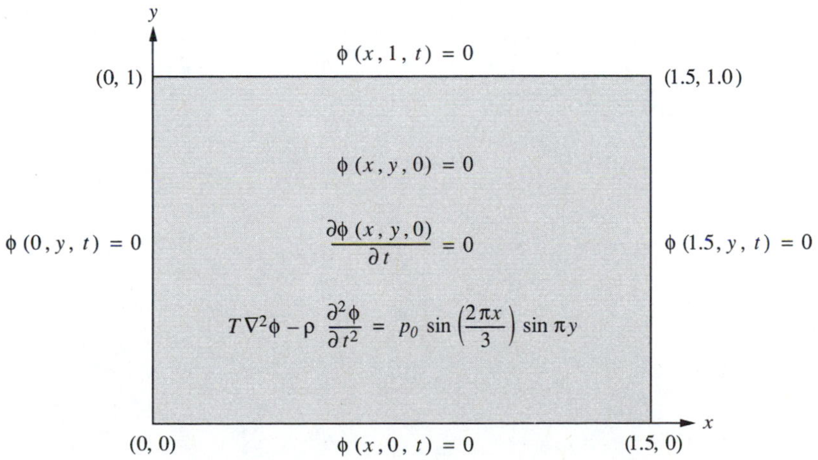

y

$\phi(x, 1, t) = 0$

(0, 1) (1.5, 1.0)

$\phi(x, y, 0) = 0$

$\phi(0, y, t) = 0$ $\quad \dfrac{\partial \phi(x, y, 0)}{\partial t} = 0$ $\quad \phi(1.5, y, t) = 0$

$$T\nabla^2\phi - \rho\frac{\partial^2 \phi}{\partial t^2} = p_0 \sin\left(\frac{2\pi x}{3}\right)\sin \pi y$$

x

(0, 0) $\quad \phi(x, 0, t) = 0$ $\quad$ (1.5, 0)

a. Set up the finite element model using six equal-size square elements as indicated. Use the results developed in Chapter 3 for the rectangular element.

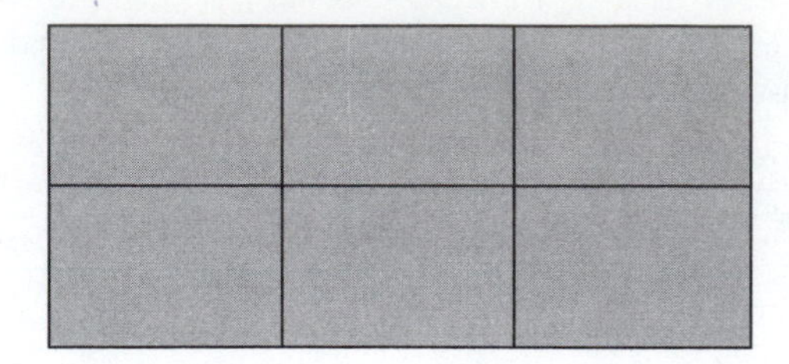

b. Enforce the constraints, solve the equation analytically, and compare the results to the exact solution

$$\phi(x, y, t) = \left(\frac{p_0}{\rho\omega^2}\right)(\cos\omega t - 1)\sin\left(\frac{2\pi x}{3}\right)\sin\pi y$$

where $\omega^2 = 13\pi^2 T/9\rho$.

4.33. Repeat Exercise 4.32 using the lumped mass matrix of Exercise 4.24.

4.34. Repeat Exercise 4.32 using the weighted mass matrix of Exercise 4.25.

4.35–4.37. Repeat the integration of each of 4.32–4.34 using the central difference algorithm with $h = h_{cr}/10$. What is h_{cr} in each case?

4.38–4.40. Repeat the integration of each of 4.32–4.34 using the Newmark algorithm with $h = h_{cr}/10$. h_{cr} refers to the central difference algorithm.

4.41–4.43. Repeat the integration of each of 4.32–4.34 using the Newmark algorithm with $h = h_{cr}$. h_{cr} refers to the central difference algorithm.

CHAPTER 5

Elasticity

Chapter Contents

5.1 Introduction
  5.1.1 Kinetics
  5.1.2 Kinematics
  5.1.3 Constitution
  5.1.4 Combination—boundary value problems
    5.1.4.1 Plane stress
    5.1.4.2 Plane strain
5.2 Weak formulation and variational principles
5.3 The Ritz finite element model
5.4 Evaluation of elemental matrices
5.5 Applications
  5.5.1 Uniform tension of a thin rectangular plate
  5.5.2 Thin rectangular plate—moment loading
  5.5.3 Thin rectangular plate—moment loading, meshes 2 and 3
  5.5.4 Stress concentration—plate with circular hole
5.6 Rectangular elements
  5.6.1 Thin rectangular plate—moment loading
  5.6.2 Thin rectangular plate—moment loading, meshes 2 and 3
5.7 Axisymmetric problems
  5.7.1 The Ritz finite element models—triangular elements
  5.7.2 Evaluation of elemental matrices
    5.7.2.1 Thick-walled pressure vessel
    5.7.2.2 Notched shaft under tension
5.8 Isoparametric formulations
  5.8.1 Applications
    5.8.1.1 Thin rectangular plate—moment loading
    5.8.1.2 Stress concentration—plate with circular hole
5.9 Closure
References
Computer projects
Exercises

5.1 INTRODUCTION

There are many important classes of physical problems to which the finite element method has been applied. One of the most important and successful of these is the area of solid mechanics, which encompasses a wide variety of different theories. In particular, there are many physical problems of interest where the linear, small displacement theory of elasticity serves as an excellent model for predicting the static and dynamic response of linearly elastic solids to mechanical and thermal inputs.

The governing equations of elasticity are generally vector equations with the displacements as dependent variables. Exact solutions are known for relatively few problems of practical interest. Finite element models for the general elasticity problem are easily constructed using the same basic ideas we have discussed in the previous chapters. The specific considerations necessary for treating the equations of elasticity are generally valid for other multivariable problems, that is, situations where there are coupled partial differential equations with more than one dependent variable. In the area of elasticity, these partial differential equations and accompanying boundary conditions are frequently expressed in terms of the displacements. For a two-dimensional theory the displacements can be taken as $u(x,y)$, $v(x,y)$, which are the displacements in the x and y directions, respectively.

The basic ideas associated with the development of any theory in solid mechanics are

1. Kinetics or balance of momentum.
2. Kinematics or strain-displacement.
3. Constitution or stress-strain-temperature.

Each of these ideas will be discussed briefly in connection with their use in developing the two-dimensional small displacement theory of elasticity. The specific theories, known as plane stress, plane strain, and the two-dimensional axisymmetric problem, will then be presented. Readers not familiar with the details of these theories should consult the references Boresi and Chong [1], Timoshenko and Goodier [2], and Love [3].

5.1.1 Kinetics

Kinetics is the basic ingredient enforcing the physical principles of (a) balance of force (equations of equilibrium) for a static situation, or (b) balance of momentum (equations of motion) for a dynamic problem. This amounts to making sure that any governing equations have embodied Newton's laws appropriately.

The equations to be used assume that the displacements, rotations, and strains are small. A two-dimensional Cartesian coordinate system is assumed. On the basis of a balance of linear momentum, the resulting equations of motion can be stated as

$$\frac{\partial \sigma_x}{\partial x} + \frac{\partial \tau_{xy}}{\partial y} + X = \rho \ddot{u}$$
$$\frac{\partial \tau_{yx}}{\partial x} + \frac{\partial \sigma_y}{\partial y} + Y = \rho \ddot{v} \tag{5.1}$$

where σ_x, σ_y, τ_{xy}, and τ_{yx} are stresses defined as shown in Fig. 5–1. The mass density is ρ, with $X(x,y,t)$ and $Y(x,y,t)$ the x and y components of body force. The x and y components of displacement are $u(x,y,t)$ and $v(x,y,t)$, with double dots denoting partial derivatives with respect to time. All of the variables are indicated in Fig. 5–1.

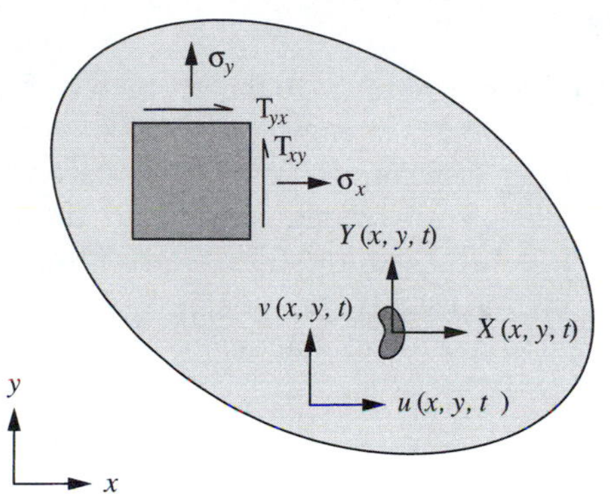

FIGURE 5–1 Stresses, displacements, and body forces

A balance of angular momentum produces the result

$$\tau_{xy} = \tau_{yx}$$

or in other words, the two-dimensional stress tensor

$$\boldsymbol{\tau} = \begin{bmatrix} \sigma_x & \tau_{xy} \\ \tau_{yx} & \sigma_y \end{bmatrix} = \begin{bmatrix} \sigma_x & \tau_{xy} \\ \tau_{xy} & \sigma_y \end{bmatrix}$$

is symmetric. The normal component of the stress tensor is the stress vector given by

$$\begin{aligned} \mathbf{t_n} = \mathbf{n} \cdot \boldsymbol{\tau} &= l(\mathbf{i}\sigma_x + \mathbf{j}\tau_{xy}) + m(\mathbf{i}\tau_{xy} + \mathbf{j}\sigma_y) \\ &= \mathbf{i}(l\sigma_x + m\tau_{xy}) + \mathbf{j}(l\tau_{xy} + m\sigma_y) \end{aligned}$$

and is frequently prescribed on a portion of the surface of the region as indicated in Fig. 5–2.

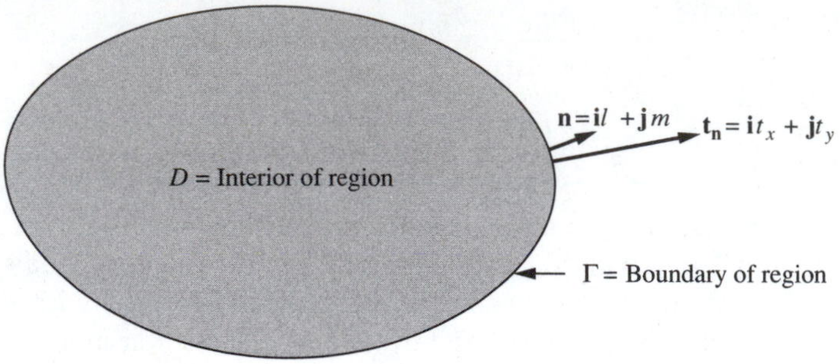

FIGURE 5–2 Prescribed tractions on the boundary

5.1.2 Kinematics

Kinematics or strain displacement refers to the geometry of deformation. Measures of changes in lengths and changes in angles, generally called strains, are important quantities that are used to assist in describing the deformation. An extensional strain is a ratio of a change in length, relative to the corresponding original length, as a result of a general deformation. Shear strains refer to distortion as a result of changes in angles usually between two originally perpendicular lines. For the two-dimensional linear theory, the appropriate linear strains are

$$\epsilon_x = \frac{\partial u}{\partial x} \qquad \epsilon_y = \frac{\partial v}{\partial y} \qquad \gamma_{xy} = \frac{\partial u}{\partial y} + \frac{\partial v}{\partial x} \tag{5.2}$$

as the extensional strains in the x and y directions and the shear strain, respectively. A quantity that is related to γ_{xy} is the tensor component ϵ_{xy}, defined as

$$\epsilon_{xy} = \frac{\gamma_{xy}}{2} = \frac{1}{2}\left(\frac{\partial u}{\partial y} + \frac{\partial v}{\partial x}\right)$$

The collection

$$\boldsymbol{\epsilon} = \begin{bmatrix} \epsilon_x & \epsilon_{xy} \\ \epsilon_{xy} & \epsilon_y \end{bmatrix}$$

is referred to as the linear strain tensor.

5.1.3 Constitution

If, in the five equations resulting from the discussion of Kinetics and Kinematics, the unknowns are listed, there results u, v, ϵ_x, ϵ_y, γ_{xy}, σ_x, σ_y, and τ_{xy}, that is, eight unknowns for only five equations. This would seem to indicate that three additional equations are required. These additional equations are usually referred to as **stress-strain equations** or **constitutive relations** in that they define the manner in which the material is constituted. The constitutive equations are necessary to specify, both qualitatively and quantitatively, the specific type of material being investigated. For the two-dimensional elasticity problem these additional equations relating the stresses to the strains and temperature changes T, assumed known, are taken to be

$$\begin{bmatrix} \sigma_x \\ \sigma_y \\ \tau_{xy} \end{bmatrix} = \begin{bmatrix} C_{11} & C_{12} & C_{13} \\ C_{12} & C_{22} & C_{23} \\ C_{13} & C_{23} & C_{33} \end{bmatrix} \begin{bmatrix} \epsilon_x \\ \epsilon_y \\ \gamma_{xy} \end{bmatrix} - \begin{bmatrix} \beta_x T \\ \beta_y T \\ 0 \end{bmatrix} \tag{5.3}$$

which can be referred to as a generalized Hooke's law. In matrix notation these relations can be written as

$$\boldsymbol{\sigma} = \mathbf{C}\boldsymbol{\epsilon} - \mathbf{f}_\mathbf{T}$$

In what follows we will deal exclusively with an orthotropic, two-dimensional elastic solid that is defined by taking $C_{13} = C_{23} = 0$. This results in the four independent elastic constants C_{11}, C_{12}, C_{22}, and C_{33}. β_x and β_y are thermal coefficients related to coefficients of thermal expansion through other elastic constants. These three additional equations provided by the stress-strain-temperature relations bring to eight the number of equations for the eight unknowns previously listed.

5.1.4 Combination—Boundary Value Problems

Eliminating the strains between Eqs. (5.2) and (5.3) and then substituting for the stresses in Eq. (5.1) results in

$$\frac{\partial\left(C_{11}\,\partial u/\partial x + C_{12}\,\partial v/\partial y\right)}{\partial x} + \frac{\partial\left(C_{33}(\partial u/\partial y + \partial v/\partial x)\right)}{\partial y} + X = \rho\ddot{u} + \frac{\partial(\beta_x T)}{\partial x}$$

$$\tag{5.4}$$

$$\frac{\partial\left(C_{33}\,(\partial u/\partial y + \partial v/\partial x)\right)}{\partial x} + \frac{\partial\left(C_{12}\,\partial u/\partial x + C_{22}\,\partial v/\partial y\right)}{\partial y} + Y = \rho\ddot{v} + \frac{\partial(\beta_y T)}{\partial y}$$

as the set of coupled partial differential equations that must be satisfied by the displacements $u(x,y,t)$ and $v(x,y,t)$. The solutions to these equations must also satisfy appropriate boundary conditions.

Figure 5–3 depicts a general two-dimensional, simply connected region in which the displacements and stresses are desired in response to the application of external body forces and surface tractions. The part of the boundary on which the displacements are prescribed is denoted as Γ_u and the part of the boundary on which the stresses are prescribed as Γ_t.

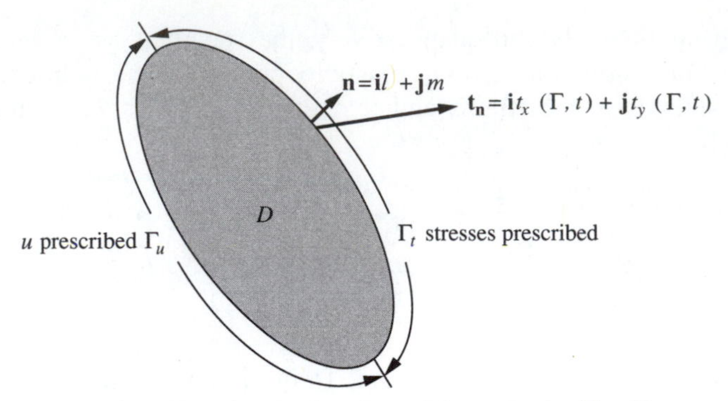

FIGURE 5–3 Region D and boundaries Γ_u, Γ_t

The formal boundary value problem that must be solved can be stated as: Determine displacements $u(x,y,t)$ and $v(x,y,t)$ that satisfy the partial differential equations

$$\frac{\partial\left(C_{11}\,\partial u/\partial x + C_{12}\,\partial v/\partial y\right)}{\partial x} + \frac{\partial\left(C_{33}\,(\partial u/\partial y + \partial v/\partial x)\right)}{\partial y} + X = \rho\ddot{u} + \frac{\partial(\beta_x T)}{\partial x} \quad \text{in } D$$

and

$$\frac{\partial\left(C_{33}(\partial u/\partial y + \partial v/\partial x)\right)}{\partial x} + \frac{\partial\left(C_{12}\,\partial u/\partial x + C_{22}\,\partial v/\partial y\right)}{\partial y} + Y = \rho\ddot{v} + \frac{\partial(\beta_y T)}{\partial y} \quad \text{in } D$$

in the interior of the region D and that also satisfy the displacement boundary conditions

$$\left.\begin{aligned} u &= u_0(\Gamma, t) \\ v &= v_0(\Gamma, t) \end{aligned}\right\} \quad \text{on } \Gamma_u$$

and the traction or stress type boundary conditions

$$\left.\begin{aligned} l\sigma_x + m\tau_{xy} &= t_x(\Gamma, t) \\ m\tau_{xy} + l\sigma_y &= t_y(\Gamma, t) \end{aligned}\right\} \quad \text{on } \Gamma_t$$

When expressed in terms of the displacements, the stress type or natural boundary conditions on Γ_t become

$$l\left(C_{11}\frac{\partial u}{\partial x} + C_{12}\frac{\partial v}{\partial y} - \beta T\right) + m\left(C_{33}\left(\frac{\partial u}{\partial y} + \frac{\partial v}{\partial x}\right)\right) = t_x(\Gamma, t) \qquad (5.5a)$$

on Γ_t

$$l\left(C_{33}\left(\frac{\partial u}{\partial y} + \frac{\partial v}{\partial x}\right)\right) + m\left(C_{12}\frac{\partial u}{\partial x} + C_{22}\frac{\partial v}{\partial y} - \beta T\right) = t_y(\Gamma, t) \qquad (5.5b)$$

Boundary conditions on Γ_u and Γ_t are essential and natural boundary conditions, respectively.

Another type of boundary condition that is sometimes prescribed is referred to as a **mixed boundary condition**. Such a mixed boundary condition is applied on a portion of the boundary Γ_m as indicated in Fig. 5–4.

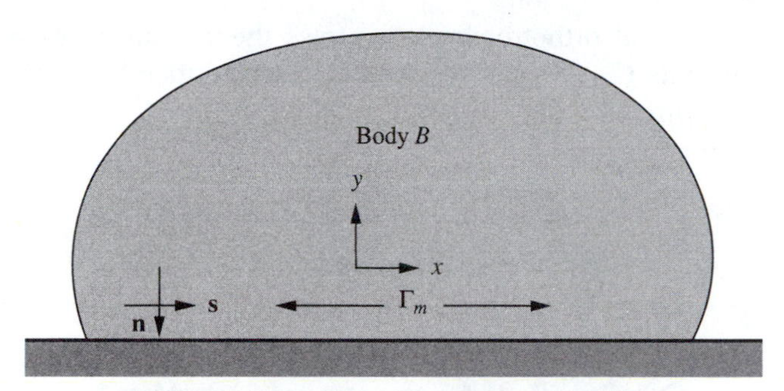

FIGURE 5–4 Mixed boundary conditions

The figure indicates a situation where the contact between the body B and the rigid base is assumed to be frictionless. In such a situation the normal component of the displacement and the tangential component of traction are prescribed. For the choice of coordinates indicated, the appropriate boundary conditions are

$$\left.\begin{aligned} \mathbf{u}\cdot\mathbf{n} &= v = 0 \\ \mathbf{t_n}\cdot\mathbf{s} &= \tau_{xy} = 0 \end{aligned}\right\} \quad \text{on } \Gamma_m$$

Mixed boundary conditions are also frequently used in situations where there are one or more planes of symmetry, in which case it can be argued that certain of the components of the displacement and stress must vanish. See Section 5.5.4 in this regard.

5.1.4.1 PLANE STRESS. A technically important special case of the general orthotropic solid described by Eq. (5.3) is known as **plane stress** or **generalized plane stress** [Boresi and Chong, 1]. This theory is appropriate for a flat thin plate loaded by surface tractions and body forces in the plane of the plate as indicated in Fig. 5–5. The elastic constants are obtained from a reduction of the three-dimensional *isotropic* case by assuming that the normal stress component $\sigma_z = 0$. The resulting elastic constant matrix, for the isotropic case, is expressed in terms of the two independent elastic constants E and ν, known as Young's modulus and Poisson's ratio, respectively. The matrix $\mathbf{C}$ is written as

$$\mathbf{C} = \frac{E}{1-\nu^2}\begin{bmatrix} 1 & \nu & 0 \\ \nu & 1 & 0 \\ 0 & 0 & \dfrac{1-\nu}{2} \end{bmatrix}$$

The thermal constants β_x and β_y are given by $E\alpha/(1-\nu)$, where α is the coefficient of thermal expansion.

Note that the general orthotropic plane stress theory can be used if the four independent constants C_{11}, C_{12}, C_{22}, and C_{33} and the thermal constants β_x and β_y have been determined.

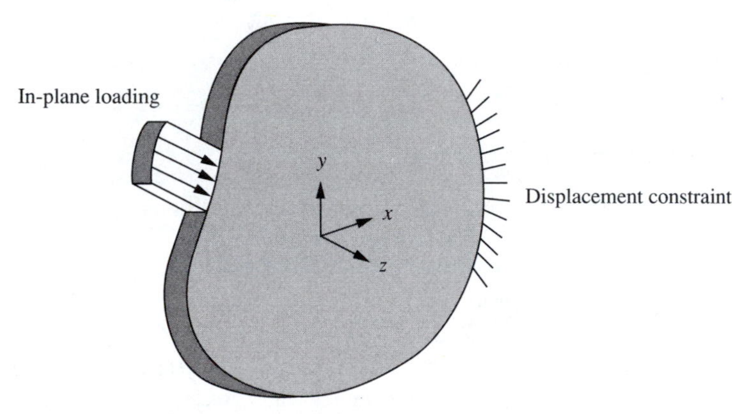

FIGURE 5–5 Region and loading for plane stress

5.1.4.2 PLANE STRAIN. A second technically important application of the two-dimensional theory of elasticity is termed **plane strain** [Boresi and Chong, 1]. As indicated in Fig. 5–6, the plane strain theory is appropriate in situations where ϵ_z (rather than σ_z) is assumed to be zero.

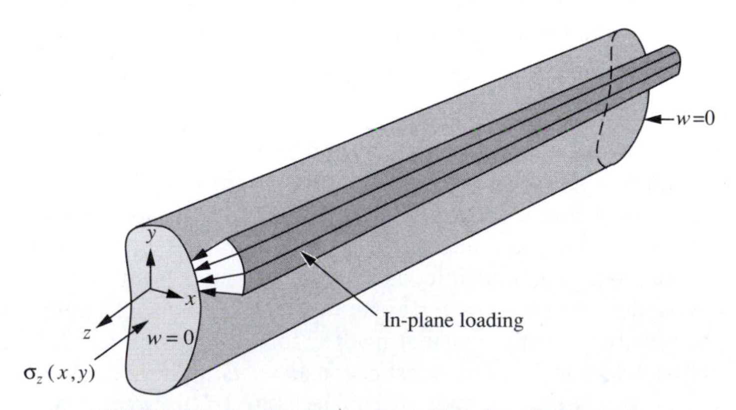

FIGURE 5–6 Region and loading for plane strain

As indicated in the figure, this state is usually obtained by enforcing $w = 0$ at the ends of the cylinder. Tractions and body forces applied along the length of the cylinder, as well as any boundary conditions that are applied, are independent of the z-coordinate. The three-dimensional isotropic case is again specialized by taking $\epsilon_z = 0$ to yield the elastic constant matrix. When expressed in terms of the two independent elastic constants E and ν, the matrix $\mathbf{C}$ is written as

$$\mathbf{C} = \frac{E(1-\nu)}{(1+\nu)(1-2\nu)} \begin{bmatrix} 1 & \dfrac{\nu}{(1-\nu)} & 0 \\ \dfrac{\nu}{(1-\nu)} & 1 & 0 \\ 0 & 0 & \dfrac{(1-2\nu)}{2(1-\nu)} \end{bmatrix}$$

The thermal constants β_x and β_y are given by $E\alpha/(1-2\nu)$ where α is the coefficient of thermal expansion. It is again possible to formulate the orthotropic two-dimensional plane strain problem if the appropriate elastic constants are known.

Note that the constraint $w = 0$ imposed at the ends of the cylinder is essentially provided by the resultant of the σ_z stresses given by

$$R_z = \iint_{\text{Area}} \sigma_z(x, y)\, dA = \iint_{\text{Area}} \nu(\sigma_x + \sigma_y)\, dA$$

5.2 WEAK FORMULATION AND VARIATIONAL PRINCIPLES

The development of the weak form for the two-dimensional elasticity problem requires a certain amount of insight in that there are now two equations involving two dependent variables, namely

$$\frac{\partial \sigma_x}{\partial x} + \frac{\partial \tau_{xy}}{\partial y} + X = \rho \ddot{u} \qquad \frac{\partial \tau_{xy}}{\partial x} + \frac{\partial \sigma_y}{\partial y} + Y = \rho \ddot{v}$$

It turns out that the appropriate initial step is to multiply the first and second equations by test functions $\phi(x, y)$ and $\psi(x, y)$ respectively, add the results, and integrate over the domain D. This results in

$$\iint_D \left(\phi \left\{ \frac{\partial \sigma_x}{\partial x} + \frac{\partial \tau_{xy}}{\partial y} + X - \rho \ddot{u} \right\} + \psi \left\{ \frac{\partial \tau_{xy}}{\partial x} + \frac{\partial \sigma_y}{\partial y} + Y - \rho \ddot{v} \right\} \right) d\,Vol = 0 \tag{5.6}$$

The test function ϕ is required to vanish at all points on the boundary where u is prescribed and, similarly, ψ is required to vanish at all points where v is prescribed. For situations where there are only boundary conditions of displacement and/or stress type (no mixed boundary conditions), this amounts to requiring that $\phi = \psi = 0$ on Γ_u.

The divergence theorem [Kaplan and Lewis, 4], which is the appropriate integration at this juncture, is used to integrate each of the terms involving derivatives of the stresses according to

$$\begin{aligned} \iint_D \frac{\partial A}{\partial x}\, d\,Vol &= \int_\Gamma lA\; d\Gamma \\ \iint_D \frac{\partial B}{\partial y}\, d\,Vol &= \int_\Gamma mB\; d\Gamma \end{aligned} \tag{5.7}$$

where (l, m) represents the exterior normal to D as indicated in Fig. 5–7.

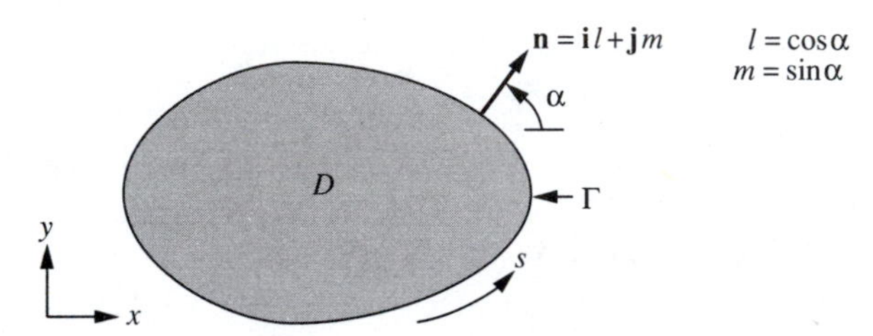

FIGURE 5–7 Domain *D*, boundary Γ, and normal *n*

For the two-dimensional case, $dVol = t\,dA$ and $d\Gamma = t\,ds$, where t is the thickness of the region and s measures arc length on the boundary as indicated.

In Eq. (5.6), the term $\phi\,\partial\sigma_x/\partial x$ is rewritten as

$$\phi\frac{\partial\sigma_x}{\partial x} = \frac{\partial(\phi\sigma_x)}{\partial x} - \sigma_x\frac{\partial\phi}{\partial x}$$

Using the first of the integral formulas (5.7) it follows that

$$\begin{aligned} \iint_D \phi\frac{\partial\sigma_x}{\partial x}\, d\,Vol &= \iint_D \left(\frac{\partial(\phi\sigma_x)}{\partial x} - \sigma_x\frac{\partial\phi}{\partial x}\right) d\,Vol \\ &= \int_\Gamma l\phi\sigma_x\, d\Gamma - \iint_D \sigma_x\frac{\partial\phi}{\partial x}\, d\,Vol \end{aligned}$$

Treating the remaining integrals involving the stress derivatives in similar fashion yields

$$\begin{aligned} &\int_\Gamma \big(\phi(l\sigma_x + m\tau_{xy}) + \psi(l\tau_{xy} + m\sigma_y)\big)\, d\Gamma \\ &\quad + \iint_D (\phi X + \psi Y)\, d\,Vol - \iint_D \rho(\phi\ddot{u} + \psi\ddot{v})\, d\,Vol \\ &\quad - \iint_D \left(\sigma_x\frac{\partial\phi}{\partial x} + \sigma_y\frac{\partial\psi}{\partial y} + \tau_{xy}\left(\frac{\partial\phi}{\partial y} + \frac{\partial\psi}{\partial x}\right)\right) d\,Vol = 0 \end{aligned}$$

Using the stress-strain and strain-displacement relations to eliminate the stresses in the last integral results in

$$\int\int_D \left(\frac{\partial\phi}{\partial x}\left\{ C_{11}\frac{\partial u}{\partial x} + C_{12}\frac{\partial v}{\partial y} \right\} + \frac{\partial\psi}{\partial y}\left\{ C_{12}\frac{\partial u}{\partial x} + C_{22}\frac{\partial v}{\partial y} \right\} \right.$$
$$+ \left\{ \frac{\partial\phi}{\partial y} + \frac{\partial\psi}{\partial x} \right\} C_{33}\left\{ \frac{\partial u}{\partial y} + \frac{\partial v}{\partial x} \right\} - T\left(\beta_x\frac{\partial\phi}{\partial x} + \beta_y\frac{\partial\psi}{\partial y} \right) - \phi X - \psi Y$$
$$\left. +\rho\{\phi\ddot{u} + \psi\ddot{v}\} \right) dVol - \int_\Gamma \big(\phi(l\sigma_x + m\tau_{xy}) + \psi(l\tau_{xy} + m\sigma_y)\big)\, d\Gamma = 0$$

Since $\Gamma = \Gamma_u + \Gamma_t$ with ϕ and ψ vanishing on Γ_u, the surface integral can be converted, using Eqs. (5.5) for the stress boundary conditions specified on Γ_t, to

$$-\int_{\Gamma_t} \big(\phi t_x(s) + \psi t_y(s)\big)\, d\Gamma$$

Finally then the weak form can be represented as

$$\int\int_D \left(\frac{\partial\phi}{\partial x}\left\{ C_{11}\frac{\partial u}{\partial x} + C_{12}\frac{\partial v}{\partial y} \right\} + \frac{\partial\psi}{\partial y}\left\{ C_{12}\frac{\partial u}{\partial x} + C_{22}\frac{\partial v}{\partial y} \right\} \right.$$
$$+ \left\{ \frac{\partial\phi}{\partial y} + \frac{\partial\psi}{\partial x} \right\} C_{33}\left\{ \frac{\partial u}{\partial y} + \frac{\partial v}{\partial x} \right\} - T\left(\beta_x\frac{\partial\phi}{\partial x} + \beta_y\frac{\partial\psi}{\partial y} \right) - \phi X - \psi Y$$
$$\left. +\rho\{\phi\ddot{u} + \psi\ddot{v}\} \right) dVol - \int_{\Gamma_t} \big(\phi t_x(s) + \psi t_y(s)\big)\, d\Gamma = 0 \qquad (5.8)$$

In what follows, the dynamic terms will be deleted and only static problems will be considered. Dynamic problems are treated in the Exercises.

In Eq. (5.8) the first terms in the integral involving the C_{ij} can be identified as a bilinear functional B written as

$$B(\phi, \psi : u, v) = \int\int_D \left(\frac{\partial\phi}{\partial x}\left\{ C_{11}\frac{\partial u}{\partial x} + C_{12}\frac{\partial v}{\partial y} \right\} + \frac{\partial\psi}{\partial y}\left\{ C_{21}\frac{\partial u}{\partial x} + C_{22}\frac{\partial v}{\partial y} \right\} \right.$$
$$\left. + \left\{ \frac{\partial\phi}{\partial y} + \frac{\partial\psi}{\partial x} \right\} C_{33}\left\{ \frac{\partial u}{\partial y} + \frac{\partial v}{\partial x} \right\} \right) dVol \qquad (5.9)$$

which can be written as

$$= \int\int_D \mathbf{\Phi}^{\mathrm{T}}\mathbf{C}\boldsymbol{\epsilon}\, dVol$$

where

$$\mathbf{\Phi}^{\mathrm{T}} = \left[\frac{\partial \phi}{\partial x} \quad \frac{\partial \psi}{\partial y} \quad \frac{\partial \psi}{\partial x} + \frac{\partial \phi}{\partial y}\right]$$

$$\boldsymbol{\epsilon}^{\mathrm{T}} = \left[\frac{\partial u}{\partial x} \quad \frac{\partial v}{\partial y} \quad \frac{\partial v}{\partial x} + \frac{\partial u}{\partial y}\right]$$

and $\mathbf{C}$ is the appropriate elastic constant matrix. The bilinear functional B is symmetric in the sense that

$$B(\phi, \psi : u, v) = B(u, v : \phi, \psi)$$

The remaining terms in Eq. (5.8) constitute the linear functional

$$l(\phi, \psi) = \int\int_D \left(T\left(\beta_x \frac{\partial \phi}{\partial x} + \beta_y \frac{\partial \psi}{\partial y}\right) + \phi X + \psi Y\right) d\,Vol + \int_{\Gamma_t} \left(\phi t_x(s) + \psi t_y(s)\right) d\Gamma \tag{5.10}$$

so that the weak form can be written as

$$B(\phi, \psi : u, v) = l(\phi, \psi) \tag{5.11}$$

This symmetric weak form can be used directly in connection with the Galerkin method to develop the finite element model for the linear two-dimensional orthotropic elasticity problem. Alternately, in view of the fact that the bilinear functional B is symmetric, the corresponding quadratic functional is given by

$$\begin{aligned} I(u, v) &= \left(\frac{1}{2}\right) B(u, v : u, v) - l(u, v) \\ &= \int\int_D \left(\left[\frac{\partial u}{\partial x}\left\{C_{11}\frac{\partial u}{\partial x} + C_{12}\frac{\partial v}{\partial y}\right\} + \frac{\partial v}{\partial y}\left\{C_{21}\frac{\partial u}{\partial x} + C_{22}\frac{\partial v}{\partial y}\right\}\right.\right. \\ &\quad \left.\left. + \left\{\frac{\partial u}{\partial y} + \frac{\partial v}{\partial x}\right\} C_{33} \left\{\frac{\partial u}{\partial y} + \frac{\partial v}{\partial x}\right\}\right]\left(\frac{1}{2}\right) - T\left(\beta_x \frac{\partial u}{\partial x} + \beta_y \frac{\partial v}{\partial y}\right) - uX - vY\right) d\,Vol \\ &\quad - \int_{\Gamma_t} \left(u t_x(s) + v t_y(s)\right) d\Gamma \\ &= \int\int_D \left(\frac{\boldsymbol{\epsilon}^{\mathrm{T}}\mathbf{C}\boldsymbol{\epsilon}}{2} - \boldsymbol{\epsilon}^{\mathrm{T}}\mathbf{f_T}\right) d\,Vol - \int\int_D \mathbf{u}\cdot\mathbf{f}\; d\,Vol - \int_{\Gamma_t} \mathbf{u}\cdot\mathbf{t_n}\; d\Gamma \end{aligned} \tag{5.12}$$

where $\mathbf{f} = [\mathrm{X}\ \mathrm{Y}]^{\mathrm{T}}$ is the body force vector. The first two terms represent the strain energy, with the third and fourth terms representing the potential energy of the body forces and surface tractions respectively. The functional $I(u,v)$ is in fact the total potential energy of the system.

Completing the circle by requiring the functional $I(u,v)$ to be stationary, namely,

$$\delta I(u, v) = 0$$

with $u = u_0(s)$ and $v = v_0(s)$ on Γ_u leads back to the equilibrium equations expressed in terms of the displacements, the essential boundary conditions on Γ_u and the natural boundary conditions on Γ_t, that is, the original statement of the nontime-dependent boundary value problem. Demonstration of this fact is left to the Exercises.

Note that by considering ϕ and ψ respectively to be δu and δv, the weak form of the problem is the principle of virtual work. The stationary value implied by $\delta I = 0$ is the principle of stationary potential energy.

A finite element model can be generated either by using the Galerkin method as applied to the weak form or by applying the method of Ritz directly to the potential energy functional. The development of the finite element model using the Ritz method will be presented in the next section.

5.3 THE RITZ FINITE ELEMENT MODEL

As developed in Section 5.2, the total potential energy can be written as

$$V = U + \Omega$$

where U and Ω are the internal and external potential energies respectively. The internal or strain energy can be written as

$$U = \int\int_D U_0 \, d\,Vol$$

where U_0 is the strain energy density [Boresi and Chong, 1] given by

$$U_0 = \left(\frac{1}{2}\right)\boldsymbol{\epsilon}^{\mathrm{T}}\mathbf{C}\boldsymbol{\epsilon} - \boldsymbol{\epsilon}^{\mathrm{T}}\mathbf{f}_{\mathrm{T}}$$

Thus

$$U = \int\int_D \left[\frac{\boldsymbol{\epsilon}^{\mathrm{T}}\mathbf{C}\boldsymbol{\epsilon}}{2} - \boldsymbol{\epsilon}^{\mathrm{T}}\mathbf{f}_{\mathrm{T}}\right] d\,Vol$$

The external potential energy consists of the potential energy of the action at a distance force $\mathbf{f} = \mathbf{i}X + \mathbf{j}Y$ given by

$$\Omega_f = -\int\int_D \mathbf{f} \cdot \mathbf{u} \, d\,Vol$$

and of the surface stresses or tractions given by

$$\Omega_t = -\int_{\Gamma_t} \mathbf{t} \cdot \mathbf{u} \, dS$$

the integral being evaluated over the Γ_t portion of the boundary on which the stresses are prescribed. The finite element model for the two-dimensional elasticity problem will be developed, following in order, the seven basic steps.

Discretization and interpolation. Nodes and elements are typically selected as indicated in Fig. 5–8, where a rectangular cantilever beam is loaded by a shear load P. Figure 5–8*a* indicates a typical mesh where linearly interpolated triangular elements have been selected; Fig. 5–8*b* depicts the same region using quadrilateral elements for the mesh.

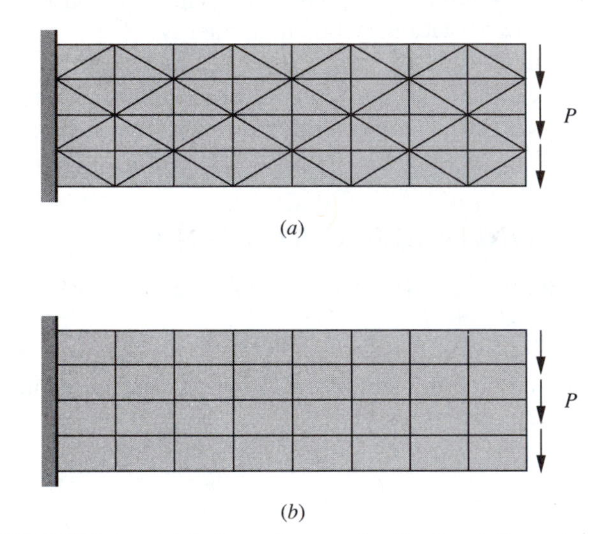

FIGURE 5–8 (*a*) Triangular mesh, (*b*) Quadrilateral mesh

For either of the two meshes indicated, each node in the mesh has two degrees of freedom, namely the components of displacement u and v. Assuming that the global coordinates are x and y in the horizontal and vertical directions respectively, the degrees of freedom consist of the u_i, v_i pairs, $i = 1, 2, \ldots, N$, where N is the number of nodes in the mesh. Thus there are $2N$ degrees of freedom for an N-node mesh.

In what follows, the linearly interpolated triangular element will be used for demonstrating the basic steps involved in generating the finite element model for the two-dimensional elasticity problem.

Consider then a typical triangular element as indicated in Fig. 5–9. The x and y components of the displacements within an element are represented as linear functions according to

$$u_e(x, y) = a_e + b_e x + c_e y$$

and

$$v_e(x, y) = d_e + f_e x + g_e y$$

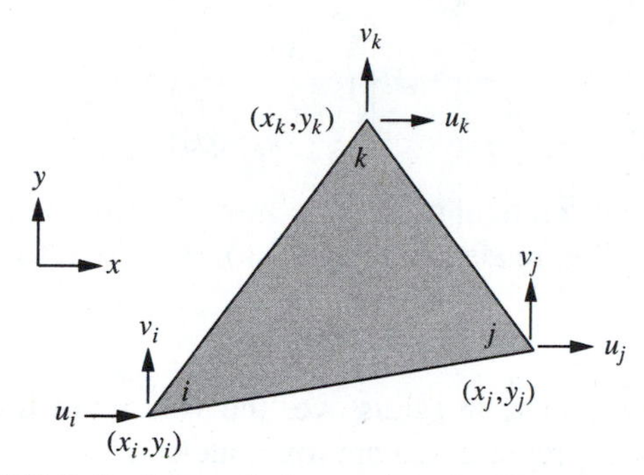

FIGURE 5–9 Linearly interpolated triangular element

The three equations for determining the coefficients a_e, b_e, and c_e are then obtained by requiring that

$$u_e(x_i, y_i) = u_i \qquad u_e(x_j, y_j) = u_j \qquad u_e(x_k, y_k) = u_k$$

Similarly, d_e, f_e, and g_e are determined from

$$v_e(x_i, y_i) = v_i \qquad v_e(x_j, y_j) = v_j \qquad v_e(x_k, y_k) = v_k$$

These three by three sets are easily solved to yield the desired coefficients.

The results can be put in the form

$$u_e(x, y) = \mathbf{N}^{\mathrm{T}}\mathbf{u_e} = \mathbf{u_e^{\mathrm{T}}}\mathbf{N} \qquad v_e(x, y) = \mathbf{N}^{\mathrm{T}}\mathbf{v_e} = \mathbf{v_e^{\mathrm{T}}}\mathbf{N} \tag{5.13}$$

where

$$\mathbf{u_e^{\mathrm{T}}} = [u_i \quad u_j \quad u_k] \qquad \mathbf{v_e^{\mathrm{T}}} = [v_i \quad v_j \quad v_k]$$

and

$$\mathbf{N}^{\mathrm{T}} = [N_i \quad N_j \quad N_k]$$

The interpolation functions N_i, N_j, N_k are exactly the same as those developed in Chapter 3, namely

$$N_i = \frac{(\alpha_i + \beta_i x + \gamma_i y)}{2A_e} \qquad i = 1, 2, 3$$

with

$$\alpha_i = x_j y_k - x_k y_j \qquad \beta_i = y_j - y_k \qquad \gamma_i = x_k - x_j$$

where in computing the α_i, β_i and γ_i, i, j, and k are to be permuted cyclically. Twice the area of the element is given by the determinant

$$2A_e = \begin{vmatrix} 1 & x_i & y_i \\ 1 & x_j & y_j \\ 1 & x_k & y_k \end{vmatrix}$$

The nodes defining any particular element must be numbered counterclockwise around the element. The reader is referred to Fig. 3–5 for sketches depicting the N_i.

Elemental formulation. The integrals over the volume of the body and over the bounding surface can be represented approximately by

$$V \approx \sum_e \int\int_{D_e} \left(\frac{\boldsymbol{\epsilon}^{\mathrm{T}} \mathbf{C} \boldsymbol{\epsilon}}{2} - \boldsymbol{\epsilon}^{\mathrm{T}} \mathbf{f}_{\mathrm{T}} \right) d\,Vol - \sum_e \int\int_{D_e} \mathbf{f} \cdot \mathbf{u} \, d\,Vol \; - \sum_e{}' \int_{\gamma_{te}} \mathbf{t}_n \cdot \mathbf{u} \, dS \tag{5.14}$$

where D_e refers to the interior of a typical element and γ_{te} to a surface element on the Γ_t portion of the boundary. The approximation arises from the fact that the straight-sided triangular elements cannot represent a curved boundary exactly.

Represent Eq. (5.14) as

$$V \approx \sum_e I_{1e} - \sum_e I_{2e} - \sum_e{}' I_{3e}$$

and consider in turn the evaluation of each of these three contributions to the total potential energy.

Evaluation of I_{1e}. In order to evaluate a typical I_{1e}, the expressions for the strains must be developed. Recall that the strain vector $\boldsymbol{\epsilon}$ is defined by

$$\boldsymbol{\epsilon}^{\mathrm{T}} = [\epsilon_x \quad \epsilon_y \quad \gamma_{xy}] = \left[\frac{\partial u}{\partial x} \quad \frac{\partial v}{\partial y} \quad \frac{\partial u}{\partial y} + \frac{\partial v}{\partial x} \right]$$

Using the representation for the displacements within the element, this can be expressed as

$$\boldsymbol{\epsilon}^{\mathrm{T}} = \left[\mathbf{u}_{\mathbf{e}}^{\mathrm{T}} \frac{\partial \mathbf{N}}{\partial x} \quad \mathbf{v}_{\mathbf{e}}^{\mathrm{T}} \frac{\partial \mathbf{N}}{\partial y} \quad \mathbf{u}_{\mathbf{e}}^{\mathrm{T}} \frac{\partial \mathbf{N}}{\partial y} + \mathbf{v}_{\mathbf{e}}^{\mathrm{T}} \frac{\partial \mathbf{N}}{\partial x} \right]$$

This can be written as

$$\boldsymbol{\epsilon} = \mathbf{B}\mathbf{U}_{\mathbf{e}} \tag{5.15}$$

where

$$\mathbf{U}_{\mathbf{e}}^{\mathrm{T}} = [u_1 \quad v_1 \quad u_2 \quad v_2 \quad u_3 \quad v_3]$$

is the elemental displacement vector, and

$$\mathbf{B} = \begin{bmatrix} \dfrac{\partial N_1}{\partial x} & 0 & \dfrac{\partial N_2}{\partial x} & 0 & \dfrac{\partial N_3}{\partial x} & 0 \\ 0 & \dfrac{\partial N_1}{\partial y} & 0 & \dfrac{\partial N_2}{\partial y} & 0 & \dfrac{\partial N_3}{\partial y} \\ \dfrac{\partial N_1}{\partial y} & \dfrac{\partial N_1}{\partial x} & \dfrac{\partial N_2}{\partial y} & \dfrac{\partial N_2}{\partial x} & \dfrac{\partial N_3}{\partial y} & \dfrac{\partial N_3}{\partial x} \end{bmatrix} \tag{5.16}$$

is referred to as the **strain matrix**. Note that in each element $\mathbf{B}$ is a constant, so that $\boldsymbol{\epsilon}$ and consequently the computed part of $\boldsymbol{\sigma}$, if $\mathbf{C}$ is a matrix of constants, is constant within an element. For this reason, this element is referred to as the *constant strain* or *constant stress* element.

Inserting the expressions for $\boldsymbol{\epsilon}$ and $\boldsymbol{\epsilon}^{\mathrm{T}}$ into the general expression for I_{1e} yields

$$\begin{aligned} I_{1e} &= \int\int_{D_e} \left(\frac{\boldsymbol{\epsilon}^{\mathrm{T}}\mathbf{C}\boldsymbol{\epsilon}}{2} - \boldsymbol{\epsilon}^{\mathrm{T}}\mathbf{f_T} \right) d\,Vol \\ &= \int\int_{D_e} \left(\frac{\mathbf{U_e^{\mathrm{T}}B^{\mathrm{T}}CBU_e}}{2} - \mathbf{U_e^{\mathrm{T}}B^{\mathrm{T}}f_T} \right) d\,Vol \\ &= \left(\frac{1}{2}\right)\mathbf{U_e^{\mathrm{T}}}\left(\int\int_{D_e} \mathbf{B^{\mathrm{T}}CB}\; d\,Vol\right)\mathbf{U_e} - \mathbf{U_e^{\mathrm{T}}}\int\int_{D_e} \mathbf{B^{\mathrm{T}}f_T}\; d\,Vol \\ &= \left(\frac{1}{2}\right)\mathbf{U_e^{\mathrm{T}}k_eU_e} - \mathbf{U_e^{\mathrm{T}}}\boldsymbol{\theta}_\mathbf{e} \end{aligned}$$

where

$$\mathbf{k_e} = \int\int_{D_e} \mathbf{B^{\mathrm{T}}CB}\; d\,Vol \tag{5.17}$$

and

$$\boldsymbol{\theta}_\mathbf{e} = \int\int_{D_e} \mathbf{B^{\mathrm{T}}f_T}\; d\,Vol \tag{5.18}$$

The size of the elemental stiffness matrix $\mathbf{k_e}$ is 6×6, and the size of the elemental load vector $\boldsymbol{\theta}_\mathbf{e}$ is 6×1. The internal potential energy can then be represented as

$$U = \left(\frac{1}{2}\right)\sum_e \mathbf{U_e^{\mathrm{T}}k_eU_e} - \sum_e \mathbf{U_e^{\mathrm{T}}}\boldsymbol{\theta}_\mathbf{e}$$

Evaluation of I_{2e}. A typical I_{2e} can be written as

$$I_{2e} = \iint_{D_e} \mathbf{f} \cdot \mathbf{u_e} \, d\,Vol = \iint_{D_e} (Xu_e + Yv_e) \, d\,Vol$$

$$= \iint_{D_e} (X\mathbf{u_e^T N} + Y\mathbf{v_e^T N}) \, d\,Vol$$

Expressed in terms of the elemental displacement vector $\mathbf{U_e}$, this can be written as

$$I_{2e} = \mathbf{U_e^T f_e} \tag{5.19}$$

where

$$\mathbf{f_e^T} = \iint_{D_e} [XN_1 \quad YN_1 \quad XN_2 \quad YN_2 \quad XN_3 \quad YN_3] \, d\,Vol \tag{5.20}$$

is the 6×1 load vector arising from the body forces.

Evaluation of I_{3e}. In evaluating the contribution to the potential energy from the stresses or tractions on the surface of the region, it is instructive to draw a typical γ_{te} portion of the boundary as shown in Fig. 5–10.

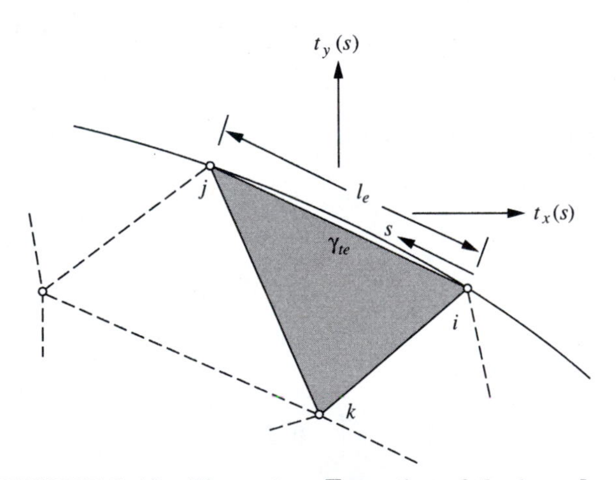

FIGURE 5–10 Element on Γ_t portion of the boundary

On a typical surface element of the boundary

$$\mathbf{t_n} \cdot \mathbf{u} = ut_x + vt_y$$

where t_x and t_y are the x and y components of the stress vector on the boundary of the region. It is often preferable to express this in terms of the normal and tangential components t_n and t_s, respectively, of the stress vector according to Fig. 5–11, where l and m are the direction cosines of the outward normal to the region.

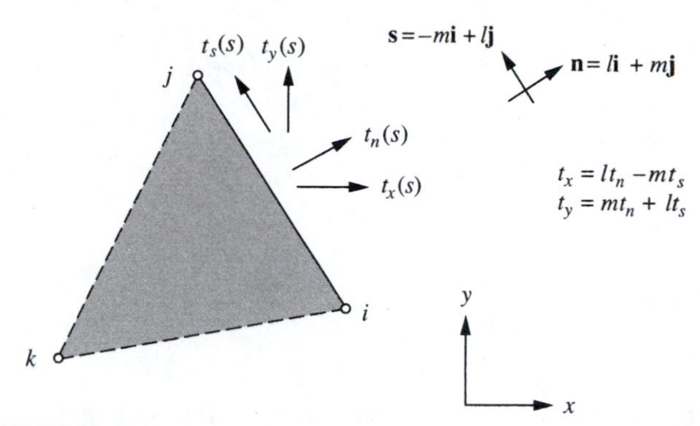

FIGURE 5–11 ***x*-*y* and *n*-*s* components of surface traction**

The integrand can then be expressed as

$$\mathbf{t_n} \cdot \mathbf{u} = ut_x + vt_y = u(lt_n - mt_s) + v(mt_n + lt_s)$$

which upon using the representation for the displacements within the element, given by Eqs. (5.13), becomes

$$\mathbf{t_n} \cdot \mathbf{u} = \mathbf{u_e^T N}(lt_n - mt_s) + \mathbf{v_e^T N}(mt_n + lt_s)$$

Denoting the two nodes on the boundary as i and j, as in Fig. 5–11, we can further simplify the expression for $\mathbf{t_n} \cdot \mathbf{u}$ to

$$\mathbf{t_n} \cdot \mathbf{u} = [u_i \quad v_i \quad u_j \quad v_j] \begin{bmatrix} N_i(lt_n - mt_s) \\ N_i(mt_n + lt_s) \\ N_j(lt_n - mt_s) \\ N_j(mt_n + lt_s) \end{bmatrix}$$

where as one would expect, only the boundary nodes of the element in question are involved. Finally then, the elemental contribution from a typical element on the boundary is

$$I_{3e} = \int_{\gamma_{te}} \mathbf{t_n} \cdot \mathbf{u}\, dS = [\, u_i \quad v_i \quad u_j \quad v_j \,]\mathbf{t_e}$$

where

$$\mathbf{t_e} = \begin{bmatrix} \int_{\gamma_{te}} N_i(lt_n - mt_s)dS \\ \int_{\gamma_{te}} N_i(mt_n + lt_s)dS \\ \int_{\gamma_{te}} N_j(lt_n - mt_s)dS \\ \int_{\gamma_{te}} N_j(mt_n + lt_s)dS \end{bmatrix} \tag{5.21}$$

Collecting all the contributions, the expression for the total potential energy can be written as

$$V = \left(\frac{1}{2}\right)\sum_e \mathbf{U_e^T k_e U_e} - \sum_e \mathbf{U_e^T[f_e + \theta_e]} - \sum_e{}' \mathbf{U_e^T t_e} \tag{5.22}$$

where the $\sum'$ indicates a sum over the elements on the Γ_t part of the boundary.

Assembly. Assembly is carried out to yield

$$V = \left(\frac{1}{2}\right)\mathbf{U_G^T K_G U_G} - \mathbf{U_G^T F_G} \tag{5.23}$$

where

$$\mathbf{K_G} = \sum_e \mathbf{k_G}$$

$$\mathbf{F_G} = \sum_e [\mathbf{f_G} + \boldsymbol{\theta}_\mathbf{G}] + \sum_e{}' \mathbf{t_G}$$

The sums again indicate the expansion of the elemental matrices to the global level.

Through the basic steps of discretization, interpolation, and elemental formulation, the total potential energy *functional* given by Eq. (5.12) has been converted into the approximate potential energy *function* given in Eq. (5.23). The independent variables for the potential energy function consist of the entire collection of (u,v) pairs of nodal displacements that are represented by the global displacement vector $\mathbf{U_G}$. Requiring the potential energy function to be stationary yields the equilibrium equations

$$\frac{\partial V}{\partial u_i} = 0 \qquad \frac{\partial V}{\partial v_i} = 0 \qquad i = 1, 2, \ldots, N$$

or collectively as

$$\frac{\partial V}{\partial \mathbf{U_G}} = 0$$

Specifically,

$$\frac{\partial V}{\partial \mathbf{U_G}} = \frac{1}{2}(\mathbf{K_G} + \mathbf{K_G^T})\mathbf{U_G} - \mathbf{F_G} = \mathbf{0}$$

or since $\mathbf{K_G}$ is symmetric,

$$\mathbf{K_G U_G} = \mathbf{F_G}$$

Constraints. In the usual manner, constraints are enforced on the assembled equations according to the specification of the displacement conditions on the Γ_u portion of the boundary.

Solution. The constrained equations are solved for the unknown displacements.

Computation of stresses. Recall from Eq. (5.3) that the stresses, strains, and temperatures are related by

$$\boldsymbol{\sigma} = \mathbf{C}\boldsymbol{\epsilon} - \mathbf{f_T} = \mathbf{CBU_e} - \mathbf{f_T}$$

The stresses can be determined per element by using the appropriate subset $\mathbf{U_e}$ of the global displacement vector, and the temperatures appropriate for the element. In the event that the temperature varies within the element, it is common to take the average of the nodal temperatures in computing the stresses. Again note that the stresses so computed, that is

$$\boldsymbol{\sigma} = \mathbf{CBU_e} - \begin{bmatrix} (\beta T)_{\text{AVG}} \\ (\beta T)_{\text{AVG}} \\ 0 \end{bmatrix}$$

are constant within a given element. Note that in the above equation $\beta_x = \beta_y = \beta$.

5.4 EVALUATION OF ELEMENTAL MATRICES

In this section the evaluation of the elemental matrices developed in Section 5.3 will be carried out. In particular, exact and approximate schemes for evaluating the elemental matrices $\mathbf{k_e}$, $\boldsymbol{\theta}_\mathbf{e}$, $\mathbf{f_e}$, and $\mathbf{t_e}$ will be discussed.

Evaluation of $\mathbf{k_e}$. Recall that the strain matrix was developed as

$$\mathbf{B} = \begin{bmatrix} \frac{\partial N_1}{\partial x} & 0 & \frac{\partial N_1}{\partial x} & 0 & \frac{\partial N_1}{\partial x} & 0 \\ 0 & \frac{\partial N_1}{\partial y} & 0 & \frac{\partial N_1}{\partial y} & 0 & \frac{\partial N_1}{\partial y} \\ \frac{\partial N_1}{\partial y} & \frac{\partial N_1}{\partial x} & \frac{\partial N_1}{\partial y} & \frac{\partial N_1}{\partial x} & \frac{\partial N_1}{\partial y} & \frac{\partial N_1}{\partial x} \end{bmatrix}$$

Using Eqs. (5.13) for the interpolation functions given in the previous section, this can be simplified to

$$\mathbf{B} = \frac{1}{2A_e} \begin{bmatrix} b_1 & 0 & b_2 & 0 & b_3 & 0 \\ 0 & c_1 & 0 & c_2 & 0 & c_3 \\ c_1 & b_1 & c_2 & b_2 & c_3 & b_3 \end{bmatrix}$$

where $b_i = y_j - y_k$, $c_i = x_k - x_j$, and i, j, and k are to be permuted cyclically. Assuming that $\mathbf{C}$ is a matrix of constants within the element in question, the fact that $\mathbf{B}$ is also a matrix of constants allows us to write

$$\begin{aligned} \mathbf{k_e} &= \int_{D_e} \mathbf{B}^{\mathrm{T}}\mathbf{C}\mathbf{B}\, d\,Vol \\ &= \mathbf{B}^{\mathrm{T}}\mathbf{C}\mathbf{B}\, Vol_e = \mathbf{B}^{\mathrm{T}}\mathbf{C}\mathbf{B}(tA_e) \end{aligned}$$

where tA_e is the volume of the element. This can be expanded to yield the general result

$$\mathbf{k_e} = \frac{t}{4A_e} \begin{bmatrix} \mathbf{A_{11}} & \mathbf{A_{12}} \\ \text{symm} & \mathbf{A_{22}} \end{bmatrix} \tag{5.24}$$

where

$$\mathbf{A_{11}} = \begin{bmatrix} C_{11}b_1^2 + C_{33}c_1^2 & (C_{12} + C_{33})b_1c_1 & C_{11}b_1b_2 + C_{33}c_1c_2 \\ & C_{22}c_1^2 + C_{33}b_1^2 & C_{12}b_2c_1 + C_{33}b_1c_2 \\ \text{symm} & & C_{11}b_2^2 + C_{33}c_2^2 \end{bmatrix}$$

$$\mathbf{A_{12}} = \begin{bmatrix} C_{12}b_1c_2 + C_{33}b_2c_1 & C_{11}b_1b_3 + C_{33}c_1c_3 & C_{12}b_1c_3 + C_{33}b_3c_1 \\ C_{22}c_1c_2 + C_{33}b_1b_2 & C_{12}b_3c_1 + C_{33}b_1c_3 & C_{22}c_1c_3 + C_{33}b_1b_3 \\ (C_{12} + C_{33})b_2c_2 & C_{11}b_2b_3 + C_{33}c_2c_3 & C_{12}b_2c_3 + C_{33}b_3c_2 \end{bmatrix}$$

$$\mathbf{A_{22}} = \begin{bmatrix} C_{22}c_2^2 + C_{33}b_2^2 & C_{12}b_3c_2 + C_{33}b_2c_3 & C_{22}c_2c_3 + C_{33}b_2b_3 \\ & C_{11}b_3^2 + C_{33}c_3^2 & (C_{12} + C_{33})b_3c_3 \\ \text{symm} & & C_{33}b_3^2 + C_{22}c_3^2 \end{bmatrix}$$

This general expression for $\mathbf{k_e}$ can be used for any of the four combinations of orthotropy or isotropy with plane stress or plane strain.

Evaluation of $\boldsymbol{\theta}_\mathbf{e}$. The general expression for the elemental matrix $\boldsymbol{\theta}_\mathbf{e}$ is

$$\boldsymbol{\theta}_\mathbf{e} = \int_{D_e} \mathbf{B}^\mathrm{T}\mathbf{f_T}\, d\,Vol$$

which with $\mathbf{f_T}$ as

$$\mathbf{f}_\mathbf{T}^\mathrm{T} = [\beta T \quad \beta T \quad 0]$$

and $\mathbf{B}$ as given above yields

$$\boldsymbol{\theta}_\mathbf{e}^\mathrm{T} = \int_{D_e} \frac{[b_1\ c_1\ b_2\ c_2\ b_3\ c_3]\beta T}{2A_e}\, d\,Vol$$

This is the general expression for $\boldsymbol{\theta}_\mathbf{e}^\mathrm{T}$. If βT is a constant, say $(\beta T)_0$, the entire integrand is constant and it follows immediately that

$$\boldsymbol{\theta}_\mathbf{e}^\mathrm{T} = \frac{[b_1\ c_1\ b_2\ c_2\ b_3\ c_3](\beta T)_0 t}{2} \tag{5.25}$$

In the event that βT is not constant, but is nevertheless a reasonably smooth function of position, the integral can be evaluated with sufficient accuracy by *approximating* the actual function by its linear interpolation according to

$$\beta T = (\beta T)_i N_i + (\beta T)_j N_j + (\beta T)_k N_k$$

Recalling from Section 3.4 that

$$\int_{A_e} N_i\, dA = \int_{A_e} N_j\, dA = \int_{A_e} N_k\, dA = \frac{A_e}{3}$$

it follows that

$$\boldsymbol{\theta}_\mathbf{e}^\mathrm{T} = \frac{[b_1 \quad c_1 \quad b_2 \quad c_2 \quad b_3 \quad c_3]\big((\beta T)_{\mathrm{AVG}}\big)t}{2} \tag{5.26}$$

where $(\beta T)_{\mathrm{AVG}} = \big((\beta T)_i + (\beta T)_j + (\beta T)_k\big)/3$. If more accuracy is required in evaluating the integrals, the integration formulas

$$\int_{A_e} N_I^a N_J^b N_K^c\, dA = \frac{a!b!c!2A_e}{(a+b+c+2)!} \tag{5.27}$$

or one of the available quadratures can be used. The nodal forces corresponding to Eq. (5.26) are indicated in Fig. 5–12.

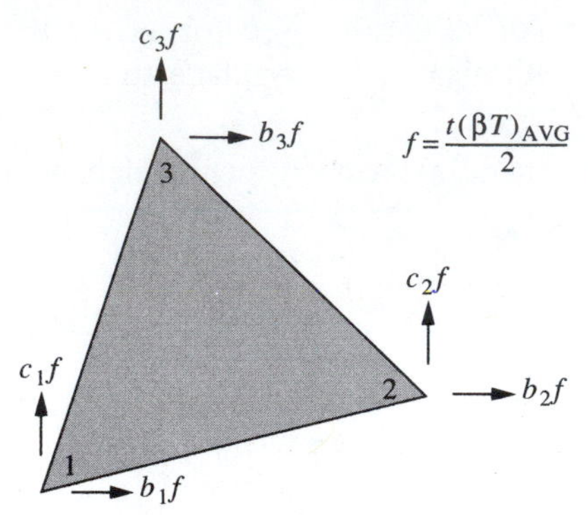

FIGURE 5–12 Thermal nodal loads

Evaluation of $\mathbf{f_e}$. From Eq. (5.20), the general expression for $\mathbf{f_e}$ is

$$\mathbf{f_e^T} = \int_{D_e} [N_1 X \quad N_1 Y \quad N_2 X \quad N_2 Y \quad N_3 X \quad N_3 Y]\, d\,Vol$$

If X and Y, the components of the body force, are constants within the element, say $X = X_0$ and $Y = Y_0$, it follows immediately using the integration results of Eq. (5.27), that

$$\mathbf{f_e^T} = \frac{[X_0 \; Y_0 \; X_0 \; Y_0 \; X_0 \; Y_0]\, A_e t}{3} \tag{5.28}$$

This result indicates that 1/3 the total of $X_0 A_e t$ is lumped at each node for the component of force in the x-direction and similarly, that 1/3 the total of $Y_0 A_e t$ is lumped at each node for the component of force in the y-direction.

If X and Y are reasonably smooth functions of position within the element, sufficient accuracy can often be obtained by *approximating* the functions X and Y respectively as

$$X = X_i N_i + X_j N_j + X_k N_k$$

$$Y = Y_i N_i + Y_j N_j + Y_k N_k$$

The integrals for $\mathbf{f_e}$ are easily evaluated using Eq. (5.27) to yield

$$\mathbf{f_e} = \begin{bmatrix} 2X_i + X_j + X_k \\ 2Y_i + Y_j + Y_k \\ X_i + 2X_j + X_k \\ Y_i + 2Y_j + Y_k \\ X_i + X_j + 2X_k \\ Y_i + Y_j + 2Y_k \end{bmatrix} \frac{A_e t}{12} = \begin{bmatrix} f_{Xi} \\ f_{Yi} \\ f_{Xj} \\ f_{Yj} \\ f_{Xk} \\ f_{Yk} \end{bmatrix} \tag{5.29}$$

If more accuracy is required in evaluating the integrals a quadrature can be used. The body force nodal forces are shown in Fig. 5–13.

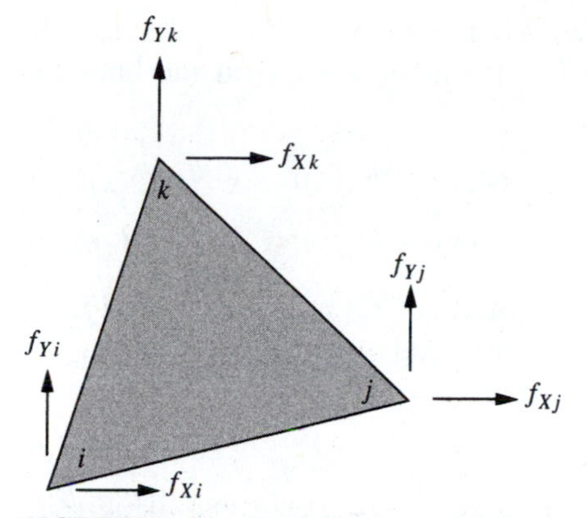

FIGURE 5–13 Body force nodal loads

Evaluation of $\mathbf{t_e}$. From Eq. (5.21), the general expression developed for $\mathbf{t_e}$ in the previous section is

$$\mathbf{t_e} = \begin{bmatrix} \int_{\gamma_{te}} N_i(lt_n - mt_s)\, dS \\ \int_{\gamma_{te}} N_i(mt_n + lt_s)\, dS \\ \int_{\gamma_{te}} N_j(lt_n - mt_s)\, dS \\ \int_{\gamma_{te}} N_j(mt_n + lt_s)\, dS \end{bmatrix}$$

where t_n, t_s, l, and m are indicated in Fig. 5–14.

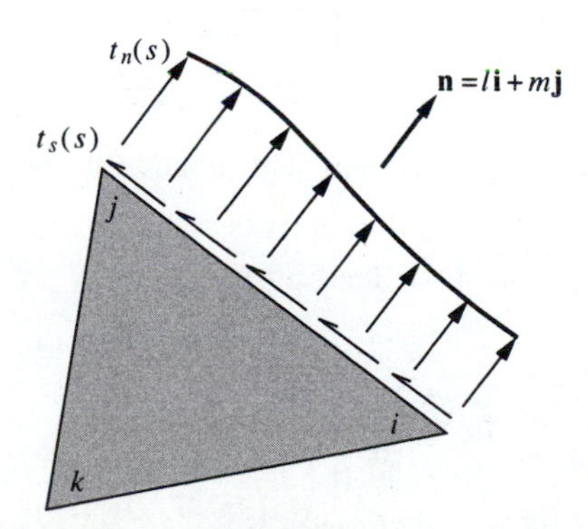

FIGURE 5–14 Definitions of t_n, t_s, l, and m

It is instructive to consider the specific form of $\mathbf{t_e}$ as follows. Assuming t_n and t_s to be reasonably smooth functions of s on the boundary, represent $t_n(s)$ and $t_s(s)$ approximately as

$$t_n(s) = N_i(s)t_{ni} + N_j(s)t_{nj}$$

$$t_s(s) = N_i(s)t_{si} + N_j(s)t_{sj}$$

where t_{ni} and t_{nj} are the values of $t_n(s)$ at nodes i and j, t_{si} and t_{sj} are the values of $t_s(s)$ at nodes i and j respectively, $N_i(s) = (1 - s/l_e)$, and $N_j(s) = s/l_e$. Using the integration formulas

$$\int N_I^a N_J^b \, dS = t \int N_I^a N_J^b \, ds = \frac{a!b!\, t l_e}{(a+b+1)!}$$

$\mathbf{t_e}$ can then be computed to be

$$\mathbf{t_e} = \frac{l_e t}{6} \begin{bmatrix} l(2t_{ni} + t_{nj}) - m(2t_{si} + t_{sj}) \\ m(2t_{ni} + t_{nj}) + l(2t_{si} + t_{sj}) \\ l(t_{ni} + 2t_{nj}) - m(t_{si} + 2t_{sj}) \\ m(t_{ni} + 2t_{nj}) + l(t_{si} + 2t_{sj}) \end{bmatrix} \tag{5.30}$$

Special cases. Consider an element for which the normal is in the x-direction as shown in Fig. 5–15.

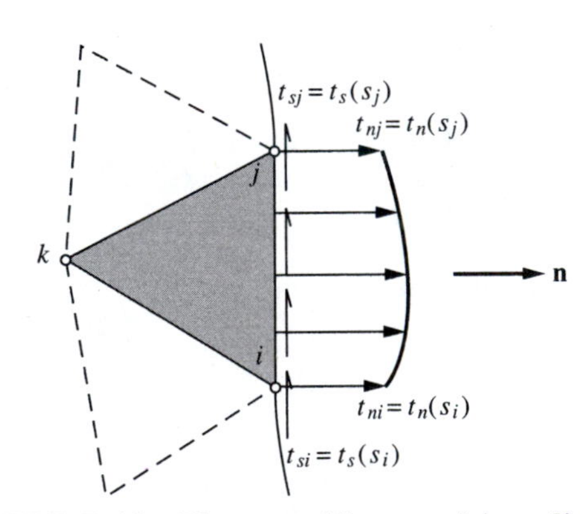

FIGURE 5–15 Element with normal in x-direction

In this case, $l = 1$ and $m = 0$, leading to

$$\mathbf{t_e} = \frac{l_e t}{6} \begin{bmatrix} (2t_{ni} + t_{nj}) \\ (2t_{si} + t_{sj}) \\ (t_{ni} + 2t_{nj}) \\ (t_{si} + 2t_{sj}) \end{bmatrix}$$

Furthermore, in case both the normal and shear components are constants on the portion of the boundary in question, with $t_{ni} = t_{nj} = \sigma_n$ and with $t_{si} = t_{sj} = \tau_n$, it follows that

$$\mathbf{t_e} = \frac{l_e t}{2} \begin{bmatrix} \sigma_n \\ \tau_n \\ \sigma_n \\ \tau_n \end{bmatrix}$$

which appears in a force sketch as in Fig. 5–16, indicating that for linear interpolation, the total elemental normal force $tl_e\sigma_n$ and the total elemental shear force $tl_e\tau_n$ are equally distributed between the nodes in question.

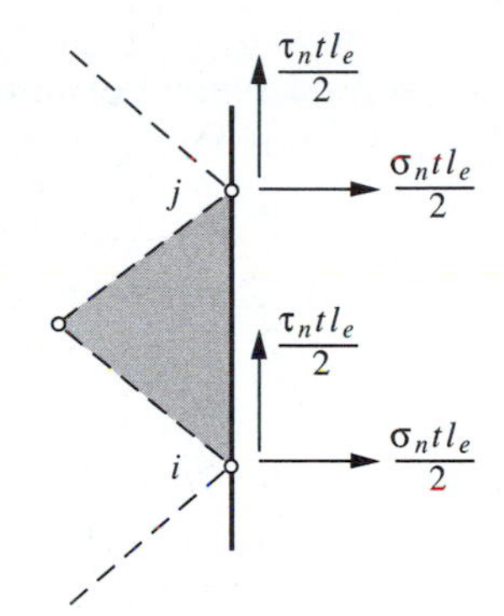

FIGURE 5–16 Force sketch

A similar development can be carried out for the case where the normal is in the y-direction, that is, $l = 0$ and $m = 1$, leading to

$$\mathbf{t_e} = \frac{l_e t}{6} \begin{bmatrix} -(2t_{si} + t_{sj}) \\ (2t_{ni} + t_{nj}) \\ -(t_{si} + 2t_{sj}) \\ (t_{ni} + 2t_{nj}) \end{bmatrix}$$

In case both the normal and shear components are constants on the portion of the boundary in question, with $t_{ni} = t_{nj} = \sigma_n$ and with $t_{si} = t_{sj} = \tau_n$, it follows that

$$\mathbf{t_e} = \frac{l_e t}{2} \begin{bmatrix} -\tau_n \\ \sigma_n \\ -\tau_n \\ \sigma_n \end{bmatrix}$$

which appears in a force sketch as in Fig. 5–17. These elemental matrices will be illustrated in the examples that follow.

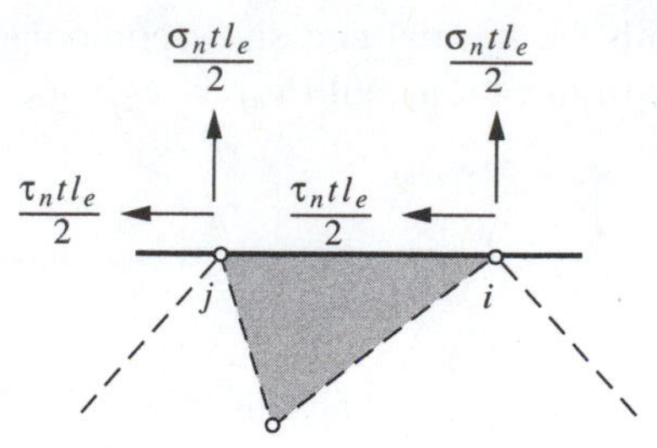

FIGURE 5–17 Force sketch

5.5 APPLICATIONS

In this section, we will discuss several typical examples of the use of the finite element model for the two-dimensional theory.

5.5.1 Uniform Tension of a Thin Rectangular Plate

Consider a thin ($t = 0.25$ in), uniform plate loaded by a uniform load on one face as shown in Fig. 5–18. The plate will be assumed to be composed of an isotropic steel with a value of Young's modulus of $E = 3 \times 10^7$ psi and a Poisson's ratio of $\nu = 0.3$. Plane stress is the appropriate theory for the type of geometry and loading involved.

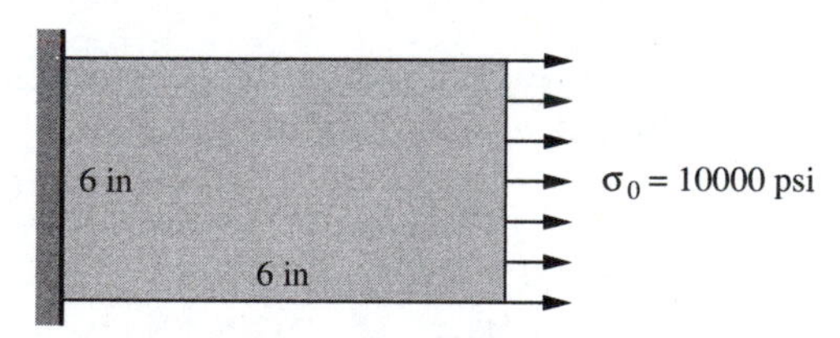

FIGURE 5–18 Uniform tensile loading of a thin rectangular plate

Discretization and interpolation. One of the simplest possible meshes using triangular elements is as indicated in Fig. 5–19.

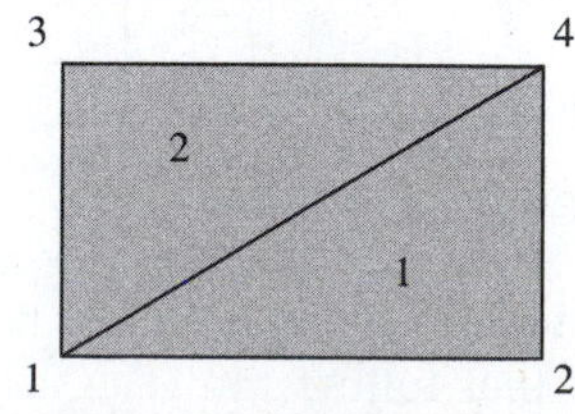

FIGURE 5–19 Two-element mesh—rectangular plate

Elemental formulation. For this example, there are only contributions to the global stiffness and load vectors, namely those from $\mathbf{k_e}$ and $\mathbf{t_e}$. Each of the $\mathbf{k_{ei}}$ will be computed, followed by the computation of the nodal loads $\mathbf{t_e}$ arising from the stresses on face 2-4 of element 1.

$\mathbf{k_{e1}}$. Each of the elemental stiffness matrices depends on the elemental geometry by virtue of the strain matrix **B**. It is convenient to form a table containing the x and y values of the nodal coordinates as indicated in Table 5.1. From this table the elements of the **B** matrix are easily constructed. The information for both elements is presented.

TABLE 5.1 Elemental and nodal data

Element	N1	N2	N3	x_1	x_2	x_3	y_1	y_2	y_3
1	1	2	4	0.0	6.0	6.0	0.0	0.0	6.0
2	1	4	3	0.0	6.0	0.0	0.0	6.0	6.0

For element 1, with node numbers 1, 2, and 4, we form the vectors $\mathbf{x_1}$, $\mathbf{y_1}$, $\mathbf{b_1}$, and $\mathbf{c_1}$ as

$$\mathbf{x_1} = [x_1 \quad x_2 \quad x_4] = [0.0 \quad 6.0 \quad 6.0]$$

$$\mathbf{y_1} = [y_1 \quad y_2 \quad y_4] = [0.0 \quad 0.0 \quad 6.0]$$

$$\mathbf{b_1} = [b_1 \quad b_2 \quad b_3] = \frac{[y_2 - y_4 \quad y_4 - y_1 \quad y_1 - y_2]}{2A_e} = \frac{[-6.0 \quad 6.0 \quad 0.0]}{36.0}$$

$$\mathbf{c_1} = [c_1 \quad c_2 \quad c_3] = \frac{[x_4 - x_2 \quad x_1 - x_4 \quad x_2 - x_1]}{2A_e} = \frac{[0.0 \quad -6.0 \quad 6.0]}{36.0}$$

The strain matrix **B** becomes

$$\mathbf{B} = \begin{bmatrix} -6.0 & 0.0 & 6.0 & 0.0 & 0.0 & 0.0 \\ 0.0 & 0.0 & 0.0 & -6.0 & 0.0 & 6.0 \\ 0.0 & -6.0 & -6.0 & 6.0 & 6.0 & 0.0 \end{bmatrix} \frac{1}{(36.0 \text{ in})}$$

The elastic coefficient matrix is

$$\mathbf{C} = \frac{3 \times 10^7}{0.91} \begin{bmatrix} 1.0 & 0.3 & 0.0 \\ 0.3 & 1.0 & 0.0 \\ 0.0 & 0.0 & 0.35 \end{bmatrix} \text{psi}$$

The elemental stiffness matrix can then be computed as

$$\mathbf{k_{e1}} = (tA_e)\mathbf{B}^{\mathrm{T}}\mathbf{C}\mathbf{B}$$

$$= \begin{array}{c} \begin{matrix} u_1 & \quad v_1 & \quad u_2 & \quad v_2 & \quad u_4 & \quad v_4 \end{matrix} \\ \begin{bmatrix} 36.0 & 0.0 & -36.0 & 10.8 & 0.0 & -10.8 \\ 0.0 & 12.6 & 12.6 & -12.6 & -12.6 & 0.0 \\ -36.0 & 12.6 & 48.6 & -23.4 & -12.6 & 10.8 \\ 10.8 & -12.6 & -23.4 & 48.6 & 12.6 & -36.0 \\ 0.0 & -12.6 & -12.6 & 12.6 & 12.6 & 0.0 \\ -10.8 & 0.0 & 10.8 & -36.0 & 0.0 & 36.0 \end{bmatrix} \end{array} (114{,}469 \text{ lbf/in})$$

The displacements above the columns indicate the positions into which the elements will be loaded or assembled in the global stiffness matrix.

$\mathbf{k_{e2}}$. Again using the information in Table 5.1, we form

$$\mathbf{x_2} = [x_1 \quad x_4 \quad x_3] = [0.0 \quad 6.0 \quad 0.0]$$

$$\mathbf{y_2} = [y_1 \quad y_4 \quad y_3] = [0.0 \quad 6.0 \quad 6.0]$$

$$\mathbf{b_2} = [b_1 \quad b_2 \quad b_3] = \frac{[y_4 - y_3 \quad y_3 - y_1 \quad y_1 - y_4]}{2A_e}$$

$$= \frac{[0.0 \quad 6.0 \quad -6.0]}{36.0}$$

$$\mathbf{c_2} = [c_1 \quad c_2 \quad c_3] = \frac{[x_3 - x_4 \quad x_1 - x_3 \quad x_4 - x_1]}{2A_e}$$

$$= \frac{[-6.0 \quad 0.0 \quad 6.0]}{36.0}$$

from which

$$\mathbf{B} = \begin{bmatrix} 0.0 & 0.0 & 6.0 & 0.0 & -6.0 & 0.0 \\ 0.0 & -6.0 & 0.0 & 0.0 & 0.0 & 6.0 \\ -6.0 & 0.0 & 0.0 & 6.0 & 6.0 & -6.0 \end{bmatrix} \frac{1}{(36.0 \text{ in})}$$

The resulting $\mathbf{k_{e2}}$ is

$$\mathbf{k_{e2}} = \begin{array}{c} \begin{matrix} u_1 & \quad v_1 & \quad u_4 & \quad v_4 & \quad u_3 & \quad v_3 \end{matrix} \\ \begin{bmatrix} 12.6 & 0.0 & 0.0 & -12.6 & -12.6 & 12.6 \\ 0.0 & 36.0 & -10.8 & 0.0 & 10.8 & -36.0 \\ 0.0 & -10.8 & 36.0 & 0.0 & -36.0 & 10.8 \\ -12.6 & 0.0 & 0.0 & 12.6 & 12.6 & -12.6 \\ -12.6 & 10.8 & -36.0 & 12.6 & 48.6 & -23.4 \\ 12.6 & -36.0 & 10.8 & -12.6 & -23.4 & 48.6 \end{bmatrix} \end{array} (114{,}469 \text{ lbf/in})$$

As mentioned previously, there are no body forces or thermal inputs so that $\mathbf{f_{e1}} = \mathbf{f_{e2}} = \boldsymbol{\theta}_{\mathbf{e1}} = \boldsymbol{\theta}_{\mathbf{e2}} = \mathbf{0}$.

$\mathbf{t_e}$. The surface tractions contribute on the 2-4 side of element 1. From the previous section, the general result for the $\mathbf{t_e}$ is

$$\mathbf{t_e} = \frac{l_e t}{6}\begin{bmatrix} l(2t_{ni} + t_{nj}) - m(2t_{si} + t_{sj}) \\ m(2t_{ni} + t_{nj}) + l(2t_{si} + t_{sj}) \\ l(t_{ni} + 2t_{nj}) - m(t_{si} + 2t_{sj}) \\ m(t_{ni} + 2t_{nj}) + l(t_{si} + 2t_{sj}) \end{bmatrix}$$

Taking $l_e = 6.0$ in, $t = 0.25$ in, $t_{ni} = t_{nj} = 10{,}000.0$ psi, $l = 1.0$, $m = 0.0$, and $t_{si} = t_{sj} = 0$, yields

$$\mathbf{t_e} = \begin{bmatrix} 7500.0 \text{ lbf} \\ 0.0 \text{ lbf} \\ 7500.0 \text{ lbf} \\ 0.0 \text{ lbf} \end{bmatrix} \begin{matrix} u_2 \\ v_2 \\ u_4 \\ v_4 \end{matrix}$$

where the positions into which the loads are to be assembled are indicated by the u_i and v_i to the right of the matrix. These loads are shown in Fig. 5–20.

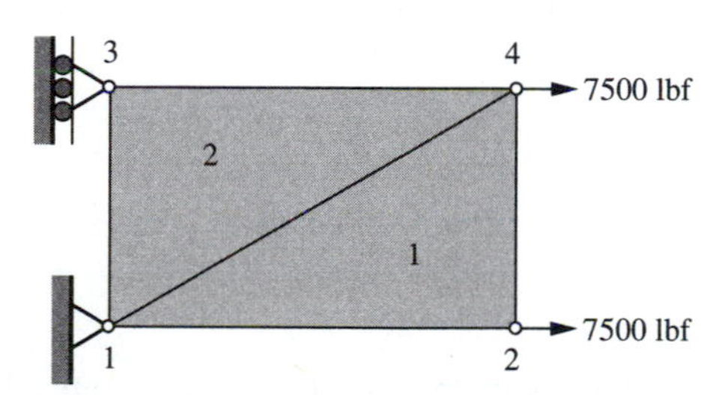

FIGURE 5–20 Boundary conditions and loading

Assembly. In order to clearly see the assembly process, rewrite the elemental stiffness matrices in partitioned form as

$$\mathbf{k_{e1}} = \begin{matrix} & \begin{matrix} 1 & 2 & 4 \end{matrix} & \\ & \begin{bmatrix} \mathbf{a}_{11} & \mathbf{a}_{12} & \mathbf{a}_{13} \\ \mathbf{a}_{21} & \mathbf{a}_{22} & \mathbf{a}_{23} \\ \mathbf{a}_{31} & \mathbf{a}_{32} & \mathbf{a}_{33} \end{bmatrix} & \begin{matrix} 1 \\ 2 \\ 4 \end{matrix} \end{matrix} \qquad \mathbf{k_{e2}} = \begin{matrix} & \begin{matrix} 1 & 4 & 3 \end{matrix} & \\ & \begin{bmatrix} \mathbf{b}_{11} & \mathbf{b}_{12} & \mathbf{b}_{13} \\ \mathbf{b}_{21} & \mathbf{b}_{22} & \mathbf{b}_{23} \\ \mathbf{b}_{31} & \mathbf{b}_{32} & \mathbf{b}_{33} \end{bmatrix} & \begin{matrix} 1 \\ 4 \\ 3 \end{matrix} \end{matrix}$$

Each of the 2×2 blocks $\mathbf{a_{ij}}$ and $\mathbf{b_{ij}}$ represents the contribution of a *u-v* pair from the node in question and is loaded into the global stiffness matrix as follows.

The global stiffness matrix is the proper sum of the elemental stiffness matrices $\mathbf{k_{e1}}$ and $\mathbf{k_{e2}}$ according to

$$\mathbf{K_G} = \begin{bmatrix} \mathbf{a_{11}} & \mathbf{a_{12}} & \mathbf{0} & \mathbf{a_{13}} \\ \mathbf{a_{21}} & \mathbf{a_{22}} & \mathbf{0} & \mathbf{a_{23}} \\ \mathbf{0} & \mathbf{0} & \mathbf{0} & \mathbf{0} \\ \mathbf{a_{31}} & \mathbf{a_{32}} & \mathbf{0} & \mathbf{a_{33}} \end{bmatrix} \begin{matrix} 1 \\ 2 \\ 3 \\ 4 \end{matrix} + \begin{bmatrix} \mathbf{b_{11}} & \mathbf{0} & \mathbf{b_{13}} & \mathbf{b_{12}} \\ \mathbf{0} & \mathbf{0} & \mathbf{0} & \mathbf{0} \\ \mathbf{b_{31}} & \mathbf{0} & \mathbf{b_{33}} & \mathbf{b_{32}} \\ \mathbf{b_{21}} & \mathbf{0} & \mathbf{b_{23}} & \mathbf{b_{22}} \end{bmatrix} \begin{matrix} 1 \\ 2 \\ 3 \\ 4 \end{matrix}$$

or

$$\mathbf{K_G} = \begin{matrix} & 1 & 2 & 3 & 4 \\ & \end{matrix}$$

$$\mathbf{K_G} = \begin{bmatrix} \mathbf{a_{11}} + \mathbf{b_{11}} & \mathbf{a_{12}} & \mathbf{b_{13}} & \mathbf{a_{13}} + \mathbf{b_{12}} \\ \mathbf{a_{21}} & \mathbf{a_{22}} & \mathbf{0} & \mathbf{a_{23}} \\ \mathbf{b_{31}} & \mathbf{0} & \mathbf{b_{33}} & \mathbf{b_{32}} \\ \mathbf{a_{31}} + \mathbf{b_{21}} & \mathbf{a_{23}} & \mathbf{b_{23}} & \mathbf{a_{33}} + \mathbf{b_{22}} \end{bmatrix} \begin{matrix} 1 \\ 2 \\ 3 \\ 4 \end{matrix}$$

As can be seen, an extra row and column must be inserted in the third position in the partitioned elemental matrix $\mathbf{k_{e1}}$ before loading into the global stiffness matrix. For $\mathbf{k_{e2}}$, an extra row and column must be inserted in the position corresponding to node 2 of the partitioned elemental stiffness matrix $\mathbf{k_{e2}}$, after which the elements in the third and fourth rows and columns of the partitioned matrices must be interchanged before loading into the global stiffness matrix. The end result is the global stiffness matrix

$$\mathbf{K_G} = \begin{bmatrix} 48.6 & 0.0 & -36.0 & 10.8 & -12.6 & 12.6 & 0.0 & -23.4 \\ & 48.6 & 12.6 & -12.6 & 10.8 & -36.0 & -23.4 & 0.0 \\ & & 48.6 & -23.4 & 0.0 & 0.0 & -12.6 & 10.8 \\ & & & 48.6 & 0.0 & 0.0 & 12.6 & -36.0 \\ & & & & 48.6 & -23.4 & -36.0 & 12.6 \\ & & & & & 48.6 & 10.8 & -12.6 \\ & \text{symm} & & & & & 48.6 & 0.0 \\ & & & & & & & 48.6 \end{bmatrix} D$$

where $D = 114{,}469$ lbf/in.

In a similar fashion, the elemental load matrix $\mathbf{t_e}$ must be expanded to the global level by inserting zeros in the positions corresponding to nodes 1 and 3. The result is the global load matrix

$$\mathbf{F_G^T} = [0.0 \quad 0.0 \quad 7500.0 \quad 0.0 \quad 0.0 \quad 0.0 \quad 7500.0 \quad 0.0] \text{ lbf}$$

Constraints. The constraints that will be used are those that correspond to constraining only the x-component of the displacement at nodes 1 and 3, and the y-component of displacement at node 1. This corresponds to the symbolic boundary conditions shown in Fig. 5–20, and allows the plate to contract freely at the left boundary so as to be able to compare the solution to an exact solution from two-dimensional isotropic plane stress theory. The y-constraint is necessary at node 1 in order to prevent rigid body motions. Thus, the three imposed constraints are $u_1 = v_1 = u_3 = 0$.

Solution. Solving the resulting constrained set of equations yields

$$\mathbf{u}_{\mathbf{G}}^{\mathrm{T}} = [0.0 \quad 0.0 \quad 0.002 \quad 0.0 \quad 0.0 \quad -0.0006 \quad 0.002 \quad -0.0006] \text{ in}$$

The nodal values

$$u_2 = 0.002 \text{ in} = u_4$$

$$v_3 = -0.0006 \text{ in} = v_4$$

coincide with the values that can be obtained from the exact solution, given by

$$u(x, y) = \frac{\sigma_0 x}{E} \qquad v(x, y) = \frac{-\nu\sigma_0 y}{E}$$

As mentioned in Section 2.8, the Ritz method has the desirable property that if within the approximate assumed Ritz solution the exact solution is contained for *some* choice of the coefficients, the Ritz method will choose the values corresponding to the exact solution. This example demonstrates that property.

Derived variables or stresses. The general expression for the stresses from Eq. (5.3) is

$$\boldsymbol{\sigma} = \mathbf{C}\boldsymbol{\epsilon} - \mathbf{f_T} = \mathbf{CBU_e} - \mathbf{f_T}$$

which must be evaluated for each element.

Element 1. For element 1, **CB** is easily computed to be

$$\mathbf{CB} = \frac{3 \times 10^7}{(0.91)(36.0)} \begin{bmatrix} -6.0 & 0.0 & 6.0 & -1.8 & 0.0 & 1.8 \\ -1.8 & 0.0 & 1.8 & -6.0 & 0.0 & 6.0 \\ 0.0 & -2.1 & -2.1 & 2.1 & 2.1 & 0.0 \end{bmatrix}$$

$\mathbf{U_e}$ is given by

$$\mathbf{U}_{\mathbf{e}}^{\mathrm{T}} = [0.0 \quad 0.0 \quad 0.002 \quad 0.0 \quad 0.002 \quad -0.0006]$$

resulting in

$$\boldsymbol{\sigma}_{\mathbf{1}}^{\mathrm{T}} = [10{,}000.0 \quad 0.0 \quad 0.0] \text{ psi}$$

These stresses are shown in Fig. 5–21. The stresses existing on the inclined segment 1-4 are also shown.

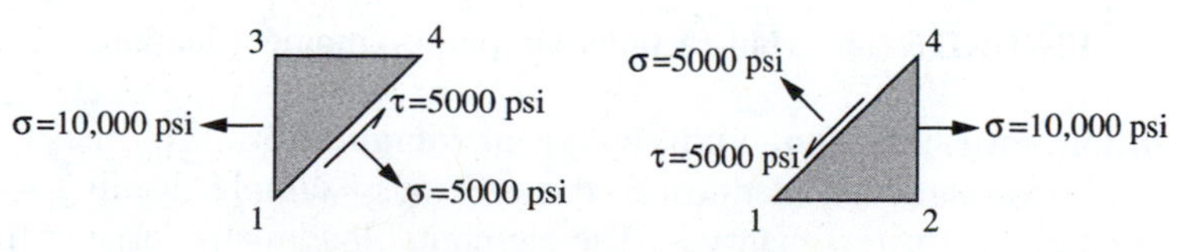

FIGURE 5–21 Elemental stresses

Element 2. For element 2, **CB** is computed as

$$\mathbf{CB} = \frac{3 \times 10^7}{(0.91)(36.0)} \begin{bmatrix} 0.0 & -1.8 & 0.0 & 0.0 & -6.0 & 1.8 \\ 0.0 & -6.0 & 1.8 & 0.0 & -1.8 & 6.0 \\ -2.1 & 0.0 & 0.0 & 2.1 & 2.1 & -2.1 \end{bmatrix}$$

$\mathbf{U_e}$ is given by

$$\mathbf{U_e^T} = [0.0 \quad 0.0 \quad 0.002 \quad -0.0006 \quad 0.0 \quad -0.0006]$$

resulting in

$$\boldsymbol{\sigma}_2^{\mathrm{T}} = [10{,}000.0 \quad 0.0 \quad 0.0] \text{ psi}$$

These stresses are also indicated in Fig. 5–21. Note that all the natural boundary conditions are satisfied, namely,

Face 1-2	$\sigma_y = \tau_{xy} = 0$	Face 2-4	$\tau_{xy} = 0$
Face 1-3	$\tau_{xy} = 0$	Face 3-4	$\sigma_y = \tau_{xy} = 0$

Note also that the stresses are continuous across the interelement boundary between nodes 1 and 4. Thus it is seen that for this particular problem the exact expressions for the displacements and stresses are generated by the Ritz finite element method.

5.5.2 Thin Rectangular Plate — Moment Loading

One main purpose for this example is to illustrate the situation that generally exists regarding interelement continuity of the derived variables (stresses); another is to investigate the satisfaction of the natural boundary conditions for problems that are modeled using the constant stress triangular element.

To these ends, consider the same thin plate as for Section 5.5.1 except that the plate is loaded by stresses with a moment resultant as indicated in Fig. 5–22.

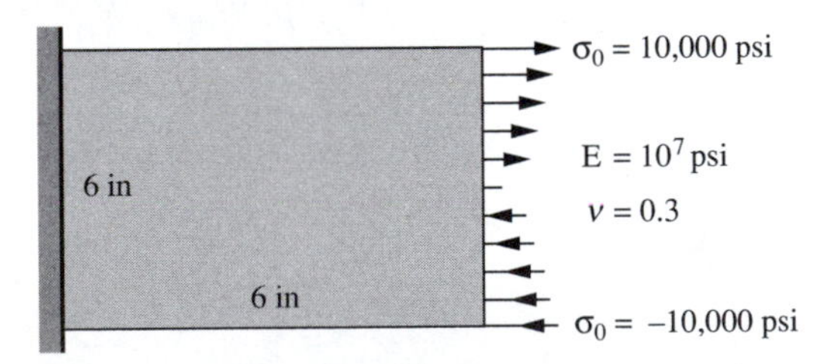

FIGURE 5–22 Thin rectangular plate — moment loading

Discretization, interpolation, and elemental formulation. The mesh, elastic, and geometric constants are identical to the previous example, leading to exactly the same elemental stiffness matrices. The elemental load matrix arising from side 2-4 of element 2 is computed from the general expression given in the previous example by taking $i = 2$, $j = 4$, $l_e = 6.0$ in, $t = 0.25$ in, $t_{si} = t_{sj} = 0.0$, $l = 1.0$, $m = 0.0$, and $t_{ni} = -10{,}000.0$ psi, and $t_{nj} = 10{,}000.0$ psi to obtain

$$\mathbf{t_e} = \begin{bmatrix} -2500.0 \text{ lbf} \\ 0.0 \text{ lbf} \\ 2500.0 \text{ lbf} \\ 0.0 \text{ lbf} \end{bmatrix} \begin{matrix} u_2 \\ v_2 \\ u_4 \\ v_4 \end{matrix}$$

where the positions into which the loads are to be assembled are indicated by the names of the displacements to the right of the matrix. Figure 5–23 shows the loads.

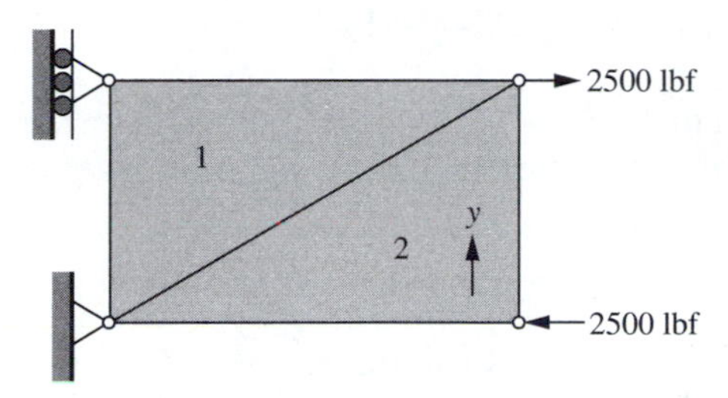

FIGURE 5–23 Nodal loads for the moment loading

Assembly, constraints, and solution. Assembling, enforcing the constraints $u_1 = v_1 = u_3 = 0$, and solving for the unknown displacements yield

$$\mathbf{u_G^T} = [\,0.0 \quad 0.0 \quad -0.000585 \quad -0.000867 \quad 0.0 \quad -0.000282 \quad 0.000585 \quad -0.000585\,] \text{ in}$$

Derived variables or stresses. Using the general expression for the stresses as in the previous example, the stresses are computed as

$$\boldsymbol{\sigma}_1^T = [2750 \quad -583.3 \quad -583.3] \text{ psi}$$

$$\boldsymbol{\sigma}_2^T = [-2750 \quad 583.3 \quad 583.3] \text{ psi}$$

At this point it is necessary to review briefly the two-dimensional stress transformation formulas. Consider a typical element or block as indicated in Fig. 5-24.

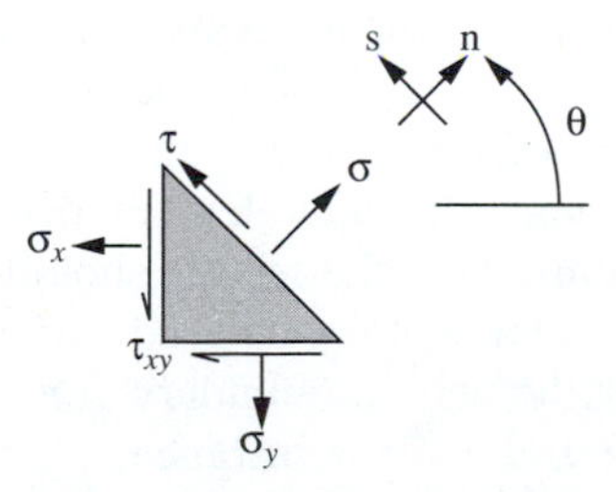

FIGURE 5–24 Stress transformation variables

The positive stress components σ_x, σ_y, and τ_{xy}, the angle θ which defines the positive normal to the face in question, and the positive values for the normal and shear stresses on the inclined face are indicated. Requiring that the block be in equilibrium ($\sum F_n = 0$ and $\sum F_s = 0$) yields the transformation equations

$$\sigma = \sigma_x \cos^2\theta + \sigma_y \sin^2\theta + 2\tau_{xy}\sin\theta\cos\theta$$

$$\tau = (\sigma_y - \sigma_x)\sin\theta\cos\theta + \tau_{xy}(\cos^2\theta - \sin^2\theta)$$

Given the state of stress σ_x, σ_y, and τ_{xy} within an element, the normal and shear stress can be determined on any inclined face or boundary.

Applying these transformation formulas to the states of stress for elements 1 and 2 of the current example yields the following.

Element 1. For element 1, $\sigma_x = 2750$ psi, $\sigma_y = -583$ psi, $\tau_{xy} = -583$ psi, and $\theta = -45°$, from which the stresses on the inclined face are $\sigma = 1667$ psi and $\tau = 1667$ psi.

Element 2. For element 2, $\sigma_x = -2750$ psi, $\sigma_y = 583$ psi, $\tau_{xy} = 583$ psi, and $\theta = 135°$, leading to $\sigma = -1667$ psi and $\tau = -1667$ psi.

The stresses for both elements are shown in Fig. 5–25.

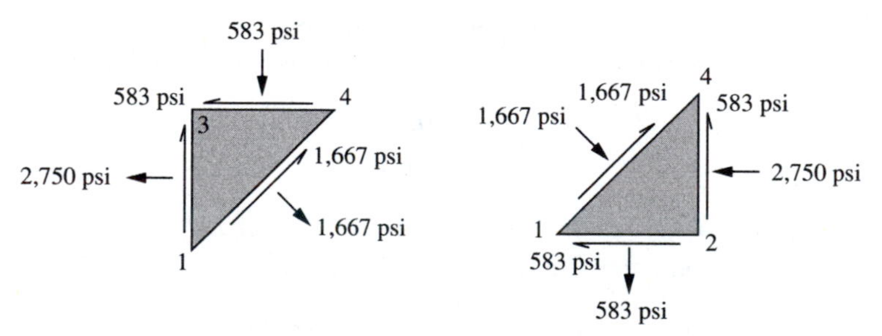

FIGURE 5–25 Elemental stresses, elements 1 and 2

Note carefully that the stresses are *not continuous* across the interelement boundary between nodes 1 and 4. This is a particularly striking characteristic of the constant stress element, that is, the presence, in general, of substantial discontinuities in both the normal and shear stresses across interelement boundaries. This condition continues to be true even for relatively fine meshes. The magnitudes of these discontinuities are a measure of the accuracy of the stresses. Until, as a result of mesh refinement, the interelement discontinuities become small compared to the stresses themselves, the results should be viewed with caution.

Another measure of the accuracy of the derived variables is the degree to which the predicted stresses satisfy the natural boundary conditions on the Γ_t portion of the boundary. On faces 1-2 and 3-4 for instance, the normal and shear stresses should all be zero. Similarly, the natural boundary conditions on the normal and shear stress on face 2-4 are clearly not satisfied, since $\sigma_x = -27{,}500$ psi and $\tau_{xy} = 5833$ psi, whereas the correct values are $\sigma_x = -10{,}000 + 20{,}000y/6$ and $\tau_{xy} = 0$, with y as defined in Fig. 5–23. The satisfaction of the natural boundary conditions on the other parts of the Γ_t portions of the boundary, that is, faces 1-2 and 3-4, are equally poor. This poor satisfaction of the natural boundary conditions is primarily due to the very crude mesh used. Nevertheless, the substantial lack of interelement continuity and nonsatisfaction of the natural boundary conditions is, for the constant stress triangular element, an ever-present condition, except for relatively fine meshes. This situation is further investigated in the next example.

5.5.3 Thin Rectangular Plate — Moment Loading, Meshes 2 and 3

Consider again the problem of Section 5.5.2: the thin rectangular plate loaded by stresses statically equivalent to a moment as shown in Fig. 5–22.

Discretization, interpolation, and elemental formulation. The mesh consists of nine nodes and eight constant stress triangular elements as indicated in Fig. 5–26.

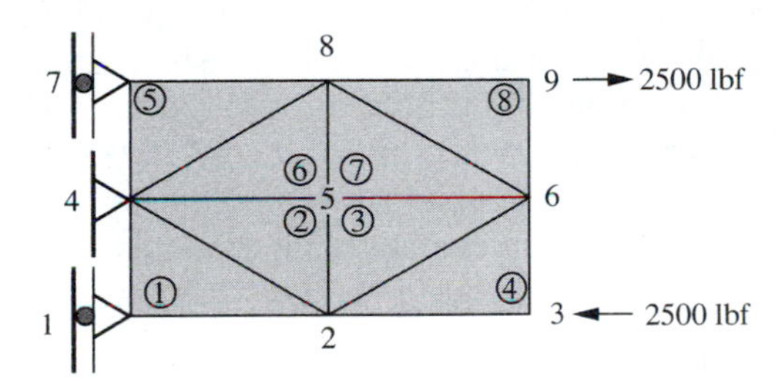

FIGURE 5–26 Mesh 2, flat plate bending

The elemental stiffness matrices are computed in precisely the same fashion as in the previous two examples. The elemental load matrices arising from the stresses prescribed on side 3-6 of element 4 and side 6-9 of element 8 are easily computed to be, respectively,

$$\mathbf{t}_{e4}^{T} = \overset{\qquad u_3 \qquad\quad v_3 \qquad\quad u_6 \qquad\quad v_6}{\begin{bmatrix} -2500 & 0.0 & -1250 & 0.0 \end{bmatrix}} \text{ lbf}$$

$$\mathbf{t}_{e8}^{T} = \overset{\qquad u_6 \qquad\quad v_6 \qquad\quad u_9 \qquad\quad v_9}{\begin{bmatrix} 1250 & 0.0 & 2500 & 0.0 \end{bmatrix}} \text{ lbf}$$

The positions occupied by the nodal forces are indicated by the names of the displacements above each of the elemental force vectors. The nodal forces are shown in Fig. 5–26.

Assembly, constraints, and solution. After assembling the load and stiffness matrices, the constraints, given by $u_1 = v_4 = u_4 = u_7 = 0$, are enforced and the solution is obtained.

Derived variables. The stresses are computed using $\sigma = \mathbf{CBU_e}$, and are displayed in Table 5.2.

TABLE 5.2 Stresses (psi), moment loading, mesh 2

Element	σ_x	σ_y	τ_{xy}
1	−6045	621	−621
2	−186	−621	621
3	−186	−621	−621
4	−6045	621	621
5	6045	−621	−621
6	186	621	621
7	186	621	−621
8	6045	−621	621

The σ_x component, often referred to as the bending stress, is positive above the midplane and negative below as expected. It is constant, however, rather than varying linearly as is prescribed on that face. This is an unavoidable consequence of having used the constant stress triangular element. The normal stresses of ± 6045 psi on faces 6-9 and 3-6, respectively, give rise to a moment of (6045)(3)(0.25)(3) = 13,601 in-lbf compared with the correct value of 15,000 in-lbf.

In order to get an idea of the degree to which the natural boundary conditions and interelement continuity conditions are being satisfied, consider free-body diagrams of elements 7 and 8 as indicated in Fig. 5–27.

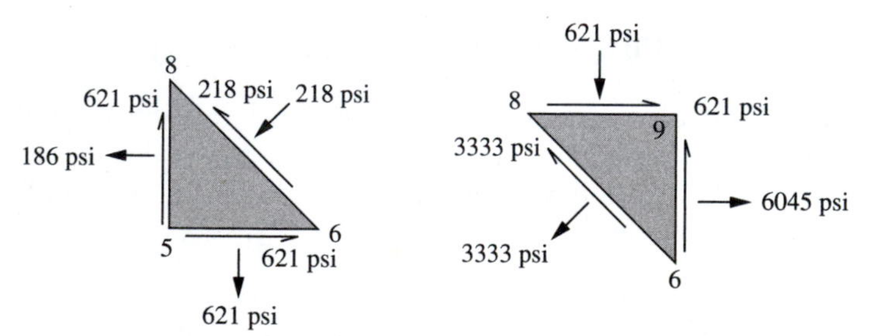

FIGURE 5–27 Boundary and interelement stresses, mesh 2

The shear stress value of 621 psi acting on faces 6-9 and 8-9, and the normal stress value of 621 psi on face 8-9, are all relatively small compared with maximum stress in element 8. The magnitudes of the discontinuities in the normal and shear stresses across the interelements boundary between nodes 6 and 8 are, however, quite large, with values of $3333 - (-218) = 3551$ psi. This mesh would have to be judged as not sufficiently refined to predict the stresses adequately.

Consider then the mesh indicated in Fig. 5–28, where the size of the elements has been halved compared with mesh 2.

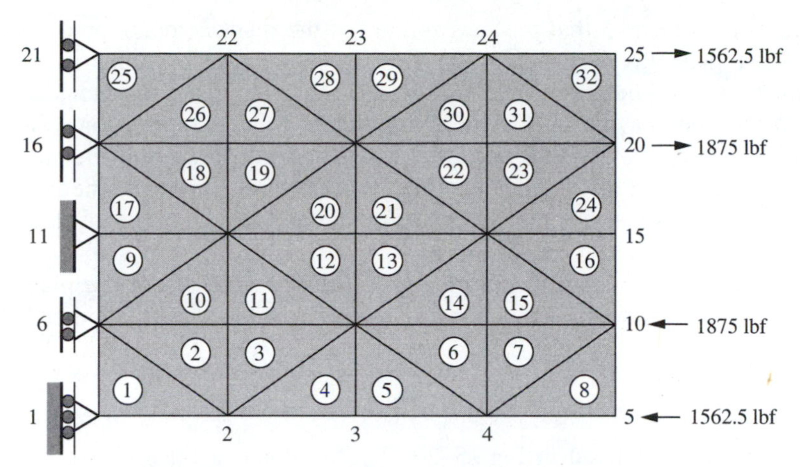

FIGURE 5–28 Bending of flat plate, mesh 3

Carrying out the basic steps outlined in detail in the previous examples leads to the displacements shown in Table 5.3. Also shown in Table 5.3 are the displacements for the nodes at the same physical locations given by the model using mesh 2.

TABLE 5.3 Comparison of displacements, meshes 2 and 3

Mesh 3			Mesh 2		
Node	***u*(in)**	***v*(in)**	**Node**	***u*(in)**	***v*(in)**
1	0.00000	−0.00039	1	0.00000	−0.00073
3	−0.00260	−0.00169	2	−0.00187	−0.00122
5	−0.00511	−0.00551	3	−0.00374	−0.00447
11	0.00000	0.00000	4	0.00000	0.00000
13	0.00000	−0.00130	5	0.00000	−0.00138
15	0.00000	−0.00520	6	0.00000	−0.00374
21	0.00000	−0.00039	7	0.00000	−0.00073
23	0.00260	−0.00169	8	0.00187	−0.00122
25	0.00511	−0.00551	9	0.00374	−0.00447

The agreement between the displacements for the two meshes is poor. In view of the fact that the displacements for the Ritz method generally converge fairly rapidly, this is particularly significant, and is a good indication that mesh 3 is probably still not adequately refined, there is still too much difference in the results given by the two models. This is borne out by the following investigation of the stresses.

Note from Table 5.3 that the magnitudes of the displacements predicted from mesh 3 are generally larger than the corresponding results from mesh 2. This is characteristic of the Ritz method: the replacement of a continuous elasticity problem with one having only a finite number of degrees of freedom effectively introduces internal constraints, or additional stiffening in the model, and results in displacements that are smaller than the exact values. Further refinement in the mesh generally results in a more flexible model and hence larger displacements, tending towards the exact values.

In this regard, the exact displacements can be obtained from the plane stress theory of elasticity, Boresi and Chong [1] and Timoshenko and Goodier [2], as

$$u(x, y) = \frac{Mxy}{EI} \qquad v(x, y) = \frac{-M(x^2 + \nu y^2)}{2EI}$$

where x and y are defined in Fig. 5–29. The resultant of the stresses prescribed on the right-hand edge of the region is M.

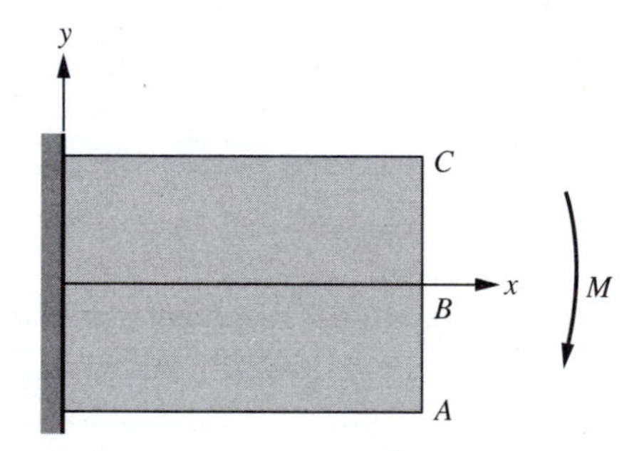

FIGURE 5–29 Coordinate system and points for comparison of displacement

Approximate displacement values from the results of mesh 3 are compared with the exact values at points A, B, and C as indicated in Fig. 5–29. These comparisons are presented in Table 5.4. For this problem, the approach from below to the exact values of the displacements can be observed on the basis of the information contained in Tables 5.3 and 5.4.

TABLE 5.4 Exact versus approximate displacements

Point	u_{ex}	u_{approx}	v_{ex}	v_{approx}
A	−0.00600	−0.00511	−0.00645	−0.00551
B	0.00000	0.00000	−0.00600	−0.00520
C	0.00600	0.00511	−0.00645	−0.00551

Stresses. The stresses computed on the basis of the procedure outlined in the previous examples are presented in Table 5.5.

TABLE 5.5 Stresses for mesh 3

Element	σ_x	σ_y	τ_{xy}
1	−8541	439	−439
2	−4490	−417	421
3	−4512	−424	−444
4	−8518	441	402
5	−8405	475	−515
6	−4628	−307	291
7	−4898	−388	−543
8	−8153	180	180
9	−128	−426	−426
10	−4230	448	443
11	−4253	442	−422
12	−120	−400	464
13	−120	−400	−336
14	−4378	526	561
15	−4647	446	−273
16	−191	−636	636
17	128	426	−426
18	4230	−448	443
19	4253	−442	−442
20	120	400	464
21	120	400	464
22	4378	−526	561
23	4647	−446	−273
24	191	636	636
25	8541	−439	−439
26	4490	417	421
27	4512	424	−444
28	8518	−441	402
29	8405	−475	−515
30	4629	307	291
31	4898	388	−543
32	8153	−180	180

Discussion. Consider, as representative of the satisfaction of the natural boundary conditions, the stresses from mesh 3 on the right-hand face as shown in Fig. 5–30.

There is poor agreement between the predicted stresses and the exact, prescribed distributions that are also indicated. The total moment supplied by the normal stresses is

$$M = 2\big(8153(1.5)(0.25)(2.25) + (191)(1.5)(0.25)(0.75)\big) = 13,866 \text{ in-lbf}$$

compared with the correct value of 15,000 in-lbf. The resultant shear force given by

$$V = 2(180 + 635)(1.5)(0.25) = 611 \text{ lbf}$$

is compared with the correct value of zero!

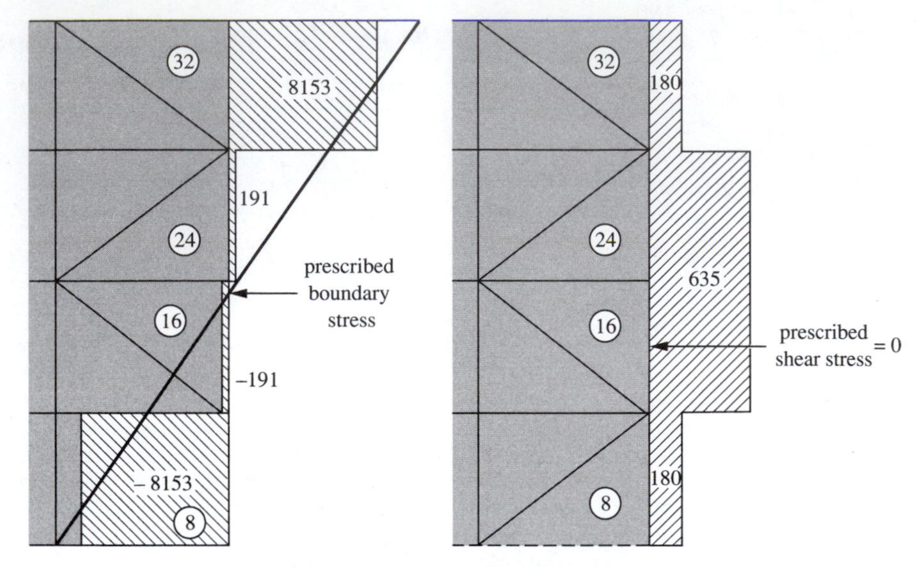

FIGURE 5–30 Boundary stresses on right face

On the top surface the stresses appear as in Fig. 5–31, again indicating a general lack of satisfaction of the natural boundary conditions of $\sigma_y = \tau_{xy} = 0$ on the top face.

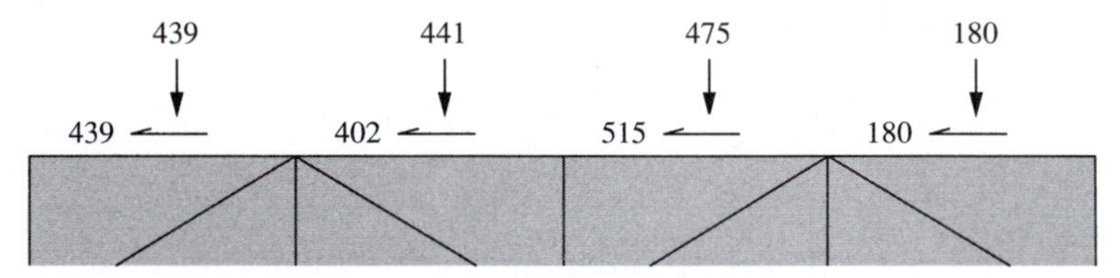

FIGURE 5–31 Stresses on top face

For mesh 3, consider the normal and shear stresses that act on a section passed through the last column of elements as shown in Fig. 5–32. The exact straight-line normal stress distribution is also shown. The exact shear stresses are, of course, zero.

The general characteristics, in terms of the normal stress increasing with distance from the neutral axis, are encouraging. However, the numerical values give only a rough idea of the correct magnitudes.

For an indication of the continuity of stresses across interelement boundaries, consider elements 31 and 32 shown in Fig. 5–33. The discontinuity in the normal stress across the interelement boundary between nodes 20 and 24 is $4167 - 2101 = 2066$ psi, with the discontinuity in the shear stress of $4167 - 2255 = 1912$ psi, both excessively large compared to the predicted bending stress of 8153 psi in element 32.

The results from mesh 3 indicate that there are still significant discontinuities in the stresses across interelement boundaries. This is a strong sign that the results are not adequate, and that further refinement is necessary. Although further refinement

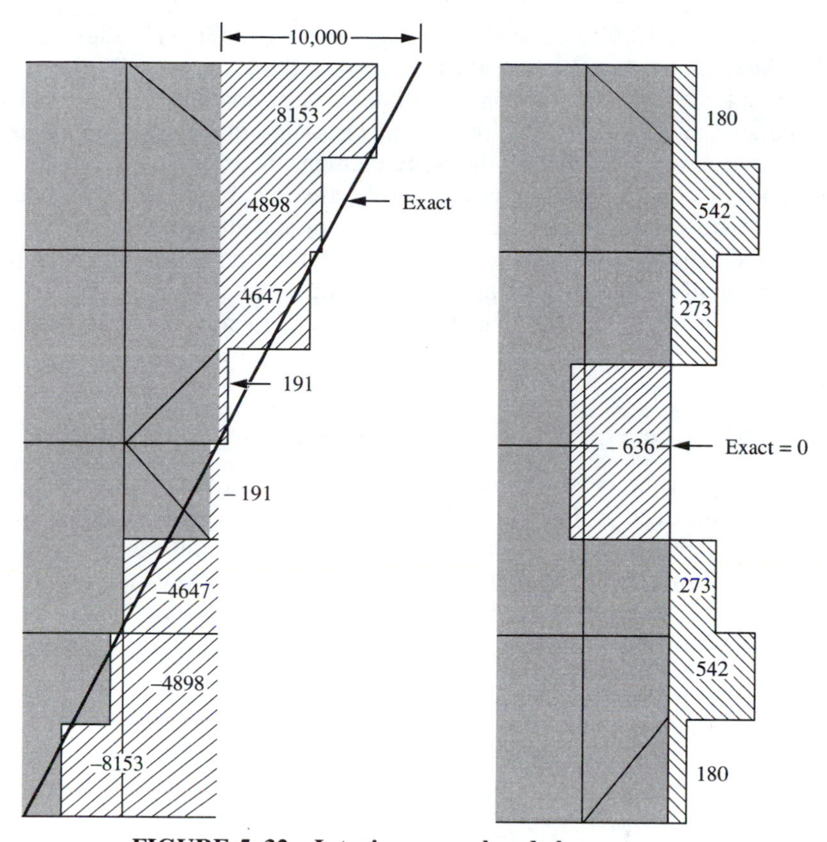

FIGURE 5–32 Interior normal and shear stresses

is certainly possible, it is somewhat futile if the constant stress triangular element is used.

Based on the results shown in Fig. 5–32, it is unlikely that, without knowledge of the exact solution, one could arrive at the conclusion that the normal stress distribution on a vertical section through the region should be linear, with a

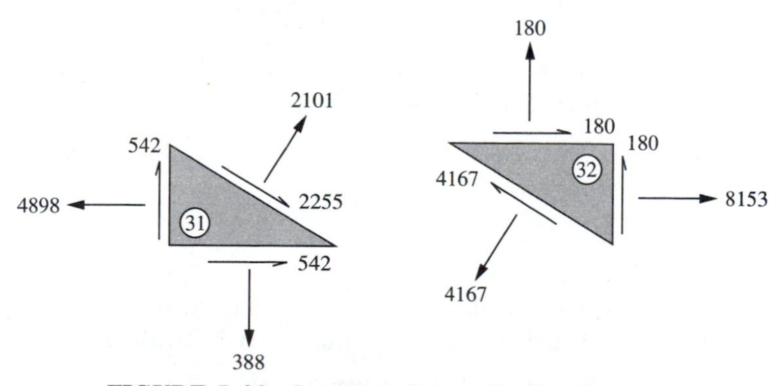

FIGURE 5–33 Interelement continuity of stresses

maximum value of 10,000 psi at the top edge and $-10{,}000$ psi at the bottom, or that the shear stress should be identically zero on the same section. For a stress analysis problem where the solution is not known, it is crucially important that the analyst be able to bring to bear some type of reasoning that permits an evaluation of the state of stress predicted by the finite element model.

The types of checks used in these examples are valuable in terms of assessing convergence from the perspectives of whether or not the stress boundary conditions are satisfied, and whether or not the interelement discontinuities in the stresses are small compared with the corresponding stresses within the elements. These are easily made checks that the student should rely on in assessing the validity of the derived variables.

5.5.4 Stress Concentration—Plate with Circular Hole

Abrupt changes in the cross-sectional area for a member that transmits an axial load such as indicated in Fig. 5–34 can result in a stress concentration; that is, the maximum stress at the section through the hole can significantly exceed the value computed on the basis of the elementary theory. The axial stress is assumed to have the uniform value σ_o at distances away from the hole as indicated. The plate has a uniform thickness t.

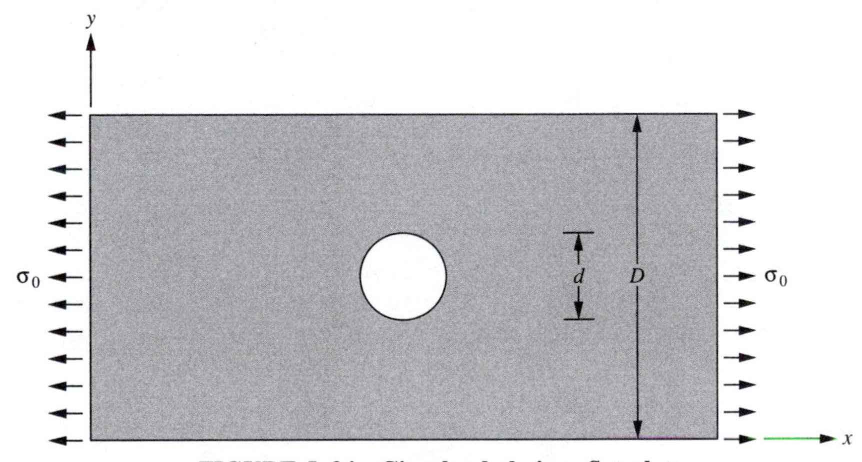

FIGURE 5–34 Circular hole in a flat plate

The average normal stress $(\sigma_x)_{AVG}$ acting on a section through the center of the hole would be computed as

$$(\sigma_x)_{AVG} = \frac{P}{A}$$

where $A = (D - d)t$ is the actual area over which the stress is distributed. The actual maximum stress, which occurs at the inner surface of the hole, is

$$(\sigma_x)_{MAX} = k_T(\sigma_x)_{AVG}$$

where k_T is the *stress concentration factor*. The stress concentration factor k_T can be determined on the basis of a finite element model using the plane stress theory outlined in Section 5.1.4.1.

From symmetry we conclude that only the lightly shaded portion of the region shown on the left in Fig. 5–35 needs to be modeled.

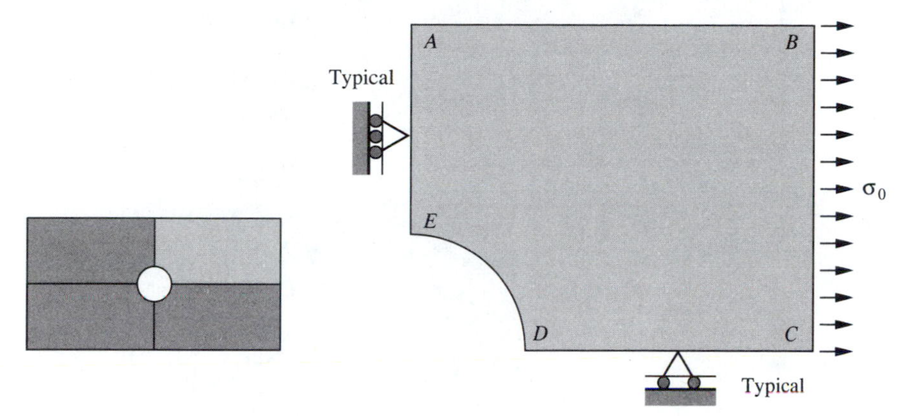

FIGURE 5–35 Quarter symmetry model

The type of boundary conditions that are appropriate on the *AE* and *CD* portions of the boundary are the *mixed type* referred to in Section 5.1. On *CD*, the vertical component of displacement v and the horizontal component of stress τ_{xy} are required to vanish due to symmetry. Similarly on *AE*, the horizontal component of displacement u and the vertical component of stress τ_{xy} must vanish. The type of support then corresponds to rollers on each of these boundaries. *AB*, *BC*, and *DE* constitute the Γ_t part of the boundary on which the stresses are prescribed. The uniform stress σ_0 on the *BC* portion of the Γ_t boundary will be converted into equivalent nodal loads for input to the loads in the global load matrix. The length of line *AB* is assumed to be large enough compared with the length of *BC* so that the stress on *BC* is the uniform value $P/A_0 = \sigma_0$ that exists in the bar away from the hole.

The displacement boundary conditions are $v = 0$ on *CD* and $u = 0$ on *EA*. The stress boundary conditions

$$\sigma_y = \tau_{xy} = 0 \quad \text{on } AB \qquad \sigma_x = \sigma_o, \quad \tau_{xy} = 0 \quad \text{on } BC$$

$$\tau_{xy} = 0 \quad \text{on } CD \qquad \sigma = \tau = 0 \quad \text{on } DE$$

$$\tau_{xy} = 0 \quad \text{on } EA$$

are all natural boundary conditions and will be satisfied only approximately by the results of the finite element method. As with any problem, the natural boundary conditions are satisfied by the finite element solution only in the limit as the mesh is sufficiently refined.

Discretization. The first mesh to be investigated is indicated in Fig. 5–36. There are 20 nodes and 24 elements.

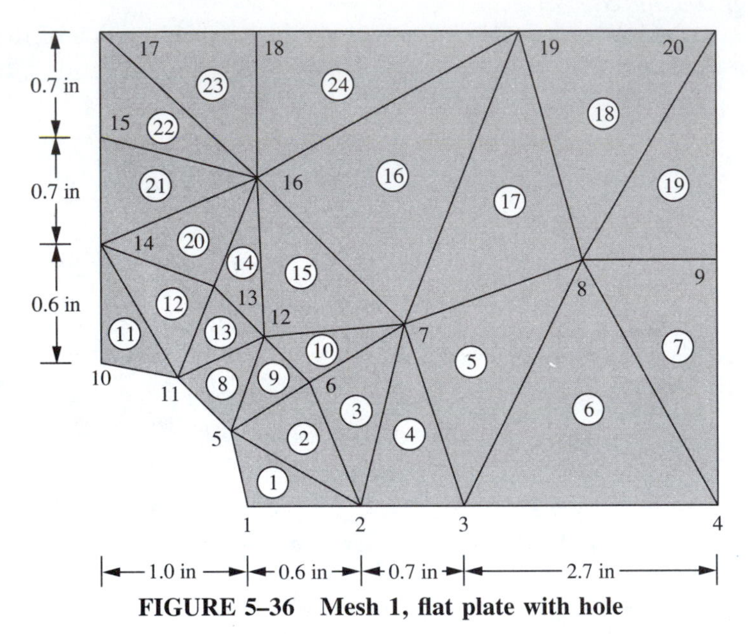

FIGURE 5–36 Mesh 1, flat plate with hole

Interpolation and elemental formulation. The linearly interpolated element discussed in Section 5.3 is used.

Assembly. Each of the 6 × 6 elemental matrices is loaded into the proper location in the global stiffness matrix as outlined in detail in the previous example.

Constraints. Enforcing the mixed boundary conditions on the segments CD and AE leads to the constraints

$$v_1 = v_2 = v_3 = v_4 = 0 \qquad \text{on } CD$$

$$u_{10} = u_{14} = u_{15} = u_{17} = 0 \qquad \text{on } AE$$

Solution. Assuming the plate to be composed of an isotropic aluminum for which $E = 10 \times 10^6$ psi, $\nu = 0.3$, and $\sigma_0 = 10^4$ psi, the constrained equations are solved for the displacements, which appear in Table 5.6. These displacements indicate generally that the plate elongates in the x-direction and contracts in the y-direction as would be expected. Note that the elongation computed for the solid plate without the hole would be

$$\Delta = \frac{PL}{AE} = \frac{\sigma L}{E} = 0.005 \text{ in}$$

which is slightly less than the 0.0057 value based on an average of the values on segment BC.

TABLE 5.6 Nodal displacements — mesh 1

Node	$u(10^5$ in)	$v(10^5$ in)
1	269	0
2	296	0
3	341	0
4	586	0
5	236	−30
6	244	−25
7	280	−31
8	373	−36
9	567	−30
10	0	−76
11	123	−78
12	180	−49
13	122	−78
14	0	−115
15	0	−132
16	145	−92
17	0	−158
18	191	−113
19	335	−84
20	546	−67

Computation of derived variables — stresses. As outlined and demonstrated in the previous example, the stresses are computed per element as

$$\boldsymbol{\sigma} = \mathbf{BCU_e} - \mathbf{f_T}$$

where for this example $\mathbf{f_T} = \mathbf{0}$. The results of these computations appear in Table 5.7.

Discussion. The mesh used for this example is fairly crude. Quantitative results for the stresses are not expected to be at all accurate. A measure of the accuracy is the extent to which the natural boundary conditions are satisfied. To this end, a sketch of the region modeled is indicated in Fig. 5–37.

The stresses in the elements on the boundary of the region are indicated. For this example, the shear stress should be zero *everywhere* on the boundary. The normal stresses on the portions *AB* and *ED* should also be zero. The normal stresses on the portion *BC* should be 10,000 psi, showing good agreement. The degree to which these natural boundary conditions are not satisfied is an indication of the deficiency of the solution. In this case, the errors are quite high; that is, $\tau = 8273$ psi on the boundary of element 8, a significant proportion of the prescribed value of 10^4 psi on side *BC*.

TABLE 5.7 Mesh 1 stresses (psi)

Element	σ_x	σ_y	τ_{xy}
1	2964	−5061	−2019
2	4072	−1287	−1131
3	8315	−434	−1325
4	6620	−798	−1362
5	10237	491	−867
6	9170	319	−806
7	10039	1032	−340
8	14961	−1584	−3233
9	8080	−2154	−3517
10	11266	−2164	−3055
11	24977	1130	−770
12	13736	1874	−1450
13	11967	238	−429
14	13736	−895	131
15	11314	−1569	−774
16	12090	−807	−360
17	10008	−195	−758
18	10545	−34	−628
19	9870	471	−408
20	14057	−112	−346
21	13016	1400	1101
22	12645	161	984
23	10926	−306	308
24	11416	64	75

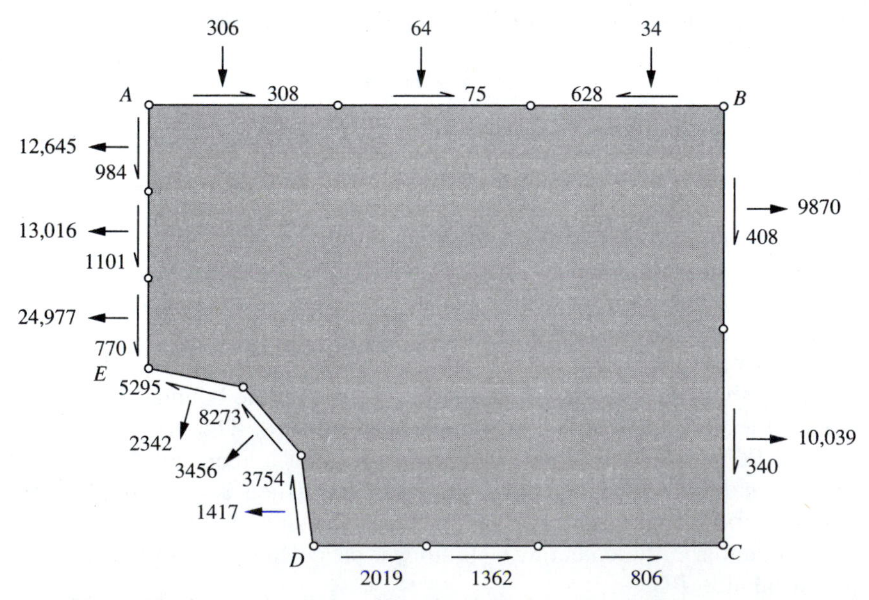

FIGURE 5–37 Satisfaction of natural boundary conditions — mesh 1

Consider also the issue of interelement continuity for the stresses in terms of the interelement continuities between elements 11, 12, 20, and 21, as shown in Fig. 5–38. Here is seen a jump of 16,703 − 8627 = 8076 psi in the normal stress between elements 11 and 12, again a significant level compared with the stresses in the two elements.

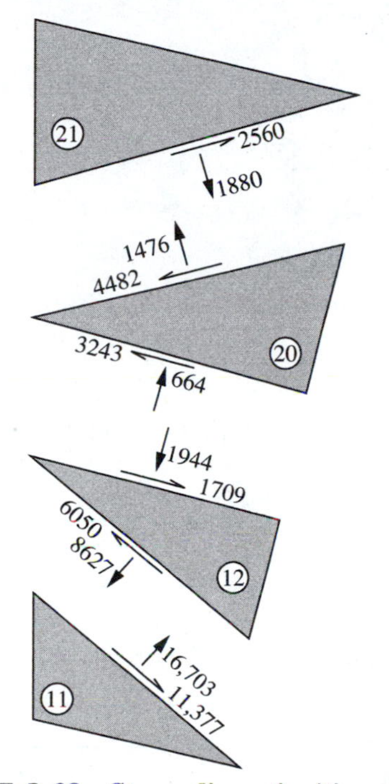

FIGURE 5–38 Stress discontinuities — mesh 1

The maximum normal stress on the section through the hole is predicted to be 24,977 psi in element 11. The average stress on this section is 15,000 psi giving rise to a predicted value of $k_T = 24,977/15,000 = 1.67$ for the stress concentration factor. This is significantly in error compared to the exact value [Cook and Young, 5] of approximately 2.25.

Consider then the more refined mesh indicated in Fig. 5–39.

Discretization and interpolation. There are 39 nodes and 54 elements. The nodes are placed uniformly around the arc of the hole. The second, third, and fourth sets of nodes are at radii of 1.3, 1.6, and 2.0 inches respectively.

Elemental formulation and assembly. These are carried out as described in the previous sections.

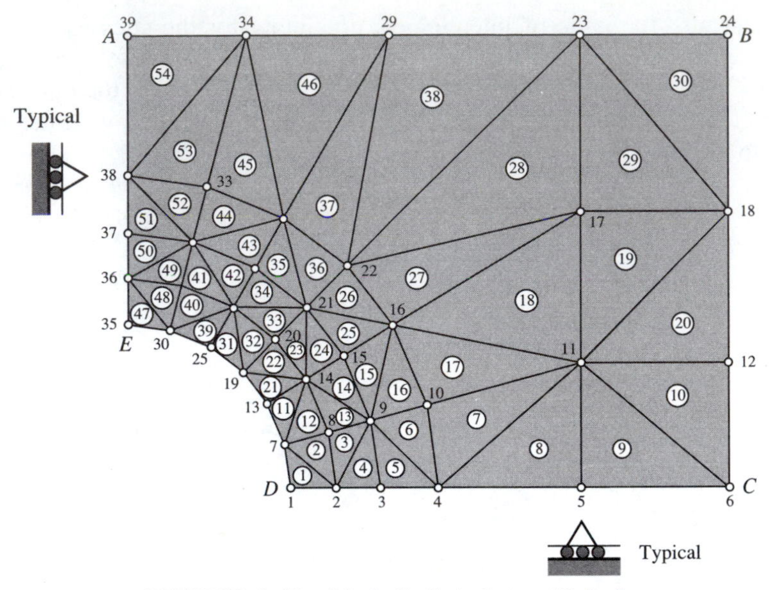

FIGURE 5–39 Mesh 2, flat plate with hole

Constraints. The constrained displacements are

$$v_1 = v_2 = v_3 = v_4 = v_5 = v_6 = 0$$

$$u_{35} = u_{36} = u_{37} = u_{38} = u_{39} = 0$$

Solution. The displacements at nodes 1, 6, 24, 35, and 39 are listed and compared with the displacements at the same physical locations from the previous model.

Mesh 2			Mesh 1		
Node	**$u(10^5$ in)**	**$v(10^5$ in)**	**Node**	**$u(10^5$ in)**	**$v(10^5$ in)**
1	309	0	1	269	0
6	600	0	4	586	0
24	553	0	21	546	0
35	0	−110	10	0	−76
39	0	−171	18	0	−113

Agreement in the overall elongation (600 versus 586 and 553 versus 546) is quite good, whereas deformations in the neighborhood of the hole differ to a larger degree. We also observe, as was pointed out in Section 5.5.1, that the finer mesh is less stiff, resulting in larger displacements.

Derived variables — stresses. The stresses in the elements forming the boundary of the region are listed in Table 5.8. The degree of success in satisfying the boundary conditions is indicated in Fig. 5–40, where the stresses acting on the boundary are shown.

TABLE 5.8 Stresses in psi — example of 5.5.4 mesh 2

Element	σ_x	σ_y	τ_{xy}
1	848	−9494	−658
4	1963	−1520	−1522
5	4411	−785	−1408
8	7329	257	−542
9	9351	864	−542
10	9754	1512	−261
11	2348	−5601	−418
20	9649	1165	−523
21	6997	−1300	−3302
30	10621	250	−394
31	15371	1263	−5151
38	11905	−704	−380
39	23180	2784	−6235
46	11613	−600	580
47	30312	1375	−1066
50	15743	3489	1064
51	15229	1777	985
54	10096	783	629

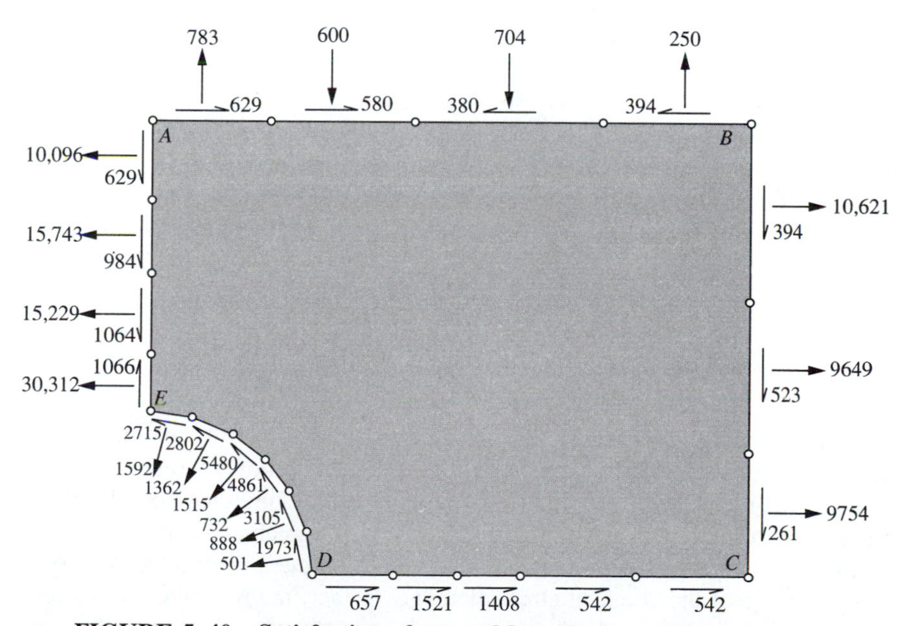

FIGURE 5–40 Satisfaction of natural boundary conditions — mesh 2

Although there is certainly still error present, the magnitudes of the stresses not satisfying the natural boundary conditions have generally decreased relative to those given by the first model in this section.

For this mesh, the maximum normal stress on the part of the boundary passing through the hole occurs in element 47 and has the value 30,312 psi. The average on this section is 15,000 psi, giving a predicted stress concentration factor $k_T = 30{,}312/15{,}000 = 2.02$, approximately 10 percent below the theoretical value of 2.25.

Consider also the interelement stress continuity for elements 47–50 as indicated in Fig. 5–41. As is clearly seen, the stress discontinuities, although reduced compared with the first mesh, are still substantial.

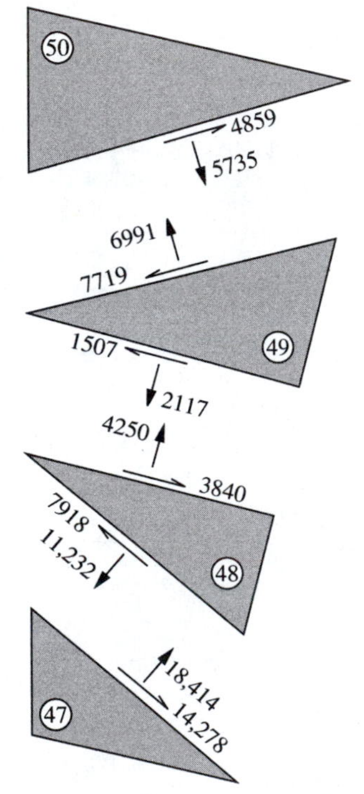

FIGURE 5–41 Interelement stress continuity—mesh 2

It cannot be overemphasized that in spite of its simplicity, the linearly interpolated triangular element or constant stress element produces results that have definite deficiencies with respect to satisfying natural boundary conditions and with respect to interelement continuity. The lack of agreement between the exact and finite element results is not due to the finite element method generally, but rather to choice of the constant stress element. In fact, many of the commercially available finite element codes have precautionary statements against the use of this particular element.

This example will be reexamined later in this chapter using other elements, which we will see provide excellent agreement with theoretical results.

5.6 RECTANGULAR ELEMENTS

From the results of Section 5.5, it is clear that the three-noded triangular or constant stress element is generally unsatisfactory in terms of accurately predicting displacements and stresses for two-dimensional elasticity. An element that significantly improves the results of the finite element method in this regard is the rectangular element discussed as follows. Extension to the more general quadrilateral element is contained in Section 5.8.

Discretization and interpolation. When using the rectangular element, the region is generally rectangular in shape and discretized as indicated in Fig. 5–42.

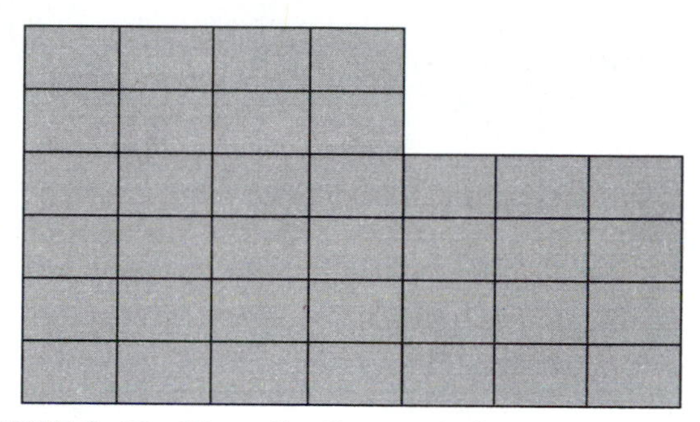

FIGURE 5–42 Discretization, typical rectangular region

When selecting the mesh, keep in mind that the aspect ratio—the ratio of the largest to the smallest side—should not exceed a value of approximately 2.5 to 3.

For a typical element, the relationship between the global coordinates (x,y) and a set of local coordinates (ξ, η) is indicated in Fig. 5–43. It turns out to be substantially easier to deal with the interpolation functions for the element in terms of a set of local coordinates such as ξ and η. Note that the relations between the global coordinates (x,y) and the local coordinates (ξ, η) is the simple translation

$$\xi = x - x_m \qquad \eta = y - y_m$$

In terms of the local coordinates ξ and η, the general form for the dependent variables u and v is expressed as

$$\begin{aligned} u_e(x,y) &= a_e + b_e\xi + c_e\eta + d_e\xi\eta \\ v_e(x,y) &= \alpha_e + \beta_e\xi + \gamma_e\eta + \delta_e\xi\eta \end{aligned} \tag{5.31}$$

The bilinear term $\xi\eta$ has been selected from among the three second-order terms ξ^2, η^2, and $\xi\eta$. The student is asked to investigate the other two terms ξ^2 and η^2 in the Exercises.

It is straightforward to conclude that the strains and hence the stresses will depend linearly upon the coordinates ξ and η within the element.

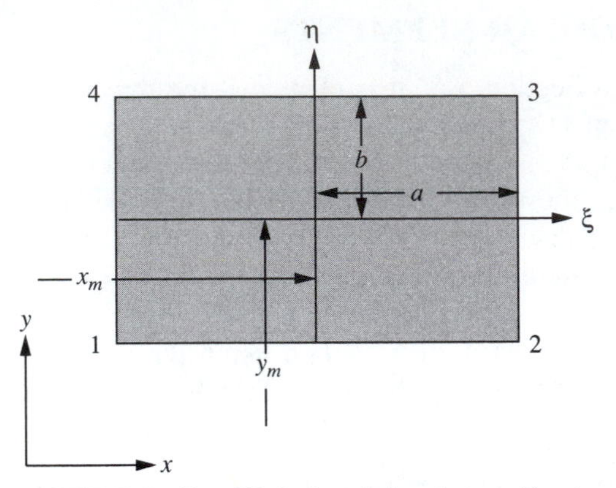

FIGURE 5–43 Global and local coordinates

Requiring that

$$\begin{aligned} u_e(x_1, y_1) &= u_1 \qquad & v_e(x_1, y_1) &= v_1 \\ u_e(x_2, y_2) &= u_2 \qquad & v_e(x_2, y_2) &= v_2 \\ u_e(x_3, y_3) &= u_3 \qquad & v_e(x_3, y_3) &= v_3 \\ u_e(x_4, y_4) &= u_4 \qquad & v_e(x_4, y_4) &= v_4 \end{aligned} \tag{5.32}$$

generates the two 4 × 4 sets of linear algebraic equations for determining the constants in Eqs. (5.31). Solving Eqs. (5.32) and reinserting the results into Eqs. (5.31) yields

$$u_e(x, y) = \sum u_i N_i(x, y) = \mathbf{u_e^T N} = \mathbf{N^T u_e}$$

$$v_e(x, y) = \sum v_i N_i(x, y) = \mathbf{v_e^T N} = \mathbf{N^T v_e}$$

with

$$N_1 = \frac{(a - \xi)(b - \eta)}{4ab} = \frac{(1 - s)(1 - t)}{4}$$

$$N_2 = \frac{(a + \xi)(b - \eta)}{4ab} = \frac{(1 + s)(1 - t)}{4}$$

$$N_3 = \frac{(a + \xi)(b + \eta)}{4ab} = \frac{(1 + s)(1 + t)}{4}$$

$$N_4 = \frac{(a - \xi)(b + \eta)}{4ab} = \frac{(1 - s)(1 + t)}{4}$$

where $s = \xi/a$ and $t = \eta/b$ are dimensionless local coordinates. See Section 3.6 for figures depicting the N_i.

Elemental formulation. The expressions for $\mathbf{k_e}$ and $\boldsymbol{\theta}_\mathbf{e}$ given by Eqs. (5.17) and (5.18) from Section 5.3 can be used in their general form,

$$\mathbf{k_e} = \int_{D_e} \mathbf{B}^{\mathrm{T}}\mathbf{C}\mathbf{B}\, d\,Vol \qquad \text{and} \qquad \boldsymbol{\theta}_\mathbf{e} = \int_{D_e} \mathbf{B}^{\mathrm{T}}\mathbf{f}_{\mathrm{T}}\, d\,Vol$$

as long as $\mathbf{B}$ is constructed keeping in mind that there are now four nodes rather than three, that is,

$$\mathbf{B} = \begin{bmatrix} \dfrac{\partial N_1}{\partial x} & 0 & \dfrac{\partial N_2}{\partial x} & 0 & \dfrac{\partial N_3}{\partial x} & 0 & \dfrac{\partial N_4}{\partial x} & 0 \\ 0 & \dfrac{\partial N_1}{\partial y} & 0 & \dfrac{\partial N_2}{\partial y} & 0 & \dfrac{\partial N_3}{\partial y} & 0 & \dfrac{\partial N_4}{\partial y} \\ \dfrac{\partial N_1}{\partial y} & \dfrac{\partial N_1}{\partial x} & \dfrac{\partial N_2}{\partial y} & \dfrac{\partial N_2}{\partial x} & \dfrac{\partial N_3}{\partial y} & \dfrac{\partial N_3}{\partial x} & \dfrac{\partial N_4}{\partial y} & \dfrac{\partial N_4}{\partial x} \end{bmatrix}$$

Similarly, the appropriate expression for the elemental load vector associated with the body force is

$$\mathbf{f}_\mathbf{e}^{\mathrm{T}} = \int_{D_e} [XN_1 \quad YN_1 \quad XN_2 \quad YN_2 \quad XN_3 \quad YN_3 \quad XN_4 \quad YN_4]\, d\,Vol \tag{5.33}$$

where the two extra entries involving N_4 clearly account for the fourth node.

Last, in view of the fact that there are two nodes on any side of the rectangular element and that the interpolation functions are linear along any edge of the element, it follows that the general form appropriate for $\mathbf{t_e}$ is given by Eq. (5.21) as

$$\mathbf{t_e} = \begin{bmatrix} \displaystyle\int_{\gamma_{te}} N_i(lt_n - mt_s)dS \\ \displaystyle\int_{\gamma_{te}} N_i(mt_n + lt_s)dS \\ \displaystyle\int_{\gamma_{te}} N_j(lt_n - mt_s)dS \\ \displaystyle\int_{\gamma_{te}} N_j(mt_n + lt_s)dS \end{bmatrix} \tag{5.34}$$

where i and j are the nodes on the boundary of the element in question as indicated in Fig. 5–44. The steps of assembly, constraints, and solution are carried out in the usual way.

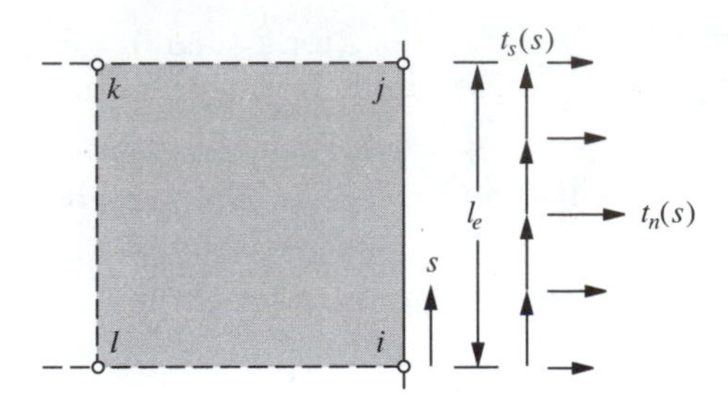

FIGURE 5–44 Boundary nodes for evaluation of $\mathbf{t_e}$

Derived variables or stresses. Consider the general expression for the stresses, namely

$$\boldsymbol{\sigma} = \mathbf{CBU_e} - \mathbf{f_T}$$

in the absence of any thermal inputs, that is, $\mathbf{f_T} = \mathbf{0}$. Assuming that $\mathbf{C}$ is constant within a particular element, there is still a dependence of the stresses within the element due to the character of the strain matrix $\mathbf{B}$. Specifically,

$$\mathbf{CB} = \begin{bmatrix} -bC_{11}(1-t) & -aC_{12}(1-s) & bC_{11}(1-t) & -aC_{12}(1+s) \\ -bC_{12}(1-t) & -aC_{22}(1-s) & bC_{12}(1-t) & -aC_{22}(1+s) \\ -aC_{33}(1-s) & -bC_{33}(1-t) & aC_{33}(1+s) & -bC_{33}(1-t) \end{bmatrix.$$
$$\begin{matrix} . & bC_{11}(1+t) & aC_{12}(1+s) & -bC_{11}(1+t) & aC_{12}(1-s) \\ & bC_{12}(1+t) & aC_{22}(1+s) & -bC_{12}(1+t) & aC_{22}(1-s) \\ & aC_{33}(1+s) & bC_{33}(1+t) & aC_{33}(1-s) & -bC_{33}(1+t) \end{bmatrix}$$

clearly indicating that all of σ_x, σ_y, and τ_{xy} are linear functions of s and t within the element. It is also apparent that in general, the equations of equilibrium are not satisfied within the element.

There are clearly several options for evaluating the stresses within an element. One approach is to compute, per element, the stresses at each of the four nodes. At a particular node these nodal stress values can then be averaged over all the elements that have that node in common. Although popular, this approach is not particularly accurate. Better values are in fact obtained if the stresses are computed at the midsides of the element and then averaged between elements.

Another approach indicated by Barlow [7] is to evaluate the stresses at the centroid ($s = t = 0$) for the rectangular element. This corresponds to the Gauss point for a one-point Gauss quadrature and represents the best point at which to evaluate the stresses for the rectangular element. "Best" refers to the fact that the error in computing the stresses is less than at other points within the element. Both of these approaches will be demonstrated and discussed in the examples that follow.

Evaluation of matrices. Both $\mathbf{k_e}$ and $\boldsymbol{\theta}_\mathbf{e}$ involve the strain matrix $\mathbf{B}$ given as before by

$$\mathbf{B} = \begin{bmatrix} \frac{\partial N_1}{\partial x} & 0 & \frac{\partial N_2}{\partial x} & 0 & \frac{\partial N_3}{\partial x} & 0 & \frac{\partial N_4}{\partial x} & 0 \\ 0 & \frac{\partial N_1}{\partial y} & 0 & \frac{\partial N_2}{\partial y} & 0 & \frac{\partial N_3}{\partial y} & 0 & \frac{\partial N_4}{\partial y} \\ \frac{\partial N_1}{\partial y} & \frac{\partial N_1}{\partial x} & \frac{\partial N_2}{\partial y} & \frac{\partial N_2}{\partial x} & \frac{\partial N_3}{\partial y} & \frac{\partial N_3}{\partial x} & \frac{\partial N_4}{\partial y} & \frac{\partial N_4}{\partial x} \end{bmatrix}$$

Note that since $\xi = x - x_m = as$ and $\eta = y - y_m = bt$, it follows that

$$\frac{\partial}{\partial x} = \left(\frac{1}{a}\right)\left(\frac{\partial}{\partial s}\right) \qquad \frac{\partial}{\partial y} = \left(\frac{1}{b}\right)\left(\frac{\partial}{\partial t}\right)$$

so that $\mathbf{B}$ may be expressed as

$$\mathbf{B}^{\mathrm{T}} = \frac{1}{4} \begin{bmatrix} -\frac{(1-t)}{a} & 0 & -\frac{(1-s)}{b} \\ 0 & -\frac{(1-s)}{b} & -\frac{(1-t)}{a} \\ \frac{(1-t)}{a} & 0 & -\frac{(1+s)}{b} \\ 0 & -\frac{(1+s)}{b} & \frac{(1-t)}{a} \\ \frac{(1+t)}{a} & 0 & \frac{(1+s)}{b} \\ 0 & \frac{(1+s)}{b} & \frac{(1+t)}{a} \\ -\frac{(1+t)}{a} & 0 & \frac{(1-s)}{b} \\ 0 & \frac{(1-s)}{b} & -\frac{(1+t)}{a} \end{bmatrix}$$

Multiplying to produce the product $\mathbf{B}^{\mathrm{T}}\mathbf{C}\mathbf{B}$ and subsequently integrating the result yields the 8×8 stiffness matrix

$$\mathbf{k_e} = \begin{bmatrix} \mathbf{A}_{11} & \mathbf{A}_{12} & \mathbf{A}_{13} \\ & \mathbf{A}_{22} & \mathbf{A}_{23} \\ \text{symm} & & \mathbf{A}_{33} \end{bmatrix} \qquad (5.35)$$

where, with $p = b/a$ and $q = 1/p$,

$$\mathbf{A}_{11} = \begin{bmatrix} 4(C_{11}p + C_{33}q) & 3(C_{12} + C_{33}) & 2(C_{33}q - 2C_{11}p) \\ & 4(C_{22}q + C_{33}p) & 3(C_{33} - C_{12}) \\ \text{symm} & & 4(C_{11}p + C_{33}q) \end{bmatrix} \frac{t}{12}$$

$$\mathbf{A_{12}} = \begin{bmatrix} 3(C_{12} - C_{33}) & -2(C_{33}q + C_{11}p) & -3(C_{12} + C_{33}) \\ 2(C_{22}q - 2C_{33}p) & -3(C_{12} + C_{33}) & -2(C_{22}q + C_{33}p) \\ -3(C_{12} + C_{33}) & 2(C_{11}p - 2C_{33}q) & 3(C_{12} - C_{33}) \end{bmatrix} \frac{t}{12}$$

$$\mathbf{A_{13}} = \begin{bmatrix} 2(C_{11}p - 2C_{33}q) & 3(C_{33} - C_{12}) \\ 3(C_{12} - C_{33}) & 2(C_{33}p - 2C_{22}q) \\ -2(C_{11}p + C_{33}q) & 3(C_{12} + C_{33}) \end{bmatrix} \frac{t}{12}$$

$$\mathbf{A_{22}} = \begin{bmatrix} 4(C_{22}q + C_{33}p) & 3(C_{33} - C_{12}) & 2(C_{33}p - 2C_{22}q) \\ & 4(C_{11}p + C_{33}q) & 3(C_{12} + C_{33}) \\ \text{symm} & & 4(C_{22}q + C_{33}p) \end{bmatrix} \frac{t}{12}$$

$$\mathbf{A_{23}} = \begin{bmatrix} 3(C_{12} + C_{33}) & -2(C_{22}q + C_{33}p) \\ 2(C_{33}q - 2C_{11}p) & 3(C_{12} - C_{33}) \\ 3(C_{33} - C_{12}) & 2(C_{22}q - 2C_{33}p) \end{bmatrix} \frac{t}{12}$$

$$\mathbf{A_{33}} = \begin{bmatrix} 4(C_{11}p + C_{33}q) & -3(C_{12} + C_{33}) \\ \text{symm} & 4(C_{22}q + C_{33}p) \end{bmatrix} \frac{t}{12}$$

The elemental stiffness matrices are loaded into the global stiffness matrix in the usual fashion.

By expanding the matrix product $\mathbf{B}^{\mathrm{T}}\mathbf{f_T}$, the elemental temperature load matrix $\boldsymbol{\theta}_{\mathbf{e}}$ can be expressed as

$$\boldsymbol{\theta}_{\mathbf{e}} = \int \beta T \begin{bmatrix} \frac{\partial N_1}{\partial x} \\ \frac{\partial N_1}{\partial y} \\ \frac{\partial N_2}{\partial x} \\ \frac{\partial N_2}{\partial y} \\ \frac{\partial N_3}{\partial x} \\ \frac{\partial N_3}{\partial y} \\ \frac{\partial N_4}{\partial x} \\ \frac{\partial N_4}{\partial y} \end{bmatrix} d\,Vol \tag{5.36}$$

If the temperature term βT is reasonably uniform within the element it can be *approximated* in terms of the interpolation functions N_i as

$$\beta T = \sum_{1}^{4} (\beta T)_i N_i \tag{5.37}$$

where $(\beta T)_i$ are suitably chosen nodal values, often taken to be the values of the function βT at the nodes. Substituting the approximate expression given by Eq. (5.37) into Eq. (5.36) and evaluating the integrals yields

$$\boldsymbol{\theta}_{\mathbf{e}} = \frac{t}{6}\begin{bmatrix} -2b & -2b & -b & -b \\ -2a & -a & -a & -2a \\ 2b & 2b & b & b \\ -a & -2a & -2a & -a \\ b & b & 2b & 2b \\ a & 2a & 2a & a \\ -b & -b & -2b & -2b \\ 2a & a & a & 2a \end{bmatrix}\begin{bmatrix} (\beta T)_1 \\ (\beta T)_2 \\ (\beta T)_3 \\ (\beta T)_4 \end{bmatrix} \tag{5.38}$$

If βT is constant or can be approximated by a constant, say $(\beta T)_0$, within the element, $\boldsymbol{\theta}_{\mathbf{e}}$ becomes

$$\boldsymbol{\theta}_{\mathbf{e}}^{\mathrm{T}} = t(\beta T)_0[-b \quad -a \quad b \quad -a \quad b \quad a \quad -b \quad a]$$

These components, which represent the nodal forces on the element due to the thermal input, are pictured in Fig. 5–45, for the case where $(\beta T)_0$ is a constant.

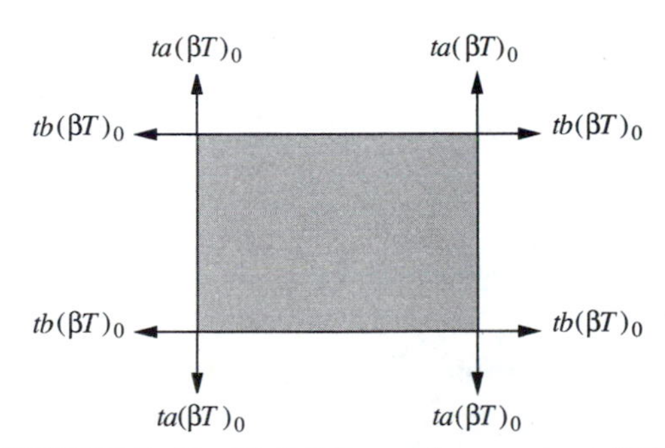

FIGURE 5–45 Nodal forces for constant $(\beta T)_0$

From Eq. (5.33), the elemental body force contribution is represented by

$$\mathbf{f}_{\mathbf{e}}^{\mathrm{T}} = \int_{D_e} [XN_1 \quad YN_1 \quad XN_2 \quad YN_2 \quad XN_3 \quad YN_3 \quad XN_4 \quad YN_4]\, d\,Vol$$

These integrals can be evaluated upon knowing the specific character of the components X and Y. If X and Y are reasonably smooth functions of position within the element, each can be *approximated* by

$$X \approx \sum_1^4 X_i N_i \qquad \text{and} \qquad Y \approx \sum_1^4 Y_i N_i$$

with the result that $\mathbf{f}_{\mathbf{e}}$ can be represented approximately as

$$
\mathbf{f_e} = \begin{bmatrix} 4X_1 + 2X_2 + X_3 + 2X_4 \\ 4Y_1 + 2Y_2 + Y_3 + 2Y_4 \\ 2X_1 + 4X_2 + 2X_3 + X_4 \\ 2Y_1 + 4Y_2 + 2Y_3 + Y_4 \\ X_1 + 2X_2 + 4X_3 + 2X_4 \\ Y_1 + 2Y_2 + 4Y_3 + 2Y_4 \\ 2X_1 + X_2 + X_3 + 4X_4 \\ 2Y_1 + Y_2 + Y_3 + 4Y_4 \end{bmatrix} \frac{abt}{36} \tag{5.39}
$$

The evaluation of the integrals for the elemental inputs arising from the stresses prescribed on the surface, namely the $\mathbf{t_e}$, is carried out in precisely the same fashion as for the triangular element in Section 5.4, to which the reader is referred.

5.6.1 Thin Rectangular Plate — Moment Loading

Consider again the plane stress problem of the thin plate loaded by a moment, as shown in Fig. 5–46.

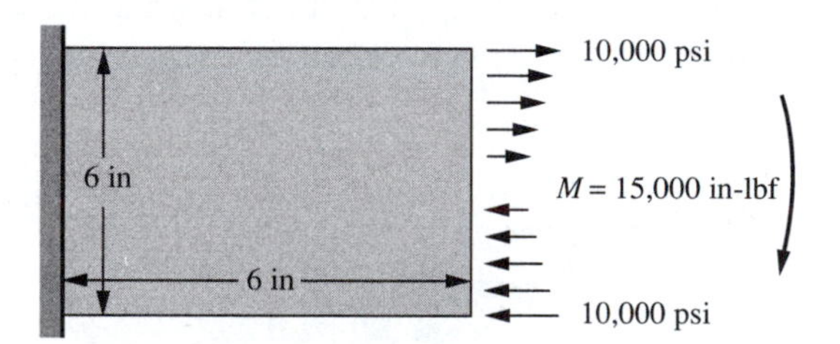

FIGURE 5–46 Moment loaded thin plate

Discretization and interpolation. The simplest possible model consists of a single element as shown in Fig. 5–47.

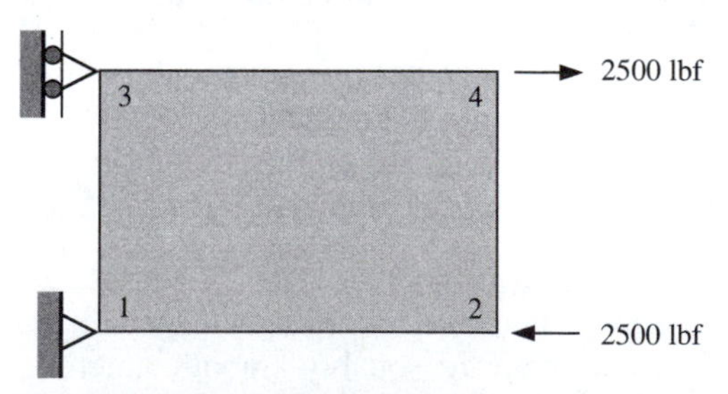

FIGURE 5–47 Mesh 1 with nodal forces and constraints

Elemental formulation and assembly. For the single element composed of an isotropic solid with $E = 10^7$ psi, $\nu = 0.3$, $a = b = 6.0$ in. $\mathbf{k_e}$ can be computed using Eq. (5.35) as

$$\begin{bmatrix} 1.2363 & 0.4464 & -0.7555 & -0.0343 & 0.1374 & 0.0343 & -0.6181 & -0.4464 \\ & 1.2363 & 0.0343 & 0.1364 & -0.0343 & -0.7555 & -0.4464 & -0.6181 \\ & & 1.2363 & -0.4464 & -0.6181 & 0.4464 & 0.1374 & -0.0343 \\ & & & 1.2363 & 0.4463 & -0.6181 & 0.0343 & -0.7555 \\ & & & & 1.2363 & -0.4464 & -0.7555 & 0.0343 \\ & & & & & 1.2363 & -0.0343 & 0.1374 \\ & \text{symm} & & & & & 1.2363 & 0.4464 \\ & & & & & & & 1.2363 \end{bmatrix} \begin{matrix} 10^6 \\ \text{lb/fin} \end{matrix}$$

which is also $\mathbf{k_G}$, since there is only one element.

Using Eq. (5.34) the nodal loads are computed to be

$$\mathbf{t_e} = \begin{matrix} & u_2 & v_2 & u_4 & v_4 & \\ [& -2500.0 & 0.0 & 2500.0 & 0.0 &]^{\mathrm{T}} \end{matrix}$$

with the expanded global load vector as

$$\mathbf{F_G} = [0.0 \quad 0.0 \quad -2500.0 \quad 0.0 \quad 0.0 \quad 0.0 \quad 2500.0 \quad 0.0]^{\mathrm{T}}$$

Constraints. From Fig. 5–47, the constraints are easily seen to be

$$u_1 = v_1 = u_3 = 0$$

with the resulting constrained equations appearing as

$$\begin{bmatrix} 1.0 & 0.0 & 0.0 & 0.0 & 0.0 & 0.0 & 0.0 & 0.0 \\ & 1.0 & 0.0 & 0.0 & 0.0 & 0.0 & 0.0 & 0.0 \\ & & 1.2363 & -0.4464 & 0.0 & 0.4464 & 0.1374 & -0.0343 \\ & & & 1.2363 & 0.0 & -0.6181 & 0.0343 & -0.7555 \\ & & & & 1.0 & 0.0000 & 0.0000 & 0.0000 \\ & & & & & 1.2363 & -0.0343 & 0.1374 \\ & \text{symm} & & & & & 1.2363 & 0.4464 \\ & & & & & & & 1.2363 \end{bmatrix} \mathbf{u_G} = \begin{bmatrix} 0.0000 \\ 0.0000 \\ -0.0025 \\ 0.0000 \\ 0.0000 \\ 0.0000 \\ 0.0025 \\ 0.0000 \end{bmatrix}$$

Solution. The above set of constrained equations is solved to yield

$$\mathbf{u_G^T} = [0.0 \quad 0.0 \quad -0.004044 \quad -0.004044 \quad 0.0 \quad 0.0 \quad 0.004044 \quad -0.004044]$$

Derived variables or stresses. As discussed in Section 5.6 for the rectangular element, the stresses computed according to

$$\boldsymbol{\sigma} = \mathbf{CBU_e}$$

are linear functions of the x and y or the s and t coordinates within the element. We will demonstrate the nodal evaluation and the centroidal evaluation of the stresses in what follows.

For the present example, the stress values computed at the nodes are indicated in Table 5.9.

The node numbers, the s and t values at the nodes, and the nodal stresses are indicated. On the basis that the stress components are linear functions of s

TABLE 5.9 Nodal stress values

Node	s	t	σ_x	σ_y	τ_{xy}
1	−1	−1	−7407	−2222	−2593
2	1	−1	−7407	−2222	2593
3	−1	1	7407	2222	−2593
4	1	1	7407	2222	2593

and t within the element, it follows that for $s \pm 1$ (faces 2-4 and 1-3 respectively), σ_x and τ_{xy} are linear functions of t. Similarly, when $t = \pm 1$ (faces 3-4 and 1-2 respectively), σ_y and τ_{xy} are linear functions of s. Using the nodal values given in Table 5.9, one can then conclude that the stresses on the faces are as shown in Fig. 5–48.

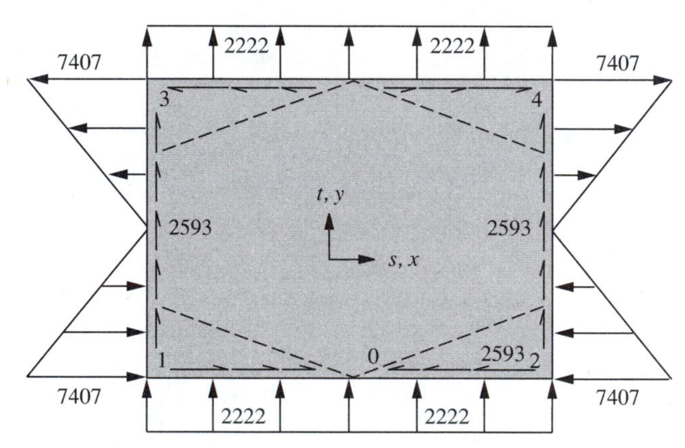

FIGURE 5–48 Stresses on the boundary—mesh 1

Note that these stresses, computed as derived variables, do not satisfy equilibrium! The sum of the forces in the y-direction is clearly not zero.

Equilibrium is satisfied, however, by the nodal forces computed on the basis of $\mathbf{F_e} = \mathbf{k_e U_e}$ for the element. For this example, these nodal forces are shown in Fig. 5–49, and clearly satisfy equilibrium.

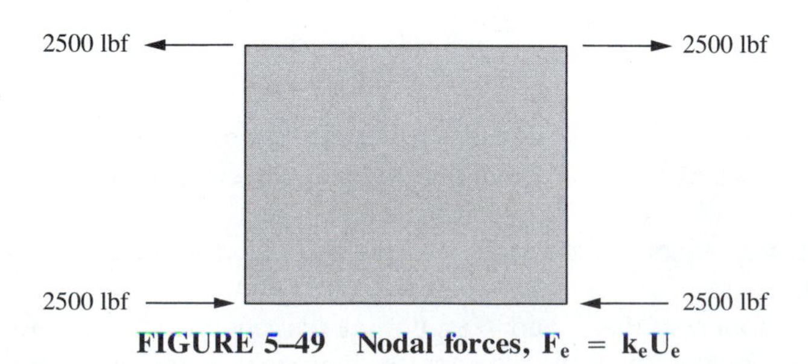

FIGURE 5–49 Nodal forces, $\mathbf{F_e} = \mathbf{k_e U_e}$

The global equilibrium equations $\mathbf{K_G U_G} = \mathbf{F_G}$ are equilibrium equations that have been assembled from elemental statements $\mathbf{F_e} = \mathbf{k_e U_e}$, each of which is an equilibrium statement. In a mesh with multiple elements, each element can be shown to be in equilibrium under the action of the nodal forces computed according to $\mathbf{F_e} = \mathbf{k_e U_e}$. The stresses, which are the derived variables for the elasticity problem, do not in general satisfy equilibrium. Considered as functions of position within the element, as for the previous one-element example, the computed stresses generally fail to satisfy the equations of equilibrium.

Evaluation of the stresses at the nodes turns out to be the least accurate, whereas stresses evaluated at the centroid of the element turn out to be the most accurate. In particular for this example, evaluating the stresses at the centroid ($s = t = 0$) of the element yields

$$(\sigma_x)_{\text{cent}} = (\sigma_y)_{\text{cent}} = (\tau_{xy})_{\text{cent}} = 0$$

These are the exact values of the stresses at the centroid. Although not particularly useful for this particular problem in terms of any resolution of the stresses within the region, the generally more accurate centroidal values are useful when the mesh is refined and a clearer picture of the spatial dependence of the stresses can be seen.

5.6.2 Thin Rectangular Plate — Moment Loading, Meshes 2 and 3

Consider once again the problem of Section 5.6.1, the moment loading of the thin plate. The reader should review the problem statement given in Section 5.6.1.

Discretization, interpolation, and elemental formulation. Discretization for meshes 2 and 3 is indicated in Fig. 5–50. The elemental stiffness matrices $\mathbf{k_e}$ each have the form given in Eq. (5.35).

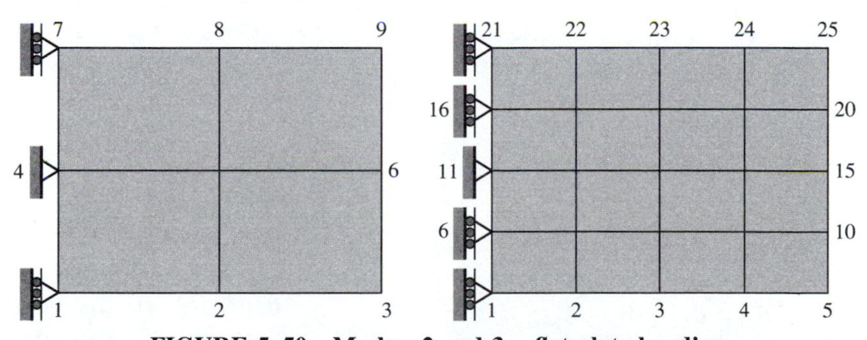

FIGURE 5–50 Meshes 2 and 3 — flat plate bending

Assembly. For mesh 2, the assembly process is indicated as follows, where each of the elemental stiffness matrices is shown before and after expansion to the global level.

Element 1.

$$\mathbf{k}_{e1} = \begin{array}{c} \begin{matrix} 1 & 2 & 5 & 4 \end{matrix} \\ \begin{bmatrix} \mathbf{a}_{11} & \mathbf{a}_{12} & \mathbf{a}_{13} & \mathbf{a}_{14} \\ \mathbf{a}_{21} & \mathbf{a}_{22} & \mathbf{a}_{23} & \mathbf{a}_{24} \\ \mathbf{a}_{31} & \mathbf{a}_{32} & \mathbf{a}_{33} & \mathbf{a}_{34} \\ \mathbf{a}_{41} & \mathbf{a}_{42} & \mathbf{a}_{43} & \mathbf{a}_{44} \end{bmatrix} \end{array} \qquad \mathbf{k}_{G1} = \begin{array}{c} \begin{matrix} 1 & 2 & 3 & 4 & 5 & 6 \end{matrix} \\ \begin{bmatrix} \mathbf{a}_{11} & \mathbf{a}_{12} & \mathbf{0} & \mathbf{a}_{14} & \mathbf{a}_{13} & \cdots \\ \mathbf{a}_{21} & \mathbf{a}_{22} & \mathbf{0} & \mathbf{a}_{24} & \mathbf{a}_{23} & \cdots \\ \mathbf{0} & \mathbf{0} & \mathbf{0} & \mathbf{0} & \mathbf{0} & \\ \mathbf{a}_{41} & \mathbf{a}_{42} & \mathbf{0} & \mathbf{a}_{44} & \mathbf{a}_{43} & \\ \mathbf{a}_{31} & \mathbf{a}_{32} & \mathbf{0} & \mathbf{a}_{34} & \mathbf{a}_{33} & \\ \vdots & \vdots & & & & \end{bmatrix} \end{array}$$

Element 2.

$$\mathbf{k}_{e2} = \begin{array}{c} \begin{matrix} 2 & 3 & 6 & 5 \end{matrix} \\ \begin{bmatrix} \mathbf{b}_{11} & \mathbf{b}_{12} & \mathbf{b}_{13} & \mathbf{b}_{14} \\ \mathbf{b}_{21} & \mathbf{b}_{22} & \mathbf{b}_{23} & \mathbf{b}_{24} \\ \mathbf{b}_{31} & \mathbf{b}_{32} & \mathbf{b}_{33} & \mathbf{b}_{34} \\ \mathbf{b}_{41} & \mathbf{b}_{42} & \mathbf{b}_{43} & \mathbf{b}_{44} \end{bmatrix} \end{array} \qquad \mathbf{k}_{G2} = \begin{array}{c} \begin{matrix} & 2 & 3 & 4 & 5 & 6 & \end{matrix} \\ \begin{bmatrix} \cdot & \cdot & \cdot & \cdot & \cdot & \cdot & \cdot \\ \cdot & \mathbf{b}_{11} & \mathbf{b}_{12} & \mathbf{0} & \mathbf{b}_{14} & \mathbf{b}_{13} & \cdot \\ \cdot & \mathbf{b}_{21} & \mathbf{b}_{22} & \mathbf{0} & \mathbf{b}_{24} & \mathbf{b}_{23} & \cdot \\ \cdot & \mathbf{0} & \mathbf{0} & \mathbf{0} & \mathbf{0} & \mathbf{0} & \cdot \\ \cdot & \mathbf{b}_{41} & \mathbf{b}_{42} & \mathbf{0} & \mathbf{b}_{44} & \mathbf{b}_{43} & \cdot \\ \cdot & \mathbf{b}_{31} & \mathbf{b}_{32} & \mathbf{0} & \mathbf{b}_{34} & \mathbf{b}_{33} & \cdot \\ \cdot & \cdot & \cdot & \cdot & \cdot & \cdot & \cdot \end{bmatrix} \end{array}$$

Element 3.

$$\mathbf{k}_{e3} = \begin{array}{c} \begin{matrix} 4 & 5 & 8 & 7 \end{matrix} \\ \begin{bmatrix} \mathbf{c}_{11} & \mathbf{c}_{12} & \mathbf{c}_{13} & \mathbf{c}_{14} \\ \mathbf{c}_{21} & \mathbf{c}_{22} & \mathbf{c}_{23} & \mathbf{c}_{24} \\ \mathbf{c}_{31} & \mathbf{c}_{32} & \mathbf{c}_{33} & \mathbf{c}_{34} \\ \mathbf{c}_{41} & \mathbf{c}_{42} & \mathbf{c}_{43} & \mathbf{c}_{44} \end{bmatrix} \end{array} \qquad \mathbf{k}_{G3} = \begin{array}{c} \begin{matrix} & 4 & 5 & 6 & 7 & 8 & \end{matrix} \\ \begin{bmatrix} \cdot & \cdot & \cdot & \cdot & \cdot & \cdot & \cdot \\ \cdot & \mathbf{c}_{11} & \mathbf{c}_{12} & \mathbf{0} & \mathbf{c}_{14} & \mathbf{c}_{13} & \cdot \\ \cdot & \mathbf{c}_{21} & \mathbf{c}_{22} & \mathbf{0} & \mathbf{c}_{24} & \mathbf{c}_{23} & \cdot \\ \cdot & \mathbf{0} & \mathbf{0} & \mathbf{0} & \mathbf{0} & \mathbf{0} & \cdot \\ \cdot & \mathbf{c}_{41} & \mathbf{c}_{42} & \mathbf{0} & \mathbf{c}_{44} & \mathbf{c}_{43} & \cdot \\ \cdot & \mathbf{c}_{31} & \mathbf{c}_{32} & \mathbf{0} & \mathbf{c}_{34} & \mathbf{c}_{33} & \cdot \\ \cdot & \cdot & \cdot & \cdot & \cdot & \cdot & \cdot \end{bmatrix} \end{array}$$

Element 4.

$$\mathbf{k}_{e4} = \begin{array}{c} \begin{matrix} 5 & 6 & 9 & 8 \end{matrix} \\ \begin{bmatrix} \mathbf{d}_{11} & \mathbf{d}_{12} & \mathbf{d}_{13} & \mathbf{d}_{14} \\ \mathbf{d}_{21} & \mathbf{d}_{22} & \mathbf{d}_{23} & \mathbf{d}_{24} \\ \mathbf{d}_{31} & \mathbf{d}_{32} & \mathbf{d}_{33} & \mathbf{d}_{34} \\ \mathbf{d}_{41} & \mathbf{d}_{42} & \mathbf{d}_{43} & \mathbf{d}_{44} \end{bmatrix} \end{array} \qquad \mathbf{k}_{G4} = \begin{array}{c} \begin{matrix} & 4 & 5 & 6 & 7 & 8 & 9 \end{matrix} \\ \begin{bmatrix} \cdot & \cdot & \cdot & \cdot & \cdot & \cdot & \cdot \\ \cdot & \cdot & \mathbf{d}_{11} & \mathbf{d}_{12} & \mathbf{0} & \mathbf{d}_{14} & \mathbf{d}_{13} \\ \cdot & \cdot & \mathbf{d}_{21} & \mathbf{d}_{22} & \mathbf{0} & \mathbf{d}_{24} & \mathbf{d}_{23} \\ \cdot & \cdot & \mathbf{0} & \mathbf{0} & \mathbf{0} & \mathbf{0} & \mathbf{0} \\ \cdot & \cdot & \mathbf{d}_{41} & \mathbf{d}_{42} & \mathbf{0} & \mathbf{d}_{44} & \mathbf{d}_{43} \\ \cdot & \cdot & \mathbf{d}_{31} & \mathbf{d}_{32} & \mathbf{0} & \mathbf{d}_{34} & \mathbf{d}_{33} \\ \cdot & \cdot & \cdot & \cdot & \cdot & \cdot & \cdot \end{bmatrix} \end{array}$$

Each of the $\mathbf{k}_{ei}$ represents the elemental stiffness matrix, whereas each of the $\mathbf{k}_{Gi}$ represents the corresponding elemental stiffness matrix expanded and ready for assembly. Note that in each case (for this particular mesh) the third and fourth rows and columns must be interchanged before the elemental matrices are expanded to the global level.

After assembly, the global stiffness matrix appears as

$$
\begin{array}{ccccccccc}
1 & 2 & 3 & 4 & 5 & 6 & 7 & 8 & 9
\end{array}
$$

$$
\begin{bmatrix}
\mathbf{a}_{11} & \mathbf{a}_{12} & \mathbf{0} & \mathbf{a}_{14} & \mathbf{a}_{13} & \mathbf{0} & \mathbf{0} & \mathbf{0} & \mathbf{0} \\
 & \mathbf{a}_{22}+\mathbf{b}_{11} & \mathbf{b}_{12} & \mathbf{a}_{24} & \mathbf{a}_{23}+\mathbf{b}_{14} & \mathbf{b}_{13} & \mathbf{0} & \mathbf{0} & \mathbf{0} \\
 & & \mathbf{b}_{22} & \mathbf{0} & \mathbf{b}_{24} & \mathbf{b}_{23} & \mathbf{0} & \mathbf{0} & \mathbf{0} \\
 & & & \mathbf{a}_{44}+\mathbf{c}_{11} & \mathbf{a}_{43}+\mathbf{c}_{12} & \mathbf{0} & \mathbf{c}_{14} & \mathbf{c}_{13} & \mathbf{0} \\
 & & & & \mathbf{k}_{55} & \mathbf{b}_{43}+\mathbf{d}_{12} & \mathbf{c}_{24} & \mathbf{c}_{23}+\mathbf{d}_{14} & \mathbf{d}_{13} \\
 & & & & & \mathbf{b}_{33}+\mathbf{d}_{22} & \mathbf{0} & \mathbf{d}_{24} & \mathbf{d}_{23} \\
 & & & & & & \mathbf{c}_{44} & \mathbf{c}_{43} & \mathbf{0} \\
 & \text{symm} & & & & & & \mathbf{c}_{33}+\mathbf{d}_{44} & \mathbf{d}_{43} \\
 & & & & & & & & \mathbf{d}_{33}
\end{bmatrix}
$$

where $\mathbf{k}_{55} = \mathbf{a}_{33} + \mathbf{b}_{44} + \mathbf{c}_{22} + \mathbf{d}_{11}$.

The elemental load vectors for elements 2 and 4 are computed using Eq. (5.34) and are shown in Fig. 5–51, resulting in the global load vector

$$
\mathbf{F}_G = [0 \quad 0 \quad 0 \quad 0 \quad -2500 \quad 0 \quad 0 \quad 0 \quad 0 \quad 0 \quad 0 \quad 0 \quad 0 \quad 0 \quad 0 \quad 0 \quad 2500 \quad 0]^T
$$

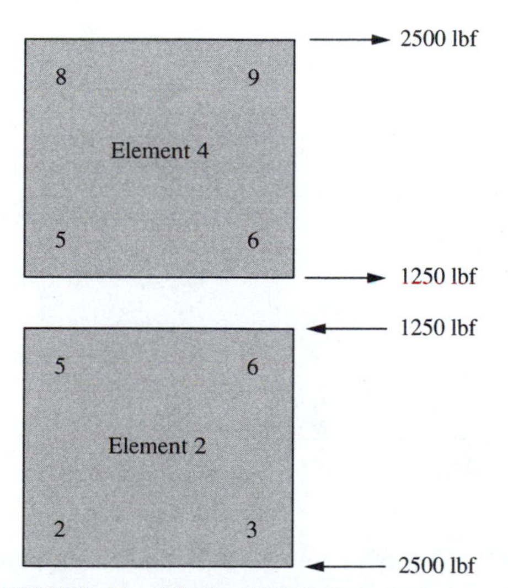

FIGURE 5–51 Elemental load vectors—mesh 2

Constraints. The constraints, which are indicated in Fig. 5–50, are

$$
u_1 = v_4 = u_4 = u_7 = 0
$$

Results. The results of evaluating the stresses at the nodes for each of the elements are presented in Table 5.10. The corresponding stresses for element 2 are indicated in Fig. 5–52. Again we see that these stresses do not satisfy equilibrium.

TABLE 5.10 Nodal stresses—mesh 2

Element 1				Element 2			
Node	σ_x	σ_y	τ_{xy}	Node	σ_x	σ_y	τ_{xy}
1	−9363	−1471	−1716	2	−9363	−1471	−1716
2	−9363	−1471	1716	3	−9363	−1471	1716
4	441	1471	−1716	5	441	1471	−1716
5	441	1471	1716	6	441	1471	1716
Element 3				**Element 4**			
Node	σ_x	σ_y	τ_{xy}	Node	σ_x	σ_y	τ_{xy}
4	−441	−1471	−1716	5	−441	−1471	−1716
5	−441	−1471	1716	6	−441	−1471	1716
7	9363	1471	−1716	8	9363	1471	−1716
8	9363	1471	1716	9	9363	1471	1716

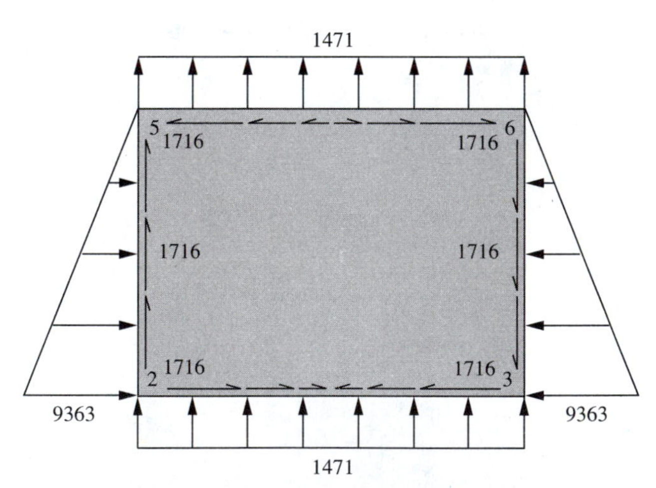

FIGURE 5–52 Boundary stresses — element 2

For this example, the results of *averaging* the stresses at the nodes as well as the centroidal values of the stresses are shown in Fig. 5–53.

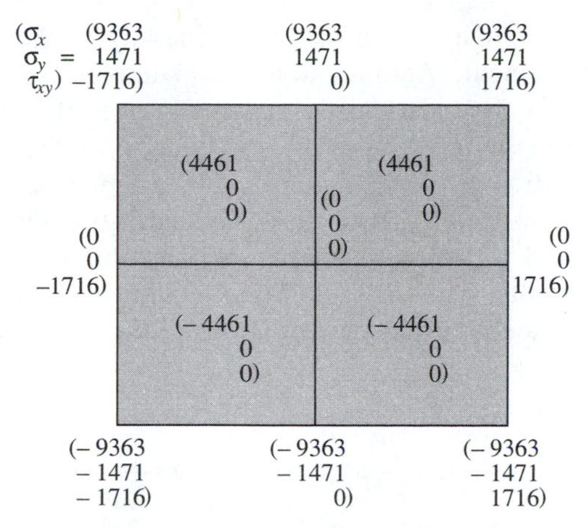

FIGURE 5–53 Averaged nodal and centroidal stresses — mesh 2

The nodally averaged values of the normal stress σ_x are quite good whereas the averaged values of the corresponding shear stress τ_{xy} and normal stress σ_y are poor. The centroidal values are quite good compared to the exact values of (±5000, 0, 0) psi, respectively, for all three of σ_x, σ_y, and τ_{xy}. Interelement continuity can be checked with the result that σ_x suffers slight discontinuities between adjacent elements with the discontinuities in σ_y and τ_{xy} relatively quite large.

For mesh 3, the centroidal values of the stresses are indicated in Fig. 5–54. These centroidal values compare well with the exact values of ±7500 and ±2500 for σ_x and 0 for both σ_y and τ_{xy}. The nodal values for the stresses (not shown here) are not as accurate as compared to the exact solution.

Nodes					Nodes
21	22	23	24		25
	(7280, 0, 0)	(7280, − 2, − 2)	(7278, − 20, − 20)	(7268, 23, − 52)	
16					20
	(2427, 0, 0)	(2428, − 4, 2)	(2443, − 23, 20)	(2509, 27, 52)	
11					15
	(− 2427, 0, 0)	(− 2428, 4, 2)	(− 2443, 23, 20)	(− 2509, − 27, 52)	
6					10
	(7280, 0, 0)	(7280, 2, − 2)	(7278, 20, − 20)	(− 7268, − 23, − 52)	
1	2	3	4		5

FIGURE 5–54 Centroidal stresses — mesh 3

Discussion. The rectangular element suffers generally in that equilibrium is not satisfied within an element. Additionally, the examples indicate that there are errors in satisfying the boundary conditions. In spite of these deficiencies, it turns out that the stresses evaluated at the centroid of the elements are quite accurate. For this particular example, the results from meshes 2 and 3 and the centroids of the elements give a pretty reasonable indication that σ_x doesn't depend upon x and that σ_y and τ_{xy} are vanishingly small, perhaps zero. Extrapolation of the centroidal σ_x results from meshes 2 and 3, respectively, are presented in Fig. 5–55, where we see good agreement between the two models and the exact results.

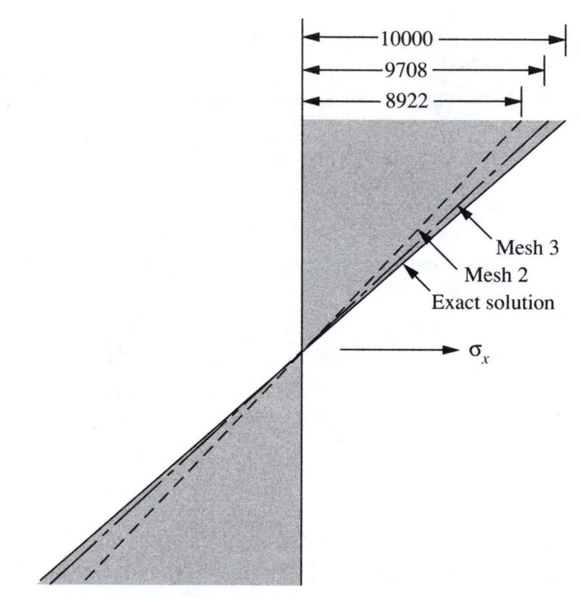

FIGURE 5–55 Extrapolated bending stress — meshes 2 and 3

When using the rectangular element it is best to compute centroidal stress values and draw conclusions and/or perform extrapolations based on the centroidal values rather than computing the stresses at other points within the element.

5.7 AXISYMMETRIC PROBLEMS

Axisymmetric problems are associated with bodies of revolution where, as indicated in Fig. 5–56, the z-axis is taken as the axis of revolution or axis of symmetry.

The two displacements are $u(r, z, t)$ and $w(r, z, t)$ in the r- and z-directions respectively. All elastic constants, body forces, surface tractions, and thermal variables are assumed to depend only on r, z, and t. For the reader unfamiliar with the details of axisymmetric elasticity, see Boresi and Chong [1] and Love [3].

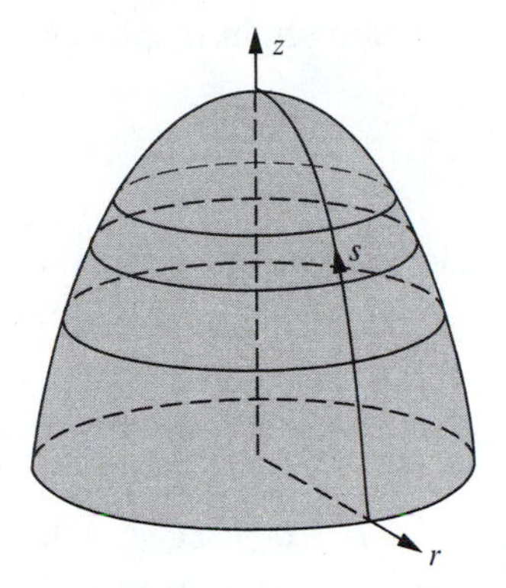

FIGURE 5–56 Axisymmetric region

Kinetics. The equations of motion for the axisymmetric problem can be written as

$$\begin{aligned} \frac{\partial \sigma_r}{\partial r} + \frac{(\sigma_r - \sigma_\theta)}{r} + \frac{\partial \tau_{rz}}{\partial z} + R &= \rho \ddot{u} \\ \frac{\partial \tau_{rz}}{\partial r} + \frac{\tau_{rz}}{r} + \frac{\partial \sigma_z}{\partial z} + Z &= \rho \ddot{w} \end{aligned} \tag{5.40}$$

where R and Z are the r and z components of the body force. The double dots denote partial derivatives with respect to time. The bounding surface S is considered to consist of a portion Γ_u on which the displacements are prescribed, and a portion Γ_t, on which the tractions or stresses are prescribed. The form of the boundary conditions on Γ_t is

$$l\sigma_r + m\tau_{rz} = t_r(S, t) \qquad l\tau_{rz} + m\sigma_z = t_z(S, t)$$

as indicated in Fig. 5–57.

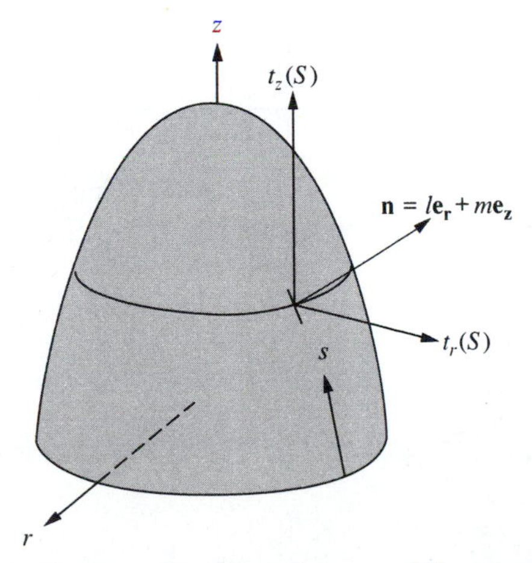

FIGURE 5–57 Boundary conditions on Γ_t

Kinematics. The small deformation strain displacement relations are

$$\epsilon_r = \frac{\partial u}{\partial r} \qquad \epsilon_\theta = \frac{u}{r} \qquad \epsilon_z = \frac{\partial w}{\partial z} \qquad \gamma_{rz} = \frac{\partial u}{\partial z} + \frac{\partial w}{\partial r} \tag{5.41}$$

Constitution. The stress-strain-temperature relations are expressed as

$$\boldsymbol{\sigma} = \mathbf{C}\boldsymbol{\epsilon} - \mathbf{f_T} \tag{5.42}$$

where

$$\mathbf{C} = \begin{bmatrix} C_{11} & C_{12} & C_{13} & 0 \\ C_{12} & C_{22} & C_{23} & 0 \\ C_{13} & C_{23} & C_{33} & 0 \\ 0 & 0 & 0 & C_{44} \end{bmatrix}$$

$$\mathbf{f_T} = [\beta T \quad \beta T \quad \beta T \quad 0]^{\mathrm{T}}$$

$$\boldsymbol{\sigma} = [\sigma_r \quad \sigma_\theta \quad \sigma_z \quad \tau_{rz}]^{\mathrm{T}}$$

and where β is a constant containing elastic constants and the coefficient of thermal expansion α. For an isotropic solid the $\mathbf{C}$ matrix can be written as

$$\mathbf{C} = \frac{E(1-\nu)}{(1+\nu)(1-2\nu)} \begin{bmatrix} 1 & \dfrac{\nu}{(1-\nu)} & \dfrac{\nu}{(1-\nu)} & 0 \\ \dfrac{\nu}{(1-\nu)} & 1 & \dfrac{\nu}{(1-\nu)} & 0 \\ \dfrac{\nu}{(1-\nu)} & \dfrac{\nu}{(1-\nu)} & 1 & 0 \\ 0 & 0 & 0 & \dfrac{(1-2\nu)}{2(1-\nu)} \end{bmatrix}$$

with

$$\mathbf{f_T} = \left[\frac{E\alpha T}{(1-2\nu)} \; \frac{E\alpha T}{(1-2\nu)} \; \frac{E\alpha T}{(1-2\nu)} \; 0 \right]$$

The strains can be eliminated between Eqs. (5.41) and (5.42), and the resulting stress-displacement relations then substituted into the equations of motion (5.40) to yield

$$\frac{\partial\big(C_{11}\partial u/\partial r + C_{12}(u/r) + C_{13}\partial w/\partial z - \beta T\big)}{\partial r} + \left((C_{11} - C_{12})\frac{\partial u}{\partial r} + (C_{12} - C_{22})\left(\frac{u}{r}\right) + (C_{13} - C_{33})\frac{\partial w}{\partial z}\right)\frac{1}{r} + \left(C_{44}\left(\frac{\partial u}{\partial z} + \frac{\partial w}{\partial r}\right)\right) + R = \rho\ddot{u}$$

and

$$\frac{\partial\big(C_{44}(\partial u/\partial z + \partial w/\partial r)\big)}{\partial r} + \frac{\big(C_{44}(\partial u/\partial z + \partial w/\partial r)\big)}{r}$$

$$+ \frac{\partial\big(C_{13}\partial u/\partial r + C_{23}(u/r) + C_{33}\partial w/\partial z\big)}{\partial z} + Z = \rho\ddot{w}$$

The basic problem is to determine functions $u(r, z, t)$ and $w(r, z, t)$ that satisfy these coupled partial differential equations in the domain D and that also satisfy the essential boundary conditions

$$\left.\begin{aligned} u(r, z, t) &= u_0(S, t) \\ w(r, z, t) &= w_0(S, t) \end{aligned}\right\} \quad \text{on } \Gamma_u$$

The natural boundary conditions

$$\left.\begin{aligned} l\sigma_r(r, z, t) + m\tau_{rz}(r, z, t) &= t_r(S, t) \\ l\tau_{rz}(r, z, t) + m\sigma_z(r, z, t) &= t_z(S, t) \end{aligned}\right\} \quad \text{on } \Gamma_t$$

must also be determined. For all but the simplest of domains D and boundary conditions on Γ_u and Γ_t, this boundary value problem is essentially unsolvable by analytical means, so that approximate and/or numerical approaches are necessary. In what follows, the weak form and the corresponding variational principle will be developed, after which the method of Ritz will be used to construct a finite element model for the axisymmetric elasticity problem.

Weak formulation. The initial step in developing the weak form for the axisymmetric problem is to multiply Eqs. (5.40) by test functions ϕ and ψ, respectively, add the two resulting equations, multiply by $d\,Vol = 2\pi r\, dr\, dz$, and integrate over the domain D to obtain

$$\int\!\!\int_D \left(\phi\left(\frac{\partial\sigma_r}{\partial r} + \frac{(\sigma_r - \sigma_\theta)}{r} + \frac{\partial\tau_{rz}}{\partial z} + R\right) + \psi\left(\frac{\partial\tau_{rz}}{\partial r} + \frac{\tau_{rz}}{r} + \frac{\partial\sigma_z}{\partial z} + Z\right)\right) r\, dr\, dz = 0 \tag{5.43}$$

where the dynamic terms have been deleted. The test functions ϕ and ψ are required to vanish on the Γ_u portion of the boundary or on the appropriate Γ_m portions of the boundary as discussed in Section 5.1. Rewrite Eq. (5.43) as

$$\int\!\!\int_D \left(\phi\left(\frac{\partial(r\sigma_r)}{\partial r} - \sigma_\theta + \frac{r\partial\tau_{rz}}{\partial z} + rR\right) + \psi\left(\frac{\partial(r\tau_{rz})}{\partial r} + \frac{r\partial\sigma_z}{\partial z} + rZ\right)\right) dr\, dz = 0 \tag{5.44}$$

Using the divergence theorem according to

$$\int\!\!\int \frac{\partial(rA)}{\partial r}\, dr\, dz = \int lrA\, ds \qquad \text{and} \qquad \int\!\!\int \frac{\partial(rB)}{\partial z}\, dr\, dz = \int mrB\, ds$$

Eq. (5.44) can be converted to

$$\int_S \Big(\phi(l\sigma_r + m\tau_{rz}) + \psi(l\tau_{rz} + m\sigma_z)\Big) r\, ds$$
$$- \int\!\!\int_D \left(\frac{\sigma_r \partial\phi}{\partial r} + \frac{\sigma_\theta \phi}{r} + \tau_{rz}\left(\frac{\partial\phi}{\partial z} + \frac{\partial\psi}{\partial r}\right) + \sigma_z \frac{\partial\psi}{\partial z}\right) r\, dr\, dz$$
$$+ \int\!\!\int_D (\phi R + \psi Z) r\, dr\, dz = 0$$

Substituting for the stresses using Eqs. (5.42) results in

$$\int_{\Gamma_t} \Big(\phi t_r(s) + \psi t_z(s)\Big) r(s)\, ds - \int\!\!\int_D \left(\frac{\partial\phi}{\partial r}\left(C_{11}\frac{\partial u}{\partial r} + C_{12}\frac{u}{r} + C_{13}\frac{\partial w}{\partial z} - \beta T\right)\right.$$
$$+ \frac{\phi}{r}\left(C_{12}\frac{\partial u}{\partial r} + C_{22}\frac{u}{r} + C_{23}\frac{\partial w}{\partial z} - \beta T\right) + \frac{\partial\psi}{\partial z}\left(C_{13}\frac{\partial u}{\partial r} + C_{23}\frac{u}{r} + C_{33}\frac{\partial w}{\partial z} - \beta T\right)$$
$$\left. + \left(\frac{\partial\phi}{\partial z} + \frac{\partial\psi}{\partial r}\right) C_{44}\left(\frac{\partial u}{\partial z} + \frac{\partial w}{\partial r}\right)\right) r\, dr\, dz + \int\!\!\int_D (\phi R + \psi Z)\, r\, dr\, dz = 0$$

where the boundary conditions on the stresses, and the fact that ϕ and ψ vanish on Γ_u, have been used in simplifying the surface integral. This equation can be written as

$$\int\!\!\int_D \mathbf{\Phi}^{\mathrm{T}}(\mathbf{C}\boldsymbol{\epsilon} - \mathbf{f}_{\mathbf{T}})\, d\,Vol = \int_{\Gamma_t} \Big(\phi t_r(s) + \psi t_z(s)\Big) r(s)\, ds + \int\!\!\int_D (\phi R + \psi Z)\, d\,Vol \tag{5.45}$$

where

$$\mathbf{\Phi}^{\mathrm{T}} = \left[\frac{\partial\phi}{\partial r} \quad \frac{\phi}{r} \quad \frac{\partial\psi}{\partial z} \quad \frac{\partial\phi}{\partial z} + \frac{\partial\psi}{\partial r}\right]$$

$$\boldsymbol{\epsilon} = \left[\frac{\partial u}{\partial r} \quad \frac{u}{r} \quad \frac{\partial w}{\partial z} \quad \frac{\partial u}{\partial z} + \frac{\partial w}{\partial r}\right]^{\mathrm{T}}$$

and $\mathbf{C}$ is the matrix of elastic constants given in Eq. (5.42). Equation (5.45) is clearly of the form

$$B(\phi, \psi : u, w) = l(\phi, \psi)$$

with the bilinear functional B being symmetric, in the sense that

$$B(\phi, \psi : u, w) = B(u, w : \phi, \psi)$$

It follows that there is a quadratic functional

$$I(u, w) = \frac{1}{2}B(u, w : u, w) - l(u, w)$$

$$= \frac{1}{2}\int\int_D \boldsymbol{\epsilon}^{\mathrm{T}}\mathbf{C}\boldsymbol{\epsilon}\, d\,Vol - \int\int_D \boldsymbol{\epsilon}^{\mathrm{T}}\mathbf{f}_{\mathrm{T}}\, d\,Vol - \int_{\Gamma_t} \mathbf{u}\cdot\mathbf{t}_{\mathbf{n}}\, dS - \int\int_D \mathbf{u}\cdot\mathbf{X}\, d\,Vol \tag{5.46}$$

which is the total potential energy. Using the principle of stationary potential energy to require that I be stationary, subject to the appropriate essential boundary conditions on Γ_u, yields the original equilibrium equations in terms of the displacement, and the natural boundary conditions on the Γ_t portion of the surface.

Geometrically, the axisymmetric problem is clearly three-dimensional. There are, however, only two independent and two dependent variables, so that from an analytical viewpoint the problem is essentially two-dimensional in character. The analysis can proceed as two-dimensional, whereas it is important to keep in mind the actual three-dimensional character of the axisymmetric problem.

5.7.1 The Ritz Finite Element Models— Triangular Elements

In this section, the potential energy functional of Eq. (5.46), namely

$$I(u, w) = \frac{1}{2}\int\int_D \boldsymbol{\epsilon}^{\mathrm{T}}\mathbf{C}\boldsymbol{\epsilon} r\, dr\, dz - \int\int_D \boldsymbol{\epsilon}^{\mathrm{T}}\mathbf{f}_{\mathrm{T}} r\, dr\, dz$$

$$- \int_{\Gamma_t} (ut_r + wt_w) r(s)\, ds - \int\int_D (uR + wZ) r\, dr\, dz$$

will be used to develop a Ritz finite element model for the axisymmetric elasticity problem. The basic steps will be discussed in order.

Discretization. An axisymmetric solid can be thought of as having been generated by revolving an area in the rz-plane around the z-axis, as indicated in Fig. 5–58. The resulting body of revolution or volume can also be considered as a collection of toroids or rings, also indicated in Fig. 5–58. In particular, for a ring that is triangular in shape, the cross section obtained by passing a plane through and parallel to the z-axis shows the corresponding triangle in the rz-plane.

Generally uniform meshes are appropriate. However, the basic $1/r$ singularity present in cylindrical coordinates suggests that if the region D includes points near or on the z-axis, a smaller element size would be appropriate near the origin. Comments made earlier regarding the limits on the geometry of a triangular element are still valid, namely, the smallest interior angle should not be smaller than approximately $\pi/8 = 22.5°$.

Interpolation. For the type of discretization indicated in Fig. 5–59, linear interpolation of the triangular element will be employed.

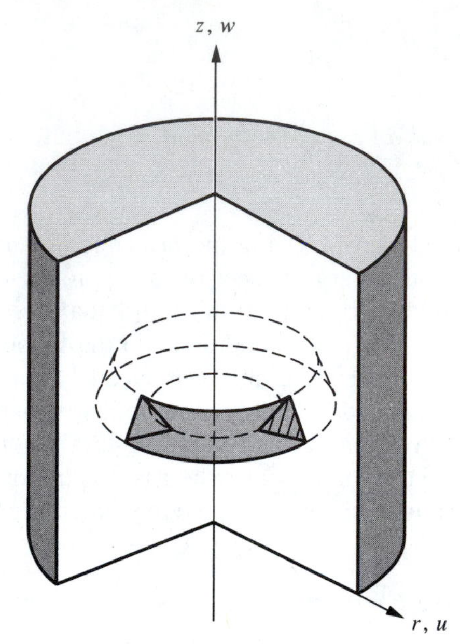

FIGURE 5–58 Ring element and triangular mesh in the rz-plane for a body of revolution

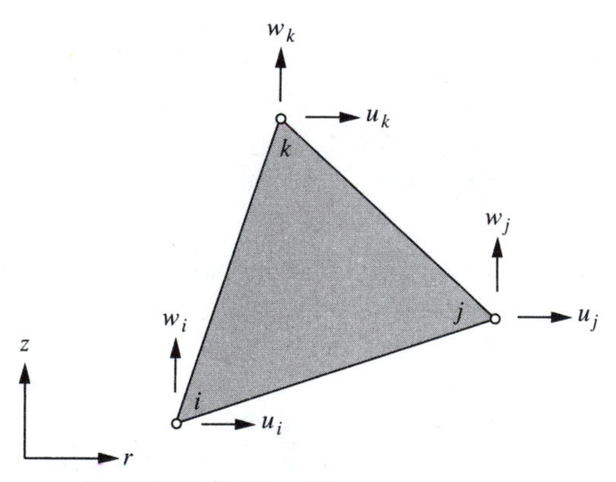

FIGURE 5–59 Element geometry

Within a typical element indicated in Fig. 5–59, the displacements are taken to be

$$\begin{aligned} u_e(r, z) &= \mathbf{u}_\mathbf{e}^\mathrm{T}\mathbf{N} = \mathbf{N}^\mathrm{T}\mathbf{u}_\mathbf{e} \\ w_e(r, z) &= \mathbf{w}_\mathbf{e}^\mathrm{T}\mathbf{N} = \mathbf{N}^\mathrm{T}\mathbf{w}_\mathbf{e} \end{aligned} \tag{5.47}$$

where $\mathbf{u}_\mathbf{e}^\mathrm{T} = [u_i\ u_j\ u_k]$, $\mathbf{w}_\mathbf{e}^\mathrm{T} = [w_i\ w_j\ w_k]$, and $\mathbf{N}$ is the interpolation vector consisting of the three N's given by

$$N_i = \frac{a_i + b_i r + c_i z}{2A_e}$$

where

$$\begin{aligned} a_i &= r_j z_k = r_k z_j \\ b_i &= z_j - z_k \\ c_i &= r_k - r_j \end{aligned}$$

and

$$2A_e = \begin{vmatrix} 1 & r_i & z_i \\ 1 & r_j & z_j \\ 1 & r_k & z_k \end{vmatrix}$$

with the j- and k-subscripted a, b, and c to be obtained by cyclic permutation of the i, j, and k. Note that these have exactly the same form as the interpolation functions presented in Section 5.3, with the x and y replaced by r and z, respectively.

Elemental formulation. The starting point for the elemental formulation is the functional given by Eq. (5.46), namely

$$I(u, w) = \frac{1}{2}\int\int_D \boldsymbol{\epsilon}^{\mathrm{T}}\mathbf{C}\boldsymbol{\epsilon} r\, dr\, dz - \int\int_D \boldsymbol{\epsilon}^{\mathrm{T}}\mathbf{f_T} r\, dr\, dz$$
$$- \int_{\Gamma_t} (ut_r + wt_z) r(s)\, ds - \int\int_D (uR + wZ) r\, dr\, dz$$

The first step is to rewrite I as the sum of the integrals over all the elements, thus reflecting the discretization. This results in

$$I \approx \frac{1}{2}\sum_e \int\int_{D_e} \boldsymbol{\epsilon}^{\mathrm{T}}\mathbf{C}\boldsymbol{\epsilon} r\, dr\, dz - \sum_e \int\int_{D_e} \boldsymbol{\epsilon}^{\mathrm{T}}\mathbf{f_T} r\, dr\, dz$$
$$- \sum_e{}' \int_{\gamma_{te}} (ut_r + wt_z) r(s)\, ds - \sum_e \int\int_{D_e} (uR + wZ) r\, dr\, dz \tag{5.48}$$

the $\approx$ reflecting the possible error in modeling a curved surface with flat-faced triangular toroids. Errors of this type tend to zero as the mesh is refined. The $\sum$ and $\sum'$ represent, respectively, sums over the triangular elements that form the region D in the rz-plane, and the straight line segments that form the Γ_t portion of the boundary. Writing Eq. (5.48) as

$$I \approx \sum_e I_{1e} - \sum_e I_{2e} - \sum_e{}' I_{3e} - \sum_e I_{4e}$$

we will consider each of the I_{ie}, $i = 1, 2, 3, 4$, in order.

Evaluation of I_{1e}. I_{1e} represents the elemental strain energy given by

$$I_{1e} = \frac{1}{2}\int\int_{D_e} \boldsymbol{\epsilon}^{\mathrm{T}}\mathbf{C}\boldsymbol{\epsilon} r\, dr\, dz$$

with

$$\boldsymbol{\epsilon}^{\mathrm{T}} = [\epsilon_r \quad \epsilon_\theta \quad \epsilon_z \quad \gamma_{rz}] = \left[\frac{\partial u}{\partial r} \quad \frac{u}{r} \quad \frac{\partial w}{\partial z} \quad \frac{\partial u}{\partial z} + \frac{\partial w}{\partial r}\right]$$

Using Eqs. (5.47) for $\mathbf{u_e}$ and $\mathbf{w_e}$, the strains $\boldsymbol{\epsilon}$ can be written as

$$\boldsymbol{\epsilon} = \mathbf{BU_e}$$

where

$$\mathbf{B} = \begin{bmatrix} \frac{\partial N_1}{\partial r} & 0 & \frac{\partial N_2}{\partial r} & 0 & \frac{\partial N_3}{\partial r} & 0 \\ \frac{N_1}{r} & 0 & \frac{N_2}{r} & 0 & \frac{N_3}{r} & 0 \\ 0 & \frac{\partial N_1}{\partial z} & 0 & \frac{\partial N_2}{\partial z} & 0 & \frac{\partial N_3}{\partial z} \\ \frac{\partial N_1}{\partial z} & \frac{\partial N_1}{\partial r} & \frac{\partial N_2}{\partial z} & \frac{\partial N_2}{\partial r} & \frac{\partial N_3}{\partial z} & \frac{\partial N_3}{\partial r} \end{bmatrix} \tag{5.49}$$

is the strain matrix, and

$$\mathbf{U_e} = [u_i \quad w_i \quad u_j \quad w_j \quad u_k \quad w_k]^{\mathrm{T}}$$

is the elemental displacement vector.

The elemental strain energy can be expressed as

$$\begin{aligned} I_{1e} &= \frac{1}{2}\int\int_{D_e} \mathbf{U_e^{\mathrm{T}}B^{\mathrm{T}}CBU_e} r\, dr\, dz \\ &= \frac{1}{2}\mathbf{U_e^{\mathrm{T}}}\left[\int\int_{D_e} \mathbf{B^{\mathrm{T}}CB} r\, dr\, dz\right]\mathbf{U_e} \\ &= \frac{1}{2}\mathbf{U_e^{\mathrm{T}}k_eU_e} \end{aligned}$$

where

$$\mathbf{k_e} = \int\int_{D_e} \mathbf{B^{\mathrm{T}}CB} r\, dr\, dz \tag{5.50}$$

is the 6×6 elemental stiffness matrix. Note that the integrand of $\mathbf{k_e}$, by virtue of $\mathbf{B}$ and the r of $r\, dr\, dz$, is not constant as was the case for the corresponding two-dimensional plane elasticity problem. The evaluation of the elemental matrices will be discussed in the next section.

I_1 then becomes

$$I_1 \approx \sum_e I_{1e} = \frac{1}{2}\sum_e \mathbf{U}_e^T \mathbf{k}_e \mathbf{U}_e$$

and represents the total internal or strain energy as the sum of the elemental internal or strain energies I_{1e}.

Evaluation of I_{2e}. The elemental thermal contribution to the total potential energy is

$$I_{2e} = \iint_{D_e} \boldsymbol{\epsilon}^T \mathbf{f}_T r\, dr\, dz$$

which on using the expression for $\boldsymbol{\epsilon}^T$ can be expressed as

$$I_{2e} = \iint_{D_e} \mathbf{U}_e^T \mathbf{B}^T \mathbf{f}_T r\, dr\, dz = \mathbf{u}_e^T \boldsymbol{\theta}_e$$

with

$$\boldsymbol{\theta}_e = \iint_{D_e} \mathbf{B}^T \mathbf{f}_T r\, dr\, dz \tag{5.51}$$

The total thermal energy is then

$$I_2 \approx \sum_e I_{2e} = \sum_e \mathbf{U}_e^T \boldsymbol{\theta}_e$$

represented as the sum of the elemental thermal energies I_{2e}.

Evaluation of I_{3e}. The I_{3e} account for the potential energy associated with the surface tractions or surface stresses, namely,

$$I_{3e} = \int_{\gamma_{te}} \big(u_e t_r(s) + w_e t_z(s)\big) r(s)\, ds$$

to be evaluated over a typical γ_{te} portion of the boundary as indicated in Fig. 5–60. Expressing the surface stresses $t_r(s)$ and $t_z(s)$ in terms of the normal and tangential stresses $t_n(s)$ and $t_s(s)$ as shown in Fig. 5–60, there results

$$I_{e3} = \int_{\gamma_{te}} \big(u_e(l t_n - m t_s) + w_e(m t_n + l t_s)\big) r(s)\, ds$$

$$= \int_{\gamma_{te}} \left[\mathbf{u}_e^T \mathbf{N}(l t_n - m t_s) + \mathbf{w}_e^T \mathbf{N}(m t_n + l t_s)\right] r(s)\, ds$$

On γ_{te}, the interpolation vectors reduce to only the two nonzero components N_j and N_k, so that I_{3e} becomes

$$I_{3e} = [u_j \quad w_j \quad u_k \quad w_k] \mathbf{t}_e$$

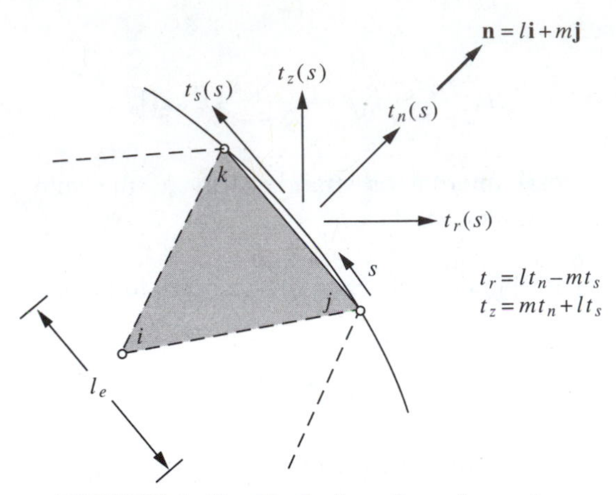

FIGURE 5–60 Typical surface element γ_{te}

where

$$\mathbf{t_e} = \int_{\gamma_{te}} \begin{bmatrix} N_j(lt_n - mt_s) \\ N_j(mt_n + lt_s) \\ N_k(lt_n - mt_s) \\ N_k(mt_n + lt_s) \end{bmatrix} r(s)\, ds \tag{5.52}$$

The total energy associated with the surface stresses is then

$$I_3 \approx \sum_e{}' I_{3e} = \sum_e{}' \mathbf{U_e^T t_e}$$

as the sum of the elemental I_{3e}, with the $\sum'$ indicating a sum over the set of boundary segments comprising Γ_t.

Evaluation of I_{4e}. The I_{4e} are the elemental contributions to the potential energy associated with the body forces and are of the form

$$I_{4e} = \iint_{D_e} (u_e R + w_e Z) r\, dr\, dz$$

Using the elemental representation for the displacements there results

$$I_{4e} = \iint_{D_e} [\mathbf{u_e^T N} R + \mathbf{w_e^T N} Z] r\, dr\, dz$$

which when expressed in terms of the elemental displacement vector $\mathbf{U_e}$ becomes

$$I_{4e} = \mathbf{u_e^T f_e}$$

where

$$\mathbf{f}_\mathbf{e}^\mathrm{T} = \int\int_{D_e} [N_1R \quad N_1Z \quad N_2R \quad N_2Z \quad N_3R \quad N_3Z] r\,dr\,dz \tag{5.53}$$

The total energy of the body forces is then

$$I_4 \approx \sum_e I_{4e} = \sum_e \mathbf{U}_\mathbf{e}^\mathrm{T}\mathbf{f}_\mathbf{e}$$

Assembly. In terms of the elemental contributions just discussed, the total potential energy function can be expressed as

$$I \approx \frac{1}{2}\sum \mathbf{U}_\mathbf{e}^\mathrm{T}\mathbf{k}_\mathbf{e}\mathbf{U}_\mathbf{e} - \sum \mathbf{U}_\mathbf{e}^\mathrm{T}\boldsymbol{\theta}_\mathbf{e} - \sum{}' \mathbf{U}_\mathbf{e}^\mathrm{T}\mathbf{t}_\mathbf{e} - \sum \mathbf{U}_\mathbf{e}^\mathrm{T}\mathbf{f}_\mathbf{e}$$

In terms of the global stiffness, load, and displacement matrices this can be written as

$$\begin{aligned} I(\mathbf{U_G}) &= I(u_1, w_1, u_2, w_2, \ldots, u_N, w_N) \\ &= \frac{1}{2}\mathbf{U}_\mathbf{G}^\mathrm{T}\mathbf{K_G}\mathbf{U_G} - \mathbf{U}_\mathbf{G}^\mathrm{T}\mathbf{F_G} \end{aligned}$$

where

$$\mathbf{K_G} = \sum_e \mathbf{k_G}$$

$$\mathbf{F_G} = \sum_e \boldsymbol{\theta}_\mathbf{G} + \sum_e \mathbf{f_G} + \sum_e{}' \mathbf{t_G}$$

with the sums indicating the assembly or loading of the elemental matrices into their proper locations in the global matrices.

The total potential energy is now a function of the collection of nodal displacements. The stationary value is obtained by requiring

$$\frac{\partial I}{\partial u_i} = 0 \qquad \frac{\partial I}{\partial w_i} = 0 \qquad i = 1, 2, \ldots, N$$

or in matrix notation

$$\frac{\partial I}{\partial \mathbf{U_G}} = \frac{1}{2}(\mathbf{K_G} + \mathbf{K}_\mathbf{G}^\mathrm{T})\mathbf{U_G} - \mathbf{F_G} = 0$$

Since $\mathbf{K_G}$ is symmetric, there results

$$\mathbf{K_G}\mathbf{U_G} = \mathbf{F_G}$$

as the set of global equilibrium equations. Constraints and solution proceed in the usual manner.

Computation of derived variables. The options for evaluating the stresses within an element will be discussed after having evaluated the elemental stiffness matrices in the next section.

5.7.2 Evaluation of Elemental Matrices

There are several different consistent schemes for evaluating the elemental matrices presented in the previous section. The simplest approach straightforwardly evaluates the integrands appearing in each of the elemental stiffness and load matrices at the centroid of the element or line segment. This simple approach is used in what follows.

Evaluation of $\mathbf{k_e}$. The general expression for the elemental stiffness matrix $\mathbf{k_e}$ is given by Eq. (5.50) as

$$\mathbf{k_e} = \int\int_{D_e} \mathbf{B}^{\mathrm{T}}\mathbf{C}\mathbf{B} r\,dr\,dz$$

with $\mathbf{B}$ given by Eq. (5.49). Note that the integrand is not a constant, as was the case for the plane elasticity problem, due to the appearance of the $1/r$ term in $\mathbf{B}$ and the r in $r\,dr\,dz$.

It is possible to evaluate the elements of $\mathbf{k_e}$ in closed form. This produces, however, undesirable $\ln r$ terms. For elements that have nodes lying on the z-axis, these terms can be handled using a limit, but they are nevertheless troublesome. Alternately, quadrature is used to avoid having to deal with the $\ln r$ terms. The simplest approach is to evaluate the integrand at the centroid of the element, namely

$$\hat{r} = \frac{(r_i + r_j + r_k)}{3} = r_{\text{cent}}$$

and

$$\hat{z} = \frac{(z_i + z_j + z_k)}{3} = z_{\text{cent}}$$

The result can be written as

$$\mathbf{k_e} = \mathbf{B}^{\mathrm{T}}(\hat{r}, \hat{z})\mathbf{C}\mathbf{B}(\hat{r}, \hat{z})\hat{r}A_e$$

or

$$\mathbf{k_e} = \hat{\mathbf{B}}^{\mathrm{T}}\hat{\mathbf{C}}\hat{\mathbf{B}}\hat{r}A_e$$

where A_e is the area of the triangle in the rz-plane. This is essentially the $N = 1$ Gauss quadrature on a triangle.

Evaluation of $\boldsymbol{\theta}_\mathbf{e}$. The general expression for the elemental thermal load $\boldsymbol{\theta}_\mathbf{e}$ is given by Eq. (5.51) as

$$\boldsymbol{\theta}_{\mathbf{e}} = \iint_{D_e} \mathbf{B}^{\mathrm{T}}\mathbf{f}_{\mathbf{T}} r \, dr \, dz$$

where the integrand $\mathbf{B}\mathbf{f}_{\mathbf{T}}r$ is clearly a function of position within the element. Again using a quadrature that simply evaluates the integrand at the centroid yields the approximate thermal contribution

$$\boldsymbol{\theta}_{\mathbf{e}} = \hat{\mathbf{B}}^{\mathrm{T}}\hat{\mathbf{f}}_{\mathbf{T}}\hat{r}A_e$$

where $\hat{\mathbf{f}}_{\mathbf{T}} = \mathbf{f}_{\mathbf{T}}(\hat{r}, \hat{z})$.

Evaluation of $\mathbf{f}_{\mathbf{e}}$. The general expression for the elemental body force vector is given by Eq. (5.53) as

$$\mathbf{f}_{\mathbf{e}} = \iint_{D_e} [N_1R \quad N_1Z \quad N_2R \quad N_2Z \quad N_3R \quad N_3Z] r \, dr \, dz$$

Evaluating the integrand at the centroid yields the approximate expression

$$\mathbf{f}_{\mathbf{e}}^{\mathrm{T}} = \frac{[\hat{R} \quad \hat{Z} \quad \hat{R} \quad \hat{Z} \quad \hat{R} \quad \hat{Z}]\hat{r}A_e}{3}$$

where $\hat{R} = R(\hat{r}, \hat{z})$ and $\hat{Z} = Z(\hat{r}, \hat{z})$.

Evaluation of $\mathbf{t}_{\mathbf{e}}$. The general expression for the elemental surface stress load is given by Eq. (5.52) as

$$\mathbf{t}_{\mathbf{e}} = \int_{\gamma_{te}} \begin{bmatrix} N_j(lt_n - mt_s) \\ N_j(mt_n + lt_s) \\ N_k(lt_n - mt_s) \\ N_k(mt_n + lt_s) \end{bmatrix} r(s) \, ds$$

These integrals are evaluated in a fashion consistent with the centroidal evaluation of the volume integrals by taking

$$\mathbf{t}_{\mathbf{e}} = \begin{bmatrix} (l\hat{t}_n - m\hat{t}_s) \\ (m\hat{t}_n + l\hat{t}_s) \\ (l\hat{t}_n - m\hat{t}_s) \\ (m\hat{t}_n + l\hat{t}_s) \end{bmatrix} \frac{\hat{r}l_e}{2}$$

where $\hat{t}_n = t_n(\hat{s})$ and $\hat{t}_s = t_s(\hat{s})$ and $\hat{s}$ and $\hat{r}$ indicate the values of s and r evaluated at the centroid of γ_{te}, respectively. This is essentially the trapezoidal rule for evaluating the line integrals.

Computation of derived variables. When linear interpolation is used for representing the displacements within the element, the best point at which to evaluate the stresses within the element is the centroid. This leads to

$$\hat{\boldsymbol{\sigma}} = \mathbf{C}\hat{\mathbf{B}}\mathbf{U}_e - \hat{\mathbf{f}}_T$$

$\mathbf{C}\hat{\mathbf{B}}$ for each element would have been computed at the time $\mathbf{k}_e$ was formed and would have been saved at that time for reuse in computing the stresses. Similarly, $\hat{\mathbf{f}}_T$ would have been computed during the process of evaluating $\mathbf{t}_e$ and would have been saved.

5.7.2.1 THICK-WALLED PRESSURE VESSEL. Consider the problem of a thick-walled pressure vessel loaded by an internal pressure as shown in Fig. 5–61. The region to be modeled is shown in Fig. 5–62, and is such that the displacements in the axial direction are prevented so as to be able to compare the results with the plane strain two-dimensional solution. The material is assumed to be isotropic.

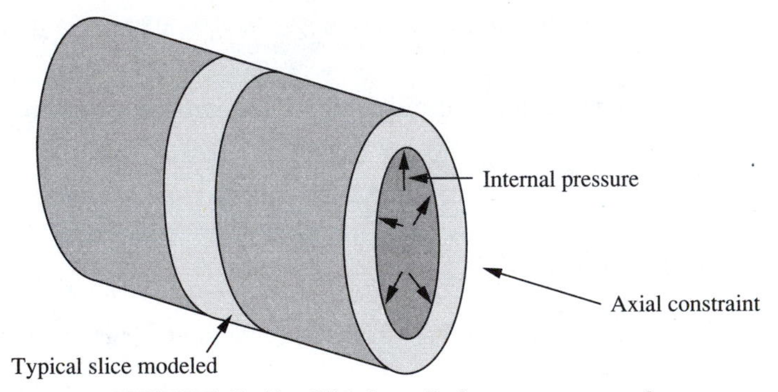

FIGURE 5–61 Thick-walled pressure vessel

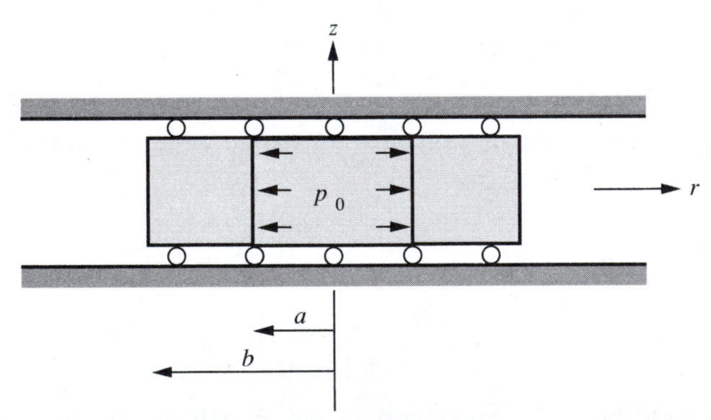

FIGURE 5–62 Equivalent plane strain problem for internal pressure

Discretization, interpolation, and elemental formulation. Using triangular elements, the first mesh to be considered is shown in Fig. 5–63. The elemental matrices are computed by evaluating at the centroid of the element all the quantities appearing in the integrands. All the $\mathbf{f}_e$ and $\boldsymbol{\theta}_e$ are zero. Assuming the inner and outer radii to be 1 in and 2 in respectively, and the internal pressure to be 1000 psi, the nodal loads $\mathbf{t_e}$ are given by

$$\mathbf{t}_{e1} = \begin{bmatrix} \overset{u_3}{100.0} & \overset{v_3}{0.0} & \overset{u_2}{100.0} & \overset{v_2}{0.0} \end{bmatrix}^T$$

$$\mathbf{t}_{e2} = \begin{bmatrix} \overset{u_2}{100.0} & \overset{v_2}{0.0} & \overset{u_1}{100.0} & \overset{v_1}{0.0} \end{bmatrix}^T$$

and are indicated in Fig. 5–63.

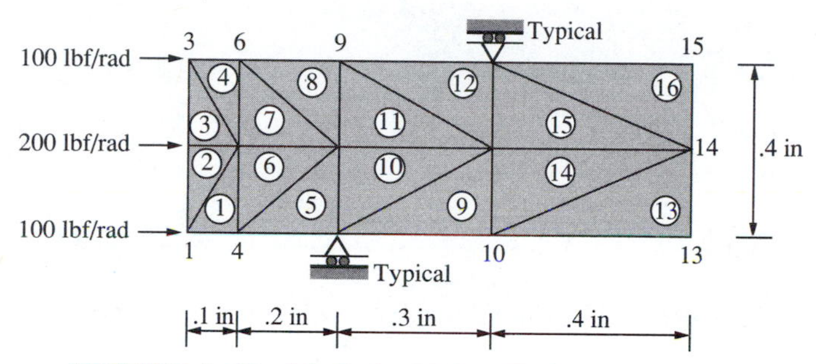

FIGURE 5–63 Mesh 1, thick-walled pressure vessel

Assembly and constraints. Assembly is carried out as usual. The resulting global stiffness and load matrices are 30 × 30 and 30 × 1, respectively. The constraints, which are appropriate for the model intended to simulate the plane strain state of stress, are indicated in Fig. 5–64. The corresponding constrained displacements are

$$w_1 = w_4 = w_7 = w_{10} = w_{13} = 0$$

and

$$w_3 = w_6 = w_9 = w_{12} = w_{15} = 0$$

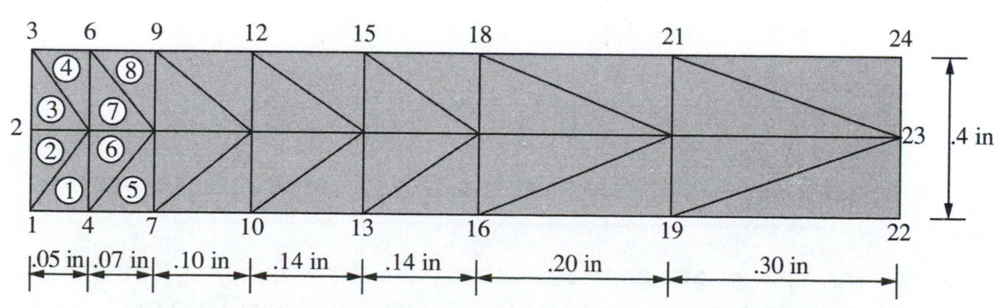

FIGURE 5–64 Mesh 2, thin-walled pressure vessel

The resulting constrained equations of equilibrium are solved for the displacements, which are plotted in Fig. 5–65.

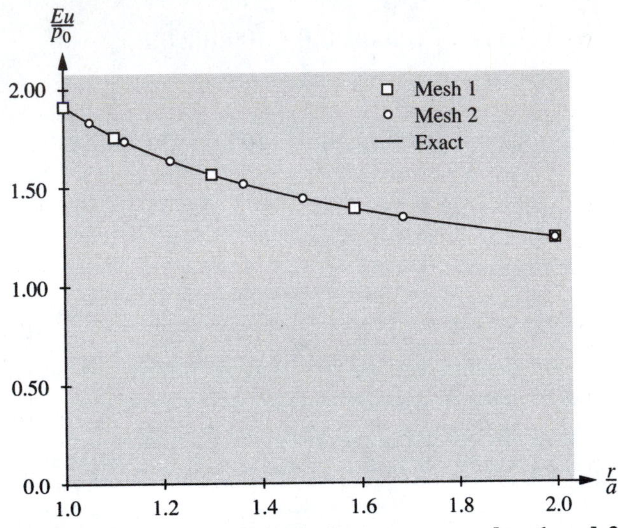

FIGURE 5–65 Radial displacement—meshes 1 and 2

Computation of derived variables. As described previously, the stresses are evaluated at the centroid using

$$\boldsymbol{\sigma} = \mathbf{CBU_e}$$

Note that **CB** is calculated during the process of evaluating $\mathbf{k_e}$ and at that time can be stored for later use in the computation of the stresses. This can result in substantial savings in computing the stresses for large problems. The corresponding stresses are presented in Table 5.11.

TABLE 5.11 Centroidal stresses—mesh 1

Element #	σ_r	σ_θ	σ_z	τ_{rz}
3	−896.7	1592	208.6	−39.4
4	−860.2	1492	189.5	−25.6
7	−592.1	1342	224.9	−25.6
8	−602.1	1187	175.6	−5.9
11	−284.3	1047	228.9	−5.9
12	−325.3	903	173.4	4.9
15	−71.9	798	217.8	4.9
16	−90.7	703	183.8	20.2

Since σ_r, σ_θ, and σ_z are symmetric about $z = 0$, with τ_{rz} asymmetric, only the stresses for elements above the $z = 0$ plane are displayed. The small centroidal shear stress values are to be compared with the exact values of $\tau_{rz} = 0$.

In Fig. 5–66, the normal stresses σ_r, σ_θ, and σ_z are plotted against the exact values [Timoshenko and Goodier, 2] given by

$$\sigma_r = -p_i\left(\frac{(b/r)^2 - 1}{(b/a)^2 - 1}\right)$$

$$\sigma_\theta = p_i\left(\frac{(b/r)^2 + 1)}{(b/a)^2 - 1}\right)$$

$$\sigma_z = \nu(\sigma_r + \sigma_\theta) = \frac{2\nu p_i}{(b/a)^2 - 1}$$

The results are seen to be quite reasonable for this relatively crude mesh.

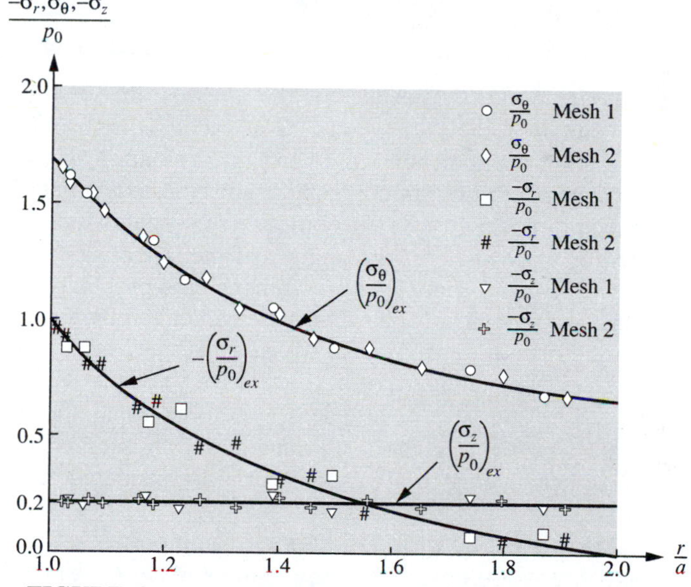

FIGURE 5–66 **Stresses σ_r, σ_θ, and σ_z—meshes 1 and 2**

Consider also the refined mesh indicated in Fig. 5–64, where again the inner and outer radii are 1 in and 2 in respectively.

The radial displacement $u(r, 0)$ resulting from this mesh is indicated in Fig. 5–65, along with the corresponding results from mesh 1 and the exact solution. The stresses are displayed in Table 5.12, and are also plotted in Fig. 5–66. Again the small centroidal values of τ_{rz} are to be compared to the correct values of zero.

Discussion. Figure 5–65 shows that there is excellent agreement between the nodal displacements and the theoretical results using either of the two meshes

TABLE 5.12 Centrodal stresses — mesh 2

Element #	σ_r	σ_θ	σ_z	τ_{rz}
3	−951	1627	203	−233
4	−921	1577	196	−19
7	−807	1499	208	−19
8	−793	1433	192	−14
11	−642	1349	212	−14
12	−642	1267	188	−7
15	−463	1181	215	−7
16	−476	1092	185	−1
19	−306	1023	215	−1
20	−333	951	185	0
23	−178	892	214	0
24	−199	821	186	3
27	−57	756	210	3
28	−59	695	191	15

considered. From Fig. 5–66, the results for the stresses are seen to be reasonably accurate for either of the two meshes. The radial component σ_r exhibits an oscillation, a known phenomenon for centroidal evaluation of the stresses in the triangular element when used in connection with the axisymmetric elasticity problem. In this regard, nodal stresses are sometimes computed and then averaged between elements for somewhat better accuracy of the stresses.

Note that this problem is essentially a one-dimensional problem in that the displacements and stresses depend only on r. For more general problems, that is, where the displacements and stresses depend on z as well as r, it is necessary when modeling the domain with triangular elements to consider a relatively fine mesh in order to get an accurate resolution of the stresses. Nevertheless, for this example, the results clearly indicate the accurate dependence of the displacements and stresses on the radial coordinate. Other z-independent axisymmetric problems are considered in the exercises.

5.7.2.2 NOTCHED SHAFT UNDER TENSION. An important technical problem is that of the stress concentration resulting from a semicircular groove in a circular shaft under tension as indicated in Fig. 5–67. The maximum tensile stress is expected to be in the z-direction at point E and is a function of the parameter ρ/D.

This is clearly a three-dimensional axisymmetric problem. The region in the rz-plane for the model to be investigated is indicated by the cross hatching in Fig. 5–67. Symmetry dictates that the z-component of displacement vanish on EA and that the radial or r-component of displacement vanish on AB. Not enforcing $u = 0$ on AB would amount to allowing the deformation to potentially create a hole at $r = 0$, clearly in violation of the continuity requirement on the radial component of displacement. Face BC is assumed to be far enough from the hole so that the normal stress can be assumed to be a constant σ_0, with the corresponding shear stress $\tau_{rz} = 0$. On the portion CD of the boundary of the modeled region, $\sigma_r = \tau_{rz} = 0$ are the natural boundary conditions. On the curved portion DE, t_r and t_z are required to vanish.

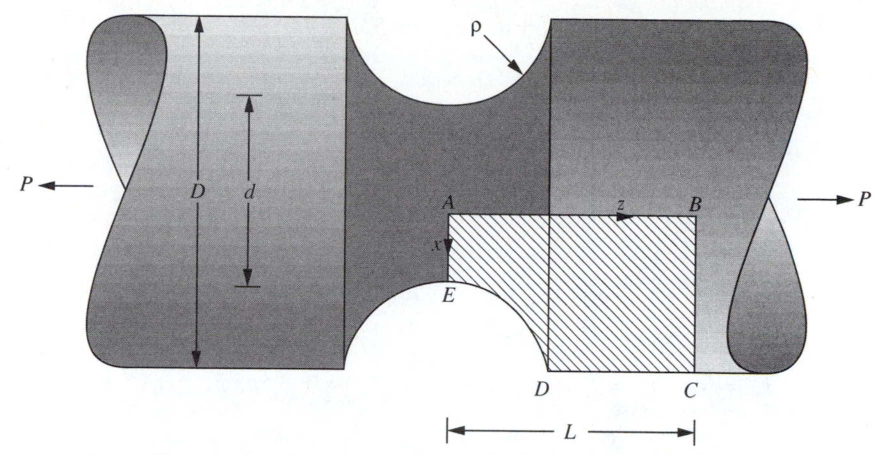

FIGURE 5–67 Notched circular shaft under tension

Discretization and interpolation. The first mesh to be investigated is indicated in Fig. 5–68. Numerical results are obtained for $D = 6$ in, $\rho = 1$ in, and $L = 5$ in.

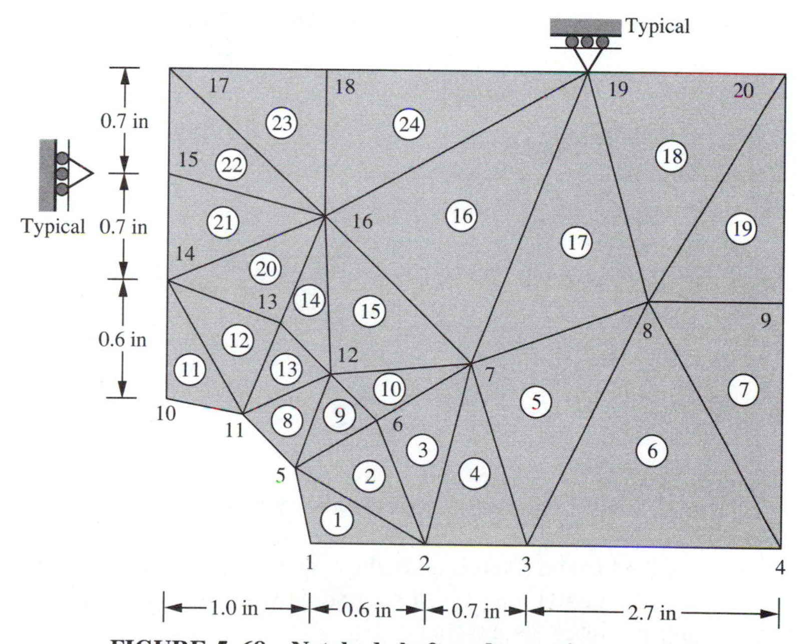

FIGURE 5–68 Notched shaft under tension—mesh 1

Elemental formulation, assembly, constraints, and solution. Using the elemental matrices described in Section 5.7.2, the global equilibrium equations are assembled in the usual fashion. Consider in particular the elemental matrices arising from the surface normal stress σ_0 on boundary BC. The $\mathbf{t}_{e1}$ and $\mathbf{t}_{e2}$ are associated with segments 4-9 and 9-20 respectively.

Using the expression developed in Section 5.7.2,

$$\mathbf{t_e} = \begin{bmatrix} lt_n - mt_s \\ mt_n + lt_s \\ lt_n - mt_s \\ mt_n + lt_s \end{bmatrix} \frac{\hat{r} l_e}{2}$$

there results, for $\mathbf{t_{e1}}$, $t_n = \sigma_0$, $t_s = 0$, $l = 0$, $m = 1$, $l_e = 1.5$ in, $\hat{r} = 2.25$ in, resulting in

$$\begin{matrix} & u_4 & w_4 & u_9 & w_9 & \\ \mathbf{t_{e1}} = [& 0 & 1.6875\sigma_0 & 0 & 1.6875\sigma_0 &]^{\mathrm{T}} \end{matrix}$$

and for $\mathbf{t_{e2}}$, $t_n = \sigma_0$, $t_s = 0$, $l = 0$, $m = 1$, $l_e = 1.5$ in, $\hat{r} = 0.75$ in, with

$$\begin{matrix} & u_9 & w_9 & u_{20} & w_{20} & \\ \mathbf{t_{e2}} = [& 0 & 0.5625\sigma_0 & 0 & 0.5625\sigma_0 &]^{\mathrm{T}} \end{matrix}$$

For each of $\mathbf{t_{e1}}$ and $\mathbf{t_{e2}}$, the degrees of freedom dictate the global positions into which the components are loaded during assembly; for example, the components of $\mathbf{t_{e1}}$ load into the 7th, 8th, 17th, and 18th positions, respectively and the components of $\mathbf{t_{e2}}$ load into the 17th, 18th, 39th, and 40th positions, respectively.

The resulting nodal loads on the face *BC* are as indicated in Fig. 5–69.

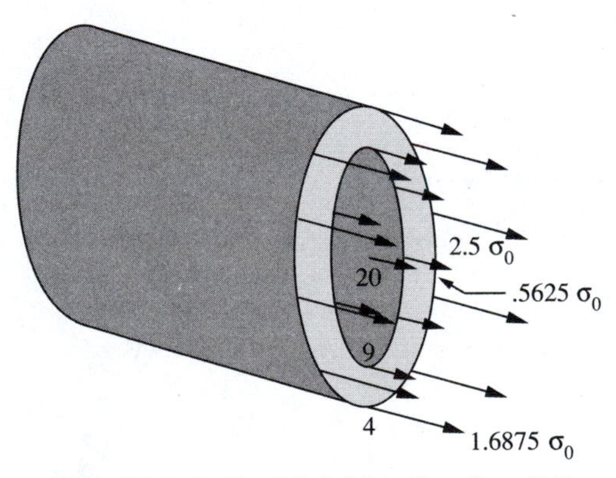

FIGURE 5–69 Nodal loads—face *BC*

The student is asked in the Exercises to show that, for this particular example, these values for the elemental load vectors coincide with those determined on the basis of evaluating the integrals exactly.

As indicated in Fig. 5–68, the constraints are

$$u_{17} = u_{18} = u_{19} = u_{20} = 0$$

$$w_{10} = w_{14} = w_{15} = w_{17} = 0$$

The nodal displacements are determined by solving the constrained equations of equilibrium.

Derived variables — stresses. The centroidal values of the stresses computed as described in the previous example are presented in Table 5.13.

TABLE 5.13 Stresses — notched shaft — mesh 1

Element #	σ_r/p_0	σ_θ/p_0	σ_z/p_0	τ_{rz}/p_0
1	0.131	−0.203	0.261	0.178
2	−0.033	−0.241	0.450	0.091
3	0.084	−0.097	0.933	0.153
4	−0.079	−0.222	0.607	0.090
5	0.102	−0.028	1.160	0.107
6	−0.019	−0.090	0.872	0.052
7	−0.007	−0.107	1.027	0.003
8	0.345	0.301	1.786	0.481
9	−0.024	−0.091	0.993	0.377
10	−0.084	−0.058	1.309	0.364
11	0.286	0.730	3.189	0.249
12	0.356	0.331	1.705	0.254
13	0.175	0.118	1.366	0.194
14	0.111	0.142	1.625	0.103
15	−0.032	−0.032	1.318	0.126
16	−0.062	−0.016	1.446	0.103
17	−0.185	−0.197	1.017	0.074
18	−0.045	−0.045	1.331	0.055
19	0.044	−0.021	1.039	−0.063
20	0.159	0.261	1.743	0.103
21	0.414	0.360	1.675	−0.059
22	0.377	0.297	1.648	−0.059
23	0.057	0.057	1.408	0.032
24	0.026	0.026	1.335	0.020

On the basis of the ratio of the areas of faces *BC* and *AE*, it is easily concluded that the average normal stress on section *AE* is $2.25\sigma_0$. The maximum normal stress on the section through the hole is seen to be $3.189\sigma_0$ at the centroid of element 11. The stress concentration factor is computed to be

$$k_T = \frac{3.189\sigma_0}{2.25\sigma_0} = 1.42$$

which is 19 percent low compared with the theoretical value [Cook and Young, 5] of approximately 1.75. The value of the stress at point *E* would logically be slightly larger than the centroidal value of $3.189\sigma_0$, resulting in some improvement in the prediction for the stress concentration factor.

For this example, consider also the refined mesh indicated in Fig. 5–70. All the basic steps are carried out as indicated in previous sections. The results indicate that the maximum axial stress σ_z occurs in element 47 with a value of $3.699\sigma_0$, from which the stress concentration factor is computed to be

$$k_T = \frac{3.699\sigma_0}{2.25\sigma_0} = 1.64$$

which is approximately 6 percent low.

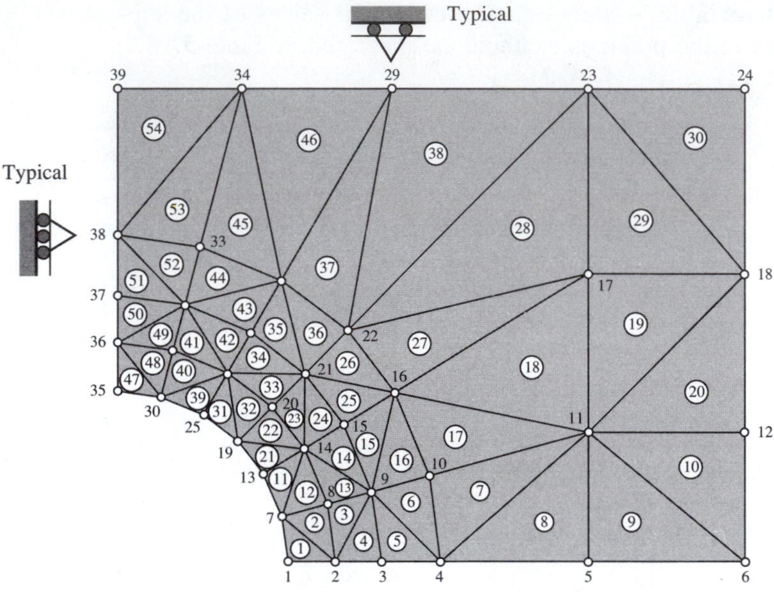

FIGURE 5–70 Notched shaft under tension—mesh 2

The value of the stress concentration factor predicted by mesh 2 is quite good. In view of the fact that the result is based on the value predicted at the centroid of the element, an extrapolation could be performed in an attempt to provide an improved estimate.

5.8 ISOPARAMETRIC FORMULATIONS

Consider a general two-dimensional region as indicated in Fig. 5–71. We will consider that the region is appropriate for either a true two-dimensional plane stress or plane strain problem or for an axisymmetric problem. If the problem is axisymmetric in nature, the region in Fig. 5–71 must either be redrawn to be symmetric about the z-axis, or be considered to represent the intersection of a $\theta =$ constant plane, with a toroidal region representing an axisymmetric problem. The general discussion presented here can be subsequently specialized to either case. We will also assume that the region has been discretized in a suitable fashion as indicated in Fig. 5–72.

Each element is assumed to be related to an appropriate parent element, as discussed in Section 3.8; that is, for suitably chosen s and t,

$$x(s,t) = \sum_{1}^{N} x_i N_i(s,t) = \mathbf{x}_\mathbf{e}^\mathrm{T}\mathbf{N} = \mathbf{N}^\mathrm{T}\mathbf{x}_\mathbf{e}$$

and

$$y(s,t) = \sum_{1}^{N} y_i N_i(s,t) = \mathbf{y}_\mathbf{e}^\mathrm{T}\mathbf{N} = \mathbf{N}^\mathrm{T}\mathbf{y}_\mathbf{e}$$

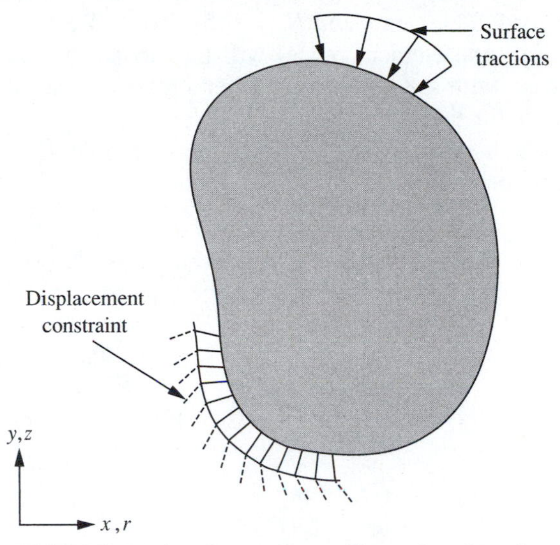

FIGURE 5–71 General two-dimensional region

define the appropriate mapping from the parent element to the actual element shape. We will further assume that the displacements are represented according to

$$u(s,t) = \sum_{1}^{N} u_i N_i(s,t) = \mathbf{u_e^T N} = \mathbf{N^T u_e}$$

and

$$v(s,t) = \sum_{1}^{N} v_i N_i(s,t) = \mathbf{v_e^T N} = \mathbf{N^T v_e}$$

In all of the functions x, y, u, and v, N represents the number of nodes that define the element so that the element is clearly isoparametric. $N = 4$ for Q4

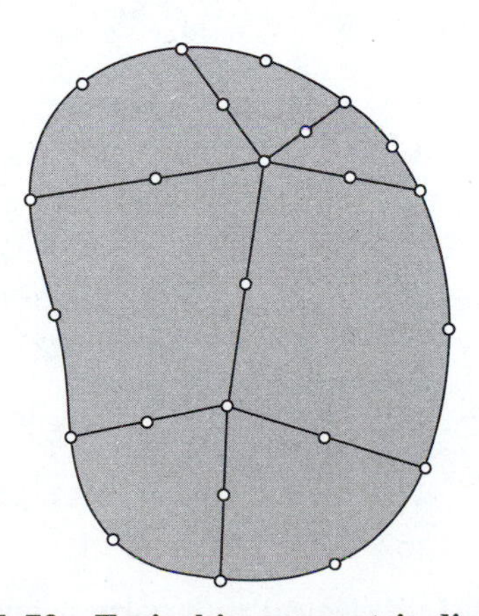

FIGURE 5–72 Typical isoparametric discretization

elements, $N = 8$ for Q8 elements and $N = 6$ for T6 elements. The development of the isoparametric finite element model will be carried through specifically for the plane stress case with Q8 elements. Other classes of problems using other elements will be left for the Exercises.

The matrices which need to be discussed are those of Eqs. (5.50), (5.51), (5.52), and (5.53), specialized for the Q8 element:

$$\mathbf{k_e} = \iint_{D_e} \mathbf{B}^{\mathrm{T}}\mathbf{C}\mathbf{B}\, d\,Vol$$

$$\boldsymbol{\theta}_\mathbf{e} = \iint_{D_e} \mathbf{B}^{\mathrm{T}}\mathbf{f_T}\, d\,Vol$$

$$\mathbf{f}_\mathbf{e}^{\mathrm{T}} = \iint_{D_e} [XN_1 \quad YN_1 \quad XN_2 \quad YN_2 \quad \ldots \quad XN_8 \quad YN_8]\, d\,Vol$$

$$\mathbf{t_e} = \int_{\gamma_{te}} \begin{bmatrix} N_i(lt_n - mt_s) \\ N_i(mt_n + lt_s) \\ N_j(lt_n - mt_s) \\ N_j(mt_n + lt_s) \\ N_k(lt_n - mt_s) \\ N_k(mt_n + lt_s) \end{bmatrix} dS$$

Evaluation of $\mathbf{t_e}$. On Γ_2, consider a typical γ_{te} portion of the boundary as indicated in Fig. 5–73. With nodes j, n, and k as defined in Fig. 5–73, $\mathbf{t_e}$ can be written explicitly as

$$\mathbf{t_e} = \int_{\gamma_{te}} \begin{bmatrix} N_j(lt_n - mt_s) \\ N_j(mt_n + lt_s) \\ N_n(lt_n - mt_s) \\ N_n(mt_n + lt_s) \\ N_k(lt_n - mt_s) \\ N_k(mt_n + lt_s) \end{bmatrix} dS$$

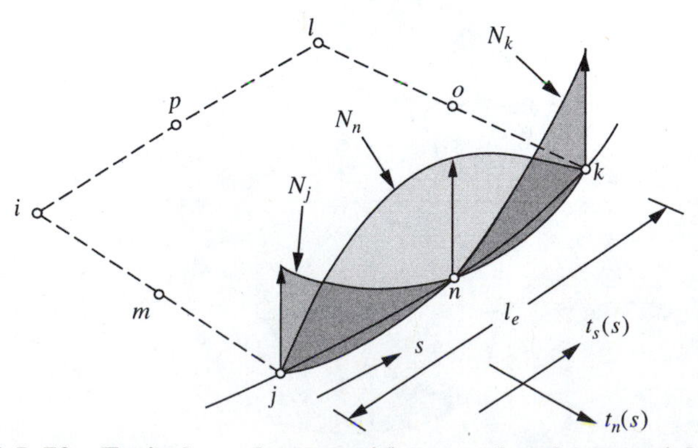

FIGURE 5–73 Typical γ_{te} element with normal and tangential tractions

with N_j, N_n, and N_k shown in Fig. 5–73. As usual, we will assume that $t_n(s)$ and $t_s(s)$ can be represented with reasonable accuracy according to

$$t_n(s) = t_{nj}N_j(s) + t_{nn}N_n(s) + t_{nk}N_k(s)$$
$$t_s(s) = t_{sj}N_j(s) + t_{sn}N_n(s) + t_{sk}N_k(s)$$

on the γ_{te} portion of the boundary. The t_{nj}, t_{nn}, and t_{nk} are generally chosen to be the nodal values of $t_n(s)$, with t_{sj}, t_{sn}, and t_{sk} the corresponding nodal values of $t_s(s)$. Evaluating the integrals yields

$$\mathbf{t_e} = \begin{bmatrix} l(4t_{nj} + 2t_{nn} - t_{nk}) - m(4t_{sj} + 2t_{sn} - t_{sk}) \\ m(4t_{nj} + 2t_{nn} - t_{nk}) + l(4t_{sj} + 2t_{sn} - t_{sk}) \\ l(2t_{nj} + 16t_{nn} + 2t_{nk}) - m(2t_{sj} + 16t_{sn} + 2t_{sk}) \\ m(2t_{nj} + 16t_{nn} + 2t_{nk}) + l(2t_{sj} + 16t_{sn} + 2t_{sk}) \\ l(-t_{nj} + 2t_{nn} + 4t_{nk}) - m(-t_{sj} + 2t_{sn} + 4t_{sk}) \\ m(-t_{nj} + 2t_{nn} + 4t_{nk}) + l(-t_{sj} + 2t_{sn} + 4t_{sk}) \end{bmatrix} \frac{hl_e}{30} \tag{5.54}$$

where l_e is the arc length of γ_{te} and can be evaluated according to the results presented in Section 3.8.2, and h represents the thickness of the element.

Note that the direction cosines l and m are actually functions of s along γ_{te} as indicated in Fig. 5–74, so that the integrals appearing in $\mathbf{t_e}$ should be evaluated with this in mind. If node n deviates significantly from the straight line connecting nodes j and k, the integrals should be evaluated using

$$l(s) = -\frac{x'(s)}{\sqrt{[x'(s)^2 + y'(s)^2]}} \qquad m(s) = \frac{y'(s)}{\sqrt{[x'(s)^2 + y'(s)^2]}}$$

where

$$x'(s) = -3x_j + 4x_n - x_k + 4\left(\frac{s}{l_e}\right)(x_j - 2x_n + x_k)$$

$$y'(s) = -3y_j + 4y_n - y_k + 4\left(\frac{s}{l_e}\right)(y_j - 2y_n + y_k)$$

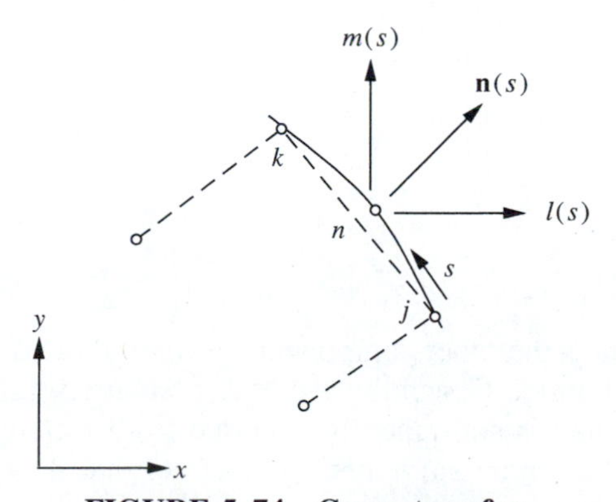

FIGURE 5–74 Geometry of γ_{te}

Otherwise, simply evaluate l and m approximately on the basis of the straight line between nodes j and k. In any event, $\mathbf{t_e}$ is clearly a 6×1 vector representing the elemental loads arising from the surface stresses. These elemental loads will contribute at assembly to the x and y loads at nodes j, n, and k of the global load vector during assembly. Note that the elemental loads are calculated to produce components in the global directions, so that no coordinate rotational transformations are necessary before the assembly process.

Evaluation of $\mathbf{f_e}$. From before, $\mathbf{f_e}$ is given by

$$\mathbf{f}_\mathbf{e}^\mathrm{T} = \int\int_{D_e} [XN_1 \quad YN_1 \quad XN_2 \quad YN_2 \quad \ldots \quad XN_8 \quad YN_8]\, d\,Vol$$
$$= [f_{x1} \quad f_{y1} \quad f_{x2} \quad f_{y2} \quad \ldots]$$

where $X = X(x, y)$ and $Y = Y(x, y)$ are the x and y components of the body force. From Section 3.8.2, the interpolation functions for the Q8 element are given by

$$N_1 = \frac{(1-s)(1-t)(-1-s-t)}{4} \qquad N_5 = \frac{(1-s^2)(1-t)}{2}$$

$$N_2 = \frac{(1+s)(1-t)(-1+s-t)}{4} \qquad N_6 = \frac{(1+s)(1-t^2)}{2}$$

$$N_3 = \frac{(1+s)(1+t)(-1+s+t)}{4} \qquad N_7 = \frac{(1-s^2)(1+t)}{2}$$

$$N_4 = \frac{(1-s)(1+t)(-1-s+t)}{4} \qquad N_8 = \frac{(1-s)(1-t^2)}{2}$$

Consider the evaluation of a typical component f_{x1} given by

$$f_{x1} = \int\int_{D_e} X(x, y)N_1\, d\,Vol$$

We choose to evaluate this over the parent element $-1 \leq s \leq 1$, $-1 \leq t \leq 1$, according to

$$f_{x1} = th \int_{-1}^{1}\int_{-1}^{1} X\big(x(s,t), y(s,t)\big) N_1(s,t)\, |\mathbf{J}|\, ds\, dt \tag{5.55}$$

with

$$x(s,t) = \mathbf{x}_\mathbf{e}^\mathrm{T}\mathbf{N} \qquad y(s,t) = \mathbf{y}_\mathbf{e}^\mathrm{T}\mathbf{N}$$

This integral, as well as the other components of $\mathbf{f_e}$, will be evaluated numerically using a Gauss quadrature. Generally a 2×2 evaluation suffices with a 3×3 necessary if the element is substantially distorted from rectangular.

If $X(x, y)$ and $Y(x, y)$ are sufficiently regular so that they can be represented according to

$$X(x, y) = \mathbf{X_e^T N} = \mathbf{N^T X_e} \qquad Y(x, y) = \mathbf{Y_e^T N} = \mathbf{N^T Y_e}$$

where

$$\mathbf{X_e^T} = [X_1 \quad X_2 \quad X_3 \quad X_4 \quad X_5 \quad X_6 \quad X_7 \quad X_8]$$

$$\mathbf{Y_e^T} = [Y_1 \quad Y_2 \quad Y_3 \quad Y_4 \quad Y_5 \quad Y_6 \quad Y_7 \quad Y_8]$$

with X_i and Y_i as the values of $X(x, y)$ and $Y(x, y)$ at node i respectively. The result for f_{x1} is

$$f_{x1} = th\left(\int_{-1}^{1}\int_{-1}^{1} \mathbf{N^T} N_1 \, |\mathbf{J}| \, ds \, dt\right)\mathbf{X_e} \tag{5.56}$$

which will also be evaluated using Gauss quadrature. The advantage of Eq. (5.56) over Eq. (5.55) is that fewer computations are involved in the evaluation of the components of $\mathbf{f_e}$ using Eq. (5.56). In either case, each of the 16 components of $\mathbf{f_e}$ is to be evaluated using a 2×2 or possibly a 3×3 Gauss quadrature, depending on the element shape. The general form of $\mathbf{f_e}$ when X and Y are approximated as before is

$$\mathbf{f_e} = th \int_{-1}^{1}\int_{-1}^{1} \left(\begin{bmatrix} N_1^2 & 0 & N_1N_2 & 0 & \cdots & & \\ 0 & N_1^2 & 0 & N_1N_2 & \cdots & & \\ N_2N_1 & 0 & N_2^2 & 0 & \cdots & & \\ 0 & N_2N_1 & 0 & N_2^2 & \cdots & & \\ \vdots & \vdots & \vdots & \vdots & & \vdots & \vdots \\ & & & & \cdots & N_8^2 & 0 \\ & & & & \cdots & 0 & N_8^2 \end{bmatrix} |\mathbf{J}| \, ds \, dt \right) \mathbf{F_e} \tag{5.57}$$

where

$$\mathbf{F_e} = [X_1 \quad Y_1 \quad X_2 \quad Y_2 \quad \ldots \quad X_8 \quad Y_8]^{\mathrm{T}}$$

The eight *xy* force pairs contribute to the appropriate positions in the global load vector during assembly. No coordinate transformations are necessary.

Evaluation of $\boldsymbol{\theta}_\mathbf{e}$ *and* $\mathbf{k_e}$. Both $\boldsymbol{\theta}_\mathbf{e}$ and $\mathbf{k_e}$ involve the strain matrix $\mathbf{B}$, which for the two-dimensional problem using Q8 elements is the 16×3 matrix

$$\mathbf{B} = \begin{bmatrix} \dfrac{\partial N_1}{\partial x} & 0 & \dfrac{\partial N_2}{\partial x} & 0 & \dfrac{\partial N_3}{\partial x} & 0 & \cdots & \dfrac{\partial N_8}{0x} & 0 \\ 0 & \dfrac{\partial N_1}{\partial y} & 0 & \dfrac{\partial N_2}{\partial y} & 0 & \dfrac{\partial N_3}{\partial y} & \cdots & 0 & \dfrac{\partial N_8}{\partial y} \\ \dfrac{\partial N_1}{\partial y} & \dfrac{\partial N_1}{\partial x} & \dfrac{\partial N_2}{\partial y} & \dfrac{\partial N_2}{\partial x} & \dfrac{\partial N_3}{\partial y} & \dfrac{\partial N_3}{\partial x} & \cdots & \dfrac{\partial N_8}{\partial y} & \dfrac{\partial N_8}{0x} \end{bmatrix}$$

In order to convert this to a form appropriate for integration over the parent element we again use the chain rule to write

$$\frac{\partial}{\partial x} = \mathbf{J}_1 \frac{\boldsymbol{\partial}}{\boldsymbol{\partial} \mathbf{s}} = J_{11}\frac{\partial}{\partial s} + J_{12}\frac{\partial}{\partial t}$$

$$\frac{\partial}{\partial y} = \mathbf{J}_2 \frac{\boldsymbol{\partial}}{\boldsymbol{\partial} \mathbf{s}} = J_{21}\frac{\partial}{\partial s} + J_{22}\frac{\partial}{\partial t}$$

where the J_{ij} are given by

$$\mathbf{J}^{-1} = \begin{bmatrix} J_{11} & J_{12} \\ J_{21} & J_{22} \end{bmatrix}$$

with $\mathbf{J}^{-1}$ the inverse of the Jacobian matrix. Then $\mathbf{B}$ can be written as

$$\mathbf{B} = \begin{bmatrix} B_{11} & B_{12} & B_{13} & B_{14} & \cdots & B_{115} & B_{116} \\ B_{21} & B_{22} & B_{23} & B_{24} & \cdots & B_{215} & B_{216} \\ B_{31} & B_{32} & B_{33} & B_{34} & \cdots & B_{315} & B_{316} \end{bmatrix} \tag{5.58}$$

where

$$\begin{aligned}
B_{11} &= J_{11}\frac{\partial N_1}{\partial s} + J_{12}\frac{\partial N_1}{\partial t} \\
B_{12} &= 0 \\
B_{13} &= J_{11}\frac{\partial N_2}{\partial s} + J_{12}\frac{\partial N_2}{\partial t} \\
B_{14} &= 0 \\
&\vdots \\
B_{115} &= J_{11}\frac{\partial N_8}{\partial s} + J_{12}\frac{\partial N_8}{\partial t} \\
B_{116} &= 0 \\
B_{21} &= 0 \\
B_{22} &= J_{21}\frac{\partial N_1}{\partial s} + J_{22}\frac{\partial N_1}{\partial t} \\
B_{23} &= 0 \\
B_{24} &= J_{21}\frac{\partial N_2}{\partial s} + J_{22}\frac{\partial N_2}{\partial t} \\
&\vdots
\end{aligned}$$

$$B_{215} = J_{21}\frac{\partial N_8}{\partial s} + J_{22}\frac{\partial N_8}{\partial t}$$

$$B_{216} = 0$$

$$B_{31} = J_{21}\frac{\partial N_1}{\partial s} + J_{22}\frac{\partial N_1}{\partial t}$$

$$B_{32} = J_{11}\frac{\partial N_1}{\partial s} + J_{12}\frac{\partial N_1}{\partial t}$$

$$B_{33} = J_{21}\frac{\partial N_2}{\partial s} + J_{22}\frac{\partial N_2}{\partial t}$$

$$B_{34} = J_{11}\frac{\partial N_2}{\partial s} + J_{12}\frac{\partial N_2}{\partial t}$$

$$\vdots$$

$$B_{315} = J_{21}\frac{\partial N_8}{\partial s} + J_{22}\frac{\partial N_8}{\partial t}$$

$$B_{316} = J_{11}\frac{\partial N_8}{\partial s} + J_{12}\frac{\partial N_8}{\partial t}$$

Then $\boldsymbol{\theta}_\mathbf{e}$ can be written as

$$\boldsymbol{\theta}_\mathbf{e} = \iint_{D_e} \mathbf{B}^\mathrm{T}\mathbf{f_T}\, d\,Vol = th \int_{-1}^{1}\int_{-1}^{1} \mathbf{B}^\mathrm{T}(s,t)\mathbf{f_T}(s,t)\,|\mathbf{J}|\, ds\, dt \tag{5.59}$$

where

$$\mathbf{f_T}(s,t) = \left[\beta T\big(x(s,t),\, y(s,t)\big) \quad \beta T\big(x(s,t),\, y(s,t)\big) \quad 0\right]$$

These elemental thermal contributions are generally evaluated using a 2×2 or if necessary a 3×3 Gauss quadrature. As usual, further simplifications are possible if βT is such that it can be represented according to

$$\beta T = \mathbf{N}^\mathrm{T}(\boldsymbol{\beta}\mathbf{T})_\mathbf{e}$$

with $(\boldsymbol{\beta}\mathbf{T})_\mathbf{e}$ an 8×1 vector of nodal values. $\boldsymbol{\theta}_\mathbf{e}$ itself is a 16×1 elemental thermal load vector whose components contribute during assembly to the global load positions corresponding to the nodes of the element.

With $\mathbf{B}$ given by Eq. (5.58), $\mathbf{k_e}$ can be expressed as

$$\mathbf{k_e} = \iint_{D_e} \mathbf{B}^\mathrm{T}\mathbf{C}\mathbf{B}\, d\,Vol = th \int_{-1}^{1}\int_{-1}^{1} \mathbf{B}^\mathrm{T}(s,t)\mathbf{C}\mathbf{B}(s,t)\,|\mathbf{J}|\, ds\, dt \tag{5.60}$$

Consider for a moment the order of integration required for the exact evaluation of $\mathbf{k_e}$. It is easily verified that the integrand of the (1,1) term of $\mathbf{k_e}$ is

$$
\begin{aligned}
I_{11} &= C_{11}B_{11}^2 + C_{33}B_{31}^2 \\
&= \left\{ C_{11}\left(\frac{J_{11}(1-t)(2s+t)}{4} + \frac{J_{12}(1-s)(s+2t)}{4}\right)^2 \right. \\
&\qquad \left. + C_{33}\left(\frac{J_{21}(1-t)(2s_t)}{4} + \frac{J_{22}(1-s)(s+2t)}{4}\right)^2 \right\} |\mathbf{J}|
\end{aligned}
$$

and is typical of any of the integrands. If the element is rectangular, J_{12} and J_{21} are zero with both J_{11} and J_{22} constant. It then follows that I_{11} is fourth degree in both s and t necessitating a 3×3 $(2 \times 3 - 1 = 5)$ Gauss quadrature for the exact evaluation of the integral. For a curvilinear Q8 shape the elements of $\mathbf{J}$ are not constant, resulting in all of the J_{ij} being rational functions (ratios of polynomials) of s and t, so that the exact evaluation of $\mathbf{k_e}$ is not possible. For general Q8 geometries a 3×3 Gaussian quadrature results in a reasonably accurate evaluation of $\mathbf{k_e}$. In any event we will refer to the 3×3 Gauss quadrature for the Q8 element as *full integration*. For the general Q4 element geometry a 2×2 Gauss quadrature is the full integration.

For convergence (in the limit as the mesh is refined) of the finite element method [Strang and Fix, 6], it is sufficient to use a quadrature rule with accuracy $O(h^{2(p-m+1)})$, where p is the order of the polynomial representation for the dependent variable(s) and $2m$ is the order of the highest derivative occurring in the governing differential equation(s). Alternatively, when an appropriate variational principle exists, m is the order of the highest derivative appearing in the energy functional. When Gauss quadrature is used for evaluating the integrals, the necessary Gauss order is $N = p - m + 1$, which for the elasticity problem is $2 - 1 + 1 = 2$. In general, p is determined on the basis of the completeness of the pth degree terms. Even though the third-order terms x^2y and xy^2 are present in the representations for the displacements, the x^3 and y^3 terms are not included, with the result that $p = 2$. The use of the 2×2 Gauss quadrature is referred to as *reduced integration* for the Q8 element, and is usually employed unless the element is substantially distorted from rectangular, or unless the mesh is coarse.

In any event, $\mathbf{k_e}$ is a 16×16 elemental stiffness matrix contributing to the global stiffness matrix in the positions corresponding to the degrees of freedom associated with the nodes of the element. After all the necessary elemental matrices have been computed and assembled according to

$$\mathbf{K_G} = \sum \mathbf{k_G}$$

$$\mathbf{F_G} = \sum(\boldsymbol{\theta}_\mathbf{G} + \mathbf{f_G}) + \sum{}' \mathbf{t_G}$$

the constraints are imposed and the solution then obtained.

Derived variables. The general expression for the stresses,

$$\boldsymbol{\sigma} = \mathbf{CBU_e} - \mathbf{f_T}$$

is a function of position within the element. There arises the obvious question as to the best locations within the element at which to evaluate the stresses. Barlow [7] showed that the stresses evaluated at the Gauss points of order $p - m + 1$ are generally more accurate than when evaluated at other points (the nodes, for instance) within the element. For the Q8 element, $p - m + 1 = 2$, as discussed previously, so that conveniently, the reduced integration to be used for evaluating the $\mathbf{k_e}$ produces the necessary information for evaluating the stresses at the best points within the element. During the process of forming the stiffness $\mathbf{k_e}$, the quantity **CB** was computed at each of the Gauss points and should, at that time, have been written to auxiliary storage along with the nodal definition for the element. For computing the stresses, this information can be retrieved and used to select the appropriate subset $\mathbf{U_e}$ (of $\mathbf{U_G}$), and then to compute the stresses at each of the Gauss points.

Frequently, however, it is desirable to have information about the stresses at the nodes. As pointed out previously, direct calculations for the nodal stresses according to

$$\boldsymbol{\sigma}(\text{nodes}) = \mathbf{CB}(\text{nodes})\mathbf{U_e} - \mathbf{f_T}(\text{nodes})$$

are generally inaccurate. A frequently used ad hoc procedure, referred to as smoothing [Hinton and Campbell, 8], uses a least-squares approach to determine nodal stresses. The end result is equivalent to an extrapolation (see Fig. 5–75) from the stresses at the Gauss points to the stresses at the nodes, outlined as follows.

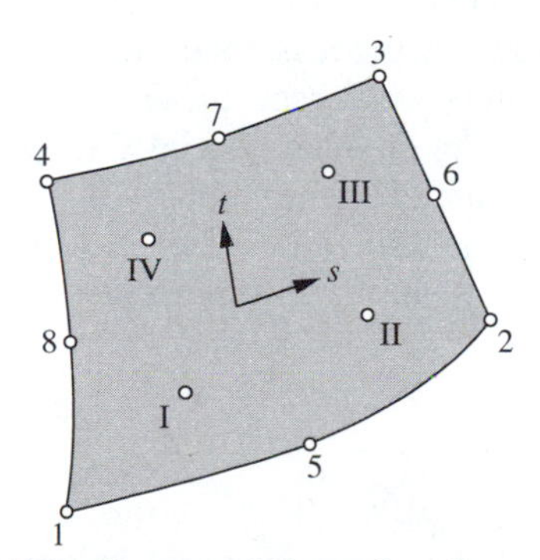

FIGURE 5–75 Gauss points and nodes—Q8 element

Using the bilinear interpolation functions

$$N_I(s,t) = \frac{(1-s)(1-t)}{4} \qquad N_{III}(s,t) = \frac{(1+s)(1+t)}{4}$$

$$N_{II}(s,t) = \frac{(1+s)(1-t)}{4} \qquad N_{IV}(s,t) = \frac{(1-s)(1+t)}{4}$$

represent a particular stress component, σ_x for instance, within the element as

$$\sigma_x = \sigma_{xI}N_I + \sigma_{xII}N_{II} + \sigma_{xIII}N_{III} + \sigma_{xIV}N_{IV}$$

This defines a bilinear σ_x within the quadrilateral region defined by the Gauss points I, II, III, and IV. Then σ_x is extrapolated to the nodes using the s and t coordinates given in Table 5.14. This same procedure is followed for each of the stress components.

TABLE 5.14 Nodal coordinates for extrapolation

Node	s	t
1	$-\sqrt{3}$	$-\sqrt{3}$
2	$+\sqrt{3}$	$-\sqrt{3}$
3	$+\sqrt{3}$	$+\sqrt{3}$
4	$-\sqrt{3}$	$+\sqrt{3}$
5	0	$-\sqrt{3}$
6	$+\sqrt{3}$	0
7	0	$+\sqrt{3}$
8	$-\sqrt{3}$	0

This process can be thought of in the following fashion. Generally for the Q8 element, the displacements are represented by quadratic functions within the element. The stresses, which are computed as derivatives of the displacements, are then roughly dependent in a linear fashion on the coordinates within the element. The extrapolation process takes the most accurate stresses, those evaluated at the Gauss points, and uses these as a basis for extrapolating linearly throughout the rest of the element. Usually these extrapolated stress values are not continuous across interelement boundaries. It is then a common practice to average a particular stress component, at a given node, over the values from all the contributing elements. As one might suspect, this is referred to as *averaging*.

5.8.1 Applications

In this section, examples that were considered in previous sections using linear interpolation will be reexamined using Q8 elements.

5.8.1.1 THIN RECTANGULAR PLATE—MOMENT LOADING. Consider again the moment-loaded 1-in thick flat plate of Section 5.5.2 shown in Fig. 5–76.

A one-element model with nodes as shown in Fig. 5–76 will be investigated. The constraints and nodal loads are shown in Fig. 5–77. Take $E = 10^7$ psi and $\nu = 0.3$.

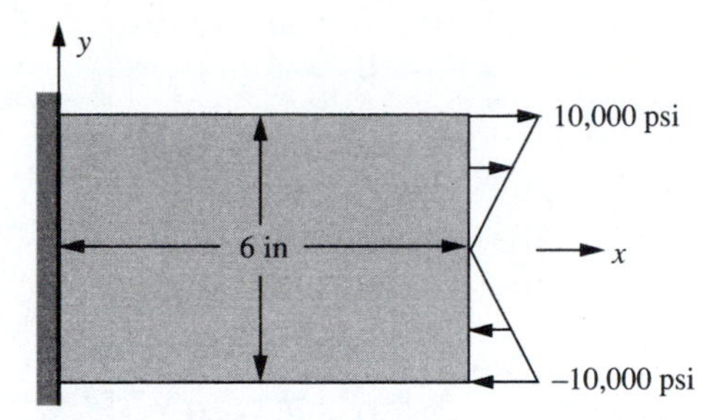

FIGURE 5–76 Moment loaded flat plate — Q8 model

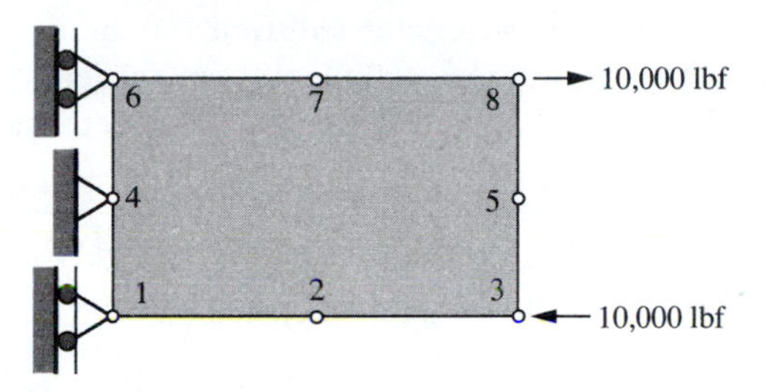

FIGURE 5–77 Constraints and nodal loads

Applying the constraints and solving the resulting equations yields the displacements shown in Table 5.15.

TABLE 5.15 Nodal displacements — moment loaded plate

Node	u (in)	u_{ex} (in)	v (in)	v_{ex} (in)
1	0.00000	0.00000	−0.00045	−0.00045
2	−0.00300	−0.00300	−0.00195	−0.00195
3	−0.00600	−0.00600	−0.00645	−0.00645
4	0.00000	0.00000	0.00000	0.00000
5	0.00000	0.00000	−0.00600	−0.00600
6	0.00000	0.00000	−0.00045	−0.00045
7	0.00300	0.00300	−0.00195	−0.00195
8	0.00600	0.00600	−0.00645	−0.00645

As seen, the results from the finite element model coincide with the exact displacements evaluated at the nodes. It is left to the Exercises to show that the displacements $u(x, y)$ and $v(x, y)$ coincide with the exact solution given in Section 5.5.3 at all points within the element.

The stresses evaluated at the Gauss points are indicated in Fig. 5–78. These stresses are also equal to those given by the exact solution evaluated at the same locations.

$$
\begin{bmatrix}
\begin{pmatrix} 7746 \\ 0 \\ 0 \end{pmatrix} & \begin{pmatrix} 7746 \\ 0 \\ 0 \end{pmatrix} & \begin{pmatrix} 7746 \\ 0 \\ 0 \end{pmatrix} \\
\begin{pmatrix} 0 \\ 0 \\ 0 \end{pmatrix} & \begin{pmatrix} 0 \\ 0 \\ 0 \end{pmatrix} & \begin{pmatrix} 0 \\ 0 \\ 0 \end{pmatrix} \\
\begin{pmatrix} -7746 \\ 0 \\ 0 \end{pmatrix} & \begin{pmatrix} -7746 \\ 0 \\ 0 \end{pmatrix} & \begin{pmatrix} -7746 \\ 0 \\ 0 \end{pmatrix}
\end{bmatrix}
$$

FIGURE 5–78 Stresses evaluated at the Gauss points

For this application, the finite element solution for the displacements and the stresses coincides with the analytical solution given in Section 5.5.3. This is to be expected in that the exact expressions for the displacements,

$$u(x, y) = \frac{Mxy}{EI} \qquad v(x, y) = -\frac{M(x^2 + \nu y^2)}{2EI}$$

are contained within the general elemental expressions

$$u_e(x, y) = c_1 + c_2x + c_3y + c_4x^2 + c_5y^2 + c_6xy + c_7x^2y + c_8xy^2$$
$$v_e(x, y) = d_1 + d_2x + d_3y + d_4x^2 + d_5y^2 + d_6xy + d_7x^2y + d_8xy^2$$

for some choice of the c_i and d_i. As pointed out several times previously, one of the properties of the Ritz method is that it will choose the exact solution if the exact solution is contained within the assumed approximate solution. This property is verified for this application.

5.8.1.2 STRESS CONCENTRATION—PLATE WITH CIRCULAR HOLE. Consider again the stress concentration example of Section 5.5.4. The reader is referred to that section for a discussion of the use of the quarter symmetry model shown in Fig. 5–79.

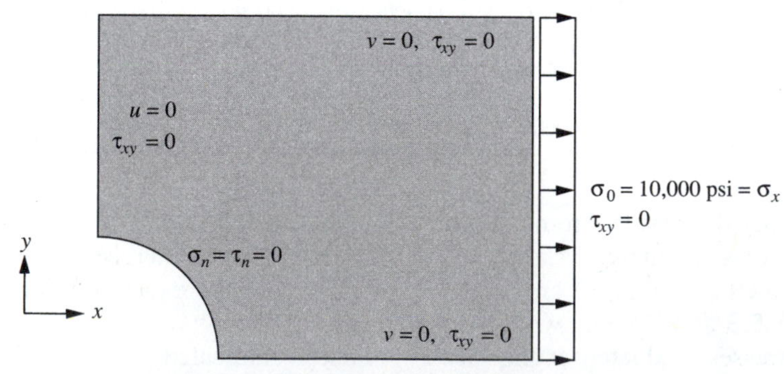

FIGURE 5–79 Quarter symmetry model

The mesh employing Q8 elements is indicated in Fig. 5–80. The mesh is chosen so as to be finer in the neighborhood of node 34, at which the stress σ_x is expected to be a maximum.

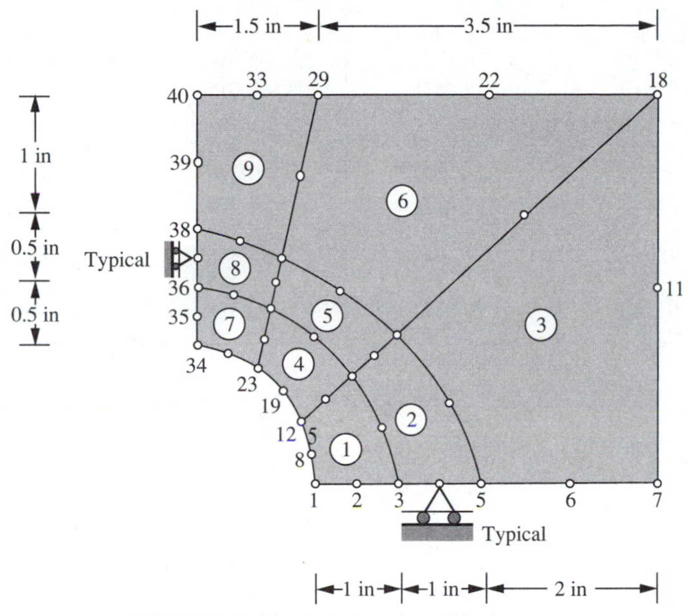

FIGURE 5–80 Mesh using Q8 elements

Numerical results for two models are presented. The first model uses reduced integration (2 × 2 Gauss) for evaluating the $\mathbf{k_e}$. The second model uses full integration (3 × 3 Gauss) for evaluating the $\mathbf{k_e}$. For each model, the global stiffness and load matrices are assembled, the typical constraints indicated in Fig. 5–80 are enforced, and the resulting equations then solved for the nodal displacements.

Figures 5–81 and 5–82 present the stress results at the $N = 2$ Gauss points $\pm 1/\sqrt{3}$, discussed in Section 5.8. Shown are the stresses in elements 4, 5, 7, and 8, which for simplicity are pictured as rectangular, with the relative positions of their nodes maintained. The three numbers in the staggered brackets represent σ_x, σ_y, and τ_{xy}, respectively, all multiplied by 10^{-4}.

The stresses presented in Fig. 5–81 are computed by reusing the **CB** matrices, which were evaluated and saved during the process of calculating the $\mathbf{k_e}$ for the reduced integration model (2 × 2 Gauss). The stresses presented in Fig. 5–82 are based on full integration (3 × 3 Gauss), so in order to compute the stresses at the best points, it is necessary to first compute the corresponding **CB** at each of the four Gauss points, an additional expense.

The two sets of results are seen to differ very little, indicating that practically, the results using reduced integration are adequate. The expense (in time) of form-

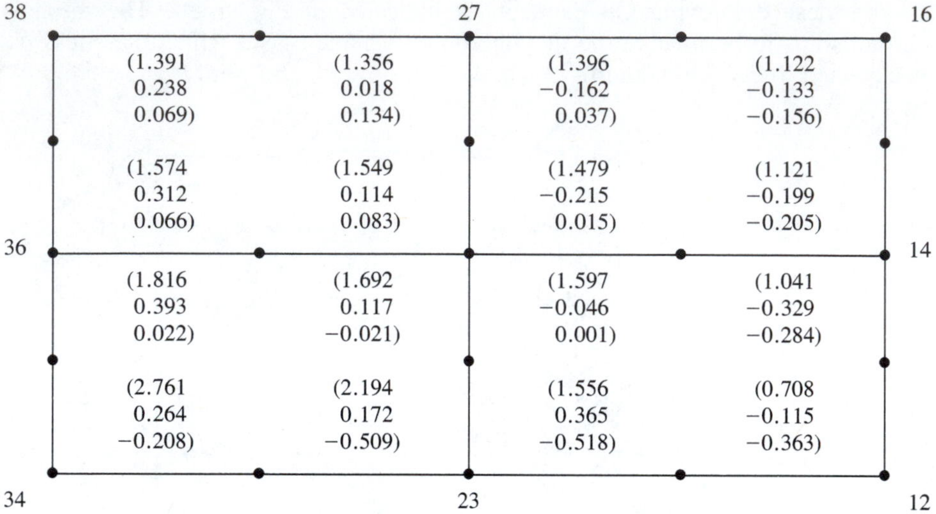
38
27
16
(1.391 0.238 0.069)
(1.356 0.018 0.134)
(1.396 −0.162 0.037)
(1.122 −0.133 −0.156)
(1.574 0.312 0.066)
(1.549 0.114 0.083)
(1.479 −0.215 0.015)
(1.121 −0.199 −0.205)
36
14
(1.816 0.393 0.022)
(1.692 0.117 −0.021)
(1.597 −0.046 0.001)
(1.041 −0.329 −0.284)
(2.761 0.264 −0.208)
(2.194 0.172 −0.509)
(1.556 0.365 −0.518)
(0.708 −0.115 −0.363)
34
23
12

FIGURE 5–81 Gauss point stresses — reduced integration

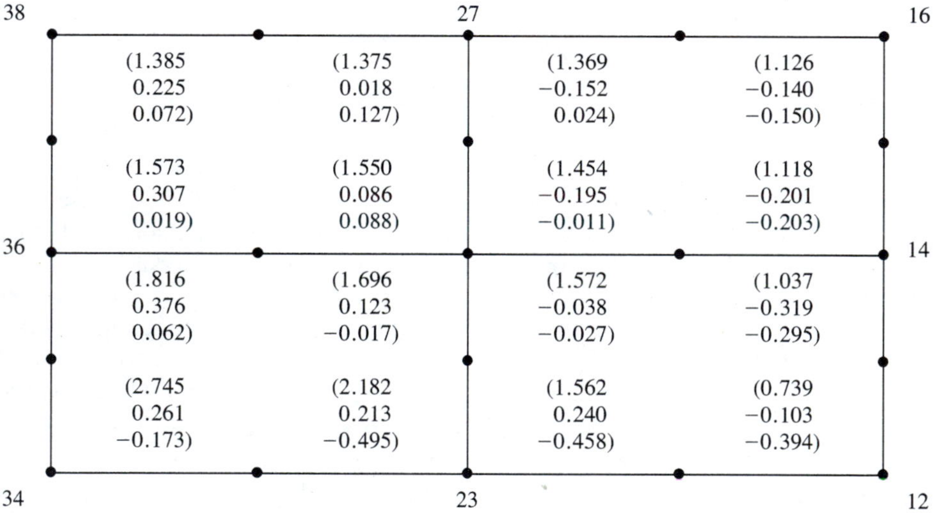
38
27
16
(1.385 0.225 0.072)
(1.375 0.018 0.127)
(1.369 −0.152 0.024)
(1.126 −0.140 −0.150)
(1.573 0.307 0.019)
(1.550 0.086 0.088)
(1.454 −0.195 −0.011)
(1.118 −0.201 −0.203)
36
14
(1.816 0.376 0.062)
(1.696 0.123 −0.017)
(1.572 −0.038 −0.027)
(1.037 −0.319 −0.295)
(2.745 0.261 −0.173)
(2.182 0.213 −0.495)
(1.562 0.240 −0.458)
(0.739 −0.103 −0.394)
34
23
12

FIGURE 5–82 Gauss point stresses—full integration

ing each of the $\mathbf{k_e}$ using reduced integration is approximately 44 percent of that using full integration. In addition, for the full integration model it is necessary to compute the **CB** before the stresses can be evaluated.

Figures 5–83 and 5–84, having used 2 × 2 and 3 × 3 Gauss, respectively, give the nodal stress extrapolated from the Gauss point stresses according to the discussion of derived variables in Section 5.8.

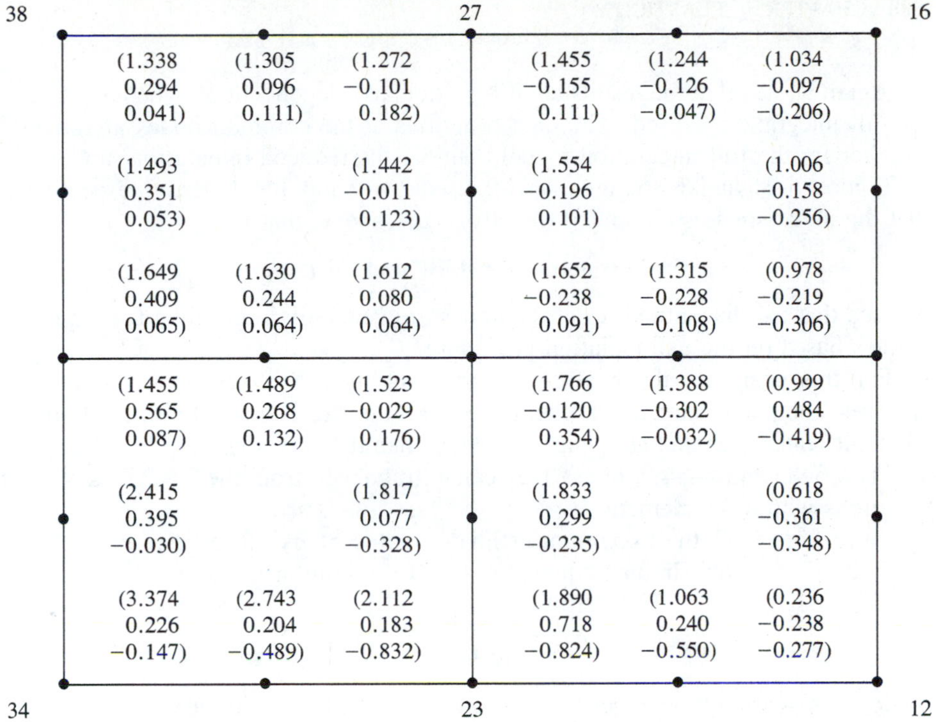

38 27 16

(1.338 0.294 0.041)	(1.305 0.096 0.111)	(1.272 −0.101 0.182)	(1.455 −0.155 0.111)	(1.244 −0.126 −0.047)	(1.034 −0.097 −0.206)
(1.493 0.351 0.053)		(1.442 −0.011 0.123)	(1.554 −0.196 0.101)		(1.006 −0.158 −0.256)
(1.649 0.409 0.065)	(1.630 0.244 0.064)	(1.612 0.080 0.064)	(1.652 −0.238 0.091)	(1.315 −0.228 −0.108)	(0.978 −0.219 −0.306)
(1.455 0.565 0.087)	(1.489 0.268 0.132)	(1.523 −0.029 0.176)	(1.766 −0.120 0.354)	(1.388 −0.302 −0.032)	(0.999 0.484 −0.419)
(2.415 0.395 −0.030)		(1.817 0.077 −0.328)	(1.833 0.299 −0.235)		(0.618 −0.361 −0.348)
(3.374 0.226 −0.147)	(2.743 0.204 −0.489)	(2.112 0.183 −0.832)	(1.890 0.718 −0.824)	(1.063 0.240 −0.550)	(0.236 −0.238 −0.277)

34 23 12

FIGURE 5–83 Reduced integration— extrapolated nodal stresses

38 27 16

(1.319 0.268 0.051)	(1.314 0.094 0.108)	(1.309 −0.080 0.165)	(1.414 −0.143 0.098)	(1.234 −0.127 −0.047)	(1.053 −0.112 −0.192)
(1.484 0.345 0.052)		(1.457 −0.027 0.080)	(1.517 −0.174 0.073)		(1.017 −0.169 −0.243)
(1.651 0.441 0.052)	(1.628 0.244 0.066)	(1.604 0.027 0.080)	(1.620 −0.206 0.049)	(1.300 −0.217 −0.123)	(0.980 −0.227 −0.294)
(1.461 0.538 0.064)	(1.497 0.548 0.124)	(1.534 −0.030 0.183)	(1.733 −0.046 0.273)	(1.361 −0.269 −0.064)	(0.988 −0.492 −0.401)
(2.406 0.374 −0.012)		(1.814 0.113 −0.321)	(1.163 0.215 −0.206)		(0.639 −0.325 −0.382)
(3.350 0.209 −0.088)	(2.722 0.232 −0.457)	(2.094 0.255 −0.826)	(1.898 0.477 −0.684)	(1.094 0.159 −0.523)	(0.290 −0.159 −0.362)

34 23 12

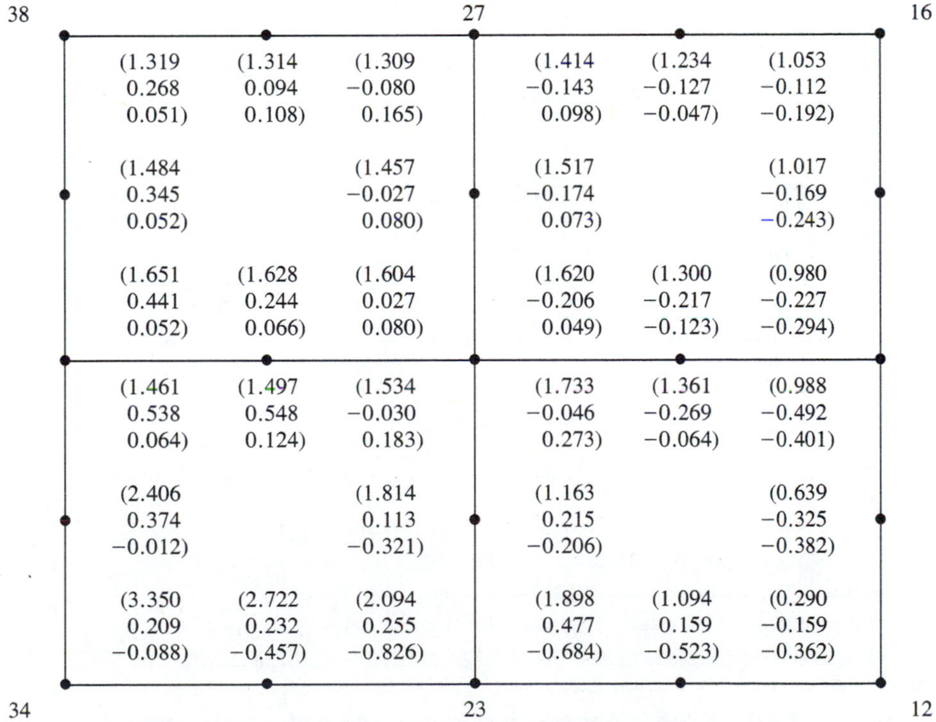

FIGURE 5–84 Full integration— extrapolated nodal stresses

Again we see that the results are not particularly dependent on whether reduced or full integration is used. The discontinuities at the common nodes are slightly smaller for the full integration model than for the reduced integration model.

Figures 5–85 and 5–86, again having used 2 × 2 and 3 × 3 Gauss, respectively, list the nodal stresses computed directly at the nodes, that is,

$$\boldsymbol{\sigma}(\text{ node }) = \mathbf{CB}(\text{ node })\mathbf{U_e}$$

we see that the discontinuities at the nodes are generally larger than for the nodal values based on the extrapolation procedure.

For this example, the results presented in Figs. 5–85 and 5–86 show that stresses computed directly at the nodes exhibit stress discontinuities at inter-element boundaries that are generally larger than the corresponding nodal stresses of Figs. 5–83 and 5–84, which have been extrapolated from the 2 × 2 Gauss point locations within the element.

From Fig. 5–83 the maximum estimated value of σ_x, located at node 34, is 33,740 psi, resulting in an estimated stress concentration factor of

$$k_T = \frac{33,740}{15,000} = 2.25$$

which is essentially the exact value quoted previously in Section 5.5.4.

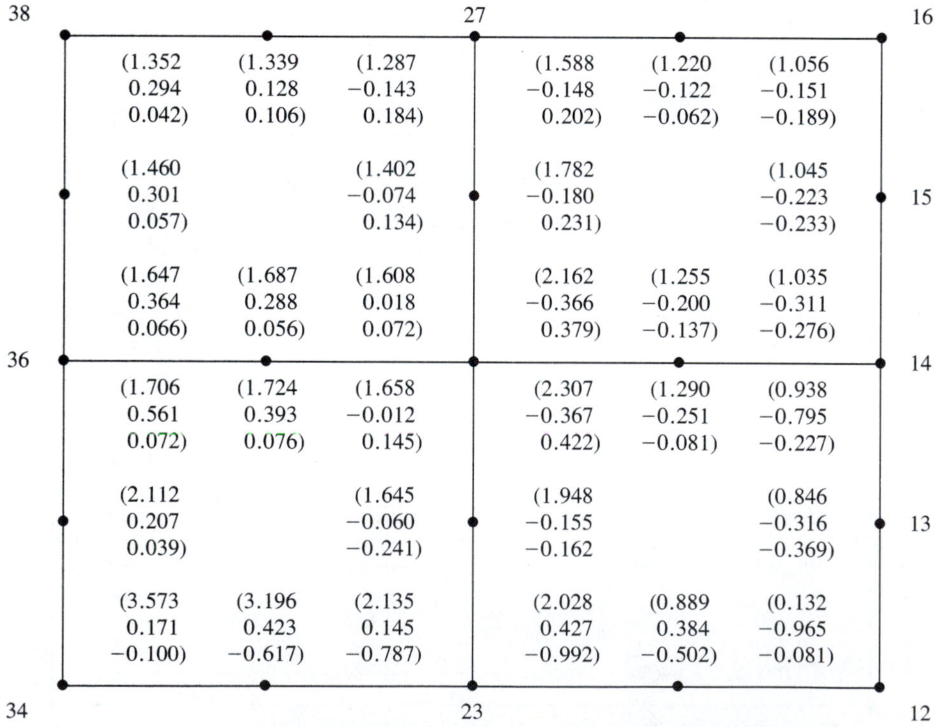

FIGURE 5–85 Reduced integration — computed nodal stresses

<table>
<tr><td>(1.330
0.221
0.052)</td><td>(1.349
0.132
0.105)</td><td>(1.321
−0.139
0.168)</td><td>(1.530
−0.139
0.178)</td><td>(1.215
−0.121
−0.062)</td><td>(1.072
−0.161
−0.175)</td></tr>
<tr><td>(1.446
0.273
0.056)</td><td></td><td>(1.415
−0.108
0.132)</td><td>(1.723
−0.161
0.185)</td><td></td><td>(1.054
−0.227
−0.221)</td></tr>
<tr><td>(1.642
0.348
0.057)</td><td>(1.687
0.281
0.059)</td><td>(1.592
0.065
0.091)</td><td>(2.115
−0.296
0.284)</td><td>(1.243
−0.193
−0.149)</td><td>(1.037
−0.305
−0.269)</td></tr>
<tr><td>(1.707
0.563
0.061)</td><td>(1.712
0.356
0.068)</td><td>(1.666
0.016
0.162)</td><td>(2.259
−0.213
0.347)</td><td>(1.263
−0.252
−0.110)</td><td>(0.959
−0.692
−0.230)</td></tr>
<tr><td>(2.134
0.248
0.057)</td><td></td><td>(1.669
−0.024
−0.227)</td><td>(1.966
−0.119
−0.114</td><td></td><td>(0.877
−0.236
−0.374)</td></tr>
<tr><td>(3.564
0.234
−0.025)</td><td>(3.132
0.403
−0.583)</td><td>(2.134
0.276
−0.772)</td><td>(2.062
0.307
−0.796)</td><td>(0.908
0.212
−0.508)</td><td>(0.272
−0.647
−0.145)</td></tr>
</table>

FIGURE 5–86 Full integration—computed nodal stresses

Discussion. For this example, the results indicate that for a mesh consisting of elements that are not unduly distorted from rectangular, reduced integration provides results that are entirely satisfactory relative to using full integration. Extrapolated nodal stresses are generally superior to stresses computed directly at the nodes. The extrapolation process yields stresses that display adequate accuracy and a lack of large interelement discontinuities. Finally, the use of the Q8 elements and the extrapolation technique is seen to provide substantially more accurate estimates of the stress concentration factor ($k_T = 2.25$) than obtained previously ($k_T = 2.02$) using linearly interpolated elements.

5.9 CLOSURE

For all of the classes of elasticity problems considered in this chapter the elemental matrices had exactly the same form:

$$\mathbf{k_e} = \iint_{D_e} \mathbf{B}^{\mathrm{T}}\mathbf{C}\mathbf{B}\, d\,Vol$$

$$\boldsymbol{\theta}_{\mathbf{e}} = \iint_{D_e} \mathbf{B}^{\mathrm{T}}\mathbf{f}_{\mathrm{T}}\, d\,Vol$$

$$\mathbf{f_e} = \iint_{D_e} [XN_1 \quad YN_1 \quad XN_2 \quad YN_2 \quad XN_3 \quad YN_3 \quad \ldots]^T dVol$$

$$\mathbf{t_e} = \begin{bmatrix} \int_{\gamma_{te}} N_i(lt_n - mt_s)dS \\ \int_{\gamma_{te}} N_i(mt_n + lt_s)dS \\ \int_{\gamma_{te}} N_j(lt_n - mt_s)dS \\ \int_{\gamma_{te}} N_j(mt_n + lt_s)dS \\ \vdots \end{bmatrix}$$

As long as the correct *number* and *type* of interpolation functions appropriate for the region and element are used, the form of the above elemental matrices is the same.

Thermal effects were discussed explicitly via the constitutive relationship

$$\boldsymbol{\sigma} = \mathbf{C}\boldsymbol{\epsilon} - \mathbf{f_T}$$

This can be generalized to include other effects according to

$$\boldsymbol{\sigma} - \boldsymbol{\sigma}_0 = \mathbf{C}(\boldsymbol{\epsilon} - \boldsymbol{\epsilon}_0) - \mathbf{f_T}$$

where $\boldsymbol{\sigma}_0$ is a prescribed or known initial stress (commonly referred to as a residual stress), and $\boldsymbol{\epsilon}_0$ is an initial strain. Rewrite this as

$$\boldsymbol{\sigma} = \mathbf{C}\boldsymbol{\epsilon} - \mathbf{f_T} - \mathbf{C}\boldsymbol{\epsilon}_0 + \boldsymbol{\sigma}_0$$

Thus, by replacing or augmenting $\mathbf{f_T}$ by $\mathbf{C}\boldsymbol{\epsilon}_0$ and/or $-\boldsymbol{\sigma}_0$, the $\boldsymbol{\theta}_e$ given above can be used to account for each of the terms $\mathbf{f_T}$, $\mathbf{C}\boldsymbol{\epsilon}_0$, and $\boldsymbol{\sigma}_0$.

Linear dynamic elasticity problems can also be handled easily with straightforward extensions of the material covered in the body of Chapter 5. These extensions are taken up in the Exercises.

Linear elasticity can be considered a logical first step in the general application of the finite element method to problems in solid mechanics. Other important areas of application include thin and thick plates, shell structures, fracture mechanics, and materially and/or geometrically nonlinear problems in solid mechanics. In addition, finite elements are finding increasing use in fluid mechanics.

As a final closing note, we observe that except for the diffusion problems of Chapter 4, all the problems considered in this text are such that a variational principle exists, enabling the finite element model to be generated using the method of Ritz. Whether or not such a variational principle exists, the Galerkin finite element method can generally be used in a wide variety of situations where it is possible to identify a differential equation or a set of differential equations together with appropriate auxiliary conditions (boundary and initial conditions). This covers essentially all of modern engineering and classical mathematical physics, a correct indication of the enormous scope of application for the finite element method.

REFERENCES

1. Boresi, A. P., and K. P. Chong: *Elasticity in Engineering Mechanics*, Elsevier-North Holland Publishing, New York, 1987.
2. Timoshenko, S. P., and J. N. Goodier: *Theory of Elasticity*, McGraw-Hill, New York, 1970.
3. Love, A. E. H.: *The Mathematical Theory of Elasticity*, Dover, New York, 1944.
4. Kaplan, W., and D. J. Lewis: *Calculus and Linear Algebra*, vol. II, John Wiley, New York, 1971.
5. Cook, R. D., and W. C. Young: *Advanced Mechanics of Materials*, Macmillan, New York, 1985.
6. Strang, G., and G. Fix: *An Analysis of the Finite Element Method*, Prentice Hall, Englewood Cliffs, New Jersey, 1973.
7. Barlow, J.: "Optimal Stress Locations in Finite Element Models," *International Journal for Numerical Methods in Engineering*, vol. 10, pp. 243–251, 1976.
8. Hinton, E., and J. S. Campbell: "Local and Global Smoothing of Discontinuous Finite Element Functions using a Least Squares Method," *International Journal for Numerical Methods in Engineering*, vol. 8, pp. 461–480, 1974.

GENERAL REFERENCES

Bathe, K. J.: *Finite Element Procedures in Engineering Analysis*, Prentice Hall, Englewood Cliffs, New Jersey, 1982.

Bathe, K. J., and E. L. Wilson: *Numerical Methods in Finite Element Analysis*, Prentice Hall, Englewood Cliffs, New Jersey, 1976.

Brebbia, C. A., and J. J. Connor: *Fundamentals of Finite Element Techniques for Structural Engineers*, Halstead Press, New York, 1974.

Cook, R. D.: *Concepts and Applications of Finite Element Analysis*, John Wiley, New York, 1981.

Desai, C. S., and J. F. Abel: *Introduction to the Finite Element Method*, Van Nostrand Reinhold, New York, 1972.

Gallagher, R. H.: *Finite Element Analysis Fundamentals*, Prentice Hall, Englewood Cliffs, New Jersey, 1975.

Grandin, H., Jr.: *Fundamentals of the Finite Element Method*, Macmillan, New York, 1986.

Hughes, T. J. R.: *The Finite Element Method, Linear Static and Dynamic Finite Element Analysis*, Prentice Hall, Englewood Cliffs, New Jersey, 1987.

Kikuchi, N.: *Finite Element Methods in Mechanics*, Cambridge University Press, Cambridge UK, 1986.

Langhaar, H. L.: *Energy Methods in Applied Mechanics*, John Wiley, New York, 1962.

Livesley, R. K.: *Finite Elements: An Introduction for Engineers*, Cambridge University Press, Cambridge UK, 1983.

Martin, H. C. and G. F. Carey: *Introduction to Finite Element Analysis—Theory and Application*, McGraw-Hill, New York, 1973.

Mikhlin, S. G.: *Variational Methods in Mathematical Physics*, Pergamon, New York, 1964.

Przemieniecki, J. S.: *Theory of Matrix Structural Analysis*, Dover, New York, 1985.

Rao, S. S.: *The Finite Element Method in Engineering*, Pergamon, Oxford, 1982.

Reddy, J. N.: *Energy and Variational Methods in Applied Mechanics with an Introduction to the Finite Element Method*, Wiley-Interscience, New York, 1984.

——: *An Introduction to the Finite Element Method*, McGraw-Hill, New York, 1984.

Rektorys, C.: *Variational Methods in Mathematics, Science, and Engineering*, Reidel, Dordrecht, Holland, 1980.

Shames, I. H., and C. L. Dym: *Energy and Finite Element Methods in Structural Mechanics*, McGraw-Hill, New York, 1985.

Tong, P., and J. N. Rossettos: *Finite Element Method: Basic Technique and Implementation*, MIT Press, Cambridge, Massachusetts, 1977.

Washizu, K.: *Variational Methods in Elasticity and Plasticity*, Pergamon, New York, 1975.

Zienkiewicz, O. C.: *The Finite Element Method,* 3rd Ed., McGraw-Hill, New York, 1977.

COMPUTER PROJECTS

Develop a computer code for the approximate solution of the general two-dimensional linear elasticity problem given as follows.

Determine displacements that satisfy the equations of equilibrium

$$\frac{\partial\left(C_{11}\,\partial u/\partial x + C_{12}\,\partial v/\partial y\right)}{\partial x} + \frac{\partial\left(C_{33}(\partial u/\partial y + \partial v/\partial x)\right)}{\partial y} + X = \partial(\beta T)/\partial x \qquad \text{in } D$$

and

$$\frac{\partial\left(C_{33}(\partial u/\partial y + \partial v/\partial x)\right)}{\partial x} + \frac{\partial\left(C_{12}\,\partial u/\partial x + C_{22}\,\partial v/\partial y\right)}{\partial y} + Y = \partial(\beta T)/\partial y \qquad \text{in } D$$

in the interior of the region D and that also satisfy the displacement boundary conditions

$$\left.\begin{aligned} u &= u_0(\Gamma, t) \\ v &= v_0(\Gamma, t) \end{aligned}\right\} \quad \text{on } \Gamma_u$$

and the traction or stress type boundary conditions

$$l\sigma_x + m\tau_{xy} = t_x(\Gamma, t)$$

and $\qquad$ on Γ_t

$$m\tau_{xy} + l\sigma_y = t_y(\Gamma, t)$$

Structure the program in roughly the following fashion.

INPUT.

1. NN $\quad$ Number of nodes.
2. NE $\quad$ Number of elements.
3. Nodal (X and Y), and body force (FX and FY) and temperature coordinates.
   X(I), I = 1, NN
   Y(I), I = 1, NN
   FX(I), I = 1, NN
   FY(I), I = 1, NN
   T(I), I = 1, NN
4. Element material properties.
   C_{11}(I), I = 1, NE
   C_{12}(I), I = 1, NE
   C_{22}(I), I = 1, NE
   C_{33}(I), I = 1, NE
   β(I), I = 1, NE
5. Element connectivity, $NI, NJ, NK, \ldots$
   NI(I), I = 1, NE
   NJ(I), I = 1, NE
   NK(I), I = 1, NE
   ⋮

The number of sets entered depends on the type of element being used, that is, four sets for the bilinearly interpolated quadrilateral element, eight sets for the Q8 element.

6. N2, the number of segments on which type two boundary conditions are prescribed.

N2S(I), I = 1 to N2	Each of these integers represents the number of nodes on the Ith segment.

Then for each segment, input:

M2(I,J) J = 1 to N2S(I)	Representing the node numbers on the Ith segment.
SN(I,J) J = 1 to N2S(I)	Each of these real numbers represents the value of the normal traction at the Jth node on the Ith segment.
SS(I,J) J = 1 to N2S(I)	Each of these real numbers represents the value of the tangential traction at the Jth node on the Ith segment.

7. NC, the number of constrained nodes.

ND(I), I = 1, NC	The node numbers at which constraints are enforced.
NCX(I), I = 1, NC	An x-direction constraint flag.
NCY(I), I = 1, NC	A y-direction constraint flag.
CX(I), I = 1, NC	The x-direction constraint value.
CY(I), I = 1, NC	The y-direction constraint value.

DATA REFLECTION. It is a good idea to print out all of the above data for a check.

ELEMENT FORMULATION AND ASSEMBLY. $\mathbf{k}_e$, $\boldsymbol{\theta}_e$, and $\mathbf{f}_e$ need to be formed and assembled for each of the NE elements. These steps are probably best handled in separate subroutines or procedures. The X, Y, FX, FY, C_{ij}, β, NI, NJ, . . . data are needed here. Also, $\mathbf{t}_e$ needs to be formed and assembled for each of the N2 segments on the boundary. The X, Y, N2, N2S, M2, SN, and SS data are required.

CONSTRAINTS. All of the constraint data must be imposed on the assembled equations. The NC, ND, NCX, NCY, CX, and CY data are required. The equations should be put into symmetric form before solving.

SOLUTION. An appropriate equation solver is employed to determine the solution for the nodal u's. See Appendix B for routines BDecomp and BSolve.

COMPUTATION OF DERIVED VARIABLES. The derived variables are the stresses, and they need to be computed for each element. Information stored during the elemental formulation should be reused at this stage. Additional routines for smoothing can be included if desired.

Specifically, the following codes are easily generated based on the material presented in the chapter:

1. Linearly interpolated triangular elements.
2. Bilinearly interpolated quadrilateral (Q4) elements.
3. A combination of 1 and 2.
4. Quadratically interpolated (T6) triangular elements.
5. Quadratically interpolated (Q8) quadrilateral elements.
6. A combination of 4 and 5.
7. Axisymmetric problems with T3, Q4, T6, or Q8 elements.

EXERCISES

Section 5.1

5.1. Draw a free-body diagram of an infinitesimal block ($d\,Vol = t\,dx\,dy$) showing the stresses and their changes from face to face together with the body forces, and hence develop the equations of motion (5.1). Also verify ($\sum M = 0$) that $\tau_{xy} = \tau_{yx}$.

5.2. For small deformations, review the definitions of extensional and shear strains and hence verify Eqs. (5.2).

5.3. Draw a free-body diagram of a portion of the region D on the boundary as shown and hence verify that $t_x = l\sigma_x + m\tau_{xy}$ and $t_y = l\tau_{yx} + m\tau_y$.

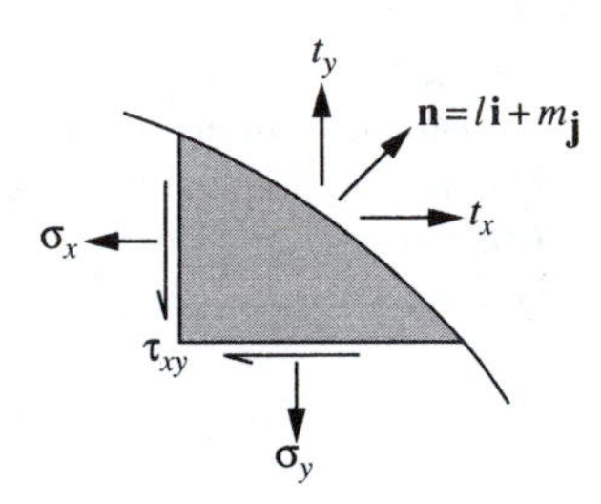

5.4. Specialize the general three-dimensional isotropic stress-strain-temperature relations

$$\epsilon_x = \frac{\sigma_x - \nu\sigma_y - \nu\sigma_z}{E} + \alpha T \qquad \gamma_{xy} = \frac{\tau_{xy}}{G}$$

$$\epsilon_y = \frac{\sigma_y - \nu\sigma_z - \nu\sigma_x}{E} + \alpha T \qquad \gamma_{yz} = \frac{\tau_{yz}}{G}$$

$$\epsilon_z = \frac{\sigma_z - \nu\sigma_x - \nu\sigma_y}{E} + \alpha T \qquad \gamma_{zx} = \frac{\tau_{zx}}{G}$$

for the plane stress assumptions, $\sigma_z = \tau_{xz} = \tau_{yz} = 0$, and hence show that the results of Section 5.1.4.1 are obtained.

5.5. Specialize the general three-dimensional isotropic stress-strain-temperature relations for the plane strain assumptions, $\epsilon_z = \gamma_{xz} = \gamma_{yz} = 0$, and hence show that the results of Section 5.1.4.2 are obtained. Show also that $\sigma_z = \nu(\sigma_x + \sigma_y)$.

Section 5.2

5.6. Verify all the details leading up to the weak form given by Eq. (5.8).

5.7. Given the potential energy functional $I(u, v)$ represented by Eq. (5.12), carry through the details of showing that requiring the potential energy functional to be stationary $\delta I(u, v) = 0$, subject to the essential boundary conditions $u = u_0(s)$, $v = v_0(s)$ on Γ_u, leads to the equilibrium equations (5.4) in terms of the displacements and the natural boundary conditions (5.5*a*) and (5.5*b*).

5.8. The strain energy density function $U_0(x, y, \epsilon_x, \epsilon_y, \gamma_{xy}, T)$ for a linear orthotropic solid is related to the stresses according to

$$\frac{\partial U_0}{\partial \epsilon_x} = \sigma_x = C_{11}\epsilon_x + C_{12}\epsilon_y - \beta T$$

$$\frac{\partial U_0}{\partial \epsilon_x} = \sigma_x = C_{21}\epsilon_x + C_{22}\epsilon_y - \beta T$$

$$\frac{\partial U_0}{\partial \gamma_{xy}} = \tau_{xy} = C_{33}\gamma_{xy}$$

Integrate these equations and hence show that U_o can be written as

$$U_0 = \frac{\boldsymbol{\epsilon}^{\mathrm{T}}\mathbf{C}\boldsymbol{\epsilon}}{2} - \boldsymbol{\epsilon}^{\mathrm{T}}\mathbf{f}_{\mathbf{T}}$$

where $\boldsymbol{\epsilon}^{\mathrm{T}} = [\epsilon_x \ \epsilon_y \ \gamma_{xy}]$ and $\mathbf{f}_{\mathbf{T}} = [\beta T \ \beta T \ 0]$. For additional reading see Boresi and Chong [1].

Section 5.3

5.9. Verify Eqs. (5.15) and (5.16) for the representation of the strains $\boldsymbol{\epsilon}$ in terms of the strain matrix $\mathbf{B}$ and the elemental displacement vector $\mathbf{U_e}$.

5.10. Verify that the elemental body force contribution is correctly represented by $\mathbf{f_e}$ of Eq. (5.20) and the elemental displacement vector $\mathbf{U_e}$.

5.11. Verify the steps leading to the elemental surface stress vector $\mathbf{t_e}$ of Eq. (5.21).

Section 5.4

5.12. Verify the expression for $\mathbf{k_e}$ given by Eq. (5.24).

5.13. Verify Eq. (5.26) for $\boldsymbol{\theta}_\mathbf{e}$ when βT is represented by $\beta T = (\beta T)_i N_i + (\beta T)_j N_j + (\beta T)_k N_k$ over the element in question.

5.14. Specialize the results of Exercise 5.13 when $\beta T = (\beta T)_0$ to verify Eq. (5.25).

5.15. Verify Eq. (5.29) for $\mathbf{f_e}$ when X and Y are represented respectively by

$$X = (X)_i N_i + (X)_j N_j + (X)_k N_k$$

$$Y = (Y)_i N_i + (X)_j N_j + (X)_k N_k$$

5.16. Specialize the results of Exercise 5.15 when $X = X_0$ and $Y = Y_0$ to verify Eq. (5.28).

5.17. When $t_n(s)$ and $t_s(s)$ are represented by linear functions of s over the surface of an element on the boundary according to

$$t_n = t_{n1}\left(1 - \frac{s}{l_e}\right) + t_{nj}\left(\frac{s}{l_e}\right) \qquad t_s = t_{s1}\left(1 - \frac{s}{l_e}\right) + t_{sj}\left(\frac{s}{l_e}\right)$$

verify the expressions given by Eq. (5.30).

5.18. Consider the special case where a portion of the boundary Γ_t on which stresses are prescribed is as indicated.

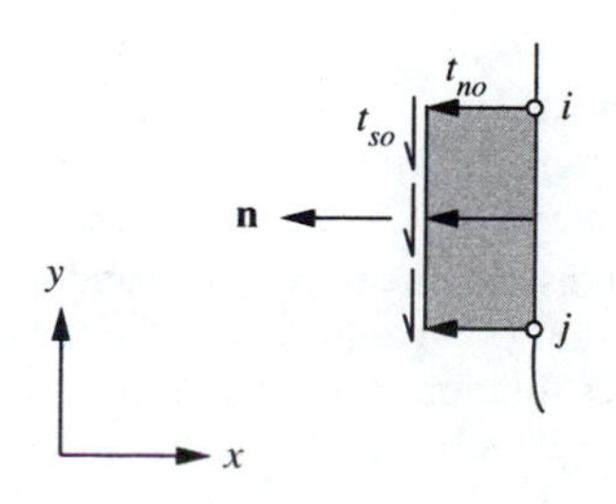

Take the normal and shear stresses to be the constants t_{no} and t_{so} positive as shown. Specialize Eqs. (5.30) to show that the magnitudes and directions of the nodal forces are given by

$$F_{xi} = \frac{-t_{no}tl_e}{2} \qquad F_{xj} = \frac{-t_{no}tl_e}{2}$$

$$F_{yi} = \frac{-t_{so}tl_e}{2} \qquad F_{yj} = \frac{-t_{so}tl_e}{2}$$

as should be the case.

5.19. Repeat Exercise 5.18 for the case shown below.

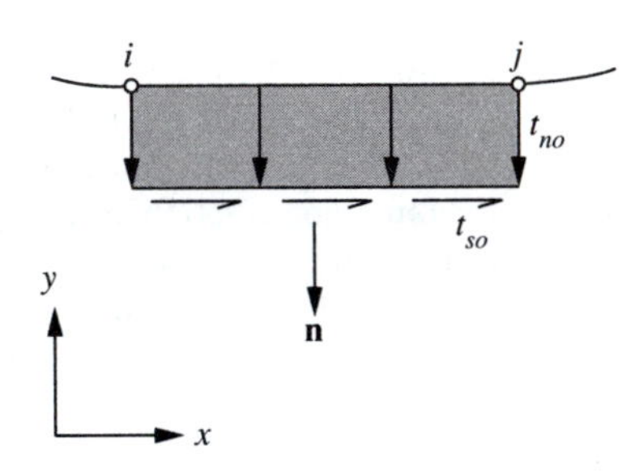

Section 5.5

5.20. By summing the forces in the n and t directions on the free-body diagram shown, verify the transformation formulas

$$\sigma_n = l^2\sigma_x + m^2\sigma_y + 2lm\tau_{xy}$$
$$\tau_n = (l^2 - m^2)\tau_{xy} + lm(\sigma_y - \sigma_x)$$

where $l = \cos\theta$ and $m = \sin\theta$.

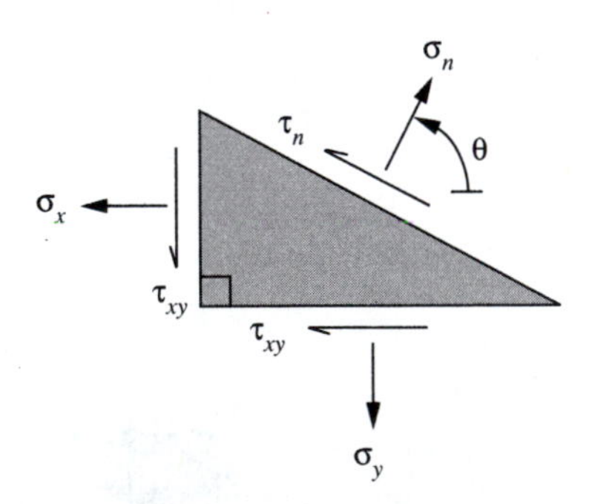

5.21. Construct the elemental displacements $u_e(x, y)$ and $v_e(x, y)$ for each of the elements of the example of Section 5.5.1 and hence show that the finite element solution coincides with the exact solution, also stated in Section 5.5.1.

Section 5.6

5.22. Perform the integrations necessary to verify at least one of the submatrices $\mathbf{A_{ij}}$ given in Eq. (5.35).

5.23. Perform the integrations necessary to verify $\boldsymbol{\theta}_e$ of Eq. (5.38).

5.24. Using the approximations $X = \sum X_i N_i$ and $Y = \sum Y_i N_i$, verify the expression given in Eq. (5.39) for $\mathbf{f}_e$.

Section 5.7

5.25. Show that by requiring the functional of Eq. (5.46) to be stationary subject to the essential boundary conditions on Γ_u, the equations of equilibrium and natural boundary conditions on Γ_t result.

5.26. Show that the nodal loads computed for mesh 1 of the example of Section 5.7.2.2 are exact.

5.27. Axisymmetric z-independent problems in linear isotropic elasticity are governed by the equations

Kinetics:
$$\frac{d\sigma_r}{dr} + \frac{(\sigma_r - \sigma_\theta)}{r} + R(r) = 0$$

Kinematics:
$$\epsilon_r = \frac{du}{dr} \qquad \epsilon_\theta = \frac{u}{r}$$

Constitution:
$$\sigma_r = \frac{E(\epsilon_r + \nu\epsilon_\theta)}{(1 - \nu^2)} - \frac{E\alpha T}{(1 - \nu)}$$

$$\sigma_0 = \frac{E(\epsilon_\theta + \nu\epsilon_r)}{(1 - \nu^2)} - \frac{E\alpha T}{(1 - \nu)}$$

for plane stress. [Note that the appropriate equation for plane strain can readily be obtained by simply replacing ν by $\nu/(1 - \nu)$, E by $E/(1 - \nu^2)$ and α by $\alpha(1 + \nu)$.] The strains and stresses can be eliminated to yield

$$u'' + \frac{u'}{r} - \frac{u}{r^2} - (1 + \nu)\alpha T' + \frac{(1 - \nu^2)R}{E} = 0$$

or

$$(ru')' - \frac{u}{r} + f = 0$$

where

$$f = -(1 + \nu)r\alpha T' + \frac{(1 - \nu^2)rR}{E}$$

and T is the temperature change and R is the body force. Boundary conditions are either of the form

u prescribed – essential boundary condition

or

$$\sigma_r = \frac{E(u' + \nu u/r)}{(1 - \nu^2)} - \frac{E\alpha T}{(1 - \nu)} \text{ prescribed}$$

which is a natural boundary condition. Alternately, the total potential energy functional can be written as

$$\frac{V}{2\pi t} = \int \left(\left(\frac{E}{2(1-\nu^2)} \right) \left(u'^2 + \frac{u^2}{r^2} + \frac{2\nu u u'}{r} \right) - \frac{E\alpha T(u' + \nu u/r)}{(1-\nu)} - rR \right) r\, dr$$

$$+ \sum_a au(a) - \sum_b bu(b)$$

where $\sum_a$ and $\sum_b$ are the prescribed tensile stresses at the inner and outer radii a and b respectively. This is clearly a Sturm-Liouville problem as studied in Chapter 2. Using four linearly interpolated elements, set up and solve the pressure vessel problem of Section 5.7.2.1. Compare your results (for the plane strain case) with those from the pressure vessel example. For additional reading see Timoshenko and Goodier [2].

5.28. Repeat Exercise 5.27 using two quadratically interpolated elements.

5.29–5.30. Repeat Exercises 5.27 and 5.28 for the problem of a rotating disk [2] with a hole. $R = \rho r \omega^2$. Assume plane stress conditions and take the outer radius to be twice the inner radius.

Section 5.8

5.31. Consider the stresses prescribed on a portion of the boundary as shown, that is, a constant normal pressure. Evaluate $\mathbf{t_e}$ assuming the direction cosines l and m to be constants, that is, using Eq. (5.54) to compute the components. Use the correct length l_e computed on the basis of the development in Section 3.8.2.2. Then evaluate the same integrals using the expressions given in Section 5.8 for $l(s)$ and $m(s)$.

5.32. Repeat Exercise 5.31 for a constant tangential distribution t_0.

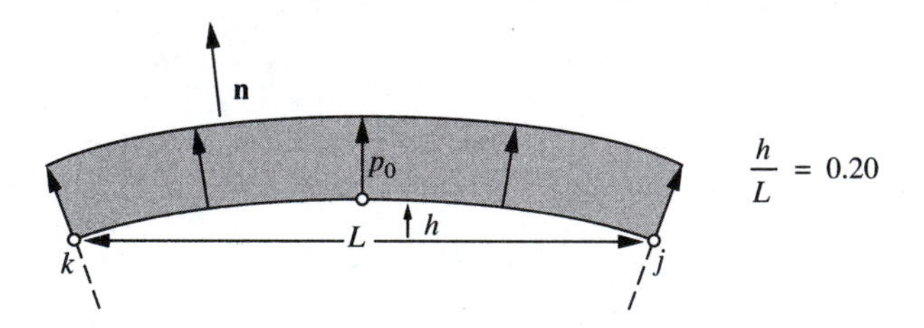

5.33. Repeat Exercise 5.31 for the linearly varying normal pressure indicated.

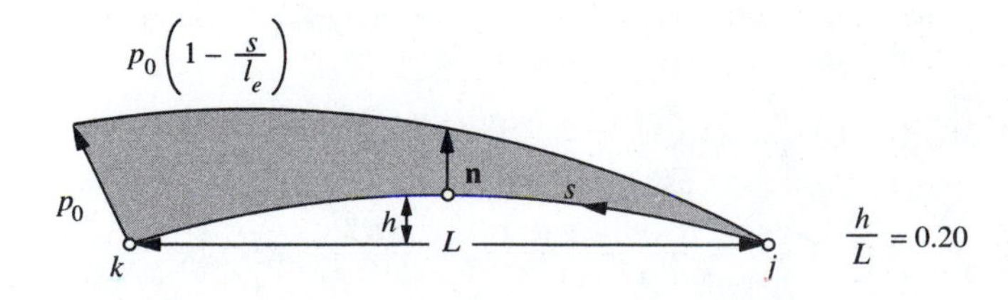

5.34. Repeat Exercise 5.31 for the linearly varying tangential pressure indicated.

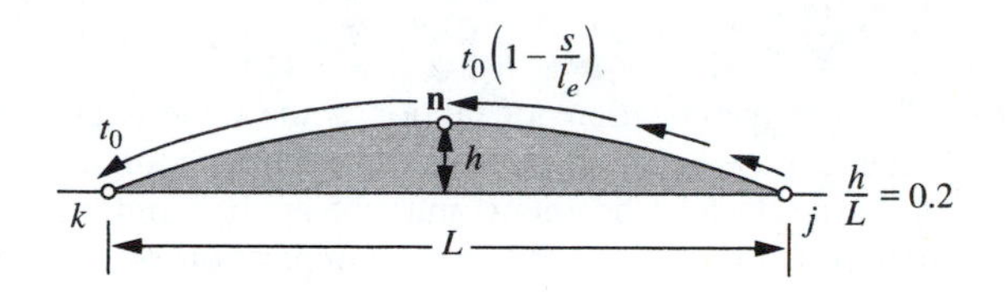

5.35. Specialize Eq. (5.57) for $\mathbf{f_e}$ to a rectangular element and hence determine the components.

5.36. Specialize the results of Exercise 5.35 to the case where $X(x, y) = X_0$ and $Y(x, y) = Y_0$.

5.37. By integrating the equations

$$\epsilon_x = \frac{\partial u}{\partial x} = \frac{\sigma_x - \nu\sigma_y}{E} = \frac{My}{EI}$$

$$\epsilon_y = \frac{\partial v}{\partial y} = \frac{\sigma_y - \nu\sigma_x}{E} = -\frac{\nu My}{EI}$$

$$\gamma_{xy} = \frac{\partial u}{\partial y} + \frac{\partial v}{\partial x} = \frac{\tau_{xy}}{G} = 0$$

subject to $u(0,0) = v(0,0) = \partial u(0,0)/\partial y - \partial v(0,0)/\partial x = 0$, determine the displacements corresponding to the pure bending problem of Section 5.8.1.1. Hence verify that the displacements predicted by the finite element model are exact throughout the region.

5.38. With $\beta T = \sum(\beta T)_i N_i$, specialize Eq. (5.59) to a rectangular element and hence evaluate the integrals to determine $\boldsymbol{\theta}_\mathbf{e}$.

5.39. Specialize the results of Exercise 5.38 when $\beta T = (\beta T)_0$.

Dynamic problems.

5.40. For the plane elasticity problem using linearly interpolated elements, carry through the details of showing that applying the semidiscrete Galerkin method to the weak form of Eq. (5.8) (minus the temperature terms) leads to the equations of motion given by

$$\mathbf{M_G\ddot{u}_G} + \mathbf{K_G u_G} = \mathbf{F_G}(t)$$

where

$$\mathbf{K_G} = \sum \mathbf{k_G} \qquad \mathbf{M_G} = \sum \mathbf{m_G}$$

$$\mathbf{F_G} = \sum \mathbf{f_G} + \sum{}' \mathbf{t_G}$$

and $\mathbf{f_e}$, $\mathbf{t_e}$, and $\mathbf{k_e}$ are as given in Section 5.3, and

$$\mathbf{m_e} = \int\int_{D_e} \cap\, \rho \cap d\,Vol$$

where

$$\cap = \begin{bmatrix} N_1 & 0 & N_2 & 0 & N_3 & 0 \\ 0 & N_1 & 0 & N_2 & 0 & N_3 \end{bmatrix}$$

[Note that the form of the equations of motion is the same regardless of the region or the type of element used. One only has to use the appropriate $\mathbf{k_e}$, $\mathbf{m_e}$, $\mathbf{t_e}$, and $\mathbf{f_e}$ for the problem in question.] Constraints must be applied and the system of second order differential equations solved. Mode decomposition or numerical integration schemes such as Newmark's method are appropriate.

5.41. For the case where there are no surface tractions, body forces, or time-dependent prescribed displacements, the constrained equations of motion reduce to

$$\mathbf{M_G \ddot{u}_G} + \mathbf{K_G u_G} = \mathbf{0}$$

Show that by taking $\mathbf{u_G} = \mathbf{v_G} \exp(i\omega t)$, there results

$$(\mathbf{K_G} - \omega^2 \mathbf{M_G})\mathbf{v_G} = 0$$

for determining the natural frequencies and mode shapes.

APPENDIX A *APPROXIMATE METHODS OF ANALYSIS*

Historically, the methods of Ritz [3] and Galerkin [4] were applied to boundary and initial-boundary value problems in a somewhat different manner than was presented in Chapters 2 through 5, where we generated finite element models. In this appendix we present and illustrate several of the more common of the **Methods of Weighted Residuals**, often abbreviated MWR's [Finlayson, 1]. The MWR's are a collection of techniques for obtaining approximate solutions to several classes of differential equations. Each of the methods uses a different approach that transforms the differential equation(s) into a set of algebraic equations; the solution of these gives information about the solution of the original differential equation. We shall be especially interested in obtaining approximate solutions to the boundary value problem studied in Chapter 2, namely,

$$L(u) = (pu')' + (\lambda\rho - q)u + f = 0 \tag{A.1}$$

$$\begin{aligned} -p(a)u'(a) + \alpha u(a) &= A \\ p(b)u'(b) + \beta u(b) &= B \end{aligned} \tag{A.2}$$

Generally, the classical MWR's use series of continuous functions satisfying the boundary conditions of the problem to represent the solution. Although the classical MWR's do not bear directly upon the finite element method, it is instructive to see how these other approximate methods are applied to the solution of differential equations.

The first step in each of the classical MWR's consists of assuming an approximate solution of the form

$$u(x) = \phi_0(x) + \sum_{1}^{N} a_n\phi_n(x) \tag{A.3}$$

where $\phi_0(x)$ is chosen so as to satisfy the nonhomogeneous boundary conditions of the problem, and each of the $\phi_n(x)$ is chosen so as to satisfy *all* the corresponding homogeneous boundary conditions. The approximate solution $u(x)$ is referred to as an N-term approximate solution. The $\phi_n(x)$ are called *admissible* functions. Substituting the approximate solution (A.3) into the differential equation (A.1) yields

$$\begin{aligned} L(u) &= (pu')' + (\lambda\rho - q)u + f \\ &= (p\phi_0')' + (\lambda\rho - q)\phi_0 + \sum \big((p\phi_n')' + (\lambda\rho - q)\phi_n\big)a_n + f \\ &= E_N(x, \mathbf{a}) \end{aligned}$$

where $E_N(x, \mathbf{a})$ is termed the **residual** or **error** of the solution. Generally each of the Methods of Weighted Residuals involves requiring that the error be orthogonal to N linearly independent weight functions $w_j(x)$ according to

$$\int_a^b w_j(x)E_N\,dx = 0 \qquad j = 1, 2, \ldots, N \tag{A.4}$$

that is, one equation is generated for each of the N linearly independent weight functions $w_j(x)$. The collection of equations (A.4) represents a system of N linear equations for the N unknowns a_n. The different MWR's use different sets of weight functions $w_j(x)$.

When the $\lambda \rho u$ term is absent, the algebraic equations are of the form

$$\mathbf{Aa} = \mathbf{b}$$

to be solved for the unknown $\mathbf{a} = [a_1 \quad a_2 \quad a_3 \quad \ldots \quad a_N]^{\mathrm{T}}$. When the $\lambda \rho u$ term is present, the algebraic equations which result from the application of one of the MWR's are of the form

$$\mathbf{Aa} - \lambda \mathbf{Ba} = \mathbf{0}$$

to be solved for the eigenvectors $\mathbf{a}$ and eigenvalues λ. Several of the different methods are described below.

Collocation. The $w_j(x)$ are chosen as

$$w_j(x) = \delta(x, x_j) \qquad j = 1, 2, \ldots, N$$

where $\delta(x, x_j)$ is the so-called **Dirac delta function** [Stakgold, 2]. The x_j are suitably chosen distinct points in the interval $a < x < b$. It is common to take the x_j so as to create N equally spaced subintervals of (a, b). One of the properties of the delta function is that for a continuous function $f(x)$,

$$\int_a^b f(x)\delta(x, x_j)\, dx = f(x_j) \qquad a < x_j < b$$

which is referred to as the **sifting** property. With the $w_j(x) = \delta(x, x_j)$ the system of equations (A.4) becomes

$$\int_a^b \delta(x, x_j) E_N(x, \mathbf{a})\, dx = E_N(x_j, \mathbf{a}) = 0 \qquad j = 1, 2, \ldots, N$$

so that the error function is simply evaluated at the N interior points x_j and then equated to zero. This results in the required N equations for the a_n.

Subdomain method. In the subdomain method, N subintervals I_j of the basic interval $a \leq x \leq b$ are chosen. The integral of the error function E_N over each of the subintervals is then required to vanish, resulting in

$$\int_{I_j} E_N(x, \mathbf{a})\, dx = 0 \qquad j = 1, 2, \ldots, N$$

again yielding N equations for the N unknowns a_n. We easily see that within the framework of the MWR's this is equivalent to choosing the $w_j(x)$ as

$$w_j(x) = 1 \qquad \text{if } x \text{ is contained in } I_j$$

$$w_j(x) = 0 \qquad \text{if } x \text{ is not contained in } I_j$$

where I_j is a subdomain or subinterval of the basic interval $[a, b]$. The intervals must be such that the N equations are independent. As an example for $N = 2$, the two intervals might be chosen as $I_1 = [a, (a + b)/2]$ and $I_2 = [(a + b)/2, b]$. The intervals do not have to be disjoint, that is, there may be overlap.

Least squares. The traditional method of least squares is based on seeking a stationary value of the quantity

$$\int_a^b E_N(x, \mathbf{a})^2 \, dx$$

with respect to the a_n. This requirement results in

$$\int_a^b E_N(x, \mathbf{a}) \frac{\partial E_N(x, \mathbf{a})}{\partial a_j} \, dx = 0 \qquad j = 1, 2, \ldots, N$$

so that $w_j(x) = \partial E_N(x, \mathbf{a})/\partial a_j$. Again an $N \times N$ system of equations results.

Galerkin method. In the Galerkin method [4] the $w_j(x)$ are chosen to be the members of the set of admissible functions $\phi_j(x)$, resulting in

$$\int_a^b E_N(x, \mathbf{a}) \phi_j(x) \, dx = 0 \qquad j = 1, 2, \ldots, N$$

These equations can be interpreted as stating that the error $E(x, \mathbf{a})$ is orthogonal to each of the admissible functions $\phi_j(x)$.

The basic underlying idea of the MWR's is that if, in the limit as N becomes large, the error function $E_N(x, \mathbf{a})$ is required to be orthogonal to each member of a set of sufficiently different (linearly independent) functions $w_j(x)$, the error itself must be small. It is then further assumed that good approximate solutions can be obtained when N is a relatively small number. That these assertions are in fact reasonable is demonstrated in the following examples.

Example A.1.

As a typical boundary value problem we consider

$$u'' + xu = 1$$
$$u(0) = 0$$
$$u(1) = 1$$

Following the procedure just presented, the approximate solution is taken to be

$$u = x + \sum_1^N a_n \sin(n\pi x)$$

where $\phi_0(x) = x$ satisfies the nonhomogeneous boundary conditions and *each* of the admissible functions $\phi_n(x) = \sin(n\pi x)$ satisfies *all* the corresponding homogeneous boundary conditions, $u(0) = u(1) = 0$. The error is easily calculated to be

$$E_N(x, \mathbf{a}) = u'' + xu - 1 = x^2 - 1 + \sum_{1}^{N} a_n \left(x - (n\pi)^2\right)\sin(n\pi x)$$

We will take $N = 3$ and generate approximate solutions using each of the above described methods. These will then be compared with the exact solution.

Collocation. Choose the three x_j values as 1/4, 1/2, and 3/4. The three equations are generated by setting

$$E_3\left(\frac{1}{4}, \mathbf{a}\right) = 0 \qquad E_3\left(\frac{2}{4}, \mathbf{a}\right) = 0 \qquad E_3\left(\frac{3}{4}, \mathbf{a}\right) = 0$$

which when expressed in augmented form are

$$\left[\begin{array}{ccc|c} 6.8021 & 39.2284 & -62.6330 & -0.9375 \\ 9.3696 & 0.0000 & -88.3264 & -0.7500 \\ 6.4485 & -38.7284 & 62.2794 & -0.4375 \end{array}\right]$$

from which $a_1 = -0.0918$, $a_2 = -0.0060$, and $a_3 = -0.0012$. The three-term approximate solution is then

$$u_C = x - 0.0918 \sin \pi x - 0.0060 \sin 2\pi x - 0.0012 \sin 3\pi x$$

Subdomain method. For this three-term solution we choose the three subdomains as the subintervals [0,1/3], [1/3,2/3], and [2/3,1] such that the three equations are

$$\int_0^{1/3} E_3\, dx = 0 \qquad \int_{1/3}^{2/3} E_3\, dx = 0 \qquad \int_{2/3}^{1} E_3\, dx = 0$$

resulting in

$$\left[\begin{array}{ccc|c} 1.5361 & 9.3763 & 18.8142 & -0.3210 \\ 2.9824 & 0.0173 & -18.7435 & -0.2469 \\ 1.4463 & -9.2345 & 18.6727 & -0.0988 \end{array}\right]$$

from which $a_1 = -0.1019$, $a_2 = -0.0114$, $a_3 = -0.0030$. The approximate solution is then

$$u_S = x - 0.1019 \sin \pi x - 0.0114 \sin 2\pi x - 0.0030 \sin 3\pi x$$

Least squares. For the least squares approach the weights are

$$w_j(x) = \frac{\partial E_3}{\partial a_j} = \left(x - (j\pi)^2\right)\sin j\pi x$$

resulting in

$$\left[\begin{array}{ccc|c} 43.911 & 4.3544 & 0.0190 & -4.2213 \\ 4.3544 & 759.69 & 12.382 & -6.3074 \\ 0.0190 & 12.382 & 3900.8 & -9.8420 \end{array}\right]$$

from which $a_1 = -0.0954$, $a_2 = -0.0077$, and $a_3 = -0.0025$. The approximate solution is then

$$u_L = x - 0.0954 \sin \pi x - 0.0077 \sin 2\pi x - 0.0025 \sin 3\pi x$$

Galerkin method. For the Galerkin method, with $w_j(x) = \phi_j(x) = \sin j\pi x$, the three equations are

$$\int_0^1 E_3 \sin \pi x \, dx = 0 \qquad \int_0^1 E_3 \sin 2\pi x \, dx = 0$$

$$\int_0^1 E_3 \sin 3\pi x \, dx = 0$$

resulting in

$$\left[\begin{array}{ccc|c} 4.6848 & 0.0901 & 0.0000 & -0.4473 \\ 0.0901 & 19.489 & 0.0973 & -0.1592 \\ 0.0000 & 0.0973 & 44.163 & -0.1109 \end{array}\right]$$

from which $a_1 = -0.0953$, $a_2 = -0.0077$, and $a_3 = -0.0025$. The approximate solution is

$$u_G = x - 0.0953 \sin \pi x - 0.0077 \sin 2\pi x - 0.0025 \sin 3\pi x$$

These different approximate solutions are compared with the exact solution u_E in Table A.1 and in Fig. A–1 that follow.

TABLE A.1 Comparison of MWR solutions

x	u_C	u_S	u_L	u_G	u_E
0.0	0.0000	0.0000	0.0000	0.0000	0.0000
0.1	0.0671	0.0594	0.0640	0.0640	0.0621
0.2	0.1392	0.1264	0.1342	0.1342	0.1341
0.3	0.2197	0.2058	0.2147	0.2148	0.2159
0.4	0.3099	0.2981	0.3062	0.3063	0.3069
0.5	0.4094	0.4011	0.4071	0.4072	0.4068
0.6	0.5169	0.5116	0.5153	0.5154	0.5145
0.7	0.6311	0.6275	0.6294	0.6295	0.6292
0.8	0.7506	0.7481	0.7489	0.7489	0.7494
0.9	0.8742	0.8728	0.8730	0.8731	0.8737
1.0	1.0000	1.0000	1.0000	1.0000	1.0000

From Table A.1, we see for this example that the numerical results generated by the least squares and Galerkin solutions are somewhat more accurate as compared to those from the collocation and subdomain methods, but that all give reasonable results considering the small number of terms taken in the series representation of the approximate solution.

Ritz method. Each of the methods of weighted residuals discussed above deals directly with the differential equation. The method of Ritz [3] on the other hand is a direct method of the calculus of variations and is based on the existence of the variational principle, when it exists, corresponding to the differential equation.

Given a variational principle of the form

$$I(u) = \int_a^b F(x, u, u') \, dx, \qquad \delta I = 0$$

$$u(a) = u_a$$

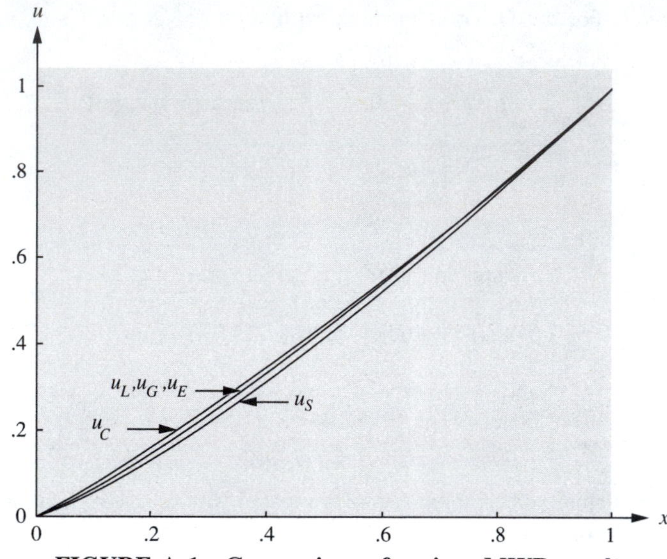

FIGURE A-1 Comparison of various MWR results

an approximate solution of the form

$$U = \phi_0(x) + \sum_{1}^{N} c_n \phi_n(x) \tag{A.5}$$

is assumed. The nonhomogeneous essential boundary conditions are satisified by $\phi_0(x)$ and each of the $\phi_n(x)$ satisfies the homogeneous essential boundary condition. The $\phi_n(x)$ are termed admissible functions. Nonessential or natural boundary conditions are not required to be satisfied by the approximate solution although accuracy and convergence are generally improved when U satisfies all boundary conditions. An approximate solution of the form (A.5) is termed an N-term Ritz solution. Note that the form of this approximate solution is similar to that used for the Galerkin method of the Methods of Weighted Residuals. The very important difference is that the admissible functions for the Ritz method are required to satisfy only the essential boundary conditions, whereas the admissible functions for the Galerkin method were required to satisfy all the boundary conditions.

Upon substitution of the approximate solution U into the functional I, the I becomes a function of the parameters c_i, $i = 1, 2, \ldots, N$. The stationary value of the function is then obtained by requiring

$$\frac{\partial I}{\partial c_i} = 0 \qquad i = 1, 2, \ldots, N$$

which constitute an algebraic set of equations. When the integrand of the functional is a quadratic function of u and u', the function I is a quadratic function of the c_i and the resulting algebraic equations are linear. This is the case for our standard boundary value problem

$$(pu')' - qu + f = 0$$
$$-p(a)u'(a) + \alpha u(a) = A$$
$$p(b)u'(b) + \beta u(b) = B$$

where the corresponding functional is

$$I(u) = \int_a^b \left(\frac{(pu'^2 + qu^2)}{2} - fu \right) dx + \frac{\alpha u(a)^2}{2} + \frac{\beta u(b)^2}{2} - Au(a) - Bu(b)$$

Example A.2.

Consider the example given above, where

$$u'' + xu = 1$$
$$u(0) = 0$$
$$u(1) = 1$$

With $p = 1$, $q = -x$, $f = -1$, $\alpha = \beta = A = B = 0$, the corresponding functional is (see Section 2.7.1)

$$I(u) = \int_0^1 \left(\frac{u'^2}{2} - \frac{xu^2}{2} + u \right) dx \tag{a}$$

For the approximate solution U we take

$$U = u_0(x) + \sum_1^N a_n \phi_n(x) = x + \sum_1^N a_n \sin(n\pi x) \tag{b}$$

where x satisfies the nonhomogeneous boundary conditions and each of the $\sin(n\pi x)$ satisfies the homogeneous boundary conditions. Substituting (b) into the functional (a) yields

$$I = \int_0^1 \left(\frac{\left(1 + \sum n\pi a_n \cos(n\pi x)\right)^2}{2} - x\frac{\left(x + \sum a_n \sin(n\pi x)\right)^2}{2} + x + \sum a_n \sin(n\pi x) \right) dx$$
$$= I(a_1, a_2, \ldots, a_N)$$

Requiring the function I to be stationary with respect to each of the a_j leads to

$$\frac{\partial I}{\partial a_j} = \int_0^1 \Bigg(j\pi \cos(j\pi x)\left(1 + \sum n\pi a_n \cos(n\pi x)\right)$$
$$- x \sin(j\pi x)\left(x + \sum a_n \sin(n\pi x)\right) + \sin(j\pi x)\ dx \Bigg) = 0$$
$$j = 1, 2, \ldots, N$$

or

$$\sum a_n \int_0^1 \Big(j\pi n\pi \cos(j\pi x)\cos(n\pi x) - x\sin(j\pi x)\sin(n\pi x)\Big)\,dx =$$

$$\int_0^1 \Big(x^2 \sin(j\pi x) - \sin(j\pi x) - j\pi\cos(j\pi x)\Big)\,dx \qquad (c)$$

$$j = 1, 2, \ldots, N$$

Taking $N = 3$, a three-term Ritz solution, the above set of linear algebraic equations becomes

$$\left[\begin{array}{ccc|c} 4.6848 & 0.0901 & 0.0000 & -0.4473 \\ 0.0901 & 19.4892 & 0.0973 & -0.1592 \\ 0.0000 & 0.0973 & 44.1632 & -0.1109 \end{array}\right]$$

which are precisely the same as those obtained using the Galerkin method. This coincidence is generally true for the class of linear boundary value problems studied in Chapter 2.

It is instructive to consider successively the solutions of (c) for $N = 1, 2$, and 3, each of which is easily obtained by retaining the upper left-hand N-square portion of the coefficient matrix and corresponding portion of the right-hand side. This yields successively the solutions

$$U_1 = x - 0.0955\sin(\pi x)$$

$$U_2 = x - 0.0953\sin(\pi x) - 0.0077\sin(2\pi x)$$

$$U_3 = x - 0.0953\sin(\pi x) - 0.0077\sin(2\pi x) - 0.0025\sin(3\pi x)$$

which are displayed in Table A.2 along with the values given by the exact solution. Figure A–2 displays the same data.

TABLE A.2 Comparison of successive Ritz solutions

x	U_1	U_2	U_3	u_E
0.0000	0.0000	0.0000	0.0000	0.0000
0.1000	0.0705	0.0660	0.0640	0.0621
0.2000	0.1439	0.1367	0.1343	0.1341
0.3000	0.2227	0.2156	0.2148	0.2159
0.4000	0.3092	0.3048	0.3063	0.3069
0.5000	0.4045	0.4047	0.4072	0.4068
0.6000	0.5092	0.5139	0.5154	0.5145
0.7000	0.6227	0.6302	0.6295	0.6292
0.8000	0.7439	0.7513	0.7489	0.7494
0.9000	0.8705	0.8751	0.8731	0.8737
1.0000	1.0000	1.0000	1.0000	1.0000

Comparison of the one-, two-, and three-term solutions provides a practical means for assessing the accuracy and convergence of the solutions. For a discussion of the convergence properties and the sense of the approximation provided by the Ritz method see Mikhlin [5] and Reddy [6].

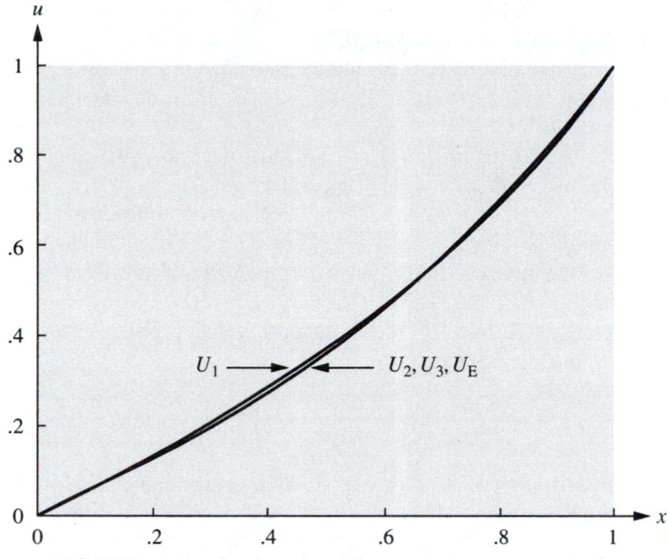

FIGURE A-2 Successive Ritz and the exact solution

REFERENCES

1. Finlayson, B. A.: *The Method of Weighted Residuals and Variational Principles*, Academic Press, New York, 1972.
2. Stakgold, I.: *Boundary Value Problems of Mathematical Physics*, vol I, Macmillan, New York, 1967.
3. Ritz, W.: *"Ueber eine neue Methode zur Losung gewisser Variationsprobleme der mathematischen Physik,"* J. Reine Angew. Math., vol 135, pp. 1–61, 1908.
4. Galerkin, B. G.: "Rods and plates. Series occurring in various questions concerning the elastic equilibrium of rods and plates," *Engineers Bulletin* (*Vestnik inshenerov*), 19, pp. 897–908, 1915.
5. Mikhlin, S. G.: *Variational Methods in Mathematical Physics*, Macmillan, New York, 1964.
6. Reddy, J. N.: *Applied Functional Analysis and Variational Methods in Engineering*, McGraw-Hill, New York, 1986.

EXERCISES

For each of the following use the method and the number of terms indicated to construct an approximate solution to the boundary value problem.

A.1. $u'' + x^2u + x = 0, u(0) = 0, u(1) = 1$. Use the collocation method with

- **a.** One term with $x_1 = 0.5$.
- **b.** Two terms with $x_1 = 1/3$ and $x_2 = 2/3$.
- **c.** Three terms with $x_1 = 1/4$, $x_2 = 2/4$, and $x_3 = 3/4$ and $u = x + \sum a_n \sin(n\pi x)$.

A.2. $u'' + x^2u + x = 0, u(0) = 0, u(1) = 1$. Use the subdomain method with
- **a.** One term with I_1 the interval (0,1).
- **b.** Two terms with $I_1 = (0, 0.5)$ and $I_2 = (0.5, 1)$.
- **c.** Three terms with $I_1 = (0, 1/3)$, $I_2 = (1/3, 2/3)$, and $I_3 = (2, 3, 1)$ and $u = x + \sum a_n \sin(n\pi x)$.

A.3. $u''+x^2u+x = 0, u(0) = 0, u(1) = 1$. Construct one-, two-, and three-term least squares solutions with $u = x + \sum a_n \sin(n\pi x)$.

A.4. $u'' + x^2u + x = 0, u(0) = 0, u(1) = 1$. Construct one-, two-, and three-term Galerkin solutions with $u = x + \sum a_n \sin(n\pi x)$. Compare each of the previous solutions on the interval (0,1). Also compare the derivatives at $x = 0$ and at $x = 1$.

A.5–8. Repeat Exercises A.1–4 for the boundary value problem $xu'' + u' + 4xu = 1, u(1) = 0, u'(2) = 0$, with

$$u = \sum_{\text{odd}} a_n \sin\left(\frac{n\pi(x-1)}{2}\right) = \sum_{\text{odd}} a_n \cos\left(\frac{n\pi x}{2}\right)$$

A.9. $u'' + \lambda xu = 0, u(0) = 0, u'(L) = 0$. Construct one-, two-, and three-term Galerkin solutions with

$$u = \sum_{\text{odd}} a_n \sin\left(\frac{n\pi x}{2L}\right)$$

A.10. $xu''+u'+\lambda xu = 0, u(L) = 0, u'(2L) = 0$. Construct one-, two-, and three-term Galerkin solutions with

$$u = \sum_{\text{odd}} a_n \cos\left(\frac{n\pi x}{2L}\right)$$

A.11. For the boundary value problem $u''+x^2u+x = 0, u(0) = 1, u(1) = 0$, construct one-, two-, and three-term Ritz solutions with

$$u = u_0(x) + \sum a_n u_n(x)$$

for
- **a.** $u_0(x) = 1 - x, u_n(x) = x^n(1 - x)$
- **b.** $u_0(x) = \cos(\pi x/2), u_n(x) = \sin(n\pi x)$

Compare the results for the solution u and the derivative u'.

A.12. For the boundary value problem $x^2u''+2xu'-2u+x = 0, u(1) = 0, u'(2) = 1$, construct one-, two-, and three-term Ritz solutions with

$$u = u_0(x) + \sum a_n u_n(x)$$

for

$$u_0(x) = x - 1 \qquad u_n(x) = \frac{\sin(2n-1)\pi(x-1)}{2}$$

Compute and plot each of the solutions and its corresponding derivative. Hence, check how well the natural boundary condition at $x = 2$ is being satisfied.

APPENDIX B *LINEAR ALGEBRAIC EQUATIONS*

For many of the problems in Chapters 1 through 5, the end result of generating the appropriate finite element model is a set of linear algebraic equations that can be written as $\mathbf{Ax} = \mathbf{b}$, where $\mathbf{A}$ is an $n \times n$ symmetric matrix, with $\mathbf{x}$ and $\mathbf{b}$ $n \times 1$ column vectors. The vector $\mathbf{b}$ represents the input, and $\mathbf{x}$ the output. For a finite element model of a one-dimensional problem with only a few elements, n may be relatively small such that a solution of the equations can be carried out by hand or on a hand calculator. The finite element models of many real-life problems involve several hundred or even several thousand unknowns, in which case a computer with appropriate software is clearly necessary.

For the time-independent problems of Chapters 1, 2, 3, and 5, there is usually a single input $\mathbf{b}$ and one desired output $\mathbf{x}$. For the time-dependent problems of Chapter 4, the time domain integration techniques require the successive solution of the set of equations

$$\mathbf{A}\mathbf{u}_{\mathbf{n+1}} = \mathbf{b}_{\mathbf{n}} \qquad n = 0, 1, 2, \ldots$$

that is, repeated solution with the same coefficient matrix $\mathbf{A}$ and a sequence of right-hand sides $\mathbf{b}$. In this Appendix we discuss techniques for solving sets of linear equations.

B.1 Gaussian Elimination

Most algorithms that are designed to solve sets of linear algebraic equations are based on Gaussian elimination, which is described as follows. Consider in scalar form the set of equations

$$\begin{aligned}
a_{11}x_1 + a_{12}x_2 + a_{13}x_3 + \cdots + a_{1n}x_n &= b_1 \\
a_{21}x_1 + a_{22}x_2 + a_{23}x_3 + \cdots + a_{2n}x_n &= b_2 \\
a_{31}x_1 + a_{32}x_2 + a_{33}x_3 + \cdots + a_{3n}x_n &= b_3 \\
&\vdots \\
a_{n1}x_1 + a_{n2}x_2 + a_{n3}x_3 + \cdots + a_{nn}x_n &= b_n
\end{aligned}$$

Assuming $a_{11} \neq 0$, solve the first equation for x_1 and substitute the result into each of the remaining equations. The result can be written as

$$\begin{aligned}
a_{11}x_1 + a_{12}x_2 + a_{13}x_3 + \cdots + a_{1n}x_n &= b_1 \\
\left(a_{22} - \frac{a_{21}a_{12}}{a_{11}}\right)x_2 + \left(a_{23} - \frac{a_{21}a_{13}}{a_{11}}\right)x_3 + \cdots &= b_2^* \\
\left(a_{32} - \frac{a_{31}a_{12}}{a_{11}}\right)x_2 + \left(a_{33} - \frac{a_{31}a_{13}}{a_{11}}\right)x_3 + \cdots &= b_3^* \\
&\vdots
\end{aligned}$$

$$\left(a_{n2} - \frac{a_{n1}a_{12}}{a_{11}}\right)x_2 + \left(a_{n3} - \frac{a_{n1}a_{13}}{a_{11}}\right)x_3 + \cdots = b_n^*$$

that is, x_1 has been eliminated from the last $n - 1$ equations. For this step a_{11} is referred to as a **pivot**. Then, assuming $a_{22}^* = a_{22} - a_{21}a_{12}/a_{11} \neq 0$, x_2 is eliminated from each of the remaining $n - 2$ equations. a_{22}^* is the pivot for this step. This process is continued until the $(n - 1)$st unknown is eliminated from the nth equation, at which time the equations appear as

$$\begin{aligned} a_{11}x_1 + a_{12}x_2 + a_{13}x_3 + \cdots + a_{1n}x_n &= b_1 \\ a_{22}^*x_2 + a_{23}^*x_3 + \cdots \quad &= b_2^* \\ a_{33}^*x_3 + \cdots \quad &= b_3^* \\ a_{nn}^*x_n &= b_n^* \end{aligned}$$

The set of operations that reduces the original square coefficient matrix **A** to the above triangular matrix **A**$*$ is referred to as the process of **forward reduction**. At the end of this process, x_n can be determined from the nth equation. The $(n - 1)$st equation then only involves the unknown x_{n-1}, which can be determined. The $(n - 2)$nd equation only involves the unknown x_{n-2}, and so forth. This set of operations is known as **back substitution**.

Example B.1.

Consider the symmetric set of equations

$$\begin{aligned} 3x_1 - x_2 + x_3 &= 3 \\ -x_1 + 3x_2 - x_3 &= 1 \\ x_1 - x_2 + 2x_3 &= 2 \end{aligned}$$

Step 1. Using the first equation, eliminate x_1. Solving the first equation for $x_1 = 1 + x_2/3 - x_3/3$ and substituting into the second and third equations yields

$$\begin{aligned} 3x_1 - x_2 + x_3 &= 3 \\ \frac{8}{3}x_2 - \frac{2}{3}x_3 &= 2 \\ -\frac{2}{3}x_2 + \frac{5}{3}x_3 &= 1 \end{aligned}$$

The student should verify that the elimination of x_1 using the first equation is equivalent to multiplying the first equation by 1/3 and then adding the first equation to the second equation, followed by the step of multiplying the first equation by $-1/3$ and then adding the first equation to the third equation. This operation of "multiplication of a row by a scalar followed by the addition of that row to another row" is called an elementary row operation (ERO) and is known to leave the solution of the set of equations unchanged. *It is exactly this operation that is implemented in an algorithm for solving* $\mathbf{Ax} = \mathbf{b}$.

Step 2. Using the second equation, eliminate x_2. Solving the second equation for $x_2 = 3/4 + x_3/4$ and substituting into the third equation yields

$$\begin{aligned} 3x_1 - x_2 + x_3 &= 3 \\ \frac{8}{3}x_2 - \frac{2}{3}x_3 &= 2 \\ \frac{9}{6}x_3 &= \frac{3}{2} \end{aligned}$$

completing the forward reduction. The corresponding elementary row operation consists of multiplying the second equation by 1/4 and adding the second equation to the third equation. The upper triangular character of the reduced equations is readily apparent.

Step 3. From the third equation $x_3 = 1$. Then from the second equation

$$x_2 = \left(\frac{3}{8}\right)\left(2 + \left(\frac{2}{3}\right)x_3\right) = \left(\frac{3}{8}\right)\left(2 + \left(\frac{2}{3}\right)1\right) = 1$$

and from the first equation

$$x_1 = \frac{(3 + x_2 - x_3)}{3} = \frac{(3 + 1 - 1)}{3} = 1$$

completing the back substitution.

Note that this set of operations is formal in that it would be possible, during the forward reduction, for one of the pivots to be zero. Unless such a possibility were anticipated and additional steps taken in the coding, trouble would certainly be in the offing for an arbitrary set of equations. Such a possibility is addressed by the technique called pivoting. The reader is referred to Strang [1] and to Noble and Daniel [2] for details.

Fortunately, the finite element models of the classes of problems covered in this text have coefficient matrices with the property that during the forward reduction process all the pivots are positive. Thus, the formal operations of forward reduction and back substitution as outlined will suffice to determine the solution.

B.2 Banded Symmetric Matrices

In many finite element applications the coefficient matrix has two desirable properties:

1. $\mathbf{A} = \mathbf{A}^T$; that is, the coefficient matrix is symmetric. This property permits the storage of only the main diagonal elements and those elements above (or below) the main diagonal, cutting the storage requirements roughly in half.

2. **A** is banded. The nonzero elements all lie within a certain distance of the main diagonal defined in Chapter 1 as the half bandwidth. This can further reduce the storage requirements.

Consider for instance the matrix shown below.

$$\left[\begin{array}{ccccccccc|c}
a_{11} & a_{12} & a_{13} & & & & & & & b_1 \\
a_{21} & a_{22} & a_{23} & & & & \text{0's} & & & b_2 \\
a_{31} & a_{32} & a_{33} & a_{34} & a_{35} & & & & & b_3 \\
 & & a_{43} & a_{44} & a_{45} & & & & & b_4 \\
 & & a_{53} & a_{54} & a_{55} & a_{56} & a_{57} & & & b_5 \\
 & & & & a_{65} & a_{66} & a_{67} & & & b_6 \\
 & & & & a_{75} & a_{76} & a_{77} & a_{78} & a_{79} & b_7 \\
 & \text{0's} & & & & & a_{87} & a_{88} & a_{89} & b_8 \\
 & & & & & & a_{97} & a_{98} & a_{99} & b_9
\end{array}\right]$$

For this set of equations, the half bandwidth defined in Chapter 1 is h = 3. All the nonzero elements lie within two positions of the main diagonal elements. Assuming the coefficient matrix to be symmetric, it is necessary to store only the main diagonal and two additional elements to the right of the main diagonal element in each row in order to retain all the information regarding the character of **A**. The necessary information for **A** can be stored in a 9×3 array with **b** stored in a 9×1 array.

$$\begin{bmatrix}
a_{11} & a_{12} & a_{13} \\
a_{22} & a_{23} & 0 \\
a_{33} & a_{34} & a_{35} \\
a_{44} & a_{45} & 0 \\
a_{55} & a_{56} & a_{57} \\
a_{66} & a_{67} & 0 \\
a_{77} & a_{78} & a_{79} \\
a_{88} & a_{89} & \underline{0} \\
a_{99} & \underline{0} & \underline{0}
\end{bmatrix}
\begin{bmatrix}
b_1 \\ b_2 \\ b_3 \\ b_4 \\ b_5 \\ b_6 \\ b_7 \\ b_8 \\ b_9
\end{bmatrix}$$

Note that the extra elements (underlined) in the last h-1 equations are taken to be zero. In general the storage requirements for **A** are hQ $\times$ h, where h is the number of equations and h is the half bandwidth. For this example we need an array of size $9 \times 3 = 27$ rather than the $9 \times 9 = 81$ for the full storage mode. A FORTRAN listing for the subroutines BDecomp and BSolve, designed to solve a symmetric banded set of equations stored in a banded format, as previously indicated, is given as follows.

```
      SUBROUTINE BDecomp(nmax, hmax, n, h, a, err)
C     ***********************************************************
C     *   This SUBROUTINE performs the forward decomposition of a     *
C     * symmetric (n × n) matrix [B] when stored in the (n × h)       *
```

```
C   * banded form [A], where h is the half-band width. The special        *
C   * form of the (n × h) matrix [A] is illustrated below for n = 7        *
C   * and h = 4:                                                           *
C   *               [11 12 13 14 00 00 00] = [11 12 13 14]                 *
C   *               [** 22 23 24 25 00 00] = [22 23 24 25]                 *
C   *               [** ** 33 34 35 36 00] = [33 34 35 36]                 *
C   *         [B] = [** ** ** 44 45 46 47] = [44 45 46 47] = [A]           *
C   *               [** ** ** ** 55 56 57] = [55 56 57 00]                 *
C   *               [** ** ** ** ** 66 67] = [66 67 00 00]                 *
C   *               [** ** ** ** ** ** 77] = [77 00 00 00]                 *
C   * where the **'s in [B] indicate symmetric elements.                   *
C   *                                                                      *
C   * INPUT ARGUMENTS:                                                     *
C   *     nmax = maximum row dimension of the (n × h) matrix [A]           *
C   *     hmax = maximum column dimension of the (n × h) matrix [A]        *
C   *     n    = actual row dimension of the (n × h) matrix [A]            *
C   *     h    = actual column dimension of the (n × h) matrix [A]         *
C   *     a    = (n × h) coefficient matrix [A]                            *
C   *                                                                      *
C   * OUTPUT ARGUMENTS:                                                    *
C   *     a   = the (n × h) decomposed coefficient matrix [A]              *
C   *     err = an error indicator with the following meaning              *
C   *         err = 0: no errors                                           *
C   *         err = i, i = 1 to n: |a(i,i)| < epsilon                      *
C   ************************************************************************

    IMPLICIT NONE

    INTEGER*2    err, h, hmax, n, nmax
    REAL*8       a(nmax,hmax)

    INTEGER*2    i, j, k, l, m
    REAL*8       eps, temp
    REAL*8       MachineEpsilon

    eps = MachineEpsilon(i)

    err = 0
    DO i = 1, n
        DO j = 2, h
            IF( a(i,j) .NE. 0.0) THEN
                IF( ABS(a(i,1)) .LT. eps) THEN
                    err = i
                    RETURN
                END IF
                temp = a(i,j) / a(i,1)
                m = 0
                l = i + j - 1
                DO k = j, h
```

```
                  m = m + 1
                  a(l,m) = a(l,m) - temp * a(i,k)
               END DO
               a(i,j) = temp
            END IF
         END DO
      END DO

      RETURN
      END

      SUBROUTINE BSolve(nmax, hmax, n, h, a, b, x)
C     ************************************************************
C     *   This SUBROUTINE performs the forward decomposition of the   *
C     * (n × 1) right-hand-side vector {b} and then uses the decomposed  *
C     * matrix [A] output from SUBROUTINE BDecomp to solve the        *
C     * system                                                       *
C     *                      [A]{x} = {b}                            *
C     * in banded form.                                              *
C     *                                                              *
C     * INPUT ARGUMENTS:                                             *
C     *     nmax = maximum row dimension of the (n × h) matrix [A]     *
C     *     hmax = maximum column dimension of the (n × h) matrix [A]  *
C     *     n    = actual row dimension of the (n × h) matrix [A]      *
C     *     h    = actual column dimension of the (n × h) matrix [A]   *
C     *     a    = (n × h) coefficient matrix [A]                      *
C     *     b    = (n × 1) right-hand-side vector {b}                  *
C     *                                                              *
C     * OUTPUT ARGUMENTS:                                            *
C     *     x = the (n × 1) solution vector {x}                        *
C     ************************************************************

      IMPLICIT NONE

      INTEGER*2     h, hmax, n, nmax
      REAL*8        a(nmax,hmax), b(nmax), x(nmax)

      INTEGER*2     i, j, k, l, m
      REAL*8        temp
      DO i = 1, n
         x(i) = b(i)
      END DO

      DO i = 1, n
         DO j = 2, h
            IF( a(i,j) .NE. 0.0 ) THEN
               l = i + j - 1
               x(l) = x(l) - a(i,j) * x(i)
            END IF
```

```
          END DO
          x(i) = x(i) / a(i,1)
     END DO

     DO m = 2, n
          i = n + 1 - m
          DO j = 2, h
               IF( a(i,j) .NE. 0.0 ) THEN
                    x(i) = x(i) - a(i,j) * x(i + j - 1)
               END IF
          END DO
      END DO

      RETURN
      END
```

Also provided is a FORTRAN listing of Decomp and Solve, which use partial pivoting in solving a system **Ax** = **b**. Decomp carries out the forward reduction using Gaussian elimination with partial pivoting. The determinant is available to check on the success of the forward elimination. If the determinant is zero, one or more of the pivots was zero and Solve should not be called; something is wrong! It may also be that the value of the determinant being smaller than some acceptable value is an indication of problems. If a more precise idea of the condition of the matrix is desired there are expanded versions of Decomp and Solve available in LINPACK [3] or in Forsythe, Malcolm, and Moler [4].

```
      SUBROUTINE Decomp(nmax, n, a, pivot, work, cond, det)
C     ********************************************************
C     *   This SUBROUTINE does the forward DECOMPosition of the    *
C     * (n × n) matrix [A] which is the coefficient matrix in the  *
C     * linear system                                              *
C     *                     [A]{x} = {b}.                          *
C     * The (n × 1) vectors {b} and {x} are not needed for the     *
C     * forward decomposition ({b} is decomposed in the SUBROUTINE *
C     * Solve). The method used is Gaussian Elimination with partial *
C     * pivoting.                                                  *
C     *                                                            *
C     * INPUT ARGUMENTS:                                           *
C     *     nmax = the maximum size of the (n × n) matrix [A]      *
C     *     n    = the actual size of the (n × n) matrix [A]       *
C     *     a    = the (n × n) matrix [A]                          *
C     *                                                            *
C     * OUTPUT ARGUMENTS:                                          *
C     *     pivot = the (n × 1) integer vector of pivot rows       *
C     *     cond  = the estimated condition number of [A]          *
C     *     det   = the determinant of [A]                         *
C     *                                                            *
```

```
C   * ADJUSTABLE ARRAY ARGUMENTS:                              *
C   *     work = an (n × 1) work vector                        *
C   ***********************************************************

      IMPLICIT NONE

C   GLOBAL VARIABLES:

      INTEGER*2    n, nmax
      INTEGER*2    pivot(nmax)
      REAL*8       cond, det
      REAL*8       a(nmax,nmax)
      REAL*8       work(nmax)

C   LOCAL VARIABLES:

      INTEGER*2    i, j, k, m
      REAL*8       anorm, ek, t, ynorm, znorm

      IF( n .EQ. 1 ) THEN
          IF( a(1,1) .NE. 0.0 ) THEN
              cond = 1.0
              det = a(1,1)
          ELSE
              cond = 1.0e+32
              det = 0.0
          END IF
          RETURN
      END IF

      anorm = 0.0
      cond  = 1.0E+32
      det   = 0.0

      DO j = 1, n
          t = 0.0
          DO i = 1, n
              t = t + ABS(a(i,j))
          END DO
          IF( t .GT. anorm ) anorm = t
      END DO

      pivot(n) = 1
      DO k = 1, n - 1
          m = k
          DO i = k + 1, n
              IF( ABS(a(i,k)) .GT. ABS(a(m,k)) ) m = i
          END DO
```

```
        pivot(k) = m
        IF( m .NE. k ) pivot(n) = -pivot(n)
        t = a(m,k)
        a(m,k) = a(k,k)
        a(k,k) = t
        IF( t .NE. 0.0 ) THEN
            DO i = k + 1, n
                a(i,k) = -a(i,k) / t
            END DO
            DO j = k + 1, n
                t = a(m,j)
                a(m,j) = a(k,j)
                a(k,j) = t
                IF( t .NE. 0.0 ) THEN
                    DO i = k + 1, n
                        a(i,j) = a(i,j) + a(i,k) * t
                    END DO
                END IF
            END DO
        END IF
    END DO

    DO k = 1, n
        t = 0.0
        IF( k .NE. 1 ) THEN
            DO i = 1, k - 1
                t = t + a(i,k) * work(i)
            END DO
        END IF
        ek = 1.0
        IF( t .LT. 0.0 ) ek = -1.0
        IF( a(k,k) .EQ. 0.0 ) THEN
            cond = 1.0e+32
            det = 0.0
            RETURN
        END IF
        work(k) = -(ek + t) / a(k,k)
    END DO
    DO k = n - 1, 1, -1
        t = 0.0
        DO i = k + 1, n
            t = t + a(i,k) * work(k)
        END DO
        work(k) = t
        m = pivot(k)
        IF( m .NE. k ) THEN
            t = work(m)
            work(m) = work(k)
```

```
            work(k) = t
         END IF
      END DO

      ynorm = 0.0
      DO i = 1, n
         ynorm = ynorm + ABS(work(i))
      END DO

      CALL Solve(nmax, n, a, work, pivot, work)

      znorm = 0.0
      DO i = 1, n
         znorm = znorm + ABS(work(i))
      END DO

      cond = anorm * znorm / ynorm
      IF( cond .LT. 1.0) cond = 1.0

      det = pivot(n)
      DO i = 1, n
         det = det * a(i,i)
      END DO

      RETURN
      END

      SUBROUTINE Solve(nmax, n, a, b, pivot, x)
C     ***********************************************************
C     *   This SUBROUTINE does the backward decomposition necessary  *
C     * to solve the system                                      *
C     *                      [A]{x} = {b}.                       *
C     * for the (n × 1) vectors {x}, given {b}. The matrix [A] is *
C     * the matrix output from the SUBROUTINE Decomp.            *
C     *                                                          *
C     * INPUT ARGUMENTS:                                         *
C     *    nmax = the maximum size of the (n × n) matrix [A]     *
C     *    n    = the actual size of the (n × n) matrix [A]      *
C     *    a    = the (n × n) matrix [A]                         *
C     *    b    = the (n × 1) right-hand-side vector             *
C     *    pivot = the (n × 1) integer vector of pivots, also from *
C     *            the SUBROUTINE Decomp                         *
C     *                                                          *
C     * OUTPUT ARGUMENTS:                                        *
C     *    x = the (n × 1) solution vector                       *
C     ***********************************************************

      IMPLICIT NONE
```

```
C GLOBAL VARIABLES:

      INTEGER*2    n, nmax
      INTEGER*2    pivot(nmax)
      REAL*8       a(nmax,nmax)
      REAL*8       b(nmax), x(nmax)

C LOCAL VARIABLES:

      INTEGER*2    i, k, kb, m
      REAL*8       t

      DO i = 1, n
         x(i) = b(i)
      END DO

      IF( n .EQ. 1 ) THEN
         x(1) = x(1) / a(1,1)
         RETURN
      END IF

      DO k = 1, n - 1
         m = pivot(k)
         t = x(m)
         x(m) = x(k)
         x(k) = t
         DO i = k + 1, n
            x(i) = x(i) + a(i,k) * t
         END DO
      END DO

      DO kb = 1, n - 1
         k = n - kb + 1
         x(k) = x(k) / a(k,k)
         t = -x(k)
         DO i = 1, k - 1
            x(i) = x(i) + a(i,k) * t
         END DO
      END DO
      x(1) = x(1) / a(1,1)

      RETURN
      END
```

For a single set of equations, Decomp and then Solve would be called with the solution returned in B. For a time-dependent situation, Decomp would be called after which Solve would be called as many times as desired.

REFERENCES

1. Strang, G.: *Linear Algebra and its Applications*, Academic Press, New York, 1980.
2. Noble, B., and J. W. Daniel: *Applied Linear Algebra*, Prentice Hall, Englewood Cliffs, New Jersey, 1977.
3. Dongarra, J. J., et al.: *LINPACK Users Guide*, SIAM, Philadelphia, 1979.
4. Forsythe, G. E., M. A. Malcolm, and C. B. Moler: *Computer Methods for Mathematical Computations*, Prentice Hall, Englewood Cliffs, New Jersey, 1977.

APPENDIX C *EIGENVALUE PROBLEMS*

Introduction

Finite element models of time-dependent problems lead generally to systems of ordinary differential equations. Whether the integration is carried out by explicitly using the eigenvalues and eigenvectors for representing the solution or implicitly using a time domain integration technique, it is clear from the developments in Chapter 4 that a knowledge of the eigenvalues and eigenvectors of the underlying problem is of primary importance. For very small systems it may be feasible to determine the eigenvalues and eigenvectors by solving the polynomial equation

$$\det(\mathbf{A} - \lambda\mathbf{B}) = 0$$

for the eigenvalues λ_i, and subsequently the eigenvectors $\mathbf{v_i}$ from the corresponding dependent set of equations

$$(\mathbf{A} - \lambda_i\mathbf{B})v_i = 0$$

Example C.1.

Consider the problem of determining the eigenvalues and eigenvectors of the matrix **A** given by

$$\mathbf{A} = \begin{bmatrix} 3 & -1 & 0 \\ -1 & 2 & -1 \\ 0 & -1 & 3 \end{bmatrix}$$

The characteristic equation is obtained by setting $\det(\mathbf{A} - \lambda\mathbf{I}) = 0$, resulting in

$$(3 - \lambda)^2(2 - \lambda) - 2(3 - \lambda) = 0$$

which has the roots $\lambda = 1$, 3, and 4. The corresponding eigenvectors are determined by back substituting each of the eigenvalues λ_i, one at a time, into the dependent set of equations $(\mathbf{A} - \lambda_i\mathbf{I})\mathbf{x}_i = 0$ to determine the $\mathbf{x_i}$. This results in

$$\lambda_1 = 1 \qquad \mathbf{x_1^T} = [1 \quad 2 \quad 1]$$

$$\lambda_2 = 3 \qquad \mathbf{x_2^T} = [1 \quad 0 \quad -1]$$

$$\lambda_3 = 4 \qquad \mathbf{x_3^T} = [1 \quad -1 \quad 1]$$

For a system of any size, however, this approach is entirely inappropriate and numerical procedures are indicated.

There are basically two distinct, general approaches for determining the eigenvalues and eigenvectors of

$$(\mathbf{A} - \lambda\mathbf{B})\mathbf{u} = \mathbf{0}$$

those that iterate in some fashion to determine the eigenvalues and eigenvectors and those that use transformations to reduce the original problem to one of simpler form from which the eigenvalues and eigenvectors are more easily determined. A general discussion of either of these two approaches is well beyond the scope of intent for this Appendix. The interested reader is referred to Bathe and Wilson [1], Strang [2], and Noble and Daniel [3].

Before embarking on a discussion of iteration techniques for the generalized problem $\mathbf{Au} - \lambda\mathbf{Bu} = \mathbf{0}$, consider the following. The statement that λ_i and $\mathbf{u_i}$ constitute an eigenvalue-eigenvector pair is equivalent to the statement

$$\mathbf{Au_i} = \lambda_i \mathbf{Bu_i}$$

Premultiply $\mathbf{u_i^T}$ and solve for λ_i to obtain

$$\lambda_i = \frac{\mathbf{u_i^T A u_i}}{\mathbf{u_i^T B u_i}}$$

valid for any eigenvalue-eigenvector pair λ_i, $\mathbf{u_i}$. The generalization to

$$\rho(\mathbf{u}) = \frac{\mathbf{u^T A u}}{\mathbf{u^T B u}}$$

is the celebrated *Rayleigh quotient*. Note that the Rayleigh quotient for the ordinary linear algebraic eigenvalue problem $\mathbf{Au} - \lambda\mathbf{u} = \mathbf{0}$ is

$$\rho(\mathbf{u}) = \frac{\mathbf{u^T A u}}{\mathbf{u^T u}}$$

In either case, the Rayleigh quotient has the property that as $\mathbf{u}$ ranges over all possible $\mathbf{u} \neq \mathbf{0}$, ρ satisfies

$$\lambda_1 \leq \rho(\mathbf{u}) \leq \lambda_N$$

where λ_1 and λ_N are the eigenvalues of least and greatest magnitude, respectively. In other words, ρ is bounded above and below by λ_1 and λ_N, respectively. It turns out that, considered as a function of $\mathbf{u}$, ρ is stationary at each of the eigenvectors with the stationary value equal to the corresponding eigenvalue.

Of primary importance to us is the fact that if an approximation $\mathbf{v_j}$ to the jth eigenvector $\mathbf{u_j}$ is available, an estimate of the corresponding eigenvalue λ_j can be obtained by computing

$$\lambda_j \approx \frac{\mathbf{v_j^T A v_j}}{\mathbf{v_j^T B v_j}}$$

A very nice property of the Rayleigh quotient is that a fairly crude approximate eigenvector can yield a fairly good estimate of the corresponding eigenvalue.

Example C.2.

From Example C.1 just considered, the eigenvector corresponding to the largest eigenvalue was $[1 \quad -1 \quad 1]^T$. Take as a fairly rough approximation, $\mathbf{v} = [1.1 \quad -0.9 \quad 1.2]^T$. Substituting into the Rayleigh quotient yields

$$\rho(\mathbf{v}) = \frac{\mathbf{v}^T\mathbf{A}\mathbf{v}}{\mathbf{v}^T\mathbf{v}} = 3.9624$$

which is obviously quite near the exact value of 4.0000. Even with errors in some of the components of the approximate eigenvector of the order of 10% to 20%, the error in the corresponding eigenvalue as computed using the Rayleigh quotient is less than 4%. The Rayleigh quotient can be used very effectively during an iteration process to obtain accurate estimates of the eigenvalue.

Iteration Methods on $\mathbf{Au} - \boldsymbol{\lambda}\mathbf{u} = \mathbf{0}$

Iteration methods are based on what is generally referred to as the power method. The power method is applied to the ordinary algebraic eigenvalue problem $\mathbf{Au} - \boldsymbol{\lambda}\mathbf{u} = \mathbf{0}$. A nonzero vector $\mathbf{u_0}$ is selected and the product $\mathbf{Au_0} = \mathbf{u_1}$ is computed. This is followed by the successive computation of $\mathbf{u_2} = \mathbf{Au_1} = \mathbf{AAu_0}$, $\mathbf{u_3} = \mathbf{Au_2} = \mathbf{AAAu_0}$, and so forth. In general,

$$\mathbf{u_{n+1}} = \mathbf{Au_n} \qquad n = 0, 1, 2, \ldots$$

It can be shown that unless $\mathbf{u_0}$ is orthogonal to the eigenvector $\mathbf{u_N}$ corresponding to the eigenvalue of largest magnitude λ_N, the process converges to $(\lambda_N)^N\mathbf{u_N}$, that is, to a vector indicating the eigenvalue of largest magnitude and the corresponding eigenvector.

Example C.3.

Consider the matrix

$$\mathbf{A} = \begin{bmatrix} 3 & -1 & 0 \\ -1 & 2 & -1 \\ 0 & -1 & 3 \end{bmatrix}$$

Take $\mathbf{u_0} = [1 \quad 0 \quad 0]^T$. It follows that

$$\begin{aligned} \mathbf{u}_1^* = \mathbf{Au}_0 &= [3 \quad -1 \quad 0]^T \\ &= 3[1 \quad -1/3 \quad 0]^T = 3\mathbf{u}_1 = \alpha_1\mathbf{u}_1 \end{aligned}$$

where $\mathbf{u}_1^*$ has been normalized with α_1 so that the first component of $\mathbf{u}_1$ is the same as the corresponding component in $\mathbf{u}_0$. This step is not necessary but serves to clarify what is actually happening. One then computes

$$\begin{aligned} \mathbf{u}_2^* = \mathbf{Au}_1 &= 3.3333[1 \quad -0.5 \quad 0.1]^T \\ &= 3.3333\mathbf{u}_2 = \alpha_2\mathbf{u}_2 \end{aligned}$$

The results of the next several steps are indicated in the table, which gives the normalization constant α_n and the corresponding vector $\mathbf{u_n}$.

n	α_n	$\mathbf{u_n}$		
3	3.5000	[1.0000	− 0.6000	0.2286]^T
4	3.6000	[1.0000	− 0.6746	0.3571]^T
5	3.6746	[1.0000	− 0.7365	0.4752]^T
6	3.7365	[1.0000	− 0.7890	0.5786]^T
20	3.9937	[1.0000	− 0.9953	0.9905]^T
21	3.9953	[1.0000	− 0.9964	0.9929]^T

It is readily apparent that α_n is heading towards $\lambda_3 = 4.0$ and that $\mathbf{u_n}$ is approaching $[1 \quad -1 \quad 1]^T$, the corresponding eigenvector. In general, the version of the power method demonstrated in this example is such that α_n converges to the eigenvalue of largest magnitude and $\mathbf{u_n}$ to the corresponding eigenvector. The process is referred to generally as *forward iteration*. By defining $\mathbf{A^{-1}} = \mathbf{B}$, and $\lambda^{-1} = \mu$, the standard linear algebraic eigenvalue problem $\mathbf{Au} - \lambda\mathbf{u} = \mathbf{0}$ can be rewritten as

$$\mathbf{Bu} - \mu\mathbf{u} = \mathbf{0}$$

The power method applied to this algebraic eigenvalue problem will converge to the largest μ, which is the smallest λ. Thus, iteration on $\mathbf{A^{-1}}$ converges to the smallest eigenvalue of $\mathbf{A}$ and the corresponding eigenvector. The inverse of $\mathbf{A}$ is easily computed to be

$$\mathbf{A}^{-1} = \begin{bmatrix} 5 & 3 & 1 \\ 3 & 9 & 3 \\ 1 & 3 & 5 \end{bmatrix} \frac{1}{12}$$

Again beginning with $\mathbf{u}_0 = [1 \quad 0 \quad 0]^T$ yields the sequence of iterates

n	α_n	$\mathbf{u_n}$		
1	0.4167	[1.0000	0.6000	0.2000]^T
2	0.5833	[1.0000	1.2857	0.5429]^T
3	0.7833	[1.0000	1.7234	0.8055]^T
4	0.9146	[1.0000	1.9067	0.9291]^T
5	0.9708	[1.0000	1.9699	0.9757]^T
6	0.9904	[1.0000	1.9903	0.9918]^T
7	0.9969	[1.0000	1.9969	0.9973]^T
8	0.9990	[1.0000	1.9990	0.9991]^T
9	0.9997	[1.0000	1.9997	0.9997]^T

clearly converging to $\lambda_1 = 1.0000$ and to $\mathbf{u}_1 = [1 \quad 2 \quad 1]^T$, that is, to the smallest eigenvalue and corresponding eigenvector. This process is referred to as *inverse iteration*. For large n, $\mathbf{A}^{-1}$ is not actually computed. Rather, a routine such as Decomp is called initially, followed repeatedly by a Solve-type algorithm.

The rate of convergence of the power method as applied to $\mathbf{A}$, that is, forward iteration, depends upon the ratio $|\lambda_{N-1}/\lambda_N|$, where λ_{N-1} and λ_N are the largest and next to largest eigenvalues, respectively, of $\mathbf{A}$. The smaller the ratio, the faster the convergence. Similarly, for inverse iteration, the rate of convergence depends upon $|\lambda_2/\lambda_1|$, where λ_1 and λ_2 are the smallest and next to smallest eigenvalues, respectively, of $\mathbf{A}$. For this example, $|\lambda_{N-1}/\lambda_N| = 3/4 = 0.75$, consistent with the relatively slow convergence of the forward iteration, whereas $|\lambda_2/\lambda_1| = 1/3 = 0.333\ldots$, corresponding to the much more rapid convergence of the inverse iteration.

Most situations encountered in practice involve the generalized linear algebraic eigenvalue problem $\mathbf{Au} - \lambda\mathbf{Bu} = \mathbf{0}$. In what follows, we will outline the corresponding forward and inverse iteration processes.

Iteration Methods on $\mathbf{Au} - \lambda\mathbf{Bu} = \mathbf{0}$

Forward iteration. Forward iteration is designed to converge to the eigenvalue of largest magnitude and to the corresponding eigenvector. With $\mathbf{u}_1$ the initial guess for the eigenvector corresponding to the largest eigenvalue, forward iteration is begun by computing $\mathbf{v}_1 = \mathbf{Au}_1$. This is followed for $k = 1, 2, \ldots,$ by the sequence of steps

1. Solve for $\mathbf{u}^*_{\mathbf{k+1}}$ from $\mathbf{Bu}^*_{\mathbf{k+1}} = \mathbf{v}_\mathbf{k}$.
2. Compute $\mathbf{v}^*_{\mathbf{k+1}} = \mathbf{Au}^*_{\mathbf{k+1}}$.
3. Compute the Rayleigh quotient according to

$$\rho(\mathbf{u}^*_{\mathbf{k+1}}) = \frac{(\mathbf{u}^*_{\mathbf{k+1}})^{\mathrm{T}}\mathbf{v}^*_{\mathbf{k+1}}}{(\mathbf{u}^*_{\mathbf{k+1}})^{\mathrm{T}}\mathbf{v}_\mathbf{k}} = \frac{(\mathbf{u}^*_{\mathbf{k+1}})^{\mathrm{T}}\mathbf{Au}^*_{\mathbf{k+1}}}{(\mathbf{u}^*_{\mathbf{k+1}})^{\mathrm{T}}\mathbf{Bu}^*_{\mathbf{k+1}}}$$

4. Last, compute

$$\mathbf{v}_{\mathbf{k+1}} = \frac{\mathbf{v}^*_{\mathbf{k+1}}}{\left((\mathbf{u}^*_{\mathbf{k+1}})^{\mathrm{T}}\mathbf{v}_\mathbf{k}\right)^{1/2}}$$

which B-normalizes $\mathbf{v}^*_{\mathbf{k+1}}$. Then return to step 1.

The Rayleigh quotient ρ converges to the eigenvalue of largest magnitude and $\mathbf{u}^*_{\mathbf{k+1}}$, usually normalized by $\rho^{-1/2}$, to the corresponding eigenvector.

Example C.4.

Take **A** and **B** as follows:

$$\mathbf{A} = \begin{bmatrix} 2 & -1 & 0 \\ -1 & 2 & -1 \\ 0 & -1 & 1 \end{bmatrix} \qquad \mathbf{B} = \begin{bmatrix} 4 & 1 & 0 \\ 1 & 4 & 1 \\ 0 & 1 & 2 \end{bmatrix}$$

and the initial guess as $\mathbf{u}_0 = [1 \;\; 0 \;\; 0]^{\mathrm{T}}$. After computing $\mathbf{v}_1 = [2.0000 \;\; -1.0000 \;\; 0\ 0000]^{\mathrm{T}}$, the results for several iterations are

$$\mathbf{u}^*_2 = [0.6154 \quad -0.4615 \quad 0.2308]^{\mathrm{T}}$$

$$\mathbf{v}^*_2 = [1.6923 \quad -1.7692 \quad 0.6923]^{\mathrm{T}}$$

$$\rho(\mathbf{u}_2^*) = 1.1923$$
$$\mathbf{v}_2 = [1.3009 \quad -1.3600 \quad 0\ \ 5322]^T$$
$$\mathbf{u}_3^* = [0.4753 \quad -0.6004 \quad 0.5663]^T$$
$$\mathbf{v}_3^* = [1.5511 \quad -2.2424 \quad 1.1667]^T$$
$$\rho(\mathbf{u}_3^*) = 1.5806$$
$$\mathbf{v}_3 = [1.1771 \quad -1.7018 \quad 0\ \ 8854]^T$$
$$\mathbf{u}_4^* = [0.4819 \quad -0.7504 \quad 0.8179]^T$$
$$\mathbf{v}_4^* = [1.7142 \quad -2.8006 \quad 1.5683]^T$$
$$\rho(\mathbf{u}_4^*) = 1.6392$$
$$\mathbf{v}_4 = [1.0696 \quad -1.7475 \quad 0\ \ 9786]^T$$
$$\mathbf{u}_5^* = [0.4600 \quad -0.7705 \quad 0.8745]^T$$
$$\mathbf{v}_5^* = [1.6906 \quad -2.8756 \quad 1.6451]^T$$
$$\rho(\mathbf{u}_5^*) = 1.6450$$
$$\mathbf{v}_5 = [1.0299 \quad -1.7519 \quad 1.0022]^T$$
$$\mathbf{u}_6^* = [0.4506 \quad -0.7725 \quad 0.8873]^T$$
$$\mathbf{v}_6^* = [1.6736 \quad -2.8828 \quad 1.6598]^T$$
$$\rho(\mathbf{u}_6^*) = 1.6455$$
$$\mathbf{v}_6 = [1.0173 \quad -1.7523 \quad 1.0089]^T$$

At the end of 10 iterations the process has converged to $\rho(\mathbf{u}_{11}^*) = 1.6456$ and $\mathbf{u}_{11}^* = [0.4461 \quad -0.7726 \quad 0.8921]^T$, which when normalized is $[0.2711 \quad -0.4695 \quad 0.5422]^T$. Note that the rate of convergence of the eigenvalue as computed by the Rayleigh quotient is faster than the rate of convergence of the corresponding eigenvector. For a general discussion in this regard see Bathe [1].

Inverse iteration. Inverse iteration is designed to converge to the eigenvalue of least magnitude and to the corresponding eigenvector. With $\mathbf{u}_1$ the initial guess for the eigenvector corresponding to the smallest eigenvalue, inverse iteration is begun by computing $\mathbf{v}_1 = \mathbf{B}\mathbf{u}_1$. This is followed for $k = 1, 2, \ldots,$ by the sequence of steps

1. Solve for $\mathbf{u}_{k+1}^*$ from $\mathbf{A}\mathbf{u}_{k+1}^* = \mathbf{v}_k$
2. Compute $v_{k+1}^* = \mathbf{B}\mathbf{u}_{k+1}^*$
3. Compute the Rayleigh quotient according to

$$\rho(\mathbf{u}_{k+1}^*) = \frac{(\mathbf{u}_{k+1}^*)^T\mathbf{v}_k}{(\mathbf{u}_{k+1}^*)^T\mathbf{v}_{k+1}^*} = \frac{(\mathbf{u}_{k+1}^*)^T\mathbf{A}\mathbf{u}_{k+1}^*}{(\mathbf{u}_{k+1}^*)^T\mathbf{B}\mathbf{u}_{k+1}^*}$$

4. Last, compute

$$\mathbf{v_{k+1}} = \frac{\mathbf{v^*_{k+1}}}{\left((\mathbf{u^*_{k+1}})^T\mathbf{v^*_{k+1}}\right)^{1/2}}$$

which B-normalizes $\mathbf{v^*_{k+1}}$. Then return to step 1.

The Rayleigh quotient ρ converges to the eigenvalue of smallest magnitude and $\mathbf{u^*_{k+1}}$, usually normalized by $\rho^{-1/2}$, converges to the corresponding eigenvector.

Example C.5.

With **A** and **B** as in the previous example and $\mathbf{u_0}$ again taken as $[1 \quad 0 \quad 0]^T$, there results after computing $\mathbf{v_1} = [4.0000 \quad 1.0000 \quad 0.0000]^T$,

$$\mathbf{u^*_2} = [5.0000 \quad 6.000 \quad 6.0000]^T$$

$$\mathbf{v^*_2} = [26.0000 \quad 35.0000 \quad 18.0000]^T$$

$$\rho(\mathbf{u^*_2}) = 0.0580$$

$$\mathbf{v_2} = [1.2284 \quad 1.6536 \quad 0.8504]^T$$

$$\mathbf{u^*_3} = [3.7324 \quad 6.2364 \quad 7.0868]^T$$

$$\mathbf{v^*_3} = [21.1660 \quad 35.7649 \quad 20.4101]^T$$

$$\rho(\mathbf{u^*_3}) = 0.0468$$

$$\mathbf{v_3} = [1.0015 \quad 1.6922 \quad 0.9657]^T$$

$$\mathbf{u^*_4} = [3.6594 \quad 6.3173 \quad 7.2830]^T$$

$$\mathbf{v^*_4} = [20.9548 \quad 36.2116 \quad 20.8833]^T$$

$$\rho(\mathbf{u^*_4}) = 0.0467$$

$$\mathbf{v_4} = [0.9797 \quad 1.6929 \quad 0.9763]^T$$

$$\mathbf{u^*_5} = [3.6489 \quad 6.3181 \quad 7.2944]^T$$

$$\mathbf{v^*_5} = [20.9136 \quad 36.2157 \quad 20.9069]^T$$

$$\rho(\mathbf{u^*_5}) = 0.0467$$

$$\mathbf{v_5} = [0.9776 \quad 1.6929 \quad 0.9773]^T$$

$$\mathbf{u^*_6} = [3.6479 \quad 6.3181 \quad 7.2954]^T$$

$$\mathbf{v^*_6} = [20.9096 \quad 36.2158 \quad 20.9090]^T$$

$$\rho(\mathbf{u^*_6}) = 0.0467$$

$$\mathbf{v_6} = [0.9774 \quad 1.6929 \quad 0.9774]^T$$

$$\mathbf{u^*_6} = [3.6478 \quad 6.3181 \quad 7.2955]^T$$

$$\mathbf{v}_6^* = [\,20.9092 \quad 36.2158 \quad 20.9092\,]^{\mathrm{T}}$$

$$\rho(\mathbf{u}_6^*) = 0.0467$$

$$\mathbf{v}_6 = [\,0.9774 \quad 1.6929 \quad 0.9774\,]^{\mathrm{T}}$$

The next iteration produces no changes so that the smallest eigenvalue is $\lambda_1 = 0.0467$ and the corresponding eigenvector $\mathbf{u}_1 = [\,3.6478 \quad 6.3181 \quad 7.2955\,]^{\mathrm{T}}$, which normalizes to $\mathbf{u}_1 = [\,0.1705 \quad 0.2953 \quad 0.3410\,]^{\mathrm{T}}$. Again note the relatively rapid convergence of the eigenvalue as compared to the convergence of the eigenvector.

Subspace Iteration

The inverse and forward iteration procedures were seen to converge to the smallest and largest $\lambda, \mathbf{u}$ pairs, respectively. More often it is the lowest several $\lambda, \mathbf{u}$ pairs that are desired. This is in view of the fact that for systems of the form $\mathbf{Au}+\mathbf{B\dot{u}} = \mathbf{f}$ or of the form $\mathbf{Au}+\mathbf{B\ddot{u}} = \mathbf{f}$, it is frequently the case that an adequate approximate solution to the system of ordinary differential equations can be determined with only a knowledge of a subset of the $\lambda, \mathbf{u}$ pairs at the lower end of the spectrum. In terms of the original physical problem, this is equivalent to the idea that in many situations, the response can adequately be represented in terms of only the lower modes. Subspace iteration, described below, is ideally suited for accomplishing the task of determining this portion of the spectrum. For a detailed discussion the reader is referred to Bathe [1].

The subspace iteration method is carried out as follows. Suppose that the first p $\lambda, \mathbf{u}$ pairs of an $n \times n$ system are desired. Select a set of $q > p$ linearly independent $n \times 1$ vectors $\mathbf{x}_1$ and form an $n \times q$ modal matrix $\mathbf{X}_1$ whose columns are the $\mathbf{x}_1$. Iterate on k as follows:

1. Solve each of the q equations indicated by

$$\mathbf{AY}_k = \mathbf{BX}_k$$

to obtain the $n \times q$ modal matrix $\mathbf{Y}_k$.

2. Form the reduced $q \times q$ eigenvalue problem

$$\mathbf{A}_k^*\mathbf{z} - \lambda\mathbf{B}_k^*\mathbf{z} = \mathbf{0}$$

where

$$\mathbf{A}_k^* = \mathbf{Y}_k^{\mathrm{T}}\mathbf{AY}_k \qquad \mathbf{B}_k^* = \mathbf{Y}_k^{\mathrm{T}}\mathbf{BY}_k$$

and determine the q eigenvalues and eigenvectors

$$\mathbf{\Lambda}_k = \mathrm{diag}\,[\,\lambda_1 \quad \lambda_2 \quad \ldots \quad \lambda_q\,] \qquad \mathbf{Q}_k = [\,\mathbf{z}_1 \quad \mathbf{z}_2 \quad \ldots \quad \mathbf{z}_q\,]$$

satisfying $\mathbf{A}_k^*\mathbf{Q}_k = \mathbf{B}_k^*\mathbf{Q}_k\mathbf{\Lambda}_k$. Then form

$$\mathbf{X}_{k+1} = \mathbf{Y}_k\mathbf{Q}_k$$

3. Check on convergence according to

$$\frac{\left|\left(\lambda_i^{(k)} - \lambda_1^{(k-1)}\right)\right|}{|\lambda_i^{(k)}|} \leq \epsilon \qquad i = 1, 2, \ldots, p$$

If these criteria are satisfied, the first p elements of $\mathbf{\Lambda_k}$ are the desired eigenvalues, with the first p columns of $\mathbf{X_{k+1}}$ the corresponding eigenvectors. If the criteria are not satisfied, return with $\mathbf{X_{k+1}}$ to step 1 and iterate again.

For best results q is to be taken to be $\min(2p, p+8)$.

Example C.6.

Consider the matrices

$$\mathbf{A} = \begin{bmatrix} 2 & -1 & 0 & 0 \\ -1 & 2 & -1 & 0 \\ 0 & -1 & 2 & -1 \\ 0 & 0 & -1 & 1 \end{bmatrix} \qquad \mathbf{B} = \begin{bmatrix} 4 & 1 & 0 & 0 \\ 1 & 4 & 1 & 0 \\ 0 & 1 & 4 & 1 \\ 0 & 0 & 1 & 2 \end{bmatrix}$$

and assume that only the first λ, $\mathbf{u}$ pair is desired. Here it would clearly be preferable to use inverse iteration, and we continue only to demonstrate the details of the subspace iteration algorithm. With one λ, $\mathbf{u}$ pair desired, the dimension of the subspace is taken to be two. The iteration is begun with $\mathbf{x}_1^T = [1 \;\; 0 \;\; 0 \;\; 0]$ and $\mathbf{x}_2^T = [0 \;\; 1 \;\; 0 \;\; 0]$. More suitable initial vectors such as $[1 \;\; 1 \;\; 1 \;\; 1]^T$ and $[1 \;\; 1 \;\; -1 \;\; -1]^T$ are clearly possible. We display below the intermediate results for the three iterations it takes for convergence.

Iteration 1. Solving the two equations $\mathbf{AY_1} = \mathbf{BX_1}$ for $\mathbf{y_1}$ and $\mathbf{y_2}$ yields

$$\mathbf{Y}_1 = \begin{bmatrix} 5.000 & 6.000 \\ 6.000 & 11.000 \\ 6.000 & 12.000 \\ 6.000 & 12.000 \end{bmatrix}$$

The first reduced eigenvalue problem is $\mathbf{A}^*\mathbf{u} - \lambda\mathbf{B}^*\mathbf{u} = \mathbf{0}$ with

$$\mathbf{A}^* = \begin{bmatrix} 26.0000 & 35.0000 \\ 35.0000 & 62.0000 \end{bmatrix} \qquad \mathbf{B}^* = \begin{bmatrix} 664.0000 & 1189.00 \\ 1189.0000 & 2176.00 \end{bmatrix}$$

The eigenvalues and eigenvectors are

$$\lambda_1 = 0.0284 \qquad \lambda_2 = 0.4376$$

and

$$\mathbf{u}_1^T = [-0.0041 \quad 0.0237] \qquad \mathbf{u}_2^T = [0.2643 \quad -0.1441]$$

respectively. The next $\mathbf{X}$ is determined from $\mathbf{X} = \mathbf{YU}$, leading to

$$\mathbf{X}_2 = \begin{bmatrix} 0.1215 & 0.4570 \\ 0.2358 & 0.0009 \\ 0.2595 & -0.1432 \\ 0.2595 & -0.1432 \end{bmatrix}$$

to start the next iteration.

Iteration 2. Solving the two equations $\mathbf{AY}_2 = \mathbf{BX}_2$ for the next two $\mathbf{y}$'s yields

$$\mathbf{Y}_2 = \begin{bmatrix} 4.3582 & 1.0013 \\ 7.9944 & 0.1737 \\ 10.306 & -0.9711 \\ 11.085 & -1.4008 \end{bmatrix}$$

The second reduced eigenvalue problem is $\mathbf{A}^*\mathbf{u} - \lambda\mathbf{B}^*\mathbf{u} = \mathbf{0}$ with

$$\mathbf{A}^* = \begin{bmatrix} 38.167 & -1.6267 \\ -1.6267 & 3.1827 \end{bmatrix} \qquad \mathbf{B}^* = \begin{bmatrix} 1465.2 & -70.494 \\ -70.494 & 14.559 \end{bmatrix}$$

The eigenvalues and eigenvectors are

$$\lambda_1 = 0.0260 \qquad \lambda_2 = 0.2789$$

and

$$\mathbf{u}_1^T = [0.0260 \quad -0.0019] \quad \mathbf{u}_2^T = [0.0146 \quad 0.2992]$$

respectively. The next $\mathbf{X}$ is determined from $\mathbf{X} = \mathbf{YU}$, leading to

$$\mathbf{X}_3 = \begin{bmatrix} 0.1115 & 0.3631 \\ 0.2078 & 0.1684 \\ 0.2702 & -0.1405 \\ 0.2913 & -0.2577 \end{bmatrix}$$

to start the next iteration.

Iteration 3. Solving the two equations $\mathbf{Ay}_1 = \mathbf{Bx}_1$ with the $\mathbf{x}_1$ as the first and second columns of $\mathbf{X}_3$, it follows that

$$\mathbf{Y}_3 = \begin{bmatrix} 4.2989 & 1.2101 \\ 7.9440 & 0.7994 \\ 10.376 & -0.5077 \\ 11.229 & -1.1636 \end{bmatrix}$$

The third reduced eigenvalue problem is $\mathbf{A}^*\mathbf{u} - \lambda\mathbf{B}^*\mathbf{u} = \mathbf{0}$ with

$$\mathbf{A}^* = \begin{bmatrix} 38.411 & -0.0335 \\ -0.0335 & 3.7717 \end{bmatrix} \qquad \mathbf{B}^* = \begin{bmatrix} 1475.4 & -1.4554 \\ -1.4554 & 14.457 \end{bmatrix}$$

The eigenvalues and eigenvectors are

$$\lambda_1 = 0.0260 \qquad \lambda_2 = 0.2609$$

and

$$\mathbf{u}_1^T = [0.0260 \quad 0.0000] \qquad \mathbf{u}_2^T = [0.0003 \quad 0.2630]$$

respectively. The next $\mathbf{X}$ is determined from $\mathbf{X} = \mathbf{YU}$, leading to

$$\mathbf{X}_4 = \begin{bmatrix} 0.1119 & 0.3194 \\ 0.2068 & 0.2123 \\ 0.2702 & -0.1308 \\ 0.2924 & -0.3031 \end{bmatrix}$$

to start the next iteration. For this example, however, λ_1 has converged for the criterion imposed, with the result that the first eigenvalue is 0.0260 with the corresponding eigenvector the first column of $\mathbf{X}_4$.

Transformation Methods

Consider the algebraic eigenvalue problem

$$\mathbf{Ax} - \lambda\mathbf{x} = \mathbf{0}$$

Set $\mathbf{x} = \mathbf{Qy}$ and premultiply by $\mathbf{Q}^T$ to obtain

$$\mathbf{Q}^T\mathbf{AQy} - \lambda\mathbf{Q}^T\mathbf{Qy} = \mathbf{0}$$

It can be shown that there is an orthogonal matrix $\mathbf{Q}$, that is, a matrix satisfying $\mathbf{Q}^T\mathbf{Q} = \mathbf{I}$, resulting in

$$\mathbf{By} - \lambda\mathbf{y} = \mathbf{0}$$

where $\mathbf{B} = \mathbf{Q}^T\mathbf{AQ} = \mathbf{\Lambda} = \text{diag}[\,\lambda_1 \quad \lambda_2 \ldots \lambda_N\,]$. The transformation $\mathbf{B} = \mathbf{Q}^T\mathbf{AQ} = \mathbf{\Lambda}$ is said to be diagonalize $\mathbf{A}$. As indicated in Chapter 4, the appropriate transformation turns out to be the modal matrix $\mathbf{X}$ consisting of the normalized eigenvectors of $\mathbf{Ax} - \lambda\mathbf{x} = \mathbf{0}$. The problem is that this matrix is not known until after the problem has been solved in some fashion. The transformation from $\mathbf{A}$ into the diagonal matrix $\mathbf{\Lambda}$ can be constructed by selecting a sequence of transformations, each of which is designed to eliminate a selected off-diagonal element of $\mathbf{A}$. After one such step with $\mathbf{A_1} = \mathbf{A}$, there results

$$\mathbf{A}_2 = \mathbf{Q}_1^T\mathbf{A}_1\mathbf{Q}_1 \qquad \text{and} \qquad \mathbf{T}_1 = \mathbf{Q}_1$$

where $\mathbf{Q_1}$ accomplishes the zeroing of a single off-diagonal element. $\mathbf{T}_1$ is the transformation at this stage. Successively, there results

$$\mathbf{A}_3 = \mathbf{Q}_2^T\mathbf{A}_2\mathbf{Q}_2 \qquad \text{and} \qquad \mathbf{T}_2 = \mathbf{Q}_1\mathbf{Q}_2$$

$$\mathbf{A}_4 = \mathbf{Q}_3^T\mathbf{A}_3\mathbf{Q}_3 \qquad \text{and} \qquad \mathbf{T}_3 = \mathbf{Q}_1\mathbf{Q}_2\mathbf{Q}_3$$

If the $\mathbf{Q_k}$ are properly chosen,

$$\lim_{k\to\infty} \mathbf{A_k} = \mathbf{\Lambda} \qquad \text{and} \qquad \lim_{k\to\infty} \mathbf{T_k} = \mathbf{X}$$

where $\mathbf{\Lambda} = \textbf{diag}\,[\,\lambda_1 \quad \lambda_2 \ldots \lambda_N\,]$, and the columns of $\mathbf{X}$ are the corresponding eigenvectors.

The basic tool of these transformation methods is the plane rotation matrix $\mathbf{R}(\theta)$ given by

$$
\mathbf{R}(\theta) = \begin{array}{c} \\ \\ \\ \\ i \\ \\ \\ \\ j \\ \\ \\ \\ \end{array}
\begin{array}{c}
\begin{array}{ccccccccccc} & & & i & & & & j & & & \end{array} \\
\left[\begin{array}{ccccccccccc}
1 & & & & & & & & & & \\
 & 1 & & & & & & & & & \\
 & & \ddots & & & & & & & & \\
 & & & \cos\theta & & & & -\sin\theta & & & \\
 & & & & 1 & & & & & & \\
 & & & & & 1 & & & & & \\
 & & & & & & \ddots & & & & \\
 & & & \sin\theta & & & & \cos\theta & & & \\
 & & & & & & & & 1 & & \\
 & & & & & & & & & 1 & \\
 & & & & & & & & & & 1
\end{array}\right]
\end{array}
$$

that is, all the main diagonal elements are unity except the ith and jth, and all the off-diagonal elements are zero except the (i, j) and (j, i) positions. It is easily shown that if $\mathbf{A}$ is premultiplied by $\mathbf{R}^{\mathrm{T}}$ and postmultiplied by $\mathbf{R}$ to form $\mathbf{B} = \mathbf{R}^{\mathrm{T}}\mathbf{A}\mathbf{R}$, the only elements that change are

$$
\begin{aligned}
b_{ik} &= b_{ki} = a_{ik}\cos\theta + a_{jk}\sin\theta \qquad && k \neq i, j \\
b_{jk} &= b_{kj} = -a_{ik}\sin\theta + a_{jk}\cos\theta \qquad && k \neq i, j \\
b_{ii} &= a_{ii}\cos^2\theta + 2a_{ij}\sin\theta\cos\theta + a_{jj}\sin^2\theta \\
b_{jj} &= a_{ii}\sin^2\theta - 2a_{ij}\sin\theta\cos\theta + a_{jj}\cos^2\theta \\
b_{ij} &= b_{ji} = a_{ij}\cos 2\theta + \left(\frac{1}{2}\right)(a_{jj} - a_{ii})\sin 2\theta
\end{aligned}
$$

Any off-diagonal element b_{ij} can be made to be zero by setting $b_{ij} = 0$, resulting in

$$
\tan 2\theta = \frac{2a_{ij}}{(a_{ii} - a_{jj})}
$$

as the angle in the rotation matrix $\mathbf{R}$. This use of the rotation matrix is associated with the method of Jacobi. There are several strategies for choosing the sequence of rotations. These include classical, cyclic, and threshold Jacobi rotations. For a description of each of these strategies, see Parlett [4].

In particular, the threshold Jacobi method involves choosing rotations to successively eliminate all of the off-diagonal elements that have an absolute value larger than some threshold value. One such set of rotations is referred to as a sweep. A sufficient number of sweeps is executed to reduce the absolute values of all the off-diagonal elements of $\mathbf{A}$ to below some threshold value.

Example C.7.

Consider

$$\mathbf{A} = \begin{bmatrix} 3 & -1 & 0 \\ -1 & 2 & -1 \\ 0 & -1 & 3 \end{bmatrix}$$

Step 1. Eliminate a_{12} by taking $\tan 2\theta_1 = 2(-1)/(3-2) = -1$, from which $\cos\theta_1 = 0.850651$ and $\sin\theta_1 = -0.525731$. It follows with $\mathbf{A}_1 = A$, that

$$\mathbf{R}_1 = \begin{bmatrix} 0.850651 & 0.525731 & 0.000000 \\ -0.525731 & 0.850651 & 0.000000 \\ 0.000000 & 0.000000 & 1.000000 \end{bmatrix} = \mathbf{T}_1$$

and that $\mathbf{R}_1^T\mathbf{A}_1\mathbf{R}_1 = \mathbf{A}_2$ is

$$\mathbf{A}_2 = \begin{bmatrix} 3.618034 & 0.000000 & 0.525731 \\ 0.000000 & 1.381966 & -0.850651 \\ 0.525731 & -0.850651 & 3.000000 \end{bmatrix}$$

Next, eliminate the a_{13} term by taking $\tan 2\theta_2 = 2(0.525731)/(3.618034 - 3.0)$, from which $\cos\theta_2 = 0.867967$ and $\sin\theta_2 = 0.496623$. It again follows that

$$\mathbf{T}_2 = \begin{bmatrix} 0.738336 & 0.525731 & -0.442452 \\ -0.456317 & 0.850651 & 0.261090 \\ 0.496623 & 0.000000 & 0.867967 \end{bmatrix} = \mathbf{R}_1\mathbf{R}_2$$

and that $\mathbf{R}_2^T\mathbf{A}\mathbf{R}_2 = \mathbf{A}_2$ is

$$\mathbf{A}_2 = \begin{bmatrix} 3.918840 & -0.422452 & 0.000000 \\ -0.422452 & 1.381966 & -0.738336 \\ 0.000000 & -0.738336 & 2.699193 \end{bmatrix}$$

Last, the A_{23} element of $\mathbf{A_2}$ is eliminated by taking $\tan 2\theta_3 = 2(-0.738336)/(1.381966 - 2.669193)$. It follows that $\cos\theta_3 = 0.910247$ and $\sin\theta_3 = 0.414067$, resulting in

$$\mathbf{T}_3 = \begin{bmatrix} 0.738336 & 0.307058 & -0.600479 \\ -0.456317 & 0.883051 & -0.109526 \\ 0.496623 & 0.354875 & 0.792104 \end{bmatrix} = \mathbf{R}_1\mathbf{R}_2\mathbf{R}_3$$

and

$$\mathbf{A}_3 = \begin{bmatrix} 3.918840 & -0.385529 & 0.172723 \\ -0.385529 & 1.051180 & 0.000000 \\ 0.172723 & 0.000000 & 3.029980 \end{bmatrix} = \mathbf{R}_3^T\mathbf{A}_2\mathbf{R}_3$$

This set of three rotations constitutes one sweep.

The main diagonal terms of $\mathbf{A}_3$ are crude approximations to the eigenvalues of $\mathbf{A}$, namely, $\lambda = 4$, 1, and 3. The three columns of $\mathbf{T}_3$ are approximations to the corresponding eigenvectors. The off-diagonal terms are still too large to assert any sort of convergence so another sweep is carried out. The results are

$$\mathbf{T}_6 = \begin{bmatrix} 0.576846 & 0.409005 & -0.707081 \\ -0.578420 & 0.815739 & -0.000025 \\ 0.576784 & 0.409005 & 0.707132 \end{bmatrix} = \mathbf{R}_1\mathbf{R}_2\ldots\mathbf{R}_6$$

and

$$\mathbf{A}_6 = \begin{bmatrix} 3.999995 & 0.003932 & 0.000044 \\ 0.003932 & 1.000005 & 0.000000 \\ 0.000044 & 0.000000 & 3.000000 \end{bmatrix} = \mathbf{R}_6^{\mathrm{T}}\mathbf{A}_5\mathbf{R}_6$$

A third sweep produces

$$\mathbf{T}_9 = \begin{bmatrix} 0.577350 & 0.408248 & -0.707107 \\ -0.577350 & 0.816497 & 0.000000 \\ 0.577350 & 0.408248 & 0.707107 \end{bmatrix} = \mathbf{X}$$

and

$$\mathbf{A}_9 = \begin{bmatrix} 4.000000 & 0.000000 & 0.000000 \\ 0.000000 & 1.000000 & 0.000000 \\ 0.000000 & 0.000000 & 3.000000 \end{bmatrix}$$

indicating convergence to the accuracy displayed. One can easily verify that the columns of $\mathbf{X}$ are the eigenvectors corresponding to the eigenvalues 4, 1, and 3, and that $\mathbf{X}^{\mathrm{T}}\mathbf{A}\mathbf{X} = \mathbf{A}_9 = \mathbf{\Lambda}$.

The problem that more frequently arises from a finite element model is the generalized problem

$$\mathbf{A}\mathbf{x} - \lambda\mathbf{B}\mathbf{x} = \mathbf{0}$$

It is possible to generalize the usual Jacobi approach for determining the eigenvalues of a matrix $\mathbf{A}$ as follows. Consider a 2×2 problem

$$\begin{bmatrix} a & b \\ b & c \end{bmatrix}\mathbf{x} - \lambda\begin{bmatrix} \alpha & \beta \\ \beta & \gamma \end{bmatrix}\mathbf{x} = \mathbf{0}$$

Setting $\mathbf{x} = \mathbf{R}\mathbf{y}$ where

$$\mathbf{R} = \begin{bmatrix} 1 & \phi \\ -\psi & 1 \end{bmatrix}$$

and premultiplying the equation by $\mathbf{R}^{\mathrm{T}}$ yields

$$\mathbf{A}^*\mathbf{y} - \lambda\mathbf{B}^*\mathbf{y} = \mathbf{0}$$

where

$$\mathbf{A}^* = \begin{bmatrix} a - 2\phi b + \phi^2 c & \psi a + (1 - \phi\psi)b - \phi c \\ \psi a + (1 - \phi\psi)b - \phi c & \psi^2 a + 2\psi b + c \end{bmatrix}$$

$$\mathbf{B}^* = \begin{bmatrix} \alpha - 2\phi\beta + \phi^2\gamma & \psi\alpha + (1 - \phi\psi)\beta - \phi\gamma \\ \psi\alpha + (1 - \phi\psi)\beta - \phi\gamma & \psi^2\alpha + 2\psi\beta + \gamma \end{bmatrix}$$

As was the case for the Jacobi method associated with $\mathbf{A}\mathbf{x} - \lambda\mathbf{x} = \mathbf{0}$, it is desired to select ϕ and ψ so as to zero the off-diagonal terms of both $\mathbf{A}^*$ and $\mathbf{B}^*$ as a result of the transformation. It is indicated in Parlett [4] that this is possible by taking $\phi = d_1/e$ and $\psi = d_2/e$ where

$$d_1 = \det\begin{bmatrix} a & b \\ \alpha & \beta \end{bmatrix} \qquad d_2 = \det\begin{bmatrix} c & b \\ \gamma & \beta \end{bmatrix}$$

with e satisfying the quadratic equation

$$e^2 - De - d_1 d_2 = 0$$

with

$$D = \det\begin{bmatrix} a & c \\ \alpha & \gamma \end{bmatrix}$$

Example C.8.

Consider

$$\mathbf{A} = \begin{bmatrix} 2 & -1 \\ -1 & 1 \end{bmatrix} \qquad \mathbf{B} = \begin{bmatrix} 4 & 1 \\ 1 & 2 \end{bmatrix}$$

it follows that $d_1 = 6,\ d_2 = 3$, and $D = 0$. The roots satisfy $e^2 - 18 = 0$, or $e = \pm\sqrt{18}$. It then follows that

$$\mathbf{R} = \begin{bmatrix} 1.0000 & 0.7071 \\ -1.4142 & 1.0000 \end{bmatrix}$$

and that

$$\mathbf{A}^* = \begin{bmatrix} 6.8284 & 0.0000 \\ 0.0000 & 0.5858 \end{bmatrix} \qquad \mathbf{B}^* = \begin{bmatrix} 5.1715 & 0.0000 \\ 0.0000 & 5.4142 \end{bmatrix}$$

The eigenvalues are given by the ratios of the corresponding main diagonal elements of $\mathbf{A}*$ and $\mathbf{B}*$, namely

$$\lambda_1 = \frac{(\mathbf{A}^*)_{11}}{(\mathbf{B}^*)_{11}} = \frac{6.8284}{5.1715} = 1.3204$$

and

$$\lambda_2 = \frac{(\mathbf{A}^*)_{22}}{(\mathbf{B}^*)_{22}} = \frac{0.5858}{5.4142} = 0.1082$$

The corresponding eigenvectors are the first and second columns of $\mathbf{R}$.

In general, with $\mathbf{A}_1 = \mathbf{A}$ and $\mathbf{B}_1 = \mathbf{B}$, a rotation matrix defined by

$$\mathbf{R}(\theta) = \begin{array}{cc} & \begin{array}{ccccccccccc} & & & i & & & & j & & & \end{array} \\ \begin{array}{c} \\ \\ \\ i \\ \\ \\ \\ j \\ \\ \\ \\ \end{array} & \left[\begin{array}{ccccccccccc} 1 &&&&&&&&&& \\ & 1 &&&&&&&&& \\ && 1 &&&&&&&& \\ &&& 1 &&&& \phi &&& \\ &&&& 1 &&&&&& \\ &&&&& 1 &&&&& \\ &&&&&& 1 &&&& \\ &&& -\psi &&&& 1 &&& \\ &&&&&&&& 1 && \\ &&&&&&&&& 1 & \\ &&&&&&&&&& 1 \end{array}\right] \end{array}$$

is used to simultaneously reduce a_{ij}, a_{ji}, b_{ij}, and b_{ji} to zero by forming

$$\mathbf{R}^T\mathbf{A}_1\mathbf{R} = \mathbf{A}_2 \qquad \text{and} \qquad \mathbf{R}^T\mathbf{B}_1\mathbf{R} = \mathbf{B}_2$$

This step is repeated successively to reduce off-diagonal elements of $\mathbf{A_k}$ and $\mathbf{B_k}$ to zero until all the off-diagonal elements have been reduced to below some threshold value, resulting in matrices $\mathbf{A}^*$ and $\mathbf{B}^*$, which are essentially diagonal. The eigenvalues are computed according to

$$\lambda_i = \frac{(\mathbf{A}^*)_{ii}}{(\mathbf{B}^*)_{ii}}$$

with the corresponding eigenvectors the columns of

$$\mathbf{R}_1\mathbf{R}_2\mathbf{R}_3 \ldots \mathbf{R}_N$$

The method is referred to as the **threshold generalized Jacobi method.**

A listing of the FORTRAN program GJacobi is given below.

```
      SUBROUTINE GJacobi (nmax, n, a, b, d, eigval, eigvec, err)

      IMPLICIT NONE

C     ***********************************************************
C     *  This SUBROUTINE uses the Jacobi threshold method to find  *
C     * all the eigenvalues and eigenvectors of the generalized eigen-  *
C     * system                                                   *
C     *                  [A] - β[B]){u} = {0}                    *
C     *                                                          *
C     * INPUT ARGUMENTS:                                         *
C     *        n = size of the (n × n) matrices [A] and [B]      *
C     *        a = (n × n) matrix [A]                            *
C     *        b = (n × n) matrix [B]                            *
C     *                                                          *
C     * OUTPUT ARGUMENTS:                                        *
C     *     eigval = (n × 1) vector of eigenvalues               *
C     *     eigvec = (n × n) matrix whose columns are the normalized  *
C     *              eigenvectors                                *
C     *     err    = error level indicator                       *
C     *              err = 0: the subroutine successfully completed  *
C     *              err = 1: [A] is NOT positive definite       *
C     *              err = 2: [B] is NOT positive definite       *
C     *              err = 3: internal check not satisfied       *
C     *              err = 4: [A] is NOT positive definite       *
C     *              err = 5: [B] is NOT positive definite       *
C     *              err = 6: maximum number of sweeps exceeded  *
C     *                                                          *
C     * ADJUSTABLE ARRAY ARGUMENTS:                              *
C     *        d = (n × 1) working vector                        *
C     ***********************************************************
```

```
C   GLOBAL VARIABLES:

      INTEGER*2   err, n, nmax
      REAL*8      a(nmax,nmax), b (nmax,nmax), eigvec(nmax,nmax)
      REAL*8      d(nmax), eigval(nmax)

C   LOCAL VARIABLES:

      INTEGER*2   i, j, k, maxsweeps, sweep
      REAL*8      ab, aj, ajj, ak, akk, bj, bk, bsqr
      REAL*8      ca, cg, check, den, dif, d1, d2, ej, ek
      REAL*8      eps, epsa, epsb, rtol, sqch, temp, tol, tola, tolb

      PARAMETER (maxsweeps = 99)

      rtol  = 1.0e-6
      tol   = 0.0
      tola  = 0.0
      tolb  = 0.0
      eps   = 0.0
      epsa  = 0.0
      epsb  = 0.0

      err   = 0

      DO i = 1, n
           IF( a(i,i) .LE. 0.0 ) THEN
                err = 1
                RETURN
           END IF
           IF( b(i,i) .LE. 0.0 ) THEN
                err = 2
                RETURN
           END IF
           d(i) = a(i,i) / b(i,i)
           eigval(i) = d(i)
      END DO

      DO i = 1, n
           DO j = 1, n
                eigvec(i,j) = 0.0
           END DO
           eigvec(i,i) = 1.0
      END DO

      IF( n .EQ. 1 ) RETURN

      sweep = 0
   10 sweep = sweep + 1
```

```
eps = (0.01**sweep)**2
DO j = 1, n - 1
     DO k = j + 1, n
          tola = (a(j,k) * a(j,k)) / (a(j,j) * a(k,k))
          tolb = (b(j,k) * b(j,k)) / (b(j,j) * b(k,k))
          IF( (tola .GE. eps) .OR. (tolb .GE. eps) ) THEN
               akk = a(k,k) * b(j,k) - b(k,k) * a(j,k)
               ajj  = a(j,j) * b(j,k) - b(j,j) * a(j,k)
               ab = a(j,j) * b(k,k) - a(k,k) * b(j,j)
               check = (ab * ab + 4.0 * akk * ajj) / 4.0
               IF( check .LT. 0.0 ) THEN
                    err = 3
                    RETURN
               END IF
               sqch = SQRT(check)
               d1 = ab / 2.0 + sqch
               d2 = ab / 2.0 - sqch
               IF( ABS(d2) .GT. ABS(d1) ) THEN
                    den = d2
               ELSE
                    den = d1
               END IF
               IF( den .EQ. 0.0) THEN
                    ca = 0.0
                    cg = -a(j,k) / a(k,k)
               ELSE
                    ca =  akk / den
                    cg = -ajj / den
               END IF
               IF( n .GT. 2 ) THEN
                    IF( j .GE. 2 ) THEN
                         DO i = 1, j - 1
                              aj  = a(i,j)
                              bj  = b(i,j)
                              ak  = a(i,k)
                              bk  = b(i,k)
                              a(i,j)  = aj + cg * ak
                              b(i,j)  = bj + cg * bk
                              a(i,k)  = ak + ca * aj
                              b(i,k)  = bk + ca * bj
                         END DO
                    END IF
                    IF( (k + 1) .LE. n ) THEN
                         DO i = k + 1, n
                              aj  = a(j,i)
                              bj  = b(j,i)
                              ak  = a(k,i)
                              bk  = b(k,i)
                              a(j,i)  = aj + cg * ak
```

```
                        b(j,i)  = bj + cg * bk
                        a(k,i)  = ak + ca * aj
                        b(k,i)  = bk + ca * bj
                    END DO
                END IF
                IF( (j + 1) .LE. (k - 1) ) THEN
                    DO i = j + 1, k - 1
                        aj  = a(j,i)
                        bj  = b(j,i)
                        ak  = a(k,i)
                        bk  = b(k,i)
                        a(j,i)  = aj + cg * ak
                        b(j,i)  = bj + cg * bk
                        a(i,k)  = ak + ca * aj
                        b(i,k)  = bk + ca * bj
                    END DO
                END IF
            END IF
            ak = a(k,k)
            bk = b(k,k)
            a(k,k)  = ak + 2.0 * ca * a(j,k) + ca * ca * a(j,j)
            b(k,k)  = bk + 2.0 * ca * b(j,k) + ca * ca * b(j,j)
            a(j,j)  = a(j,j) + 2.0 * cg * a(j,k) + cg * cg * ak
            b(j,j)  = b(j,j) + 2.0 * cg * b(j,k) + cg * cg * bk
            a(j,k)  = 0.0
            b(j,k)  = 0.0
            DO i = 1, n
                ej = eigvec(i,j)
                ek = eigvec(i,k)
                eigvec(i,j) = ej + cg * ek
                eigvec(i,k) = ek + ca * ej
            END DO
        END IF
    END DO
END DO

DO i = 1, n
    IF( ABS(a(i,i)) .LT. rtol ) a(i,i) = rtol
    IF( ABS(b(i,i)) .LT. rtol ) b(i,i) = rtol

    IF( a(i,i) .LE. 0.0 ) THEN
        err = 4
        RETURN
    END IF

    IF( b(i,i) .LE. 0.0 ) THEN
        err = 5
        RETURN
    END IF
```

```
        eigval(i) = a(i,i) / b(i,i)
     END DO

     DO i = 1, n
        tol = rtol * d(i)
        dif = ABS(eigval(i) - d(i))
        IF( dif .GT. tol ) GO TO 30
     END DO

     eps = rtol**2

     DO j = 1, n - 1
        DO k = j + 1, n
           epsa = (a(j,k) * a(j,k)) / (a(j,j) * a(k,k))
           epsb = (b(j,k) * b(j,k)) / (b(j,j) * b(k,k))
           IF( (epsa .GE. eps) .OR. (epsb .GE. eps) ) THEN
              GO TO 30
           END IF
        END DO
     END DO

20   DO i = 1, n
        DO j = 1, n
           a(j,i) = a(i,j)
           b(j,i) = b(i,j)
        END DO
     END DO

     DO j = 1, n
        bsqr = SQRT(b(j,j))
        DO k = 1, n
           eigvec(k,j) = eigvec(k,j) / bsqr
        END DO
     END DO

C    Sort the eigenvalues and eigenvectors

     DO i = 1, n - 1
        DO j = i + 1, n
           IF( eigval(i) .GT. eigval(j) ) THEN
              temp = eigval(i)
              eigval(i) = eigval(j)
              eigval(j) = temp
              DO k = 1, n
                 temp = eigvec(k,i)
                 eigvec(k,i) = eigvec(k,j)
                 eigvec(k,j) = temp
              END DO
           END IF
```

```
          END DO
      END DO

      RETURN

30    DO i = 1, n
          d(i) = eigval(i)
      END DO

      IF( sweep .LT. maxsweeps) GOTO 10

      IF( sweep .GT. maxsweeps) err = 6

      GO TO 20

      RETURN
      END
```

REFERENCES

1. Bathe, K. J.: *Finite Element Procedures in Engineering Analysis*, Prentice Hall, Englewood Cliffs, New Jersey, 1982.
2. Strang, G.: *Linear Algebra and its Applications*, Academic Press, New York, 1980.
3. Noble, B., and J.W. Daniel: *Applied Linear Algebra*, Prentice Hall, Englewood Cliffs, New Jersey, 1977.
4. Parlett, B. N.: *Symmetric Eigenvalue Problems*, Prentice Hall, Englewood Cliffs, New Jersey, 1980.

APPENDIX D *SOLVABILITY OF FINITE ELEMENT EQUATIONS*

The theory of linear algebraic equations indicates that there are three distinct possibilities for the solution to a set of equations $\mathbf{Ax} = \mathbf{b}$, namely,

1. There is a unique solution.
2. There are an infinite number of solutions.
3. There is no solution.

These can be summarized by the so-called **Fredholm alternative,** which states "Either the nonhomogeneous equation $\mathbf{Ax} = \mathbf{b}$ possesses a unique solution or there are an infinite number of solutions to the homogeneous equation $\mathbf{Ax} = \mathbf{0}$." Solutions to elliptic boundary value problems of the types studied in Chapters 2 and 3 exhibit properties that are precisely analogous. These ideas are considered below together with a discussion of the connection between the boundary value problems and the system of linear algebraic equations that results from constructing the corresponding finite element model.

One-dimensional problems. Consider the simple boundary value problem

$$u'' + f = 0 \qquad -u'(a) = A \qquad u'(b) = B \tag{D.1}$$

Integrate the differential equation from a to b to obtain

$$u'(b) - u'(a) + \int_a^b f\,dx = 0$$

or

$$B + A + \int_a^b f\,dx = 0 \tag{D.2}$$

showing that there is a relationship between A, B, and the integral of f from a to b, which must be satisfied. If the condition given by Eq. (D.2) is satisfied, there is a solution that is not unique. If the condition is not satisfied there is no solution. Note that the corresponding completely homogeneous boundary value problem

$$u'' = 0 \qquad -u'(a) = 0 \qquad u'(b) = 0$$

has the nontrivial solution u = constant.

For the more general Sturm-Liouville boundary value problem

$$\begin{aligned}(pu')' - qu + f &= 0\\ -p(a)u'(a) + \alpha u(a) &= A\\ -p(b)u'(b) + \beta u(a) &= B\end{aligned}$$

the result of integrating the differential equation over the interval is

$$B + A + \int_a^b f\,dx = \beta u(b) + \alpha u(a) + \int_a^b qu\,dx \tag{D.3}$$

which is always satisfied as long as not all of α, β, and q are zero.

Practically, one needs to be aware that *if Neumann type boundary conditions* (u') *are specified at both boundaries*, there is potentially a problem with the existence and uniqueness of the solution.

A simple physical problem useful in helping to understand this issue is the problem of the axial deformation of an elastic bar as indicated.

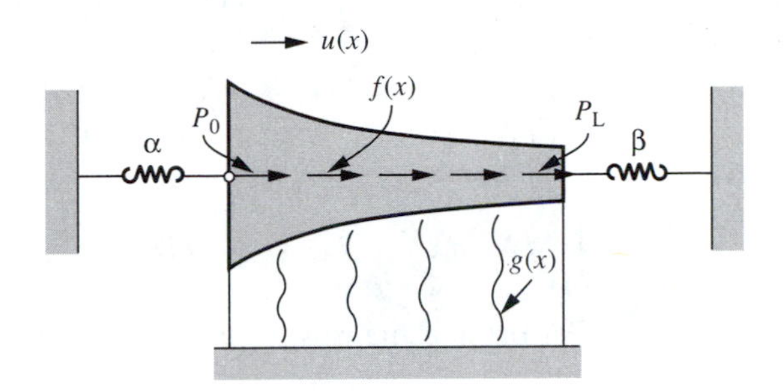

The term $f(x)$ is the distributed loading with α and β representing linear springs connecting the bar to ground at the ends, and $q(x)$ representing a linear elastic medium to which the bar is connected along its length. The elastic foundation $q(x)$ resists motion in the axial direction. The differential equation of equilibrium is

$$(AEu')' - qu + f = 0$$

with the boundary conditions

$$-A(0)E(0)u'(0) + \alpha u(0) = P_0$$

$$A(L)E(L)u'(L) + \beta u(L) = P_L$$

As long as the elastic medium or at least one of the springs is present, the resultant force F, given by

$$F = P_0 + P_L + \int_0^L f\, dx$$

can be reacted by one of the springs or the elastic medium q, resulting in a unique solution for the axial deformation $u(x)$.

If both springs and the elastic medium q are absent ($\alpha = \beta = q = 0$), there will be a solution if and only if equilibrium is satisfied, that is, if $F = 0$. When equilibrium is satisfied, a rigid body motion (u = constant) can be added to the internal deformations resulting from P_0, P_L, and f. This rigid body motion can be chosen as any value and imposed at either end of the bar, resulting in a unique solution.

To see how all of this manifests itself in the finite element model, consider the model corresponding to Eq. (D.1). Using three equal-length linearly interpolated elements, the assembled equations appear as

$$\left[\begin{array}{cccc|c} c & -c & 0 & 0 & A+f_{11} \\ -c & 2c & -c & 0 & f_{12}+f_{21} \\ 0 & -c & 2c & -c & f_{22}+f_{31} \\ 0 & 0 & -c & c & f_{32}+B \end{array}\right]$$

where

$$f_{11} = \int_a^{x_2} f N_1 \, dx \qquad f_{12} = \int_a^{x_2} f N_2 \, dx$$

$$f_{21} = \int_{x_2}^{x_3} f N_1 \, dx \qquad f_{22} = \int_{x_2}^{x_3} f N_2 \, dx$$

$$f_{31} = \int_{x_3}^{b} f N_1 \, dx \qquad f_{32} = \int_{x_3}^{b} f N_2 \, dx$$

and where $c = 3/(b - a)$. Performing the row reduction leads to

$$\left[\begin{array}{cccc|l} c & -c & 0 & 0 & A+f_{11} \\ 0 & c & -c & 0 & A+f_{11}+f_{12}+f_{21} \\ 0 & 0 & c & -c & A+f_{11}+f_{12}+f_{21}f_{22}+f_{31} \\ 0 & 0 & 0 & 0 & A+f_{11}+f_{12}+f_{21}+f_{22}+f_{31}f_{32}+B \end{array}\right]$$

showing that the coefficient matrix has a rank of 3. There is a solution that is not unique if and only if the rank of the augmented matrix is 3, that is, if

$$A + f_{11} + f_{12} + f_{21} + f_{22} + f_{31} f_{32} + B = 0$$

With a little algebra this can be expressed as

$$B + A + \int_a^b f \, dx = 0$$

which is clearly the same condition as resulted from the boundary value problem.

The point is that one needs to be aware that there are boundary value problems that have no solution or an infinity of solutions and that the algebraic equations of the corresponding finite element models will possess these same characteristics. Hopefully the analyst will recognize any such situations and formulate the problem so that the final system of equations possesses a unique solution.

In any event, the equation solver used should be capable of detecting when the coefficient matrix is singular and warn the user. This would usually be accomplished by using a code that employs partial pivoting that checks to see whether the coefficient matrix is numerically singular.

In summary, for the general Sturm-Liouville problem with two natural boundary conditions, the algebraic equations representing the finite element model will possess a unique solution as long as not all of q, α, and β are zero. If all of α, β, and q are zero and the integrability condition

$$A + B + \int_a^b f \, dx = 0$$

is satisfied, a unique solution can be obtained by constraining the degree of freedom at either end. On the other hand, if at least one boundary condition is essential, the resulting constraint will give rise to a nonsingular coefficient matrix and a unique solution.

Two-dimensional problems. The two-dimensional boundary value problem that is the analogue of problem (D.1) is

$$\nabla^2 u + f = 0 \qquad \text{in } D$$

$$\frac{\partial u}{\partial n} = g \qquad \text{on } \Gamma$$

and is known as the Neumann problem. Integrating over the domain D and using the divergence theorem stating that

$$\int_D \nabla^2 u \, dA = \int_\Gamma \frac{\partial u}{\partial n} \, ds$$

it follows that

$$\int_D f \, dA + \int_\Gamma \frac{\partial u}{\partial n} \, ds = 0$$

or

$$\int_D f \, dA + \int_\Gamma g \, ds = 0$$

that is, an additional condition relating the functions f and g. Unless this condition is satisfied there is no solution. If the condition is satisfied the solution is not unique. Again, the corresponding homogeneous boundary value problem

$$\nabla^2 u = 0 \qquad \text{in } D$$

$$\frac{\partial u}{\partial n} = 0 \qquad \text{on } \Gamma$$

clearly always has the solution u = constant.

For the more general problem

$$\nabla^2 u + f = 0 \qquad \text{in } D$$

$$\frac{\partial u}{\partial n} + \alpha u = g \qquad \text{on } \Gamma$$

there is a unique solution unless $\alpha = 0$. Similarly, the boundary value problem

$$\nabla^2 u + f = 0 \qquad \text{in } D$$

$$\frac{\partial u}{\partial n} = g \qquad \text{on } \Gamma_2$$

$$u = u_0 \qquad \text{on } \Gamma_1$$

possesses a unique solution unless $\Gamma_1 = 0$. The completely general problem

$$\nabla^2 T + f = 0 \qquad \text{in } D$$

$$\frac{\partial u}{\partial n} + \alpha u = g \qquad \text{on } \Gamma_2$$

$$u = u_0 \qquad \text{on } \Gamma_1$$

possesses a unique solution as long as at least one of α or Γ_1 is not identically zero.

A physical problem that is useful in understanding these ideas as they pertain to the two-dimensional boundary value problems is steady-state heat conduction. The general boundary value problem can be stated as

$$\nabla^2 T + f = 0 \qquad \text{in } D$$

with

$$T = T_0 \qquad \text{on } \Gamma_1$$

and

$$\frac{\partial T}{\partial n} + \alpha T = g \qquad \text{on } \Gamma_2$$

If Γ_1 and α are both zero, it follows from the development above that

$$\int_D f\, dA + \int_\Gamma g\, ds = 0$$

which is a global energy balance equation. The first term represents the amount of energy generated by the internal sources and the second term, the total energy passing through the boundary. Unless this condition is satisfied, energy is not balanced, the basic physical principle is violated, and there is clearly no solution to the boundary value problem. If the condition is satisfied there is a nonunique temperature distribution determined by specifying the temperature at one point in the region.

If there is a portion of the boundary on which essential boundary conditions are prescribed, energy can be transferred across that part of the boundary with a resulting global energy balance. Similarly, energy can be transferred by convection through the Γ_2 portion of the boundary when α is not zero. Either situation results in a set of algebraic equations possessing a unique solution.

In summary, for the general elliptic boundary value problem, the algebraic equations representing the finite element model will possess a unique solution as long as both α and Γ_1 are not zero. If both α and Γ_1 are zero and the integrability condition

$$\int_D f\, dA + \int_\Gamma g\, ds = 0$$

is satisfied, a unique solution can be obtained by specifying the degree of freedom at one node. If both α and Γ_1 are zero and the integrability condition is not satisfied, there is no solution.

Index

A

Accuracy of solution, 272, 286
Admissible functions, 121, 601, 606
Algebraic equations, systems of, 606, 611
Approximations of functions, 107, 146, 255, 256, 277, 307, 330, 354
Arc length, 331, 400
Area coordinates, 249, 346
Aspect ratio, 305
Assembly, 15, 18, 23, 26, 30, 64, 71, 77, 163, 227, 247, 261, 289, 412, 502
Average acceleration method, 455
Axial deformation, 76, 146, 434
Axisymmetric problems, 364
  elasticity, 550

B

Bandwidth, 31, 376
Barlow points, 538, 581
Best approximation property, 147–8
Body force vector, 506, 541
Body forces, 485
Boundary conditions
  essential, 76, 125, 128, 411, 489
  mixed, 489, 528
  natural, 125, 129, 155, 489
  satisfaction of, 266, 271, 286, 321, 343, 363, 518, 524, 530, 533
Boundary value problems, 216, 364, 487

C

Calculus of variations, 119, 193
Central difference method, 448
Chain rule, 299, 351
Collocation, 602
Completeness of interpolation functions, 132, 159, 297
Condition number, 74
Conforming elements, 376
Constitutive relations, 50, 51, 53, 486
Constraints, 14, 33, 71, 74, 84, 88, 92, 95, 147, 152, 164, 247, 265, 270, 290, 413, 467, 517, 565
Convergence, 168, 171, 266, 267, 272, 283, 286, 297, 321, 363, 521, 580
Coordinates
  global, 27, 30, 98, 101, 160, 171
  intrinsic or natural, 100, 102, 141, 152
  local, 27, 30, 83, 98, 101, 160, 171, 249
Counterclockwise labeling, 222
Crank-Nicolson method, 426
Curved boundaries, 220, 321, 344

D

Degrees of freedom, 7
Derivatives of interpolation functions, 63, 142, 243, 253
Derived variables, 75, 79, 85, 89, 110, 113, 138, 147, 151, 163, 248, 262, 270, 321, 363
Differential equations, 57, 219
Diffusion, 408, 464
Dimensional homogeneity, 36
Dirac delta function, 196, 602
Directional derivative, 305
Dirichlet boundary conditions, 219, 365
Discretization, 5, 62, 76, 83, 87, 220, 241, 295, 371, 410, 436, 448, 496
Distributed forces, 9
Divergence theorem, 225, 365, 465, 555

E

Eigenfunctions, 95, 157, 288
Eigenvalue problems, 56, 93, 286

Eigenvalues, 95, 157, 294
Eigenvectors, 93, 157, 294
Elasticity, 484
Element shapes, 221, 242, 305, 327, 352
Elemental formulations, 64, 71, 76, 83, 87, 161, 224, 244, 369, 411, 414, 437
Elemental matrices
  bar, 29, 117
  beam, 158
  diffusion problems, 408, 464
  discrete problems, 4
  elasticity, 484
  heat transfer, 18
  isoparametric elements, 295
  pipe flow, 23
  Sturm-Liouville problems, 69, 175
  truss, 26
  wave equation problems, 434, 467
Elliptic equations, 216
Energy, 50, 53, 91
Energy equations, 91, 409
Equilibrium equations, 12, 51
  checks, 15, 35, 81, 518, 520, 525, 531, 533, 545, 548, 549
Errors, 65, 601
  in derivatives, 169, 170
  in discretization, 220, 424
  in solutions, 168, 283, 364
Euler equations, 123, 128, 195
  method, 423
Examples
  elasticity, 510, 516, 519, 526, 542, 545, 564, 568
  fluid mechanics, 3, 23, 51, 82, 148, 386
  heat transfer, 4, 18, 53, 86, 152, 267, 283, 317, 341, 362, 408
  solid mechanics, 50, 76, 144, 257, 279, 314, 341, 362
  vibrations, 55, 86, 156
Extremals, 71

F

Finite differences
  backward, 430
  central, 427, 448
  forward, 423
Fixed endpoint problem, 121
Fluid flow problems, 51, 82, 148
Forward iteration, 629
Fourier's Law, 19, 54
Full integration, 580, 585, 589
Functionals, 120
  bilinear, 116, 240, 366, 494, 554
  linear, 116, 240, 366
  quadratic, 117, 240, 366, 494, 555

G

Galerkin method, 65, 224, 603
Gauss elimination, 611
Gauss quadrature. *See* Numerical integration
Global matrices
  stiffness matrices, 13, 33, 94, 147, 237
  load matrices, 119, 158

H

Half bandwidth, 30, 377, 413, 439
Heat transfer. *See also* Examples, heat transfer, 18, 86, 152
  conduction, 218, 267, 283
  convection, 218, 267, 283
Helmholtz's equation, 218, 288, 397, 398, 399
Higher-order interpolations
  one-dimensional, 138
  two-dimensional, 321
Hyperbolic problems, 434

I

Improved Euler method, 426
Incompatible elements, 377
Initial conditions, 409, 414, 435, 440
Initial-boundary value problems, 286
Integration by parts, 115, 119, 224, 239
Interelement boundaries, 220
Interelement continuity, 244, 375
Interpolation, 62, 138, 221, 273, 298, 367, 410, 436, 496
Interpolation functions
  bilinear, 272
  isoparametric, 295, 322, 344
  linear, 63, 131, 221
  quadratic, 138, 344
Inverse iteration, 630

Isoparametric elements
Q4 elements, 297
Q8 elements, 321, 574
T3 elements, 296
T6 elements, 344

J

Jacobian determinant, 100, 253, 275
Jacobian matrix, 252, 278, 299, 334, 578

K

Kinematic viscosity, 23, 52
Kinematics, 51, 486, 552
Kinetics, 484, 551
Kronecker delta, 67

L

Laplace's equation, 216, 364
Least squares, 603

M

Mass matrix
consistent, 289, 415, 438
lumped, 416, 441
weighted, 441
Membrane
analogy, 258
vibration, 290, 291
Mesh refinement by quartering, 271, 283, 285
Method of Weighted Residuals, 601
Moan points, 75, 144, 306, 329, 353
Modal matrix, 421, 430, 447, 461

N

Natural frequency
axial deformation, 443
membrane, 290
torsional, 56, 93, 156
Neumann boundary conditions, 219, 365
Newmark's method, 455, 461
Nodal averaging, 585
Nodal interpolation functions, 63 223, 381, 411, 436, 465
Nodes, 62, 139, 224,
numbering, 31, 377
Numerical integration, 100
Gauss quadrature, 103, 105, 109, 144, 309, 335, 356, 583
Newton-Cotes, 101
rules for triangles, 356
Simpson's rule, 102, 106
special integration, 106, 110, 144, 190
trapezoidal rule, 101, 105
Numerical oscillations, 434

O

One-dimensional problems, 50
Operator, 174, 370
Ordinary differential equations, 52, 57
Orthotropic, 487
Outward normal, 225, 486

P

Parabolic problems, 408, 464
Parent element, 298, 302, 303, 325, 349
Partial differential equations, 216, 218, 408, 434
Penalty method for constraints, 72
Piecewise continuous, 62
Piping networks, 23
Plane strain, 490
Plane stress, 489
Poisson's equation, 216
Potential energy, 16, 118, 161, 494
Pressure, 23
Pressure vessel, 564

Q

Quadrature. *See* Numerical integration
Quadrilateral elements
Q4 elements, 297
Q8 elements, 321, 574

R

Rayleigh quotient, 200, 289, 397
Rectangular elements, 272, 394, 535
Reduced integration, 580, 585
Residuals, 601
Richardson's extrapolation, 170
Ring elements, 367, 556
Ritz method, 129, 237, 495, 590
Robins boundary condition, 219, 365

Romberg integration, 170
Rotation matrix, 28

S

Sampling point, 103
Self-adjoint, 58
Semidiscretization, 411, 436
Shape function. *See also* Interpolation functions
  Q4 elements, 297
  Q8 elements, 321
  T6 elements, 344
Shear modulus, 56
Shear strain, 56
Shear stress, 56
Singular matrix, 231
Slope, 159
Smoothing, 142, 329, 353, 403, 581
Solution, 74, 248, 367
  homogeneous, 417, 441
  particular, 418, 442
Springs, 7, 11, 16, 27
Stability, 424, 453, 462
  conditional, 424, 449, 456, 463, 471
  unconditional, 427, 431, 456, 463, 471
Stationary potential energy
  functionals, 120, 194, 494
  functions, 16, 137, 247
Stationary value, 16, 120, 137, 383, 502, 561
Stiffness matrices. *See* Elemental matrices
Strain, 486, 552
Strain energy, 494
Strain matrix, 499, 558
Stress, 217, 485
  computation of, 503, 515, 529, 548, 581
  concentration of, 315, 339, 360, 526, 568, 584
  continuity of, 515, 516, 518, 520, 525, 531
Stress-strain relations, 486, 552
Sturm-Liouville problems, 58
Subdoman method, 602
Subparametric, 296
Superparametric, 297
Surface tractions, 488, 501
  loads, 502, 507, 513, 517, 543, 547, 560
Symmetry conditions, 489, 527, 584
  exploitation of, 258

T

Temperature, 19, 410, 487
Test function, 115, 237, 289, 365, 411, 437, 465, 468
Thermal conductivity, 19
Thermal loads, 505, 541, 562
Time dependent problems, 407
  hyperbolic problems, 434, 467
  parabolic problems, 408, 464
Time domain integration
  first-order systems, 422
  second-order systems, 447
Time step, critical value of, 424, 426, 432, 449, 451, 463
Torsion, 215, 216, 278, 314, 339, 360
Total potential energy, 17, 118, 127, 161, 494, 555
Transverse displacement, 127, 158
Triangular elements, 221, 241, 344, 497, 555
Truss, 26
Two-dimensional problems
  elliptic boundary value problems, 216
  hyperbolic initial-boundary value problems, 434, 467
  parabolic initial-boundary value problems, 408, 464

V

Variational principles, 17, 114, 237
Vibration problems, 55, 93, 156
Virtual work, 118, 495

W

Wave equation, 434, 466, 467, 477, 481
Weak formulation, 115, 119, 237, 365, 411, 437, 465, 468, 493, 555
Weighted residuals, 601
Weighting functions, 57, 601, 602
Well-posed problem, 221, 406

Y

Young's Modulus, 27, 51, 432, 489

Conversion of an elliptic boundary value problem into the algebraic equations of the corresponding finite element model

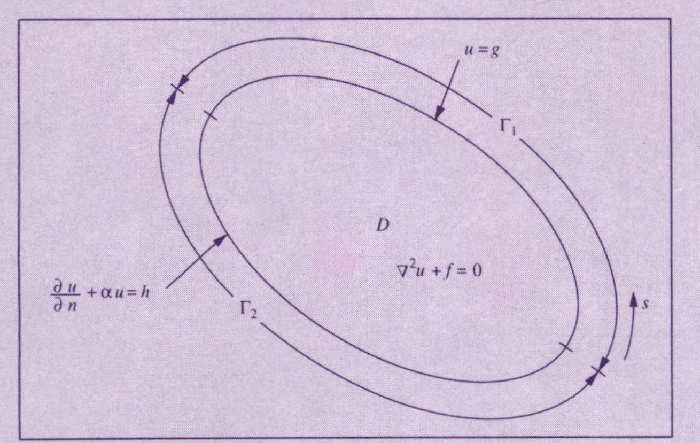

Typical region showing D, Γ_1, and Γ_2

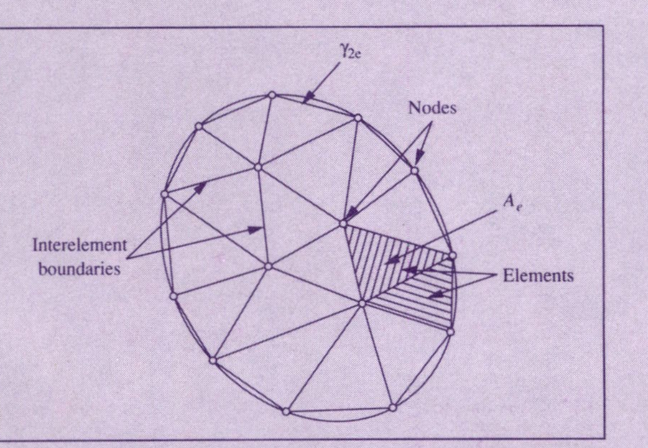

Typical mesh for the two-dimensional boundary value problem using linearly interpolated triangular elements

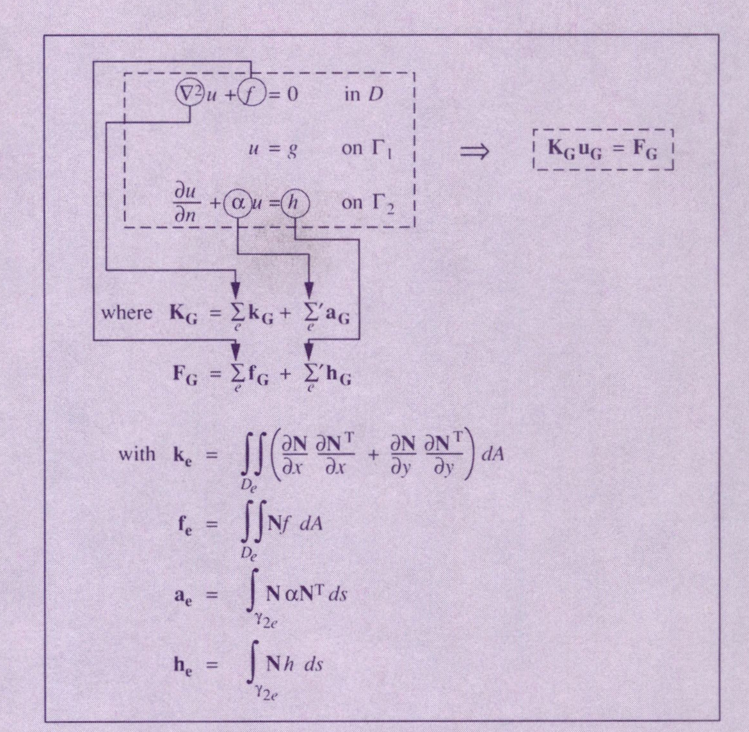

Statement of the boundary value problem and the corresponding finite element model